Plant Genome Diversity Volume 2

Ilia J. Leitch
Editor-in-chief

Johann Greilhuber · Jaroslav Doležel ·
Jonathan F. Wendel
Editors

Plant Genome Diversity Volume 2

Physical Structure, Behaviour and Evolution of Plant Genomes

Springer

Editor-in-chief
Ilia J. Leitch
Jodrell Laboratory
Royal Botanic Gardens,
Kew Richmond, Surrey
United Kingdom

Editors
Johann Greilhuber
Department of Systematic and Evolutionary Botany
Faculty of Life Sciences
University of Vienna
Vienna
Austria

Jaroslav Doležel
Institute of Experimental Botany ASCR
Centre of the Region Hana for
Biotechnological and Agricultural
Research
Olomouc
Czech Republic

Jonathan F. Wendel
Department of Botany
Iowa State University
Ames, Iowa
USA

ISBN 978-3-7091-1673-9 ISBN 978-3-7091-1160-4 (eBook)
DOI 10.1007/978-3-7091-1160-4
Springer Wien Heidelberg New York Dordrecht London

Printed on acid-free paper

Springer is part of Springer Science+Business Media (www.springer.com)

Preface

Ever since the origin of life, the evolution of living organisms and their hereditary information has been accompanied by the development of genetic machinery capable of storing, utilizing, and transmitting this information between generations. Importantly, this machinery has had to be flexible, able to respond to the environment and evolve. A characteristic feature of the genetic machinery in eukaryotes is the partitioning of the hereditary information into smaller portions—chromosomes. Indeed, the appearance of linear chromosomes was one of the great evolutionary inventions and paved the way for the formation of large and complex genomes, in plants as well as in animals. Any consideration of plant genome structure, evolution, and function is thus incomplete if it does not take into account its higher-order structure and the behaviour of its principal units—the chromosomes.

The chromosome theory of heredity, which linked the behaviour of Mendel's "factors" (units of inheritance) with that of chromosomes, was coined by Walter S. Sutton more than a century ago. This was followed, only a few decades later by Cyril D. Darlington's demonstration that the behaviour of chromosomes and meiotic crossing over in particular, was the main force behind evolution as opposed to single gene mutations and deletions. This set in motion the quest to understand the nature of inheritance, leading to the discovery of the structure of DNA and the advent of molecular biology and genomics. At this point the goal seemed clear—all that was needed was to establish the sequence of bases in the DNA. However, as increasing amounts of DNA sequences were generated, it became obvious that there was still a lot to discover about how DNA was organized within chromosomes and how the DNA sequence information was interpreted, processed, and utilized in the nuclear and cellular environments. The days when the DNA sequence itself was considered a holy grail are over and we now know that things are considerably more complicated.

Luckily, progress in genomics has been complemented by advances in understanding the dynamic structure of chromatin, the organization of interphase nuclei, and the behaviour of chromosomes during mitosis and meiosis. The latter includes novel insights into modified cell cycles, which may lead to chromosomes with more than two chromatids. Impressive progress has also been made in understanding the origin and function of specialized chromosomes (e.g., B chromosomes and sex chromosomes) and in appreciating the extent and significance of polyploidy in plant evolution. Although the frequent occurrence of polyploidy has been known for a long time based on chromosome counts and behaviour, the advent of DNA sequencing and comparative genomics has been instrumental in uncovering evidence of further rounds of polyploidy buried within the genome and now no longer visible at the chromosome level. Such studies have reinforced and extended our understanding of the significance of this mechanism as one of the main forces underlying the evolution and large diversity of many plant genomes. The effect has been multiplied by the extensive structural chromosome changes, which, together with alterations in chromosome number and genome size, can accompany plant speciation.

To fully understand and appreciate the diversity, functioning, and evolution of plant genomes, a holistic knowledge of the current status in each of these individual areas is vital, yet there is no single accessible source of information currently available. Thus, we felt it timely to fill this gap.

This book is the second volume of a two-volume set on Plant Genome Diversity. Our aim is to assist students and researchers by providing as complete an overview as possible of each respective area of research. We have succeeded in engaging leading experts in each field who describe the current state-of-the-art knowledge without overwhelming the reader with details that can be found elsewhere. What we offer in the present volume are 20 chapters whose topics have been chosen carefully to provide a complete picture. Each chapter can stand on its own and thus the reader does not need to read all chapters if he/she is only interested in a specific area. We sincerely hope that this model serves our readers well. It is up to them to decide if we have succeeded.

The 20 chapters deal with individual aspects of plant genome structure, function, and evolution and they are divided into five informal sections.

Evolutionary Framework for Studying the Diversity of Plant Genomes

Although we do not necessarily expect our readers to read all chapters, we do recommend that those interested in the evolution of plant genomes read the first chapter by Soltis and Soltis (Chap. 1) who provide an overview of plant phylogeny, with an emphasis on angiosperms. Among other things, they highlight research projects that have deposited phylogenetic trees in public databases and can be downloaded for analysis.

Architecture and Dynamics of the Plant Cell Nucleus

In nondiving cells, the chromosomes are organized within the nucleus, although the structural and functional complexity of this organization is still poorly understood. One can hardly imagine the intricacy of interactions of DNA with various molecules necessary to control tens of thousands of genes and process transcripts of genic and non-genic DNA. In addition, this is all taking place in a tightly packed nuclear environment, which also harbors structures needed for DNA synthesis and repair, chromosome reduplication, posttranscriptional modifications, and synthesis of ribosomal subunits, to name but a few. Jones and Langdon (Chap. 2) review nuclear organization and discuss the consequences of interspecific hybridization, which results in two different genomes being accommodated within a single nucleus. This cohabitation may not be peaceful and can result in dramatic structural and epigenetic reorganizations in subsequent generations.

The way DNA is organized and packaged into chromatin, particularly at the higher-order level has never been entirely clear although numerous models have been proposed. However, recent discoveries question even the existence of the 30-nm fibre, which traditionally has been considered to originate by folding the 11 nm nucleosome fibre. In Chap. 3, Takata et al. address this topic by describing the composition of chromatin in relation to chromosome condensation and DNA packing. Moreover, they present a novel model for chromosome structure, which suggests that the nucleosome fibres exist in a highly disordered state and do not form 30-nm chromatin fibres at all.

In addition to separating the nuclear environment from the cytoplasm, the nuclear envelope performs many important functions; one of which is the control of molecular traffic between both cellular compartments. Kiseleva et al. (Chap. 4) describe the composition of the nuclear envelope, the nuclear pore complexes, and their assembly and function and discuss possible interactions of the envelope with the cytoplasmic and nucleoplasmic components.

Nuclei are known to contain a variety of nuclear bodies, but only the nucleolus can be easily identified by optical light microcopy. It is where the cell produces ribosomes, which are required by the cell in large numbers. Shaw (Chap. 5) summarizes the current state of knowledge of the nucleolus, which is formed on nucleolar organizing regions of chromosomes.

For plants to grow and reproduce, the cells must divide either through mitosis or meiosis. The aim is that the hereditary material is faithfully transmitted to the daughter cells. Not only must the chromosomes be fully reduplicated but their chromatids must separate at the right moment and move in the right direction to form daughter nuclei. In addition, the nuclear envelope breaks down during cell division and this represents an additional major challenge for the genetic apparatus. Magyar et al. (Chap. 6) provide an insightful review on molecular events underlying the mitotic cell cycle and mitosis itself.

Following cell division, cell and tissue differentiation is often accompanied by modified cell cycles in which the mitosis step is omitted and the nuclear envelope does not break down. Maluszynska et al. (Chap. 7) outline these different types of endopolyploidy and describe the molecular pathways involved in switching from the mitotic to the endopolyploidization cycle and how the number of endocycles are regulated. They also review the occurrence of endopolyploidy, its biological significance, and the structure of endopolyploid nuclei.

The production of gametes provides an important means to generate genetic variation via recombination of parental chromatids and their random segregation. Given the complexity of the process, it is not surprising that it is unclear exactly how the mitotic machinery is modified for the purpose of meiosis. Nevertheless, Jenczewski et al. (Chap. 8) describe the current knowledge in this area, covering chromosome dynamics during meiosis, initiation of meiotic recombination, regulation of double strand break repair, crossover formation and interference, genetic control of crossing-over formation, and its distribution in polyploids.

Karyotype Diversity Across Plants and Trends in Evolution

One of the ways in which the diversity of plant genomes is manifested is through a wide range of chromosome numbers. Lysák and Schubert (Chap. 9) explain that in many cases this originates via chromosome rearrangements. The authors outline in detail the mechanisms of chromosome rearrangements detectable by microscopic techniques and highlight those that have had an impact on the alteration of chromosome number and structure during evolution and thus may have played a role in speciation.

The compartmentalization of genomes into chromosomes has provided opportunities for the development of specialized chromosomes. One such example is the B chromosome (often called supernumerary chromosome), and Houben et al. (Chap. 10) describe its structure, DNA composition, and evolution. The authors explain peculiarities in the behaviour of Bs during mitosis and meiosis and list various drive mechanisms responsible for retaining Bs in the population.

Sex chromosomes are another classic example of specialized chromosomes, and Janoušek et al. (Chap. 11) review sex determination systems in various plant groups and, based on taxonomic distribution, argue that dioecy has originated independently many times during evolution. The authors introduce the genus *Silene* as an excellent system to study the evolution of sex chromosomes and present the first ever evidence of sex dimorphism in dioecious plants.

Bureš et al. (Chap. 12) describe holocentric chromosomes which differ from the more common monocentric chromosomes by the way in which spindle microtubules attach along the whole chromosome length through kinetochores that cover a substantial part of their poleward surfaces during mitosis. They review the occurrence of holocentric chromosomes in plants and describe their chromatin structure and behaviour during mitosis and meiosis and the evolutionary processes that have contributed to the diversity of holocentric karyotypes.

The remaining three chapters in this section analyse karyotype diversity in three different groups of plants. Weiss-Schneeweiss and Schneeweiss (Chap. 13) provide a comprehensive account of karyotype diversity and evolutionary trends in angiosperms. They discuss in detail how changes in chromosome number, including dysploidy and aneuploidy, as well as changes in chromosome morphology contribute to the karyotype diversity observed. They also outline various cytogenetic methods which can be used to characterize chromosomes in a karyotype and study their changes during evolution and speciation. Murray (Chap. 14) presents a survey of chromosome numbers and size variation in gymnosperms, the sister group to angiosperms, and describes the methods used to analyse karyotype diversity in this group of seed plants. After reviewing available data, the author concludes that in contrast to angiosperms, gymnosperms are characterized by much greater uniformity in chromosome number and karyotype. The third chapter in this block is by Barker (Chap. 15) who focuses on karyotype and genome evolution in pteridophytes (monilophytes and lycophytes). He draws attention to the high chromosome numbers typical of many ferns, particularly the homosporous species, which on average contain over three-fold more chromosomes than the average flowering plant. Interestingly, there is currently no conclusive answer as to why this should be so although it is expected that the availability of complete genome sequences will contribute to solving this long-standing mystery.

Generative Polyploidy

The three chapters in this section evaluate various features of generative polyploidy, which is widespread in land plants. Husband et al. (Chap. 16) examine patterns of polyploid occurrence, such as the variation among taxonomic groups at or above the species level, intraspecific variation, variation in mechanisms of formation, geographic and ecological patterns of polyploid incidence, and associations between ploidy and reproduction.

Thanks to the advances in DNA sequencing and genomics there is now evidence to suggest that most seed plants have undergone at least one episode of poylploidization. Thus, one cannot consider the evolution of land plants without understanding polyploidy. Fawcett et al. (Chap. 17) explain how the episodes of ancient polyploidization can be identified and dated and describe the immediate consequences of polyploidization to genes and genomes. They also discuss changes within polyploid genomes during evolution and the contribution of polyploidization to the evolutionary success of descendant lineages.

While the majority of studies on polyploidy have focused on angiosperms, Rensing et al. (Chap. 18) consider the importance of polyploidy in haploid-dominant land plants, the bryophytes. Here, polyploidy may play an even more essential evolutionary role than in other evolutionary lineages, rendering a bryophyte more robust against somatic mutations, while changes in chromosome number through polyploidy can lead to changes in the sexual system. The need for more genomic data and model species is paramount and the sequencing of the genome of the moss *Physcomitrella patens* together with the eagerly anticipated genome sequences from other moss species and the liverwort *Marchantia polymorpha* in the near future should shed further light on genome evolution and the role of polyploidy in these haploid-dominant land plants.

Genome Size Diversity and Consequences

The book closes with chapters that consider the whole genome in bulk. Leitch and Leitch (Chap. 19) take advantage of the recent increase in the number of species with genome size data and provide a comprehensive review on diversity of genome sizes across all groups of land plants. Evaluation of individual groups suggests that most plant genomes are rather small, probably due to strong selection pressure to limit genome size. Importantly, the chapter also considers how the diversity in genome size might have evolved. The last chapter of this volume by Greilhuber and Leitch (Chap. 20) examines the phenotypic correlates of variation in genome size, which include cell size and cell division rate. It also discusses the theories to explain the causality behind this variation observed, considers alternative views, and puts important studies into focus.

There is no doubt that it was an ambitious goal to cover the broad range of biological phenomena related to the structure, function, and evolution of plant genomes. However, we were motivated by the lack of a single resource, which is so needed in this era of rapid DNA sequence data generation. The chapters included in this volume deliver exciting facts from the history and life of plant genomes and present unanswered questions and hypotheses. We hope that the readers will find that the time spent with the book is both enjoyable and stimulating.

This volume would not exist without the contribution of the authors of individual chapters. Busy leaders in their areas of research, they spared precious time to share with us their knowledge and visions. We cannot be grateful enough for this and we appreciate their efforts and patience when responding to our requests for revisions. The only reward for them may be a response from the readers. So why not contact them? Sincere thanks go to the publisher, Springer-Verlag, Vienna and New York, who initiated and accompanied this project and made the publication of this volume possible. We appreciate the careful and professional management of the project.

RBG Kew, United Kingdom
Vienna, Austria
Olomouc, Czech Republic
Ames, Iowa, USA
March, 2012

Ilia. J. Leitch
Johann Greilhuber
Jaroslav Doležel
Jonathan F. Wendel

Contents

Contributors

Dr. Sarah J. Baldwin Department of Integrative Biology, University of Guelph, Guelph, ON, Canada

Michael S. Barker Department of Ecology & Evolutionary Biology, University of Arizona, Tucson, USA, msbarker@email.arizona.edu

Dr. Anna K. Beike Plant Biotechnology, Faculty of Biology, University of Freiburg, Freiburg, Germany

Dr. Pavla Binarová Institute of Microbiology, ASCR, Prague 4, Czech Republic

Prof. Laszlo Bogre Royal Holloway, University of London, Centre for Systems and Synthetic Biology, Egham, UK, l.bogre@rhul.ac.uk

Prof. RNDr. Petr Bureš Department of Botany and Zoology, Faculty of Science, Masaryk University, Brno, Czech Republic, bures@sci.muni.cz

Dr. Jeffrey A. Fawcett Graduate University for Advanced Studies, Hayama, Kanagawa, Japan

Dr. Jindriska Fiserova Department of Biological and Biomedical Sciences, Durham University, Durham, UK

Dr. Martin W. Goldberg Department of Biological and Biomedical Sciences, Durham University, Durham, UK

Prof. Johann Greilhuber Department of Systematic and Evolutionary Botany, Faculty of Life Sciences, University of Vienna, Vienna, Austria

Dr. Roman Hobza Institute of Biophysics, Academy of Sciences of the Czech Republic, Brno, Czech Republic

Dr. Andreas Houben Leibniz Institute of Plant Genetics and Crop Plant Research (IPK), Gatersleben, Germany, houben@ipk-gatersleben.de

Prof. Dr. Brian C. Husband Department of Integrative Biology, Science Complex, University of Guelph, Guelph, ON, Canada, bhusband@uoguelph.ca

Dr. Masaki Ito Graduate School of Bioagricultural Sciences, Nagoya University, Nagoya, Japan

Dr. Bohuslav Janoušek Institute of Biophysics, Academy of Sciences of the Czech Republic, Brno, Czech Republic

Dr. Eric Jenczewski Institut Jean-Pierre Bourgin, Institut National de Recherche Agronomique, cedex, France

Prof. R. Neil Jones Institute of Biological, Environmental and Rural Sciences, Aberystwyth University, Wales, UK, rnj@aber.ac.uk

Dr. Elena Kiseleva Laboratory of Morphology and Function of Cell Structure, Institute of Cytology and Genetics, Novosibirsk, Russia, elka@bionet.nsc.ru

Dr. Sonja Klemme Leibniz Institute of Plant Genetics and Crop Plant Research (IPK), Chromosome Structure and Function Laboratory, Gatersleben, Germany

Dr. Bozena Kolano Department of Plant Anatomy and Cytology, University of Silesia, Katowice, Poland

Dr. Daniel Lang Plant Biotechnology, Faculty of Biology, University of Freiburg, Freiburg, Germany

Dr. Tim Langdon Institute of Biological, Environmental and Rural Sciences, Aberystwyth University, Wales, UK

Prof. Andrew R. Leitch School of Biological and Chemical Sciences, Queen Mary, University of London, London, UK

Dr. Ilia J. Leitch Jodrell Laboratory Royal Botanic Gardens, Kew, Richmond, Surrey, UK, I.Leitch@kew.org

Dr. Martin Lysák Department of Experimental Biology, Faculty of Science, Masaryk University, Brno, Czech Republic, lysak@sci.muni.cz

Dr. Nicolas Macaisne Institut Jean-Pierre Bourgin, Institut National de Recherche Agronomique, Versailles, cedex, France

Prof. Steven Maere Department of Plant Systems Biology, VIB, Ghent, Belgium, stmae@psb.vib-ugent.be

Dr. Kazuhiro Maeshima Laboratory for Biological Macromolecules, Structural Biology Center, National Institute of Genetics, Mishima, Shizuoka, Japan, kmaeshim@lab.nig.ac.jp

Dr. Zoltán Magyar Institute of Plant Biology, Biological Research Centre, Szeged, Hungary

Prof. Jolanta Maluszynska Department of Plant Anatomy and Cytology, University of Silesia, Katowice, Poland, jolanta.maluszynska@us.edu.pl

Dr. Michaela Marková Department of Botany and Zoology, Faculty of Science, Masaryk University, Brno, Czech Republic

Dr. Sachihiro Matsunaga Department of Biotechnology, Graduate School of Engineering, Osaka University, Suita, Osaka, Japan

Dr. Raphael Mercier Institut Jean-Pierre Bourgin, Institut National de Recherche Agronomique, Versailles, cedex, France

Prof. Christine Mézard Station de Génétique et d'Amélioration des Plantes, Versailles, France, Christine.Mezard@versailles.inra.fr

Dr. Binish Mohamed Royal Holloway, University of London, Centre for Systems and Synthetic Biology, Egham, UK

Dr. Ali Mohammad Banaei Moghaddam Leibniz Institute of Plant Genetics and Crop Plant Research (IPK), Gatersleben, Germany

Prof. Brian G. Murray School of Biological Sciences, University of Auckland, Auckland, New Zealand, b.murray@auckland.ac.nz

Dr. Stefan A. Rensing FRISYS, Faculty of Biology, University of Freiburg, Freiburg, Germany, stefan.rensing@biologie.uni-freiburg.de

Prof. Ingo Schubert Leibniz Institute of Plant Genetics and Crop Plant Research (IPK), Gatersleben, Germany

Dr. Hanna Sas-Nowosielska Department of Plant Anatomy and Cytology, University of Silesia, Katowice, Poland

Prof. Gerald M. Schneeweiss Department of Systematic and Evolutionary Botany, Faculty Center Botany, University of Vienna, Vienna, Austria

Dr. Peter Shaw Cell and Developmental Biology Department, John Innes Centre, Norwich, UK, peter.shaw@bbsrc.ac.uk

Prof. Douglas E. Soltis Department of Biology and the Genetics Institute, University of Florida, Gainesville, FL, USA, dsoltis@botany.ufl.edu

Prof. Pamela S. Soltis Laboratory of Molecular Systematics and Evolutionary Genetics, Florida Museum of Natural History, University of Florida, Gainesville, FL, USA, psoltis@flmnh.ufl.edu

Prof. Jan Suda Department of Botany, Faculty of Science, Charles University in Prague, Prague, Czech Republic

Dr. Hideaki Takata Biological Macromolecules Laboratory, Structural Biology Center, National Institute of Genetics, Mishima, Shizuoka, Japan

Prof. Yves Van de Peer Department of Plant Systems Biology, VIB, Ghent, Belgium, yves.vandepeer@psb.vib-ugent.be

Prof. Boris Vyskot Institute of Biophysics, Czech Academy of Sciences, Laboratory of Plant Developmental Genetics, Brno, Czech Republic, vyskot@ibp.cz

Prof. Hanna Weiss-Schneeweiss Department of Systematic and Evolutionary Botany, Faculty Center Botany, University of Vienna, Vienna, Austria, Hanna.Weiss@univie.ac.at

Dr. František Zedek Department of Botany and Zoology, Faculty of Science, Masaryk University, Brno, Czech Republic

Angiosperm Phylogeny: A Framework for Studies of Genome Evolution

1

Pamela S. Soltis and Douglas E. Soltis

Contents

1.1 Introduction

The angiosperms—or flowering plants—comprise an estimated 260,000 (Takhtajan 1997)–400,000 (Raven in Jarvis 2007) extant species and occupy nearly all habitats on Earth except the coldest arctic and polar regions and the deepest oceans. Their diversification has occurred over a relatively short timespan, with the fossil record placing the earliest angiosperms in the early Cretaceous, approximately 132 million years ago. Molecular clock estimates suggest that the angiosperms are perhaps older, dating to the Jurassic (e.g., Sanderson et al. 2004; Bell et al. 2005; Bell et al. 2010) or even the Triassic (Magallon 2010; Smith et al. 2010).

Our understanding of the phylogeny of angiosperms has improved dramatically in recent years through large-scale collaborative analyses (e.g., Chase et al. 1993; Soltis et al. 1999; Soltis et al. 2000; Hilu et al. 2003; Soltis et al. 2011) and the application of molecular data, from single genes to entire plastid genomes (e.g., Jansen et al. 2007; Moore et al. 2007; Moore et al. 2010). Likewise, many clade-specific analyses have clarified relationships within some of the largest groups of angiosperms: e.g., *Monocotyledoneae* (monocots *sensu* Cantino et al. 2007; subsequent italicized names refer to phylogenetically defined clades in Cantino et al. 2007), Chase et al. (2006); Caryophyllales, Brockington et al. (2009); *Eudicotyledoneae* (eudicots), Moore et al. (2010); *Campanulideae* (campanulids), Tank and Donoghue (2010). In less than 20 years time, our view of angiosperm phylogeny has been transformed from a nebulous series of possible transitions to a well-supported and well-resolved tree of explicit sister-group relationships (summarized in Fig. 1.1). The stability of this tree is reflected in the modest changes to the classification of the Angiosperm Phylogeny Group (APG) over the past decade (1998, 2003, 2009; summarized by Stevens 2001 onward) and in the development of a phylogenetic nomenclature for angiosperms (Cantino et al. 2007).

P.S. Soltis (✉)
Florida Museum of Natural History and the Genetics Institute, University of Florida, Gainesville, FL 32611, USA
e-mail: psoltis@flmnh.ufl.edu

I.J. Leitch et al. (eds.), *Plant Genome Diversity Volume 2*,
DOI 10.1007/978-3-7091-1160-4_1, © Springer-Verlag Wien 2013

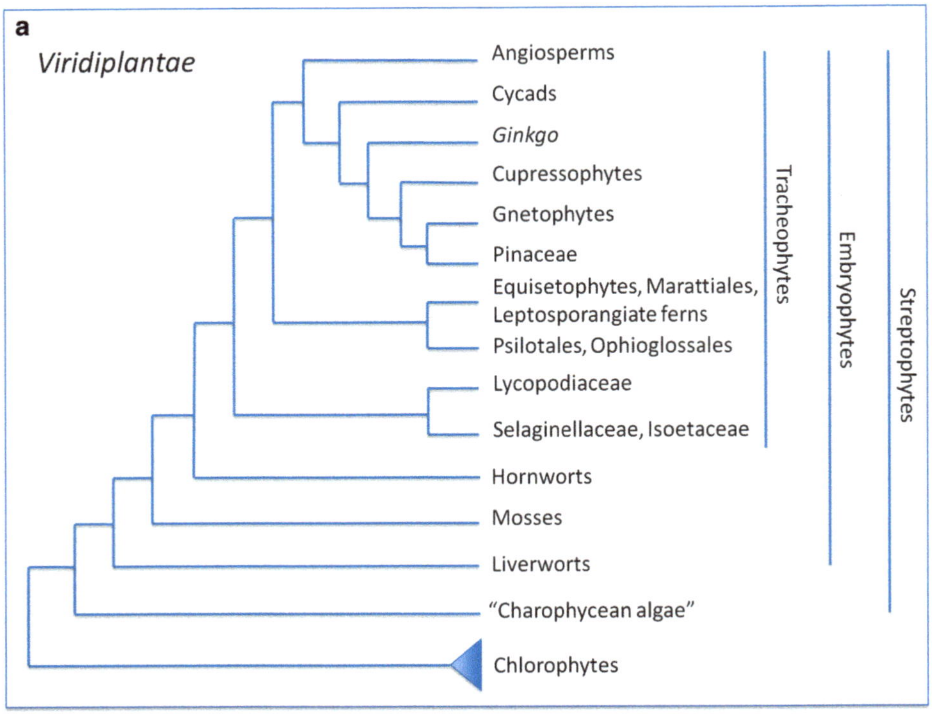

Fig. 1.1 (continued)

Phylogenetic trees of angiosperms have been used to address a range of evolutionary and ecological questions, such as the causes of diversification (Davies et al. 2004), the evolution of reproductive systems (e.g., Culley et al. 2002), the evolution of syncarpy and its role in pollination (Armbruster et al. 2002), and the relationship among phylogeny, biogeography, and biodiversity (Donoghue 2008). In addition, trees for which internal nodes have been dated (e.g., Wikstrom et al. 2001; Bell et al. 2005; Bell et al. 2010) have supplied a framework for many additional studies (Slingsby and Verboom 2006; Vamosi et al. 2006; Edwards et al. 2007; Webb et al. 2008).

Recent advances in angiosperm phylogenetics have also played a significant role in selecting taxa for genetic analysis and genome sequencing (e.g., Pryer et al. 2002; Soltis et al. 2008). For example, studies of gene family evolution have focused on representatives of basal angiosperm clades, basal eudicots, and selected monocots and eudicots (e.g., Kim et al. 2004; Kim et al. 2005; Zahn et al. 2005) to investigate patterns of gene duplication and loss. The results yield complex patterns of gene family dynamics—patterns that are not apparent through analysis of model systems alone. Likewise, genomic resources (e.g., BAC libraries) have been developed for a set of phylogenetically important plant species in anticipation of eventual genomic analysis and sequencing. Most recently, genome sequencing of *Amborella trichopoda*, the sister to all other extant angiosperms (e.g., Soltis et al. 1999; Soltis et al. 2000; Hilu et al. 2003; Leebens-Mack et al. 2005; Jansen et al. 2007), has been initiated, to provide an evolutionary reference for genome analysis within the angiosperms and across all green plants (Soltis et al. 2008; Chamala et al. 2011). The genome sequence of *Aquilegia* of Ranunculales, the sister to all other eudicots, will similarly provide an evolutionary reference for eudicots and a further point of comparison among the genomes of *Amborella*, monocots, and model eudicots.

Here we provide an overview of plant phylogeny, with an emphasis on angiosperms, based on the past two decades of research, to serve as the basis for investigating patterns of genome evolution. We give a summary, as well as many original citations, with an emphasis on those analyses that have deposited trees in public databases, such as TreeBASE, where they are available for download and analysis.

1.2 Methods of Phylogenetic Analysis: A Primer

The development of phylogenetic methods during the past decade has produced a perhaps baffling array of approaches, algorithms, and software. The state of the art a mere decade ago was maximum parsimony, with numerous options, e.g., TNT (Goloboff 1999), parsimony ratchet (Nixon 1999), to allow for analysis of perhaps several hundred taxa to a few thousand (Kallersjo et al. 1999) and one or a handful of genes. Concerns that sufficient tree space was searched were paramount, given the restrictions in memory and speed of most computers at the time. Maximum likelihood analyses were possible for only tens of taxa. In the early 2000s, major shifts occurred to model-based approaches as Bayesian methods (MrBayes, Huelsenbeck and Ronquist 2001; Huelsenbeck et al. 2001; and then BEAST, Drummond and Rambaut

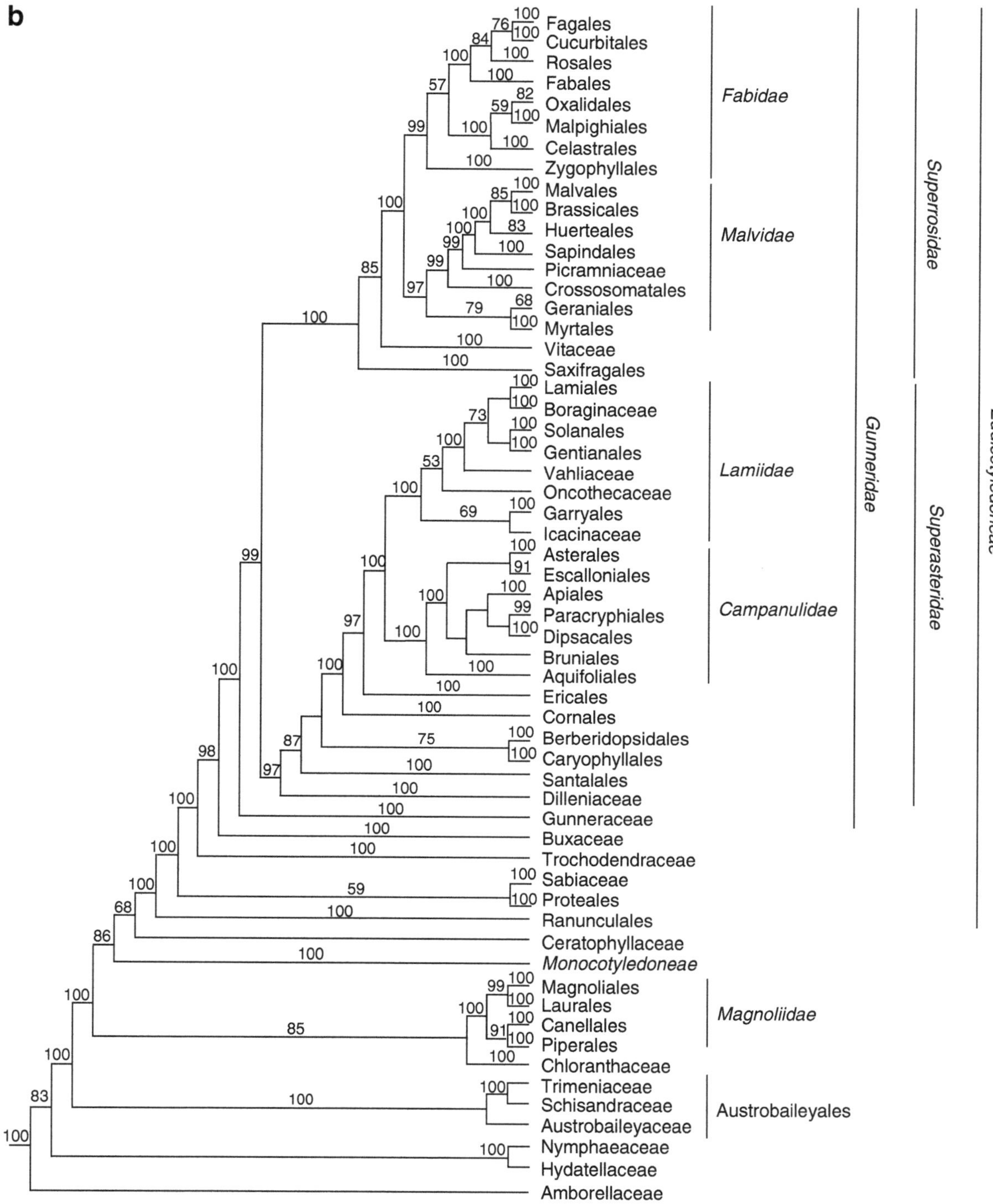

Fig. 1.1 Summary of phylogenetic relationships among major clades of green plants (*Viridiplantae*). (**a**) Overview, based on consensus of many studies. (**b**) Overview of angiosperm phylogeny, based on maximum likelihood analysis, with bootstrap values, redrawn from Soltis et al. (2011)

2007), along with maximum likelihood approaches using new algorithms (e.g., genetic algorithm, GARLI, Zwickl 2006; RAxML, Stamatakis 2006; Stamatakis et al. 2008), made it possible to reconstruct large trees (hundreds of taxa) with confidence scores (posterior probabilities or bootstrap values, respectively). Parallelization has helped to reduce run times dramatically for large problems, but has not been universally implemented to date (although dividing bootstrap analyses among an array of processors in a cluster is a form of parallel analysis that can considerably shorten run times).

Our assessment is that most projects today employ both parsimony and at least one model-based approach (typically RAxML or MrBayes).

Whereas many analyses of phylogeny reconstruction have embraced model-based methods, most analyses of character evolution continue to rely on parsimony, despite the implementation of both maximum likelihood and Bayesian methods for inferring ancestral states and mapping character variation. Although the reason for this bias is unclear, it may be that researchers are more comfortable applying statistical methods to tree selection than to character mapping, in which parsimony has an intuitive appeal. Although as in tree selection, likelihood, Bayesian, and parsimony methods typically produce similar patterns of character evolution, parsimony results may differ from likelihood and Bayesian reconstructions, particularly when branch lengths are short. We encourage expanded use of likelihood and/or Bayesian methods for character reconstructions, at least for comparison with parsimony results.

Most analyses of plant phylogeny to date have focused on plastid genes, with an emphasis at deep levels on *rbc*L, *atp*B, *ndh*F, and to some extent *mat*K. The former two have similar rates of evolution and are easily alignable, *ndh*F tends to evolve slightly more rapidly and is longer (although only part of the gene is sometimes used), and *mat*K has a higher rate of both nucleotide substitution and indels, leading to more difficult alignment. Of course, plastid genes provide only the evolutionary history of the plastid, and although this may not be a concern at the deepest levels of plant phylogeny, one must consider how well a plastid gene tree may reflect the organismal tree. To date, at deep levels, the only nuclear genes that have been used widely are the 18S and 26S ribosomal RNA genes. A number of MADS-box genes have been shown to track angiosperm phylogeny (Litt and Irish 2003; Kim et al. 2004; Kim et al. 2005; Zahn et al. 2005), but these genes have not yet been applied solely for the purpose of phylogeny reconstrution. However, a number of other nuclear genes (or their introns) have been used at more shallow levels: *LEAFY, APETALA3, PISTILLATA, ALCOHOL DEHYDROGENASE, GLYCERALDEHYDE 3-PHOSPHATE DEHYDROGENASE, CHALCONE SYNTHASE*, and *WAXY*, to name a few. All of this latter set of genes tend to have regions of 1,000 bp (plus or minus a few hundred) and are generally fairly easy to amplify with standard primers. However, the use of nuclear genes carries its own concerns, most notably issues of orthology, allelic diversity, and recombination, often requiring extensive cloning, sequencing of clones, and analyses of recombination prior to phylogenetic analysis. A set of mitochondrial genes (e.g., *mat*R, *atp*1, *nad*5, *rps*3) has also been applied to plant phylogeny. These genes tend to evolve more slowly than either plastid or nuclear genes used to date and can supply characters that are useful deep in plant phylogeny. However, mitochondrial-based trees have shown evidence of horizontal transfer of mitochondrial genes (e.g., Won and Renner 2003; Bergthorsson et al. 2004; Davis and Wurdack 2004) and should therefore be used in conjunction with other markers, particularly in groups that contain parasites.

1.3 The Phylogeny of Embryophytes: An Abbreviated Overview

A thorough summary of plant phylogeny is beyond the scope of this chapter; instead, we present a simple overview to set the stage for further discussion of genome evolution in plants and the phylogenetic placement of angiosperms, the focus of this chapter. Although important for understanding major patterns of plant evolution, especially in morphological and anatomical characters, the fossil record plays a less crucial role in understanding genome evolution, and we have therefore largely confined our discussion of plant phylogeny to extant taxa. However, the fossil record provides a requisite timeframe on our interpretation of phylogeny, and the phylogenetic placement of fossil groups may affect the topology of extant groups; we therefore introduce data from fossils as needed in this brief overview but we recognize that our treatment is incomplete.

The "green plants" (sometimes referred to as *Viridiplantae* or viridophytes) are a clade of at least half a million species with a fossil record that extends back nearly one billion years. They share a common cyanobacterial endosymbiotic event with red algae and glaucophytes and can be diagnosed by chlorophyll *b*, starch as the storage product for photosynthesis, and a stellate flagellar structure (e.g., Judd et al. 2008). A basal split in the green plants produced two clades, the chlorophytes (mostly marine "green algae") and streptophytes (which include freshwater "green algae" and embryophytes). *Mesostigma*, a freshwater "alga", has been identified as the sister to all other streptophytes. Subsequently branching lineages include the Klebsormidiales, the Zygnematales, and the Coleochaetales and Charales, the limits of which are not completely clear. *Chara* (and relatives) and Coleochaetales seem to be the sister group(s) of the embryophytes, although some recent studies place Zygnematales in this position (e.g., Timme et al. 2012). Embryophytes, or land plants, trace their history to at least the Ordovician and began to diversify extensively in the Silurian and Devonian. Morphological and anatomical synapomorphies of the embryophytes are a multicellular sporangium, thick-walled spores, multicellular gametangia, an embryo, and a cuticle.

Within the embryophytes, the phylogeny of the major clades is not fully resolved (Fig. 1.1a). For example, although traditionally recognized as a single taxonomic group, the bryophytes (consisting of mosses, liverworts, and hornworts) are paraphyletic, and the branching order of these clades relative to the tracheophytes is not yet clear. All possible branching orders have been proposed. Most morphological and molecular data support liverworts as sister to all other embryophytes, and the major remaining disagreement is between the topology of (liverworts, (hornworts, (mosses + tracheophytes))), which is supported by the shared feature of a sporophyte apical meristem in mosses and tracheophytes, and (liverworts, (mosses, (hornworts + tracheophytes))), which is supported by the persistently green sporophyte in hornworts and tracheophytes (reviewed in Judd et al. 2008).

The tracheophytes (*Tracheophyta*) comprise two major clades, lycophytes (*Lycopodiophyta*) and euphyllophytes (*Euphyllophyta*). *Lycopodiophyta* comprises *Isoetes*, *Selaginella*, and Lycopodiaceae and forms a clade with some of the most prominent early vascular plants: *Cooksonia*, zosterophytes, and Lepidodendrales (Judd et al. 2008). Extant *Euphyllophyta* is united by a plastid genome inversion, multicellular sperm, overtopping, and terminal sporangia on lateral branches. The clade contains two major clades, *Monilophyta* and *Lignophyta*. The former is composed of Psilotales, Ophioglossales, Equisetales, Marattiales, and the leptosporangiate ferns (*Leptosporangiatae*). This assemblage of monilophytes was recognized by Kenrick and Crane (1997) on the basis of stem anatomy and was later supported by molecular data as well (Pryer et al. 2001; Pryer et al. 2004). The *Lignophyta* comprises several fossil lineages and the *Spermatophyta*, the seed plants.

Relationships among seed plants perhaps remain the most challenging in plant phylogeny. Extensive extinction within this clade has undoubtedly contributed to the difficulty of phylogeny reconstruction. The "gymnosperms" as typically recognized are paraphyletic and include several clades with extant members (cycads, *Ginkgo*, conifers, and gnetophytes) and several other groups that are only found in the fossil record (*Medullosa*, seed ferns, glossopterids, *Caytonia*, Bennetittales). However, the paraphyly of the "gymnosperms" is not apparent in molecular-based trees that typically (but not always) recover reciprocally monophyletic gymnosperms and angiosperms. When fossils are included in phylogenetic analyses of seed plants, disagreements exist with regard to both the placement of many of the non-flowering seed plants and in the sister group of the angiosperms, and there is little consensus on the overall phylogeny of all seed plants. It appears that glossopterids, *Caytonia*, and Bennetittales are more closely related to angiosperms than to other "gymnosperms" but beyond that, there is little resolution. When only extant seed plants are considered, disagreement still abounds. Most recent studies have found a topology in which extant gymnosperms and angiosperms are sister groups, with cycads either sister to all other extant gymnosperms or sister to *Ginkgo*, and with conifers and gnetophytes either sister to each other, or more often, gnetophytes nested within conifers or within Pinaceae. Despite extensive study using many sources of data and modes of analysis, the phylogenetic relationships among extant seed plants remain unresolved. For analyses that seek to examine patterns of evolution among angiosperms, this frustrating result precludes identification of the appropriate outgroup for comparative studies.

1.4 The Phylogeny of Angiosperms: An Overview

1.4.1 Major Clades

Angiosperm phylogeny has been studied extensively in recent decades, from the perspective of deep-level branching patterns to clades of closely related species. Ultimately and ideally, these results will be linked, either through supertree methods that combine published trees via shared taxa or through new supermatrix analyses that combine all data into a single matrix for analysis (see below). Here we will provide only an overview of the major clades and their interrelationships, followed by further discussion on some of the emergent patterns from analyses conducted to date.

Nearly all molecular-based analyses of the past decade have identified *Amborella* as the sister to all other extant angiosperms, most often alone or occasionally with Nymphaeales (Fig. 1.1b). All analyses are consistent in then placing Austrobaileyales (comprising *Austobaileya*, *Trimenia*, *Illicium*, and Schisandraceae) as the sister group to all other extant angiosperms. This large remainder, the *Mesangiospermae*, comprises *Magnoliidae* + Chloranthaceae as sister to *Monocotyledoneae* + *Eudicotyledoneae* + *Ceratophyllum* (Moore et al. 2007). Although long recognized as an ancient group, the placement of Chloranthaceae has been elusive, but recent analyses place them as sister to *Magnoliidae*. The relationship among magnoliids, monocots, and eudicots has been very difficult to disentangle, but plastid genome sequences support the sister-group relationship of monocots and eudicots (+ *Ceratophyllum*).

Relationships among major clades of monocots are now clear, but they do not follow traditional taxonomic circumscriptions. Acorales are sister to all other extant monocots, and a grade that includes Alismatales, followed by Petrosaviaceae, subtends a clade comprising the majority of monocot species diversity. Pandanales + Dioscoreales are sister to a clade of (Liliales, (Asparagales + *Commelinidae*)). One of the most substantial reorganizations of monocot classification is based on new understanding of relationships of the former Liliaceae. Although dismantling of this large family was proposed many years ago, the placements of its components have not always been clear. Progress has been substantial, but questions remain. Likewise, relationships within Asparagales have been difficult to resolve, and although recent analyses have done much to resolve phylogeny, few morphological characters have been idenitifed to diagnose the component clades. The *Commelinidae* are diagnosed by starchy pollen, UV-fluorescent ferulic and coumaric acids in the cell walls, and *Strelitzia*-type epicuticular wax. The clade is large, comprising over 25,000 species, with diversity spanning grasses to palms. Component clades are Commelinales, Zingiberales, Arecales, and Poales (+ Dasypogonaceae).

Within eudicots, a basal grade consisting of Ranunculales, Proteales, Sabiales, Trochodendrales, and Buxales subtends the "core eudicots", or *Gunneridae*. Gunnerales are sister to the remaining *Gunneridae*, the *Pentapetalae*, which fall into two major clades (Moore et al. 2010; Soltis et al. 2011): *Superrosidae* and *Superasteridae* (*sensu* Soltis et al. 2011). *Superrosidae* comprises Saxifragales, Vitaceae, and *Rosidae*, whereas *Superasteridae* contains Santalales, Berberidopsidales, Caryophyllales, and *Asteridae*. Dilleniaceae, which has been associated with Caryophyllales in some previous analyses and unplaced in many others, remains unplaced, with alternative placements in *Superrosidae* and *Superasteridae* (see below). The positions of most major clades of core eudicots (*Gunneridae*) remain unchanged relative to earlier studies (e.g., Soltis et al. 1999; Soltis et al. 2000; Soltis et al. 2005), although additional resolution has been obtained in both the *Rosidae* and *Asteridae* (see Soltis et al. 2011, and references therein).

Phylogenetic analyses of the angiosperms at these deep levels have resulted in new classifications, such as the Angiosperm Phylogeny Group's (APG) (1998, 2003, 2009) classifications at the familial and ordinal levels, with rank-free names assigned to clades corresponding to groups larger than recognized orders. An alternative rank-free classification has also emerged (Cantino et al. 2007), with phylogenetic definitions provided for many of the clades that correspond to those recognized at the ordinal level and above in the APG system.

1.4.2 Repeated Radiations

A prominent pattern apparent in the angiosperm phylogenetic trees is a series of polytomies interspersed by regions of well-resolved relationships (see Soltis et al. 2005; Soltis et al. 2008; Wang et al. 2009; Soltis et al. 2010). Although polytomies may be due to insufficient data or taxon sampling, they may also represent real radiations. Within the angiosperms, several apparent radiations have persisted through the addition of new data and more taxa, suggesting that in fact these radiations are real.

The origin of the angiosperms themselves has often been considered a rapid radiation, based on the fossil record and Darwin's words themselves: *"The rapid rise and early diversification of angiosperms is an abominable mystery. . ."* (Darwin 1903). However, phylogenetic reconstructions suggest instead that the angiosperms radiated, not immediately upon their origin, but a few nodes subsequent to the common ancestor of all extant angiosperms (Mathews and Donoghue 1999; Soltis et al. 1999; Soltis et al. 2005). This radiation corresponds to the diversification of the *Mesangiospermae* (*sensu* Cantino et al. 2007), the clade comprising magnoliids, Chloranthaceae, monocots, Ceratophyllaceae, and eudicots—in other words, all angiosperms except *Amborella*, Nymphaeales, and Austrobaileyales (see Moore et al. 2007). Subsequent radiations appear to follow the origin and early diversification of many large clades of angiosperms: for example, within the eudicots (*Eudicotyledoneae*), within the core eudicots (*Gunneridae*), within *Rosidae* (and within the fabid and malvid clades, *Fabidae* and *Malvidae*, of *Rosidae*), within the *Asteridae*, and within clades of monocots, to name a few.

Attempts to find causes for these apparent radiations have met with mixed success. One of the most extensive analyses of possible factors associated with radiations addressed both the early radiation of the angiosperms themselves and subsequent radiations (Davies et al. 2004). Davies et al. (2004) tested a range of traits reflecting prominent hypotheses for the "success" of the angiosperms on rates of diversification and found no significant association at any level of the tree. Despite the lack of significance of specific features, radiations within the angiosperms may be explained by biotic factors and cospeciation with other clades. For example, modern ferns diversified alongside angiosperms, suggesting either that angiosperms provided new habitats for ferns or that the same causal factors allowed diversification of both clades (Schneider et al. 2004). Moreover, within the angiosperms, the radiation of the rosids is associated with radiations in several other clades, such as ants, amphibians, and even primates (see Wang et al. 2009, for review). Other possible biotic interactions, such as those with mycorrhizal fungi, may also have contributed to radiations of angiosperms.

Recent observations of gene duplications at or near nodes associated with radiations are suggestive of a causal role of genetic or genomic factors in these radiations themselves. For example, coincident gene duplications in multiple subfamilies of the MADS-box gene family prior to the origin of the angiosperms raised hypotheses about the role that these duplications in genes important in the specification of floral organ identity and other features of the flower may have played in the early evolution of angiosperms (Buzgo et al. 2005; De Bodt et al. 2005; Zahn et al. 2005). Likewise, similar patterns of duplication appear to be associated with the early evolution of the eudicots, again suggesting a causal role in the floral changes that occurred at that point in angiosperm phylogeny and a further role in diversification (for reviews see Soltis et al. 2006; Soltis et al. 2009b). Finally, duplications in the *CYCLOIDEA* gene family suggest possible roles in both floral and species diversification in asterids (Howarth and Donoghue 2006).

Coincident gene duplications at specific nodes are suggestive of whole-genome duplications (WGD; see Buzgo et al. 2005; De Bodt et al. 2005; Zahn et al. 2005; reviewed in Soltis et al. 2009a; Soltis et al. 2009b), and it may be that WGD rather than duplications of specific floral genes triggered radiation. Whole-genome duplication (polyploidy) has, in fact, been suggested as the impetus for angiosperm success following the K-T boundary (Fawcett et al. 2009). Genomic data have revealed unsuspected episodes of WGD throughout green plants (see below; Blanc and Wolfe 2004; Cui et al. 2006; Soltis et al. 2009a; Jiao et al. 2011 for review). In several instances, the clade marked by WGD is more species-rich than the sister clade that lacks the duplication; however, genomic data are lacking for a sufficient number of species to allow for thorough statistical analyses of heterogeneity of diversification rates (Soltis et al. 2009a). Additional data for more species, so that WGD events can be plotted more accurately on a phylogenetic tree, are needed.

1.4.3 Unresolved Relationships

Given the recent progress in angiosperm phylogenetics, few major issues of deep-level relationships remain, although problems abound within clades recognized as "orders" and "families" *sensu* APG. Among deep-level problems, one of the most perplexing placements is that of Dilleniaceae, which occupies different positions depending on the data set and analysis, from sister to *Superrosidae* (*sensu* Soltis et al. 2011), to sister to *Superasteridae* (*sensu* Soltis et al. 2011), to sister to *Superasteridae* + *Superrosidae* (see Soltis et al. 2011, for discussion). Other prominent areas requiring further analysis include: (1) the branching order among basal eudicots; (2) relationships among major clades of rosids; (3) relationships within Malpighiales; (4) Lamiales. Most of these regions can most likely be resolved with additional taxa and DNA sequence data, although problem areas such as Malpighiales have recently received substantial attention, with at least some progress (Wurdack and Davis 2009).

1.4.4 "Big Trees"

Until recently, most tree reconstruction algorithms could not handle data sets of 1,000 or more terminals. However, recent modifications have yielded trees with many thousands of species (e.g., Goloboff et al. 2009; Smith et al. 2009; Smith et al. 2011). The most recent tree generated by Smith et al. (2011)—with 55,000 taxa!—used a newly adapted version of RAxML, demonstrating the ability to use model-based approaches for "big tree" reconstruction. The concern with such large trees, however, is their accuracy: with so many terminals, the thoroughness of the searches is reduced, raising the question of the accuracy of the results. The overall structure of the Smith et al. (2011) 55,000-taxon tree is quite similar to trees based on far fewer taxa (such as the 640-taxon tree of Soltis et al. 2011), suggesting that for many purposes, this very large tree will be very useful. However, without further diagnostics on the performance of tree reconstruction at this large scale, many close relationships may require cautious acceptance. And this may be an issue for studies aimed at reconstructing the evolution of genes or specific genomic traits at a fine scale. Nevertheless, breakthroughs in tree reconstruction will undoubtedly lead to new and exciting opportunities for learning about the evolution of plant genomes.

An alternative to the supermatrix approach described above is the construction of a supertree from smaller trees with overlapping taxa (see Sanderson et al. 1998; Davies et al. 2004). Supertree methods take advantage of vast amounts of data collected and analysed in the past to produce a summary of phylogenetic inferences contained in published trees. Although not without their own problems, supertrees offer a solution to generating large phylogenies. Recent methods that combine elements of supermatrix and supertree approaches show particular promise.

1.5 Studies of Genome Evolution in Angiosperms

Hypotheses on patterns of chromosomal evolution in the angiosperms abound. Classical perspectives were based on integrated inferences on ancient and recent polyploidy, processes of chromosomal fission and fusion, and relative "advancement" of a taxonomic group. More recently, it has been possible to test hypotheses of increases and decreases in chromosome number and genome size by mapping these characteristics across an explicit phylogenetic tree. Longstanding hypotheses of chromosomal evolution have suggested that the ancestral chromosome number for angiosperms ranged from $x = 6$–9, with $x = 7$ a commonly proposed base number (e.g., Stebbins 1950; Ehrendorfer et al. 1968; Stebbins 1971; Raven 1975; Grant 1981). Chromosome numbers in angiosperms vary dramatically, from $2n = 4$ (e.g., *Haplopappus gracilis*, Asteraceae) to $2n =$ c. 640 (*Sedum suaveolens*, Crassulaceae), a 160-fold difference (Uhl 1978). However, it is clear that genome size varies independently of chromosome number, with genome size ranging from 1C = 0.065 pg to 1C = 152.23 pg, a c. 2,400-fold difference (Greilhuber et al. 2006; Pellicer et al. 2010; Leitch and Leitch, 2013 this volume). Previous reconstructions of genome size across angiosperms found that the ancestral genome was "very small", with 1C $\leq$ 1.4 pg (Leitch et al. 1998; Soltis et al. 2003), with multiple increases and decreases in genome size from this ancestral condition.

Another clear attribute of angiosperm genomes is polyploidy, and events of whole-genome duplication have also been mapped across an angiosperm tree (Soltis et al. 2009a). Whereas polyploidy has long been recognized as an important speciation mechanism in angiosperms, events of genome duplication were long considered to be confined to the tips of the tree, with a few putative cases of ancient polyploidy, e.g., Magnoliaceae, Lauraceae, Salicaceae (Stebbins 1950; Stebbins 1971). Remarkably, genome sequencing studies have revealed multiple rounds of genome duplication throughout the evolutionary history of angiosperms. The surprising finding that the very small genome of *Arabidopsis thaliana* has undergone multiple rounds of duplication set the stage for investigations of other unsuspected events of genome duplication, and all other angiosperms sequenced to date likewise exhibit signatures of ancient duplication (see Soltis et al. 2009a; Fawcett et al. 2013, this volume). As more genome sequences are generated, it will be possible to map these duplication events more clearly onto a phylogenetic tree. From there, hypotheses of possible causal effects of genome duplication may be tested.

1.6 Future Prospects

1.6.1 New Scope, New Tools

Despite the progress in reconstructing the major relationships within the angiosperms, much remains to be done to produce a comprehensive phylogeny. One approach to increasing taxon sampling is to include all species that are represented in GenBank, regardless of gene. Methods such as those explored by Driskell et al. (2004) have proven that even very sparse matrices can yield reasonable phylogenetic trees. A modification of this approach that includes only species for which any of a small, specified set of genes has been sequenced (Smith et al. 2009; Smith et al. 2011) has resulted in a 55,000-taxon tree, generated using maximum likelihood (via RAxML, Stamatakis 2006), that has been used to address the evolution of various traits (Smith et al. 2011). The ability to generate trees of this size is a recent breakthrough and sets the stage for future large-scale analyses. Whereas until recently, the limitation in phylogenetics was computational power, the bottleneck very soon will be a lack of data. Tools being developed by the iPlant Collaborative (iplantcollaborative.org) will soon be available for large-scale tree reconstruction and post-tree analyses.

An alternative approach to using data fortuitously available from GenBank is the deliberate generation of new data for taxa not represented in GenBank. For example, only a fraction of the estimated 15,000 genera of angiosperms are included in GenBank. Although most genera are likely not monophyletic (Judd et al. 2008), generic-level classification reflects a loose assessment of diversity and therefore a framework for ongoing phylogenetic analysis. However, substantial specimen collecting, including samples for DNA analysis, is needed to conduct such a study. Thus, given new developments in phylogenetic software, perhaps the major limitation to a comprehensive angiosperm phylogenetic tree is the lack of material for molecular analysis.

1.6.2 Improved Access to Data, Trees, and Tools

Until now, most phylogenetic reconstructions of morphological, physiological, ecological, or other characters have involved, initially, sharing of trees and data sets by systematists and, more recently, downloading published trees and data from public databases such as TreeBASE (treebase.org) and Dryad (datadryad.org). These analyses have been necessarily limited to those taxa included in prior phylogenetic trees, without representation of those taxa that might be of greater interest from the perspective of morphology or other traits. A solution is to reconstruct a new tree that includes such taxa, but large phylogenetic analyses may not be feasible for those interested in reconstructing patterns of character evolution. New approaches, such as an automatically generated tree with each new GenBank release (every 2 months) and the availability of data matrices and computational resources for tree estimation, are being developed by the iPlant Collaborative. Implementation of these tools will facilitate customized phylogenetic analyses and lead to "democratization" of angiosperm phylogenetics. Continued investment in phylogenetic cyberinfrastructure to address such issues as reticulation, horizontal gene transfer, and patterns of character evolution will lead to further advancements in our understanding of angiosperm evolution.

Acknowledgments This work was supported in part by the US National Science Foundation (grants EF-0431266 and PGR-0638595) and the NSF-funded iPlant Collaborative.

References

Angiosperm Phylogeny Group (1998) An ordinal classification for the families of flowering plants. Ann Mo Bot Gard 85:531–553

Angiosperm Phylogeny Group (2003) An update of the angiosperm phylogeny group classification for the orders and families of flowering plants. Bot J Linn Soc 141:399–436

APG, III [The Angiosperm Phylogeny Group] (2009) An update of the angiosperm phylogeny group classification for the orders and families of flowering plants: APG III. Bot J Linn Soc 161:105–121

Armbruster WS, Debevec EM, Willson MF (2002) Evolution of syncarpy in angiosperms: theoretical and phylogenetic analyses of the effects of carpel fusion on offspring quantity and quality. J Evol Biol 15:657–672

Bell CD, Soltis DE, Soltis PS (2005) The age of the angiosperms: a molecular time-scale without a clock. Evolution 59:1245–1258

Bell CD, Soltis DE, Soltis PS (2010) The age and diversification of the angiosperms re-revisited. Am J Bot 97:1296–1313

Bergthorsson U, Richardson AO, Young GJ, Goertzen LR, Palmer JD (2004) Massive horizontal transfer of mitochondrial genes from diverse land plant donors to the basal angiosperm *Amborella*. Proc Natl Acad Sci USA 101:17747–17752

Blanc G, Wolfe KH (2004) Widespread paleopolyploidy in model plant species inferred from age distributions of duplicate genes. Plant Cell 16:1667–1678

Brockington SF, Alexandre R, Ramdial J, Moore MJ, Crawley S, Dhingra A, Hilu K, Soltis DE, Soltis PS (2009) Phylogeny of Caryophyllales and patterns of floral evolution. Int J Plant Sci 170:627–643

Buzgo M, Soltis PS, Kim S, Soltis DE (2005) The making of the flower. Biologist 52:149–154

Cantino P, Doyle J, Graham S, Judd W, Olmstead R, Soltis D, Soltis P, Donoghue M (2007) Towards a phylogenetic nomenclature of Tracheophyta. Taxon 56:822–846

Chamala S, Walts B, Albert V, dePamphilis C, Der J, Estill J, Leebens-Mack J, Lee S, Ma H, Rounsley S, Schuster S, Soltis D, Soltis P, Tomsho L, Wessler S, Wing R, Yu Y, Barbazuk B (2011) The *Amborella* genome project: generating a reference sequence for angiosperm evolutionary analysis. PAG abstract

Chase MW, Soltis DE, Olmstead RG, Morgan D, Les DH, Mishler BD, Duvall MR, Price RA, Hills HG, Qiu Y-L et al (1993) Phylogenetics of seed plants: an analysis of nucleotide sequences from the plastid gene *rbc*L. Ann Mo Bot Gard 80:528–580

Chase MW, Fay MF, Devey D, Maurin O, Rønsted N, Davies J, Pillon Y, Petersen G, Seberg O, Tamura MN, Asmussen CB, Hilu K, Borsch T, Davis JI, Stevenson DW, Pires JC, Givnish TJ, Sytsma KJ, McPherson MA, Graham SW, Rai HS (2006) Multigene analyses of monocot relationships: a summary. In: Columbus JT, Friar EA, Porter JM, Prince LM, Simpson MG (eds) Monocots: comparative biology and evolution (excluding Poales), vol 22. Aliso, pp 63–75

Cui L, Wall PK, Leebens-Mack J, Lindsay BG, Soltis D, Doyle JJ, Soltis P, Carlson J, Arumuganathan A, Barakat A, Albert V, Ma H, dePamphilis CW (2006) Widespread genome duplications throughout the history of flowering plants. Genome Res 16:738–749

Culley TM, Weller SG, Sakai AK (2002) The evolution of wind pollination in angiosperms. Trends Ecol Evol 17:361–369

Darwin C (1903) Letter to J. D. Hooker. In: Darwin F, Seward AC (eds) More letters of Charles Darwin, vol 2. John Murray, London

Davies TJ, Barraclough TG, Chase MW, Soltis PS, Soltis DE, Savolainen V (2004) Darwin's abominable mystery: insights from a supertree of the angiosperms. Proc Natl Acad Sci USA 101:1904–1909

Davis CC, Wurdack KJ (2004) Host-to-parasite gene transfer in flowering plants: phylogenetic evidence from Malpighiales. Science 305:676–678

De Bodt S, Maere S, Van de Peer Y (2005) Genome duplication and the origin of angiosperms. Trends Ecol Evol 20:591–597

Donoghue MJ (2008) A phylogenetic perspective on the distribution of plant diversity. Proc Natl Acad Sci USA 105(suppl 1):11549–11555

Driskell AC, Ané C, Burleigh JG, McMahon MM, O'Meara BC, Sanderson MJ (2004) Phylogenetic utility of large sequence databases for building the tree of life. Science 306:1172–1174

Drummond AJ, Rambaut A (2007) BEAST: Bayesian evolutionary analysis by sampling trees. BMC Evol Biol 7:214

Edwards EJ, Stills CJ, Donoghue MJ (2007) The relevance of phylogeny to the studies of global change. Trends Ecol Evol 22:243–249

Ehrendorfer F, Krendl F, Habeler E, Sauer W (1968) Chromosome numbers and evolution in primitive angiosperms. Taxon 17:337–468

Fawcett JA, Maere S, Van de Peer Y (2009) Plants with double genomes might have had a better chance to survive the cretaceous-tertiary extinction event. Proc Natl Acad Sci USA 106:5737–5742

Fawcett JA, Van de Peer Y, Maere S (2013) Significance and biological consequences of polyploidization in land plants. In: Leitch IJ, Greilhuber J, Doležel J, Wendel JF (eds) Plant genome diversity, vol 2, Physical structure, behaviour and evolution of plant genomes. Springer-Verlag, Wien, pp 277–293

Goloboff PA (1999) Analyzing large data sets in reasonable times: solutions for composite optima. Cladistics 15:415–428

Goloboff PA, Catalano SA, Mirande JM, Szumik CA, Arias JS, Kallersjo M, Farris JS (2009) Phylogenetic analysis of 73060 taxa corroborates major eukaryotic groups. Cladistics 25:211–230

Grant V (1981) Plant speciation. Columbia University Press, New York

Greilhuber J, Borsch T, Müller K, Worberg A, Porembski S, Barthlott W (2006) Smallest angiosperm genomes found in Lentibulariaceae with chromosomes of bacterial size. Plant Biol 8:770–777

Hilu KW, Borsch T, Müller K, Soltis DE, Soltis PS, Savolainen V, Chase MW, Powell MP, Alice LA, Evens R, Sauquet H, Neinhuis C, Slotta TAB, Rohwer JG, Campbell CS, Chatrou LW (2003) Angiosperm phylogeny based on *mat*K sequence information. Am J Bot 90:1758–1776

Howarth DG, Donoghue MJ (2006) Phylogenetic analyses of the "ECE" (CYC/TB1) clade reveal duplications that predate the core eudicots. Proc Natl Acad Sci USA 103:9101–9106

Huelsenbeck JP, Ronquist F (2001) MrBayes: Bayesian inference of phylogeny. Bioinformatics 17:754–755

Huelsenbeck JP, Ronquist F, Nielsen R, Bollback J (2001) Bayesian inference of phylogeny and its impact on evolutionary biology. Science 294:2310–2314

Jansen RK, Cai Z, Raubeson LA, Daniell H, dePamphilis CW, Leebens-Mack J, Müller KF, Guisinger-Bellian M, Haberle RC, Hansen AK, Chumley TW, Lee S-B, Peery R, McNeal J, Kuehl JV, Boore JL (2007) Analysis of 81 genes from 64 chloroplast genomes resolves relationships in angiosperms and identifies genome-scale evolutionary patterns. Proc Natl Acad Sci USA 104:19369–19374

Jarvis C (2007) Order out of chaos: Linnaean plant names and their types. Linnean Society of London and the Natural History Museum, London

Jiao Y, Wickett NJ, Ayyampalayam S, Chanderbali AS, Landherr L, Ralph PE, Tomsho LP, Hu Y, Liang H, Soltis PS, Soltis DE, Clifton SW, Schlarbaum SE, Schuster SC, Ma H, Leebens-Mack J, dePamphilis CW (2011) Ancestral polyploidy in seed plants and angiosperms. Nature 473:97–100

Judd WS, Campbell CS, Kellogg EA, Stevens PF, Donoghue MJ (2008) Plant systematics—a phylogenetic approach, 3rd edn. Sinauer, Sunderland

Kallersjo M, Albert V, Farris J (1999) Homoplasy increases phylogenetic structure. Cladistics 15:91–93
Kenrick P, Crane PR (1997) The origin and early diversification of land plants. Smithsonian Institution Press, Washington
Kim S, Soltis DE, Albert V, Yoo MJ, Farris JS, Soltis PS, Soltis DE (2004) Phylogeny and diversification of B-function MADS-box genes in angiosperms: evolutionary and functional implications of a 260-million-year-old duplication. Am J Bot 91:2102–2118
Kim S, Koh J, Yoo M-J, Kong H, Hu Y, Ma H, Soltis PS, Soltis DE (2005) Expression of floral MADS-box genes in basal angiosperms: implications for the evolution of floral regulators. Plant J 43:724–744
Leebens-Mack J, Raubeson LA, Cui L, Kuehl JV, Fourcade MH, Chumley TW, Boore JL, Jansen RK, dePamphilis CW (2005) Identifying the basal angiosperm node in chloroplast genome phylogenies: sampling one's way out of the Felsenstein zone. Mol Biol Evol 22:1948–1963
Leitch IJ, Chase MW, Bennett MD (1998) Phylogenetic analysis of DNA C-values provides evidence for a small ancestral genome size in flowering plants. Ann Bot 82:85–94
Leitch IJ, Leitch AR (2013) Genome size diversity and evolution in land plants. In: Leitch IJ, Greilhuber J, Doležel J, Wendel JF (eds) Plant genome diversity, vol 2, Physical structure, behaviour and evolution of plant genomes. Springer-Verlag, Wien, pp 307–322
Litt A, Irish VF (2003) Duplication and diversification in the *APETALA1/FRUITFULL* floral homeotic gene lineage: implications for the evolution of floral development. Genetics 165:821–833
Magallon S (2010) Using fossils to break long branches in molecular dating: a comparison of relaxed clocks applied to the origin of angiosperms. Syst Biol 59:384–389
Mathews S, Donoghue M (1999) The root of angiosperm phylogeny inferred from duplicate phytochrome genes. Science 286:947–950
Moore MJ, Bell CD, Soltis PS, Soltis DE (2007) Using plastid genomic-scale data to resolve enigmatic relationships among basal angiosperms. Proc Natl Acad Sci USA 104:19363–19368
Moore MJ, Soltis PS, Bell CD, Burleigh JG, Soltis DE (2010) Phylogenetic analysis of 83 plastid genes further resolves the early diversification of eudicots. Proc Natl Acad Sci USA 107:4623–4628
Nixon KC (1999) The parsimony ratchet, a new method for rapid parsimony analysis. Cladistics 15:407–414
Pellicer J, Fay MF, Leitch IJ (2010) The largest eukaryotic genome of them all? Bot J Linn Soc 164:10–15
Pryer KM, Schneider H, Smith AR, Cranfill R, Wolf PG, Hunt JS, Sipes SD (2001) Horsetails and ferns are a monophyletic group and the closest living relatives to seed plants. Nature 409:618–622
Pryer KM, Schneider H, Zimmer EA, Banks J (2002) Deciding among green plants for whole genome studies. Trends Plant Sci 7:550–554
Pryer KM, Schuettpelz ME, Wolf PG, Schneider H, Smith AR, Cranfill R (2004) Phylogeny and evolution of ferns (monilophytes) with a focus on the early leptosporangiate divergences. Am J Bot 91:1582–1598
Raven PH (1975) The bases of angiosperm phylogeny: cytology. Ann Mo Bot Gard 62:724–764
Sanderson MJ, Purvis A, Henze C (1998) Phylogenetic supertrees: assembling the trees of life. Trends Ecol Evol 13:105–109
Sanderson MJ, Thorne JL, Wikstrom N, Bremer K (2004) Molecular evidence on plant divergence times. Am J Bot 91:1656–1665
Schneider H, Schuettpelz E, Pryer KM, Cranfill R, Magallon S, Lupia R (2004) Ferns diversified in the shadow of angiosperms. Nature 428:553–557
Slingsby JA, Verboom GA (2006) Phylogenetic relatedness limits co-occurrence at fine spatial scales: evidence from the schoenoid sedges (Cyperaceae: Schoeneae) of the Cape Floristic Region, South Africa. Am Nat 168:14–27
Smith SA, Beaulieu JM, Donoghue MJ (2009) Mega-phylogeny approach for comparative biology: an alternative to supertree and supermatrix approaches. BMC Evol Biol 9:37
Smith SA, Beaulieu JM, Donoghue MJ (2010) An uncorrelated relaxed-clock analysis suggests an earlier origin for flowering plants. Proc Natl Acad Sci USA 107:5897–5902
Smith SA, Beaulieu JM, Stamatakis A, Donoghue MJ (2011) Understanding angiosperm diversification using small and large phylogenetic trees. Am J Bot 98:404–414
Soltis PS, Soltis DE, Chase MW (1999) Angiosperm phylogeny inferred from multiple genes as a tool for comparative biology. Nature 402:402–404
Soltis DE, Soltis PS, Chase MW, Mort M, Albach D, Zanis M, Savolainen V, Hahn W, Hoot S, Fay M, Axtell M, Swensen S, Nixon K, Farris J (2000) Angiosperm phylogeny inferred from a combined data set of 18 S rDNA, rbcL and atpB sequences. Bot J Linn Soc 133:381–461
Soltis DE, Soltis PS, Bennett MD, Leitch IJ (2003) Evolution of genome size in the angiosperms. Am J Bot 90:1596–1603
Soltis DE, Soltis PS, Endress PK, Chase MW (2005) Phylogeny and evolution of the angiosperms. Sinauer, Sunderland
Soltis PS, Soltis DE, Kim S, Chanderbali A, Buzgo M (2006) Expression of floral regulators in basal angiosperms and the origin and evolution of the ABC model. Adv Bot Res 44:483–506
Soltis DE, Bell CD, Kim S, Soltis PS (2008) Origin and early evolution of angiosperms. Ann NY Acad Sci 1133:3–25
Soltis DE, Albert VA, Leebens-Mack J, Bell CD, Paterson A, Zheng C, Sankoff D, Wall PK, Soltis PS (2009a) Polyploidy and angiosperm diversification. Am J Bot 96:336–348
Soltis PS, Brockington SF, Yoo MJ, Piedrahita A, Latvis M, Moore MJ, Chanderbali AS, Soltis DE (2009b) Floral variation and floral genetics in basal angiosperms. Am J Bot 96:110–128
Soltis PS, Burleigh JG, Chanderbali AS, Yoo M-J, Soltis DE (2010) Gene and genome duplication in plants. In: Dittmar K, Liberles DA (eds) Evolution after genome duplication. Wiley-VCH, Weinheim, pp 369–398
Soltis DE, Smith SA, Cellinese N, Wurdack KJ, Tank DC, Brockington SF, Refulio-Rodriguez NF, Walker JB, Moore MJ, Carlsward BS, Bell CD, Latvis M, Crawley S, Black C, Diouf D, Xi Z, Rushworth CA, Gitzendanner MA, Sytsma KJ, Qiu Y-L, Hilu KW, Davis CC, Sanderson MJ, Beaman RS, Olmstead RG, Judd WS, Donoghue MJ, Soltis PS (2011) Angiosperm phylogeny: 17 genes, 640 taxa. Am J Bot 98:704–730
Stamatakis A (2006) RAxML-VI-HPC: maximum likelihood-based phylogenetic analyses with thousands of taxa and mixed models. Bioinformatics 22:2688–2690
Stamatakis A, Hoover AP, Rougemont J (2008) A fast bootstrapping algorithm for the RAxML web-servers. Syst Biol 57:758–771
Stebbins GL (1950) Variation and evolution in plants. Columbia University Press, New York
Stebbins GL (1971) Chromosomal evolution in higher plants. Addison-Wesley, London
Stevens PF (2001 onward) Angiosperm phylogeny website. http://www.mobot.org/MOBOT/research/APweb/
Takhtajan A (1997) Diversity and classification of flowering plants. Columbia University Press, New York
Tank DC, Donoghue MJ (2010) Phylogeny and phylogenetic nomenclature of the Campanulidae based on an expanded sample of genes and taxa. Syst Bot 35:425–441
Timme RE, Bachvaroff TR, Delwiche CF (2012) Broad phylogenomic sampling and the sister lineage of land plants. PLoS One 7:e29696. doi:10.1371/journal.pone.0029696
Uhl C (1978) Chromosomes of Mexican *Sedum* II. Section *Pachysedum*. Rhodora 80:491–512

Vamosi JC, Knight TM, Steets JA, Mazer SJ, Burd M, Ashman TL (2006) Pollination decays in biodiversity hotspots. Proc Natl Acad Sci USA 103:956–961

Wang H, Moore MJ, Soltis PS, Bell CD, Brockington SF, Alexandre R, Davis CC, Latvis M, Manchester SR, Soltis DE (2009) Rosid diversification and the rapid rise of angiosperm-dominated forests. Proc Natl Acad Sci USA 106:3853–3858

Webb CO, Cannon CH, Davies SJ (2008) Ecological organization, biogeography, and the phylogenetic structure of tropical forest tree communities. In: Carson WP, Schnitzer SA (eds) Tropical forest community ecology. Wiley-Blackwell, Oxford, pp 79–97

Wikström N, Savolainen V, Chase MW (2001) Evolution of the angiosperms: calibrating the family tree. Proc Roy Soc Lond B Bio 268:2211–2220

Won H, Renner SS (2003) Horizontal gene transfer from flowering plants to *Gnetum*. Proc Natl Acad Sci USA 100: 10824–10829

Wurdack KJ, Davis CC (2009) Malpighiales phylogenetics: gaining ground on one of the most recalcitrant clades in the angiosperm tree of life. Am J Bot 96:1551–1570

Zahn LM, Kong H, Leebens-Mack JH, Kim S, Soltis PS, Landherr LL, Soltis DE, dePamphilis CW, Ma H (2005) The evolution of the *SEPALLATA* subfamily of MADS-box genes: a pre-angiosperm origin with multiple duplications throughout angiosperm history. Genetics 169:2209–2223

Zwickl DJ (2006) GARLI. Genetic algorithm for rapid likelihood inference. Software available at http://www.bio.utexas.edu/grad/zwickl/web/garli.html

The Plant Nucleus at War and Peace: Genome Organization in the Interphase Nucleus

2

R. Neil Jones and Tim Langdon

Contents

R.N. Jones (✉)
Institute of Biological, Environmental and Rural Sciences, Aberystwyth University, Edward Llwyd Building, Penglais Campus, Aberystwyth, Wales SY23 3DA, UK
e-mail: rnj@aber.ac.uk

I.J. Leitch et al. (eds.), *Plant Genome Diversity Volume 2*,
DOI 10.1007/978-3-7091-1160-4_2,

2.1 Introduction

The organization of the plant nucleus has been seen as a key to understanding the workings of plants themselves for the past two centuries. This aspiration seems to be being finally realised as new sequencing technologies are helping to draw together old strands of research as well as supporting novel approaches to bridge the gap between cytological and molecular scales of description; and recent advances in understanding epigenetic processes and mechanisms are providing a fresh perspective on studies of interphase organization. The inevitable caveat is that new questions may need to be answered, as the comparative detail now available throws up additional complexities which were previously hidden. Chief among these is the extent to which the genome is not constant within a species. Limited surveys of genes and genomic regions in maize revealed some years ago that there was more genetic diversity within this one species than between humans and chimpanzees; and a more recent genome wide survey indicates that if anything this was an underestimate (Gore et al. 2009). More generally in both animals and plants, intragenic copy number variation in non-repetitive sequences is being found to rival that seen in heterochromatic repeats; 10% or more of coding sequences show copy number variation in maize and rice (Ding et al. 2007; Springer et al. 2009). Nuclear organization in many plants must therefore be capable of accommodating a wide range of structural and nucleotide polymorphisms without suffering detrimental effects; indeed, maize itself demonstrates that the hybridization of divergent parents may be strongly advantageous (Shull 1948). This plasticity may explain the ability of many wide crosses to eventually generate stable derivatives, which in turn is likely to underpin many of the numerous examples of reticulate evolution being found. Nevertheless, interspecific hybridizations have often been shown to trigger genome wide reactions, frequently followed by structural and epigenetic reorganization in subsequent generations, which can be seen as a war between parental genomes. Here we review some of the aspects of nuclear organization that may underlie

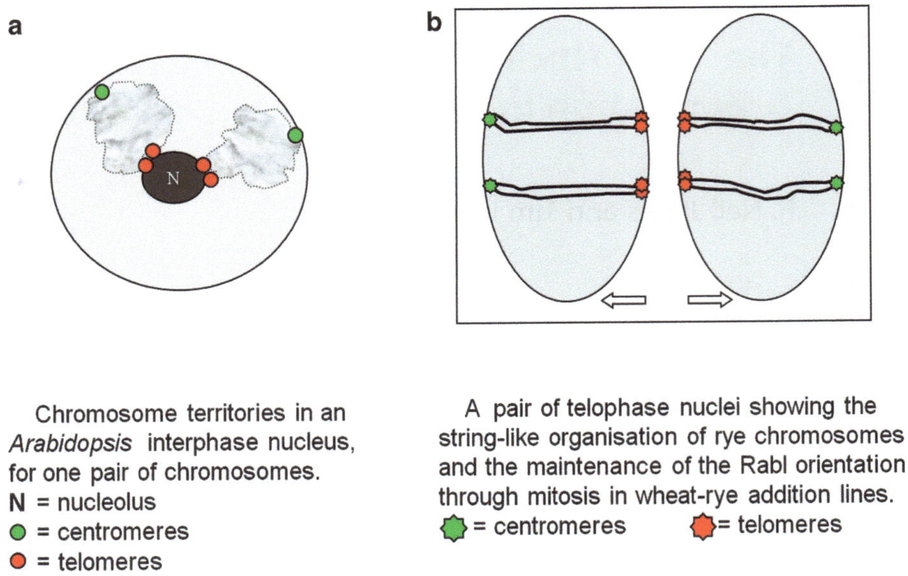

Fig. 2.1 Highly schematic representation of the disposition of chromosomes in species with a small (**a**) *(Arabidopsis)* and a large (**b**) (hexaploid wheat) genome. Probes identify the centromeres (*red*) and telomeres (*green*). The chromosome territories in *Arabidopsis* are visualized with chromosome-specific probes from pools of BAC contigs; whereas in wheat individual chromosomes, present as a wheat/rye addition line 1R, can be seen by probing with whole genome DNA which discriminates between the dispersed repeats of wheat and rye chromatin

incompatibilities, and indicate some of the changes that may be required to restore if not peace, then at least a truce.

2.2 General Aspects of Organization in the Peaceful Nucleus

A number of significant discoveries on the structure of the nucleus were made early on, at the limits of what could then be observed with the light microscope. In 1885 Rabl first described the arrangement of the chromosomes of salamanders in terms of the orientation of their centromeres and telomeres. He explained how the anaphase configuration, with the telomeres at the nuclear periphery of adjacent daughter cells and the centromeres at the relic poles, was maintained through to the following interphase (Rabl 1885). Laibach (1907) later made the remarkable finding that the number of chromocentres in *Arabidopsis* corresponded to the number of chromosomes, and Heitz (1928) demonstrated the continuity of the structure of chromosomes throughout the cell cycle in *Pellia endiviifolia*, using blocks of constitutive heterochromatin as markers for the identity of individual chromosomes. An early inference was that the interphase nucleus would be made up of discrete chromosomal territories (CTs) (Boveri 1909), and each subsequent development of microscopic techniques has improved our understanding of the nuclear architecture of CTs and its impact on gene expression (reviewed in Rouquette et al. 2010). Most recently, direct 3D analysis of living cells using green fluorescent protein (GFP) tags and BACs, and of accurately fixed material by confocal methods, has been extended by indirect analysis via sequencing of DNA retrieved from common pools of cross linked proteins (3C, 4C, 5C and HiC methods), which is providing resolution beyond the limits of microscopy (Rajapakse and Groudine 2011). Nevertheless, the role and even the extent of CTs remains unclear, and it appears that this is as much because of inherent variability as because of technical limitations.

This variability is well illustrated by the inconsistent occurrence even of Rabl arrangements. They do appear to be a fixture of plant species with relatively large genomes (such as wheat, oat, barley, *Vicia faba* and *Allium cepa*, all with genomes in excess of 1C = 5,000 Mb), where probes identifying alien segments can be used to show chromosome arms assuming a string-like form running between the centromeric and telomeric nuclear poles (Abranches et al. 1998) (Fig. 2.1). On the other hand, plant species with small genomes (such as sorghum, rice, *Arabidopsis*, and brassicas, with genomes of less than 1C = 1,000 Mb) generally lack the Rabl arrangement. This can be observed, for example, by 3D reconstruction of living cells of *A. thaliana* after tagging with a GFP probe, where the centromeres are predominantly dispersed around the nuclear periphery in different cell types (Fang and Spector 2005) although the telomeres are associated within the nucleus core around the nucleolus (Armstrong et al. 2001). As a result individual *Arabidopsis* chromosomes appear as radial euchromatic loops emanating from the heterochromatic centromeric chromocentres when painted with specific BAC markers (Fransz et al. 2002). Maize, with 1C = 3,000 Mb of

DNA, is intermediate in size, and cannot be classified as either having a Rabl arrangement or not (Cowan et al. 2001) although rice (1C = 430 Mb) does display a tissue-specific Rabl arrangement in endopolyploid differentiating cells in the xylem (Santos and Shaw 2004). Dong and Jiang (1998) suggested that chromosome, rather than genome, size might be a crucial factor; and Fransz and de Jong (2011) have suggested 500 Mb as a threshold value. However, while it might be reasoned that the larger the chromosome, the greater the logistical problem of interphase chromatin organization, and hence the longer persistence of anaphase organization (Cowan et al. 2001), this is not the full answer, as yeasts may show a Rabl configuration while larger genome mammals do not (discussed further in Schubert and Shaw 2011).

2.2.1 Chromosome Territories (CT)

The enigmatic nature of Rabl arrangements also applies to lower levels of nuclear organization, despite extraordinarily detailed analyses of interphase CTs in a variety of animal and plant cell types. As with Rabl, there are examples of highly ordered structures, which logic suggests must play an important functional role; however, again as with Rabl, there are examples where these structures do not occur or are disrupted without apparent ill effect. These conflicting observations are described in a number of excellent reviews (Cremer and Cremer 2010; Woodcock and Ghosh 2010) and will not be repeated in detail here. At present studies on human cells support a model of regions of increased gene expression (ridges) and similar sized domains enriched for genes with low expression (anti-ridges). Originally defined by mapping transcriptome data onto linear chromosome maps, which found one to two orders of magnitude difference in expression levels between ridges and anti-ridges (Caron et al. 2001), the patterns have been visualized in three dimensions using confocal microscopy, which found anti-ridges to be more compact and regularly organized than ridges, and to be closer to the nuclear envelope, where they may be anchored (Goetze et al. 2007). Large cell-to-cell variation in this organization was seen and interpreted as structural plasticity rather than experimental error. This is consistent with global patterns of chromatin organization revealed indirectly by the HiC procedure, which supports a fractal globule model where self-organizing domains ‘crumple’ in such a way that the resulting structures do not penetrate into other crumples (Lieberman-Aiden et al. 2009). Accordingly, there are no knots in the structures to prevent dynamic looping out, whether by diffusion or in response to specific cues, but defined CTs are expected to persist over the period of the cell cycle as crumples reform. The HiC data support a division into two spatial sub-compartments, similar to the ridge/anti-ridge model, and indicate frequent interchromosomal contact (Lieberman-Aiden et al. 2009).

Studies on plant CTs are not yet as detailed as for humans but do not conflict with the fractal globule model. Significant patterns of co-expressed intrachromosomal neighbourhoods have been found (Williams and Bowles 2004; Riley et al. 2007) and it is likely that chromosome conformation capture methods will soon allow these to be examined in three dimensions. Clear demonstrations have been made, however, that CT orthologues do not co-localize at high frequency during interphase (reviewed in Schubert and Shaw 2011) and transcription appears to occur at sites distributed throughout the nucleus, rather than concentrated in regions expected to be gene-rich on the basis of, for example, Rabl disposition. Strong interchromosomal associations may normally be limited to heterochromatic chromocentres and nucleolar organizers. Reorganization of chromatin in response to a number of cues has been followed, including large scale reversible decondensation of heterochromatin in *Arabidopsis* (Tessadori et al. 2007). Reassembly of the *Arabidopsis* chromocentres is sequential with an order which appears to be determined simply by weak interactions whose strength is proportional to array length. Similarly, treatment of wheat seedlings with chemicals which disrupt chromatin silencing, at levels too low to significantly reduce growth, was found to cause decondensation of chromosome arms and remodelling of CTs, but did not alter the Rabl organization of heterochromatic blocks (Santos et al. 2002).

The disposition of chromosome pairs at interphase in *Arabidopsis* has been studied by chromosome painting with pools of chromosome-specific BAC contigs, and it appears that association patterns on homologous and heterologous chromosomes is largely random, and that only chromosomes bearing nucleolus organizing regions (NORs) show that homologs associate more frequently than random (Pecinka et al. 2004). It further transpires that the cohesion between BAC-labelled sister chromatids is lacking, and that the strict parallel alignment between them is greater in differentiated and in endopolyploid cells (Schubert et al. 2006).

Overall, then, CTs in both animals and plants behave as if they are largely self-organized and independent, influenced by weak preferences rather than dependent on absolute configurations. This is perhaps to be expected given the high levels of copy number variation being found.

2.2.2 Modifications of Chromatin and DNA

One area where plants differ markedly from animal models of chromatin organization is in the use of the modified histones which discriminate transcriptionally active regions from those that are silenced, particularly heterochromatin. Surprisingly the same histone code is not maintained between animals and plants, or even between plant species (Fuchs et al. 2006; Shi and Dawe 2006; Braszewska-Zalewska et al. 2009). Soppe et al. (2002) noted that the diffuse presence of H3K9 modifications of histone H3 in euchromatin was characteristic of large genomes, such as maize, while it was found only concentrated in heterochromatin in *Arabidopsis*. The ability to redefine such a fundamental aspect of chromatin organization may reflect a divergent approach to genome surveillance by plants compared to animals. Plants lack a dedicated germline and the bulk of silencing signals is maintained at all times, unlike animals which reset the genome at least twice per generation (plants do have 'facultative' silencing, involved in processes such as vernalisation and parental imprinting, e.g., Hsieh et al. 2011). Plant genes are also relatively more compact and intergenic conserved non-coding sequences (CNS) are relatively rare. This combination suggests that meristem integrity may have been protected by enhancing mechanisms which monitor intergenic regions in order to restrict mobile element colonisation. In large part this monitoring is carried out via a small RNA based silencing pathway, involving the plant-specific RNA polymerases IV and V (RNAP IV, RNAP V) (Herr et al. 2005; Onodera et al. 2005; Zhang et al. 2007; Matzke et al. 2009). This represents a significant extension to the more universal eukaryotic use of different classes of small RNA to carry out a variety of roles in modulating chromatin structure and gene expression, for example the use of miRNA to co-ordinate developmental switches in animals and plants (reviewed in Kaufmann et al. 2010).

The central role of siRNAs in the control of plant genome stability has been emerging over the last 10 years but details about mechanisms are still being revealed. Briefly, it appears that all eukaryotes are able to detect 'alien' transcripts through the recognition of aberrant double stranded RNA (dsRNA) structures; the 'aliens' are typically transposable elements (TEs) or viruses but may include transgenes or even some host genes. Once recognised, the dsRNA is cut into small fragment(s), destroying the original transcript but in most cases retaining a short guide which can be used to recognise complementary RNA or DNA templates. The ways in which guides are exploited to modify DNA can vary greatly; the most extreme use is shown by some fungi, in which a series of steps leads to mutation of all cognate sequences (Selker 2002). Plants use a family of Dicer-like (DCL) enzymes to generate small RNAs from a variety of dsRNAs which may then be bound by ARGONAUTE (AGO) proteins and used to direct DNA methylation (RNA-directed DNA methylation, RdDM) and histone modification at relevant chromatin sites. Different dsRNA classes are processed by different, partially redundant, pathways but the majority of plant small RNAs are 24 nt long, individually at low abundance and derived from transcripts generated by RNA polymerases IV and V. These enzymes supplement the universal RNA polymerase, Pol II, and produce templates primarily from intergenic and heterochromatic regions (Zheng et al. 2009) so that the targets of the 24 nt silencing RNAs (siRNAs) are predominantly mobile elements, tandem repeats and associated debris. In addition to transcripts from promoters in the repetitive elements themselves, there are a range of host driven transcripts, including for long intergenic non-coding RNAs (lincRNAs) which may result in almost all of the genome being transcribed at some point, and hence potentially able to generate siRNAs. The unusually high level of retained introns in mature transcripts in plants (Filichkin et al. 2010) may further add to the pool of potential silencing signals. The importance of siRNAs is demonstrated by the presence of a range of additional DCL, AGO and other associated components, including plant-specific pathways to generate dsRNA from single-stranded RNA templates, and to methylate asymmetrical DNA targets (with the consensus sequence CHH). Plants have also evolved mechanism(s) to transfer silencing between cells and tissues (Molnar et al. 2010).

Once symmetrical methylation (at CG and CHG sites) has been established, it may be maintained at DNA replication by a hemimethylation recognition mechanism. The 24 nt siRNAs are required to direct fresh methylation at the asymmetrical CHH sites, however. This illustrates one of the features of epigenetic control, that there tend to be multiple mechanisms which may reinforce each other, but may all be required for full effect. In combination this may result in epigenetic haplotypes ('epialleles') which are stable for generations (Johannes et al. 2009; Reinders et al. 2009; Baubec et al. 2010) but, intriguingly, global responses to stress may drive some of the initial epigenetic responses, creating heritable phenotypic variation which may be adaptive and allow selection for further epigenetic changes (reviewed in Boyko and Kovalchuk 2011; Verhoeven et al. 2010; Mirouze and Paszkowski 2011). An interesting model has suggested that the diversity of DCLs may facilitate the gradual domestication of particularly adaptive siRNA sources and their integration into developmental pathway regulation (Vazquez et al. 2008).

Since methylation and histone modification have longer range impacts on recombination frequencies and patterns of DNA replication, and since TE-directed epigenetic marks may spread to have impacts on adjacent host genes (Gehring et al. 2009; He and Dooner 2009; Martin et al. 2009), it can be seen that even in a mature genome, with low levels of both genic polymorphism and endogenous TE activity, growing in a relatively stable environment, there is much scope for divergence of individual lineages. These constrained conditions rarely apply. Moreover, the prediction that parental conflict over resource allocation in endosperm would result in parent-specific gene expression in that tissue (Haig and Westoby 1989) is being borne out in surveys of the distribution of imprinted genes and release of maternal silencing, while there is also some evidence for recurring conflicts between divergent centromeric sequences during meiosis (Malik and Henikoff 2009). There is therefore intrinsic potential for intragenomic 'squabbles' between both conventional and epigenetic polymorphisms to accumulate over time. In the next section we consider how frequently evidence for these squabbles can be seen.

2.3 The Nucleus Under Strain

Some indication of the tolerance required by plant nuclear organization is provided as soon as a rigorous definition of a plant species is required. In the extreme this should depend on strict interfertility criteria but this approach soon runs into difficulty, both in the inclusion of phenotypically distinct hybrids, particularly allopolyploids, which suggests the definition is too broad, and in the exclusion of some clearly related populations or lines, which suggests the definition is too narrow (discussed in Soltis and Soltis 2009). Sequence data do not always resolve these issues (Petersena et al. 2011), and indeed are blurring species boundaries as it becomes possible to track the history of introgression and recombination of contrasting haplotypes over million year periods of time (Wicker et al. 2009). The breadth of intraspecific phenotypes in the shape of karyotype variants or crossing incompatibilities has, of course, been known for many decades. Here we consider some of the genomic polymorphisms accomodated within a species as generally understood (Rieseberg et al. 2006), which appears increasingly likely to overlap the variation provided by interspecific hybridization.

2.3.1 Plasticity of Gene Family Organization

The best defined category of variants could be described as potential speciation genes (Rieseberg and Blackman 2010) where polymorphisms at a very limited number of loci are sufficiently antagonistic that they have a major impact on fertility. The classic example is hybrid necrosis, where allele-specific interactions between two loci trigger a disease-like phenotype leading in the most extreme cases to the death of progeny. Hybrid necrosis is a problem for most if not all crop breeding programmes, but is also seen in wild species, occurring in 2% of *Arabidopsis* intraspecific crosses, for example (Bomblies et al. 2007). Interactions with a similar maladaptive epistatic allele-specific basis, known as Dobzhansky-Muller incompatibilities, include hybrid male sterility in rice (Ouyang et al. 2010), and appear to be relatively more common in plants than animals although whether this reflects a genuine increase in the number of candidate genes or an ascertainment bias remains to be established (Rieseberg and Blackman 2010). At the simplest level, these interactions may appear to be idiosyncratic, with little direct relevance to nuclear organization. However, it is useful to consider hybrid necrosis in some detail as it appears to be an almost inevitable by-product of the amplification and diversification of microbial pathogen resistance genes (R-genes) (Bomblies 2009). Pathogens provide one of the major selective pressures on plants and models of R-gene evolution are helping to illustrate the ways in which sessile organisms can generate the plasticity required within their nuclear organization to match environmental challenges. It is likely that this plasticity is required for many other aspects of adaptation, and will be seen across the nucleus.

There are many R-genes (>150 in *Arabidopsis*, >400 in rice) and they are found both singly and in clusters across genomes. Although they appear to be particularly polymorphic (Zeller et al. 2008) and to operate under particularly strong selection (Chen et al. 2010), the evolutionary dynamics of their tandemly arranged clusters are similar to those of other, less well characterised, gene clusters such as those for F-box proteins (Jain et al. 2007). Tandem duplications of these genes follow a complex birth-and-death pattern of evolution, which contrasts with the concerted evolution of more regular arrays such as rDNA genes (Nei and Rooney 2005). This may be promoted by the recombinogenic effect of the unpaired additional copy at meiosis (Sun et al. 2008), enhancing not only ectopic recombination between related sequences but also using microhomologies to create unusual chimeras such as the MAF2 alleles affecting *Arabidopsis* flowering time (Rosloski et al. 2010). At R-gene clusters this inherent instability helps to generate the range of defense options necessary to counter microbial pathogens. Insertion-deletion polymorphisms (indels) are also common, found at eleven times the frequency in R-genes as housekeeping genes in rice and resulting in some 20% of R-genes displaying a presence or absence variation (PAV) in both rice and *Arabidopsis* (Shen et al. 2006). Surprisingly, those loci showing PAV appear to represent a distinct class which does not have the high SNP content of other R-gene loci. Instead they are maintained over relatively much longer periods, presumably by balancing selection across the population (Shen et al. 2006).

Further R-gene complexity is being uncovered in the role of siRNAs, for example at the *Arabidopsis* RPP5 region, which may attenuate expression under normal conditions but allow release under stress or when pathogens compromise silencing mechanisms (Yi and Richards 2007).

An epigenetic component is likely to be a general feature of R-gene expression, allowing modulation by other biotic and abiotic stresses (Alcazar et al. 2009), and the development of immunity (Alvarez et al. 2010) and transgenerational effects on locus methylation and stability (Boyko et al. 2007). The need for adaptable regulation is demonstrated by the wide range of maladaptive phentypes potentially caused by R-gene mutation, including developmental abnormalities (reviewed in Uchida and Tasaka 2010) while, more generally, enhanced disease resistance frequently comes at the cost of other fitness parameters (e.g., Todesco et al. 2010), leaving the genome exposed to conflicting pressures. For example, lines exhibiting hybrid necrosis symptoms at one temperature may be more stable at another, and may even show greater disease resistance than sister lines. It is highly likely, therefore, that plant nuclear organization has had to evolve to permit some chromosomal regions the ability to create novel gene organization, as has been suggested to occur for 'gene nurseries' in some animal systems (Nahon 2003; Bailey and Eichler 2006). However, gene nurseries occur at particular chromosomal contexts and while there may be analagous regions in plants (Tian et al. 2011), R-gene clusters rarely persist at syntenic locations, resulting in at least one locus being dubbed 'nomadic' (David et al. 2009). Given these variables, the frequency of maladaptive necrotic interactions could be considered rather low!

Another aspect of R-gene evolution that must be accomodated is the asymmetrical nature of the loci, that is high levels of PAV and copy number variation (CNV). This is a pervasive phenomenon in plant genomes (Ding et al. 2007; Springer et al. 2009; DeBolt 2010). The number of functional genes involved is difficult to quantify accurately using indirect methods, such as array hybridization, as a variety of mobile elements have been found to transpose genic fragments. In maize, fragments of at least 376 different genes have been duplicated by helitrons (Du et al. 2009) while in rice fragments of 1,500 genes have been duplicated by PACK-Mules (Hanada et al. 2009). A very small proportion of these fragments are likely to retain coding function. Nevertheless, the scale of CNV identified in the most recent screen of maize (Swanson-Wagner et al. 2010) supports observations made on smaller sequence-based data sets (Fu and Dooner 2002; Wang and Dooner 2006) and is beyond levels expected to be due to genic 'debris'. Thus, over 10% of ~32,500 genes surveyed showed CNV, the majority being absent in one or more lines despite having Gene Ontology annotations and orthologs in other species. Most surprisingly, PAV patterns appeared to have an ancient origin in the maize progenitor, teosinte, and to have persisted despite domestication and movement into new environments. It is possible that this represents the presence of many alternative haplotypes generated as the ancient palaeoploidy event identified in the maize lineage (Gaut and Doebley 1997) is resolved by gradual diploidisation, and less redundant diploid species may not be able to support such high levels of PAV. Nevertheless, significant levels of stable CNVs could be generated by relatively mild stress in *Arabidopsis* over five generations (DeBolt 2010), affecting over 1% of total genes. Consistent with the expectation of tandem gene instability, just over half of genes affected were in tandemly duplicated regions.

2.3.2 Limits to Expression Neighbourhoods

It is therefore clear that intraspecific crosses must almost inevitably generate asymmetric CT orthologues. As a consequence it is difficult to envisage mechanisms by which orthologous loci could be provided with conserved nuclear environments in which subtle regulatory cues could be consistently supplied. Do individual genes have a high degree of autonomy in their regulation, to allow them to overcome shifting levels of neighbourhood-wide expression? Such autonomy should be encoded within the gene itself (*cis*-regulation) but may require more or less strong contributions from external network-specific factors (*trans*-regulation). There has been much recent interest in characterising this balance and identifying relevant sequences, a field described as expression or eQTL mapping. A number of studies have used array hybridization or specific PCR to look at asymmetrical expression of alleles across the transcriptome in a variety of plant species. Differences from parental mid-point values are frequently found, for example at rates ranging from 4% to 32% of genes in comparisons between *Arabidopsis thaliana* ecotypes (Zhang et al. 2008a). However in *Arabidopsis* over-dominant expression was seen in ~9% of transcripts examined (Vuylsteke et al. 2005) while a wider maize study found that the great majority of alleles were expressed at parental or intermediate levels (Stupar and Springer 2006). In particular both individual maize parents frequently showed similar expression levels for a locus; in all these cases the same level was maintained in the hybrid. The authors of the maize study concluded that *trans*-regulation alone was rarely responsible for asymmetric expression, although *cis*-regulation frequently was (Stupar and Springer 2006). This fits an alternative view of expressed neighbourhoods, in which they are created because of 'ripple effects' of read-through transcription outside the target locus (Ebisuya et al. 2008; Yanai and Hunter 2009). Genes moved into novel neighbourhoods will adopt the expression pattern of that neighbourhood to some extent, but *cis*-acting elements may allow locus-specific regulation to prevail. Presumably it is efficient to maintain neighbourhoods of housekeeping genes, but equally it might be presumed dangerous to fix the expression patterns of loci responsible for, for example, stress or developmental responses.

2.3.3 Hybrid Vigour

One of the goals of understanding how nuclear organization ensures optimal plant performance is to improve crop breeding strategies. A particular target is understanding the basis of hybrid vigour. This occurs when crosses between certain subsets of a species generate progeny which have far greater adaptive potential than the additive value of their parents (Chen 2010). Two contrasting models have been proposed to account for this. In the dominance model, deleterious alleles in one parent are complemented by superior alleles from the other; in the over-dominance model novel interactions between different parental alleles create phenotypes not possible from the use of either allele alone. Despite many decades of use in maize, it has not proved possible to breed out the putative deleterious alleles of the dominance model, although there has been a recent claim that vigour has been dissected into a small number of complementing QTL (Schön et al. 2010). Now, a recent maize resequencing project has provided evidence that absent PAV loci are being complemented by PAV present loci from the other parent (Lai et al. 2010 - note that the resequencing approach provides a more conservative estimate of ~300 genes present in the B73 reference but missing in other lines). This is consistent with the lack of significant over-dominant expression patterns in hybrids, although some component of vigour could still be contributed by such interactions (Stupar et al. 2008).

2.3.4 Mobile Elements

Brief mention should also be made of the very large contribution that mobile or transposable elements (TEs) make to intraspecific variability. Variation in mobile element profiles across, and even within, populations has been well documented and has frequently provided the basis for characterising crop genetics and domestication histories (Kalendar et al. 2011). As an example of the extent of TE content variation, two sequenced maize genomes differ in size by ~20%, mostly ascribed to differences in TE content (Vielle-Calzada et al. 2009). As an example of TE links to host phenotype, variation in the *BARE*-1 retrotransposon correlated with adaptation to local environmental conditions in *Hordeum spontaneum* (Kalendar et al. 2000). Across a 300 m transect, *BARE*-1 copy number varied three-fold. Perhaps most surprisingly, plants in the most arid and inhospitable conditions exhibited both the highest numbers of *BARE*-1 elements and the greatest retention of *BARE*-1 sequences, despite the naive expectation that these properties would reduce fitness. TE profiles do not, therefore, automatically conform to intuitive expectations although a general correlation of expression with stress responses is found, coupled with a propensity to act as nucleation points for epigenetic effects which spread into adjacent host genes. Ecological analogies for TE behaviour appear to be apt and descriptions of the opportunistic nature of niche exploitation by specific element types or sub-families abound (see Kejnovsky et al. 2012; Slotkin et al. 2012). A crucial point, however, is that horizontal transfer of TEs is not sufficiently widespread in plants to provide a reliable escape mechanism such as may be argued to occur, for example, in *Drosophila* (Bartolome et al. 2009); the fates of plant TEs are therefore tied to those of their hosts and in the long-term their behaviour cannot afford to be as destructive as, say, viral retroelements. As part of the restraining process acting on TE ‘selfishness’, non-autonomous elements frequently mediate the most significant phenotypic effects of TE families on their hosts but also attenuate the activity of rare active autonomous elements and help to disperse their impact across populations. The variability created by TE activity is not restricted to direct impact at insertion sites. A good recent example of their general ability to remodel expression demonstrated that TE proximity reduced *Arabidopsis* gene expression (Hollister et al. 2011); the strength of this effect depended on both the host species and the frequency with which the TE was targeted by siRNAs.

2.4 The Nucleus at War

The previous section outlined the sometimes surprising levels of structural variation that may be found between individuals of the same plant species. Not all variation within a species may be accommodated in any single cross, and not all intraspecific incompatibilities may be ascribed to such relatively simple causes as hybrid necrosis (e.g., Zhao et al. 2010). However, by definition, it might be expected that the great majority of genomic tensions would be held in check within a species. As McClintock (1984) pointed out ‘*species crosses are a potent source of genomic modification. .. the alterations produced when the genomes of two species are combined reflect their basic incompatibilities*’. New molecular tools are now providing us with the means to investigate what happens when the genomes of different species, as F1 hybrids or allopolyploids, come together for the first time within a single nucleus (Hegarty and Hiscock 2005) and in the common cytoplasm from the maternal parent (Fig. 2.2). What happens, then, once there is no longer a need for

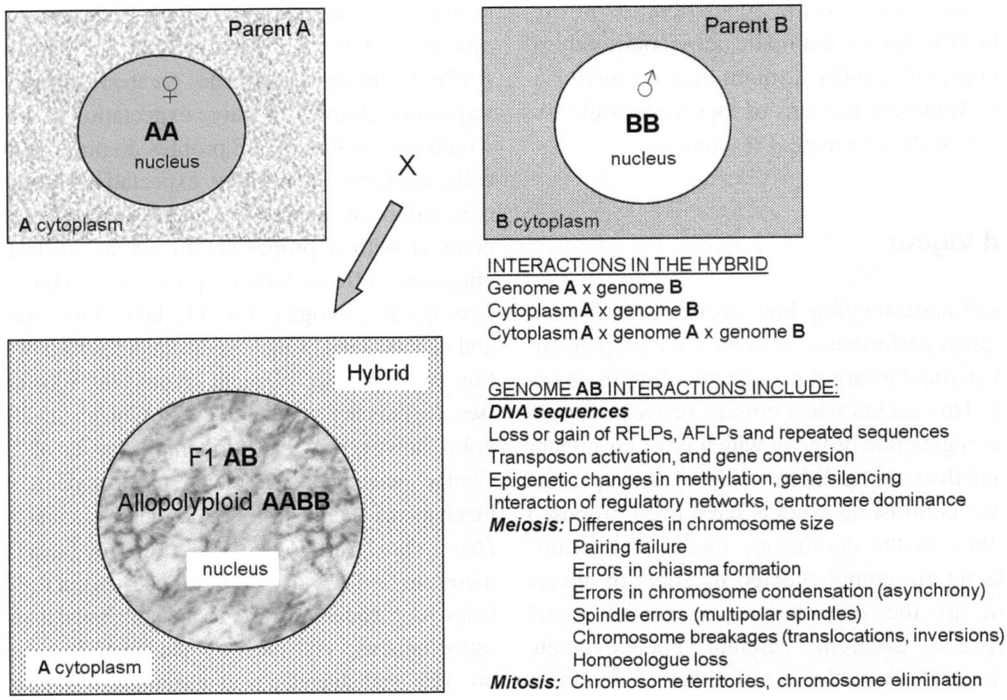

Fig. 2.2 Diagrammatic representation of the plethora of interactions which can occur when genomes of different species are combined together within a single nucleus, and within the cytoplasm of the maternal parent (partly based on Gill 1991)

genomes to be compatible in shared germplasm? Do interspecific hybrids require extreme chance or strong selection for divergent genomes to cohabit, or will the forces constraining normal sibling jostling continue to maintain common standards of organization and expression in the absence of routine requirements for interfertility? In this section we review the adjustments found to occur in interspecific hybrids, both where genomes are sufficiently similar to allow homoploid formation, and the far better characterised situation where each parental genome persists separately as a component of a new polyploid. Sometimes hybridization appears to proceed relatively smoothly, while in other cases one genome appears to fall victim to another. We are far from understanding the processes responsible for these differences, although some common patterns appear to be emerging.

2.4.1 Ploidy Levels

Homoploid hybridization generates novel mosaic genomes, in which some proportion of each parent must be lost. The gene content and expression patterns of each parental genome should therefore be relatively well conserved. It might also be expected that parental chromosomes must show extensive colinearity to allow reciprocal recombination. Consequently there is likely to be a limit to the amount of divergence that can be successfully tolerated for the homoploid generation. The parameters of this limit have been vigorously discussed (Buggs et al. 2009; Paun et al. 2009). Very detailed work has been carried out on a number of young homoploid hybrids (Baack and Rieseberg 2007), but higher sequencing throughput is likely to throw up multiple examples of paleohomoploidy, with signs of suspect ancestry already showing in model systems, including an *Arabidopsis* relative (Wang et al. 2010) and *Brachypodium distachyon* (Wolny et al. 2011).

Polyploidy requires an additional genome doubling step compared to homoploid hybridization but allows cohabitation of far more divergent genomes. It not only retains all parental material at initial stages, so that parent-specific processes may potentially be maintained, but it also provides redundancy such that maladaptive interactions may subsequently be overcome by silencing or deletion rather than requiring specific adaptation. From the previous sections it can be seen that plant nuclear organization is already primed to accomodate 'alien' CTs and unstable gene neighbourhoods, and to identify and suppress loci producing aberrant transcripts, so it is not surprising that successful polyploidisation events occur frequently. There is evidence for them having occurred in more than 70% of species at some stage during their evolution (Levin 2002; Fawcett et al. 2009), and it has been suggested that polyploidy may even be ubiquitous among angiosperms (Soltis et al. 2009; Jiao et al. 2011).

Autopolyploids have multiple sets of chromosomes from a single species, and are of interest because they allow examination of the simple effect of increasing CT numbers in the nucleus. As might be expected, only relatively subtle effects on expression are usually seen (Stupar et al. 2007; Parisod et al. 2010) although in the longer term differentiation of the chromosomes occurs, mostly by loss of material and presumably to reduce irregularities at meiosis. The majority of polyploids are formed between divergent species (allopolyploids) and conflicts might therefore be expected at hybridization due to differences in genome size, genome composition, regulatory mechanisms, cell cycle duration, genetic and epigenetic modifications and indeed all of the aspects that contribute to harmony in the physiology of the diploid or autopolyploid nucleus. A large number of potential conflicts must be resolved rapidly, typically within a generation or two, a scale which dictates extensive use of epigenetic regulation. This phase of polyploidisation has been called 'revolutionary', in contrast with the much longer 'evolutionary' phase during which the new species progresses back towards diploidy (Levy and Feldman 2002); this phase appears to still be ongoing in maize at least five million years since a tetraploidy event (Schnable et al. 2011).

2.4.2 Changes at the Gene Level

Newly synthesised allopolyploids and F1 hybrids are subject to new combinations of regulatory networks, as well as rapid genetic modifications which can result in altered patterns or levels of transcription. The extent of these interactions will depend on the degree of divergence between the progenitors. Recent advances in molecular techniques have now enabled researchers to directly compare new hybrids and their parents; and the impact of hybridization and genome duplication at the genic level can involve sequence loss, changes in transposable elements, altered patterns of gene expression and epigenetic modifications. These changes have been dealt with in detail in Jones and Hegarty (2009), and are considered here only briefly.

2.4.2.1 Sequence Loss

Over time polyploids are known to lose redundant parental sequences, a process known as fractionation or diploidisation. However, sequence loss can occur surprisingly rapidly, as was first reported for newly formed allopolyploids in *Brassica* (Song et al. 1995), followed soon after by *Triticum aestivum* (Feldman et al. 1997). These early reports used restriction analysis and Southern blotting to detect changes in low-copy sequences. There were some differences between the species, with *Brassica* instabilities continuing beyond the fifth generation while wheat stabilised by the second or third allopolyploid generation (Ozkan et al. 2001; Shaked et al. 2001). Strikingly, there was evidence both for preferential loss of material from specific genomes, and for recapitulation of at least some of the changes found to have occurred in natural polyploids (Shaked et al. 2001). Genome-wide characterisation of both newly synthesised and ancient allopolyploids is now possible at the sequence level and it is apparent that preferential loss is a general phenomenon. In maize, there is evidence that one parental genome was not only the preferential target for initial sequence loss, but that this preference has persisted over millions of years of subsequent evolution (Schnable et al. 2011). There is also evidence for preferential loss of particular TE sub-families (Kraitshtein et al. 2010; see below). The mechanism(s) responsible for this loss are unknown but a recent review of *Triticeae* polyploids found evidence that the largest genome in the hybrid was most likely to suffer loss (Bento et al. 2011). However, experimental conditions or intraspecific variation among parents may have a strong influence on this behaviour, as Mestiri et al. (2010) found no evidence for any rearrangement in a number of synthetic wheat allohexaploids.

2.4.2.2 Changes in Transposable Elements (TE)

Widespread changes to the activity of TEs were proposed by McClintock (1984) as a likely consequence of the 'genome shock' induced by the merger of two divergent genomes in a single nucleus; and experimental studies have now confirmed that proliferation of TEs occurs in newly formed diploid hybrids of *Helianthus* (Ungerer et al. 2007) and contributes to genome size expansion in these taxa. The activation of normally silenced retrotransposons has also been demonstrated in synthetic allotetraploid wheat (Kashkush et al. 2002), but in this case no accompanying burst of transposition was observed (Kashkush et al. 2003). Surprisingly, analysis of the history of retrotransposon colonisation of the wheat genomes also found that proliferation rates were neither enhanced nor repressed by polyploidisation (Charles et al. 2008). A more recent paper from the Kashkush group (Kraitshtein et al. 2010) examined the behaviour of a specific TE sub-family, the *Veju* TRIM retrotransposons, in a new allohexaploid. *Veju* was chosen as a high copy functional family, expected to be particularly sensitive to epigenetic changes. Over 50% of *Veju* sites showed altered methylation, in most cases a reduction, in early generations, returning to normal or hypermethylated levels by S4. Unexpectedly there was widespread loss of the elements in the first generation, partly compensated by a subsequent increase in copy number. As with sequence loss (above), TE modification may be strongly dependent on the genetic backgrounds or ploidies of the hybrids used; the Kashkush group found hypermethylation of class II TEs (DNA transposons) in synthetic allohexaploids, but hypomethylation in synthetic allotetraploids (reviewed in Yaakov and Kashkush 2011).

2.4.2.3 Altered Gene Expression in Allopolyploids

Changes in gene expression appear to occur during both revolutionary and evolutionary phases of polyploidy. While the immediate (revolutionary) selective pressures might be to minimise deleterious expression, in the long term the presence of duplicate loci may permit the evolution of homoeologs with different functions. Some of the most detailed transcriptome work has been carried out in *Gossypium,* which includes allotetraploids combining Old and New World lineages in hybridization events some one to two million years ago. Gene deletion is very rare but there appears to be a constant bias for the polyploids to recover the transcription pattern of the New World lineage (D genome) whether this be for higher or lower levels of expression than seen in the Old World lineage (A genome) (Flagel and Wendel 2010). Synthetic *Gossypium* allopolyploids suggest that bias is particularly high during the 'revolutionary' phase (Rapp et al. 2009) and that it occurs even in an F1 hybrid where no change in ploidy is involved (Flagel and Wendel 2010). The 'evolutionary' phase of the natural allopolyploids therefore appears to have rather equalised the contribution of each parent, partially reversing the initial bias (Flagel and Wendel 2010). As predicted, there is good evidence that at least some homoeologous loci have evolved differential expression patterns (Dong and Adams 2011).

Other key work on altered patterns of gene expression in allopolyploids has come from studies on *Arabidopsis.* Synthetic lines of allotetraploid *Arabidopsis suecica,* using oligonucleotide arrays, showed non-additive changes to the expression of approximately 5% of genes represented on the array (Wang et al. 2006); and around 56% of these genes were affected in a similar way in two independently generated lines. Wang et al. (2006) showed that the affected genes did not appear to be limited to any particular chromosomal region, and belonged to a broad range of functional categories. An excellent recent paper (Chang et al. 2010) may have resolved these complexities and provides a model against which to gauge other systems. A tiling array with over 400k features, confirmed by resequencing, was used exhaustively to uncover a pattern in which interacting networks of genes derived from a single parent appeared to have been selected. Heterologous networks were significantly under-represented and the loci from the *Arabidopsis thaliana* lineage also tended to be expressed less frequently than those of the *A. arenosa* lineage. The authors propose a number of explanations for these findings, which are not mutually exclusive. Firstly, the *A. thaliana* genome may be disadvantaged both by its selfing history and by the fact that its native range is to the south of both *A. arenosa* and the hybrid, exacerbated by the possibly fortuitous predominance of *A. arenosa*-derived components in the hybrid's transcription machinery. Second, selection for networks derived from a single parent would minimise the expression of Dobzhansky-Muller incompatibilities, such as hybrid necrosis as discussed above. Such selection could also follow from disruption of expression neighbourhoods containing genes in the same network. Observations from a number of systems suggest that fractionation tends to be clustered, which might simply represent elimination of neighbourhoods damaged by a random initial event.

Similar analysis could be applied to data reported in other allopolyploid systems such as *Senecio* (Hegarty et al. 2006; Hegarty et al. 2008), and wheat (Pumphrey et al. 2009). This topic is explored in much greater depth in Jones and Hegarty (2009).

2.4.2.4 Epigenetic Changes

Much of the initial research into gene silencing in allopolyploids was performed using synthetic lines of *Arabidopsis suecica*, which displayed considerable phenotypic variation in the F2 generation and several cases of unstable phenotypes compared with the F1 (Comai et al. 2000). cDNA-AFLP analysis showed that a number of genes were potentially silenced in the allotetraploid lineages; and Southern blot assays using methylation-sensitive restriction enzymes indicated that these genes were methylated in the hybrids. It appeared, however, that there were no gross changes to the overall level of cytosine methylation in the allotetraploid genome (Madlung et al. 2002); but the use of methylation-sensitive AFLP (MSAP) showed that 8.3% of loci scored showed differential methylation between the parents and the F3 synthetic allotetraploids. The majority of these (76.9%) were consistent across the four allotetraploid individuals studied, with 62.5% of this group showing demethylation and the remainder displaying an increase in methylation (Madlung et al. 2002). These results suggest that the methylation pattern of allopolyploids is subject to widespread modification in a concerted manner. The role of small RNAs in allopolyploid systems is only now coming to light, although theories suggest that incompatibilities between divergent parental genomes could result in three specific effects in polyploids. The interaction of the two (or more) genomes could result in either (1) changes to the accumulation of siRNAs and miRNAs, or (2) changes in the efficiency of the siRNA/miRNA biosynthetic machinery, or (3) alterations to the specificity of the targets (Chen and Ni 2006). In the first such scenario, transcription will be affected differently depending upon the fidelity of the small RNAs for their targets. Low fidelity, where small RNA accumulation is increased (due to ploidy change), will result in silencing/downregulation of both parental transcripts, but high fidelity will affect only targets from one parent. This topic is also covered in greater detail in Jones and Hegarty (2009).

2.4.2.5 Non-destabilised Genomes

Finally it should be noted that not all detailed studies find major genome shocks to occur on hybridization (e.g., Ammiraju et al. 2010). A screen of ~22,0,000 loci in cotton allopolyploids failed to find evidence for sequence or methylation changes, nor were gross changes in TE content detected (Liu et al. 2001), suggesting that the expression changes described above could have been mediated by 'normal' pathway interactions. In support of this, a recent comparison of domesticated and wild *Gossypium* species suggests that domestication alone has led to alterations in expression of almost a quarter of transcripts in fibre cells, without recourse to 'genomic shock' (Rapp et al. 2010).

2.4.3 Adjustments at the Chromosome Level

Insights into genome organization and evolution at the sequence level are now based on comprehensive data sets, with a flood of comparative information anticipated over the next few years. Polyploids remain difficult for current technologies but clear pictures of the extent of sequence change induced by hybridization, as opposed to normal wear and tear, can be expected. Possibly the greatest challenge for the future, however, will be to understand how nuclear organization at the chromosome level is maintained and adapts in the face of sequence change, particularly the large structural variations and dynamic shifts in gene content and activity described above. These descriptions did not extend to other aspects of chromosome organization which are also subject to selection, such as variable recombination rates, maintenance of synteny, replication origins and so forth. As discussed for CTs, a strong self-organizing potential must be involved, as well as a high degree of plasticity. These properties seem to hold within a species, and intraspecific chromosomal polymorphisms are usually rather limited, in contrast to the sequence-level 'strains' outlined in Sect. 2.3. Hybridisation can lead to wholescale and often rapid reorganization, however. In the final section we will discuss how chromosomes adapt (or not) when they fall out with the new neighbours in the nucleus.

The process of adjustment leading to the establishment of stable polyploids can involve random translocations as well as species-specific rearrangements in particular chromosomes in all polyploid populations of a species (Leitch and Bennett 1997; Lim et al. 2008; Wright et al. 2009). Studies in newly formed or experimentally induced allopolyploids give us insights into the scope and the timing of these events, be they stochastic or otherwise. The pairing of chromosomes at meiosis can be compromised, especially in F1 hybrids, sudden changes in histone codes can determine how DNA is organized and how sequences are expressed in chromatin, and the way in which DNA is embedded in chromatin, and these conflicts in turn must impact on the way in which genomes interact and accommodate to sharing a nucleus (Jones and Pašakinskienė 2005). There may also be problems for the spatial separation of genomes and the arrangements of nuclear territories. Centromeres may be under separate genetic control, and those from one genome can suffer suppression and silencing by the other. This may lead to chromosome instabilities, even at a tissue-specific level, with the whole genome or part of the genome being eliminated as the response. The balance of chromatin may also change, with one species coming to sequentially represent a larger component of the genome than its partner over successive generations. Intragenomic recombination may also remodel a hybrid, as explained below.

2.4.3.1 Chromosome Changes in F1 Hybrids

F1 hybrids have serious problems in dealing with chromosome pairing. There may be differences in chromosome number as well as chromosome size, and even where there is some balance between the species involved there is frequent formation of univalents and lack of viable gametes. Nothwithstanding these conflicts there are some striking examples of their success in dealing with their hybridity. Asexual reproduction, as in the Egyptian tree onion, $2n = 2x = 16$, *Allium cepa* × *A. fistulosum* (Hanelt 1990; Jones 1991), is a prime example. Bulbils form in place of flowers. Apomixis is also a common means by which F1 hybrids can persist (Rieseberg 1997).

A large difference in genome size, based on the measurement of nuclear DNA amounts, as in the FI hybrid, is not necessarily a factor in creating genome conflict. The hybrid ryegrass, *Lolium temulentum* × *L. perenne* ($2n = 2x = 14$), for example, has two sets of chromosomes which are structurally and genetically dissimilar, differing in DNA amounts by about 50%, and yet they form a stable F1 hybrid with regular bivalents at meiosis and with chiasma frequencies which are similar to those of the parents, although the fertility is low. This hybrid has the capacity to resolve differences in terms of its synaptonemal complexes and to produce homoeologous bivalents with functional and morphological integrity (Jenkins and White 1990). This raises the question – if genome or chromosome size is not a critical factor then what is it that makes for conflict? The answer it seems is that structural differences between chromosomes are not necessarily size-dependent. Hybrid systems are known where translocations or tranversions may be the main cause of loss of fertility in an F1 hybrid (Rieseberg 1997).

2.4.3.2 Chromosome Elimination in F1 Hybrids

The extreme of the conflict in F1 hybrids is where one parental genome is rapidly eliminated by the other due to mitotic processes. This often happens in the first few nuclear divisions of the zygote; although in some cases where perenniality is involved the elimination may be incomplete. The phenomenon was first correctly interpreted in *Hordeum vulgare* × *H. bulbosum* by Kasha and Kao (1970), and is now known in many other interspecies and intergeneric crosses.

The essential components of the system can be briefly summarised as follows:

(1) Species vary in the extent of the elimination involved. In some hybrids stable F1 plants may be established, while in other genotypes there is elimination of one of the parental genomes, and when it does occur the extent varies, so that some plants can remain as mosaics for long periods of time, e.g., *H. vulgare* × *H. bulbosum* (Thomas and Pickering 1983; Riera-Lizarazu et al. 1996).

(2) Eliminated chromosomes often fail to congress on to the metaphase plate or to reach the anaphase poles during the early divisions of the zygote when the elimination is taking place, as in *H. vulgare* × *H. bulbosum* (Bennett et al. 1976). Failure to congress could be due to a lack of efficiency in attachment to spindle microtubules; and as later studies in wheat × maize crosses demonstrated, all of the much smaller maize genome chromosomes were lost during the first three cell divisions (Laurie and Bennett 1989), or early embryogenesis (Mochida et al. 2004) in most of the embryos. The maize centromeres were either tiny or non-visible and without affinity for spindle attachment, and the maize nucleolar organizing regions (NORs) were also suppressed (Laurie and Bennett 1989). The same is true in wheat × sorghum crosses (Laurie and Bennett 1988). In barley × maize hybrids the maize chromosomes in the zygote had well defined centromeres, but even so they were still eliminated (Laurie and Bennett 1988).

(3) In stable hybrids the genomes may show some level of spatial separation, as in *H. vulgare* cv. Tuleen 346 × *H. bulbosum* (Anamthawat-Jonsson et al. 1993), and the differential behaviour of the parents indicated that their centromere activity must be under separate genetic control. In later studies on the same hybrid Schwarzacher et al. (1992) confirmed genome separation, as well as the near identical size of the two genomes. Centromere-associated structures of *H. vulgare* were larger than those of the more peripheral *H. bulbosum* chromosomes which have 'weaker' (smaller) centromeres. Genome separation has also been found in several other hybrids (Finch et al. 1981; Leitch et al. 1990; Leitch et al. 1991).

(4) In the hybrid *H. vulgare* cv. Tuleen 346 × *H. marinum* elimination of the Tuleen 346 genome occurs in the endosperm, while that of *H. marinum* is lost in the embryo; i.e., there is tissue-specific 'alternative elimination' of parental genomes (Finch 1983; Finch and Bennett 1983). The eliminated chromosomes again had smaller centromeres, and tended to occupy more peripheral positions. The authors suggested tissue-specific suppression of genes for centromere function. Interestingly, the eliminated chromosomes also showed suppression of their NORs.

(5) The most compelling evidence we have to date that the centromere underlies chromosome instabilities in hybrids comes from the application of DNA fibre-FISH to addition lines of individual maize chromosomes added to oat (Jin et al. 2004). The centromeric DNA (CEN-DNA) is intermingled with centromere-specific retrotransposons (CRM), and these two components make up a range of sizes varying between ~ 300 and 2,800 kb for individual maize centromeres. The point of interest is that in addition lines with two different genes coding for the CENH3 histone, it is the oat gene that is dominant and the oat CENH3 that becomes incorporated into the maize centromeres. The maize CENH3 gene is silenced, and the oat CENH3 organizes the kinetochore on the maize chromosomes. Interestingly, in oat × maize hybrids it is the maize chromosomes which are eliminated and the maize centromeres which are impaired in the hybrids. In the oat × maize crosses the elimination of the maize chromosomes is more gradual than it is in crosses such as wheat × maize and barley × maize (Riera-Lizarazu et al. 1996), but nonetheless it does occur, and the least we can now say is that the centromeres find themselves compromised in hybrids and are subject to suppression and silencing together with the NOR ribosomal RNA genes.

In wheat × pearl millet hybrids Gernand et al. (2005) have shown how the pearl millet chromosomes in young embryos occupy a distinct interphase territory, mainly at the periphery of the nucleus. The pearl millet chromatin is then sequentially eliminated, over successive mitoses, by the formation of micronuclei which bud-off from the wheat nucleus. At the same time there is failure of segregation of pearl millet chromosomes at mitosis itself as well as obvious changes to their structural integrity. Thus there are two processes going on which gradually eliminate all the pearl millet chromatin, which is then degraded and lost, leaving the haploid wheat nucleus. The authors further speculate that the elimination process involves differences between wheat and pearl millet in post-translational histone modifcations; but such differences were not found in *Hordeum marinum* × *H. bulbosum* hybrids and their parents (Sanei et al. 2010).

New data from Ishii et al. (2010), using *in situ* hybridization to probe both genomic DNA and centromere-specific repeats, confirm that pearl millet chromosomes are eliminated in crosses with several members of the Triticeae.

The elimination process showed several of the features previously described by Gernand et al. (2005), including chromosome breakage, nondisjunction, micronuclei formation and failure of chromosome segregation at mitosis. Ishii et al. (2010) further revealed that the cause of elimination cannot be explained by malfunction of kinetochores binding to spindles, but proposed that it could be due to differences in the properties of cohesins between sister chromatids in the wheat and pearl millet chromosomes. In hybrids with oat however, all seven of the pearl millet chromosomes were retained, except for a small percentage of cells with decreased (6.1%) or increased (2.4%) numbers of pearl millet chromosomes, and carried centromere signals of the pearl millet chromosomes.

2.4.3.3 Chromosome Changes in Allopolyploids

New hybrids arising by allopolyploidisation, either through genome duplication of an existing F1 or through the union of unreduced gametes, are immediately subject to a strong form of reproductive isolation, and allopolyploidy is therefore a direct route to instant speciation (reviewed in Hegarty and Hiscock 2008). The allopolyploid 'marriage' however is not harmonious: it results in an irreversible burst of reorganization and modification of the genomes involved. There are immediate and genome-wide changes at the visible level of chromosomes and phenotype, as well as cryptic modifications involving gene expression mediated by genetic and epigenetic mechanisms caused by alterations in gene regulatory networks (Matzke et al. 1999; Riddle and Birchler 2003).

Changes at the whole genome/chromosome level can be visualized using genomic *in situ* hybridization (GISH) (Pašakinskienė and Jones 2005); and this differential 'painting' of chromosome sets is a useful cytological resource for studying genome conflict in hybrids and allopolyploids. An interesting and novel case involves a hybrid between *Lolium multiflorum* and the allopolyploid *Festuca arundinacea:* itself a natural allohexaploid originating as a hybrid between *F. pratensis* and *F. glaucescens* (genome composition *FpFpFgFgFgFg,* Humphreys et al. 1995). In colchicine-doubled F1 C0 octoploids of *L. multiflorum* × *F. arundinacea* ($2n = 8x = 56$) some genotypes, as revealed by GISH, restructured themselves as 'novel diploids' by diploidisation (i.e., becoming diploid, $2n = 14$) and somatic recombination (Pašakinskienė et al. 1997). Using GISH these new genomic variants were shown to contain components derived from *F. pratensis*, *L. multiflorum* and *F. glaucescens*, with genomes of the three species being present in different proportions and as variable patterns, and with *F. pratensis* chromatin as their genomic basis.

Similar events were recently found in a selected F1 C1 hybrid from the same population of plants. A hexaploid genotype F2 3–18 ($2n = 6x = 42$) was characterised as a 'super-recombinant', with a high number of recombinant chromosomes. Some of these new chromosomes were constructs composed of chromatin from the three component species *L. multiflorum – F. pratensis – F. glaucescens*. This genotype was observed for a number of years, and was found to have high vigour complemented by a good level of fertility. Its instability became apparent in phenotypic segregation, during vegetative multiplication. In a few cases the initial hexaploid gave rise to somatic segregants with a diploid chromosome number of $2n = 2x = 14$ (Pašakinskienė and Jones 2005).

The instabilities described above in the *L. multiflorum* × *F. arundinacea* hybrids are genotype-specific, and the plants which became diploid ($2n = 2x = 14$) are constructed *de novo* from the tri-specific genomic conflict involving *L. multiflorum – F. pratensis – F. glaucescens.* In most cases the chromosomes of the novel diploid are very similar to the chromosomes of pure *F. pratensis* diploid, but in some cases the *F. pratensis* genome has gained various-sized blocks of *L. mulitflorum* chromatin, and sometimes *F. glaucescens* is present but not in all of the segregants. In any case the reconstructed chromosomes are different from those existing in the *F. pratensis* genome within *F. arundinacea*, according to their GISH-banding pattern (Pašakinskienė et al. 1998). It is assumed that the 'novel diploids' could have resulted from concerted translocations, where at some stage the entire newly made allopolyploid genome was a ferment of rearrangements of its constituent species-specific parts. The centromeres have most likely also played an important role in the formation of these 'novel diploids'. We speculate that the centromeres could be 'novel' as well, built on the basis of the centromeric components of the parental species involved in the chromosome set of the hybrid genome. A similar story to *Lolium – Festuca*, but involving meiotic as well as mitotic recombination, has recently been reported for a derivative reconstructed 'Zebra' chromosome from an *Elymus trachycaulus* × *Triticum aestivum* hybrid (Zhang et al. 2008b).

2.4.3.4 Genome Drift in *Festulolium*

Another manifestation of 'genomes at war' is revealed by changes to 'genome balance', or 'genome drift' in otherwise stable *Lolium-Festuca* allopolyploids. In the F8 population of the tetraploid hybrid 'Prior', *L. perenne* × *F. pratensis,* meiosis was observed to be stable in the early generations, with a high level of bivalent formation. However, a GISH study revealed that extensive recombination had taken place between homoeologs of the two genomes, and that the balance of chromatin was not equal (Canter et al. 1999). The substitution of *Festuca*-origin chromosomes by those of *Lolium*-origin resulted in a mean of 17.9 *Lolium* and 9.7 *Festuca* chromosomes per genotype. These results are similar to those of an earlier study looking at another hybrid *L. multiflorum* × *F. pratensis* (Zwierzykowski et al. 1998).

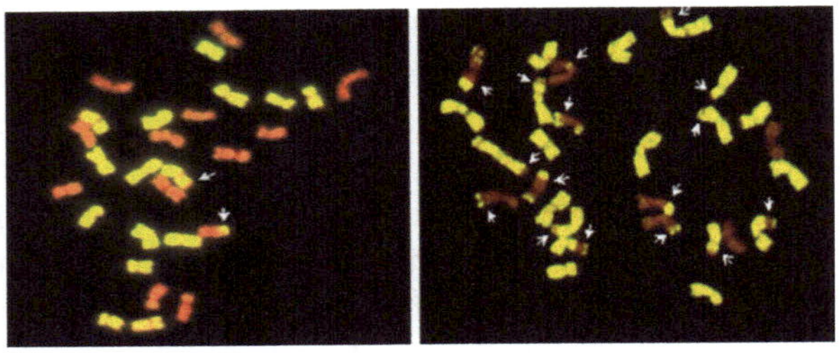

Fig. 2.3 Changes in the proportion of parental chromatin in the allotetraploid *Festuca pratensis* × *Lolium perenne* between generation F2 (*left*) and F6 (*right*) using genomic *in situ* hybridization. *L. perenne* chromosomes are labelled yellow whereas those originating from *F. pratensis* appear orange. Recombinant chromosomes are indicated by *arrows*, and the genomes are virtually identical in size

Here it was shown that the proportion of the genome occupied by *L. multiflorum* chromatin ranged from 49.2% to 66.7%, and this likewise confirms the balance in favour of *Lolium* over *Festuca*. Kopecký et al. (2006) have now shown that the phenomenon is widespread within the *Festulolium* species complex. They used GISH to study the genomic constitution of more than 600 plants from virtually all commercially available *Festulolium* cultivars, and found a large range of variation in the proportions of parental genomes and the level of intergenomic recombination, including several cultivars which were at the F9–F10 generation.

The reasons for the dominance of *Lolium* over *Festuca* in this way are not understood. Various theories have been proposed, such as gametic competition, pollination effects or selection for vigour in the early stages of seedling growth, but no definitive answers have yet emerged. In the light of recent knowledge of the centromere organization and function we could conjecture that the *Lolium* centromeres are more competitive than those of *Festuca*, and this may account for the predominance of *Lolium* chromatin in these hybrids, although any convincing mechanism still awaits discovery.

A similar story to that of *Festulolium* was earlier described by Anamthawat-Jonsson (1999). Using GISH she recovered a unique set of chromosomes in *Triticum* ($2n = 4x = 28$, AABB) × *Leymus* ($2n = 4x = 28$, NNXX) hybrids. The allopolyploids stabilised over a number of years as a set of six pairs of *Leymus* and 15 pairs of wheat chromosomes (Anamthawat-Jonsson 1999). The unique composition probably resulted from the stable replacement of one pair of *Leymus* chromosomes by the addition of one pair from wheat, together with the selective elimination of eight pairs from *Leymus*. It is clear the conflicts and chromosome instabilities in allopolyploids have a higher degree of complexity and a larger variety of outcomes. There are more conflicts to resolve and more ways in which resolutions can occur.

Recent studies involving allotetraploids of *Festuca pratensis* × *Lolium perenne* have now tracked changes in the balance of chromatin in somatic cells over six successive generations of open pollination (Zwierzykowski et al. 2006), and the outcome is the same (Fig. 2.3.). The proportion of total genome length contributed by *L. perenne* chromatin increased from about 50% in the F2 to about 60% in the F6, raising speculation about how far this chromatin replacement could go over further cycles of open pollination. The mechanism therefore appears to be progressive by small incremental shifts at each generation.

A key question in all of these studies revolves around the use of GISH, the level of resolution and how to interpret the observations. The sensitivity of GISH is limited to large megabase-sized chromosome segments (Lukaszewski et al. 2005), which could explain the imbalance in genome proportions of the hybrids, but this still leaves open the mechanism by which GISH megabases change. A new approach to dealing with this question has been to up the level of resolution of genome structure using Diversity Array Technology with polymorphic DArT markers, which allows for the screening of several thousands of markers at a single pass (Kopecký et al. 2009; Kopecký et al. 2011). Species-specific markers were identified which yielded results at variance with the GISH work, but the study also confirmed the enhanced sensitivity of the method and its potential for future analysis of 'genome drift'.

2.5 Concluding Remarks

The story of order and chaos in the plant nucleus is incomplete and open-ended. The knowledge that we have is based on only a handful of species, including model organisms with small genomes and a few plants of economic significance such as maize, rice and wheat. Nonetheless, certain observations have emerged which are both surprising and confusing. There is no normal or fixed order for the structure of the interphase plant nucleus; it depends on genome size which has a seamless array of values with no cut-off point between what is a small and large genome, and what happens to the order of the nucleus across this continuum of sizes. Nature has dealt with the advent of allopolyploidy, and experimental studies on nascent allopolyploids confirm the strong barriers that have had to be rapidly overcome in the process. There is a rich harvest of new discoveries yet to be made on order and chaos in the plant nucleus, and in the process of doing so we still need to bear in mind recent findings (e.g., maize and *Arabidopsis*) that many diploids, if not all of them, have a history of having passed through earlier cycles of ploidy events, and still bear the duplications as evidence. As far as hybrids are concerned, our new discoveries on genome readjustment have implications for the fundamental understanding of genome change in evolution, as well as presenting opportunities for the release of new forms of genetic and epigenetic variation in crop plants.

References

Abranches R, Beven AF, Aragón-Alcaide L, Shaw PJ (1998) Transcription sites are not correlated with chromosome territories in wheat nuclei. J Cell Biol 143:5–12

Alcazar R, Garcia AV, Parker JE, Reymond M (2009) Incremental steps toward incompatibility revealed by *Arabidopsis* epistatic interactions modulating salicylic acid pathway activation. Proc Natl Acad Sci USA 106:334–339

Alvarez ME, Nota F, Cambiagno DA (2010) Epigenetic control of plant immunity. Mol Plant P 11:563–576

Ammiraju JS, Fan C, Yu Y, Song X, Cranston KA, Pontaroli AC, Lu F, Sanyal A, Jiang N, Rambo T, Currie J, Collura K, Talag J, Bennetzen JL, Chen M, Jackson S, Wing RA (2010) Spatio-temporal patterns of genome evolution in allotetraploid species of the genus *Oryza*. Plant J 63:430–442

Anamthawat-Jonsson K (1999) Variable genome composition in *Triticum* × *Leymus* amphiploids. Theor Appl Genet 99:1087–1093

Anamthawat-Jonsson K, Schwarzacher T, Heslop-Harrison JS (1993) Behaviour of parental genomes in the hybrid *Hordeum vulgare* × *H. bulbosum*. J Hered 84:78–82

Armstrong SJ, Franklin FCH, Jones GH (2001) Nucleolus-associated telomere clustering precedes meiotic chromosome synapsis in *Arabidopsis thaliana*. J Cell Sci 114:4207–4217

Baack EJ, Rieseberg LH (2007) A genomic view of introgression and hybrid speciation. Curr Opin Genet Dev 17:513–518

Bailey JA, Eichler EE (2006) Primate segmental duplications: crucibles of evolution diversity and disease. Nat Rev Genet 7:552–564

Bartolome C, Bello X, Maside X (2009) Widespread evidence for horizontal transfer of transposable elements across *Drosophila* genomes. Genome Biol 10:R22

Baubec T, Dinh HQ, Pecinka A, Rakic B, Rozhon W, Wohlrab B, von Haeseler A, Mittelsten Scheid O (2010) Cooperation of multiple chromatin modifications can generate unanticipated stability of epigenetic states in *Arabidopsis*. Plant Cell 22:34–47

Bennett MD, Finch RA, Barclay IR (1976) The time, rate and mechanism of chromosome elimination in *Hordeum* hybrids. Chromosoma 54:175–200

Bento M, Gustafson JP, Viegas W, Silva M (2011) Size matters in *Triticeae* polyploids: larger genomes have higher remodelling. Genome 54:175–183

Bomblies K (2009) Too much of a good thing? Hybrid necrosis as a by-product of plant immune system diversification. Botany-Botanique 87:1013–1022

Bomblies K, Lempe J, Epple P, Warthmann N, Lanz C, Dangl JL, Weigel D (2007) Autoimmune response as a mechanism for a Dobzhansky-Muller-type incompatibility syndrome in plants. PLoS Biol 5:1962–1972

Boveri T (1909) Die Blastomerenkerne von *Ascaris megalocephala* und die Theorie der Chromosomenindividualität. Arch Zellforsch 3:181–268

Boyko A, Kathiria P, Zemp FJ, Yao Y, Pogribny I, Kovalchuk I (2007) Transgenerational changes in the genome stability and methylation in pathogen-infected plants (virus-induced plant genome instability). Nucleic Acids Res 35:1714–1725

Boyko A, Kovalchuk I (2011) Genome instability and epigenetic modification - heritable responses to environmental stress? Curr Op Plant Biol 14:260–266

Braszewska-Zalewska A, Bernas T, Maluszynska J (2009) Epigenetic chromatin modifications in *Brassica* genomes. Genome 53:203–210

Buggs RJA, Soltis PS, Soltis DE (2009) Does hybridization between divergent progenitors drive whole-genome duplication? Mol Ecol 18:3334–3339

Canter PH, Pašakinskienė I, Jones RN, Humphreys MW (1999) Chromosome substitutions and recombination in the amphiploid *Lolium perenne* × *Festuca pratensis* cv. Prior ($2n = 4x = 28$). Theor Appl Genet 98:809–814

Caron H, van Schaik B, van der Mee M, Baas F, Riggins G, van Sluis P, Hermus MC, van Asperen R, Boon K, Voûte PA, Heisterkamp S, van Kampen A, Versteeg R (2001) The human transcriptome map: clustering of highly expressed genes in chromosomal domains. Science 291:1289–1292

Chang PL, Dilkes BP, McMahon M, Comai L, Nuzhdin CL (2010) Homoeolog-specific retention and use in allotetraploid *Arabidopsis suecica* depends on parent of origin and network partners. Genome Biol 11:R125

Charles M, Belcram H, Just J, Huneau C, Viollet A, Couloux A, Segurens B, Carter M, Huteau V, Coriton O, Appels R, Samain S, Chalhoub B (2008) Dynamics and differential proliferation of transposable elements during the evolution of the B and A genomes of wheat. Genetics 180:1071–1086

Chen ZJ (2010) Molecular mechanisms of polyploidy and hybrid vigor. Trends Plant Sci 15:57–71

Chen ZJ, Ni Z (2006) Mechanisms of genomic rearrangements and gene expression changes in plant polyploids. Bioessays 28:240–252

Chen Q, Han Z, Jiang H, Tian D, Yang S (2010) Strong positive selection drives rapid diversification of R-genes in *Arabidopsis* relatives. J Mol Evol 70:137–148

Comai L, Tyagi AP, Winter K, Holmes-Davis R, Reynolds SH, Stevens Y, Byers B (2000) Phenotypic instability and rapid gene silencing in newly formed *Arabidopsis* allotetraploids. Plant Cell 12:1551–1567

Cowan CR, Carlton PM, Cande WZ (2001) The polar arrangement of telomeres in interphase and meiosis. Rabl organisation and the bouquet. Plant Physiol 125:532–538

Cremer T, Cremer M (2010) Chromosome territories. Cold Spring Harb Perspect Biol 2:a00388

David P, Chen NW, Pedrosa-Harand A, Thareau V, Sévignac M, Cannon SB, Debouck D, Langin T, Geffroy V (2009) A nomadic subtelomeric disease resistance gene cluster in common bean. Plant Physiol 151:1048–1065

DeBolt S (2010) Copy number variation shapes genome diversity in *Arabidopsis* over immediate family generational scales. Genome Biol Evol 2:441–453

Ding J, Araki H, Wang Q, Zhang P, Yang S, Chen JQ, Tian D (2007) Highly asymmetric rice genomes. BMC Genomics 8:154

Dong S, Adams K (2011) Differential contributions to the transcriptome of duplicated genes in response to abiotic stresses in natural and synthetic polyploids. New Phytol 190:1045–1057

Dong F, Jiang J (1998) Non-Rabl patterns of centromere and telomere distribution in the interphase nuclei of plant cells. Chromosome Res 6:551–558

Du C, Fefelova N, Caronna J, He L, Dooner HK (2009) The polychromatic Helitron landscape of the maize genome. Proc Natl Acad Sci USA 106:19916–19921

Ebisuya M, Yamamoto T, Nakajima M, Nishida E (2008) Ripples from neighbouring transcription. Nat Cell Biol 10:1106–1113

Fang Y, Spector DL (2005) Centromere positioning and dynamics in living *Arabidopsis* plants. Mol Biol Cell 16:5710–5718

Fawcett JA, Maere S, Van de Peer Y (2009) Plants with double genomes might have had chance to survive the Cretaceous – Tertiary extinction event. Proc Natl Acad Sci USA 106:5737–5742

Feldman M, Liu B, Segal G, Abbo S, Levy AA, Vega JM (1997) Rapid elimination of low-copy DNA sequences in polyploid wheat: a possible mechanism for differentiation of homoeologous chromosomes. Genetics 147:1381–1387

Filichkin SA, Priest HD, Givan SA, Shen R, Bryant DW, Fox SE, Wong WK, Mockler TC (2010) Genome-wide mapping of alternative splicing in *Arabidopsis thaliana*. Genome Res 20:45–58

Finch RA (1983) Tissue-specific elimination of alternative whole parental genomes in one barley hybrid. Chromosoma 88:386–393
Finch RA, Bennett MD (1983) The mechanism of somatic chromosome elimination in *Hordeum*. In: Brandham PE, Bennett MD (eds) Kew chromosome conference II. Allen and Unwin, London, pp 147–154
Finch RA, Smith JB, Bennett MD (1981) *Hordeum* and *Secale* mitotic genomes lie apart in a hybrid. J Cell Sci 52:391–403
Flagel LE, Wendel JF (2010) Evolutionary rate variation, genomic dominance and duplicate gene expression evolution during allotetraploid cotton speciation. New Phytol 186:184–193
Fransz P, de Jong H (2011) From nucleosome to chromosome: a dynamic organization of genetic information. Plant J 66:4–17
Fransz P, De Jong H, Lysák M, Castiglione MF, Schubert I (2002) Interphase chromosomes in *Arabidopsis* are organised as well defined chromocenters from which euchromatin loops emanate. Proc Natl Acad Sci USA 99:14584–14589
Fu HH, Dooner HK (2002) Intraspecific violation of genetic colinearity and its implications in maize. Proc Natl Acad Sci USA 99:9573–9578
Fuchs J, Demidov D, Houben A, Schubert I (2006) Chromosomal histone modification patterns — from conservation to diversity. Trends Plant Sci 11:199–208
Gaut BS, Doebley JF (1997) DNA sequence evidence for the segmental allotetraploid origin of maize. Proc Natl Acad Sci USA 94:6809–6814
Gehring M, Bubb KL, Henikoff S (2009) Extensive demethylation of repetitive elements during seed development underlies gene imprinting. Science 324:1447–1451
Gernand D, Rutten T, Varshney A, Rubtsova M, Prodanovic S, Brüss C, Kumlehn J, Matzk F, Houben A (2005) Uniparental chromosome elimination at mitosis and interphase in wheat and pearl millet crosses involves micronucleus formation, progressive heterochromatinization, and DNA fragmentation. Plant Cell 17:2431–2438
Gill BS (1991) Nucleo-cytoplasmic interaction (NCI) hypothesis of genome evolution and speciation in polyploid plants. In: Sasakuma T, Kinoshita T (eds) pp 48–53, Proceedings of the Kihara memorial symposium on cytoplasmic engineering in wheat. Yokohama, Japan
Goetze S, Mateos-Langerak J, Gierman HJ, de Leeuw W, Giromus O, Indemans MH, Koster J, Ondrej V, Versteeg R, van Driel R (2007) The three-dimensional structure of human interphase chromosomes is related to the transcriptome map. Mol Cell Biol 27: 4475–4487
Gore MA, Chia JM, Elshire RJ, Sun Q, Ersoz ES, Hurwitz BL, Peiffer JA, McMullen MD, Grills GS, Ross-Ibarra J, Ware DH, Buckler ES (2009) A first-generation haplotype map of maize. Science 326: 1115–1117
Haig D, Westoby M (1989) Parent-specific gene-expression and the triploid endosperm. Am Nat 134:147–155
Hanada K, Vallejo V, Nobuta K, Slotkin RK, Lisch D, Meyers BC, Shiu SH, Jiang N (2009) The functional role of pack-MULEs in rice inferred from purifying selection and expression profile. Plant Cell 21:25–38
Hanelt P (1990) Taxonomy, evolution, and history. In: Rabinowitch HD, Brewster JL (eds) Onions and allied crops, vol I, Botany, physiology and genetics. CRC Press, Boca Raton, pp 1–24
He LM, Dooner HK (2009) Haplotype structure strongly affects recombination in a maize genetic interval polymorphic for Helitron and retrotransposon insertions. Proc Natl Acad Sci USA 106:8410–8416
Hegarty MJ, Hiscock SJ (2005) Hybrid speciation in plants: new insights from molecular studies. New Phytol 165:411–423
Hegarty MJ, Hiscock SJ (2008) Genomic clues to the evolutionary success of polyploid plants. Curr Biol 18:R435–R444
Hegarty MJ, Barker GL, Wilson ID, Abbott RJ, Edwards KJ, Hiscock SJ (2006) Transcriptome shock after interspecific hybridisation in *Senecio* is ameliorated by genome duplication. Curr Biol 16:1652–1659
Hegarty MJ, Barker GL, Brennan AC, Abbott RJ, Edwards KJ, Hiscock SJ (2008) Changes to gene expression associated with hybrid speciation in plants: further insights from transcriptomic studies in *Senecio*. Philos Trans Roy Soc Lond B Bio 363:3055–3069
Heitz E (1928) Das Heterochromatin der Moose. I. Jahrb Wiss Bot 69:762–818
Herr AJ, Jensen MB, Dalmay T, Baulcombe DC (2005) RNA polymerase IV directs silencing of endogenous DNA. Science 308:118–120
Hollister JD, Smith LM, Guo YL, Ott F, Weigel D, Gaut BS (2011) Transposable elements and small RNAs contribute to gene expression divergence between *Arabidopsis thaliana* and *Arabidopsis lyrata*. Proc Natl Acad Sci USA 108:2322–2327
Hsieh TF, Shin J, Uzawa R, Silva P, Cohen S, Bauer MJ, Hashimoto M, Kirkbride RC, Harada JJ, Zilberman D, Fischer RL (2011) Regulation of imprinted gene expression in *Arabidopsis* endosperm. Proc Natl Acad Sci USA 108:1755–1762
Humphreys MW, Thomas HM, Morgan WG, Meredith MR, Harper JA, Thomas H, Zwierzykowski Z, Ghesquiére M (1995) Discriminating the ancestral progenitors of hexaploid *Festuca arundinacea* using genomic *in situ* hybridisation. Heredity 75:171–174
Ishii T, Ueda T, Tanaka H, Tsujimoto H (2010) Chromosome elimination by wide hybridisation between *Triticeae* or oat plant and pearl millet: pearl millet chromosome dynamics in hybrid embryo cells. Chromosome Res 18:821–831
Jain M, Nijhawan A, Arora R, Agarwal P, Ray S, Sharma P, Kapoor S, Tyagi AK, Khurana JP (2007) F-box proteins in rice. Genome-wide analysis classification temporal and spatial gene expression during panicle and seed development and regulation by light and abiotic stress. Plant Physiol 143:1467–1483
Jenkins G, White J (1990) Elimination of synaptonemal complex irregularities in a *Lolium* hybrid. Heredity 64:45–53
Jiao Y, Wickett NJ, Ayyampalayam S, Chanderbali AS, Landherr L, Ralph PE, Tomsho LP, Hu Y, Liang H, Soltis PS, Soltis DE, Clifton SW, Schlarbaum SE, Schuster SC, Ma H, Leebens-Mack J, dePamphilis CW (2011) Ancestral polyploidy in seed plants and angiosperms. Nature 473:97–100
Jin W, Melo JR, Nagaki K, Talbert PB, Henikoff S, Dawe RK, Jiang J (2004) Maize centromeres: organization and functional adaptation in the genetic background of oat. Plant Cell 16:571–581
Johannes F, Porcher E, Teixeira FK, Saliba-Colombani V, Simon M, Agier N, Bulski A, Albuisson J, Heredia F, Audigier P, Bouchez D, Dillmann C, Guerche P, Hospital F, Colot V (2009) Assessing the impact of transgenerational epigenetic variation on complex traits. PLoS Genet 5:e1000530
Jones RN (1991) Cytogenetics of alliums. In: Tsuchiya T, Gupta PK (eds) Chromosome engineering in plants: genetics, breeding, evolution, Part B. Elsevier, Amsterdam, pp 215–227
Jones RN, Hegarty M (2009) Order out of chaos in the hybrid plant nucleus. Cytogenet Genome Res 126:376–389
Jones RN, Pašakinskienė I (2005) Genome conflict in the Gramineae. New Phytol 165:391–410
Kalendar R, Tanskanen J, Immonen S, Nevo E, Schulman AH (2000) Genome evolution of wild barley (*Hordeum spontaneum*) by *BARE*-1 retrotransposon dynamics in response to sharp microclimatic divergence. Proc Natl Acad Sci USA 97:6603–6607
Kalendar R, Flavell AJ, Ellis TH, Sjakste T, Moisy C, Schulman AH (2011) Analysis of plant diversity with retrotransposon-based molecular markers. Heredity 106:520–530
Kasha KJ, Kao KN (1970) High frequency haploid production in barley (*Hordeum vulgare* L.). Nature 225:874–876
Kashkush K, Feldman M, Levy AA (2002) Gene loss, silencing and activation in a newly synthesised wheat allotetraploid. Genetics 160:1651–1659
Kashkush K, Feldman M, Levy AA (2003) Transcriptional activation of retrotransposons alters the expression of adjacent genes in wheat. Nat Genet 33:102–106

Kaufmann K, Pajoro A, Angenent GC (2010) Regulation of transcription in plants: mechanisms controlling developmental switches. Nat Rev Genet 11:830–842

Kejnovsky E, Hawkins JS, Feschotte C (2012) Plant transposable elements: biology and evolution. In: Wendel JF, Greilhuber J, Doležel J, Leitch IJ (eds) Plant genome diversity, vol 1, Plant genomes, their residents, and their evolutionary dynamics. Springer-Verlag, Wien, pp 17–34

Kopecký D, Loureiro J, Zwierzykowski Z, Ghesquière M, Doležel J (2006) Genome constitution and evolution in *Lolium* × *Festuca* hybrid cultivars (*Festulolium*). Theor Appl Genet 113:731–742

Kopecký D, Bartoš J, Lukaszewski AJ, Baird JH, Černoch V, Kölliker R, Rognli OA, Blois H, Craig V, Lübberstedt T, Studer B, Shaw B, Dolezel J, Kilian A (2009) Development and mapping of DArT markers within the *Festuca-Lolium* complex. BMC Genomics 10:473

Kopecký D, Bartoš J, Christelová P, Černoch V, Kilian A, Doležel J (2011) Genomic constitution of *Festuca* × *Lolium* hybrids revealed by the DArTFest array. Theor Appl Genet 122:355–363

Kraitshtein Z, Yaakov B, Khasdan V, Kashkush K (2010) Genetic and epigenetic dynamics of a retrotransposon after allopolyploidization of wheat. Genetics 186:801–812

Lai J, Li R, Xu X, Jin W, Xu M, Zhao H, Xiang Z, Song W, Ying K, Zhang M, Jiao Y, Ni P, Zhang J, Li D, Guo X, Ye K, Jian M, Wang B, Zheng H, Liang H, Zhang X, Wang S, Chen S, Li J, Fu Y, Springer NM, Yang H, Wang J, Dai J, Schnable PS, Wang J (2010) Genome-wide patterns of genetic variation among elite maize inbred lines. Nat Genet 42:1027–1030

Laibach F (1907) Zur Frage nach der Individualität der Chromosomen im Pflanzenreich. Beih Bot Centralbl 22:191–210

Laurie DA, Bennett MD (1988) Chromosome behaviour in wheat × maize, wheat × sorghum and barley × maize crosses. In: Brandham PE (ed) Kew chromosome conference III. HMSO, London, pp 167–177

Laurie DA, Bennett MD (1989) The timing of chromosome elimination in hexaploid wheat × maize crosses. Genome 32:953–961

Leitch IJ, Bennett MD (1997) Polyploidy in angiosperms. Trends Plant Sci 2:470–476

Leitch AR, Mosgoller W, Schwarzacher T, Bennett MD, Heslop-Harrison JS (1990) Genomic *in situ* hybridisation to sectioned nuclei shows chromosome domains in grass hybrids. J Cell Sci 95:335–341

Leitch AR, Schwarzacher T, Mosgoller W, Bennett MD, Heslop-Harrison JS (1991) Parental genomes are separated throughout the cell cycle in a plant hybrid. Chromosoma 101:206–213

Levin DA (2002) The role of chromosomal change in plant evolution. Oxford University Press, New York

Levy AA, Feldman M (2002) The impact of polyploidy on grass genome evolution. Plant Physiol 130:1587–1593

Lieberman-Aiden E, van Berkum NL, Williams L, Imakaev M, Ragoczy T, Telling A, Amit I, Lajoie BR, Sabo PJ, Dorschner MO, Sandstrom R, Bernstein B, Bender MA, Groudine M, Gnirke A, Stamatoyannopoulos J, Mirny LA, Lander ES, Dekker J (2009) Comprehensive mapping of long-range interactions reveals folding principles of the human genome. Science 326:289–293

Lim KY, Soltis DE, Soltis PS, Tate J, Matyasek R, Srubarova H, Kovarik A, Pires JC, Xiong Z, Leitch AR (2008) Rapid chromosome evolution in recently formed polyploids in *Tragopogon* (Asteraceae). PLoS One 3:e3353

Liu B, Brubaker CL, Mergeai G, Cronn RC, Wendel JF (2001) Polyploid formation in cotton is not accompanied by rapid genomic changes. Genome 44:321–330

Lukaszewski AJ, Lapiński B, Rybka K (2005) Limitations of *in situ* hybridization with total genomic DNA in routine screening for alien introgression in wheat. Cytogenet Genome Res 109:373–377

Madlung A, Masuelli RW, Watson B, Reynolds SH, Davison J, Comai L (2002) Remodeling of DNA methylation and phenotypic and transcriptional changes in synthetic *Arabidopsis* allotetraploids. Plant Physiol 129:733–746

Malik HS, Henikoff S (2009) Major evolutionary transitions in centromere complexity. Cell 138:1067–1082

Martin A, Troadec C, Boualem A, Rajab M, Fernandez R, Morin H, Pitrat M, Dogimont C, Bendahmane A (2009) A transposon-induced epigenetic change leads to sex determination in melon. Nature 461:1135–1138

Matzke MA, Mittelsten Scheid O, Matzke AJM (1999) Rapid stuctural and epigenetic changes in polyploid and aneuploid genomes. Bioessays 21:761–767

Matzke M, Kanno T, Daxinger L, Huettel B, Matzke AJ (2009) RNA-mediated chromatin-based silencing in plants. Curr Opin Cell Biol 21:367–376

McClintock B (1984) The significance of responses of the genome to challenge. Science 226:792–801

Mestiri I, Chagué V, Tanguy AM, Huneau C, Huteau V, Belcram H, Coriton O, Chalhoub B, Jahier J (2010) Newly synthesized wheat allohexaploids display progenitor-dependent meiotic stability and aneuploidy but structural genomic additivity. New Phytol 186:86–101

Mirouze M, Paszkowski J (2011) Epigenetic contribution to stress adaptation in plants. Curr Opin Plant Biol 14:1–8

Mochida K, Tsujimoto H, Sasakuma T (2004) Confocal analysis of chromosome behaviour in wheat × maize zygotes. Genome 47:199–205

Molnar A, Melnyk CW, Bassett A, Hardcastle TJ, Dunn R, Baulcombe DC (2010) Small silencing RNAs in plants are mobile and direct epigenetic modification in recipient cells. Science 328:872–875

Nahon JL (2003) Birth of 'human-specific' genes during primate evolution. Genetica 118:193–208

Nei M, Rooney AP (2005) Concerted and birth-and-death evolution of multigene families. Annu Rev Genet 39:121–152

Onodera Y, Haag JR, Ream T, Costa Nunes P, Pontes O, Pikaard CS (2005) Plant nuclear RNA polymerase IV mediates siRNA and DNA methylation-dependent heterochromatin formation. Cell 120:613–622

Ouyang YD, Liu YG, Zhang QF (2010) Hybrid sterility in plant: stories from rice. Curr Opin Plant Biol 13:186–192

Ozkan H, Levy AA, Feldman M (2001) Allopolyploidy-induced rapid genome evolution in the wheat (*Aegilops* – *Triticum*) group. Plant Cell 13:1735–1747

Parisod C, Holderegger R, Brochmann C (2010) Evolutionary consequences of autopolyploidy. New Phytol 186:5–17

Pašakinskienė I, Anamthawat-Jonsson K, Humphreys MW, Jones RN (1997) Novel diploids following chromosome elimination and somatic recombination in *Lolium multiflorum* × *Festuca arundinacea* hybrids. Heredity 78:464–469

Pašakinskienė I, Anamthawat-Jonsson K, Humphreys MW, Paplauskiene V, Jones RN (1998) New molecular evidence on genome relationships and chromosome identification in *Festuca* and *Lolium*. Heredity 81:659–665

Pašakinskienė I, Jones RN (2005) A decade of 'chromosome painting' in *Lolium* and *Festuca*: species relationships and genome adjustment in hybrids. Cytogenet Genome Res 109:393–399

Paun O, Forest F, Fay MF, Chase MW (2009) Hybrid speciation in angiosperms: parental divergence drives ploidy. New Phytol 182:507–518

Pecinka A, Schubert V, Meister A, Kreth G, Klatte M, Lysák MA, Fuchs J, Schubert I (2004) Chromosome territory arrangement and homologous pairing in nuclei of *Arabidopsis thaliana* are predominantly random except for NOR-bearing chromosomes. Chromosoma 113:258–269

Petersena G, Aagesenb L, Seberga O, Larsen IH (2011) When is enough, enough in phylogenetics? A case in point from *Hordeum* (Poaceae). Cladistics 27:1–19

Pumphrey M, Bai J, Laudencia-Chingcuanco D, Anderson O, Gill BS (2009) Nonadditive expression of homoeologous genes is established upon polyploidization in hexaploid wheat. Genetics 181:1147–1157

Rabl C (1885) Über Zelltheilung. Morphol Jahrb 10:214–330

Rajapakse I, Groudine M (2011) On emerging nuclear order. J Cell Biol 192:711–721

Rapp RA, UdalL JA, Wendel JF (2009) Genomic expression dominance in allopolyploids. BMC Biol 7:18

Rapp RA, Haigler CH, Flagel L, Hovav RH, Udall JA, Wendel JF (2010) Gene expression in developing fibers of Upland cotton (*Gossypium hirsutum* L.) was massively altered by domestication. BMC Biol 8:139

Reinders J, Wulff BB, Mirouze M, Marí-Ordóñez A, Dapp M, Rozhon W, Bucher E, Theiler G, Paszkowski J (2009) Compromised stability of DNA methylation and transposon immobilization in mosaic *Arabidopsis* epigenomes. Genes Dev 23:939–950

Riddle NC, Birchler JA (2003) Effects of reunited diverged regulatory hierarchies in allopolyploids and species hybrids. Trends Genet 19:597–600

Riera-Lizarazu O, Rines HW, Phillips RL (1996) Cytological and molecular characterisation of oat × maize partial hybrids. Theor Appl Genet 93:123–135

Rieseberg LH (1997) Hybrid origins of plant species. Annu Rev Ecol Syst 28:359–389

Rieseberg LH, Blackman BK (2010) Speciation genes in plants. Ann Bot 106:439–455

Rieseberg LH, Wood TE, Baack EJ (2006) The nature of plant species. Nature 440:524–527

Riley MC, Clare A, King R (2007) Locational distribution of gene functional classes in *Arabidopsis thaliana*. BMC Bioinformatics 8:112

Rosloski SM, Jali SS, Balasubramanian S, Weigel D, Grbic V (2010) Natural diversity in flowering responses of *Arabidopsis thaliana* caused by variation in a tandem gene array. Genetics 186:263–276

Rouquette J, Cremer C, Cremer T, Fakan S (2010) Functional nuclear architecture studied by microscopy: present and future. Int Rev Cell Mol Biol 282:1–90

Sanei M, Pickering R, Fuchs J, Banaei Moghaddam AM, Dziurlikowska A, Houben DA (2010) Interspecific hybrids of *Hordeum marinum* ssp. *marinum* × *H. bulbosum* are mitotically stable and reveal no gross alterations in chromatin properties. Cytogenet Genome Res 129:110–116

Santos AP, Shaw P (2004) Interphase chromosomes and the Rabl configuration: does genome size matter? J Microsc 214:201–206

Santos AP, Abranches R, Stroger E, Beven A, Viegas W, Shaw PJ (2002) The architecture of interphase chromosomes and gene positioning are altered by changes in DNA methylation and histone acetylation. J Cell Sci 115:4597–4605

Schnable JC, Springer NM, Freeling M (2011) Differentiation of the maize subgenomes by genome dominance and both ancient and ongoing gene loss. Proc Natl Acad Sci USA 108:4069–4074

Schön CC, Dhillon BS, Utz HF, Melchinger AE (2010) High congruency of QTL positions for heterosis of grain yield in three crosses of maize. Theor Appl Genet 120:321–332

Schubert I, Shaw P (2011) Organization and dynamics of plant interphase chromosomes. Trends Plant Sci 16:273–281

Schubert V, Klatte M, Pecinka A, Meister A, Jasencakova Z, Schubert I (2006) Sister chromatids are often incompletely aligned in meristematic and endopolyploid interphase nuclei of *Arabidopsis thaliana*. Genetics 172:467–475

Schwarzacher T, Heslop-Harrison JS, Anamthawat-Jonsson K (1992) Parental genome separation in reconstruction of somatic and premeiotic metaphases of *Hordeum vulgare* × *H. bulbosum*. J Cell Sci 101:13–24

Selker EU (2002) Repeat-induced gene silencing in fungi. Adv Genet 46:439–450

Shaked H, Kashkush K, Ozkan H, Feldman M, Levy AA (2001) Sequence elimination and cytosine methylation are rapid and reproducible responses of the genome to wide hybridisation and allopolyploidy in wheat. Plant Cell 13:1749–1759

Shen J, Araki H, Chen L, Chen JQ, Tian D (2006) Unique evolutionary mechanism in R-genes under the presence/absence polymorphism in *Arabidopsis thaliana*. Genetics 172:1243–1250

Shi JH, Dawe RK (2006) Partitioning of the maize epigenome by the number of methyl groups on histone H3 lysines 9 and 27. Genetics 173:1571–1583

Shull GH (1948) What is "heterosis"? Genetics 33:439–446

Slotkin RK, Nuthikattu S, Jiang N (2012) The impact of transposable elements on gene and genome evolution. In: Wendel JF, Greilhuber J, Doležel J, Leitch IJ (eds) Plant genome diversity, vol 1, Plant genomes, their residents, and their evolutionary dynamics. Springer-Verlag, Wien, pp 35–58

Soltis PS, Soltis DE (2009) The role of hybridization in plant speciation. Annu Rev Plant Biol 60:561–588

Soltis DE, Albert VA, Leebens-Mack J, Bell CD, Paterson AH, Zheng C, Sankoff D, Depamphilis CW, Wall PK, Soltis PS (2009) Polyploidy and angiosperm diversification. Am J Bot 96:336–348

Song KM, Lu P, Tang KL, Osborn TC (1995) Rapid genome change in synthetic polyploids of Brassica and its implications for polyploid evolution. Proc Natl Acad Sci USA 92:7719–7723

Soppe WJ, Jasencakova Z, Houben A, Kakutani T, Meister A, Huang MS, Jacobsen SE, Schubert I, Fransz PF (2002) DNA methylation controls histone H3 lysine 9 methylation and heterochromatin assembly in *Arabidopsis*. EMBO J 21:6549–6559

Springer NM, Ying K, Fu Y, Ji T, Yeh CT, Jia Y, Wu W, Richmond T, Kitzman J, Rosenbaum H, Iniguez AL, Barbazuk WB, Jeddeloh JA, Nettleton D, Schnable PS (2009) Maize inbreds exhibit high levels of copy number variation (CNV) and presence/absence variation (PAV) in genome content. PLoS Genet 5:e1000734

Stupar RM, Springer NM (2006) Cis-transcriptional variation in maize inbred lines B73 and Mo17 leads to additive expression patterns in the F-1 hybrid. Genetics 173:2199–2210

Stupar RM, Bhaskar PB, Yandell BS, Rensink WA, Hart AL, Ouyang S, Veilleux RE, Busse JS, Erhardt RJ, Buell CR, Jiang J (2007) Phenotypic and transcriptomic changes associated with potato autopolyploidization. Genetics 176:2055–2067

Stupar RM, Gardiner JM, Oldre AG, Haun WJ, Chandler VL, Springer NM (2008) Gene expression analyses in maize inbreds and hybrids with varying levels of heterosis. BMC Plant Biol 8:33

Sun X, Zhang Y, Yang S, Chen JQ, Hohn B, Tian D (2008) Insertion DNA promotes ectopic recombination during meiosis in *Arabidopsis*. Mol Biol Evol 25:2079–2083

Swanson-Wagner RA, Eichten SR, Kumari S, Tiffin P, Stein JC, Ware D, Springer NM (2010) Pervasive gene content variation and copy number variation in maize and its undomesticated progenitor. Genome Res 20:1689–1699

Tessadori F, Schulkes RK, van Driel R, Fransz P (2007) Light-regulated large-scale reorganization of chromatin during the floral transition in *Arabidopsis*. Plant J 50:848–857

Thomas HM, Pickering RA (1983) Chromosome elimination in *Hordeum vulgare* × *H. bulbosum* hybrids. I. Comparisons of stable and unstable amphiploids. Theor Appl Genet 66:135–140

Tian Z, Yu Y, Lin F, Yu Y, Sanmiguel PJ, Wing RA, McCouch SR, Ma J, Jackson SA (2011) Exceptional lability of a genomic complex in rice and its close relatives revealed by interspecific and intraspecific comparison and population analysis. BMC Genomics 12:142

Todesco M, Balasubramanian S, Hu TT, Traw MB, Horton M, Epple P, Kuhns C, Sureshkumar S, Schwartz C, Lanz C, Laitinen RA, Huang Y, Chory J, Lipka V, Borevitz JO, Dangl JL, Bergelson J, Nordborg M, Weigel D (2010) Natural allelic variation underlying a major fitness trade-off in *Arabidopsis thaliana*. Nature 465:632–636

Uchida N, Tasaka M (2010) Intersections between immune responses and morphological regulation in plants. J Exp Bot 61:2539–2547

Ungerer MC, Strakosh SC, Zhen Y (2007) Genome expansion in three hybrid sunflower species is associated with retrotransposon proliferation. Curr Biol 16:R872–R873

Vazquez F, Blevins T, Ailhas J, Boller T, Meins F Jr (2008) Evolution of *Arabidopsis* MIR genes generates novel microRNA classes. Nucleic Acids Res 36:6429–6438

Verhoeven KJ, Jansen JJ, van Dijk PJ, Biere A (2010) Stress-induced DNA methylation changes and their heritability in asexual dandelions. New Phytol 185:1108–1118

Vielle-Calzada JP, Martínez de la Vega O, Hernández-Guzmán G, Ibarra-Laclette E, Alvarez-Mejía C, Vega-Arreguín JC, Jiménez-Moraila B, Fernández-Cortés A, Corona-Armenta G, Herrera-Estrella L, Herrera-Estrella A (2009) The Palomero genome suggests metal effects on domestication. Science 326:1078

Vuylsteke M, van Eeuwijk F, Van Hummelen P, Kuiper M, Zabeau M (2005) Genetic analysis of variation in gene expression in *Arabidopsis thaliana*. Genetics 171:1267–1275

Wang Q, Dooner HK (2006) Remarkable variation in maize genome structure inferred from haplotype diversity at the bz locus. Proc Natl Acad Sci USA 103:17644–17649

Wang J, Tian L, Lee HS, Wei NE, Jiang H, Watson B, Madlung A, Osborn TC, Doerge RW, Comai L, Chen ZJ (2006) Genome-wide nonadditive genes regulation in *Arabidopsis* allotetraploids. Genetics 17:507–517

Wang WK, Ho CW, Hung KH, Wang KH, Huang CC, Araki H, Hwang CC, Hsu TW, Osada N, Chiang TY (2010) Multilocus analysis of genetic divergence between outcrossing *Arabidopsis* species: evidence of genome-wide admixture. New Phytol 188:488–500

Wicker T, Krattinger SG, Lagudah ES, Komatsuda T, Pourkheirandish M, Matsumoto T, Cloutier S, Reiser L, Kanamori H, Sato K, Perovic D, Stein N, Keller B (2009) Analysis of intraspecies diversity in wheat and barley genomes identifies breakpoints of ancient haplotypes and provides insight into the structure of diploid and hexaploid *Triticeae* gene pools. Plant Physiol 149:258–270

Williams EJB, Bowles DJ (2004) Coexpression of neighbouring genes in the genome of *Arabidopisis thaliana*. Genome Res 14:1060–1067

Wolny E, Lesniewska K, Hasterok R, Langdon T (2011) Compact genomes and complex evolution in the genus *Brachypodium*. Chromosoma 120:199–212

Woodcock CL, Ghosh RP (2010) Chromatin higher-order structure and dynamics. Cold Spring Harbor Perspect Biol 2:a000596

Wright KM, Pires C, Madlung A (2009) Mitotic instability in resynthesised and natural polyploids of the genus *Arabidopsis* (Brassicaceae). Am J Bot 96:1656–1664

Yaakov B, Kashkush K (2011) Methylation transcription and rearrangements of transposable elements in synthetic allopolyploids. Int J Plant Genomics. Article ID 569826. doi:10.1155/2011/569826

Yanai I, Hunter CP (2009) Comparison of diverse developmental transcriptomes reveals that coexpression of gene neighbors is not evolutionarily conserved. Genome Res 19:2214–2220

Yi H, Richards EJ (2007) A cluster of disease resistance genes in *Arabidopsis* is coordinately regulated by transcriptional activation and RNA silencing. Plant Cell 19:2929–2939

Zeller G, Clark RM, Schneeberger K, Bohlen A, Weigel D, Rätsch G (2008) Detecting polymorphic regions in *Arabidopsis thaliana* with resequencing microarrays. Genome Res 18:918–929

Zhang X, Henderson IR, Lu C, Green PJ, Jacobsen SE (2007) Role of RNA polymerase IV in plant small RNA metabolism. Proc Natl Acad Sci USA 104:4536–4541

Zhang X, Byrnes JK, Gal TS, Li WH, Borevitz JO (2008a) Whole genome transcriptome polymorphisms in *Arabidopsis thaliana*. Genome Biol 9:R165

Zhang P, Li W, Friebe B, Gill BS (2008b) The origin of a 'Zebra' chromosome in wheat suggests nonhomologous recombination as a novel mechanism for new chromosome evolution and step changes in chromosome number. Genetics 179:1169–1177

Zhao XX, Chai Y, Wang HY, Zhao N, Liu B (2010) Rapid genomic alteration in an 'incompatible' pair of maize reciprocal F(1) hybrids – a possible cause for the accumulation of inter-strain genetic diversity. Hereditas 147:1–9

Zheng B, Wang Z, Li S, Yu B, Liu JY, Chen X (2009) Intergenic transcription by RNA Polymerase II coordinates Pol IV and Pol V in siRNA-directed transcriptional gene silencing in *Arabidopsis*. Genes Dev 23:2850–2860

Zwierzykowski Z, Tayyar R, Brunell M, Lukaszewski AJ (1998) Genome recombination in intergeneric hybrids between tetraploid *Festuca pratensis* and *Lolium multiflorum*. J Hered 89: 324–328

Zwierzykowski Z, Kosmala A, Zwierzykowska E, Jones RN, Jokś W, Bocianowski J (2006) Genome balance in six successive generations of the allotetraploid *Festuca pratensis* × *Lolium perenne*. Theor Appl Genet 113:539–547

The Organization of Genomic DNA in Mitotic Chromosomes: A Novel View

3

Hideaki Takata, Sachihiro Matsunaga, and Kazuhiro Maeshima

Contents

K. Maeshima (✉)
Biological Macromolecules Laboratory, Structural Biology Center, National Institute of Genetics, Mishima, Shizuoka 411-8540, Japan

Department of Genetics, School of Life Science, Graduate University for Advanced Studies (Sokendai), Mishima, Shizuoka, Japan
e-mail: kmaeshim@lab.nig.ac.jp

3.1 Introduction

In 1878, W. Flemming discovered a nuclear substance that was visible on staining under the light microscope and named it 'chromatin', which is the basic unit of genomic DNA organization (Olins and Olins 2003). During cell division, chromatin forms a condensed structure, called a 'chromosome', which ensures transmission of the duplicated genomic DNA. The term 'chromosome' is derived from the Greek for 'coloured body', reflecting the observation that a condensed chromosome is clearly visible with dyes. Yet long before the discovery of DNA as a genetic material, the mitotic chromosome fascinated biologists as being a candidate structure involved in heredity. The basic mitotic chromosome structure is well conserved among eukaryotes, with some minor differences. In this chapter, we will focus on the mitotic chromosome structure in mammals with a historical background. We will also highlight the structural differences between mammalian and plant chromosomes where data are available.

3.2 Chromatin Structure

3.2.1 Nucleosome Structure

DNA has a negatively charged phosphate backbone that produces electrostatic repulsion between adjacent DNA regions, making it difficult for DNA to fold upon itself (Bloomfield 1996; Yoshikawa and Yoshikawa 2002). For the first level of folding, the long negatively charged DNA is wrapped around a basic protein complex called a core histone octamer, which consists of two sets of histones, H2A and H2B, and H3 and H4 proteins, and forms a nucleosome (Fig. 3.1) (Kornberg and Lorch 1999). The structural details of the nucleosome core are now known at a resolution of 1.9 Å (Davey et al. 2002). In the core particle, 147 base pairs (bp) of DNA are wrapped in 1.7 left-handed superhelical turns

I.J. Leitch et al. (eds.), *Plant Genome Diversity Volume 2*,
DOI 10.1007/978-3-7091-1160-4_3,

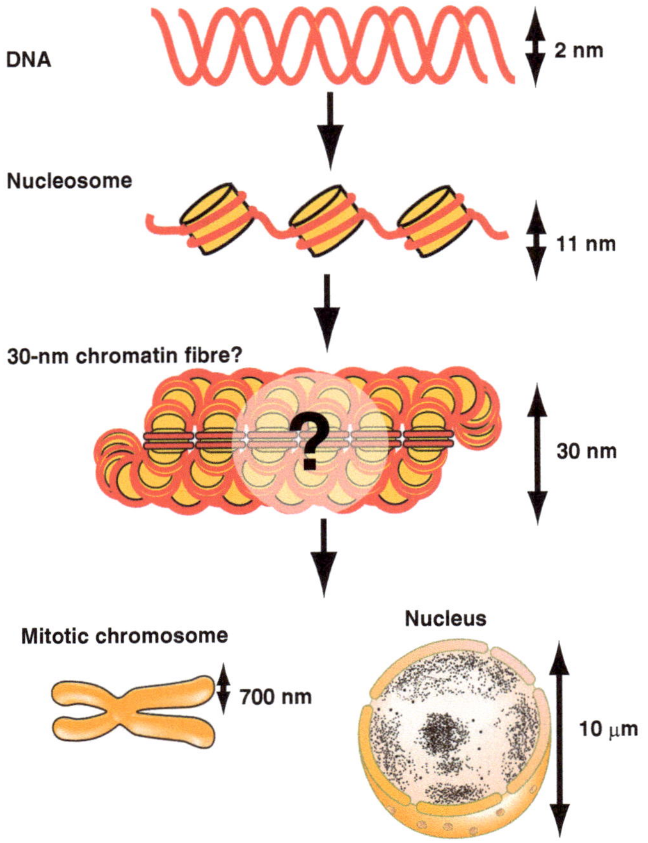

Fig. 3.1 A long DNA molecule with a diameter of 2 nm is wrapped around a core histone octamer that consists of histone H2A, H2B, H3 and H4 proteins, and it forms a nucleosome with a diameter of 11 nm. The nucleosome has long been assumed to be folded into 30-nm chromatin fibres (Robinson and Rhodes 2006). How the nucleosome or 30-nm fibre is compacted into the chromatid with a diameter of approximately 0.7 μm and interphase nucleus with a diameter of approximately 10 μm, however, remains unclear. Note that such 30-nm chromatin fibres in native chromosomes are not detected by cryo-electron microscopy (see Sect. 3.4.1)

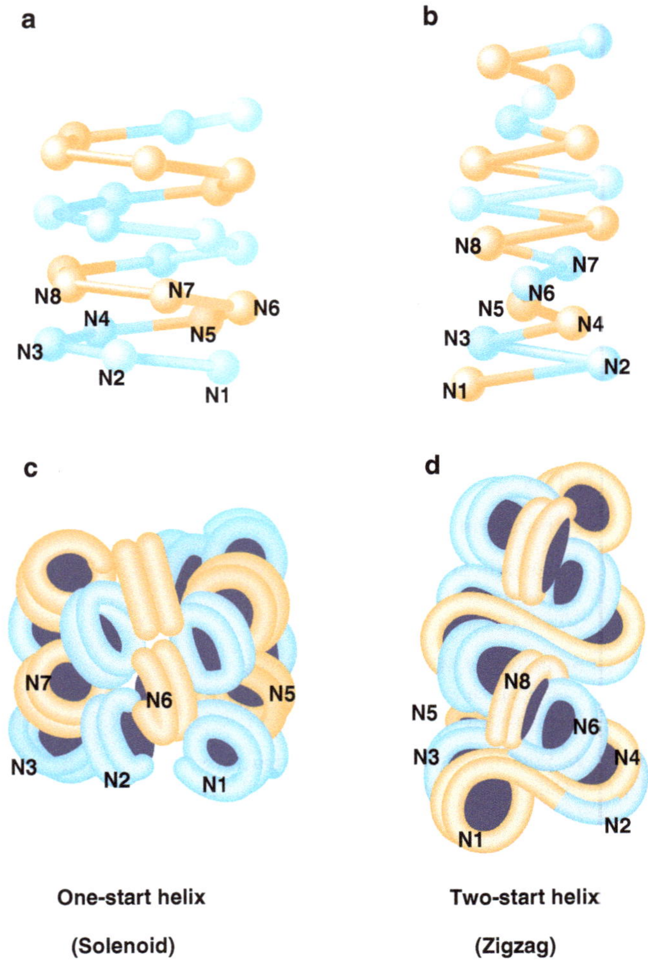

Fig. 3.2 Models of a 30-nm chromatin fibre. Two well-known structural models exist for 30-nm chromatin fibres: one-start helix (*solenoid*) (**a**) and two-start helix (*zigzag*) (**b**). Positions from the first (N1) to eighth (N8) nucleosome are labelled as in Robinson and Rhodes (2006). (**c**) In the one-start helix model proposed by Robinson and Rhodes, the 30-nm chromatin fibre is an interdigitated solenoid. Essentially, a nucleosome in the fibre interacts with its fifth and sixth neighbours. (**d**) In the two-start model proposed by Richmond and co-workers (Schalch et al. 2005), nucleosomes are essentially arranged in a zigzag manner such that alternate nucleosomes form interacting partners. A nucleosome in the fibre binds to the second neighbour nucleosome. Alternate nucleosome pairs are coloured *blue* and *orange* (The images are reproduced from Maeshima et al. (2010) with the permission of Elsevier)

around the histone octamer. Each nucleosome core is connected by approximately 200 bp of 'linker DNA' to make repetitive motifs. Accordingly, the nucleosome fibre was originally described as 'beads on a string' (Olins and Olins 2003).

3.2.2 30-nm Chromatin Fibre

In 1976, Finch and Klug first proposed that the nucleosome, with linker histone H1 or Mg^{2+} ions, was folded into "30-nm chromatin fibres" (Fig. 3.1) (Finch and Klug 1976). In fact, isolated nucleosomes looked like 30-nm-diameter fibres under transmission electron microscopy (EM). In their model called the 'solenoid', consecutive nucleosomes are located next to each other in the fibre, folding into a simple one-start helix (Figs. 3.1 and 3.2a, c). Subsequently, a second model of the 'two-start helix' was proposed based on microscopic observations of isolated nucleosomes (Fig. 3.2b, d) (Woodcock et al. 1984). Although some variations exist in this model (Bassett et al. 2009), nucleosomes are arranged essentially in a zigzag manner, such that a nucleosome in the fibre is bound to the second neighbour but not the first one (Fig. 3.2b, d).

In 2004, Richmond and colleagues found that their cross-linking study on nucleosomal arrays (12-nucleosome repeats) was in good agreement with the zigzag conformation of the two-start helix (Dorigo et al. 2004). Furthermore, they succeeded in resolving the crystal structure of a tetranucleosome (four nucleosome cores) at a resolution of 9 Å

(Fig. 3.2d) (Schalch et al. 2005). Although the structure resolution is relatively low, they defined the positions of the linker DNA and nucleosomes in the fibre by replacing the coarse core region with the fine atomic structure of a nucleosome core particle (Davey et al. 2002). Their results support the zigzag two-start helix model of the 30-nm chromatin fibre (Fig. 3.2d).

However, more recently, Rhodes et al. reconstituted long, regular chromatin fibres from a range of nucleosomal repeats with chicken erythrocyte linker histone H5 (Robinson and Rhodes 2006; Robinson et al. 2006). The *in vitro* assembled chromatin fibres were observed using cryo-electron microscopy (cryo-EM), which is based on vitrification (frozen hydrated) by rapid cooling, which ensures immobilisation of all molecules in a sample to be in a close-to-native state (for a review, see Frank 2006). Using the obtained data, they proposed that the 30-nm chromatin fibre is an interdigitated solenoid (Fig. 3.2a, c); their model shows that a nucleosome in the fibre contacts with the fifth and sixth neighbours along the DNA path (Fig. 3.2a, c) (Robinson et al. 2006), whereas it interacts with its first neighbour in the classic solenoid (Fig. 3.1) (Finch and Klug 1976) . Moreover, recent single-molecule studies of long nucleosome arrays to probe their mechanical properties also support the one-start helix topology underlying the 30-nm fibre (Kruithof et al. 2009).

What caused this difference? Further work from Rhodes' group suggested that the solenoid or zigzag mode of compaction is defined by the length of the nucleosomal linker DNA (Routh et al. 2008). Recently, Grigoryev et al. showed that the two-start zigzag and one-start solenoid modes may be present simultaneously in a 30-nm chromatin fibre under certain conditions (Grigoryev et al. 2009). Consequently, the structural details of the 30-nm chromatin fibre remain controversial. Nonetheless, the nucleosome chain has long been assumed to form a 30-nm chromatin fibre before the higher-order organization of mitotic chromosomes and also interphase nuclei.

3.3 Structure of Mitotic Chromosomes

3.3.1 Classical Views

In the late 1970s, Laemmli and colleagues proposed a hypothesis stating that chromosome structure arises from a set of non-histone proteins that fold the chromatin fibres into loops. They isolated histone-depleted chromosomes (Fig. 3.3a) (Paulson and Laemmli 1977) by gently removing the histones from isolated mitotic chromosomes via competition with an excess of the polyanions dextran sulphate and heparin. This approach was effective for separating the structural contributions of the histones from those of certain non-histone proteins. After removing the histones, they found that the DNA remained highly organized by a residue of non-histone proteins whose structure retained the size and shape of the original chromosomes (Fig. 3.3a) (Paulson and Laemmli 1977). Therefore, Laemmli called the central axial structure made from these non-histone proteins the 'chromosome scaffold' (Fig. 3.3a). The scaffold consists of a subset of non-histone proteins that includes two major high-molecular-weight proteins, Sc1 (170 kDa) and Sc2 (135 kDa), and several minor proteins (Lewis and Laemmli 1982). Sc1 was later identified as topoisomerase II (Earnshaw et al. 1985; Gasser et al. 1986), an evolutionary conserved protein that can untangle DNA and relax the interwound supercoils in a DNA molecule by passing one DNA molecule through a transient double-stranded break in another (Wang 2002). The requirement of topoisomerase II for chromosome condensation was clearly demonstrated by using fission yeast genetics (Uemura et al. 1987) and an *in vitro* system in which chromosomes were assembled by adding sperm nuclei to extracts prepared from *Xenopus laevis* eggs (Adachi et al. 1991; Hirano and Mitchison 1993).

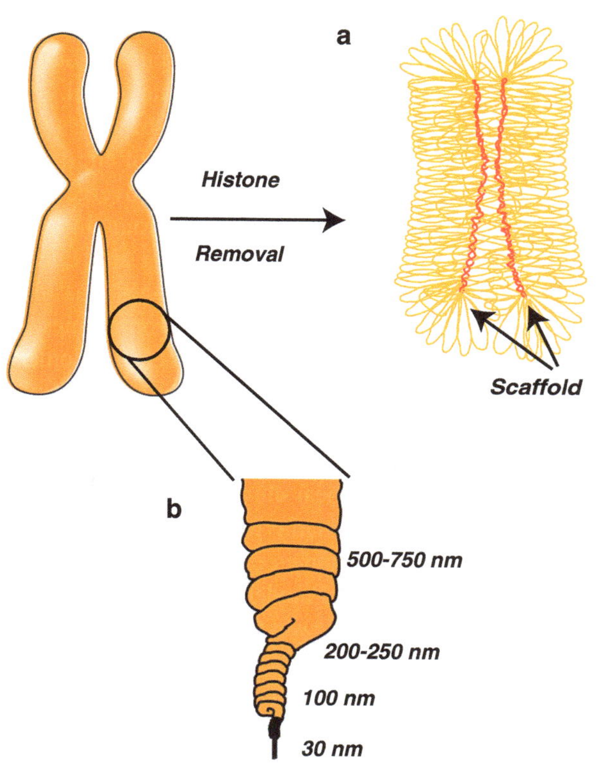

Fig. 3.3 Classical models of mitotic chromosome structure. (**a**) Schematic representation of a histone-depleted chromosome. The histones were removed from the isolated mitotic chromosome gently via competition with an excess of polyanionic dextran sulphate and heparin. After removing the histones, the DNA remained highly organized by the chromosome scaffold (drawn in *red*), maintaining the size and shape of the original chromosomes. (**b**) Hierarchical helical folding model. This model assumes that 30-nm chromatin fibres are folded into 100-nm fibres and then progressively into 200–250 nm fibres that coil to form the final mitotic chromosomes

Such chromatin loop observations were subsequently obtained from several EM studies (Marsden and Laemmli 1979; Earnshaw and Laemmli 1983; Maeshima et al. 2005). As we describe later, isolated chromosomes become swollen in low-salt buffer. Cross-sectional images of the swollen chromosomes appear star-like; the fibres converge on the central axis, suggesting tethering of the chromatin fibre to the axial element, forming loops (Marsden and Laemmli 1979; Earnshaw and Laemmli 1983; Maeshima et al. 2005). This loop structure of mitotic chromosomes may be analogous to that of the lampbrush chromosomes in amphibian oocytes (Morgan 2002) or the meiotic prophase chromosomes in various organisms (Kleckner 2006) that are organized into an enormous number of large chromatin loops emanating from a linear chromosome axis. Evidence clearly implies that chromatin loops are a fundamental organizing unit of chromosomes.

In 1978, another classical model, the so-called 'hierarchical helical folding model', was proposed from extensive observations using high-voltage electron microscopes (Sedat and Manuelidis 1978). This model postulates that the 30-nm chromatin fibres are folded into 100-nm fibres and then progressively into 200- to 250-nm fibres that coil to form the final mitotic chromosomes (Fig. 3.3b), i.e., a hierarchy in the chromosomes based on regular helical structures. Indeed, EM observations showed that chromosomes appeared to be assembled from chromatin fibres of various diameters (Belmont et al. 1987). Furthermore, analysis using engineered chromosomes with large copy numbers of *lac* operator repeats revealed 250-nm-diameter coiling domains in the chromosome, supporting this model (Strukov et al. 2003).

Although these two models appear incompatible, both structures may co-exist in chromosomes. Chromosomes often become X-shaped with helically-folded chromatids after prolonged treatment with microtubule-depolymerising drugs such as nocodazol or colcemid (Boy de la Tour and Laemmli 1988; Maeshima and Laemmli 2003). Such chromosomes show helical coiling of a fibre that is composed of radial loops (Rattner and Lin 1985).

3.3.2 Chromosome Condensation Factors

3.3.2.1 Discovery of Condensins

In 1994, three groups independently made a landmark discovery. Hirano and Mitchison (1994) identified a series of chromosome associated polypeptides (CAPs) in *Xenopus* egg extracts. Of these, the two abundant proteins, CAP-C and CAP-E, had sequence similarity with a family of proteins in budding yeast, which was later called the 'structural maintenance of chromosomes' (SMC) family (Strunnikov et al. 1995). Almost simultaneously, Earnshaw's group showed that a major chromosome scaffold component, Sc2, is a chicken homolog of a SMC protein (Saitoh et al. 1994). Yanagida's group also demonstrated that fission yeast had two SMC homologs Cut3 and Cut14, and that both were involved in chromosome condensation (Saka et al. 1994).

Subsequent characterisation of the CAPs led Hirano's group to the discovery of a protein complex called condensin, which consists of five different subunits, including a heterodimer of CAP-C (SMC4) and CAP-E (SMC2) (Hirano et al. 1997). When condensin is depleted from *Xenopus* extracts using a specific antibody, mitotic chromosome condensation is defective, and chromatin forms swollen puffs, not a compact structure. When purified condensin complex is added back to the depleted extracts, chromosome condensation recovers, implying that the complex has a key role in the chromosome condensation process.

The condensin family was discovered to be conserved from bacteria to mammals (Nasmyth and Haering 2005; Hirano 2006), suggesting a universal role of this family in compacting the genome. Although exactly how the complex functions in the condensation process is unknown, interesting clues came from the discovery that condensin can introduce positive supercoils into closed circular DNA (Kimura and Hirano 1997), depending on ATP-hydrolysis and mitosis-specific phosphorylation of the complex (Kimura et al. 1998; Takemoto et al. 2004).

Further biochemical and cytological analyses of the chromosome scaffold demonstrated that it is composed predominantly of a condensin complex and topoisomerase IIα (Maeshima and Laemmli 2003). Both components have an axial distribution with a diameter of about 200 nm at the centre of each chromatid in swollen and compact chromosomes (Saitoh et al. 1994; Coelho et al. 2003; Maeshima and Laemmli 2003; Kireeva et al. 2004), and even in the chromosomes of living cells (Tavormina et al. 2002; Hirota et al. 2004). Although an argument was made that the scaffold structure was an artefact resulting from the non-specific aggregation of non-histone proteins in the histone-depleted chromosomes (Okada and Comings 1980), these findings support the physiological relevance of the chromosome scaffold structure (Maeshima and Laemmli 2003; Maeshima and Eltsov 2008).

Ten years after the condensin discovery in vertebrates, condensin II was isolated as a second condensin complex, composed of the common heterodimer of SMC2 and SMC4 and three non-SMC subunits related to, but distinct from, those in condensin I (Ono et al. 2003; Yeong et al. 2003). Condensins I and II show an axial localization in chromosomes in a rather complementary manner (Ono et al. 2003). Experiments knocking down condensin I or II separately or together have revealed their distinct functions in part (Hirota et al. 2004; Ono et al. 2004). Condensin I is localized in the cytoplasm until the nuclear envelope breaks down, whereas condensin II is located in the nuclei during interphase. Consistent with this observation, depletion of condensin I does not affect prophase chromosome condensation

(Hirota et al. 2004). By contrast, condensin II knockdown significantly delays the initiation of prophase chromosome condensation (Hirota et al. 2004). The chromosomal binding of condensin II but not condensin I seems to be regulated by phosphatase PP2A (Takemoto et al. 2009).

In the plant *Arabidopsis thaliana*, there are two SMC2 genes (AtCAP-E1 and AtCAP-E2) (Liu Cm et al. 2002; Siddiqui et al. 2003) and one SMC4 gene (AtCAP-C). AtCAP-E1 comprises more than 85 % of the total SMC2 transcripts. The non-SMC subunits (AtCAP-H and AtCAP-H2) of condensin I and II have also been identified in *A. thaliana* (Fujimoto et al. 2005). AtCAP-H and AtCAP-H2 are localized on mitotic chromosomes from prometaphase to telophase; however, their localizations during interphase are different, as in mammalian cells. While AtCAP-H is localized in the cytoplasm, AtCAP-H2 resides in the nucleus, particularly at the nucleolus. These different localizations during interphase are indicative of functional differentiation between condensins I and II in *A. thaliana*.

3.3.2.2 A Paradox Involving Condensin and Topoisomerase II

No doubt exists of the universal roles of condensin and topoisomerase II in genome compaction. However, available genetic and RNAi data from various organisms (Steffensen et al. 2001; Hagstrom et al. 2002; Hudson et al. 2003) indicate that condensin mutants mainly have a segregation defect that does not cause dramatic abnormalities in chromosome morphology. Earnshaw and colleagues showed that metaphase chromosomes in condensin-knockout chicken DT40 cells could still condense normally, as in wild-type cells, although there were frequent chromosomal bridges during anaphase (Vagnarelli et al. 2006). Thus, they inferred a novel chromosome condensation component, regulator of chromosome architecture (RCA), which enabled metaphase chromosome condensation without condensins (Vagnarelli et al. 2006). These results suggest that condensins are required for chromosome structural integrity but not for chromosome condensation. As well as condensins, knockdown of the topoisomerase II in fly or human cells also results in chromosome segregation defects, but no prominent condensation defect (Chang et al. 2003; Carpenter and Porter 2004; Sakaguchi and Kikuchi 2004), whereas it is essential for chromosome condensation in some other systems (Uemura et al. 1987; Adachi et al. 1991; Hirano and Mitchison 1993). In higher eukaryotic organisms, which have more complex systems, proteins might acquire more specialized functions or functional redundancy to ensure their long-term survival, although one hypothesis states that a residual condensin or topoisomerase II after their depletion could still create the chromosomal shape.

In plants there is also no direct evidence so far to indicate that plant condensins are involved in chromosome condensation. Embryo lethality caused by the double mutation of SMC2 genes suggests that plant condensins are involved in *A. thaliana* development. A mutant of a SMC4 homolog, AtCAP-C, displays phenotypes similar to those exhibited by the AtCAP-E double mutant, including altered segregation of alleles during gametogenesis and embryo lethality (Siddiqui et al. 2006). Moreover, knockout plants of two condensin II subunits, AtCAP-G2 and AtCAP-H2, have no defects in chromosome condensation (Sakamoto et al. 2011).

3.3.2.3 Other Non-Histone Chromosomal Proteins

Little information is available regarding the global protein composition of mitotic chromosomes, except for major non-histone chromosomal proteins such as topoisomerase II and condensin proteins. Several attempts have been made to identify non-histone chromosomal proteins using a proteomic approach. Some chromosome passenger proteins required for mitotic chromosome alignment and segregation, and also proteins located at the chromosome periphery, have been identified (Morrison et al. 2002; Uchiyama et al. 2005; Takata et al. 2007). The number of identified proteins including uncharacterised ones is more than 200. Additionally, several proteins specifically located at centromeric and telomeric regions have also been reported (Foltz et al. 2006; Okada et al. 2006; Dejardin and Kingston 2009). These studies suggest that complex protein networks might exist that organize mitotic chromosome structure, which might explain why depletion of condensin or topoisomerase II mainly results in a defect in chromosome segregation but not condensation, as described above.

3.3.2.4 Histone Phosphorylation

Histone tails are subjected to various post-translational modifications and could be related to chromatin structure and function. The most striking modification of histone tails in mitosis is the histone H3 phosphorylation at serine 10 (Guo et al. 1995) and at serine 28 (Goto et al. 1999). A causal relationship between histone H3 phosphorylation and chromosome condensation has been identified in *Tetrahymena* (Wei et al. 1999). A mutant in which histone H3 cannot be phosphorylated exhibits abnormal chromosome condensation and segregation, demonstrating that phosphorylation of histone H3, at least, at serine 10 is required for proper chromosome dynamics. However, histone H3 phosphorylation is not required for chromosome condensation in many other species, such as yeast (Hsu et al. 2000), *Xenopus* (MacCallum et al. 2002), *Drosophila* (Adams et al. 2001) and mammals (Van Hooser et al. 1998). In plants, several studies have suggested that histone H3 phosphorylation is involved in sister chromatid cohesion but not in chromosome condensation (Gernand et al. 2003). Moreover, histone H3 phosphorylation in plant chromosomes is strictly limited to centromeric regions in contrast to the chromosome-wide distribution in mammalian chromosomes (Kurihara et al. 2006).

3.3.2.5 Cations

Other than protein factors, such as condensins or topoisomerase II, cations seem to be essential participants in chromosome condensation. As mentioned above, mitotic chromosomes become very swollen following the depletion of Ca^{2+} or Mg^{2+} and this process is completely reversible. Almost 40 years ago, Cole demonstrated that the repeated removal and addition of Mg^{2+} results in cycles of chromosome swelling and compaction (Cole 1967). DNA has a negatively charged phosphate backbone that produces electrostatic repulsion. In the presence of cations, DNA condensation results from charge neutralisation, as the binding of cations specifically to the DNA phosphates decreases the overall electrostatic repulsion between adjacent DNA regions (Bloomfield 1996; Yoshikawa and Yoshikawa 2002). In the case of chromatin, the negative DNA charges are about 60 % neutralised by core histones, which have tails with positively charged lysine and arginine residues (Strick et al. 2001). Therefore, the remaining approximately 40 % of the DNA charge must be neutralised by other factors such as histone H1, non-histone proteins and cations. In fact, using secondary ion mass spectrometry, Strick et al. reported that Ca^{2+}, Mg^{2+}, Na^{+} and K^{+} are highly enriched in mitotic chromosomes, compared with interphase nuclei, suggesting a potentially important role in chromosome condensation (Strick et al. 2001). Chromatin 'opening' and 'closing' with DNA-charge neutralisation may also be involved in global gene regulation as a result of histone tail modifications, such as phosphorylation and acetylation (Kornberg and Lorch 1999).

3.3.2.6 Repetitive Sequences

As more 'complete' genome sequences become available it seems clear that many genomes consist mainly of repetitive DNA sequences. For example, in humans satellite DNA repeats are clustered in discrete areas, such as the centromere, whereas other repeats such as short interspersed nuclear elements (SINEs) and long interspersed nuclear elements (LINEs) are dispersed throughout the genome. Plant genomes are also composed predominantly of repetitive DNA although retrotransposons with long terminal repeats (LTR retrotransposons) are generally the predominant repeat type (see Kejnovsky et al. 2012). Are these really "junk"? Do they have any function? One of proposed functions is that repetitive sequences could be involved in genome compaction. Indeed, the genomic regions, which are abundant with satellite DNA, are condensed throughout the cell cycle (e.g. centromere). Double-stranded DNA has the property of self-assembly in aqueous solutions containing physiological concentrations of divalent cations. Ohyama and his colleagues found that DNA molecules preferentially interact with molecules with an identical sequence and length, even in a solution containing heterogeneous DNA species (Inoue et al. 2007). Therefore, postulating that this attractive force due to repetitive sequences functions in chromosome condensation is tempting. Furthermore, evidence implicates LINEs (L1s) in X chromosome condensation (Lyon 2003). One of the female X chromosomes is well known to be highly condensed as a Barr body (Lyon 2003). Bailey et al. showed that the level of LINEs in X chromosomal DNA was roughly double that in autosomes (Bailey et al. 2000). They also found that the level was highest in a region near the X-inactivation centre and lowest in a short arm region containing numerous genes that escape inactivation. These data are consistent with the postulate that the density of LINEs correlates with the extent of inactivation or condensation. Moreover, LINEs are concentrated mainly in the dark G-bands in autosomes, where chromatin is condensed as heterochromatin.

'Knob', a classic cytogenetic marker of plant chromosomes, has a chromosome structure that can be identified microscopically as being darkly stained compared to the surrounding region. The knob was first reported in maize (McClintock 1930). Such a lump-like structure in maize chromosomes is a large block of heterochromatin that contains largely tandem repetitive sequences and retrotransposons with very few transcribed genes. Maize knobs are mainly composed of a 180 bp repeat and a 350 bp repeat, which are organized in tandem arrays (Peacock et al. 1981; Ananiev et al. 1998). The knob satellite is found near the ends of all or nearly all maize chromosomes (Lamb et al. 2007). The structure of tandem repeats in such a knob could be associated with heterochromatic domains. At the very least, structural conservation demonstrates the importance of tandem repeats of repetitive sequences in heterochromatin formation, suggesting that repetitive sequences influence chromosome condensation. Although chromosomes cannot be condensed without the help of protein factors such as condensins, the large amounts of repetitive sequences may well contribute to genome compaction and have been under selection during evolution.

3.4 *In Vivo* Mitotic Chromosome Structure

In the classical EM views, chromosomes seem to consist of radial 30-nm chromatin loops that are somehow tethered centrally by scaffold proteins such as condensin or topoisomerase II. Although this is a rather classical view of chromosome organization, one must keep in mind that the samples are fixed chemically, dehydrated and embedded in plastic. In this section, we will introduce a recent view of chromosome structure based on observations made in nearly intact cell conditions.

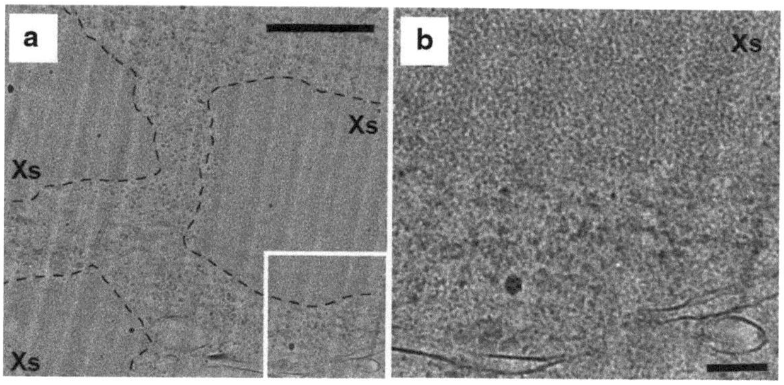

Fig. 3.4 (**a**) Cryo-electron micrograph image of a vitrified (frozen-hydrated) section of mitotic HeLa cells. The compact areas, outlined by the *dashed line*, are cross sections of mitotic chromosomes (Xs) that are surrounded by a cytoplasm full of electron-dense ribosomes and other particles. The scale bar indicates 500 nm. (**b**) Magnification of selected area of (**a**). Note the grainy, homogeneous texture of the chromosome (Xs). No higher-order or periodic structures, such as 30-nm chromatin fibres, are recognised. The scale bar indicates 100 nm (The images are reproduced from Maeshima and Eltsov (2008) with the permission of Oxford University Press)

3.4.1 Inside Mitotic Chromosomes: Does the 30-nm Fibre Exist *In Vivo*?

What do mitotic chromosomes look like in living cells? One of the best ways to address this question is by using cryo-EM of vitreous sections (CEMOVIS). In CEMOVIS, after the frozen hydrated (vitrified) cells are sectioned, the thin sections are observed directly under a cryo-EM with no chemical fixation or staining. This approach enables direct high-resolution imaging of cell structures in a close-to-native state.

More than 20 years ago, Dubochet and his colleagues first observed vitrified sections of mammalian mitotic cells using CEMOVIS (McDowall et al. 1986). Unexpectedly, they found that the chromosomes showed a homogeneous and grainy texture with a c. 11 nm spacing. Neither higher-order nor periodic structures, including 30-nm chromatin fibres, were recognised. Consequently, Dubochet et al. proposed that the basic structure of the chromosome was a liquid-like compact aggregation of 11 nm nucleosome fibres (Dubochet et al. 1988). We also investigated human mitotic cells using CEMOVIS and a subsequent image processing analysis and concluded that the 30-nm fibres are essentially absent in human mitotic chromosomes (Fig. 3.4) (Eltsov et al. 2008).

These observations are consistent with a report on the three-dimensional structure of *Xenopus* chromosomes assembled *in vitro* using electron tomography (König et al. 2007). In the cryo-substituted chromosomes, König et al. detected no continuous fibre-like structures, such as 30-nm fibres. No other regular ultrastructural organization was observed. This conclusion is supported by the finding that chromosome assembly in *Xenopus* extracts proceeds without the linker histone H1, which supposedly stabilises the 30-nm fibres (Ohsumi et al. 1993). Furthermore, histone H1 is highly mobile in the chromosomes of live cells (Chen et al. 2005). These data again support the absence of 'static' continuous 30-nm chromatin fibres in native chromosomes.

However, why can we sometimes see the 30-nm chromatin fibres? As described above, the original concept of the 30-nm chromatin fibre was derived from studies using conventional transmission or scanning EM: fibres of approximately 30 nm in diameter were observed under certain conditions involving isolated nucleosomes, nuclei and mitotic chromosomes. To answer this question, one should emphasise that the observation of 30-nm chromatin fibres required a strict cationic environment, namely a low-salt buffer containing 1–2 mM Mg^{2+}; under such conditions, isolated nuclei or chromosomes become swollen. Accordingly, the local nucleosome concentration decreases, favouring intra-fibre nucleosome associations, leading to the formation of 30-nm chromatin fibres (Fig. 3.5a, b). Furthermore, in conventional EM observations, the formation of 30-nm chromatin fibres might be stabilised through chemical cross-linking (e.g. glutaraldehyde fixation) of intra-fibre nucleosome associations and further shrinkage with alcohol dehydration during sample preparation (Belmont 2006; Maeshima and Eltsov 2008).

However, at the high nucleosome concentrations that occur *in vivo*, inter-fibre nucleosome interactions become increasingly dominant (Fig. 3.5c, d). Nucleosome fibres are forced to interdigitate with one another. This interferes with the formation and maintenance of the 30-nm chromatin fibre, leading to the 'polymer melt' state (Fig. 3.5c, d).

3.4.2 A Novel Model of Chromosome Structure

By its nature, chromatin forms aggregates in the presence of proper concentrations of divalent or multivalent cations (de Frutos et al. 2001). Again, how is long genomic DNA

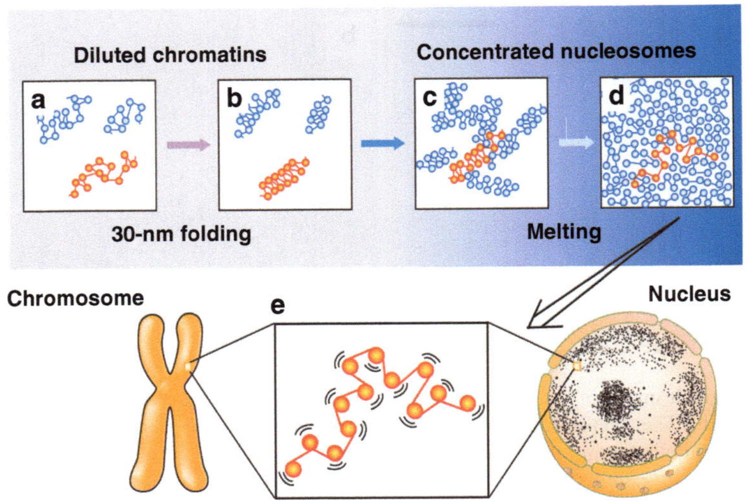

Fig. 3.5 (**a, b**) Under the diluted condition, the flexible nucleosome fibres compact through selective close-neighbour associations, forming the 30-nm chromatin fibres. An increase in nucleosome concentration results in inter-fibre nucleosomal contacts, which interfere with the intra-fibre bonds (**c**). The nucleosomes of adjacent fibres interdigitate and intermix. This disrupts the 30-nm folding, and the nucleosomal fibres progress to a state of 'polymer melt' (**d**). (**e**) The concept of polymer melt implies dynamic polymer chains; i.e., nucleosome fibres may be moving and rearranging constantly. This may provide several advantages during chromosome condensation and segregation of mitosis and the transcription and DNA replication processes during interphase (The images are reproduced with minor modifications from Maeshima et al. (2010) with the permission of Elsevier)

packaged into compact mitotic chromosomes? When isolated chromosomes become swollen under low-salt conditions, they seem to consist of radial chromatin loops that are somehow tethered centrally by scaffold proteins. Considering a structural analogy with meiotic and lampbrush chromosomes, chromatin loops may be the fundamental organizing unit of chromosomes. In our cryo-EM observations, chromosomes show a homogenous, grainy texture with no higher-order or periodic structures (Fig. 3.4a, b) (McDowall et al. 1986; Eltsov et al. 2008). As the intracellular cations increase during mitosis, the loop structures would fold irregularly toward the chromosome centre, which contains abundant condensins. This loop collapse process might be enhanced by the attractive force of repetitive sequences dispersed throughout the genome. Therefore, the compact native chromosome would be made up primarily of an irregular chromatin network further cross-linked by condensins (Fig. 3.6).

3.4.3 Organization of Genomic DNA in Interphase Nuclei

The compartmentalisation of DNA within the nucleus accompanies genome organization. A long time ago, Rabl suggested that interphase nuclei have a preferentially polarised organization of centromeres at the apical side and telomeres at the basal side in salamander nuclei (Rabl 1885). This 'Rabl organization' was also observed in *Drosophila* (Wilkie et al. 1999), onion, wheat and barley (Dong and Jiang 1998). However, such organization is not found in mammalian nuclei. In the case of plants, nuclei of small genome species exhibit diffuse chromosome territories without Rabl organization, whereas those of large genome species have Rabl organization (Dong and Jiang 1998; Fujimoto et al. 2005). The Rabl orientation might be an effective way to arrange the huge chromosomes, at least in plants.

We next focus on the genome organization in interphase nuclei. Unexpectedly, interphase nuclei in most higher-eukaryote cell types examined by cryo-EM do not contain a regular 30-nm chromatin fibre as the underlying structure (Dubochet et al. 1988; Bouchet-Marquis et al. 2006; Fakan and van Driel 2007). As Woodcock has stated (Woodcock 1994), this phenomenon could not be due to insufficient contrast of cryo-EM because 30-nm fibres are not visible in the majority of interphase nuclei after conventional chemical fixation and plastic embedding (Horowitz et al. 1990). Although cryo-EM analysis of specific chromatin domains, for example, chromocentres, inactive X chromosomes and senescence-associated heterochromatin foci, have not been reported, typical heterochromatin regions in plant or mammalian nuclei that have been visualized by cryo-EM look very similar to mitotic chromosomes (Bouchet-Marquis et al. 2006; Fakan and van Driel 2007). Consequently, the melt may represent the predominant state of compacted chromatin *in vivo*. Consistent with this, using a combination of chromosome conformation capture (3C) technique and polymer modelling, Dekker found that chromatin in a transcriptionally-active domain in yeast did not form a compact 30-nm fibre, but instead noted that the chromatin was

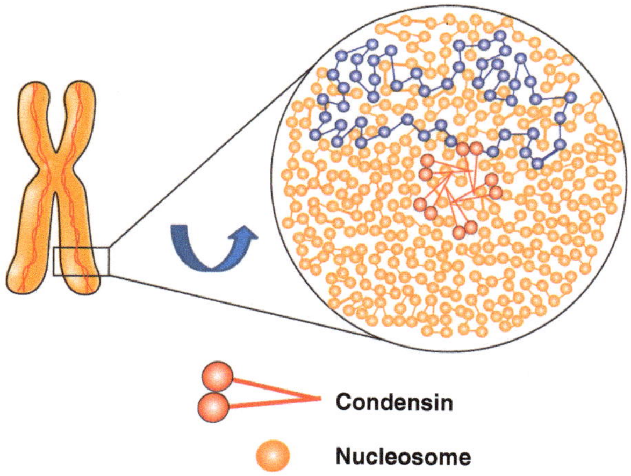

Fig. 3.6 Novel structural model of mitotic chromosomes based on the polymer melt concept. We postulate that chromatin loops are the fundamental organizing unit of chromosomes. We propose that with increased nucleosome concentration, presumably due to increases in intracellular cation concentration during mitosis, the loop structures fold irregularly toward the chromosome centre (the *blue* nucleosome fibre loop), which contains abundant condensins. Note that no continuous 30-nm chromatin fibres are visible in the native chromosomes

extended with a rather loose arrangement of nucleosomes (Dekker 2008). The chromatin in the neighbouring domain was more compact, but the mass density was still well below that of a canonical 30-nm fibre (Dekker 2008).

3.4.4 Dynamic Nature of the Melted Polymer

The concept of polymer melt implies dynamic polymer chains: nucleosome fibres may be constantly moving and rearranging at the local level (Fig. 3.5e). During mitosis, this dynamic irregular folding of nucleosome fibres could be a driving force for chromosome condensation and segregation. Recently, the Jun group proposed that the entropic force of a dynamic flexible polymer chain can drive bacterial chromosome segregation (Jun and Wright 2010). Because folding of the nucleosome fibre determines DNA accessibility (Tremethick 2007), these dynamics may have several advantages in a template-directed biological process, i.e., transcriptional regulation and DNA replication in interphase nuclei. In the case of transcriptional regulation, the dynamic movement of nucleosome fibres will contribute to the targeting of transcription factors to the *cis*-elements along DNA and the forming of complexes because target sequences are exposed more often from the folding structure. Moreover, dynamic irregular folding can easily form loops, facilitating interaction between promoter and enhancer sequences at the three-dimensional level.

Conclusions

We have discussed the structural aspects of mitotic chromosomes in mammals and plants and proposed a novel model. In the model, the nucleosome fibres in the bulk of mitotic chromosomes do not form 30-nm chromatin fibres but exist in a highly disordered state, which is locally similar to a polymer melt (Figs. 3.5 and 3.6). A similar state exists in the majority of active interphase nuclei. The concept of the polymer melt implies dynamic polymer chains (Fig. 3.5e). This may have several advantages during chromosome condensation and segregation during mitosis and the transcription and DNA replication processes during interphase.

3.4.5 Added Note

In this chapter, we proposed a novel chromosome structure model, mainly from cryo-EM observations. Recently, using synchrotron X-ray scattering analysis, we obtained conclusive evidence that human mitotic HeLa chromosomes are predominantly composed of irregularly folded nucleosome fibres rather than 30-nm chromatin fibres (Nishino et al. 2012). The scattering pattern implied a fractal nature in the chromosomes, which permits a more dynamic and flexible genome organization. We also found a similar scattering pattern in interphase nuclei of HeLa cells in the range up to ~275 nm (Joti et al. 2012). Our findings suggest a common structural feature in interphase and mitotic chromatin: compact and irregular folding of nucleosome fibres occurs without a 30-nm chromatin structure (Joti et al. 2012).

Acknowledgments We are grateful to Dr. Eltosv, Prof. Dubochet, and Prof. Frangakis for collaboration with KM We would like to thank Prof. Laemmli, Prof. Fukui, Prof. Yoshikawa, Dr. Uchiyama and Ms. Hihara for exciting discussions. KM was supported by a MEXT grant-in-aid and JST CREST. HT is a research fellow of the Japan Society for the Promotion of Science.

References

Adachi Y, Luke M, Laemmli UK (1991) Chromosome assembly in vitro: topoisomerase II is required for condensation. Cell 64:137–148

Adams RR, Maiato H, Earnshaw WC, Carmena M (2001) Essential roles of *Drosophila* inner centromere protein (INCENP) and aurora B in histone H3 phosphorylation, metaphase chromosome alignment, kinetochore disjunction, and chromosome segregation. J Cell Biol 153:865–880

Ananiev EV, Phillips RL, Rines HW (1998) A knob-associated tandem repeat in maize capable of forming fold-back DNA segments: are chromosome knobs megatransposons? Proc Natl Acad Sci USA 95:10785–10790

Bailey JA, Carrel L, Chakravarti A, Eichler EE (2000) Molecular evidence for a relationship between LINE-1 elements and X chromosome inactivation: the Lyon repeat hypothesis. Proc Natl Acad Sci USA 97:6634–6639

Bassett A, Cooper S, Wu C, Travers A (2009) The folding and unfolding of eukaryotic chromatin. Curr Opin Genet Dev 19:159–165

Belmont AS (2006) Mitotic chromosome structure and condensation. Curr Opin Cell Biol 18:632–638

Belmont AS, Sedat JW, Agard DA (1987) A three-dimensional approach to mitotic chromosome structure: evidence for a complex hierarchical organization. J Cell Biol 105:77–92

Bloomfield VA (1996) DNA condensation. Curr Opin Struct Biol 6:334–341

Bouchet-Marquis C, Dubochet J, Fakan S (2006) Cryoelectron microscopy of vitrified sections: a new challenge for the analysis of functional nuclear architecture. Histochem Cell Biol 125:43–51

Boy de la Tour E, Laemmli UK (1988) The metaphase scaffold is helically folded: sister chromatids have predominantly opposite helical handedness. Cell 55:937–944

Carpenter AJ, Porter AC (2004) Construction, characterization, and complementation of a conditional-lethal DNA topoisomerase IIalpha mutant human cell line. Mol Biol Cell 15:5700–5711

Chang CJ, Goulding S, Earnshaw WC, Carmena M (2003) RNAi analysis reveals an unexpected role for topoisomerase II in chromosome arm congression to a metaphase plate. J Cell Sci 116: 4715–4726

Chen D, Dundr M, Wang C, Leung A, Lamond A, Misteli T, Huang S (2005) Condensed mitotic chromatin is accessible to transcription factors and chromatin structural proteins. J Cell Biol 168:41–54

Coelho PA, Queiroz-Machado J, Sunkel CE (2003) Condensin-dependent localisation of topoisomerase II to an axial chromosomal structure is required for sister chromatid resolution during mitosis. J Cell Sci 116:4763–4776

Cole A (1967) Chromosome structure. Theor Biophys 1:305–375

Davey CA, Sargent DF, Luger K, Maeder AW, Richmond TJ (2002) Solvent mediated interactions in the structure of the nucleosome core particle at 1.9 a resolution. J Mol Biol 319:1097–1113

de Frutos M, Raspaud E, Leforestier A, Livolant F (2001) Aggregation of nucleosomes by divalent cations. Biophys J 81:1127–1132

Dejardin J, Kingston RE (2009) Purification of proteins associated with specific genomic loci. Cell 136:175–186

Dekker J (2008) Mapping in vivo chromatin interactions in yeast suggests an extended chromatin fiber with regional variation in compaction. J Biol Chem 283:34532–34540

Dong F, Jiang J (1998) Non-Rabl patterns of centromere and telomere distribution in the interphase nuclei of plant cells. Chromosome Res 6:551–558

Dorigo B, Schalch T, Kulangara A, Duda S, Schroeder RR, Richmond TJ (2004) Nucleosome arrays reveal the two-start organization of the chromatin fiber. Science 306:1571–1573

Dubochet J, Adrian M, Chang JJ, Homo JC, Lepault J, McDowall AW, Schultz P (1988) Cryo-electron microscopy of vitrified specimens. Q Rev Biophys 21:129–228

Earnshaw WC, Laemmli UK (1983) Architecture of metaphase chromosomes and chromosome scaffolds. J Cell Biol 96:84–93

Earnshaw WC, Halligan B, Cooke CA, Heck MM, Liu LF (1985) Topoisomerase II is a structural component of mitotic chromosome scaffolds. J Cell Biol 100:1706–1715

Eltsov M, Maclellan KM, Maeshima K, Frangakis AS, Dubochet J (2008) Analysis of cryo-electron microscopy images does not support the existence of 30-nm chromatin fibers in mitotic chromosomes in situ. Proc Natl Acad Sci USA 105:19732–19737

Fakan S, van Driel R (2007) The perichromatin region: a functional compartment in the nucleus that determines large-scale chromatin folding. Semin Cell Dev Biol 18:676–681

Finch JT, Klug A (1976) Solenoidal model for superstructure in chromatin. Proc Natl Acad Sci USA 73:1897–1901

Foltz DR, Jansen LE, Black BE, Bailey AO, Yates JR 3rd, Cleveland DW (2006) The human CENP-A centromeric nucleosome-associated complex. Nat Cell Biol 8:458–469

Frank J (2006) Three-dimensional electron microscopy of macromolecular assemblies: visualization of biological molecules in their native state, 2nd edn. Oxford University Press, New York

Fujimoto S, Ito M, Matsunaga S, Fukui K (2005) An upper limit of the ratio of DNA volume to nuclear volume exists in plants. Genes Genet Syst 80:345–350

Gasser SM, Laroche T, Falquet J, Boy de la Tour E, Laemmli UK (1986) Metaphase chromosome structure: involvement of topoisomerase II. J Mol Biol 188:613–629

Gernand D, Demidov D, Houben A (2003) The temporal and spatial pattern of histone H3 phosphorylation at serine 28 and serine 10 is similar in plants but differs between mono- and polycentric chromosomes. Cytogenet Genome Res 101:172–176

Goto H, Tomono Y, Ajiro K, Kosako H, Fujita M, Sakurai M, Okawa K, Iwamatsu A, Okigaki T, Takahashi T, Inagaki M (1999) Identification of a novel phosphorylation site on histone H3 coupled with mitotic chromosome condensation. J Biol Chem 274:25543–25549

Grigoryev SA, Arya G, Correll S, Woodcock CL, Schlick T (2009) Evidence for heteromorphic chromatin fibers from analysis of nucleosome interactions. Proc Natl Acad Sci USA 106:13317–13322

Guo XW, Th'ng JP, Swank RA, Anderson HJ, Tudan C, Bradbury EM, Roberge M (1995) Chromosome condensation induced by fostriecin does not require p34cdc2 kinase activity and histone H1 hyperphosphorylation, but is associated with enhanced histone H2A and H3 phosphorylation. EMBO J 14:976–985

Hagstrom KA, Holmes VF, Cozzarelli NR, Meyer BJ (2002) *C. elegans* condensin promotes mitotic chromosome architecture, centromere organization, and sister chromatid segregation during mitosis and meiosis. Genes Dev 16:729–742

Hirano T (2006) At the heart of the chromosome: SMC proteins in action. Nat Rev Mol Cell Biol 7:311–322

Hirano T, Mitchison TJ (1993) Topoisomerase II does not play a scaffolding role in the organization of mitotic chromosomes assembled in *Xenopus* egg extracts. J Cell Biol 120:601–612

Hirano T, Mitchison TJ (1994) A heterodimeric coiled-coil protein required for mitotic chromosome condensation in vitro. Cell 79:449–458

Hirano T, Kobayashi R, Hirano M (1997) Condensins, chromosome condensation protein complexes containing XCAP-C, XCAP-E and a *Xenopus* homolog of the *Drosophila* Barren protein. Cell 89:511–521

Hirota T, Gerlich D, Koch B, Ellenberg J, Peters JM (2004) Distinct functions of condensin I and II in mitotic chromosome assembly. J Cell Sci 117:6435–6445

Horowitz RA, Giannasca PJ, Woodcock CL (1990) Ultrastructural preservation of nuclei and chromatin: improvement with low-temperature methods. J Microsc 157:205–224

Hsu JY, Sun ZW, Li X, Reuben M, Tatchell K, Bishop DK, Grushcow JM, Brame CJ, Caldwell JA, Hunt DF, Lin R, Smith MM, Allis CD (2000) Mitotic phosphorylation of histone H3 is governed by Ipl1/aurora kinase and Glc7/PP1 phosphatase in budding yeast and nematodes. Cell 102:279–291

Hudson DF, Vagnarelli P, Gassmann R, Earnshaw WC (2003) Condensin is required for nonhistone protein assembly and structural integrity of vertebrate mitotic chromosomes. Dev Cell 5:323–336

Inoue S, Sugiyama S, Travers AA, Ohyama T (2007) Self-assembly of double-stranded DNA molecules at nanomolar concentrations. Biochemistry 46:164–171

Joti Y, Hikima T, Nishino Y, kamada F, Hihara S, Takata H, Ishikawa T, Maeshima K (2012) Chromosomes without a 30-nm chromatin fibre. Nucleus (in press)

Jun S, Wright A (2010) Entropy as the driver of chromosome segregation. Nat Rev Microbiol 8:600–607

Kejnovsky E, Hawkins JS, Feschotte C (2012) Plant transposable elements: biology and evolution. In: Wendel JF, Greilhuber J, Doležel J, Leitch IJ (eds) Plant genome diversity, vol 1, Plant genomes, their residents, and their evolutionary dynamics. Springer-Verlag, Wien, pp 17–34

Kimura K, Hirano T (1997) ATP-dependent positive supercoiling of DNA by 13S condensin: a biochemical implication for chromosome condensation. Cell 90:625–634

Kimura K, Hirano M, Kobayashi R, Hirano T (1998) Phosphorylation and activation of 13S condensin by Cdc2 in vitro. Science 282:487–490

Kireeva N, Lakonishok M, Kireev I, Hirano T, Belmont AS (2004) Visualization of early chromosome condensation: a hierarchical folding, axial glue model of chromosome structure. J Cell Biol 166:775–785

Kleckner N (2006) Chiasma formation: chromatin/axis interplay and the role(s) of the synaptonemal complex. Chromosoma 115:175–194

König P, Braunfeld MB, Sedat JW, Agard DA (2007) The three-dimensional structure of in vitro reconstituted *Xenopus laevis* chromosomes by EM tomography. Chromosoma 116:349–372

Kornberg RD, Lorch Y (1999) Twenty-five years of the nucleosome, fundamental particle of the eukaryote chromosome. Cell 98:285–294

Kruithof M, Chien FT, Routh A, Logie C, Rhodes D, van Noort J (2009) Single-molecule force spectroscopy reveals a highly compliant helical folding for the 30-nm chromatin fiber. Nat Struct Mol Biol 16:534–540

Kurihara D, Matsunaga S, Kawabe A, Fujimoto S, Noda M, Uchiyama S, Fukui K (2006) Aurora kinase is required for chromosome segregation in tobacco BY-2 cells. Plant J 48:572–580

Lamb JC, Meyer JM, Corcoran B, Kato A, Han F, Birchler JA (2007) Distinct chromosomal distributions of highly repetitive sequences in maize. Chromosome Res 15:33–49

Lewis CD, Laemmli UK (1982) Higher order metaphase chromosome structure: evidence for metalloprotein interactions. Cell 29:171–181

Liu Cm CM, McElver J, Tzafrir I, Joosen R, Wittich P, Patton D, Van Lammeren AA, Meinke D (2002) Condensin and cohesin knockouts in *Arabidopsis* exhibit a titan seed phenotype. Plant J 29:405–415

Lyon MF (2003) The Lyon and the LINE hypothesis. Semin Cell Dev Biol 14:313–318

MacCallum DE, Losada A, Kobayashi R, Hirano T (2002) ISWI remodeling complexes in *Xenopus* egg extracts: identification as major chromosomal components that are regulated by INCENP-aurora B. Mol Biol Cell 13:25–39

Maeshima K, Eltsov M (2008) Packaging the genome: the structure of mitotic chromosomes. J Biochem 143:145–153

Maeshima K, Laemmli UK (2003) A two-step scaffolding model for mitotic chromosome assembly. Dev Cell 4:467–480

Maeshima K, Eltsov M, Laemmli UK (2005) Chromosome structure: improved immunolabeling for electron microscopy. Chromosoma 114:365–375

Maeshima K, Hihara S, Eltsov M (2010) Chromatin structure: does the 30-nm fibre exist in vivo? Curr Opin Cell Biol 22:291–297

Marsden MP, Laemmli UK (1979) Metaphase chromosome structure: evidence for a radial loop model. Cell 17:849–858

McClintock B (1930) A cytological demonstration of the location of an interchange between two non-homologous chromosomes of *Zea mays*. Proc Natl Acad Sci USA 16:791–796

McDowall AW, Smith JM, Dubochet J (1986) Cryo-electron microscopy of vitrified chromosomes in situ. EMBO J 5:1395–1402

Morgan GT (2002) Lampbrush chromosomes and associated bodies: new insights into principles of nuclear structure and function. Chromosome Res 10:177–200

Morrison C, Henzing AJ, Jensen ON, Osheroff N, Dodson H, Kandels-Lewis SE, Adams RR, Earnshaw WC (2002) Proteomic analysis of human metaphase chromosomes reveals topoisomerase II alpha as an Aurora B substrate. Nucleic Acids Res 30:5318–5327

Nasmyth K, Haering CH (2005) The structure and function of SMC and kleisin complexes. Annu Rev Biochem 74:595–648

Nishino Y, Eltsov M, Joti Y, Ito K, Takata H, Takahashi Y, Hihara S, Frangakis AS, Imamoto N, Ishikawa T, Maeshima K (2012) Human mitotic chromosomes consist predominantly of irregularly folded nucleosome fibres without a 30-nm chromatin structure. EMBO J 31:1644–1653

Ohsumi K, Katagiri C, Kishimoto T (1993) Chromosome condensation in *Xenopus* mitotic extracts without histone H1. Science 262: 2033–2035

Okada TA, Comings DE (1980) A search for protein cores in chromosomes: is the scaffold an artifact? Am J Hum Genet 32:814–832

Okada M, Cheeseman IM, Hori T, Okawa K, McLeod IX, Yates JR 3rd, Desai A, Fukagawa T (2006) The CENP-H-I complex is required for the efficient incorporation of newly synthesized CENP-A into centromeres. Nat Cell Biol 8:446–457

Olins DE, Olins AL (2003) Chromatin history: our view from the bridge. Nat Rev Mol Cell Biol 4:809–814

Ono T, Losada A, Hirano M, Myers MP, Neuwald AF, Hirano T (2003) Differential contributions of condensin I and condensin II to mitotic chromosome architecture in vertebrate cells. Cell 115:109–121

Ono T, Fang Y, Spector DL, Hirano T (2004) Spatial and temporal regulation of Condensins I and II in mitotic chromosome assembly in human cells. Mol Biol Cell 15:3296–3308

Paulson JR, Laemmli UK (1977) The structure of histone-depleted metaphase chromosomes. Cell 12:817–828

Peacock WJ, Dennis ES, Rhoades MM, Pryor AJ (1981) Highly repeated DNA sequence limited to knob heterochromatin in maize. Proc Natl Acad Sci USA 78:4490–4494

Rabl C (1885) Über Zelltheilung. Morphol Jahrb 10:214–330

Rattner JB, Lin CC (1985) Radial loops and helical coils coexist in metaphase chromosomes. Cell 42:291–296

Robinson PJ, Rhodes D (2006) Structure of the ‚'30 nm' chromatin fibre: a key role for the linker histone. Curr Opin Struct Biol 16:336–343

Robinson PJ, Fairall L, Huynh VA, Rhodes D (2006) EM measurements define the dimensions of the "30-nm" chromatin fiber: evidence for a compact, interdigitated structure. Proc Natl Acad Sci USA 103:6506–6511

Routh A, Sandin S, Rhodes D (2008) Nucleosome repeat length and linker histone stoichiometry determine chromatin fiber structure. Proc Natl Acad Sci USA 105:8872–8877

Saitoh N, Goldberg IG, Wood ER, Earnshaw WC (1994) ScII: an abundant chromosome scaffold protein is a member of a family of

putative ATPases with an unusual predicted tertiary structure. J Cell Biol 127:303–318
Saka Y, Sutani T, Yamashita Y, Saitoh S, Takeuchi M, Nakaseko Y, Yanagida M (1994) Fission yeast cut3 and cut14, members of a ubiquitous protein family, are required for chromosome condensation and segregation in mitosis. EMBO J 13:4938–4952
Sakaguchi A, Kikuchi A (2004) Functional compatibility between isoform alpha and beta of type II DNA topoisomerase. J Cell Sci 117:1047–1054
Sakamoto T, Tsujimoto-Inui Y, Uraguchi S, Yoshizumi T, Matsunaga S, Mastui M, Umeda M, Fukui K, Fujiwara T (2011) Condensin II alleviates DNA damage and is essential for tolerance of B overload stress in Arabidopsis thaliana. Plant Cell 23:3533–3546
Schalch T, Duda S, Sargent DF, Richmond TJ (2005) X-ray structure of a tetranucleosome and its implications for the chromatin fibre. Nature 436:138–141
Sedat J, Manuelidis L (1978) A direct approach to the structure of eukaryotic chromosomes. Cold Spring Harb Symp Quant Biol 42(Pt 1):331–350
Siddiqui NU, Stronghill PE, Dengler RE, Hasenkampf CA, Riggs CD (2003) Mutations in *Arabidopsis* condensin genes disrupt embryogenesis, meristem organization and segregation of homologous chromosomes during meiosis. Development 130:3283–3295
Siddiqui NU, Rusyniak S, Hasenkampf CA, Riggs CD (2006) Disruption of the *Arabidopsis* SMC4 gene, AtCAP-C, compromises gametogenesis and embryogenesis. Planta 223:990–997
Steffensen S, Coelho PA, Cobbe N, Vass S, Costa M, Hassan B, Prokopenko SN, Bellen H, Heck MM, Sunkel CE (2001) A role for *Drosophila* SMC4 in the resolution of sister chromatids in mitosis. Curr Biol 11:295–307
Strick R, Strissel PL, Gavrilov K, Levi-Setti R (2001) Cation-chromatin binding as shown by ion microscopy is essential for the structural integrity of chromosomes. J Cell Biol 155:899–910
Strukov YG, Wang Y, Belmont AS (2003) Engineered chromosome regions with altered sequence composition demonstrate hierarchical large-scale folding within metaphase chromosomes. J Cell Biol 162:23–35
Strunnikov AV, Hogan E, Koshland D (1995) SMC2, a *Saccharomyces cerevisiae* gene essential for chromosome segregation and condensation, defines a subgroup within the SMC family. Genes Dev 9:587–599
Takata H, Uchiyama S, Nakamura N, Nakashima S, Kobayashi S, Sone T, Kimura S, Lahmers S, Granzier H, Labeit S, Matsunaga S, Fukui K (2007) A comparative proteome analysis of human metaphase chromosomes isolated from two different cell lines reveals a set of conserved chromosome-associated proteins. Genes Cells 12:269–284
Takemoto A, Kimura K, Yokoyama S, Hanaoka F (2004) Cell cycle-dependent phosphorylation, nuclear localization, and activation of human condensin. J Biol Chem 279:4551–4559
Takemoto A, Maeshima K, Ikehara T, Yamaguchi K, Murayama A, Imamura S, Imamoto N, Yokoyama S, Hirano T, Watanabe Y, Hanaoka F, Yanagisawa J, Kimura K (2009) The chromosomal association of condensin II is regulated by a noncatalytic function of PP2A. Nat Struct Mol Biol 16:1302–1308
Tavormina PA, Come MG, Hudson JR, Mo YY, Beck WT, Gorbsky GJ (2002) Rapid exchange of mammalian topoisomerase II alpha at kinetochores and chromosome arms in mitosis. J Cell Biol 158:23–29
Tremethick DJ (2007) Higher-order structures of chromatin: the elusive 30 nm fiber. Cell 128:651–654
Uchiyama S, Kobayashi S, Takata H, Ishihara T, Hori N, Higashi T, Hayashihara K, Sone T, Higo D, Nirasawa T, Takao T, Matsunaga S, Fukui K (2005) Proteome analysis of human metaphase chromosomes. J Biol Chem 280:16994–17004
Uemura T, Ohkura H, Adachi Y, Morino K, Shiozaki K, Yanagida M (1987) DNA topoisomerase II is required for condensation and separation of mitotic chromosomes in *S. pombe*. Cell 50:917–925
Vagnarelli P, Hudson DF, Ribeiro SA, Trinkle-Mulcahy L, Spence JM, Lai F, Farr CJ, Lamond AI, Earnshaw WC (2006) Condensin and Repo-Man-PP1 co-operate in the regulation of chromosome architecture during mitosis. Nat Cell Biol 8:1133–1142
Van Hooser A, Goodrich DW, Allis CD, Brinkley BR, Mancini MA (1998) Histone H3 phosphorylation is required for the initiation, but not maintenance, of mammalian chromosome condensation. J Cell Sci 111(Pt 23):3497–3506
Wang JC (2002) Cellular roles of DNA topoisomerases: a molecular perspective. Nat Rev Mol Cell Biol 3:430–440
Wei Y, Yu L, Bowen J, Gorovsky MA, Allis CD (1999) Phosphorylation of histone H3 is required for proper chromosome condensation and segregation. Cell 97:99–109
Wilkie GS, Shermoen AW, O'Farrell PH, Davis I (1999) Transcribed genes are localized according to chromosomal position within polarized *Drosophila* embryonic nuclei. Curr Biol 9:1263–1266
Woodcock CL (1994) Chromatin fibers observed in situ in frozen hydrated sections. Native fiber diameter is not correlated with nucleosome repeat length. J Cell Biol 125:11–19
Woodcock CL, Frado LL, Rattner JB (1984) The higher-order structure of chromatin: evidence for a helical ribbon arrangement. J Cell Biol 99:42–52
Yeong FM, Hombauer H, Wendt KS, Hirota T, Mudrak I, Mechtler K, Loregger T, Marchler-Bauer A, Tanaka K, Peters JM, Ogris E (2003) Identification of a subunit of a novel Kleisin-beta/SMC complex as a potential substrate of protein phosphatase 2A. Curr Biol 13:2058–2064
Yoshikawa K, Yoshikawa Y (2002) Compaction and condensation of DNA. In: Mahato RI, Kim SW (eds) Pharmaceutical perspectives of nucleic acid-based therapeutics. Taylor and Francis, New York, pp 137–163

Structural Organization of the Plant Nucleus: Nuclear Envelope, Pore Complexes and Nucleoskeleton

4

Elena Kiseleva, Jindriska Fiserova, and Martin W. Goldberg

Contents

4.1 Introduction

Plants, like all other eukaryotes, contain their genome within the nuclear compartment. The purpose of this compartment is to separate the transcriptional machinery from the sites of protein synthesis. There is therefore an impermeable barrier between the nuclear and cytoplasmic compartments, the nuclear envelope (NE). However the NE is permeated with large protein channels, the nuclear pore complexes (NPCs). Trafficking through the NPCs can therefore be used to control gene expression at several levels. The outer and inner membranes both contain distinct and complex sets of proteins, which link to the cytoskeleton as well as the nuclear interior. The nucleus also therefore has important functions in organizing both the genome and the cytoplasm. Here we describe the architecture and dynamics of the structural components of the nucleus and discuss how the plant nucleus appears to differ in important ways from animals and fungi, while maintaining many similarities. First we will focus on the NE of plants and their specific protein composition, structure and function compared to animal systems. We will then discuss the role of the NE as a barrier and interface between the cytoplasm and nucleoplasm, before focussing on communication across the NE and finally discussing how the nuclear interior is structurally organized by the nucleoskeleton. Although the proteins and functions of the NE and nucleoskeleton appear to overlap, we discuss them separately so as not to confuse the distinct functions that do clearly exist.

4.2 The Nuclear Envelope

The eukaryotic nucleus is surrounded by a double phospholipid bilayer called the nuclear envelope (NE). The NE of plants, animals or fungi is organized in a similar manner: chemically distinct inner and outer nuclear membranes (INM and ONM) encircle the nucleus and enclose the

E. Kiseleva (✉)
Laboratory of Morphology and Function of Cell Structure, Institute of Cytology and Genetics, Russian Academy of Science, Novosibirsk 630090, Russia
e-mail: elka@bionet.nsc.ru

I.J. Leitch et al. (eds.), *Plant Genome Diversity Volume 2*,
DOI 10.1007/978-3-7091-1160-4_4,

perinuclear space between them. The fusion of the two membranes occurs at the insertion sites of the nuclear pore complexes (NPCs), the only gates for regulated protein transport. Inner and outer membranes of the NE differ not only by chemical composition of their lipid components but, importantly, by the type and number of the associated proteins. A particular set of proteins essentially defines the specific structural and functional characteristics of the NEs of different cells, tissues and species.

4.2.1 Proteins of the Plant Nuclear Envelope

In animals and yeast, a combination of approaches has led to the identification of multiple NE proteins (for review see Dauer and Worman 2009; Hetzer 2010). In comparison, the plant field is less advanced and only a few NE proteins have been identified in the last decade. The following text will concentrate on the structure and function of plant-specific proteins putatively linked to the inner nuclear membrane (lamin-like proteins) (1), proteins of the inner (2) and outer (3) nuclear membrane and ion channels (4).

4.2.2 Lamin-Like Proteins in Plants

The nuclear lamina represents a multifunctional, highly organized protein layer closely attached to the INM in vertebrates. While biochemical, immunological and structural evidence for a lamina-like structure in plants have been presented (McNulty and Saunders 1992; Minguez and Moreno Diaz de la Espina 1993; Fiserova et al. 2009), the lack of any gene coding for a plant protein homolog of a lamin raises the question as to which proteins make up the detected structures. Animal lamins are ~66 kDa in size, contain a central coil-coiled region to enable dimerisation and N- and C-terminal globular head and tail domains. In vertebrates, lamins are encoded by three genes, LMNA, LMNB1 and LMNB2. The three genes encode at least seven different polypeptides resulting from splice or post-translational modifications (for review see Gruenbaum et al. 2003; Hetzer 2010). The nuclear lamina was shown to be essential in many processes including chromatin organization, transcription, NPC anchoring or communication between the nucleus and the cytoplasm. Therefore, it is likely that some plant-specific proteins are required in plants to fulfil at least some of the proposed lamin roles.

To date, the best candidate for a lamin-like protein in plants is LINC1 (Little nuclei 1) in *Arabidopsis* alias NMCP1 (Nuclear Matrix Constituent Protein 1) in carrot (Masuda et al. 1997; Dittmer et al. 2007). About twice the size of lamin (LINC1 is about 134 kD), the LINC1 protein contains a coil-coiled domain, localizes at the NE and mutations cause altered nuclear morphology (Dittmer et al. 2007). LINC1 belongs to a family of four LINC proteins (LINC1-4) with varying functions and localizations within the cell. LINC2 and LINC3 were identified as nuclear proteins with no specificity for the NE, while LINC4 was shown to be a plastid protein (Kleffmann et al. 2006; Dittmer et al. 2007). It is not known, however, whether LINC1 can form filaments *in vitro* or *in vivo*, what proteins it binds or how it is involved in maintaining nuclear morphology.

Other putative lamin-like proteins in plants include a family of seven filament-like proteins identified from *Arabidopsis* (AtFPP1-7 for filament-like plant protein of *Arabidopsis thaliana*). Homologs have been identified in tomato (LeFPP) and rice (OsFPP) (Gindullis et al. 2002). In yeast two-hybrid screens, these proteins associate with another nuclear envelope protein MAF1 (MFP1 associated factor1). However, their more specific subcellular localisation or function is uncertain.

To complete the list of plant-specific coiled-coil NE proteins, MAR binding filament-like protein 1 (MFP1) was found to specifically bind to matrix attachment regions (MARs) of DNA at the NE in tomato. MFP1 contains a coil-coiled domain, an N-terminal hydrophobic region responsible for targeting the protein to the NE but no globular head or tail. It is located at the NE and was predicted to be a component of the plant nuclear matrix and to connect the nuclear matrix to the NE (Gindullis and Meier 1999; Gindullis et al. 1999).

4.2.3 Integral Membrane Proteins of the Inner Nuclear Membrane

Most INM proteins characterised in vertebrates do not have homologs in plants (for review see Meier 2007; Meier and Brkljacic 2009b; Graumann and Evans 2010). The only proteins identified to date to reside at the plant INM are two *Arabidopsis* SUN-domain (Spindle architecture defective 1/UNC84 homology) containing proteins (AtSun1 and AtSun2) (Graumann et al. 2009). Smaller than animal SUN proteins (~50 kDa vs. ~85 kDa) they contain the highly conserved C-terminal SUN domain and share structural features with animal and fungal SUN-domain proteins including a functional coiled-coil domain and nuclear localization signal. The proteins localize to the INM and are expressed in various tissues and cell types, but most abundantly in proliferating tissues (Graumann et al. 2009). Similarity searches have revealed the presence of the two SUN proteins in other plant species including *Oryza sativa*, *Vitis vinifera* and *Zea mays*. In maize, five different SUN genes were identified (Murphy et al. 2010), which fall into two classes. Whereas ZmSun1 and ZmSun2 represent structural homologs of animal SUNs, ZmSun3-5 include a novel structural variant of SUN-domain proteins with likely plant-specific roles at the NE (Murphy et al. 2010).

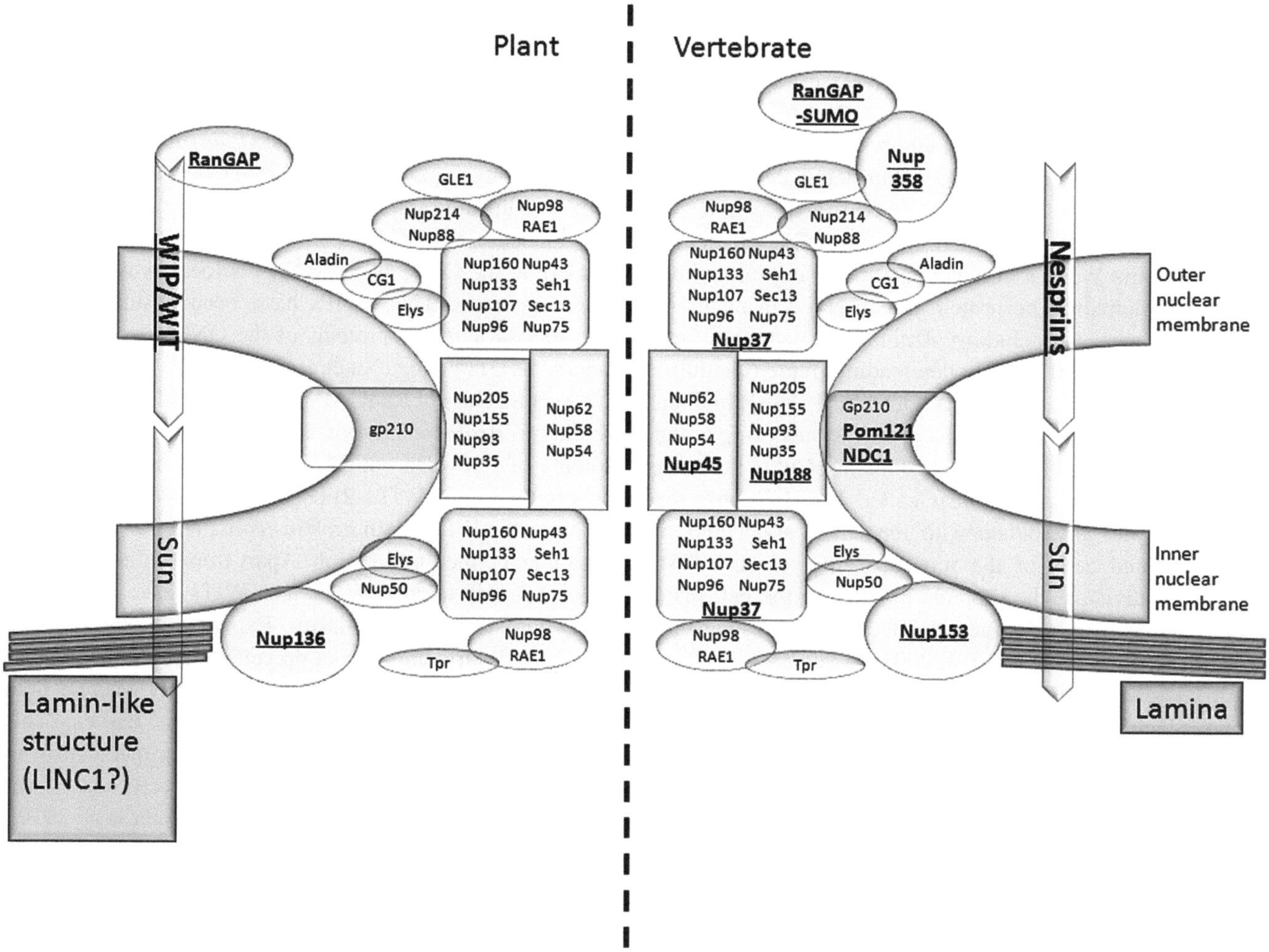

Fig. 4.1 Comparison of plant and animal nucleoporins and associated proteins, showing approximate relative positions within the NPC. Proteins labelled in *bold* are those that appear to differ between the two kingdoms

In vertebrates, SUN proteins interact in the intramembrane space via their SUN-domain with the KASH domain of KASH-domain proteins (Klarsicht/ANC-1/SYNE1 homology, also referred to as Nesprins) located at the ONM. Thus, they form a bridge (the bridging complexes are also known as the LINC complexes for LInker of Nucleoskeleton and Cytoskeleton, which should not be confused with the LINC1-4 proteins) spanning the two membranes of the NE and physically connecting the nuclear interior with the cytoskeleton (for review see Starr and Fridolfsson 2010; Starr 2011). Thus, LINC complexes connect microtubules, centrosomes, actin filaments, or intermediate filaments to the surface of the nucleus. They are essential in cell cycle control, nuclear import, and apoptosis or affect global cytoskeletal organization (reviewed in Starr 2011). Until recently, the other partner of the LINC complex, KASH-domain proteins of the ONM, were not identified in plants. Interestingly, WIP proteins (Fig. 4.1), which have a role in localizing the GTPase activating protein for the small GTPase nuclear transport regulator, Ran, to the outer nuclear membrane, have been shown to contain sequences similar to the KASH domain. Most importantly, these proteins also bind to the SUN proteins and so appear to be functionally equivalent to animal KASH-domain proteins (Zhou et al. 2012).

4.2.4 Proteins of the Outer Nuclear Membrane

In vertebrates, a number of spectrin repeat containing nuclear envelope proteins (Syne/Nesprins) have been identified at the ONM (reviewed in Mellad et al. 2011). Nesprins are characterised, among other things, by the C-terminal KASH domain (Klarsicht/ANC-1/SYNE1 homology) that physically links to the SUN-domain of SUN proteins in the intramembrane space. Nesprins thus form molecular bridges between the cytoplasm and the nucleus termed LINC complexes (reviewed in Starr 2011). These proteins have important roles in nuclear positioning and migration, focal adhesions and cell motility. Whether WIPs perform such functions is unknown, but it seems likely that other candidates will also be discovered to fulfil such roles in plant-specific environments.

On the other hand, other plant-specific ONM proteins have been found and studied in greater detail. MFP attachment factor 1 (MAF1) is a small (~16 kDa) globular NE associated protein first identified in tomato (Gindullis et al. 1999). Like the WIPs it contains a tryptophan/proline/proline (WPP) domain. The protein is conserved in plants as other plant species including *Arabidopsis*, soybean and maize contain the full MAF1 open reading frame (Gindullis et al. 1999). The protein interacts with MFP1 (MAR binding filament-like protein 1) which may provide anchoring to the NE via its N-terminal hydrophobic domains. Out of the three MAF1 homologs in *Arabidopsis* called WPP1-3, only the WPP1 and WPP2 associate with the ONM and NPCs in undifferentiated cells of the root tip (Patel et al. 2004). During cell division, the proteins relocate to the cell plate and RNAi leads to developmental defects in roots suggesting a role in cytokinesis (Patel et al. 2004).

An interesting difference between plants and animals is the way plants use a family of ONM proteins to orientate the Ran gradient which is required for the directionality of nuclear transport. RanGAP is a GTPase activating protein for Ran and is essential for nuclear transport as well as cell division. In yeast it is located in the cytosol, whereas in animals it is also associated with the outer surface of the NPCs. Four RanGAPs have been identified in plants to date; two from *Arabidopsis* and one each from rice and alfalfa (Pay et al. 2002). The proteins are ~60 kDa in size and their N-terminal domain shows significant homology to the WPP domain of plant MAF1 protein. In interphase cells, AtRanGAP1 localizes at the NE while it is associated with the preprophase band, spindle and phragmoplast during mitosis. The detected localization pattern corresponds with the postulated function of plant RanGAPs in the regulation of nuclear import during interphase and suggests a role as a continuous protein marker of the plant cell division plane (Xu et al. 2008).

RanGAP is targeted to the ONM in a plant-specific manner. Animal RanGAP targeting to the NPCs requires interaction of its SUMOylated C-terminus with the nucleoporin Nup358/RanBP2 (which is, interestingly, lacking in plants), whereas AtRanGAP1 associates with the NE via its N-terminal plant-specific WPP domain (Rose and Meier 2001; Pay et al. 2002). Two classes of WPP-domain interacting proteins have been further identified. Firstly, a family of three WPP-domain interacting proteins (WIPs) of ~70 kDa in size were identified from *Arabidopsis* in yeast two hybrid screens to interact with AtRanGAP1. Similarity searches showed that the protein family is conserved in many plant species including tomato (*Solanum lycopersicum*), grape (*Vitis vinifera*), barrel medic (*Medicago truncatula*), poplar (*Populus deltoides*), wheat (*Triticum aestivum*), rice (*Oryza sativa*), maize (*Zea mays*), barley (*Hordeum vulgare*), sorghum (*Sorghum bicolor*), sugarcane (*Saccharum officinarum*), and pine (*Pinus taeda*) (Xu et al. 2007). The proteins contain a predicted C-terminal transmembrane domain and central coiled-coil domain that is necessary and sufficient for RanGAP interaction. WIPs co-localize with RanGAP at the ONM as shown by confocal as well as electron microscopy. Thus, WIPs have been postulated to be plant-specific intrinsic proteins of the ONM responsible for RanGAP NE targeting (Xu et al. 2007).

A second class of RanGAP interacting and NE associated proteins in *Arabidopsis* root tip cells is constituted of two proteins of ~90 kDa called WPP-domain interacting tail-anchored proteins (WIT1-2) (Zhao et al. 2008). Structurally similar to the WIPs, both proteins contain a transmembrane domain and coiled-coil region. Apart from the homology in the coiled-coil region, WIPs and WITs have no further amino acid sequence similarity. Experiments showed that RanGAP NE targeting in root tip cells, required the activity of at least one member of each protein family whereas both families were dispensable in other plant tissues. This suggested a complex RanGAP NE targeting mechanism in higher plants contrasting to both animal cells and lower eukaryotes (Zhao et al. 2008, reviewed in Meier et al. 2010).

4.2.5 Ion Channels at the Nuclear Envelope

In animals, the NE hosts a vast number of ion channels and transporters (Schirmer et al. 2003; Batrakou et al. 2009). Whether these transporters reside exclusively at the NE is not yet certain in most cases and many are likely to be present in other cellular membranes as well (Foster et al. 2006; for review see Matzke et al. 2010). Several nuclear membrane ion channels have been identified in plants by nuclear patch clamping and through genetic screens (reviewed in Matzke et al. 2010). SERCA-type calcium ATPases were found in tomato and *Medicago truncatula* and termed LCA and MCA8, respectively (Downie et al. 1998; Capoen et al. 2011). LCA is ~116 kDa and seems to be present only as a single copy gene in tomato. Several putative SERCA homologs have been identified in *Arabidopsis* (Evans and Williams 1998). This suggests that, similar to animals, there may be at least three SERCA homologs in plants located at the NE or endoplasmic reticulum (Downie et al. 1998; Evans and Williams 1998). The pump locates at the ONM and likely functions to replenish Ca^{2+} stores in the perinuclear space (Bunney et al. 2000; Mazars et al. 2011). MCA8 belongs to the large family of calcium ATPases consisting of ten members in *Medicago*, only MCA8, however, is NE targeted (Capoen et al. 2011).

One ortholog containing a nuclear localization sequence was found in *Lotus japonicus*. Orthologs of *Arabidopsis* or soybean however lack this nuclear localization signal (Capoen et al. 2011). MCA8 locates on both the INM and ONM in *Medicago*, which is in contrast to the nuclear calcium ATPases in animals that appear strictly localized to the ONM (Humbert et al. 1996; Capoen et al. 2011). The localization at both the INM and ONM suggests a role in recapture of calcium released into the nucleus as well as reloading the perinuclear space (Capoen et al. 2011).

In plants, most attention has been paid to the calcium signalling during nodulation in legumes. Perinuclear Ca^{2+} oscillations are produced in response to root bacterial infections and result in the development of a specialized organ, the root nodule. Genetic screens for nodulation-defective mutants identified several related proteins essential for the Ca^{2+} oscillations: DMI1-DMI3 (Does not Make Infections 1–3) were identified from *Medicago truncatula* and shown to be required for both nodulation and mycorrhization (Ane et al. 2004; Riely et al. 2007). DMI1 was studied in greater detail and shown to preferentially locate at the INM. The protein is similar to the bacterial calcium-activated potassium channel (MthK). A conserved motif for the potassium selective filter is, however, missing leaving open the possibility that the protein is permeable to other ions. Its role in calcium regulation is likely when considering the DMI1 requirement for calcium spikes in response to rhizobium signalling and nodule development (Peiter et al. 2007; Mazars et al. 2009). DMI1 homologs have also been identified in other plant species including CASTOR and POLUX from *Lotus japonicus* (Imaizumi-Anraku et al. 2005; Charpentier et al. 2008) and SYMB8 (Symbiosis8) from *Pisum sativum* (Edwards et al. 2007). It has been hypothesised that CASTOR is present in the ONM and POLUX in both the ONM and INM (Charpentier et al. 2008). Accordingly, both channels play a role in regulating NE membrane potential (Charpentier et al. 2008).

Nuclear calcium signalling as well as the role of the NE in calcium signalling is a recent, strongly developing field of plant biology and a full consideration goes beyond the scope of this chapter. Therefore, we refer the reader to recent reviews (Matzke et al. 2009; Mazars et al. 2009, 2011).

4.2.6 Functions of the Nuclear Envelope Proteins

The NE represents a highly organized barrier separating the nucleoplasm from the cytoplasm. As such it allows the segregation of the processes involving genetic material storage, expression and replication from the metabolic processes taking place in the cytoplasm. Many roles of the NE closely relate to the roles of the nuclear lamina and proteins associated with or incorporated inside the NE. In this respect the roles of the specific proteins and the functioning of the NE are indistinguishable. With the assistance of the nuclear lamina, for instance, the NE provides shape, mechanical stability and plasticity for the nucleus (Houben et al. 2007). Growing evidence also suggests roles for lamins in maintaining the mechanical properties of the entire cell (Broers et al. 2006; Lee et al. 2007). In addition, lamins are essential for chromatin organization: gene-poor regions of chromatin have been shown to localize at the periphery whereas gene-rich regions have been located in the nuclear interior (Shimi et al. 2008; reviewed in Cremer et al. 2004). Finally, lamins have been implicated to play a role in the regulation of gene expression and DNA replication (Reddy and Singh 2008; reviewed in Kumaran et al. 2008).

Importantly, the NE anchors NPCs and thus enables the actively regulated transport of molecules between the nucleus and the cytoplasm (Dechat et al. 2009). Communication of the nucleus with cytoskeletal networks is further facilitated via transmembrane proteins and intermembrane bridges such as the LINC complex. Microtubules, actin filaments or intermediate filaments are connected to the surface of the nucleus and via these bridging complexes communicate with the nuclear interior. Thus, the two distinct compartments are in close contact (Starr 2011) and may affect each other according to actual needs. Details of these interactions, though, have not been described yet (Broers et al. 2004; Lee et al. 2007).

In animals, the importance of the NE proteins in development and differentiation is clearly demonstrated by a fascinating set of genetically inherited diseases referred to as nuclear envelopathies or laminopathies. These disorders affect muscle, adipose, bone, nerve and skin cells and range from muscular dystrophies to accelerated aging (reviewed in Mattout et al. 2006; Dauer and Worman 2009).

The NE of plants has received far less attention in comparison to animals and thus, many specific roles of the plant NE may be only guessed. However, in complex multicellular organisms with organized large genomes like those in some higher plants (nuclei of many plant species contain 10–100 times more chromatin than animal nuclei) (Shaw et al. 2002; Flowers and Purugganan 2008; Leitch and Leitch 2013, this volume), similar complex roles of the NE are very likely. Clearly, the NE serves as a barrier between the nucleoplasm and the cytoplasm, and apparently provides a platform for communication in a similar manner to that of animals: via NPCs or via proteins physically traversing the NE such as SUNs. The limited knowledge of the protein composition of the INM and clear differences from animal systems may indicate that plant-specific mechanisms exist for the involvement of the nuclear periphery in nuclear processes.

Limited evidence from mutant studies of NE proteins in plants further confirms that, like animals, NE proteins function in plant growth and development: mutant plants display not only altered nuclear morphologies but also varying developmental phenotypes including dwarfing and leaf curling (LINC1 mutants, Dittmer et al. 2007), root shortening and reduced numbers of lateral roots (WPP mutants, Patel et al. 2004). However, a combination of approaches will be necessary to confirm the NE functions established in animals as well as to understand plant-specific NE roles.

One plant-specific role identified for the NE is its ability to nucleate microtubules during interphase, as well as during cell division. Microtubules of the mitotic spindle are formed with the aid of microtubule organizing centres (MTOCs). To allow access of the mitotic spindle to the chromatin, different organisms have evolved various mechanisms. These range between the two extremes from "closed" mitosis (where the NE stays intact and MTOCs are either a constant part of the NE or are inserted into the NE during the entry of the cell into mitosis, and the mitotic spindle is formed inside the nucleus) to "open" mitosis (where the NE is completely disassembled as the cell enters mitosis and the mitotic spindle forms from MTOCs in the cytoplasm) (reviewed in Guttinger et al. 2009). Open mitosis is typical for higher eukaryotes including vertebrates and plants. While centrosomes serve as the MTOCs in vertebrates, plants do not posses any centrosomes or other microtubule organizing structures in the cytoplasm and, instead, the NE has the ability to organize microtubules and act as the MTOC (Stoppin et al. 1994; Magyar et al. 2013, this volume). This plant-specific NE function remains far from understood and only fragmented information is available on the molecular nature of the NE-localized MTOC. For detailed information on plant MTOCs as well as on the dynamics of the NE during the cell cycle we refer the interested reader to recent reviews (Rose et al. 2004; Meier and Brkljacic 2009b; Evans et al. 2011).

4.2.7 The Nuclear Envelope of Plants is Similar but Distinct from Animals

Structurally, the plant NE resembles that of animals or yeast. However, its molecular composition gives rises to a common theme when comparing plants and animals: similar in basic characteristics but clearly distinct in others. The absence of plant homologs of animal NE proteins like lamins, LBR (lamin B receptor), emerin, MAN1 or Nesprins (Graumann and Evans 2010; Hetzer 2010), the presence of plant-specific NE proteins such as FPP or MAF (Gindullis et al. 1999, 2002; Meier and Brkljacic 2009b) and the unique NE targeting mechanisms of RanGAP in plants and vertebrates (Meier et al. 2008) suggest that NE structure and composition in plants and animals were selected with respect to the specific needs of the two different kingdoms for NE function. However, much remains to be resolved to understand not only the structural details of the NE in plants but also the functional connections of the NE to the surrounding environment and within the cell. Thus, this field awaits yet a further bloom of exciting discoveries.

4.3 Plant Nuclear Pore Complex (NPC)

The NPC is a specialized multiprotein complex and is the sole gateway for macromolecular exchange between the nucleus and cytoplasm (Brohawn et al. 2009; Hoelz et al. 2011). Much attention has been paid to the structural and molecular details of the NPCs and their dynamics and function in animals and yeast (Hoelz et al. 2011; Kahms et al. 2011; Onischenko and Weis 2011). In contrast, until the last decade, little was known about the plant NPC (Heese-Peck and Raikhel 1998; Meier and Somers 2011). Here, aspects of the functional and structural dynamics of the plant NPC are reviewed.

4.3.1 Structural Organization of the Plant and Mammalian NPC is Similar

The ultrastructure of higher plant NPCs was first described during the analysis of sections by transmission electron microscopy (TEM) (Yoo and Bayley 1967; Roberts and Northcote 1970). Plant NPCs appeared similar to those observed in animal cells (Feldherr 1965; Stevens and Swift 1966; Hinshaw et al. 1992; Akey and Radermacher 1993). Detailed structural organization of individual components of NPCs (Goldberg and Allen 1992, 1996) as well as conservation of morphology between higher and lower eukaryotes, has been demonstrated by field emission in-lens scanning electron microscopy (feSEM) (Kiseleva et al. 1998, 2004). These studies have shown that vertebrate NPCs have a three-layered ring structure with the spoke ring complex sandwiched between the nuclear and cytoplasmic rings (Fig. 4.2) (Allen et al. 2000; Vasu and Forbes 2001). *Xenopus* oocyte NPCs have a diameter of ~110 nm (Goldberg and Allen 1995, 1996; Stoffler et al. 1999; Allen et al. 2000; Kiseleva et al. 2000) and possess eight-fold (top view) and two-fold (side view) symmetry (Unwin and Milligan 1982; Hinshaw et al. 1992; Akey and Radermacher 1993). The inner spoke ring (central aperture ~40 nm) and central transporter (which, although controversial, appears to alter conformation during mRNP transport) (Akey 1990; Kiseleva et al. 1996, 1998) are within the pore. Eight spokes emanate from the inner spoke ring, penetrate the pore membrane and join together via radial arms in the NE lumen to form the lumenal ring. The cytoplasmic part of the NPC contains the star ring, which is embedded in the outer membrane. The cytoplasmic ring lies on top and is

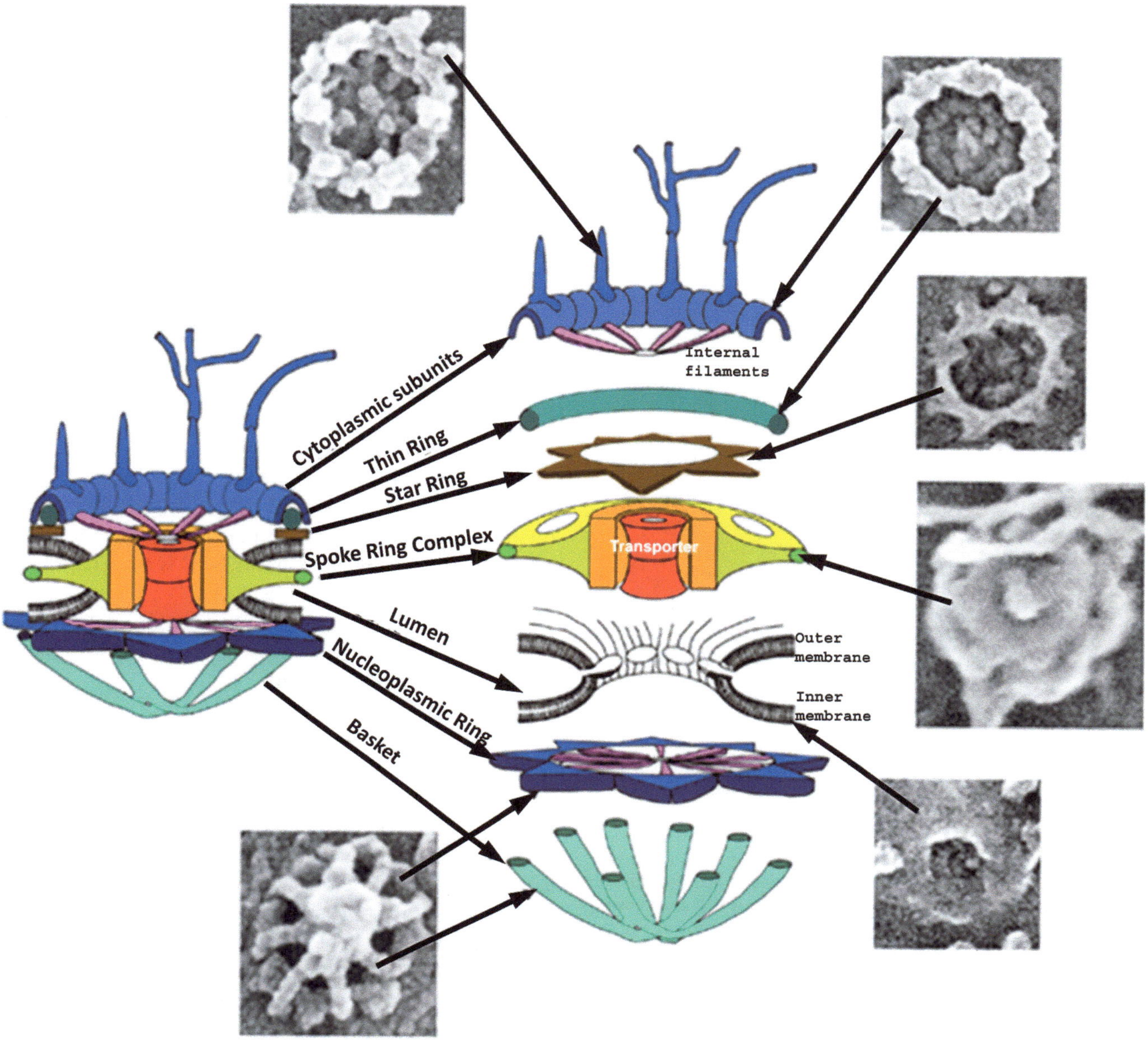

Fig. 4.2 Modular structural components of the NPC based on feSEM images of NEs from *Xenopus* and plants (Taken from Allen et al. (2000))

composed of a thin ring with eight subunits. Rod-like particles extend into the cytoplasm and play a role in cargo import (Panté and Aebi 1996; Rutherford et al. 1997). A basket-like structure attached to the nucleoplasmic ring (Goldberg and Allen 1996; Ris 1997) contains the distal basket ring which opens and closes during mRNP export (Kiseleva et al. 1996). Attached to the cytoplasmic and nucleoplasmic rings are internal filaments which have a spoke-like appearance (Goldberg and Allen 1996; Kiseleva et al. 1998).

Recent feSEM investigations of both sides of the NE from tobacco BY-2 cells as well as the cytoplasmic side of the NE from onion root cells have shown that the NPCs of flowering plants have a conserved structure which is related to vertebrate NPCs (Fig. 4.3) (Fiserova et al. 2009). NPCs in plants appear a little smaller (diameter ~105 nm) than in *Xenopus* oocytes (diameter 110–120 nm) (Goldberg and Allen 1996), but they are larger than in yeast (diameter ~95 nm) (Kiseleva et al. 2004), as observed by feSEM. The plant NPC also consists of a cytoplasmic ring formed by eight subunits with cytoplasmic filaments, the star ring and a central particle of 24–28 nm which is smaller than those identified in *Xenopus* oocytes and *Chironomus* salivary glands (outer diameter 33–38 nm; Goldberg and Allen 1996; Kiseleva et al. 1998). Fracturing of BY-2 cells revealed well-preserved structures of the inner nuclear membrane showing the existence of plant NPC baskets, similar to those of *Xenopus* oocytes. Plant baskets consist of ~6 nm filaments anchored to the nucleoplasmic ring and

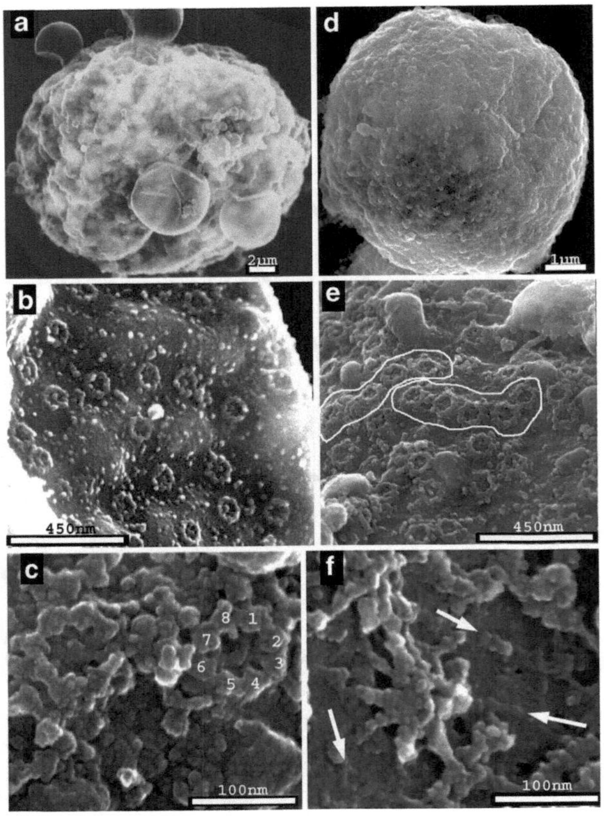

Fig. 4.3 Nuclei isolated from onion root tip (**a**-**c**) or tobacco BY-2 cells (**d**-**f**) and visualized with feSEM at low (**a** and **d**) and high (**b**-**f**) magnification. Large endoplasmic reticulum membranes are fused to the outer nuclear membrane (**a**). Many NPCs appear distributed over the nuclear envelope in onion root tips (**b**) but are arranged in rows in 7 day BY-2 cultures (as indicated in **e**), suggesting some underlying organization. Such an organization could be a lamina-like structure on the nucleoplasmic side, indicated by *arrows* in (**f**), which also shows a BY-2 NPC basket. The eight-fold rotational repeating structure of the plant cytoplasmic NPC ring is indicated in (**c**)

thicker filaments (~10 nm) extending from the distal ring of the baskets towards the nuclear interior (Fiserova et al. 2009). It was suggested that these could have structural roles in connecting the NPCs to each other or could act as docking sites for transport factors. It is relevant that similar but typically distinct inter-basket filament connections were observed in *Xenopus* and bird oocytes (Jarnik and Aebi 1991; Goldberg and Allen 1992; Goldberg et al. 1997).

The NPC density over the nuclear surface varies, depending on species, nuclear size and activity (Lim et al. 2008). It can range from 2 to 4 NPCs/μm^2 in chick erythrocytes or mammalian lymphocytes (Maul et al. 1971) to over 60 NPCs/μm^2 in *Acetabularia* or mature *Xenopus* oocytes (Gerace and Burke 1988; Görlich and Kutay 1999). The mean density of NPCs in plants varies from over 25 NPCs/μm^2 in onion root cells to 40–50/μm^2 in tobacco BY-2 cells (Fiserova et al. 2009). The NPCs are not distributed in a random manner over the nuclear surface, but are aligned in rows (Aaronson and Blobel 1975; Goldberg and Allen 1992; Maeshima et al. 2006). The arrangement of NPCs in plants into compact linear rows suggests that a lamina-like structure could play a role in the organization of the NPCs in plants (Fiserova and Goldberg 2010) and other eukaryotes (mammalian cells, Maeshima et al. 2006; *Caenorhabditis*, Liu et al. 2000; or in *Drosophila* cells, Lenz-Böhme et al. 1997). Loss of A-type lamins in mammals, or lamin Dmo in *Drosophila* correlates with lower NPC densities or deviant NPC spatial organization (Sullivan et al. 1999; Maeshima et al. 2006). Moreover yeast does not have a nuclear lamina and NPCs are arranged as mobile clusters (Belgareh and Doye 1997; Bucci and Wente 1997; Winey et al. 1997). NPC clustering was also noted in *Caenorhabditis* mutants whose lamin content was reduced (Liu et al. 2000). Overall, feSEM together with other approaches has revealed further information on the structural organization of NPCs in plants and demonstrated that the general morphology of the NPC is conserved from yeast to humans and plants (Fig. 4.4).

4.3.2 Plant Nucleoporins

The NPC represents the largest multiprotein channel in the cell and is a fundamental component of all eukaryotes. Therefore information about the proteins (nucleoporins or Nups) that it is composed of, is essential for understanding the mechanism of molecular trafficking. Although yeast and mammalian NPCs differ in size, their components are similarly arranged. Proteome analysis of animal and yeast NPCs demonstrated that both contain ~30 different Nups (Cronshaw et al. 2002; Brohawn and Schwartz 2009; Elad et al. 2009). Although recent investigations have suggested that the plant NPC also consists of ~30 different Nups (Meier and Brkljacic 2009a; Meier and Somers 2011), we know much less about the transport machinery in plants than in vertebrates and yeast. Moreover, a complete plant NPC proteome has not been established.

In all eukaryotes, including plants, NPC proteins form several subcomplexes. In vertebrates, three subcomplexes have been described: the Nup62 subcomplex (Nup62, Nup58, Nup54, and Nup45); the Nup107-160 subcomplex (Nup160, Nup133, Nup107, Nup96, Nup75, Nup43, Nup37, Seh1, Sec13, and ALADIN); and the Nup93 subcomplex (Nup205, Nup188, Nup155, Nup93, and Nup35) (Tran and Wente 2006). Other investigators have separated Nups into three categories, depending on their location in the different compartments within the pore membrane: Nups integrated into the pore membrane itself; scaffold Nups involved in the formation of the NPC framework; and barrier Nups critical for the selective permeability of the NPC (Fig. 4.1)

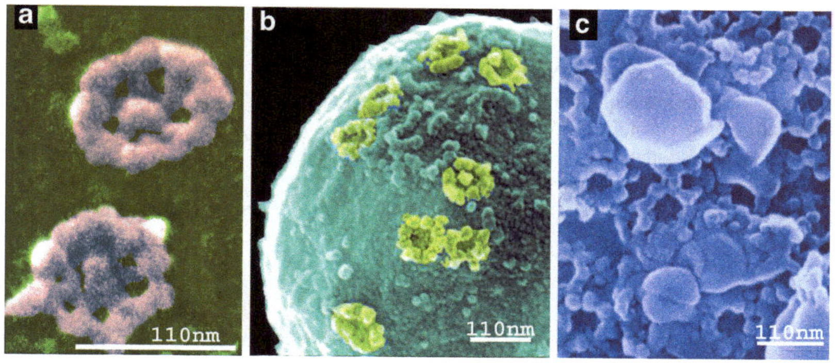

Fig. 4.4 Evolutionary conservation of the structural organization of NPCs from different organisms, showing NPCs from onion (**a**), yeast (**b**) and *Drosophila* embryos (**c**)

(Onischenko and Weis 2011; Tamura and Hara-Nishimura 2011).

The first layer consists of membrane Nups that facilite the anchoring of the neighbouring scaffold layer Nups. This layer in turn attaches barrier Nups that form the wall of the central channel and the asymmetric NPC extensions (Figs. 4.1 and 4.2). Barrier Nups contain multiple phenylalanine-glycine (FG) repeats giving rise to either a hydrogel based on hydrophobic interactions between the FG repeats (Ribbeck and Görlich 2001) or a brush border arrangement of the intrinsically disordered FG domains (Rout et al. 2000). The permeability properties of these molecular organizations allow selective access of the transport complexes while inhibiting the entry of other proteins of comparable size. Nups localize on both sides of a symmetrical axis in the NE plane.

A number of plant Nups that are similar to the animal and yeast Nups have been found using large scale genetic and reverse genetic approaches (Boruc et al. 2012). Sequence similarity searches have failed to identify plant homologs for about half of the nucleoporins identified in metazoan NPCs (Xu and Meier 2008). Nevertheless, proteome analysis of the nuclear pore composition in *Arabidopsis* using transgenic *Arabidopsis* expressing GFP-tagged RAE1 (RNA export factor 1) and mass spectrometry led to the identification of a total of 30 *Arabidopsis* Nups (Tamura et al. 2010). Most of these proteins exhibited sequence homology to vertebrate Nups however no homologs in the *Arabidopsis* genome were found for the human Nup358, Nup153, Nup45, Nup37, NDC1, or POM121 (indicated in bold in Fig. 4.1) (Xu et al. 2007; Tamura et al. 2010). A candidate for *Arabidopsis* NDC1, which was localized at the NE in *Arabidopsis* root tip cells, was proposed (Stavru et al. 2006). Nup136 is thought to be a functional homolog to animal Nup153, although they have no sequence homology (Tamura et al. 2010). Interestingly overexpression of Nup136 affects nuclear morphology. Plant Nup136 (Tamura et al. 2010), also named Nup1 (Lu et al. 2010), is a FG-Nup.

Plant Nups are important in diverse processes such as the resistance response and hormone signaling (Parry et al. 2006), cold-stress tolerance (Dong et al. 2006), or rhizobial and fungal symbiosis (Kanamori et al. 2006; Saito et al. 2007; Boruc et al. 2012).

4.3.3 Nucleocytoplasmic Transport Through the NPC

Small molecules and ions diffuse freely through the NPC, whereas the transport of large molecules such as proteins and RNA is a regulated process (Wälde and Kehlenbach 2010; Zhao and Meier 2011). The mechanism of nucleocytoplasmic transport through the mammalian NPC has been thoroughly studied. Most actively transported proteins contain a NLS (nuclear localization signal) or NES (nuclear export signal), or indeed both. These motifs have also been found in many plant proteins (Merkle 2001). Selective transport is regulated by many soluble factors. The most important carrier proteins belong to the superfamily of importin β-like proteins (also referred to as "karyopherins" or importins, exportins and transportins) (Fried and Kutay 2003; Xylourgidis and Fornerod 2009). Karyopherins are conserved eukaryotic proteins but in plants they do reveal some unique features (Meier 2007).

Another soluble factor, the small GTPase Ran, exists in either the GTP- or GDP-bound form. RanGTP and its conversion to RanGDP crucially control the interactions between karyopherins and cargoes (Merkle 2001; Meier 2007; Xu et al. 2007). Three genes encoding Ran have been identified in *Arabidopsis* (Haizel et al. 1997). Transformation of RanGTP to RanGDP takes place in the cytoplasm and is under the control of Ran GTPase Activating Protein (RanGAP) and Ran Binding Proteins (RanBPs). The chromatin-bound Ran nucleotide exchange factor RCC1 (regulator of chromosome condensation) converts RanGDP to RanGTP in the nucleus. These nuclear transport factors and the import factor for RanGDP (NTF2) maintain the nucleocytoplasmic gradient between RanGTP within the nucleus and RanGDP in the cytoplasm which determines directionality of nucleocytoplasmic transport (Wälde and Kehlenbach 2010). Most components of the Ran cycle

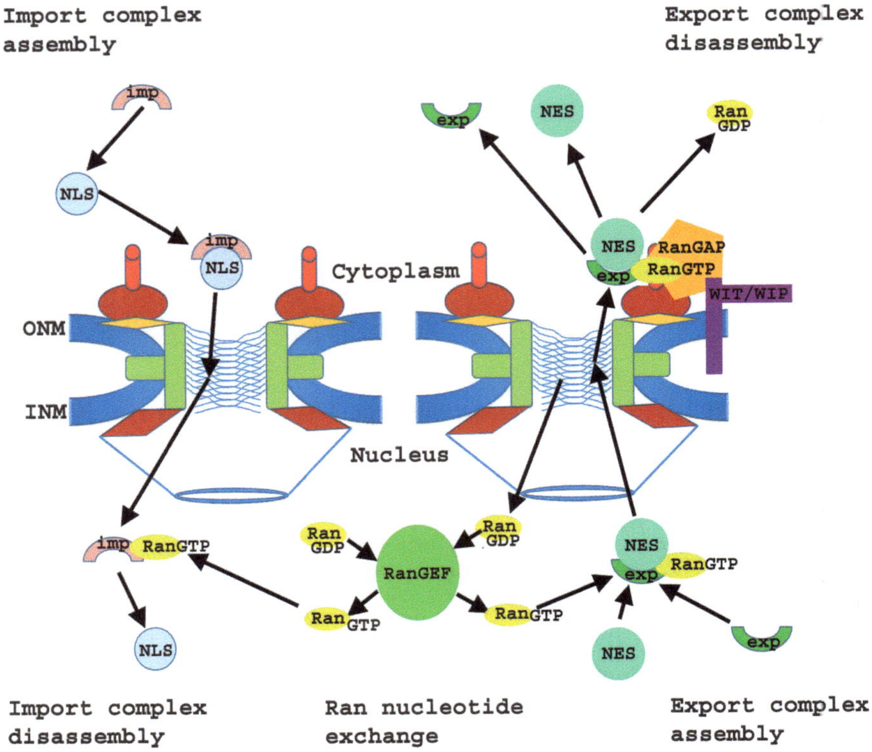

Fig. 4.5 Summary of the steps of protein import and export in plants. *Imp* import receptor, *NLS* nuclear localization sequence containing cargo protein, *NES* nuclear export sequence containing protien, *exp* export receptor (for details see text)

have now been identified in different species of plants including RanGAP, RanBPs, and two homologs of NTF2 (Xu et al. 2007; Meier and Brkljacic 2009b).

The classical import process in animals and plants starts with the binding of proteins with NLSs to importin α which then binds to importin β. Importin β mediates the translocation of the complex across the NPC via interaction with FG-domains. Figure 4.5 illustrates the steps of nucleocytoplasmic transport of molecules across the NPC governed by Ran. It has been suggested that FG-Nups may play a dual role in nuclear transport: they may create a permeability barrier for the Nups, and they may transiently interact with the soluble nuclear transport receptors (NTRs) mediating the translocation across the NPC. Several importin α and 17 importin β-like proteins have been identified in *Arabidopsis* (Bollman et al. 2003; Meier 2007). In contrast to animals, the plant importin At-IMPα is capable of mediating nuclear import in the absence of importin β in *in vitro* experiments (Hubner et al. 1999). Proteins transported to the nucleus dissociate from the cargo after binding of RanGTP to importin β, and this is followed by recycling of the importin β-RanGTP complex back to the cytoplasm.

Proteins containing a NES interact with the nuclear export receptor, CRM1, also known as exportin-1, which is the major karyopherin that exports many cellular proteins from the nucleus to the cytoplasm. The homologs of different export karyopherins such as exportin 1 (AtXPO1), exportin 5 (HASTY) and exportin t (PAUSED, involved in tRNA export) were determined and functionally characterized in plants (Meier 2007). For protein export, RanGTP associates broadly with the nuclear export receptor/cargo and this complex translocates across the NPC to the cytoplasm. The dissociation of the transport complex is triggered in the cytoplasm by GTP-hydrolysis on Ran under the action of the RanBP1 and RanGAP. This is followed by recycling of the free exportin back to the nucleus. Export of mRNA to the cytoplasm is controlled by different mechanisms that are not dependent on karyopherins nor on the Ran system (Kelly and Corbett 2009). At present there are several models that have been developed to explain how macromolecules pass through the NPC central channel (Wälde and Kehlenbach 2010).

Despite considerable similarities between the nuclear transport machineries of plants and animals (reviewed in Meier 2000; Xu et al. 2007) there are clear-cut differences between the Ran components in cells from these two kingdoms. The Ran-specific guanine nucleotide exchange factor RCC1 has not so far been described in plants, although this function must exist. Plant RanGAP1 contains a unique WPP domain at the N-terminal end which has not been found in vertebrates (Rose and Meier 2001). Two novel proteins (WIP and WIT) which control RanGAP1 binding to the NE have been described (Meier and Brkljacic 2009b). The vital importance of plant transport factors is suggested by studies showing that defects in their expression have major effects on plant development (Zhao et al. 2008; Meier and Brkljacic 2009a). In particular, the functional role of RanGAP in plant cytokinesis has been demonstrated (Xu et al. 2008).

4.3.4 Cell Cycle Dynamics of the Nuclear Envelope

The cell cycle in all eukaryotes is controlled by cyclin-dependent kinases (CDKs), see Magyar et al. 2013; this volume. Mitosis equally partitions replicated chromosomes as well as duplicated cytoplasmic organelles into the daughter cells (Prunuske and Ullman 2006; Costas et al. 2011). CDKs are also central in controlling the steps of interphase (Dultz and Ellenberg 2010; Talamas and Hetzer 2011). Cell cycle transitions depend on protein phosphorylation and dephosphorylation, which is controlled by CDKs and cyclins, as well as phosphatases. The higher-plant cell cycle differs from other higher eukaryotes because plants do not have centrosomes, as discussed above. Instead the NE plays the role of the microtubule organizing center (MTOC) (Stoppin et al. 1994). Additionally, in contrast to metazoans, mitotic nuclear membranes remain in close proximity to the chromosomes and NE reformation is spatially organized in dividing plant cells (Graumann and Evans 2011). Before preprophase, plant CDKs and cyclin B move from the cytoplasm to the nucleus and accumulate at the NE, initiating its disruption and chromosome condensation (Boruc et al. 2012). Mitotic NE disassembly in metazoans involves the disintegration of NPCs, lamina depolymerization and retraction of nuclear membranes into mitotic endoplasmic reticulum (ER). The mechanism of this well orchestrated basic process is not completely understood (Maeshima et al. 2011). There are three types of mitosis: open (higher plants and animals), closed (lower eukaryotes such as yeast) and semi-closed (e.g., *Drosophila* early embryos) (Kiseleva et al. 2001). NPCs are assembled in the re-forming nuclear envelope (NE) during the exit from mitosis as well as into intact, expanding NEs during interphase. During open mitosis, NPCs, membranes and the nuclear lamina are broken down in prophase and reassembled around the daughter chromosomes in telophase (Fernandez-Martinez and Rout 2009). During interphase, new NPCs are embedded in the intact NE and double the NPC number before the nucleus starts to divide. On the basis of biochemical and genetic studies, *in vitro* assembly assays and Nup dynamics during the assembly, two models for NE and NPC reformation have been developed (Hetzer et al. 2005; Doucet et al. 2010). One model proposes that fusion of the outer and inner membrane creates a pore for NPC assembly. In the other model specific Nups bind to chromatin to form "pre-pores" and the stepwise assembly of NPC structures then follows. At the final step, the membrane encloses newly formed NPCs and peripheral nucleoporins are added.

4.3.5 NPC Disassembly

While we have significant information about NPC composition, our understanding of assembly and especially disassembly of this multiprotein complex is not completely clear (Laurell et al. 2011). In higher eukaryotes, including plants, the NPCs disassemble in late prophase. This is controlled by the family of evolutionary conserved CDKs, complexed with their regulatory cyclin B and NIMA (Onischenko et al. 2005; Mühlhäusser and Kutay 2007; Magyar et al. 2013, this volume), resulting in phosphorylation of many Nups (Macaulay et al. 1995; Laurell et al. 2011). NPC disassembly is a quick process (about 1 min) and starts with disassembly of the peripheral NPC components such as the cytoplasmic ring and filaments, together with the star ring, then the central transporter and finally the scaffold and transmembrane nucleoporins (Kiseleva et al. 2001; Lénárt et al. 2003). In metazoans the NE loses its integrity during the transition from prophase to prometaphase (Guttinger et al. 2009) and the pore membrane proteins diffuse into the ER network. In vertebrates during mitosis the GLFG (glycine-leucine-phenylalanine-glycine) nucleoporin Nup98 localizes to both sides of the NPC with its partner Rae1 and the cytoplasmic filament protein Nup358/RanBP2. Nup98 is phosphorylated earlier than other Nups such as gp210, Nup214, Nup153, the Nup62 subcomplex, the Nup53/93 subcomplex, and the Nup107-160 subcomplex (Glavy et al. 2007; Dultz et al. 2008). It has been demonstrated that phosphorylation of Nup98 by multiple kinases is crucial for NPC disassembly during entry into mitosis (Laurell et al. 2011). Disassembly and reassembly of the NPC during mitosis is also regulated by a dynamic equilibrium between the activities of CDK1 and okadaic acid-sensitive protein phosphatases (Onischenko et al. 2005). Disassembly of NPCs in plants has not been investigated to the same extent as in animals. Mitotic protein phosphorylation releases Nups from the plant NPCs and they become distributed on the spindles or soluble in the cytoplasm at late preprophase (Boruc et al. 2012). It has been shown in plants that Nup136/Nup1 disperses into the cytosol during NE breakdown and then reassociates with the NPCs as the NE reforms in the daughter cells (Xu et al. 2007). NUA (Nuclear Pore Anchor, the *Arabidopsis* homolog of Tpr/Mlp1/Mlp2/Megator) and Rae1 relocate during mitosis (Lee et al. 2009).

Because plants have no centrosomes there is a preprophase period during the G2 phase of the cell cycle when the NE plays the role of MTOC (Stoppin et al. 1994; see also Magyar et al. 2012, this volume). Before this stage, plant CDKs and cyclin B move from the cytoplasm to the nucleus and accumulate at the NE initiating its disruption and chromosome condensation (Boruc et al. 2012).

Apart from mediating nucleocytoplasmic transport in higher eukaryotes, Nups play an important role in the regulation of mitosis (Meier and Brkljacic 2009a). After phosphorylation and disassembly most animal Nups are soluble and disperse in the cytoplasm. Some Nup complexes (Nup62, Nup88 and Nup107) are associated with spindles or kinetochores in metaphase. For example, it has been found that Nup98 participates in the regulation of the anaphase promoting complex during early mitosis and spindle assembly (Laurell et al. 2011). The Nup107-160 complex is involved in mitotic microtubule (MT) assembly and together with Nup358 in stabilization of MT/kinetochore attachment (Zuccolo et al. 2007). Nup358 (RanBP2), which is located to the cytoplasmic filaments in animals, plays an additional role in chromosome segregation.

Nup133 is required for efficient anchoring of tubulin associated proteins, the dynein–dynactin complex, to the NE. Tpr and Nup153 are active players in the SAC (spindle assembly check point) regulation and Nup153 together with Nup62 are critical for cytokinesis (Mackay et al. 2009).

Plant Nups and related factors might also play a role in cell division and cytokinesis (Meier and Brkljacic 2009a). It was shown that *Arabidopsis* NUA (a plant Nup with a role in control of flowering time and development) can influence gene expression (Xu and Meier 2008). Mutations in *Arabidopsis* RanBP1 cause mitotic arrest in metaphase/anaphase and cell death in the root meristematic zone (Kim et al. 2001). Plant Nup136, which is probably a functional homolog of animal Nup153, is involved in regulation of nuclear morphology (Tamura et al. 2010). The quantity of Nup136 on the NE determines nuclear shape in *Arabidopsis* (Tamura et al. 2010). Mutations in the plant Nup107–160 complex lead to mitotic as well as nucleocytoplasmic transport defects (Xu and Meier 2008).

4.3.6 NPC Assembly at the End of Mitosis

NPC assembly at the end of mitosis requires dephosphorylation of Nups and the correct recruitment of all the dispersed components back into the multiprotein NPC. The steps of this process were first investigated in detail *in vitro*. Studies have shown that NPCs assemble after the double nuclear membrane forms, a process that starts with the binding of membrane vesicles to chromatin (Lohka and Masui 1984; Salpingidou et al. 2008). Plant NE reassembly commences in anaphase and appears to be completed concomitant with early cell-plate formation (Xu et al. 2007). In late anaphase these membranes and Nups are recruited to the surface of decondensing chromatin and rapidly form NE fragments in which NPCs are assembled (Dultz et al. 2008). In animals, the Nup107-160 complex is recruited first to the reforming NE and this step is mediated by DNA-interacting Nup, ELYS (Belgareh et al. 2001; Rasala et al. 2008). This is followed by the incorporation of membrane vesicles containing transmembrane Nups, POM121 and NDC1, which in turn leads to the incorporation of other Nups (Onischenko et al. 2009). Nup133 and Nup62, then other Nups are recruited in a stepwise manner relatively late in the assembly process (Bodoor et al. 1999; Dultz et al. 2008). The sequentially formed structural intermediates of the NPC (dimples, pores, star-rings, and thin rings) have been identified on growing NEs assembled in *Xenopus* egg extracts (Goldberg et al. 1997), and also in the NEs of actively dividing nuclei from early *Drosophila* embryos by feSEM (Kiseleva et al. 2001). Similar NPC intermediates have recently been demonstrated (Fig. 4.6) during investigations of plant nuclei from proliferating tobacco BY-2 cells (Fiserova et al. 2009). This supports the conclusions about the conserved mechanism of NPC assembly through intermediates in higher eukaryotes.

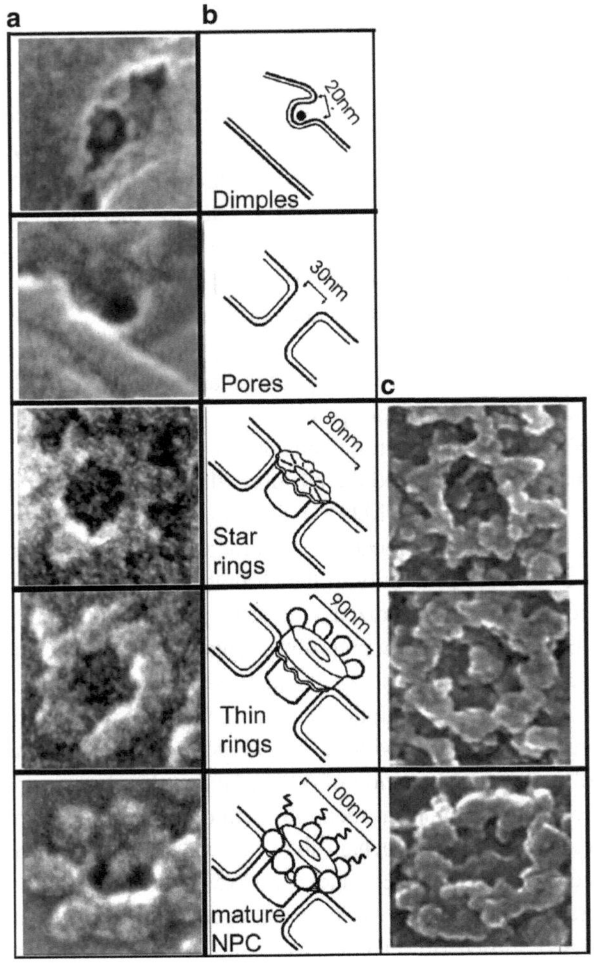

Fig. 4.6 Structural intermediates visualized during NPC assembly and the corresponding model of the steps. (**a**) NPC intermediates in the nuclear envelope of a telophase nucleus isolated from *Drosophila* early embryo; (**b**) schematic representation of the steps of NPC formation; (**c**) NPC intermediates in the nuclear envelope of the nucleus isolated from tobacco cell culture

4.3.7 NPC Assembly During Interphase

NPC biogenesis during interphase may differ from mitosis because NPCs must be inserted into a fully formed NE lined by the nuclear lamina (Maul et al. 1971; Goldberg et al. 1997; Winey et al. 1997; Dultz and Ellenberg 2010; Talamas and Hetzer 2011). The membrane curvature-inducing protein reticulon 4a (Nogo A) is involved in local fusion of the inner and outer NE membranes at the site of the forming pore (Kiseleva et al. 2007; Dawson et al. 2009). It has been suggested that the lipid composition of the membrane is critical for the early steps of NPC assembly before the incorporation of Nups (Fichtman et al. 2010). It was shown that POM121, rather than the Nup107–160 complex, is recruited first to the site of the new pore and plays a role in the early steps of NPC assembly (Funakoshi et al. 2011; Talamas and Hetzer 2011). Sun1 is an INM-specific protein, found in plants, animals and fungi. It interacts with the outer membrane nesprins in animals, linking the inner membrane to the actin network in the cytoplasm and possibly the nucleus. In plants it binds WIP/WIT proteins, but connections to the cytoskeleton or nucleoskeleton have not yet been demonstrated. In animals, Sun1 is present with POM121 at the forming NPC and may promote membrane fusion (Talamas and Hetzer 2011). There is evidence from vertebrates that the Ran GTPase cycle, the membrane layer Nup53 and the DNA-binding nucleoporin, Mel28/ELYS are necessary for the early steps of NPC biogenesis (Onischenko and Weis 2011). Currently nothing is known about plant NPC assembly during interphase.

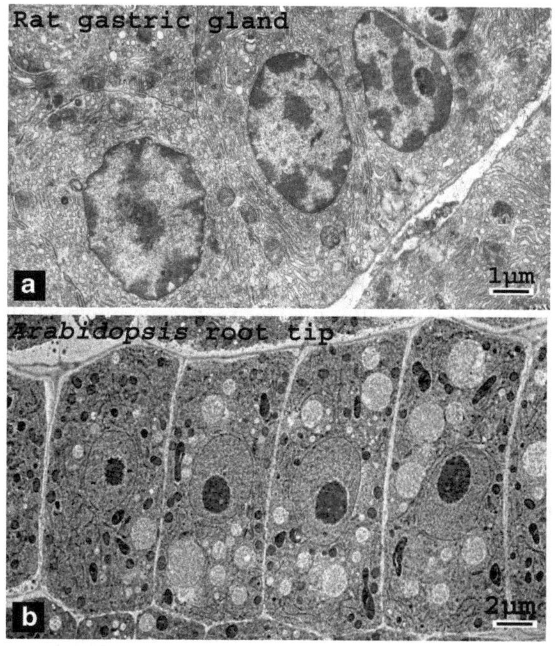

Fig. 4.7 Thin section TEM comparing *Arabidopsis thaliana* root tip cells (**b**) to rat gastric gland cells (**a**), showing the highly organized and consistent nature of plant nuclei, like in many animal tissues

4.3.8 The NPC in Plants is Similar but Distinct from Other Organisms

Significant progress has been achieved in unravelling the structural organization and biochemical composition of the plant NPCs. Plant and animal NPCs are similar in many respects supporting the notion that NPC organization is conserved in evolution. However our knowledge is far from complete and it is likely that not all plant Nups or nuclear transport proteins have been discovered. Important remaining questions are as follows: how are NPCs assembled during interphase? how do Nups contribute to different cellular functions in addition to nuclear transport? to what extent are Nup complexes in plants and animals similar? More detailed analysis of Nup dynamics and the role of specific FG domains should increase our understanding of the mechanisms of nucleoplasmic transport in plants which may reveal some unique features. Further work should focus on the connection between apparently plant-specific nuclear pore components and their fate and function during plant-specific aspects of the cell cycle.

4.4 The Plant Nucleoskeleton and Intra-nuclear Organization

The enclosure of the genome into a compartment that separated the transcription process from the translation machinery was a pivotal event in the evolution of complex eukaryotes. It enabled an exquisite control of gene expression by regulating the access of proteins such as transcription factors to the genome and exit of RNAs from the nucleus. As discussed, the nuclear envelope provides the impermeable wall to the compartment and the NPCs are the selective and controllable gates.

The evolution of the nuclear compartment also allowed the genome to be dynamically organized on several levels. For instance, the location of chromosomes, chromosomal domains and even individual genes can be controlled within the nuclear space. The control of such localization may be developmental, leading to epigenetic control, where housekeeping genes and tissue-specific genes are located to active euchromatic regions whereas genes that are not required are located to inactive heterochromatin. Alternatively, recent evidence from budding yeast (Taddei 2007) has shown that gene activity can be switched on, or enhanced, by relocating specific genes from the nuclear interior to the NPCs on a short timescale. On top of this, higher plant genomes can be very large, with multiple gene duplications and differentially expressed variants as well as vestigial genes.

Plants in particular, therefore, have a requirement to organise the genome into regions of different activity which can be controlled either during development or in response to specific signals. It also must organize other processes such as DNA replication and RNA processing. In addition, examination of electron micrographs of plant tissue, as well as animals, gives the immediate impression of a highly regulated and consistent organization (Fig. 4.7).

How this organization is achieved is not fully understood, especially in plants. A basic requirement for a spatial organizing system is a structural framework. Such a framework can be used to provide a semi-rigid set of spatial co-ordinates, onto which specific regions of the genome, or other functional centres, can be attached in specific domains. For instance in animals, a subset of heterochromatin is located to the peripheral nuclear lamina (Towbin et al. 2009). This structural framework, which includes the nuclear lamina, is known as the nuclear matrix or nucleoskeleton. The nuclear matrix was discovered using electron microscopy ~35 years ago in rat liver cells (Berezney and Coffey 1974, 1977), but partly because of the highly extracted nature of this structure it remains controversial to this day.

This latter point was partially addressed by Jackson and Cook (1988), who observed an intermediate filament-like network throughout the nucleus after removal of chromatin using "physiological" buffers. This was termed the "nucleoskeleton" and interestingly was shown to retain enzymes involved in transcription and DNA replication (Hozák et al. 1993), suggesting that the framework has a direct role in organizing biochemical processes of the nucleus.

Although the protein composition of the nucleoskeleton has not been fully elucidated, antibodies raised against animal nuclear matrix preparations appear to react with plant nuclei (Chaly et al. 1986). A number of studies subjected plant cells to nuclear matrix preparation methods and observed filamentous structures similar to the animal nuclear matrix/nucleoskeleton (Frederick et al. 1992; McNulty and Saunders 1992; Mínguez et al. 1994; Wang et al. 1996; Blumenthal et al. 2004). Therefore it can probably be concluded that if this is a *bone fide* structure in animals, it probably also exists in plants.

4.4.1 Intermediate Filament-Like Proteins of the Nucleoskeleton

The nucleoskeleton consists of two structurally distinct components: the nuclear lamina and the intra-nuclear network. All multicellular animal species studied so far, from hydra to humans, possess lamins (Erber et al. 1999; Burke and Stewart 2006; Melcer et al. 2007; Mencarelli et al. 2011). Lamins constitute the archetypal intermediate filament network that lines the nucleoplasmic face of the inner nuclear membrane. The nuclear lamina provides the outer structural shell to the nucleus and is clearly involved in the structural rigidity of the nucleus (Schäpe et al. 2009). In humans and in animal models, certain lamin mutations result in easily damaged nuclei (Shimi et al. 2010). Interestingly, the study of disease causing mutations in humans has revealed that lamins also have important roles in organizing chromatin and the nucleus in general (Shimi et al. 2010) and can therefore be considered as an integral, or even a major structural and functional component of the nucleoskeleton. Furthermore lamins are also found in intra-nuclear structures (Markiewicz et al. 2002), and may be an integral component of the internal nucleoskeleton. However we do not know if internal lamins form intermediate filaments and indeed they appear to be more easily extracted than the peripheral lamina and other nuclear matrix proteins (Markiewicz et al. 2002).

There is both structural and immuno-histochemical evidence for a nuclear lamina-like structure in plants (McNulty and Saunders 1992; Minguez and Moreno Diaz de la Espina 1993; Fiserova et al. 2009). However, the protein composition of this structure has not been characterised. Lamins are coiled-coil containing intermediate filament proteins and there are coiled-coil proteins that localize to the NE (e.g., NMCP1, LINC1 and MFP1, as discussed above). These are therefore candidate lamin-like proteins. However, of these, only LINC1 is exclusively located to the NE when expressed as a YFP-chimera (Dittmer et al. 2007). LINC2-YFP, on the other hand, is found throughout the nucleoplasm (Dittmer et al. 2007) and therefore could perform similar functions to the intra-nuclear lamins or other filamentous nuclear matrix components. Like the intra-nuclear lamins however, it is not known if LINC2 assembles into filaments. Interestingly, mutation of both LINC1 and LINC2 in *A. thaliana* results in stunted plants with small, misshapen nuclei, reminiscent of the "laminopathy" phenotypes in humans (Worman et al. 2009). It is likely therefore that LINC1 and LINC2, as well as LINC3, are important constituents of the different nucleoskeletal components: the lamina and the intra-nuclear network respectively.

Much of the evidence for a plant nucleoskeleton comes from staining nuclear matrix preparations with antibodies against intermediate filaments in general and lamins in particular (Frederick et al. 1992; McNulty and Saunders 1992; Mínguez et al. 1994; Wang et al. 1996; Blumenthal et al. 2004). Usually such antibodies stain throughout the nuclear matrix preparations. This shows that there are lamin/intermediate filament-like epitopes throughout the plant nucleus. Despite this, no sequences within plant genomes have been recognised that code for any protein that conforms to the properties of any intermediate filament protein. Plants do possess long coiled-coil proteins (Rose et al. 2004, 2005), but they do not have other clear intermediate filament structural properties.

4.4.2 A DNA Binding Integral Membrane Protein

MAR binding filament like protein 1 (MFP1) is an interesting long coiled-coil protein that locates to the nuclear periphery. It is thought to be membrane bound due to the predicted transmembrane domain. This is in contrast to lamins which become membrane associated via a lipid modification. MFP1 also has DNA binding activity (Meier et al. 1996) and appears to have a preference for DNA sequences known as Matrix Attachment Regions (MARs). MARs (also known as scaffold attachment regions or SARs) are ~200 bp long AT-rich DNA sequences which are substrates for topoisomerase II and are found tightly bound to nuclear matrix preparations after extensive nuclease digestion of the DNA and biochemical extraction (Wang et al. 2010). MARs therefore are thought to anchor DNA to the nuclear matrix.

In some studies MFP1 is located in a speckled pattern to the nuclear periphery (Gindullis and Meier 1999), but others (Samaniego et al. 2006) indicate that it is intra-nuclear. The latter localization would be unusual if MFP1 was truly an integral membrane protein, as predicted from the sequence, because as far as we know there are no intra-nuclear membranes in plant cells. It is however possible that MFP1 exists as a membrane bound or soluble protein depending on its functional state. Unfortunately the picture for MFP1 is far from clear: we know little about its function and there is evidence that it may also be associated with chloroplasts and Golgi as well as or instead of the nucleus (Jeong et al. 2003; Samaniego et al. 2006).

An MFP1 binding protein has been discovered called MFP1 associated factor (MAF1) (Gindullis et al. 1999). MAF1 contains a WPP (tryptophan-proline-proline) domain which is an interaction domain also found in RanGAP. In RanGAP the WPP domain binds to WPP domain-interacting proteins (WIPs) and WPP domain-interacting tail-anchored proteins (WITs) (Zhao et al. 2008; Brkljacic et al. 2009) in order to locate it to the nuclear envelope (Meier et al. 2008). Recently (Zhou et al. 2012) it was shown that WIPs may in fact be the elusive equivalent to metazoan KASH-domain proteins, which link the outer nuclear membrane to the nuclear interior via the inner nuclear membrane SUN proteins. It is possible therefore that the function of MFP1 is to locate other proteins, such as MAF1, to the nucleoskeleton, possibly bringing them into close association with specific regions of the genome. However, as yet, we do not know the function of MAF1.

4.4.3 DNA Sequences of the Nucleoskeleton and Proteins That Bind Them

The identification of MARs and the discovery that some genes that code for proteins involved in the same pathway are clustered (Loc and Strätling 1988), together with the demonstration of a loop-like organization of chromatin, suggested that one function of the nucleoskeleton could be to organize chromatin into structural domains. MARs clearly exist in plants (Hall et al. 1991; Breyne et al. 1992) and there is evidence that incorporation of MARs into transgenes can be used to enhance their expression (Maximova et al. 2003; Van der Geest et al. 2004; Wang et al. 2007). Topoisomerase II is a component of the nuclear matrix in animals and MARs contain topoisomerase II consensus sequences (Berrios et al. 1985). It is uncertain however whether plant MARs are topoisomerase II targets. Treatment of plant nuclei with topoisomerase II inhibitors, which cause double stranded breaks in the DNA, result in fragmentation of the genome into consistent sized fragments (Razin et al. 1991). This suggests a role for topoisomerase II in determining chromatin domains. However, it is uncertain whether topoisomerase II actually cleaves at the MARs in *Arabidopsis* (Makarevitch and Somers 2006).

4.4.4 NuMA

Another protein identified as a consistent constituent of various nuclear matrix preparations is Nuclear Mitotic Apparatus (NuMA) (Radulescu and Cleveland 2010). Nuclear matrix preparations from onion cells contain a large protein (similar to human NuMA) that is recognised by anti-NuMA antibodies (Yu et al. 1999). These antibodies also label nuclear matrix filaments as shown by immuno-electron microscopy. Although this protein shares similar properties with mammalian NuMA, no NuMA homolog has been clearly identified in plants. Mammalian NuMA is a large coiled-coil protein that tethers microtubules to the spindle poles during mitosis but is also associated with the nuclear matrix during interphase (Kallajoki et al. 1991). When over-expressed, intra-nuclear filaments are observed (Gueth-Hallonet et al. 1998), suggesting it could constitute a structural component of the nucleoskeleton. Therefore like several nucleoskeletal components, there is likely to be a functional equivalent of NuMA, which may be structurally similar enough to react with antibodies raised against the mammalian protein, but without enough sequence similarity to be recognised by sequence comparison algorithms.

Conclusion

A more detailed understanding of the animal nucleoskeleton, NE and NPCs as well as identification of further components will clearly help inform the plant field. On the other hand there appear to be significant differences. These differences are not only important for understanding the structural and functional organization of the plant nucleus, but could also provide important clues to understanding this immensely complex organelle in all other eukaryotes.

Acknowledgements This work was supported by grants from the Biotechnology and Biological Sciences Research Council, UK, grant numbers BB/E015735/1 and BB/G011818/1 (MWG and JF), and grants for EK by Russian Federation for Basic Research and the Program of the RAS Presidium "Molecular and Cell Biology". Thanks to Christine Richardson (Durham University) for Fig. 4.7.

References

Aaronson RP, Blobel G (1975) Isolation of nuclear pore complexes in association with a lamina. Proc Natl Acad Sci U S A 72:1007–1011

Akey CW (1990) Visualization of transport-related configurations of the nuclear pore transporter. Biophys J 58:341–355

Akey CW, Radermacher M (1993) Architecture of the *Xenopus* nuclear pore complex revealed by three-dimensional cryo-electron microscopy. J Cell Biol 22:1–19

Allen TD, Cronshaw JM, Bagley S, Kiseleva E, Goldberg MW (2000) The nuclear pore complex: mediator of translocation between nucleus and cytoplasm. J Cell Sci 113:1651–1659

Ane JM, Kiss GB, Riely BK, Penmetsa RV, Oldroyd GE, Ayax C, Levy J, Debelle F, Baek JM, Kalo P, Rosenberg C, Roe BA, Long SR, Denarie J, Cook DR (2004) *Medicago truncatula* DMI1 required for bacterial and fungal symbioses in legumes. Science 303:1364–1367

Batrakou DG, Kerr AR, Schirmer EC (2009) Comparative proteomic analyses of the nuclear envelope and pore complex suggests a wide range of heretofore unexpected functions. J Proteomics 72:56–70

Belgareh N, Doye V (1997) Dynamics of nuclear pore distribution in nucleoporin mutant yeast cells. J Cell Biol 136:747–759

Belgareh N, Rabut G, Bai SW, van Overbeek M, Beaudouin J, Daigle N, Zatsepina OV, Pasteau F, Labas V, Fromont-Racine M, Ellenberg J, Doye V (2001) An evolutionarily conserved NPC subcomplex, which redistributes in part to kinetochores in mammalian cells. J Cell Biol 154:1147–1160

Berezney R, Coffey DS (1974) Identification of a nuclear protein matrix. Biochem Biophys Res Commun 60:1410–1417

Berezney R, Coffey DS (1977) Nuclear matrix. Isolation and characterization of a framework structure from rat liver nuclei. J Cell Biol 73:616–637

Berrios M, Osheroff N, Fisher PA (1985) In situ localization of DNA topoisomerase II, a major polypeptide component of the *Drosophila* nuclear matrix fraction. Proc Natl Acad Sci U S A 82:4142–4146

Blumenthal SS, Clark GB, Roux SJ (2004) Biochemical and immunological characterization of pea nuclear intermediate filament proteins. Planta 218:965–975

Bodoor K, Shaikh S, Salina D, Raharjo WH, Bastos R, Lohka M, Burke B (1999) Sequential recruitment of NPC proteins to the nuclear periphery at the end of mitosis. J Cell Sci 112:2253–2264

Bollman KM, Aukerman MJ, Park MY, Hunter C, Berardini TZ, Poethig RS (2003) HASTY, the *Arabidopsis* ortholog of exportin 5/MSN5, regulates phase change and morphogenesis. Development 130:1493–1504

Boruc J, Zhou X, Meier I (2012) Dynamics of the plant nuclear envelope and nuclear pore. Plant Physiol 158:78–86

Breyne P, van Montagu M, Depicker N, Gheysen G (1992) Characterization of a plant scaffold attachment region in a DNA fragment that normalizes transgene expression in tobacco. Plant Cell 4:463–471

Brkljacic J, Zhao Q, Meier I (2009) WPP-domain proteins mimic the activity of the HSC70-1 chaperone in preventing mistargeting of RanGAP1-anchoring protein WIT1. Plant Physiol 151:142–154

Broers JL, Peeters EA, Kuijpers HJ, Endert J, Bouten CV, Oomens CW, Baaijens FP, Ramaekers FC (2004) Decreased mechanical stiffness in LMNA-/- cells is caused by defective nucleo-cytoskeletal integrity: implications for the development of laminopathies. Hum Mol Genet 13:2567–2580

Broers JL, Ramaekers FC, Bonne G, Yaou RB, Hutchison CJ (2006) Nuclear lamins: laminopathies and their role in premature ageing. Physiol Rev 86:967–1008

Brohawn SG, Schwartz TU (2009) A lattice model of the nuclear pore complex. Commun Integr Biol 2:205–207

Brohawn SG, Partridge JR, Whittle JR, Schwartz TU (2009) The nuclear pore complex has entered the atomic age. Structure 17:1156–1168

Bucci M, Wente SR (1997) In vivo dynamics of nuclear pore complexes in yeast. J Cell Biol 136:1185–1199

Bunney TD, Shaw PJ, Watkins PA, Taylor JP, Beven AF, Wells B, Calder GM, Drobak BK (2000) ATP-dependent regulation of nuclear Ca(2+) levels in plant cells. FEBS Lett 476:145–149

Burke B, Stewart CL (2006) The laminopathies: the functional architecture of the nucleus and its contribution to disease. Annu Rev Genomics Hum Genet 7:369–405

Capoen W, Sun J, Wysham D, Otegui MS, Venkateshwaran M, Hirsch S, Miwa H, Downie JA, Morris RJ, Ane JM, Oldroyd GE (2011) Nuclear membranes control symbiotic calcium signaling of legumes. Proc Natl Acad Sci U S A 108:14348–14353

Chaly N, Sabour MP, Silver JC, Aitchison WA, Little JE, Brown DL (1986) Monoclonal antibodies against nuclear matrix detect nuclear antigens in mammalian, insect and plant cells: an immunofluorescence study. Cell Biol Int Rep 10:421–428

Charpentier M, Bredemeier R, Wanner G, Takeda N, Schleiff E, Parniske M (2008) *Lotus japonicus* CASTOR and POLLUX are ion channels essential for perinuclear calcium spiking in legume root endosymbiosis. Plant Cell 20:3467–3479

Costas C, Desvoyes B, Gutierrez C (2011) A chromatin perspective of plant cell cycle progression. Biochim Biophys Acta 1809:379–387

Cremer T, Kupper K, Dietzel S, Fakan S (2004) Higher order chromatin architecture in the cell nucleus: on the way from structure to function. Biol Cell 96:555–567

Cronshaw JM, Krutchinsky AN, Zhang W, Chait BT, Matunis MJ (2002) Proteomic analysis of the mammalian nuclear pore complex. J Cell Biol 158:915–927

Dauer WT, Worman HJ (2009) The nuclear envelope as a signaling node in development and disease. Dev Cell 17:626–638

Dawson TR, Lazarus MD, Hetzer MW, Wente SR (2009) ER membrane-bending proteins are necessary for de novo nuclear pore formation. J Cell Biol 184:659–675

Dechat T, Adam SA, Goldman RD (2009) Nuclear lamins and chromatin: when structure meets function. Adv Enzyme Regul 49:157–166

Dittmer TA, Stacey NJ, Sugimoto-Shirasu K, Richards EJ (2007) LITTLE NUCLEI genes affecting nuclear morphology in *Arabidopsis thaliana*. Plant Cell 19:2793–2803

Dong CH, Hu X, Tang W, Zheng X, Kim YS, Lee BH, Zhu JK (2006) A putative *Arabidopsis* nucleoporin, AtNUP160, is critical for RNA export and required for plant tolerance to cold stress. Mol Cell Biol 26:9533–9543

Doucet CM, Talamas JA, Hetzer MW (2010) Cell cycle-dependent differences in nuclear pore complex assembly in metazoa. Cell 141:1030–1041

Downie L, Priddle J, Hawes C, Evans DE (1998) A calcium pump at the higher plant nuclear envelope? FEBS Lett 429:44–48

Dultz E, Ellenberg J (2010) Live imaging of single nuclear pores reveals unique assembly kinetics and mechanism in interphase. J Cell Biol 191:15–22

Dultz E, Zanin E, Wurzenberger C, Braun M, Rabut G, Sironi L, Ellenberg J (2008) Systematic kinetic analysis of mitotic dis- and reassembly of the nuclear pore in living cells. J Cell Biol 180:857–865

Edwards A, Heckmann AB, Yousafzai F, Duc G, Downie JA (2007) Structural implications of mutations in the pea SYM8 symbiosis gene, the DMI1 ortholog, encoding a predicted ion channel. Mol Plant Microbe Interact 20:1183–1191

Elad N, Maimon T, Frenkiel-Krispin D, Lim RY, Medalia O (2009) Structural analysis of the nuclear pore complex by integrated approaches. Curr Opin Struct Biol 19:226–232

Erber A, Riemer D, Hofemeister H, Bovenschulte M, Stick R, Panopoulou G, Lehrach H, Weber K (1999) Characterization of the Hydra lamin and its gene: a molecular phylogeny of metazoan lamins. J Mol Evol 49:260–271

Evans DE, Williams LE (1998) P-type calcium ATPases in higher plants—biochemical, molecular and functional properties. Biochim Biophys Acta 1376:1–25

Evans DE, Shvedunova M, Graumann K (2011) The nuclear envelope in the plant cell cycle: structure, function and regulation. Ann Bot 107:1111–1118

Feldherr CM (1965) The effect of the electron-opaque pore material on exchanges through the nuclear annuli. J Cell Biol 25:43–53

Fernandez-Martinez J, Rout MP (2009) Nuclear pore complex biogenesis. Curr Opin Cell Biol 21:603–612

Fichtman B, Ramos C, Rasala B, Harel A, Forbes DJ (2010) Inner/outer nuclear membrane fusion in nuclear pore assembly: biochemical demonstration and molecular analysis. Mol Biol Cell 21:4197–4211

Fiserova J, Goldberg MW (2010) Relationships at the nuclear envelope: lamins and nuclear pore complexes in animals and plants. Biochem Soc Trans 38:829–831

Fiserova J, Kiseleva E, Goldberg MW (2009) Nuclear envelope and nuclear pore complex structure and organization in tobacco BY-2 cells. Plant J 59:243–255

Flowers JM, Purugganan MD (2008) The evolution of plant genomes: scaling up from a population perspective. Curr Opin Genet Dev 18:565–570

Foster LJ, de Hoog CL, Zhang Y, Zhang Y, Xie X, Mootha VK, Mann M (2006) A mammalian organelle map by protein correlation profiling. Cell 125:187–199

Frederick SE, Mangan ME, Carey JB, Gruber PJ (1992) Intermediate filament antigens of 60 and 65 kDa in the nuclear matrix of plants: their detection and localization. Exp Cell Res 199:213–222

Fried H, Kutay U (2003) Nucleocytoplasmic transport: taking an inventory. Cell Mol Life Sci 60:1659–1688

Funakoshi T, Clever M, Watanabe A, Imamoto N (2011) Localization of Pom121 to the inner nuclear membrane is required for an early step of interphase nuclear pore complex assembly. Mol Biol Cell 22:1058–1069

Gerace L, Burke B (1988) Functional organization of the nuclear envelope. Annu Rev Cell Biol 4:335–374

Gindullis F, Meier I (1999) Matrix attachment region binding protein MFP1 is localized in discrete domains at the nuclear envelope. Plant Cell 11:1117–1128

Gindullis F, Peffer NJ, Meier I (1999) MAF1, a novel plant protein interacting with matrix attachment region binding protein MFP1, is located at the nuclear envelope. Plant Cell 11:1755–1768

Gindullis F, Rose A, Patel S, Meier I (2002) Four signature motifs define the first class of structurally related large coiled-coil proteins in plants. BMC Genomics 3:9

Glavy JS, Krutchinsky AN, Cristea IM, Berke IC, Boehmer T, Blobel G, Chait BT (2007) Cell-cycle-dependent phosphorylation of the nuclear pore Nup107-160 subcomplex. Proc Natl Acad Sci U S A 104:3811–3816

Goldberg MW, Allen TD (1992) High resolution scanning electron microscopy of the nuclear envelope: demonstration of a new, regular, fibrous lattice attached to the baskets of the nucleoplasmic face of the nuclear pores. J Cell Biol 119:1429–1440

Goldberg MW, Allen TD (1995) Structural and functional organization of the nuclear envelope. Curr Opin Cell Biol 7:301–309

Goldberg MW, Allen TD (1996) The nuclear pore complex and lamina: three-dimensional structures and interactions determined by field emission in-lens scanning electron microscopy. J Mol Biol 257:848–865

Goldberg MW, Wiese C, Allen TD, Wilson KL (1997) Dimples, pores, star-rings, and thin rings on growing nuclear envelopes: evidence for structural intermediates in nuclear pore complex assembly. J Cell Sci 110:409–420

Görlich D, Kutay U (1999) Transport between the cell nucleus and the cytoplasm. Annu Rev Cell Dev Biol 15:607–660

Graumann K, Evans DE (2010) The plant nuclear envelope in focus. Biochem Soc Trans 38:307–311

Graumann K, Evans DE (2011) Nuclear envelope dynamics during plant cell division suggest common mechanisms between kingdoms. Biochem J 435:661–667

Graumann K, Runions J, Evans DE (2009) Characterization of SUN-domain proteins at the higher plant nuclear envelope. Plant J 61:134–144

Gruenbaum Y, Goldman RD, Meyuhas R, Mills E, Margalit A, Fridkin A, Dayani Y, Prokocimer M, Enosh A (2003) The nuclear lamina and its functions in the nucleus. Int Rev Cytol 226:1–62

Gueth-Hallonet C, Wang J, Harborth J, Weber K, Osborn M (1998) Induction of a regular nuclear lattice by overexpression of NuMA. Exp Cell Res 243:434–452

Guttinger S, Laurell E, Kutay U (2009) Orchestrating nuclear envelope disassembly and reassembly during mitosis. Nat Rev Mol Cell Biol 10:178–191

Haizel T, Merkle T, Pay A, Fejes E, Nagy F (1997) Characterization of proteins that interact with the GTP-bound form of the regulatory GTPase Ran in *Arabidopsis*. Plant J 11:93–103

Hall G Jr, Allen GC, Loer DS, Thompson WF, Spiker S (1991) Nuclear scaffolds and scaffold-attachment regions in higher plants. Proc Natl Acad Sci U S A 88:9320–9324

Heese-Peck A, Raikhel NV (1998) The nuclear pore complex. Plant Mol Biol 38:145–162

Hetzer MW (2010) The nuclear envelope. Cold Spring Harb Perspect Biol 2:a000539

Hetzer MW, Walther TC, Mattaj IW (2005) Pushing the envelope: structure, function, and dynamics of the nuclear periphery. Annu Rev Cell Dev Biol 21:347–380

Hinshaw JE, Carragher BO, Milligan RA (1992) Architecture and design of the nuclear pore complex. Cell 69:1133–1141

Hoelz A, Debler EW, Blobel G (2011) The structure of the nuclear pore complex. Annu Rev Biochem 80:613–643

Houben F, Ramaekers FC, Snoeckx LH, Broers JL (2007) Role of nuclear lamina-cytoskeleton interactions in the maintenance of cellular strength. Biochim Biophys Acta 1773:675–686

Hozák P, Hassan AB, Jackson DA, Cook PR (1993) Visualization of replication factories attached to nucleoskeleton. Cell 73:361–373

Hubner S, Smith HMS, Hu W, Chan CK, Rihs HP, Paschal BM, Raikhel NV, Jans DA (1999) Plant importin a binds nuclear localization sequences with high affinity and can mediate nuclear import independent of importin b. J Biol Chem 274:22610–22617

Humbert JP, Matter N, Artault JC, Koppler P, Malviya AN (1996) Inositol 1,4,5-trisphosphate receptor is located to the inner nuclear membrane vindicating regulation of nuclear calcium signaling by inositol 1,4,5-trisphosphate. Discrete distribution of inositol phosphate receptors to inner and outer nuclear membranes. J Biol Chem 271:478–485

Imaizumi-Anraku H, Takeda N, Charpentier M, Perry J, Miwa H, Umehara Y, Kouchi H, Murakami Y, Mulder L, Vickers K, Pike J, Downie JA, Wang T, Sato S, Asamizu E, Tabata S, Yoshikawa M, Murooka Y, Wu GJ, Kawaguchi M, Kawasaki S, Parniske M, Hayashi M (2005) Plastid proteins crucial for symbiotic fungal and bacterial entry into plant roots. Nature 433:527–531

Jackson DA, Cook PR (1988) Visualization of a filamentous nucleoskeleton with a 23 nm axial repeat. EMBO J 7:3667–3677

Jarnik M, Aebi U (1991) Toward a more complete 3-D structure of the nuclear pore complex. J Struct Biol 107:291–308

Jeong SY, Rose A, Meier I (2003) MFP1 is a thylakoid-associated, nucleoid-binding protein with a coiled-coil structure. Nucleic Acids Res 31:5175–5185

Kahms M, Hüve J, Wesselmann R, Farr JC, Baumgärtel V, Peters R (2011) Lighting up the nuclear pore complex. Eur J Cell Biol 90:751–758

Kallajoki M, Weber K, Osborn M (1991) A 210 kDa nuclear matrix protein is a functional part of the mitotic spindle; a microinjection study using SPN monoclonal antibodies. EMBO J 10:3351–3362

Kanamori N, Madsen LH, Radutoiu S, Frantescu M, Quistgaard EM, Miwa H, Downie JA, James EK, Felle HH, Haaning LL, Jensen TH, Sato S, Nakamura Y, Tabata S, Sandal N, Stougaard J (2006) A nucleoporin is required for induction of Ca2+ spiking in legume nodule development and essential for rhizobial and fungal symbiosis. Proc Natl Acad Sci U S A 103:359–364

Kelly SM, Corbett AH (2009) Messenger RNA export from the nucleus: a series of molecular wardrobe changes. Traffic 10:1199–1208

Kim SH, Arnold D, Lloyd A, Roux SJ (2001) Antisense expression of an *Arabidopsis* Ran binding protein renders transgenic roots hypersensitive to auxin and alters auxin-induced root growth and development by arresting mitotic progress. Plant Cell 13:2619–2630

Kiseleva E, Goldberg MW, Daneholt B, Allen TD (1996) RNP export is mediated by structural reorganization of the nuclear pore basket. J Mol Biol 260:304–311

Kiseleva E, Goldberg MW, Allen TD, Akey CW (1998) Active nuclear pore complexes in *Chironomus*: visualization of transporter configurations related to mRNP export. J Cell Sci 111:223–236

Kiseleva E, Goldberg MW, Cronshaw J, Allen TD (2000) The nuclear pore complex: structure, function, and dynamics. Crit Rev Eukaryot Gene Expr 10:101–112

Kiseleva E, Rutherford S, Cotter LM, Allen TD, Goldberg MW (2001) Steps of nuclear pore complex disassembly and reassembly during mitosis in early *Drosophila* embryos. J Cell Sci 114:3607–3618

Kiseleva E, Allen TD, Rutherford S, Bucci M, Wente SR, Goldberg MW (2004) Yeast nuclear pore complexes have a cytoplasmic ring and internal filaments. J Struct Biol 145:272–288

Kiseleva E, Morozova KN, Voeltz GK, Allen TD, Goldberg MW (2007) Reticulon 4a/NogoA locates to regions of high membrane curvature and may have a role in nuclear envelope growth. J Struct Biol 160:224–235

Kleffmann T, Hirsch-Hoffmann M, Gruissem W, Baginsky S (2006) plprot: a comprehensive proteome database for different plastid types. Plant Cell Physiol 47:432–436

Kumaran RI, Thakar R, Spector DL (2008) Chromatin dynamics and gene positioning. Cell 132:929–934

Laurell E, Beck K, Krupina K, Theerthagiri G, Bodenmiller B, Horvath P, Aebersold R, Antonin W, Kutay U (2011) Phosphorylation of Nup98 by multiple kinases is crucial for NPC disassembly during mitotic entry. Cell 144:539–550

Lee JS, Hale CM, Panorchan P, Khatau SB, George JP, Tseng Y, Stewart CL, Hodzic D, Wirtz D (2007) Nuclear lamin A/C deficiency induces defects in cell mechanics, polarization, and migration. Biophys J 93:2542–2552

Lee JS, Wang S, Sritubtim S, Chen JG, Ellis BE (2009) *Arabidopsis* mitogen-activated protein kinase MPK12 interacts with the MAPK phosphatase IBR5 and regulates auxin signaling. Plant J 57:975–985

Leitch AR, Leitch IJ (2012) Ecological and genetic factors linked to contrasting genome dynamics in seed plants. New Phytol. 194:629–646

Leitch IJ, Leitch AR (2013) Genome size diversity and evolution in land plants. In: Leitch IJ, Greilhuber J, Doležel J, Wendel JF (eds) Plant genome diversity, vol 2, Physical structure, behaviour and evolution of plant genomes. Springer-Verlag, Wien, pp 307–322

Lénárt P, Rabut G, Daigle N, Hand AR, Terasaki M, Ellenberg J (2003) Nuclear envelope breakdown in starfish oocytes proceeds by partial NPC disassembly followed by a rapidly spreading fenestration of nuclear membranes. J Cell Biol 160:1055–1068

Lenz-Böhme B, Wismar J, Fuchs S, Reifegerste R, Buchner E, Betz H, Schmitt B (1997) Insertional mutation of the *Drosophila* nuclear lamin Dm0 gene results in defective nuclear envelopes, clustering of nuclear pore complexes, and accumulation of annulate lamellae. J Cell Biol 137:1001–1006

Lim RY, Aebi U, Fahrenkrog B (2008) Towards reconciling structure and function in the nuclear pore complex. Histochem Cell Biol 129:105–116

Liu J, Rolef Ben-Shahar T, Riemer D, Treinin M, Spann P, Weber K, Fire A, Gruenbaum Y (2000) Essential roles for *Caenorhabditis elegans* lamin gene in nuclear organization, cell cycle progression, and spatial organization of nuclear pore complexes. Mol Biol Cell 11:3937–3947

Loc PV, Strätling WH (1988) The matrix attachment regions of the chicken lysozyme gene co-map with the boundaries of the chromatin domain. EMBO J 7:655–664

Lohka MJ, Masui Y (1984) Roles of cytosol and cytoplasmic particles in nuclear envelope assembly and sperm pronuclear formation in cell-free preparations from amphibian eggs. J Cell Biol 98:1222–1230

Lu Q, Tang X, Tian G, Wang F, Liu K, Nguyen V, Kohalmi SE, Keller WA, Tsang EW, Harada JJ, Rothstein SJ, Cui Y (2010) *Arabidopsis* homolog of the yeast TREX-2 mRNA export complex: components and anchoring nucleoporin. Plant J 61:259–270

Macaulay C, Meier E, Forbes DJ (1995) Differential mitotic phosphorylation of proteins of the nuclear pore complex. J Biol Chem 270:254–262

Mackay DR, Elgort SW, Ullman KS (2009) The nucleoporin Nup153 has separable roles in both early mitotic progression and the resolution of mitosis. Mol Biol Cell 20:1652–1660

Maeshima K, Yahata K, Sasaki Y, Nakatomi R, Tachibana T, Hashikawa T, Imamoto F, Imamoto N (2006) Cell-cycle-dependent dynamics of nuclear pores: pore-free islands and lamins. J Cell Sci 119:4442–4451

Maeshima K, Iino H, Hihara S, Imamoto N (2011) Nuclear size, nuclear pore number and cell cycle. Nucleus 2:113–118

Magyar Z, Ito M, Binarová P, Mohamed B, Bogre L (2013) Cell cycle modules in plants for entry into proliferation and for mitosis. In: Leitch IJ, Greilhuber J, Doležel J, Wendel JF (eds) Plant genome diversity, vol 2, Physical structure, behaviour and evolution of plant genomes. Springer-Verlag, Wien, pp 77–97

Makarevitch I, Somers DA (2006) Association of *Arabidopsis* topoisomerase IIA cleavage sites with functional genomic elements and T-DNA loci. Plant J 48:697–709

Markiewicz E, Venables R, Alvarez-Reyes M, Quinlan R, Dorobek M, Hausmanowa-Petrucewicz I, Hutchison C (2002) Increased solubility of lamins and redistribution of lamin C in X-linked Emery-Dreifuss muscular dystrophy fibroblasts. J Struct Biol 140:241–253

Masuda K, Xu ZJ, Takahashi S, Ito A, Ono M, Nomura K, Inoue M (1997) Peripheral framework of carrot cell nucleus contains a novel protein predicted to exhibit a long alpha-helical domain. Exp Cell Res 232:173–181

Mattout A, Dechat T, Adam SA, Goldman RD, Gruenbaum Y (2006) Nuclear lamins, diseases and aging. Curr Opin Cell Biol 18:335–341

Matzke M, Weiger TM, Papp I, Matzke AJ (2009) Nuclear membrane ion channels mediate root nodule development. Trends Plant Sci 14:295–298

Matzke AJ, Weiger TM, Matzke M (2010) Ion channels at the nucleus: electrophysiology meets the genome. Mol Plant 3:642–652

Maul GG, Price JW, Lieberman MW (1971) Formation and distribution of nuclear pore complexes in interphase. J Cell Biol 51:405–418

Maximova S, Miller C, Antúnez de Mayolo G, Pishak S, Young A, Guiltinan MJ (2003) Stable transformation of *Theobroma cacao* L. and influence of matrix attachment regions on GFP expression. Plant Cell Rep 21:872–883
Mazars C, Bourque S, Mithofer A, Pugin A, Ranjeva R (2009) Calcium homeostasis in plant cell nuclei. New Phytol 181:261–274
Mazars C, Briere C, Bourque S, Thuleau P (2011) Nuclear calcium signaling: an emerging topic in plants. Biochimie 93:2068–2074
McNulty AK, Saunders MJ (1992) Purification and immunological detection of pea nuclear intermediate filaments: evidence for plant nuclear lamins. J Cell Sci 103(Pt 2):407–414
Meier I (2000) A novel link between Ran signal transductionand nuclear envelope proteins in plants. Plant Physiol 124:1507–1510
Meier I (2007) Composition of the plant nuclear envelope: theme and variations. J Exp Bot 58:27–34
Meier I, Brkljacic J (2009a) The nuclear pore and plant development. Curr Opin Plant Biol 12:87–95
Meier I, Brkljacic J (2009b) Adding pieces to the puzzling plant nuclear envelope. Curr Opin Plant Biol 12:752–759
Meier I, Somers DE (2011) Regulation of nucleocytoplasmic trafficking in plants. Curr Opin Plant Biol 14:538–546
Meier I, Phelan T, Gruissem W, Spiker S, Schneider D (1996) MFP1, a novel plant filament-like protein with affinity for matrix attachment region DNA. Plant Cell 8:2105–2115
Meier I, Xu XM, Brkljacic J, Zhao Q, Wang HJ (2008) Going green: plants' alternative way to position the Ran gradient. J Microsc 231:225–233
Meier I, Zhou X, Brkljacic J, Rose A, Zhao Q, Xu XM (2010) Targeting proteins to the plant nuclear envelope. Biochem Soc Trans 38:733–740
Melcer S, Gruenbaum Y, Krohne G (2007) Invertebrate lamins. Exp Cell Res 313:2157–2166
Mellad JA, Warren DT, Shanahan CM (2011) Nesprins LINC the nucleus and cytoskeleton. Curr Opin Cell Biol 23:47–54
Mencarelli C, Ciolfi S, Caroti D, Lupetti P, Dallai R (2011) Isomin: a novel cytoplasmic intermediate filament protein from an arthropod species. BMC Biol 28:9–17
Merkle T (2001) Nuclear import and export of proteins in plants: a tool for the regulation of signaling. Planta 213:499–517
Minguez A, Moreno Diaz de la Espina S (1993) Immunological characterization of lamins in the nuclear matrix of onion cells. J Cell Sci 106(Pt 1):431–439
Mínguez A, Franca S, Díaz M, de la Espina S (1994) Dinoflagellates have a eukaryotic nuclear matrix with lamin-like proteins and topoisomerase II. J Cell Sci 107:2861–2873
Mühlhäusser P, Kutay U (2007) An in vitro nuclear disassembly system reveals a role for the RanGTPase system and microtubule-dependent steps in nuclear envelope breakdown. J Cell Biol 178:595–610
Murphy SP, Simmons CR, Bass HW (2010) Structure and expression of the maize (*Zea mays* L.) SUN-domain protein gene family: evidence for the existence of two divergent classes of SUN proteins in plants. BMC Plant Biol 10:269
Onischenko E, Weis K (2011) Nuclear pore complex—a coat specifically tailored for the nuclear envelope. Curr Opin Cell Biol 23:293–301
Onischenko EA, Gubanova NV, Kiseleva EV, Hallberg E (2005) Cdk1 and okadaic acid-sensitive phosphatases control assembly of nuclear pore complexes in *Drosophila* embryos. Mol Biol Cell 16:5152–5162
Onischenko E, Stanton LH, Madrid AS, Kieselbach T, Weis K (2009) Role of the Ndc1 interaction network in yeast nuclear pore complex assembly and maintenance. J Cell Biol 185:475–491
Panté N, Aebi U (1996) Toward the molecular dissection of protein import into nuclei. Curr Opin Cell Biol 8:397–406
Parry G, Ward S, Cernac A, Dharmasiri S, Estelle M (2006) The *Arabidopsis* supressor of auxin resistance proteins are nucleoporins with an important role in hormone signaling and development. Plant Cell 18:1590–1603
Patel S, Rose A, Meulia T, Dixit R, Cyr RJ, Meier I (2004) *Arabidopsis* WPP-domain proteins are developmentally associated with the nuclear envelope and promote cell division. Plant Cell 16:3260–3273
Pay A, Resch K, Frohnmeyer H, Fejes E, Nagy F, Nick P (2002) Plant RanGAPs are localized at the nuclear envelope in interphase and associated with microtubules in mitotic cells. Plant J 30:699–709
Peiter E, Sun J, Heckmann AB, Venkateshwaran M, Riely BK, Otegui MS, Edwards A, Freshour G, Hahn MG, Cook DR, Sanders D, Oldroyd GE, Downie JA, Ane JM (2007) The *Medicago truncatula* DMI1 protein modulates cytosolic calcium signaling. Plant Physiol 145:192–203
Prunuske AJ, Ullman KS (2006) The nuclear envelope: form and reformation. Curr Opin Cell Biol 18:108–116
Radulescu AE, Cleveland DW (2010) NuMA after 30 years: the matrix revisited. Trends Cell Biol 20:214–222
Rasala BA, Ramos C, Harel A, Forbes DJ (2008) Capture of AT-rich chromatin by ELYS recruits POM121 and NDC1 to initiate nuclear pore assembly. Mol Biol Cell 19:3982–3996
Razin SV, Vassetzky YS, Hancock R (1991) Nuclear matrix attachment regions and topoisomerase II binding and reaction sites in the vicinity of a chicken DNA replication origin. Biochem Biophys Res Commun 177:265–270
Reddy KL, Singh H (2008) Using molecular tethering to analyze the role of nuclear compartmentalization in the regulation of mammalian gene activity. Methods 45:242–251
Ribbeck K, Görlich D (2001) Kinetic analysis of translocation through nuclear pore complexes. EMBO J 20:1320–1330
Riely BK, Lougnon G, Ane JM, Cook DR (2007) The symbiotic ion channel homolog DMI1 is localized in the nuclear membrane of *Medicago truncatula* roots. Plant J 49:208–216
Ris H (1997) High-resolution field-emission scanning electron microscopy of nuclear pore complex. Scanning 19:368–375
Roberts K, Northcote DH (1970) Structure of the nuclear pore in higher plants. Nature 228:385–386
Rose A, Meier I (2001) A domain unique to plant RanGAP is responsible for its targeting to the plant nuclear rim. Proc Natl Acad Sci U S A 98:15377–15382
Rose A, Patel S, Meier I (2004) Plant nuclear envelope proteins. Symp Soc Exp Biol 56:69–88
Rose A, Schraegle SJ, Stahlberg EA, Meier I (2005) Coiled-coil protein composition of 22 proteomes—differences and common themes in subcellular infrastructure and traffic control. BMC Evol Biol 5:66
Rout MP, Aitchison JD, Suprapto A, Hjertaas K, Zhao Y, Chait BT (2000) The yeast nuclear pore complex: composition, architecture, and transport mechanism. J Cell Biol 148:635–651
Rutherford SA, Goldberg MW, Allen TD (1997) Three-dimensional visualization of the route of protein import: the role of nuclear pore complex substructures. Exp Cell Res 232:146–160
Saito K, Yoshikawa M, Yano K, Miwa H, Uchida H, Asamizu E, Sato S, Tabata S, Imaizumi-Anraku H, Umehara Y, Kouchi H, Murooka Y, Szczyglowski K, Downie JA, Parniske M, Hayashi M, Kawaguchi M (2007) Nucleoporin 85 is required for calcium spiking, fungal and bacterial symbioses, and seed production in *Lotus japonicus*. Plant Cell 19:610–624
Salpingidou G, Rzepecki R, Kiseleva E, Lyon C, Lane B, Fusiek K, Golebiewska A, Drummond S, Allen T, Ellis JA, Smythe C, Goldberg MW, Hutchison CJ (2008) NEP-A and NEP-B both contribute to nuclear pore formation in *Xenopus* eggs and oocytes. J Cell Sci 121:706–716
Samaniego R, Jeong SY, Meier I, de la Espina SM (2006) Dual location of MAR-binding, filament-like protein 1 in *Arabidopsis*, tobacco, and tomato. Planta 223:1201–1206

Schäpe J, Prausse S, Radmacher M, Stick R (2009) Influence of lamin A on the mechanical properties of amphibian oocyte nuclei measured by atomic force microscopy. Biophys J 96:4319–4325
Schirmer EC, Florens L, Guan T, Yates Jr III, Gerace L (2003) Nuclear membrane proteins with potential disease links found by subtractive proteomics. Science 301:1380–1382
Shaw PJ, Abranches R, Paula SA, Beven AF, Stoger E, Wegel E, Gonzalez-Melendi P (2002) The architecture of interphase chromosomes and nucleolar transcription sites in plants. J Struct Biol 140:31–38
Shimi T, Pfleghaar K, Kojima S, Pack CG, Solovei I, Goldman AE, Adam SA, Shumaker DK, Kinjo M, Cremer T, Goldman RD (2008) The A- and B-type nuclear lamin networks: microdomains involved in chromatin organization and transcription. Genes Dev 22:3409–3421
Shimi T, Butin-Israeli V, Adam SA, Goldman RD (2010) Nuclear lamins in cell regulation and disease. Cold Spring Harb Symp Quant Biol 75:525–531
Starr DA (2011) KASH and SUN proteins. Curr Biol 21:R414–R415
Starr DA, Fridolfsson HN (2010) Interactions between nuclei and the cytoskeleton are mediated by SUN-KASH nuclear-envelope bridges. Annu Rev Cell Dev Biol 26:421–444
Stavru F, Hulsmann BB, Spang A, Hartmann E, Cordes VC, Gorlich D (2006) NDC1: a crucial membrane-integral nucleoporin of metazoan nuclear pore complexes. J Cell Biol 173:509–519
Stevens BJ, Swift H (1966) RNA transport from nucleus to cytoplasm in *Chironomus* salivary glands. J Cell Biol 31:55–77
Stoffler D, Fahrenkrog B, Aebi U (1999) The nuclear pore complex: from molecular architecture to functional dynamics. Curr Opin Cell Biol 11:391–401
Stoppin V, Vantard M, Schmit AC, Lambert AM (1994) Isolated plant nuclei nucleate microtubule assembly: the nuclear surface in higher plants has centrosome-like activity. Plant Cell 6:1099–1106
Sullivan T, Escalante-Alcalde D, Bhatt H, Anver M, Bhat N, Nagashima K, Stewart CL, Burke B (1999) Loss of A-type lamin expression compromises nuclear envelope integrity leading to muscular dystrophy. J Cell Biol 147:913–920
Taddei A (2007) Active genes at the nuclear pore complex. Curr Opin Cell Biol 19:305–310
Talamas JA, Hetzer MW (2011) POM121 and Sun1 play a role in early steps of interphase NPC assembly. J Cell Biol 194:27–37
Tamura K, Hara-Nishimura I (2011) Involvement of the nuclear pore complex in morphology of the plant nucleus. Nucleus 2:168–172
Tamura K, Fukao Y, Iwamoto M, Haraguchi T, Hara-Nishimura I (2010) Identification and characterization of nuclear pore complex components in *Arabidopsis thaliana*. Plant Cell 22:4084–4097
Towbin BD, Meister P, Gasser SM (2009) The nuclear envelope—a scaffold for silencing? Curr Opin Genet Dev 19:180–186
Tran EJ, Wente SR (2006) Dynamic nuclear pore complexes: life on the edge. Cell 125:1041–1053
Unwin PN, Milligan RA (1982) A large particle associated with the perimeter of the nuclear pore complex. J Cell Biol 93:63–75
Van der Geest AH, Welter ME, Woosley AT, Pareddy DR, Pavelko SE, Skokut M, Ainley WM (2004) A short synthetic MAR positively affects transgene expression in rice and *Arabidopsis*. Plant Biotechnol J 2:13–26
Vasu SK, Forbes DJ (2001) Nuclear pores and nuclear assembly. Curr Opin Cell Biol 13:363–375
Wälde S, Kehlenbach RH (2010) The part and the whole: functions of nucleoporins in nucleocytoplasmic transport. Trends Cell Biol 20:461–469
Wang J, Yang C, Hans R, Zhai Z (1996) Nuclear lamina in plant cells. Sci China C Life Sci 39:449–457
Wang T, Xue L, Hou W, Yang B, Chai Y, Ji X, Wang Y (2007) Increased expression of transgene in stably transformed cells of *Dunaliella salina* by matrix attachment regions. Appl Microbiol Biotechnol 76:651–657
Wang TY, Han ZM, Chai YR, Zhang JH (2010) A mini review of MAR-binding proteins. Mol Biol Rep 37:3553–3560
Winey M, Yarar D, Giddings TH Jr, Mastronarde DN (1997) Nuclear pore complex number and distribution throughout the *Saccharomyces cerevisiae* cell cycle by three-dimensional reconstruction from electron micrographs of nuclear envelopes. Mol Biol Cell 8:2119–2132
Worman HJ, Fong LG, Muchir A, Young SG (2009) Laminopathies and the long strange trip from basic cell biology to therapy. J Clin Invest 119:1825–1836
Xu XM, Meier I (2008) The nuclear pore comes to the fore. Trends Plant Sci 13:20–27
Xu XM, Meulia T, Meier I (2007) Anchorage of plant RanGAP to the nuclear envelope nuclear-pore-associated involves novel proteins. Curr Biol 17:1157–1163
Xu XM, Zhao Q, Rodrigo-Peiris T, Brkljacic J, He CS, Muller S, Meier I (2008) RanGAP1 is a continuous marker of the *Arabidopsis* cell division plane. Proc Natl Acad Sci U S A 105:18637–18642
Xylourgidis N, Fornerod M (2009) Acting out of character: regulatory roles of nuclear pore complex proteins. Dev Cell 17:617–625
Yoo BY, Bayley ST (1967) Structure of pores in isolated pea nuclei. J Ultrastruct Res 18:651–660
Yu W, Díaz M, de la Espina S (1999) The plant nucleoskeleton: ultrastructural organization and identification of NuMA homologues in the nuclear matrix and mitotic spindle of plant cells. Exp Cell Res 246:516–526
Zhao Q, Meier I (2011) Identification and characterization of the *Arabidopsis* FG-repeat nucleoporin Nup62. Plant Signal Behav 6:330–334
Zhao Q, Brkljacic J, Meier I (2008) Two distinct interacting classes of nuclear envelope-associated coiled-coil proteins are required for the tissue-specific nuclear envelope targeting of *Arabidopsis* RanGAP. Plant Cell 20:1639–1651
Zhou X, Graumann K, Evans DE, Meier I (2012) Novel plant SUN-KASH bridges are involved in RanGAP anchoring and nuclear shape determination. J Cell Biol 196:203–211
Zuccolo M, Alves A, Galy V, Bolhy S, Formstecher E, Racine V, Sibarita JB, Fukagawa T, Shiekhattar R, Yen T, Doye V (2007) The human Nup107-160 nuclear pore subcomplex contributes to proper kinetochore functions. EMBO J 26:1853–1864

The Plant Nucleolus

5

Peter Shaw

Contents

5.1 Introduction and History

The nucleolus is where the cell produces ribosomes and ribosomes are required by the cell in prodigious numbers. For example, a yeast cell contains about 200,000 ribosomes, and has a generation time of about 100 min; thus a rapidly dividing cell must make 2,000 ribosomes per minute. This means that 60% of the cell's total RNA transcription is of the ribosomal DNA (rDNA) alone. Additionally, about 80 ribosomal proteins must be synthesised and imported into the nucleolus for each ribosome made. Typically yeast nuclei have about 150 nuclear pores, so each pore must import about 1,000 ribosomal proteins and export about 25 ribosomal subunits per minute (Warner 1999). Similar considerations apply to other eukaryotic cells, but with a large plant cell probably requiring several million ribosomes. Thus the majority of an actively dividing cell's metabolic activity is devoted to ribosome biogenesis, and most traffic in and out of the nucleus is targeted to or from the nucleolus. It is therefore not surprising that the nucleolus is the most prominent and easily observable structure within the nucleus (Figs. 5.1 and 5.2). It has been studied for more than 200 years, is still an active subject of research and is still generating surprising discoveries.

The first mention of the nucleolus was by Fontana (1781), but the name 'nucleolus' was coined by Valentin (1839); it means literally little nucleus, and he described most cells as having a nucleus within the nucleus. Heitz (1931), in a study of plants, was the first to show the correlation between the number of secondary constrictions in the chromosomes – regions that appear as gaps in metaphase chromosomes and where DNA was not detected by the Feulgen stain – and the number of nucleoli that reappear immediately after cell division. McClintock (1934), studying maize, then showed that this region alone was sufficient to generate a nucleolus and proposed that the chromatin at the secondary constriction was the genetic element that organized the nucleolus, now called the nucleolar organizing region or NOR. In the

P. Shaw (✉)
Cell and Developmental Biology Department, John Innes Centre, Colney, Norwich NR4 7UH, UK
e-mail: peter.shaw@jic.ac.uk

I.J. Leitch et al. (eds.), *Plant Genome Diversity Volume 2*,
DOI 10.1007/978-3-7091-1160-4_5, © Springer-Verlag Wien 2013

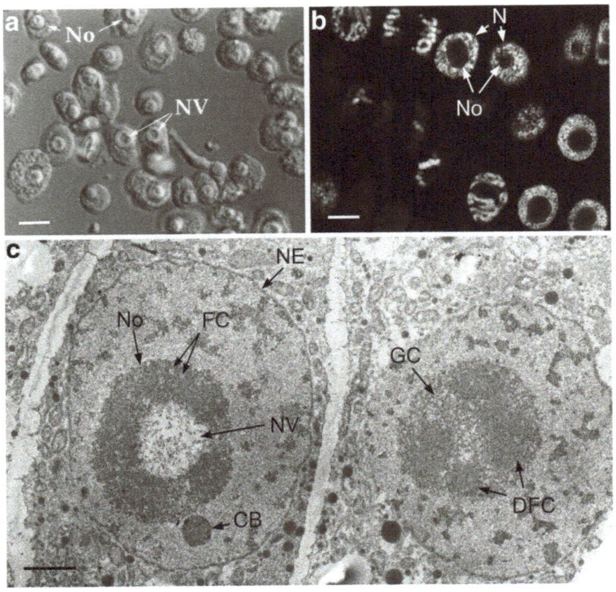

Fig. 5.1 Different microscopical views of plant nucleoli. (**a**) Differential interference contrast image of isolated tobacco nuclei clearly shows the nucleoli and nucleolar vacuoles or cavities within most nucleoli. *Bar* = 10 μm. (**b**) A single confocal optical section through pea root tissue stained with the DNA dye DAPI shows the heterochromatin in the nucleoplasm. The much more decondensed, highly transcribed DNA in the nucleoli stains at a very low level with DAPI giving the nucleoli the appearance of voids in the nuclei. *Bar* = 5 μm. (**c**) Ultrathin section transmission electron micrograph of pea root tissue. The nucleolus appears more electron dense than most of the nucleoplasm, with the nucleolar vacuole or cavity staining lightly. The dense fibrillar component and granular component show a different texture, the GC being more open. *Bar* = 2 μm. *No* nucleolus, *N* nucleus, *NV* nucleolar vacuole, *NE* nuclear envelope, *FC* fibrillar centre, *DFC* dense fibrillar component, *GC* granular component, *CB* Cajal body

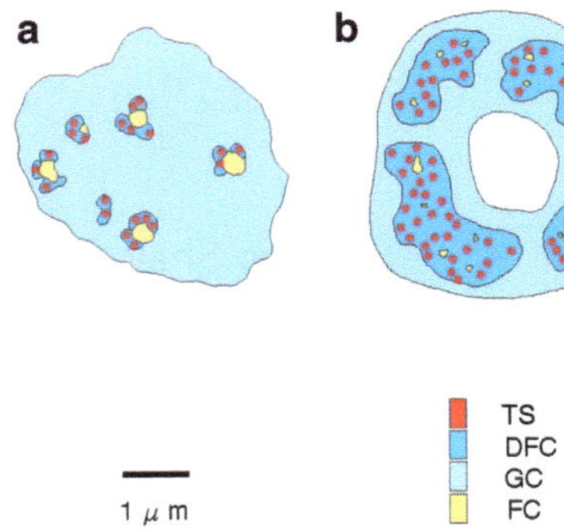

Fig. 5.2 Diagrammatic comparison of typical mammalian (**a**) and plant (**b**) nucleoli. The DFC is much smaller and more densely staining in mammalian nucleoli, whereas it generally constitutes a large proportion of the volume of plant nucleoli. Transcription sites are scattered through the DFC in both cases. *Bar* = 1 μm. *TS* transcription site, *DFC* dense fibrillar component, *GC* granular component, *FC* fibrillar centre

early 1960s it was established by a number of groups that the nucleolus is the site of ribosomal RNA transcription and ribosome biosynthesis; one of the first demonstrations was by Birnstiel et al. (1963) in pea nucleoli. Later in the decade Miller and Beatty (1969) developed a spreading technique for electron microscopy to produce beautiful images of the rDNA genes (= Miller spreads), each gene showing up to 100 attached RNA polymerases, and the RNA transcripts increasing in length as the polymerases progressed along the genes (see Fig. 5.3b). The resulting images have appeared in reviews and text books ever since, and have often been likened to 'Christmas trees'. The next 20 years saw a relative decline of interest in the nucleolus as increasingly powerful molecular biology techniques made it possible to study single copy genes; in fact the multiple tandem repeats in NORs still present significant problems for modern molecular biology techniques. From the mid-1990s on there has been a resurgence of interest in the nucleolus prompted initially by radical improvements in cell biology and imaging techniques, and still more recently by coupling cell biology to mass spectrometry-based proteomics (Andersen et al. 2002; Pendle et al. 2005). Live cell imaging approaches, particularly using green fluorescent protein (GFP), have shown that the nucleolus, in common with other subnuclear structures, is far more dynamic than had been previously appreciated (Phair and Misteli 2000).

In this chapter we shall summarize the current state of knowledge of the nucleolus, with particular reference to recent developments. Where appropriate we shall concentrate on work from plants. In some respects there seem to be significant differences between plant nucleoli and those of other kingdoms, but our knowledge is still so incomplete that it is impossible to tell whether the differences are fundamental or merely apparent. The key reviews of the nucleolus in the 'classical era' are by Busch and Smetana (1970) and Hadjiolov (1985). Recent reviews include Raska et al. (2006a, b).

5.2 Functional Organization of the Nucleolus

The nucleolus has a substantially different biochemical composition from the rest of the nucleus, giving it a different refractive index. Nucleoli are thus easily visible by phase or differential interference contrast microscopy. In many plant cells the nucleolus is almost spherical in shape, and often has a central, internal region called the nucleolar vacuole or cavity, also visible by optical microscopy (Fig. 5.1a). Using DNA-specific dyes such as DAPI and epifluorescence microscopy, the nucleolus usually appears as a dark region, suggesting it lacks DNA (Fig. 5.1b). In fact the nucleolus is the most transcriptionally active region of the nucleus, and therefore must contain substantial numbers of active genes. The relatively low level of DNA labelling shows that these active genes are highly decondensed; what is mainly seen with DNA stains is condensed, mostly inactive DNA.

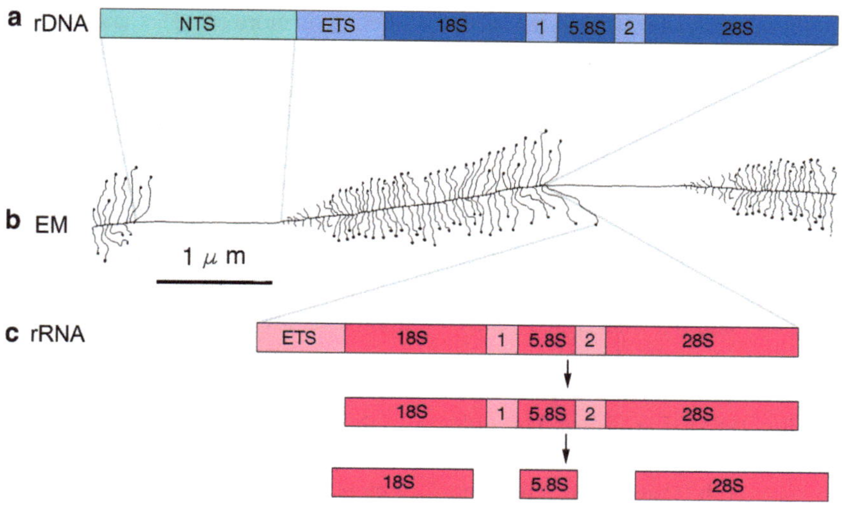

Fig. 5.3 rDNA and rRNA organization. (**a**) Organization of a single rDNA repeat. (**b**) Typical EM spread image, showing the path of the gene, the increasing length of attached transcripts along the gene and the terminal knobs. *Bar* = 1 μm. (**c**) Organization of the initial pre-rRNA transcript and its processing to remove the transcribed spacers. *NTS* non-transcribed spacer, *ETS* external transcribed spacer, *1* internal transcribed spacer 1, *2* internal transcribed spacer 2

Electron microscopy (EM) of intact nucleoli has proved fairly uninformative about their functional organization, since the structures seen cannot easily be interpreted in molecular terms. This is disappointing, as spread preparations clearly show that the structures of active transcription units are within the resolution achievable by EM (Miller and Beatty 1969). The problem in resolving functional units within nucleoli must lie with the great level of compaction of the structures *in vivo*. When ultra-thin sections of mammalian and various other animal cells are stained with standard EM stains (osmium tetroxide, uranyl acetate, lead citrate etc.) a characteristic nucleolar substructure is often seen (Shaw and Jordan 1995). This consists of one or more lightly staining structures, often with a fibrous appearance, called fibrillar centres (FC), surrounded by a layer of densely staining material called the dense fibrillar component (DFC). The rest of the nucleolus consists of granules about the size of ribosomes, and is termed the granular component (GC). There has been a tendency to interpret all nucleoli in terms of this so-called tripartite structure. However in reality there is great variety in the EM ultrastructure between different animals and even between different cell types and metabolic state of cells within the same species. In plants, structures resembling fibrillar centres are often, but not always, seen. They are embedded in a region of the nucleolus which has a somewhat fibrillar texture, assumed to be the DFC, but which does not generally stain intensely as does mammalian DFC, and which can often only be distinguished from the enveloping granular component by a difference in texture (Fig. 5.1c). The DFC region of plant nucleoli is typically a much larger fraction of the total nucleolar volume than in mammalian cells (Shaw and Jordan 1995) (Fig. 5.2).

5.2.1 Organization of rDNA

Three out of the four eukaryotic ribosomal RNAs (18S, 5.8S and 28S) are transcribed by RNA polymerase I (pol 1) from the tandem rDNA repeats in the nucleolus. The fourth ribosomal RNA, 5S, is transcribed by RNA polymerase III from tandem repeats elsewhere in the nucleus and imported into the nucleolus (Highett et al. 1993a). Given a transcription rate estimated to be about 40 nt/s (Kos and Tollervey 2010), it is clear that a single rDNA copy could not provide enough primary transcripts for the cell's ribosome requirements. Thus all eukaryotes have multiple copies of the rRNA genes. In certain specialized cells, like amphibian oocytes, extrachromosomal amplification of rDNA occurs. In virtually all eukaryotes the rDNA copies occur as tandem repeats (Hadjiolov 1985); the reason for this is unknown, but it is tempting to speculate that tandem repeats are more likely to produce high local concentrations of the various factors necessary for transcription and subsequent transcript processing and ribosome assembly – and that this is essentially what constitutes a nucleolus (Melese and Xue 1995). In fact multiple tandem repeats, a visible nucleolus and even pol I transcription are not strictly necessary for ribosome biosynthesis, since a pol I deficient yeast strain in which the rDNA is transcribed from a plasmid by pol II have been created (Oakes et al. 1993). In these mutants the typical crescent shaped yeast nucleolus was absent and instead a number of bodies termed mini-nucleolar bodies were observed. In order to form a normal nucleolar structure, however, it seems that pol I transcription of repeated rDNA copies is required.

The rDNA repeat contains a transcribed region that gives rise to a 45S pre-rRNA transcript (35S in yeast) and an intergenic region, often called the non-transcribed spacer (NTS), that contains promoter and enhancer elements. In a number of species it has been shown that a second upstream promoter can produce low levels of a transcript that includes the sequence of the major promoter. The pre-rRNA transcript contains a leader sequence, the 5′ external transcribed spacer (5′-ETS), which is removed after transcription, the small subunit s-RNA, 18S, two internal transcribed spacers (ITS1 and ITS2), which flank the 5.8S RNA, and finally the large subunit l-RNA 28S followed by a short 3′ external transcribed spacer (Hadjiolov 1985) (Fig. 5.3a). The order of the RNA transcripts and their sequences are highly conserved, but the spacers, both transcribed and non-transcribed, are very variable even between closely related species. The NTS is about 2–3 kb in plants, but much longer—20–30 kb—in vertebrates. The transcribed spacers are also longer in vertebrates, particularly in birds, than in plants. The primary 45S rRNA transcript is processed to remove leader, tail and intergenic sequences in an ordered process (Fig. 5.3c).

The number of copies of the rDNA is highly variable throughout the eukaryotes. Mammals typically have a couple of hundred copies, whereas most plants have several thousand copies. One study estimated that only about 5% of these copies were actively transcribed in pea root cells, and the reason for such large numbers of repeats is unknown (Gonzalez-Melendi et al. 2001). There is evidence that in the human genome some rDNA repeat copies are inverted and may not be functional (Caburet et al. 2005), but this has not been fully confirmed as yet. Fibre fluorescence *in situ* hybridization (FISH) of NORs in rice has suggested a regular pattern of rDNA repeats, apparently without obvious inversions or rearrangements (Mizuno et al. 2008). In fact, it is a glaring omission that the NORs have not been fully sequenced in any organism, due to the difficulties of sequencing large repetitive regions with current technology, so we have no real idea what proportion of the rDNA genes in any plant or animal are functional or what other sequences might be hidden in the intergenic regions.

FISH has been used extensively to examine the location of the rDNA in well-preserved, fixed tissue. In plants this generally shows a few dense ‘knobs’ or concentrations of rDNA around the periphery of the nucleolus together with some fainter labelling within the nucleolus. The knobs correspond to the inactive rDNA copies which remain condensed as heterochromatin, their number usually corresponding to the number of NORs, while the active copies are decondensed within the body of the nucleolus. In some species, such as the diploid species rye, the internal path of the decondensed rDNA can be clearly seen, whereas in the closely related species hexaploid wheat, the internal labelling is more complex and may contain small condensed regions of rDNA, while some NORs remain inactive and unassociated with the nucleolus (Leitch et al. 1992). In pea, the four NORs all contribute to the nucleolus, and the size of the knobs varies inversely with the size and presumed activity of the nucleolus, showing that increased nucleolar activity causes more of the rDNA copies to decondense and become active (Highett et al. 1993b).

Only the NORs that were active during the previous interphase produce secondary constrictions on mitotic chromosomes, and these NORs are also stained by silver salts in the so-called Ag-NOR labelling. In animals it has been established that active rDNA copies are associated with binding of upstream binding factor (UBF), a DNA binding protein containing a number of high mobility group (HMG) protein motifs, which are the DNA binding domains. Mais et al. (2005) integrated arrays of ectopic UBF heterologous binding sequences into human chromosomes and showed that they bound UBF and some pol I components. These pseudo-NORs gave silver-positive secondary constrictions in the metaphase chromosomes, showing that UBF binding is responsible for generating the decondensed rDNA seen in the secondary constrictions and for their Ag-NOR labelling. No UBF homolog has yet been identified in plants, but the equivalent behaviour of plant NORs strongly indicates that such a homolog must exist.

5.2.2 rDNA Transcription and Ribosome Biogenesis

The location within the nucleolus of the actively transcribed genes has been a matter of intense debate over about 25 years; see Raska et al. (2006b) for a recent summary. Most of this debate has centred around their location with respect to the EM ultrastructure, with some groups maintaining that all transcription takes place in the FCs and others that it is within the DFC. Early studies using radioactive tritiated uridine labelling to locate incorporation into nascent RNA in the nucleolus showed predominant labelling of the DFC. However, Scheer and Rose (1984) showed by immunogold labelling that the FCs contained concentrations of RNA pol I, and that little was detected elsewhere in the nucleolus. This was followed by various immunogold studies that showed DNA in the FCs; see Scheer and Weisenberger (1994) and Shaw and Jordan (1995) for summaries. The problem with all these latter studies is that most rDNA and most pol I is inactive at any given time. It is only a small proportion, perhaps just a few percent, that is active. In order to locate the active genes, nascent rRNA must be localized. Unfortunately the tritiated uridine method lacked both resolution and sensitivity. With the introduction of bromo-uridine as a marker for nascent RNA, supplied to

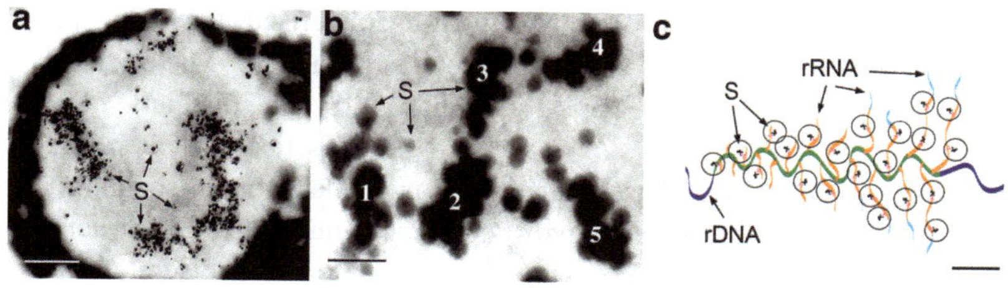

Fig. 5.4 Silver-enhanced 1 nm gold labelling of BrU in nascent transcripts in pea root tissue. (**a**) View of an entire nucleolus showing dense labelling of transcript sites within the DFC. *Bar* = 1 μm. (**b**) Higher magnification electron micrograph showing five clusters of 1 nm gold particles. The clusters are approximately conical in shape and contain 20–30 particles. *Bar* = 100 nm. (**c**) Diagram of proposed interpretation of the 1 nm gold labelling of nascent transcripts, drawn to scale (see Gonzalez-Melendi et al. 2001). *Bar* = 50 nm. *S* examples of silver-enhanced 1 nm gold particles

the cell as BrUTP, it became possible to examine the nucleolar transcription sites with much greater sensitivity and resolution (Dundr and Raska 1993; Hozak et al. 1993; Wansink et al. 1993). This labelling showed many foci within the DFC region of the nucleolus which sometimes contacted the periphery of the FCs (Hozak et al. 1994). Similar results were obtained in plants, where the more extensive and less dense DFC made the results even more unequivocal (Melcak et al. 1996; Thompson et al. 1997). Double FISH labelling of the NTS and transcribed rDNA region in peas and comparison with BrUTP transcript labelling showed that the transcribed DNA overlapped well with the transcript labelling as expected, but that the intergenic NTS labelling had very little overlap with the transcribed region. This suggests that the transcribed genes are in an extended conformation (Thompson et al. 1997). In a subsequent study in pea roots, Gonzalez-Melendi et al. (2001) showed by thin section EM that 1 nm gold labelling of BrU consisted of discrete elongated clusters of label, about 300 nm in length (see Fig. 5.4). The clusters were often approximately conical in shape, and the authors suggested these corresponded to individual transcription units – condensed Christmas trees – compacted by a factor of 5–8 compared to Miller spreads. Similar conclusions were reached by Koberna et al. (2002) in animal nucleoli.

The external and internal spacers are removed from the pre-rRNA in an ordered series of cleavage and trimming steps, which has been well studied in yeast, but less studied in other species. The rRNAs are also modified at numerous sites by 2′-O-ribose methylation and pseudouridylation. The reason for this is unclear, but the majority of the changes are in the ribosome active site, and are thought to improve ribosome efficiency. The site of each modification and cleavage is specified by a cognate guide small nucleolar RNA (snoRNA) about 60–150 nt in length, which contains a complementary sequence to the target sequence in the rRNA. The initial cleavage of the pre-rRNA involves U3, U14, MRP, snR10 and snr30 and the resulting cleaved products are cleaved by specific exonucleases Rat1p, Xrn1p and the exosome (Fatica and Tollervey 2002, 2003). The methylations and pseudouridylations are catalysed by fibrillarin and dyskerin respectively (Nop1p and Cbf5p in yeast). Methylations are guided by box CD snoRNAs (containing RUGAUGA and CUGA elements) and pseudouridylations by box H/ACA snoRNAs (containing ANANNA and ACA elements) (Kiss 2002). An EM structure for a box CD snoRNP has been determined recently for an archeon by Bleichert et al. (2009). Many of these snoRNAs and cleavage intermediates have been identified and localized in plant nucleoli (Brown et al. 2003; Kim et al. 2010). The early stages of processing, in which the 5′-ETS was present, were found closely enveloping the transcription foci in the DFC, whereas the later stages of processing, where the ITS sequences were still present, were found further away from the transcription sites, in regions broadly corresponding to the GC (Shaw et al. 1995; Beven et al. 1996; Brown and Shaw 1998, 2008). Thus the current model is of a vectorial distribution of processing steps with early steps close to transcription sites and successive steps displaced further outwards from them. Little analysis has been carried out on the later biochemistry of ribosome subunit assembly and export in plants, but the principles are assumed to be the same as in animals and in yeast, which has been analysed in the most detail. The RNA cleavages, in yeast at least, are begun co-transcriptionally (Kos and Tollervey 2010). The s-RNA is processed into the small ribosomal subunit. The terminal knobs that are seen in Miller spreads initially are about 15 nm in size, but become larger – about 40 nm – and at this stage represent the small subunit processome complexes (Dragon et al. 2002; Bernstein et al. 2004; Osheim et al. 2004). Little is known about the corresponding processing complex for the large subunit although a pre-60S particle has been imaged at high resolution in the EM (Nissan et al. 2004). The large and small ribosomal subunits are exported independently to the cytoplasm.

5.3 Assembly and Dynamics of the Nucleolus

The nucleolus disassembles at the end of the G2 phase of the cell cycle as most transcription ceases and the nuclear envelope breaks down and reassembles with the onset of rDNA transcription at the beginning of G1. The GC components are lost from the disassembling nucleolus first, followed by the DFC (Gautier et al. 1992). Subunits of pol I and other DNA binding factors such as UBF remain with the rDNA arrays; the presence of UBF alone is sufficient to produce a secondary constriction in the mitotic chromosome (Prieto and McStay 2008). Some nucleolar components diffuse throughout the mitotic cytoplasm, whereas others, such as the protein B23, associate with the periphery of the mitotic chromosomes as chromosomal 'passengers' (Hernandez-Verdun and Gautier 1994). When rDNA transcription is halted during mitosis, unprocessed pre-rRNA transcripts persist through the mitotic cell, demonstrating that pre-rRNA transcript processing is also halted.

At the end of mitosis, the nucleolus reforms. First, small round bodies, called pre-nucleolar bodies are formed (Gimenez-Martin et al. 1974; Angelier et al. 2005; Hernandez-Verdun 2006). When transcription of the rDNA is reinitiated, the pre-nucleolar bodies (PNBs) disappear as new nucleoli are formed. Originally it was assumed that the PNBs condensed onto the active rDNA, but recent data obtained using the optical technique FRAP (fluorescence recovery after photobleaching) suggest rather that pre-rRNA processing complexes (and also unprocessed pre-rRNA) are preassembled in the PNB, and then diffuse out of the PNBs, associating with the reforming nucleoli. Where more than one active NOR is present in the nucleus, separate nucleoli generally initially form at each active NOR in early G1. These small nucleoli then have a tendency, especially in plants, to fuse together to a single nucleolus as interphase progresses (Shaw and Jordan 1995).

Recent dynamic studies using a variety of nuclear and nucleolar proteins marked by GFP have led to a complete reassessment of the interpretation of nucleolar and nuclear structure. For many years cell biologists have visualized fixed cells and have thus had a tendency to regard the structures seen as stationary and long-lived. However live cell imaging studies have revealed a much more dynamic picture. First, sub-nuclear structures themselves move and rearrange themselves within the nucleus, and the nucleus itself moves and changes shape. For example, Boudonck et al. (1999) showed in *Arabidopsis* that Cajal bodies move and fuse together, changing their positions and number. At the molecular level, all nucleolar and nuclear proteins are in constant flux, exchanging between the nucleolus and cytoplasm, and the mean nucleolar residence time of even well-characterised 'nucleolar' proteins is only a few tens of seconds (Phair and Misteli 2000; Misteli 2001; Olson and Dundr 2005). The distinction between 'nuclear' and 'nucleolar' proteins lies in their residence time in the nucleolus, with nucleolar proteins spending a greater proportion of their time in the nucleolus. The structure and even the existence of the nucleolus as a discrete structure must depend on the rDNA nucleating a small sub-population of proteins that then form a structure on which all the other proteins assemble and disassemble dynamically. The nucleolus (and other nuclear bodies such as Cajal bodies) thus represent a steady state flux of proteins in rapid equilibrium with the surrounding nucleoplasm (Raska et al. 2006a). It is even possible that the DNA in the nucleolus is in dynamic equilibrium with the rDNA at the nucleolar periphery; this has yet to be tested in living cells.

5.4 Epigenetics and Nucleolar Dominance

Not all rRNA genes are necessarily transcriptionally active in a given nucleolus. In fact in most plants, the vast majority are not transcribed. We assume that all rDNA copies are identical and potentially transcribable (but as mentioned above, this is by no means certain). Current evidence, however, is that rRNA genes may be in one of three states: (1) inactive and condensed into heterochromatin, which in plants is mostly seen as knobs of heterochromatin at the nucleolar periphery, but also within the nucleolus; (2) active and transcribed, in an extended conformation within the nucleolus; finally (3) an 'open', poised conformation – potentiated and available for transcription, but not currently transcribed (Huang et al. 2006; McKeown and Shaw 2009). The balance between these states, and ultimately the level of pol I loading, may depend on DNA methylation, differences in the histone variants associated with the DNA, remodelling of the DNA, particularly the promoter regions, and the presence of histone modifications (Grummt and Pikaard 2003). rDNA can retain the level of pol I association through mitosis, and so the chromatin state of rDNA can be epigenetically inherited.

Nucleolar dominance is a particularly striking effect that is seen in hybrid organisms, where it is often found that the NORs of one parental genome are silent while those of the other are active (see Tucker et al. (2010) for a recent review). This is usually ascribed to the suppression of the under-dominant genome by the dominant one, but recent evidence from a wheat-rye hybrid has suggested that the NORs from the dominant genome are also up-regulated (Silva et al. 2008). Nucleolar dominance has been observed in many plants and animals, but has been most thoroughly studied in plants, probably because inter-species hybrids are much easier to study in plants than animals. In a given hybrid cross, the same member of the pair is always under-dominant or dominant, irrespective of which parent provides which genome. This means that nucleolar dominance is not mediated by an equivalent mechanism to parental imprinting, nor to X chromosome inactivation, in which a random choice of inactive chromosome is made. In *Brassica* hybrids, it has been demonstrated that nucleolar dominance is disrupted by aza-deoxycytidine, which reduces DNA methylation, and by Trichostatin A, a histone deacetylase inhibitor which leads to an increase in histone acetylation.

Significant progress in understanding nucleolar dominance has been made using *Arabidopsis suecica*, a hybrid of *A. thaliana* and *A. arenosa*. In young seedlings of *A. suecica*, the rRNA genes from both genomes are highly expressed, but as the plant grows, the *A. thaliana*-derived rRNA genes become silenced. This suggests that nucleolar dominance may be an aspect of active gene dosage control mechanisms, where different levels of rRNA are required at different stages of development (Tucker et al. 2010). RNAi has been used systematically to determine which histone modifying enzymes are required for gene silencing in nucleolar dominance. This approach has pinpointed the histone deacetylases HDT1 and HDA6, the *de novo* DNA methyltransferase DRM2, and the methylcytosine binding domain proteins MBD6 and MBD10 (Preuss et al. 2008). MBD6 is presumed, by analogy with animal studies, to participate in the formation of heterochromatin, but this has not yet been formally shown in plants. DRM2 is part of the RNA-directed DNA methylation pathway, in which double stranded templates are formed from pol IV RNA transcripts by the RNA-dependent RNA polymerase, RDR2, diced into 24 nt siRNAs, which then guide DRM2 to methylate the homologous DNA sequences. In confirmation that this pathway is indeed involved in nucleolar dominance in plants, knockdown of RDR2, DCL3 as well as DRM2 disrupted the rRNA silencing of the *A. thaliana* derived rRNA genes in *A. suecica* (Preuss et al. 2008).

There is also evidence for RNA-mediated silencing of rRNA genes in mammals, where rRNA genes are silenced by the nucleolar remodelling complex, NoRC, which is recruited to a subset of rRNA genes by 200–300 nt RNA species, termed pRNA, which themselves derive from intergenic regions of rDNA (Mayer et al. 2008; Santoro et al. 2010). Thus although the detailed mechanisms may differ, in both plants and animals control of rRNA gene expression depends on RNA-mediated silencing by sequences derived from the intergenic rDNA, which is presumably expressed from the minor intergenic rDNA promoter. As yet, however, little is known about how the subset of genes to be silenced is chosen, and why one particular genome is dominant or under-dominant. In hybrids, it is tempting to speculate that this may be due to the relative strength of interactions between the silencing RNAs and the rDNA of the two genomes.

5.5 Non-conventional Nucleolar Functions

5.5.1 Proteomics

High throughput proteomics has now been applied to purified nucleoli in a number of studies including humans (Andersen et al. 2002, 2005; Scherl et al. 2002) and the plant *A. thaliana* (Pendle et al. 2005). These studies have uncovered an enormous range of several hundred proteins as nucleolar constituents, and has added a new dimension to the previous observations indicating that the nucleolus is the site of many other functions than ribosome biogenesis (Pederson 1998). In this type of proteomic analysis of complex mixtures, the question of possible contaminants immediately arises. Pendle et al. (2005) answered this question by localizing GFP fusions *in vivo* of a substantial, randomly chosen set of proteins identified in the nucleolar fractions (see Fig. 5.5). The vast majority (87%) were indeed located in the nucleolus, but most were also seen in other parts of the nucleus or cytoplasm, as would be expected. In fact since the dynamic studies mentioned above have shown that virtually all nuclear proteins at least visit the nucleolus, there is a real question of what actually constitutes a nucleolar protein. The best approach is to compare the nucleolar protein profile quantitatively with the nuclear and cytoplasmic protein profiles, and thus arrive at a nucleolar partition ratio for each protein. This has been done for human cell culture nucleoli using stable isotopic labelling of the different fractions – SILAC – prior to mass spectrometry analysis (Boisvert et al. 2009). Given the rapid diffusion of most proteins in and out of the nucleolus, it is fair to ask how nucleoli can be purified at all. The answer to this is not clear, but it is presumably because the breakage of the cell and nuclear membranes that precedes nucleolar purification must also alter the solution conditions to prevent most proteins from diffusing away from the nucleoli. However the caveat that proteins may be selectively lost during nucleolar isolation is important, and shows the need to complement proteomics approaches by *in vivo* studies of specific proteins.

SILAC methods have also been used to analyse the dynamics of nucleolar proteins after treatment with specific drugs or stresses (Lamond and Sleeman 2003; Andersen et al. 2005), during the cell cycle (Leung and Lamond 2003), and after viral infection (Emmott et al. 2010; Hiscox et al. 2010). As an example, proteomic analysis of nucleoli after treatment with various inhibitors of the proteasome showed a large accumulation of ribosomal proteins. Photobleaching experiments of individual GFP-tagged ribosomal proteins showed that these proteins are synthesized and imported to the nucleolus very rapidly. The number of rRNA molecules needs to be balanced with the number of ribosomal proteins since they are required in stoichiometric amounts (Rudra and Warner 2004). These experiments suggest that this balance is

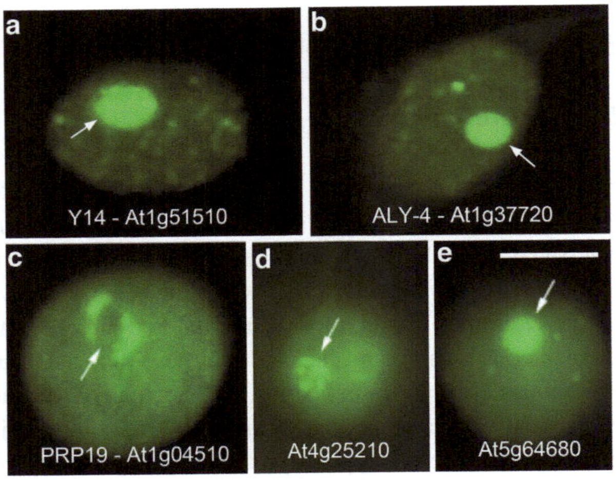

Fig. 5.5 Examples of expression of GFP-fusion constructs for proteins found in proteomic analyses of *Arabidopsis* nucleoli. Transient expression in *Arabidopsis* culture cells (see Pendle et al. 2005). (**a**) EJC component Y14. (**b**) EJC/export factor ALY/REF. (**c**) Splicing factor PRP19 shows a perinucleolar distribution. (**d**) Protein of unknown function localized to nucleolar substructures. (**e**) Protein of unknown function localized to nucleolus and other nuclear bodies. *Bar* = 5 μm

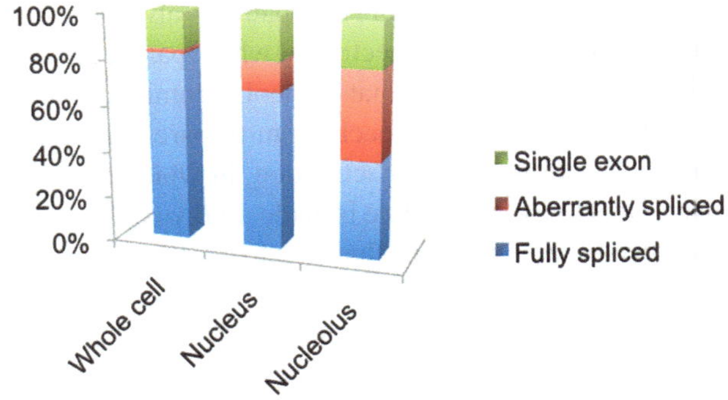

Fig. 5.6 Relative amounts of single exon, aberrantly spliced and fully spliced mRNAs in polyA + libraries made from whole cells, nuclear extracts and nucleolar extracts respectively. Whereas the single exon mRNAs are present in about the same proportion in each library (15–20%), the percentage of aberrantly spliced mRNAs increases from a very low level in whole cells, mainly from cytoplasmic mRNA (2%), to an intermediate level in nuclear extracts (13%), and the highest level in nucleolar extracts (38%). Thus the aberrant mRNAs purify predominantly with the nucleolar fraction (see Kim et al. 2009)

achieved by making an excess of the r-proteins and ubiquitinating and degrading any that remain unincorporated into ribosomes (Andersen et al. 2005).

5.5.2 mRNAs and Nonsense-Mediated mRNA Decay (NMD)

A detailed analysis of the nucleolar proteome from *Arabidopsis* showed that many proteins involved in mRNA splicing and translation were present in the nucleolus. In particular, it was striking that almost all the known components of the post-splicing exon-junction complex (EJC) were detected, and were subsequently shown by GFP fusions to indeed be associated with the nucleolus (Pendle et al. 2005). In contrast, although EJC components were detected in the human nucleolar proteome, localization studies have not so far confirmed their localization (Custodio et al. 2004), suggesting possible differences between plants and animals. The EJC is a multi-protein complex that is deposited 20–30 nucleotides upstream of splice junctions in mRNAs. The complex contains residual spliceosomal proteins, as well as factors involved in mRNA export and translation. The complex remains in place until the pioneer round of translation (Lejeune et al. 2004). The EJC also mediates nonsense-mediate mRNA decay (NMD) (Lejeune et al. 2004). In the most studied mechanism, if the ribosome during the initial round of translation encounters a stop codon upstream of an EJC complex, the stop codon is identified as a premature stop codon. The mRNA is thus marked as aberrant, the NMD factors upf1, upf2 and upf3 are recruited to the EJC complex and the mRNA is degraded, probably mostly through cytoplasmic P bodies (Parker and Sheth 2007; Xu and Chua 2009).

The observation of EJC components in the nucleolus in plants suggests that mRNAs may also be located there. This has been confirmed by constructing cDNA libraries from polyA + RNA extracted from purified nucleoli (Kim et al. 2009). Many individual clones were sequenced from the nucleolar library and compared with similar libraries made from purified nuclei and entire cell extracts respectively. Remarkably, this analysis showed that mis-spliced and otherwise aberrant mRNAs were greatly enriched in the nucleolar extract – about ten-fold compared to the entire cell extract, which would be expected to be mainly cytoplasmic (see Fig. 5.6). Single exon transcripts, which do not undergo splicing, were found in the same amounts in all three libraries, showing that contamination could not explain these results. The aberrant transcripts contained all sorts of splicing errors, including intron retention and splice boundary mis-sensing of all types. It might be argued that many of these species were alternatively spliced rather than mis-spliced. Since alternative splicing has been little studied in plants, this is difficult to assess. However, the spliced variants were at odds with the standard gene models from the genome sequence, most of which have been verified by EST and other mRNA sequences. Thus it is likely that most of these species should be considered as mis-spliced rather than alternatively spliced. About 90% of the aberrant transcripts fulfilled the conditions for targeting for NMD, at least according to the mammalian NMD criteria. Finally Kim et al. (2009) showed by GFP fusion analysis that the NMD factors upf3 and upf2 were localized to the nucleolus, although upf1 was not. These results strongly argue that the nucleolus is involved in mRNA surveillance and export.

In fact the nucleolus has been previously implicated in mRNA export in experiments going back to the 1960s (Pederson 1998). Harris (1967) showed that in heterokaryons between chicken erythrocytes and human HeLa cells that no proteins of chicken origin were produced until the previously inactive chicken nucleus reformed a nucleolus, suggesting that the lack of a functional nucleolus impaired mRNA export from the chicken nucleus. More recently, the nucleolus in transport-defective yeast mutants has been shown to be disrupted (Schneiter et al. 1995), and heat shock or mutation of nucleolar proteins lead to accumulation of polyA + RNA in the nucleolus (Kadowaki et al. 1995). Ideue et al. (2004) have shown that a subset of poly A + mRNA associated transiently with the nucleolus during export, and an intron-containing transcript accumulated in the nucleolus in export-deficient mutants, whereas transcripts from the intronless cDNA did not.

5.5.3 Nucleolar Translation?

The finding that mis-spliced mRNAs are preferentially concentrated in the nucleolus, whether they are degraded there or simply pass through on their way to the cytoplasm, raises the interesting question of how these RNA species are identified. The best studied mechanism for NMD requires a ribosome to detect a premature termination codon during the initial pioneer round of translation. If such transcripts are identified before nuclear export and sent to or preferentially retained in the nucleolus as the current data imply, this mechanism would require at least the pioneer round of translation for some mRNAs to take place in the nucleus or nucleolus or both. This is a controversial idea, but one that, surprisingly, has some experimental support. There is evidence for the presence of amino-acylated tRNAs in the nucleolus of yeast (Steiner-Mosonyi and Mangroo 2004), and in the nucleus of *Xenopus* oocytes (Lund and Dahlberg 1998). The nucleolus is also full of ribosomes, some of which could be competent for translation. Evidence for actual protein translation in pea nucleoli was first published during the 1960s (Birnstiel et al. 1961; Birnstiel and Hyde 1963). This work was subsequently assumed to be due to cytoplasmic contamination after it was shown that translation occurred in the cytoplasm. However the idea was revisited using modern cell biological methods by Iborra et al. (2001), who allowed cells to incorporate labelled amino-acyl tRNA and then detected the incorporated labelled amino acid residues by fluorescence microscopy. This showed 9–15% of the labelling within the nucleus, with prominent nucleolar labelling. Nathanson et al. (2003) questioned these experiments, showing that in their hands only about 1% of the labelling was intranuclear, and suggested that the results of Iborra et al. (2001) were due to cytoplasmic contamination. This controversy has yet to be satisfactorily resolved (Iborra et al. 2004).

5.5.4 Other RNA Species

There is emerging evidence for a number of other non-conventional roles for the nucleolus (Table 5.1) (Pederson 1998; Olson et al. 2002; Raska et al. 2006a). For example, tRNA genes have been shown to be preferentially located in or at the periphery of the nucleolus, and this location is dependent on their transcription (Thompson et al. 2003). The resulting transcripts are processed by trimming at 5′ and 3′ ends by RNAse P, an RNA-containing enzyme, which is also found in the nucleolus (and Cajal bodies) (Jarrous et al. 1999). Similarly Highett et al. (1993a) showed a preferential location of 5S genes at the nucleolar periphery by FISH. Telomerase (both RNA and protein components) has also been found in the nucleolus, either as part of its biosynthesis or sequestered there as a control mechanism (Wong et al. 2002). Many RNAs, including snoRNAs, tRNAs and telomerase RNA are modified by pseudo-uridylation, which is catalyzed by cbf5p/dyskerin, which is found in the nucleolus in plants, animals and yeast, and by 2′-O-ribose methylation, catalysed by fibrillarin, also found in the nucleolus (and Cajal bodies). Yet another RNA complex which has been associated with the nucleolus is the signal recognition particle (SRP). This is an RNA-containing complex that targets the translation of certain proteins to the endoplasmic reticulum (ER) by first blocking and then releasing translation on binding to the SRP receptor in the ER. Stages in the assembly of the SRP have been shown to occur in the nucleolus by *in situ* hybridization, biochemical fractionation and live cell microinjection studies (Chen et al. 1998; Jacobson and Pederson 1998; Politz et al. 2002). Finally, the various components required for both rDNA silencing (see above) and heterochromatic silencing co-localize with the siRNAs themselves in the nucleoli and in Cajal bodies (Li et al. 2006; Pontes et al. 2006). Thus the nucleolus (and Cajal bodies) are involved in siRNA production and assembly of silencing complexes both for rDNA and for other genes.

Why should all these other activities and complexes be associated with the nucleolus? One clue is that they all have the post-transcriptional processing of RNA species and association of the resulting RNAs with multiple proteins in common with ribosome biosynthesis. In all cases, the RNA protein assembly pathways also probably require a complex series of steps involving various chaperones and accessory factors. Clearly in some way concentrating all the factors and processes needed to make ribosomes together in a specialized region of the nucleus—the nucleolus—has important benefits, and the same may apply to the biosynthesis of many other multi-component RNA complexes. In addition, some of the activities and factors necessary to make the various RNA machines are shared between many different processes, such as pseudo-uridylation, RNA trimming, ribose 2′-O-methylation. Concentrating these factors

Table 5.1 Non-conventional functions of the nucleolus in RNA metabolism and other cell processes

Partial assembly of telomerase RNP
Partial assembly of Signal Recognition Particle
5′ and 3′ processing of some pre-tRNAs by RNAse P
Processing and assembly of RNAse P
Processing of polycistronic pre-snoRNAs in plants
Nucleotide modifications in snoRNAs and snRNAs
Production of heterochromatin siRNAs in plants
Nucleolar phase for some mRNAs
Concentration of aberrant mRNAs/ NMD (plants)
Nucleolar trafficking of some animal and plant virus proteins
Sensor of cell stress
Sequestration of various factors – cdc14, p53/MDM2/ARF, telomerase

in the nucleolus for ribosome biosynthesis may have the side effect of locating other processes that require some of the same factors in the nucleolus as well.

Conclusion

Over the past 15–20 years the importance of the nucleolus has grown with increasing understanding of the range of activities located in this nuclear region. It is clear that it is the major centre in the nucleus for RNA transcription and processing and for the assembly of a wide variety of RNP complexes. It is involved with the products of all the DNA-dependent RNA polymerases in one way or another, and recent evidence from plants shows that it is likely to be involved in mRNA export and surveillance, as well as RNAi silencing mechanisms. The extent of this involvement and the detailed mechanisms underlying it are the subject of active research.

The activity of the nucleolus underpins most of the activity of the cell, and it is therefore not unexpected that responses to growth conditions and to stresses involve modulation and responses in the nucleolus. In human pathology, nucleolar morphology has been used in tumour diagnosis and grading for prognosis over many years and an enigmatic body called the perinucleolar compartment has emerged as closely linked to malignant transformation (Kopp and Huang 2005). One of the major mediators of cellular stress responses and genome damage in mammalian cells is the p53 transcription factor; more than 50% of human cancers have impaired p53 pathways, which has made its regulation the subject of intense study. Rubbi and Milner (2003) have shown that disruption of the integrity of the nucleolus, by targeted UV irradiation, drug treatment or specific antibodies, can induce activation of the p53 pathway, suggesting that the nucleolus itself is the upstream stress sensor for DNA damage and other stresses. It is not known whether similar mechanisms operate in plants, but recent evidence has implicated relocation of the RNA binding factor and EJC component eiF4a-III to the nucleolus and nuclear granules in response to hypoxia and other stresses (Koroleva et al. 2009a, b), suggesting that the nucleolus is indeed involved in stress responses in plants.

Research on the nucleolus has led the way in a number of areas of cell and molecular biology. Its fundamental importance to all eukaryotic cells means that it can provide paradigms for many activities and mechanisms. It is also involved in biosynthesis of some of the most ancient cellular machinery involving RNA. In the light of this, the evolution of the nucleolus in the first eukaryotes and its relation to the organization of equivalent processes in archaea, from which the informational processes in eukaryotes are thought to have developed, is likely to be a fruitful field for future study.

References

Andersen JS, Lyon CE, Fox AH, Leung AK, Lam YW, Steen H, Mann M, Lamond AI (2002) Directed proteomic analysis of the human nucleolus. Curr Biol 12:1–11

Andersen JS, Lam YW, Leung AK, Ong SE, Lyon CE, Lamond AI, Mann M (2005) Nucleolar proteome dynamics. Nature 433:77–83

Angelier N, Tramier M, Louvet E, Coppey-Moisan M, Savino TM, De Mey JR, Hernandez-Verdun D (2005) Tracking the interactions of rRNA processing proteins during nucleolar assembly in living cells. Mol Biol Cell 16:2862–2871

Bernstein KA, Gallagher JE, Mitchell BM, Granneman S, Baserga SJ (2004) The small-subunit processome is a ribosome assembly intermediate. Eukaryot Cell 3:1619–1626

Beven AF, Lee R, Razaz M, Leader DJ, Brown JW, Shaw PJ (1996) The organization of ribosomal RNA processing correlates with the distribution of nucleolar snRNAs. J Cell Sci 109:1241–1251

Birnstiel ML, Hyde BB (1963) Protein synthesis by isolated pea nucleoli. J Cell Biol 18:41–50

Birnstiel ML, Chipchase M, Bonner J (1961) Incorporation of leucine-h3 into subnuclear components of isolated pea nuclei. Biochem Biophys Res Commun 6:161–166

Birnstiel ML, Chipchase MI, Hyde BB (1963) The nucleolus, a source of ribosomes. Biochim Biophys Acta 76:454–462

Bleichert F, Gagnon KT, Brown BA II, Maxwell ES, Leschziner AE, Unger VM, Baserga SJ (2009) A dimeric structure for archaeal box c/d small ribonucleoproteins. Science 325:1384–1387

Boisvert FM, Lam YW, Lamont D, Lamond AI (2009) A quantitative proteomics analysis of subcellular proteome localization and changes induced by DNA damage. Mol Cell Proteomics 9:457–470

Boudonck K, Dolan L, Shaw PJ (1999) The movement of coiled bodies visualized in living plant cells by the green fluorescent protein. Mol Biol Cell 10:2297–2307

Brown JWS, Shaw PJ (1998) Small nucleolar RNAs and pre-rRNA processing in plants. Plant Cell 10:649–657

Brown JW, Shaw PJ (2008) The role of the plant nucleolus in pre-mRNA processing. Curr Top Microbiol Immunol 326:291–311

Brown JW, Echeverria M, Qu LH (2003) Plant snoRNAs: functional evolution and new modes of gene expression. Trends Plant Sci 8:42–49

Busch H, Smetana K (1970) The nucleolus. Academic, New York

Caburet S, Conti C, Schurra C, Lebofsky R, Edelstein SJ, Bensimon A (2005) Human ribosomal RNA gene arrays display a broad range of palindromic structures. Genome Res 15:1079–1085

Chen Y, Sinha K, Perumal K, Gu J, Reddy R (1998) Accurate 3′ end processing and adenylation of human signal recognition particle RNA and alu RNA in vitro. J Biol Chem 273:35023–35031

Custodio N, Carvalho C, Condado I, Antoniou M, Blencowe BJ, Carmo-Fonseca M (2004) In vivo recruitment of exon junction complex proteins to transcription sites in mammalian cell nuclei. RNA 10:622–633

Dragon F, Gallagher JE, Compagnone-Post PA, Mitchell BM, Porwancher KA, Wehner KA, Wormsley S, Settlage RE, Shabanowitz J, Osheim Y, Beyer AL, Hunt DF, Baserga SJ (2002) A large nucleolar u3 ribonucleoprotein required for 18S ribosomal RNA biogenesis. Nature 417:967–970

Dundr M, Raska I (1993) Nonisotopic ultrastructural mapping of transcription sites within the nucleolus. Exp Cell Res 208:275–281

Emmott E, Rodgers M, Macdonald A, McCrory S, Ajuh P, Hiscox JA (2010) Quantitative proteomics using stable isotope labeling with amino acids in cell culture (SILAC) reveals changes in the cytoplasmic, nuclear and nucleolar proteomes in Vero cells infected with the coronavirus infectious bronchitis virus. Mol Cell Proteomics 9:1920–1936

Fatica A, Tollervey D (2002) Making ribosomes. Curr Opin Cell Biol 14:313–318

Fatica A, Tollervey D (2003) Insights into the structure and function of a guide rnp. Nat Struct Biol 10:237–239

Fontana F (1781) Traite sur le venin de la viper, sur les poisons americains, sur le laurier-cerise et sur quelques autres poisons vegetaux. Gibelin, Florence

Gautier T, Robert-Nicoud M, Guilly MN, Hernandez-Verdun D (1992) Relocation of nucleolar proteins around chromosomes at mitosis – a study by confocal laser scanning microscopy. J Cell Sci 102:729–737

Gimenez-Martin G, De la Torre C, Fernandez-Gomez ME, Gonzalez-Fernandez A (1974) Experimental analysis of nucleolar reorganization. J Cell Biol 60:502–507

Gonzalez-Melendi P, Wells B, Beven AF, Shaw PJ (2001) Single ribosomal transcription units are linear, compacted christmas trees in plant nucleoli. Plant J 27:223–233

Grummt I, Pikaard CS (2003) Epigenetic silencing of RNA polymerase I transcription. Nat Rev Mol Cell Biol 4:641–649

Hadjiolov AA (1985) The nucleolus and ribosome biogenesis, vol 12, Cell biology monographs. Springer, Wien

Harris H (1967) The reactivation of the red cell nucleus. J Cell Sci 2:23–32

Heitz E (1931) Die Ursache der gesetzmäßigen Zahl, Lage, Form und Größe pflanzlicher Nukleolen. Planta 12:775–844

Hernandez-Verdun D (2006) Nucleolus: from structure to dynamics. Histochem Cell Biol 125:127–137

Hernandez-Verdun D, Gautier T (1994) The chromosome periphery during mitosis. Bioessays 16:179–185

Highett MI, Beven AF, Shaw PJ (1993a) Localization of 5S genes and transcripts in *Pisum sativum* nuclei. J Cell Sci 105:1151–1158

Highett MI, Rawlins DJ, Shaw PJ (1993b) Different patterns of rDNA distribution in *Pisum sativum* nucleoli correlate with different levels of nucleolar activity. J Cell Sci 104:843–852

Hiscox JA, Whitehouse A, Matthews DA (2010) Nucleolar proteomics and viral infection. Proteomics 10:4077–4086

Hozak P, Hassan AB, Jackson DA, Cook PR (1993) Visualization of replication factories attached to a nucleoskeleton. Cell 73:361–373

Hozak P, Cook PR, Schofer C, Mosgoller W, Wachtler F (1994) Site of transcription of ribosomal RNA and intranucleolar structure in HeLa cells. J Cell Sci 107:639–648

Huang S, Rothblum LI, Chen D (2006) Ribosomal chromatin organization. Biochem Cell Biol 84:444–449

Iborra FJ, Jackson DA, Cook PR (2001) Coupled transcription and translation within nuclei of mammalian cells. Science 293:1139–1142

Iborra FJ, Jackson DA, Cook PR (2004) The case for nuclear translation. J Cell Sci 117:5713–5720

Ideue T, Azad AK, Yoshida J, Matsusaka T, Yanagida M, Ohshima Y, Tani T (2004) The nucleolus is involved in mRNA export from the nucleus in fission yeast. J Cell Sci 117:2887–2895

Jacobson MR, Pederson T (1998) Localization of signal recognition particle RNA in the nucleolus of mammalian cells. Proc Natl Acad Sci USA 95:7981–7986

Jarrous N, Wolenski JS, Wesolowski D, Lee C, Altman S (1999) Localization in the nucleolus and coiled bodies of protein subunits of the ribonucleoprotein ribonuclease P. J Cell Biol 146:559–572

Kadowaki T, Schneiter R, Hitomi M, Tartakoff AM (1995) Mutations in nucleolar proteins lead to nucleolar accumulation of polyA + RNA in *Saccharomyces cerevisiae*. Mol Biol Cell 6:1103–1110

Kim SH, Koroleva OA, Lewandowska D, Pendle AF, Clark GP, Simpson CG, Shaw PJ, Brown JW (2009) Aberrant mRNA transcripts and the nonsense-mediated decay proteins upf2 and upf3 are enriched in the *Arabidopsis* nucleolus. Plant Cell 21:2045–2057

Kim SH, Spensley M, Choi SK, Calixto CP, Pendle AF, Koroleva O, Shaw PJ, Brown JW (2010) Plant u13 orthologues and orphan snoRNAs identified by RNomics of RNA from *Arabidopsis* nucleoli. Nucleic Acids Res 38:3054–3067

Kiss T (2002) Small nucleolar RNAs: an abundant group of noncoding RNAs with diverse cellular functions. Cell 109:145–148

Koberna K, Malinsky J, Pliss A, Masata M, Vecerova J, Fialova M, Bednar J, Raska I (2002) Ribosomal genes in focus: new transcripts label the dense fibrillar components and form clusters indicative of "Christmas trees" in situ. J Cell Biol 157:743–748

Kopp K, Huang S (2005) Perinucleolar compartment and transformation. J Cell Biochem 95:217–225

Koroleva OA, Brown JW, Shaw PJ (2009a) Localization of eif4a-III in the nucleolus and splicing speckles is an indicator of plant stress. Plant Signal Behav 4:1148–1151

Koroleva OA, Calder G, Pendle AF, Kim SH, Lewandowska D, Simpson CG, Jones IM, Brown JW, Shaw PJ (2009b) Dynamic behavior of *Arabidopsis* eif4a-III, putative core protein of exon junction complex: fast relocation to nucleolus and splicing speckles under hypoxia. Plant Cell 21:1592–1606

Kos M, Tollervey D (2010) Yeast pre-rRNA processing and modification occur cotranscriptionally. Mol Cell 37:809–820

Lamond AI, Sleeman JE (2003) Nuclear substructure and dynamics. Curr Biol 13:R825–R828

Leitch IJ, Leitch AR (2013) Genome size diversity and evolution in land plants. In: Leitch IJ, Greilhuber J, Doležel J, Wendel JF (eds) Plant genome diversity, vol 2, Physical structure, behaviour and evolution of plant genomes. Springer-Verlag, Wien, pp 307–322

Leitch AR, Mosgoller W, Shi M, Heslop-Harrison JS (1992) Different patterns of rDNA organization at interphase in nuclei of wheat and rye. J Cell Sci 101:751–757

Lejeune F, Ranganathan AC, Maquat LE (2004) Eif4g is required for the pioneer round of translation in mammalian cells. Nat Struct Mol Biol 11:992–1000

Leung AK, Lamond AI (2003) The dynamics of the nucleolus. Crit Rev Eukaryot Gene Expr 13:39–54

Li CF, Pontes O, El-Shami M, Henderson IR, Bernatavichute YV, Chan SW, Lagrange T, Pikaard CS, Jacobsen SE (2006) An argonaute4-containing nuclear processing center colocalized with Cajal bodies in *Arabidopsis thaliana*. Cell 126:93–106

Lund E, Dahlberg JE (1998) Proofreading and aminoacylation of tRNAs before export from the nucleus. Science 282:2082–2085

Mais C, Wright JE, Prieto JL, Raggett SL, McStay B (2005) Ubf-binding site arrays form pseudo-NORs and sequester the RNA polymerase I transcription machinery. Genes Dev 19:50–64

Mayer C, Neubert M, Grummt I (2008) The structure of NORC-associated RNA is crucial for targeting the chromatin remodelling complex NORC to the nucleolus. EMBO Rep 9:774–780

McClintock B (1934) The relation of a particular chromosomal element to the development of the nucleoli in *Zea mays*. Z Zellforsch Mikrosk Anat 21:294–328

McKeown PC, Shaw PJ (2009) Chromatin: linking structure and function in the nucleolus. Chromosoma 118:11–23

Melcak I, Risueno MC, Raska I (1996) Ultrastructural nonisotopic mapping of nucleolar transcription sites in onion protoplasts. J Struct Biol 116:253–263

Melese T, Xue Z (1995) The nucleolus—an organelle formed by the act of building a ribosome. Curr Opin Cell Biol 7:319–324

Miller OLJ, Beatty RR (1969) Visualization of nucleolar genes. Science 164:955–957

Misteli T (2001) Protein dynamics: implications for nuclear architecture and gene expression. Science 291:843–847

Mizuno H, Sasaki T, Matsumoto T (2008) Characterization of internal structure of the nucleolar organizing region in rice (*Oryza sativa* L.). Cytogenet Genome Res 121:282–285

Nathanson L, Xia T, Deutscher MP (2003) Nuclear protein synthesis: a re-evaluation. RNA 9:9–13

Nissan TA, Galani K, Maco B, Tollervey D, Aebi U, Hurt E (2004) A pre-ribosome with a tadpole-like structure functions in ATP-dependent maturation of 60S subunits. Mol Cell 15:295–301

Oakes M, Nogi Y, Clark MW, Nomura M (1993) Structural alterations of the nucleolus in mutants of *Saccharomyces cerevisiae* defective in RNA polymerase-I. Mol Cell Biol 13:2441–2455

Olson MO, Dundr M (2005) The moving parts of the nucleolus. Histochem Cell Biol 123:203–216

Olson MO, Hingorani K, Szebeni A (2002) Conventional and nonconventional roles of the nucleolus. Int Rev Cytol 219:199–266

Osheim YN, French SL, Keck KM, Champion EA, Spasov K, Dragon F, Baserga SJ, Beyer AL (2004) Pre-18S ribosomal RNA is structurally compacted into the SSU processome prior to being cleaved from nascent transcripts in *Saccharomyces cerevisiae*. Mol Cell 16:943–954

Parker R, Sheth U (2007) P bodies and the control of mRNA translation and degradation. Mol Cell 25:635–646

Pederson T (1998) The plurifunctional nucleolus. Nucleic Acids Res 26:3871–3876

Pendle AF, Clark GP, Boon R, Lewandowska D, Lam YW, Andersen J, Mann M, Lamond AI, Brown JW, Shaw PJ (2005) Proteomic analysis of the *Arabidopsis* nucleolus suggests novel nucleolar functions. Mol Biol Cell 16:260–269

Phair RD, Misteli T (2000) High mobility of proteins in the mammalian cell nucleus. Nature 404:604–609

Politz JC, Lewandowski LB, Pederson T (2002) Signal recognition particle RNA localization within the nucleolus differs from the classical sites of ribosome synthesis. J Cell Biol 159:411–418

Pontes O, Li CF, Nunes PC, Haag J, Ream T, Vitins A, Jacobsen SE, Pikaard CS (2006) The *Arabidopsis* chromatin-modifying nuclear siRNA pathway involves a nucleolar RNA processing center. Cell 126:79–92

Preuss SB, Costa-Nunes P, Tucker S, Pontes O, Lawrence RJ, Mosher R, Kasschau KD, Carrington JC, Baulcombe DC, Viegas W, Pikaard CS (2008) Multimegabase silencing in nucleolar dominance involves siRNA-directed DNA methylation and specific methylcytosine-binding proteins. Mol Cell 32:673–684

Prieto JL, McStay B (2008) Pseudo-NORs: a novel model for studying nucleoli. Biochim Biophys Acta 1783:2116–2123

Raska I, Shaw PJ, Cmarko D (2006a) New insights into nucleolar architecture and activity. Int Rev Cytol 255:177–235

Raska I, Shaw PJ, Cmarko D (2006b) Structure and function of the nucleolus in the spotlight. Curr Opin Cell Biol 18:325–334

Rubbi CP, Milner J (2003) Disruption of the nucleolus mediates stabilization of p53 in response to DNA damage and other stresses. EMBO J 22:6068–6077

Rudra D, Warner JR (2004) What better measure than ribosome synthesis? Genes Dev 18:2431–2436

Santoro R, Schmitz KM, Sandoval J, Grummt I (2010) Intergenic transcripts originating from a subclass of ribosomal DNA repeats silence ribosomal RNA genes in trans. EMBO Rep 11:52–58

Scheer U, Rose KM (1984) Localization of RNA polymerase-I in interphase cells and mitotic chromosomes by light and electron-microscopic immunocytochemstry. Proc Natl Acad Sci USA 81:1431–1435

Scheer U, Weisenberger D (1994) The nucleolus. Curr Opin Cell Biol 6:354–359

Scherl A, Coute Y, Deon C, Calle A, Kindbeiter K, Sanchez JC, Greco A, Hochstrasser D, Diaz JJ (2002) Functional proteomic analysis of human nucleolus. Mol Biol Cell 13:4100–4109

Schneiter R, Kadowaki T, Tartakoff AM (1995) mRNA transport in yeast: time to reinvestigate the functions of the nucleolus. Mol Biol Cell 6:357–370

Shaw PJ, Jordan EG (1995) The nucleolus. Annu Rev Cell Dev Biol 11:93–121

Shaw PJ, Highett MI, Beven AF, Jordan EG (1995) The nucleolar architecture of polymerase I transcription and processing. EMBO J 14:2896–2906

Silva M, Pereira HS, Bento M, Santos AP, Shaw P, Delgado M, Neves N, Viegas W (2008) Interplay of ribosomal DNA loci in nucleolar dominance: dominant NORs are up-regulated by chromatin dynamics in the wheat-rye system. PLoS One 3:e3824

Steiner-Mosonyi M, Mangroo D (2004) The nuclear tRNA aminoacylation-dependent pathway may be the principal route used to export tRNA from the nucleus in *Saccharomyces cerevisiae*. Biochem J 378:809–816

Thompson WF, Beven AF, Wells B, Shaw PJ (1997) Sites of rDNA transcription are widely dispersed through the nucleolus in *Pisum sativum* and can comprise single genes. Plant J 12:571–581

Thompson M, Haeusler RA, Good PD, Engelke DR (2003) Nucleolar clustering of dispersed tRNA genes. Science 302:1399–1401

Tucker S, Vitins A, Pikaard CS (2010) Nucleolar dominance and ribosomal RNA gene silencing. Curr Opin Cell Biol 22:351–356

Valentin G (1839) Repertorium für Anatomie und Physiologie, vol 4. Huber und Co., Bern/St. Gallen

Wansink DG, Schul W, van der Kraan I, van Steensel B, van Driel R, de Jong L (1993) Fluorescent labeling of nascent RNA reveals transcription by RNA polymerase II in domains scattered throughout the nucleus. J Cell Biol 122:283–293

Warner JR (1999) The economics of ribosome biosynthesis in yeast. Trends Biochem Sci 24:437–440

Wong JM, Kusdra L, Collins K (2002) Subnuclear shuttling of human telomerase induced by transformation and DNA damage. Nat Cell Biol 4:731–736

Xu J, Chua NH (2009) *Arabidopsis* decapping 5 is required for mRNA decapping, P-body formation, and translational repression during postembryonic development. Plant Cell 21:3270–3279

6 Cell Cycle Modules in Plants for Entry into Proliferation and for Mitosis

Zoltán Magyar, Masaki Ito, Pavla Binarová, Binish Mohamed, and Laszlo Bogre

Contents

L. Bogre (✉)
Centre for Systems and Synthetic Biology, Royal Holloway, University of London, Egham Hill, Egham TW20 0EX, UK
e-mail: l.bogre@rhul.ac.uk

6.1 Choreography of the Cell Division Cycle

The cell cycle is an orderly progression through a series of events culminating in the duplication of chromosomes during S-phase and in the segregation of chromosomes into daughter cells during mitosis. The S-and M-phases are the so called active phases, and each is preceded by a preparatory gap phase, (known as G1 and G2 respectively) when regulatory inputs are pereceived; (Fig. 6.1). Chromosomes during G1-, S and G2-phases (interphase) are decondensed but these intephase chromosomes are still known to occupy distinct territories within the nucleus and this organization is important for replication, transcription, repair and recombination processes (Schubert and Shaw 2011). Mitosis is a cell cycle period packed with morphological events subdivided into pro-, meta-, ana- and telophase followed by cytokinesis. Within this chapter we will focus on the central regulators of the cell cycle phase transitions, but before we get onto the regulators in detail we first summarise what is being regulated during these cell cycle phases with an aim to pinpoint conserved and plant-specific mechanisms.

DNA synthesis is largely conserved in eukaryotes. It involves (i) assembly of the prereplicative complex at multiple replication origins during G1 (ii) origin activation by firing of these prereplicative complexes and (iii) formation of the preinitiation complex that leads to DNA synthesis. The re-replication of DNA during G2 is blocked by inhibiting the re-assembly of the prereplicative complexes. This is done by the so called licensing factor, which links to a cell cycle stage with high activity of the central regulator, cyclin-dependent kinase (CDK, see later). These prereplicative complexes can therefore only form after mitosis in G1 when CDK activity is low, thus ensuring that the synthesis and segregation of sister chromatids alternate. Both the components of the prereplicative and preinitiation complexes and the regulatory protein kinases (CDK and Cdc7) are conserved in yeasts, animals and plants (Costas et al. 2011a). Genome wide mapping studies have identified plant replication origins and chromatin marks which

I.J. Leitch et al. (eds.), *Plant Genome Diversity Volume 2*,
DOI 10.1007/978-3-7091-1160-4_6,

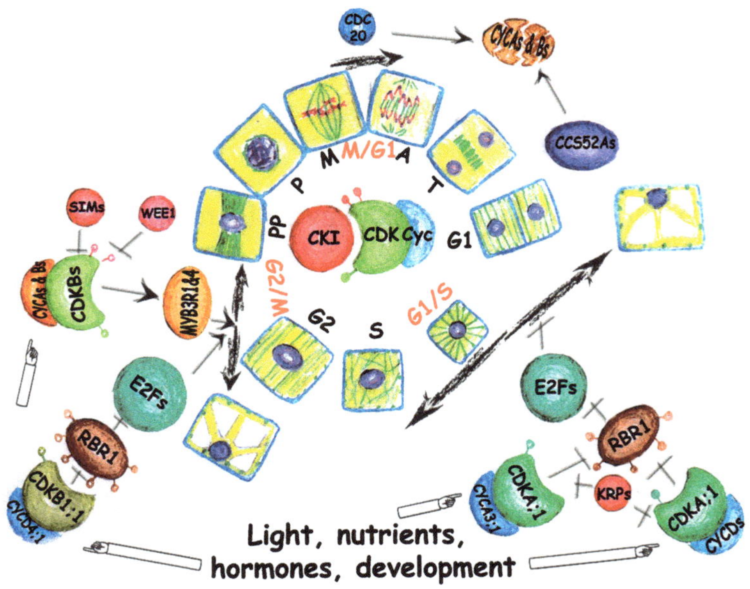

Fig. 6.1 Regulatory components of the three main cell cycle transitions, G1/S, G2/M and M/G1. At the centre of the conserved cell cycle regulation is cyclin-dependent kinase (CDK) in complex with the phase-specific cyclins, and opposed by CDK inhibitors (CKI) and further regulated by positive (P-161) and inhibitory (P-T14; P-Tyr15) phosphorylations. At the G1/S transition, D-type cyclins (CYDs) preferentially interact with CDKA;1, and these complexes win against the opposing CDK inhibitors (KRPs). The main target of CYCD-CDKA;1 complex is RBR1, which is inactivated through CDKA;1 phosphorylation leading to the release of E2F transcription factors (primarily E2FB) to activate genes for G1/S transition. CYCA3;1 cyclin has also been shown to interact with CDKA;1, phosphorylate RBR1 and be involved in the G1/S transition. G2/M is preferentially regulated by B-type CDKs (*CDKB*s) in complex with A- and B-type cyclins (CYCA and CYCB). These mitotic CDKs are opposed by the plant-specific CDK inhibitor, SIM. The role of inhibitory phosphorylation on CDKB is not well understood. Interestingly, CYCD4;1 can associate with mitotic CDKB1;1 and has been shown to trigger mitosis, possibly through RBR1 phosphorylation. Exit from mitosis at the meta- to anaphase transition is trigerred by the degradation of CYCA and CYCB through the activation of the anaphase promoting complex (APC) by CDC20 during the M/G1 transition and later in G1. CYCA and CYCB levels are kept low by the activation of the APC by CCS52 proteins. In plants, cells can exit from proliferation, enter into a G0 state and differentiate both at G1/S and G2/M transitions. Both these transitions are regulated by external signals such as light, nutrient availability, hormones and developmental cues. *Arrows* indicate activation, *hammers* repression, *pointing fingers* show regulatory inputs to the cell cycle

act as important regulatory elements (Costas et al. 2011b, c). Incomplete DNA synthesis and DNA damage is sensed and signalled to the cell cycle through the DNA damage checkpoint and this halts cell cycle progression to allow time to repair the DNA (Cools and De Veylder 2009; Cross et al. 2011). Some but not all the components of the DNA integrity checkpoint are conserved among eukaryotic organisms.

Newly synthesised sister chromatids are held together by the cohesin ring complex. This is resolved by the release of active separase which cleaves cohesins at the meta- to anaphase transition of mitosis, leading to perfect timing of sister chromatid separation. During meiosis I sister chromatids are held together by meiosis-specific cohesins. Cohesins and associated proteins also help to establish the bi-orientation of kinetochores during mitosis, while the meiosis-specific cohesion at kinetochores establishes the mono-orientation of sister chromatids. The molecular players and mechanisms of sister chromatid separation during mitosis and meiosis are largely conserved in eukaryotes, including plants (Yuan et al. 2011). Following the onset of mitosis, chromatin gradually becomes condensed from prophase to metaphase driven by histone modifications, primarily histone phosphorylation and deacetylation as well as molecular motors called condensins (Costas et al. 2011b).

The segregation of chromosomes into daughter cells is driven by the mitotic spindle. In higher plants microtubules are nucleated at dispersed sites on existing microtubules, membranes or chromatin (see later). In contrast to animals, there are no microtubule organizing centres (MTOC) such as centrosomes to control the nucleation and organization of spindle microtubules. Instead, chromatin-mediated microtubule nucleation and organization is the major mechanism to build mitotic spindles (Karsenti and Vernos 2001). The mitotic checkpoint monitors the full alignment of chromosomes at the metaphase plate through the tension exerted by the kinetochore microtubules on the sister chromatids. This signals to the anaphase promoting complex (APC) and triggers the dissolution of cohesion molecules through the proteolysis of securin, an inhibitor of separase. Dissolution of cohesion between sister chromatids allows the chromosomes to separate by the pulling force of the anaphase spindle (Morgan 2007). The regulatory mechanisms for the mitotic checkpoint are also conserved in eukaryotic organisms (Caillaud et al. 2009).

Plant cells are encased in a cell wall, and therefore building the septum inbetween two daughter cells during cytokinesis is an inside out process through the expansion of a disk like structure, called the phragmoplast. Microtubules provide the central skeleton for this structure as well as the highway for the delivery of vesicles to the cell plate (Lloyd 2011). The timing when phragmoplast forms and its spatial organization is controlled by cytokinetic MAPK signalling pathways (Bogre 2011).

6.2 The Core Cell Cycle Regulators, Can One Do the Job for All?

The cell cycle is an orderly progression through a series of events leading to the duplication of chromosomes during S-phase and in the segregation of chromosomes into daughter cells during mitosis. There are three main control points on this road; (1) the entry into S-phase, (2) the entry into mitosis and (3) the exit from mitosis back into G1 phase (Fig. 6.1). In simple eukaryotes, such as yeasts, there is a single controller; the cyclin-dependent kinase (CDK) that governs all three transitions. A central question is; how a single regulator can perform such diverse tasks. In fact in more complex animals and higher plants there has been an expansion of the CDK and cyclin families. For example, in *Arabidopsis*, there are 12 members of the CDK gene family falling into 8 groups, CDKA-F (Vandepoele et al. 2002). This grouping appears conserved as it has also been found in other plants, such as rice (Guo et al. 2007), lower plants including the lycophyte *Selaginella* (Banks et al. 2011) and in algae (Bisova et al. 2005). Not all of these CDKs are directly involved in cell cycle regulation; the two main groups of CDKs with cell cycle functions are CDKA, of which there is only a single gene in all flowering plants analysed, and CDKBs with several members (De Veylder et al. 2007). CDKA can functionally complement the yeast *cdc2* mutant, and is thought to control multiple cell cycle transitions throughout the cell cycle (Hirt et al. 1991). However, evidence is accumulating that in plants CDKA is predominantly specialized towards a singly target, RBR1 for transcriptional control of cell cycle transitions (Van Leene et al. 2011). The plant-specific CDKB1 and CDKB2 groups are expressed from late S-phase through G2 to M-phases and have functions during these phases. To what degree plant CDKA and CDKBs are specialized or carry out redundant functions is an interesting but not fully resolved question. Pollen mitosis was found to be delayed or absent in heterozygous *cdka;1* mutants, resulting in mature pollen with only a single sperm cell (Iwakawa et al. 2006; Nowack et al. 2006). However, when plants were rescued through this stage, the *cdka;1* mutant was surprisingly viable, indicating that CDKA;1 is not absolutely required for cell cycle progression. Nevertheless, with only a single sperm, the double fertilization fails because only the female egg cell is fertilized, and not the central cell. The result is seed abortion due to the lack of endosperm (Nowack et al. 2007; Aw et al. 2010). CDKB is a larger gene family (Table 6.1) than A, and therefore it is not unexpected that a *cdkb2* mutant in *Arabidopsis* is not fully compromised in cell division, although tissue organization and stem cell maintenance in meristems are affected (Andersen et al. 2008). The level of CDKA activity is also an important determinant for meristem organization in *Arabidopsis* (Gaamouche et al. 2010) and reprogramming of cell differentiation in the moss *Physcomitrella* (Ishikawa et al. 2011).

It is assumed that cell cycle phase specificity is achieved by the interchange of the cyclin regulatory subunits of CDKs, and correspondingly, in all organisms there are multiple cyclins with cell cycle phase-specific expression patterns and roles (Hochegger et al. 2008). In higher plants, there has been a large expansion of the cyclin family with around 30 members in *Arabidopsis* (Vandepoele et al. 2002; Menges et al. 2007). These can largely be matched to the animal A-, B-, and D-type cyclins. In agreement with the phase-specific role of cyclin subunits, the expression of cyclins is also tightly regulated; either linked to cell cycle phases, or coupled to external signals, hormones and developmental programs (Menges et al. 2003). The sequence similarity-based classification of cyclins can be matched across the higher plant species *Arabidopsis*, rice and maize (Vandepoele et al. 2002; Guo et al. 2007; Hu et al. 2010; Buendia-Monreal et al. 2011), as well as to lower plants and algae (Bisova et al. 2005; Banks et al. 2011). The transcriptional regulation of cell cycle genes is also similar across species such as *Arabidopsis* and rice (Menges et al. 2003; Guo et al. 2007). How these core cell cycle genes have evolved in the plant lineage is becoming apparent by comparative analysis of genome sequences from algae, moss and *Selaginella* (Bisova et al. 2005; Huysman et al. 2010; Banks et al. 2011). Moreover, having the genome sequence of eukaryotic organisms such as *Naegleria*, a free living amoeba, that is considered to be close to the base of the tree separating plants and fungi-animal kingdoms, allows the evolution of the conserved regulatory mechanism of the cell cycle to be unravelled (Cross et al. 2011). This approach holds great promise, suggesting that knowledge gained from studying model organisms such as yeasts, human cells or *Arabidopsis* can be used to infer cell cycle regulation in other species or crops.

Is such complexity in CDK and cyclin numbers an essential feature of the core cell cycle machinery for regulating cell cycle transitions? This was recently tested in fission yeast, where the cell cycle control system is somewhat simpler with a single CDK, CDC2, and a couple of G1 and G2 cyclins. The Nurse group first replaced CDC2, and the

Table 6.1 List of core cell cycle genes in representative sequenced plant genomes

Core cell cycle genes	Description	*Chlamydomonas*	*Physcomitrella*	*Selaginella*	*Oryza sativa*	*Arabidopsis*
CDKA CDKA;1	Canonical cyclin dependent kinase (CDK) in plants complementing the yeast *Saccharomyces pombe cdc2* mutant. Primarily interacts with D-type cyclins and KRPs and therefore one of the main functions is to inactivate RBR	1	2	1	3	1
CDKB CDKB1 CDKB2	Plant-specific CDK functions in G2- and M-phases, the two major subclasses in flowering plants are expressed during S/G2 (B1) or G2/M (B2)	1	7	2	2	4
CYCD CYCD1-7	Closely related to animal D-type cyclins, can complement yeast G1 cyclins. There are seven subclasses in flowering plants. Their expression can be cell cycle phase specific (both G1/S and G2/M) or coupled to external stimuli, such as phytohormones, sucrose, or to developmental cues	3	2	3	12	10
CYCA CYCA1 CYCA2 CYCA3	Related to animal A-type cyclins, there are three main subclasses in flowering plants; CYCA3s are upregulated at the G1/S transition and can interact with CDKA;1. CYCA1 and CYCA2 members are expressed from G2- to M-phase, associate with CDKBs, and negatively regulate endocycle	1	8	3	7	10
CYCB CYCB1 CYCB2 CYCB3	Related to animal B-type cyclins, divided into three subclasses: B1, B2 and B3; transcriptionally controlled by M-specific activator promoter elements (called MSA) and are destroyed by APC ubiquitin ligase	1	2	1	5	11
KRP KRP1-7	KIP Related Proteins (KRP) are functionally related to yeast and animal CDK inhibitors, such as the $p27^{KIP}$. Interact with CDKA;1 and D-type cyclins, KRP1/2 are phosphorylated by CDKA;1 and CDKB1;1 preceding their destruction	–	1	3	6	7
SIM/ **SMR**	Plant-specific CDK inhibitors, comprising SIAMESE (SIM) and SIAMESE-related (SMR) proteins. Appear to specifically interact with and oppose the activities of G2- and M-phase specific CDKB/CYCA or CDKB/CYCB complexes. Activators of endocycle	?	?	?	?	13
WEE1	Putatively involved in the inhibitory phosphorylation of Tyr residues in CDKs, though in CDKA;1 this phosphorylation appears to be indispensible. Controls cell cycle arrest in response to activation of the DNA integrity checkpoint	1	1	2	1	1
CDC25	The sole catalytic domain of CDC25 phosphatase can be identified in plants, but the N-terminal regulatory domain is missing. It is debated but has not yet been fully ruled out whether these CDC25 genes indeed function to activate CDKs through dephosphorylation of inhibitory Tyr residues	3	1	1	2	1

(continued)

Table 6.1 (continued)

Core cell cycle genes	Description	*Chlamydomonas*	*Physcomitrella*	*Selaginella*	*Oryza sativa*	*Arabidopsis*
CDC20	The anaphase promoting complex (APC) is responsible for the destruction of mitotic substrates, it is activated at the metaphase/anaphase transition by CDC20	1	5	2	3	6
CCS52 CCS52A1 CCS52A2 CCS52B	The CDH1 homologs are known as CCS52 genes in plants, they are required to establish the G1-phase state of low CDK levels by activating the APC in telo- and G1-phases. They positively regulate the endocycle	1	4	2	2	3
CKS	CDK-interacting protein related to yeast SUC1; responsible for substrate binding. *Arabidopsis* CKS is known to interact both with CDKA and CDKB	1	1	2	1	2
RBR	Related to animal retinoblastoma (RB), a tumour suppressor which functions as a transcriptional repressor for a broad spectrum of genes. RBR is phosphorylated and inactivated by the CDK/CYCD complexes. RBR1 forms a co-repressor complex involving other transcription factors and chromatin modifying enzymes, e.g. histone deacetylases. RBR is not only involved in controlling proliferation, but also in stem cell maintenance, cell differentiation and imprinting; controls the arrest of unfertilized gametophytes	1	3	2	2	1
E2F E2FA E2FB E2FC	Related to animal E2F transcription factors, pivotal targets for RB	1	3	2	4	3
DP DPA DPB	Dimerization partners of E2Fs, required for their DNA binding	1	2	1	3	2
DEL DEL1-3	DP/E2F-like proteins; can bind E2F elements as a monomer owing to their tandem DNA binding domain. They do not bind RBR. In *Arabidopsis* DEL1 inhibits the endocycle by blocking the expression of CCS52A genes	1	3	1	2	3
R1R2R3-MYB	The R1R2R3-MYB type transcription factors (TF) are conserved regulators of the cell cycle both in animal and plant cells. While in animals these TFs are known for their role in G1/S, in plants they have been shown to directly bind to the mitosis-specific MSA element, and activate or repress a suit of G2/M-specific genes	1	2	1	5	5

The name of core cell cycle genes are given in bold followed by the nomenclature of *Arabidopsis* subgroups. The number of homologous genes in representative sequenced genomes is indicated. For more information and references see the main body of text as well as Van Leene et al. (2011). *Chlamydomonas* genes are based on Bisova et al. (2005), *Physcomitrella, Selaginella, Oryza sativa, Arabidopsis* genes are taken from Banks et al. (2011). R1R2R3-MYBs are defined in Haga et al. (2007) and Ito (2005). Missing orthologues were identified using the PLAZA resource for comparative plant genomics

main mitotic cyclin, CDC13, with a single fusion protein and showed that this chimera molecule could carry out all CDK functions by deleting all the other remaining cyclins in this organism (Coudreuse and Nurse 2010). At the heart of the cell cycle regulatory system is the oscillation of CDK activity. This is mainly driven by the accumulation and degradation of the cyclin subunits. To experimentally demonstrate this, they further engineered the CDC2-CDC13 fusion molecule to become sensitive to inhibition by a "bulky" ATP analogue, and showed that the absolute level of CDK activity was the only essential feature necessary to drive the three key cell cycle transitions: a low level of CDK activity triggers entry into S-phase, a medium level blocks the repeated S-phase in G2 cells, while a burst of activity is required for entry into mitosis. Whether in plants a single CDK could carry out all cell cycle functions or the diverse CDK and cyclin repertoire is indeed required to carry out specialized functions has not yet been fully investigated.

To tune CDK activity so that it becomes switch-like to flip the three main transition points, there are additional layers of regulators incorporated into the central regulatory system (Tyson and Novak 2008). Firstly, CDK activity can be inhibited by a family of proteins that are related to the animal CDK inhibitor protein KIP, named inhibitor of CDK or KIP-related proteins (ICK/KRP). There are seven ICK/KRP in *Arabidopsis* (De Veylder et al. 2007), but apparently these proteins are missing in algae (Bisova et al. 2005; Banks et al. 2011). It could be that only a small sequence stretch is conserved in CDK inhibitors, and that might not be recognised (Cross et al. 2011). In yeast and animal cells these CDK inhibitory proteins mostly function in the G1 to S-phase transition to build a bistable regulatory switch at the transition point called start in yeast and restriction point in animal cells (Tyson and Novak 2008). However, in plants it has been proposed that ICK/KRP proteins might also function to establish the switch-like behaviour of the CDK oscillator during the G2 to M-phase transition (Dissmeyer et al. 2010). This idea mostly came in to explain the void of CDC25, the identity of which remains ambiguous in plants (Dissmeyer et al. 2009). CDC25 is a phosphatase that activates CDK by dephosphorylating conserved Thr14; Tyr15 sites, its activity opposes WEE1 kinase, which inhibits CDK activity through phosphorylation of the same sites. Though these sites are conserved in plant CDKA proteins, and WEE1 kinase is also present, the role of this phosphorylation is unclear during normal cell cycle progression, as the *wee1* mutant is indistinguishable from wild type, and replacing wild type CDKA with a mutant form mimicking the dephosphorylated state of CDKA has no effect either on plant growth or development (Dissmeyer et al. 2009). The authors therefore concluded that cell cycle control is efficiently carried out without the reversible CDK phosphorylation in plants. To compensate for the missing negative inhibitory loop built around CDK phosphorylation by WEE1/CDC25 in animal and yeast cells, it has therefore been suggested that the CDK inhibitors perform a similar role to build the G2 to M-phase switch (Dissmeyer et al. 2010).

Based on mathematical modelling of the cell cycle regulatory system it was proposed that these additional layers of negative regulators are fundamental to render unidirectional progression through all three main cell cycle control points (Tyson and Novak 2008). This idea was also tested in fission yeast by deleting the additional layers of CDK regulation. Surprisingly, they could delete CDC25 and WEE1 without a major effect on the yeast cell cycle progression (Coudreuse and Nurse 2010). It is therefore not too surprising that in plants the CDK phosphorylation is also dispensable for normal cell cycle progression. Furthermore, they have also deleted the yeast CDK inhibitor, RUM1 and shown that it is dispensable for normal cell cycle progression. These results demonstrate that a single monomolecular CDK molecule without additional layers of regulators can sustain an effective mitotic cycle and thus constitutes the minimal architecture of the eukaryotic cell cycle (Coudreuse and Nurse 2010). The expansion of CDK and cyclin families in animals and plants could provide a more elaborate regulatory system to meet more complex demands in multicellular organisms (Fig. 6.1), but it appears that CDK subunits are also redundant in metazoans (Hochegger et al. 2008).

How can we dissect the large complexity of cell cycle regulatory systems in plants? There are 99 putative core cell cycle genes in *Arabidopsis*, with CDKs and cyclins being the largest families of proteins. Though this number is somewhat lower in rice, 41, the number is still daunting and makes the understanding of cell cycle regulation in higher plants a complex task. To systematically map the regulatory interactions among these core cell cycle regulators in *Arabidopsis*, three independent methods have been used in two studies, the yeast two hybrid (Y2H) assay, the bimolecular fluorescence complementation (BiFC) assay (Boruc et al. 2010), and affinity purification coupled to mass spectrometry-based identification of proteins in these complexes (AP–MS). Amalgamation of experimental data on the binary links among the 58 core cell cycle proteins generated by these three independent methods has uncovered 416 interactions (Boruc et al. 2010; Van Leene et al. 2010; Van Leene et al. 2011). This dataset has been further extended by literature-curated interactions and by those computationally predicted based on the integration of multiple sources of data such as sequence similarities, gene order on chromosomes, and gene expression (Van Leene et al. 2011). Collectively, these works have established a protein-protein interaction (PPI) network of the plant cell cycle consisting of 506 binary interactions among 58 proteins. Besides being a valuable source of information to help interpret existing experimental data and formulate new hypotheses of the cell cycle PPI network, it has highlighted the complex combinatorial interactions among CDKs, cyclins and ICK/KRP proteins (Fig. 6.1). We can also make some firm conclusions from the topology of the network. CDKA is shown to make the most connections within the network, emphasizing its central regulatory role, although surprisingly this is almost exclusively restricted to

D-type cyclins and ICK/KRP proteins. The only exception to this rule is the CDKA interaction with CYCA3-type cyclins. D-type cyclins in association with CDKA have a predominant role in regulating entry into cell proliferation by phosphorylation of the retinoblastoma related (RBR) protein. This protein controls the expression of a large battery of cell cycle genes through the E2F transcription factors. Interestingly, CYCA3-type cyclins share some of these properties with D-type cyclins, such as the peak in expression at the G1 to S-phase boundary, and in association with CDKA they promote the phosphorylation of RBR (Takahashi et al. 2010). Thus in plants CDKA seems to have become specialized for regulating the entry into cell proliferation in complex with CYCDs, CYCA3 and ICK/KRPs.

While in animals novel CDKs have evolved to fulfil G1 - specific roles, plant lineage-specific CDKs, the B-type CDKs, are functioning in the G2 to M-phases (De Veylder et al. 2007). Consistent with a distinct role, the network around B-type CDKs are clearly separated from those of CDKA. B-type CDKs are associated with the mitotic cyclins, CYCA1s, CYCA2s and CYCBs (Van Leene et al. 2010). The only exception to this rule is the interaction of CDKB1;1 with CYCD4;1 which, when co-expressed, can lead to ectopic cell division in epidermal cells (Kono et al. 2007) and initiation of lateral root formation (Nieuwland et al. 2009). Since cyclins within the CDK complex have important roles in determining substrate specificities, CYCD4 might have retained broad substrate specificities towards both the G1 to S and G2 to M targets. The SUC1 related CDK adaptor molecules, CKS1 and CKS2 also connect to both CDKA;1 and CDKBs, further indicating some possible share of substrates for these two modules.

If all KRPs are dedicated to control the CDKA;1/CYCD complexes, which inhibitors regulate the mitotic CDKB complexes? A prime candidate is the plant-specific CDK inhibitor, SIAMESE (SIM), which was shown to antagonize with CYCB and cooperate with the CYCB degradation factor, CCS52A1 (Kasili et al. 2010). A key decision to make within the G2-phase of the cell cycle is whether to commit to mitosis or engage in cycles of repeated DNA sysnthesis, called endoreduplication or endocycle (see Maluszynska et al. 2013; this volume). In yeast it was shown that blocking of repeated DNA synthesis relies on the presence of CDK activity in G2. In plants the onset of endocycle also involves the selective inactivation of M-phase-promoting factors, such as the B1-type CDK (CDKB1;1), through proteolytic destruction of its cyclin partner, CYCA2;3 (Boudolf et al. 2009). In agreement, *Arabidopsis* relatives of the animal fizzy-related activators of the anaphase promoting complex (APC), CCS52A1 and CCS52A2, also promote the switch from mitosis to endocycle (Larson-Rabin et al. 2009; Vanstraelen et al. 2009). The anaphase-promoting complex/cyclosome (APC/C) triggers the degradation of mitotic cyclins leading to the exit from mitosis (Fulop et al. 2005).

Research into the regulation of the plant cell cycle is largely motivated by its role in determining plant organ size and potentially crop yield (Bogre et al. 2008). This principally relies on the number of cells produced during meristematic cell proliferation, and cell expansion after cells have exited proliferation, a decision point that is tightly controlled (Andriankaja et al. 2012). Can sequence variation in cell cycle genes account for size differences among plants? A large scale population genetics study of polymorphisms in 61 core cell cycle genes across 30 natural *Arabidopsis* accessions revealed signs of purifying selection in central regulators, CDK subunits, WEE1 and RBR, large effect mutations in CDKB1;1, adaptive protein evolution in KRP6 and departure from equilibrium for CYCA3;3. These data suggest that the robustness of the cell cycle regulatory network is more due to functional redundancy than high selective constraints (Sterken et al. 2009).

6.3 The Entry Module into the Cell Cycle

In plants growth is restricted to meristems, and thus a key factor is the duration of cell proliferation and the timing of the exit from proliferation to cell expansion and differentiation (Doonan and Sablowski 2010). The current view is that these events are controlled by an evolutionary-conserved transcriptional regulatory switch, the E2F-RB pathway (Inze and De Veylder 2006; Magyar 2008). The *Retinoblastoma* (RB) was the first tumour suppressor gene cloned from mammalian cells, while the first adenovirus E2 binding transcription factor (E2F) was identified based on its ability to form a complex with the RB protein (van den Heuvel and Dyson 2008). Originally E2F function was linked to cell cycle control, but further studies revealed that E2F functions extend beyond the control of cell proliferation. On the basis of most recent animal studies E2F-RB may provide a broadly utilised transcriptional switch for genes and thereby co-ordinate the temporal expression of genes involved in the regulation of cell cycle and differentiation (Korenjak and Brehm 2005).

In principle, E2F works as a heterodimeric transcription factor, forming a complex with a dimerization partner (DP) protein prior to binding to the DNA and activating the expression of target genes required for entry into the cell cycle. In cells leaving mitosis, RB binds to E2F at the carboxyl terminal RB binding motif and inhibits their activities. This is converted, upon mitogen stimulation, by hyperphosphorylation of RB by specific CDK-Cyclin D complexes, leading to the activation of genes required for DNA synthesis (Korenjak and Brehm 2005). In addition to this repression function of RB through direct binding and inhibition of E2Fs, RB can also function together with E2Fs and actively repress transcription. This is because RB, and its pocket protein relatives in mammalian cells p107 and p130, are able to simultaneously bind to E2Fs and chromatin remodelling enzymes such as histone deacetylases (HDACs) (Rayman et al. 2002); or histone methyl transferases (e.g. SUV39H1) (Liu et al. 2005). Animal E2Fs have been classified as activators and repressors based on whether they transactivate genes which are repressed by RB or form a co-repressor complex with RB, respectively (van den Heuvel and Dyson 2008). However, recent *in vivo* studies have clearly demonstrated that activator E2Fs can also work as repressors and, vice versa, repressor E2Fs can also function as activators in certain developmental contexts and in a tissue-specific manner (Danielian et al. 2008; Infante et al. 2008; Tsai et al. 2008; Chen et al. 2009; Chong et al. 2009; Wenzel et al. 2011).

Initially E2F function was linked to the control of DNA synthesis through regulating the expression of critical cell cycle genes required for S-phase progression. However, in *Drosophila* it was shown that dE2F and RB (called RBF1-2 in the fly) proteins can control replication through a mechanism which is distinct from their effects on gene expression (Cayirlioglu et al. 2001, 2003). Although the exact mechanism is still unclear, dE2F and RB proteins were found to associate with origin recognition complex (ORC) proteins and they were co-localized at amplifying loci (Royzman et al. 1999; Bosco et al. 2001).

Several studies have revealed that animal E2Fs function not only in the control of the G1/S transition but are also involved in the regulation of the G2/M transition. A number of critical mitotic genes in animal cells have been identified as E2F targets including cyclin B1, the regulatory subunit of CDK1 in mammalian cells (Zhu et al. 2005), CDC25 phosphatase, an activator of the mitotic CDK1 in *Drosophila* (Neufeld et al. 1998), and MAD2, a mitotic checkpoint component gene in mouse (Hernando et al. 2004). Interestingly, the regulation of MAD2 expression by RB/E2F transcriptional regulators was found to play important roles in the control of chromosome stability (Sage and Straight 2010). Therefore it is suggested that E2F-RB may mediate both the accurate and timely replication of DNA and the accurate distribution of chromosomes during mitosis (Bosco 2010).

Moreover, studies in animal cells have demonstrated that the E2F-RB repressor complexes are present in actively dividing cells and regulate both the expression of genes involved in differentiation and the G2- and M-phase transition of the cell cycle (Dimova et al. 2003; van den Heuvel and Dyson 2008). These newly identified E2F-RB complexes are regulated differently from the classical E2F-RB complexes since they are insensitive to the traditional RB-kinases consisting of cyclin D and CDK proteins.

The E2F-RB network is remarkably conserved between the plant and animal kingdoms. E2F and RB homologs have been identified from the unicellular green alga *Chlamydomonas reinhardtii* to the higher plants including the model plant *Arabidopsis thaliana* (Inze and De Veylder 2006). Interestingly, nearly 25 % of the *Arabidopsis* genes in total contain consensus E2F binding element(s) within their promoter region, indicating that E2Fs could play a fundamental role in regulating many aspects of plant development including cell proliferation and differentiation (Vandepoele et al. 2005, 2009). *Arabidopsis* contains a family of six E2F-related proteins (Vandepoele et al. 2002). Three of them, E2FA, E2FB and E2FC, are structurally related to the canonical animal E2Fs since they all have the conserved domains characteristic for the animal E2F1-3, including the DNA-binding, dimerization, transactivation and RB binding domains. The three other E2F-related members in *Arabidopsis* are DEL1, DEL2 and DEL3 (*DP-E2F-like* proteins). These are structurally related to mammalian E2F7 and E2F8 that lack all the E2F-specific domains except the DNA-binding domain, which is present in tandem duplication. *Arabidopsis* DELs represent a subgroup of the E2F family which does not require heterodimer formation binding to the DNA with the two known dimerization partner proteins, DPA and DPB, and their function is not directly regulated by RBR1 protein since they lack the RB-binding motif. In analogy to animal systems, plant E2Fs have also been classified as transcriptional activators (E2FA and E2FB), or transcriptional repressors (E2FC), although most of these data were derived from over-expression studies (Magyar 2008). Genome wide expression profiling has shown that ectopic co-expression of E2FA with the dimerization partner A (DPA) not only leads to transcriptional activation but also to the transcriptional repression of a large number of genes, perhaps indicating that the same E2F/DP complex can be an activator of some genes and a repressor of others (Vandepoele et al. 2005).

Functional characterization of the individual members of *Arabidopsis* E2Fs has already revealed differences among them: ectopic expression of E2FA with DPA resulted in strong activation of both the mitotic cell cycle and endocycle (De Veylder et al. 2002; Kosugi and Ohashi 2003), overexpression of E2FB was also able to activate mitosis but it repressed the endocycle (Magyar et al. 2005; Sozzani et al. 2006), whereas reduction in the level of E2FC confirmed its negative regulatory function in mitosis but a positive one in endoreduplication (del Pozo et al. 2006).

Thus, E2FB and E2FC are antagonistic transcription factors, while E2FA can regulate both proliferation and endocycle. The target gene specificities of E2FB and E2FC are not fully understood, but the regulation of the mitotic CYCB1;1 gene is clearly opposite for E2FB and E2FC (del Pozo et al. 2006; Sozzani et al. 2006). Lowering the amount of E2FC by RNAi markedly upregulated CYCB1;1 expression, much more than the other E2F target genes such as CDC6, EXP3, which have clear E2F binding elements, while CYCB1;1 does not. This suggests that CYCB1;1 might be a direct target for E2FC-dependent repression. Curiously,

E2FC-RNAi only had an effect on CYCB1;1 expression in mature leaves but not in young leaves, suggesting that E2FC forms a repressor complex in differentiated cells. The endocycle was also compromised in the leaves of E2FC-RNAi plants, suggesting that the repression of mitotic genes such as cyclin B1;1, by E2FC is part of the switch from mitotic cell cycle to endocycle. Curiously, E2FC appears to repress mitosis in mature leaves but it is most abundant in proliferating cells (del Pozo et al. 2002, 2006). In accordance with these data from *Arabidopsis*, in *Drosophila* embryos the expression of cyclin B1 and B3 genes are also repressed as cells enter endocycle (Edgar and Orr-Weaver 2001). Trichome development is linked with endocycle, but by targetting the expression of the mitotic cyclin B1;2 to trichomes, the endocycle is converted into mitosis, leading to multicellular trichomes (Schnittger et al. 2002). E2FC might function as a transcriptional repressor on E2F target genes. This is supported also by the over-expression of a stabilized mutant ΔE2FC lacking the amino terminal in *Arabidopsis* which in dark grown plants led to the repression of CDC6, a gene involved in S-phase (del Pozo et al. 2002). In mouse, there are two alternative splicing variants of E2F3, a full length form (E2F3a) and an amino terminally deleted (E2F3b) form. These two variants have opposite functions; E2F3a is an activator while E2F3b is a repressor (Leone et al. 2000; Aslanian et al. 2004). The function of the N-terminal part of E2F3 is unknown but in E2F1 it is required for ubiquitin-mediated degradation (Marti et al. 1999). In plants the N-terminal extension of E2Fs has numerous CDK phosphorylation sites. For E2FC this N-terminal part was shown to be required for ubiquitin-mediated degradation. Deletion of the N-terminal part led to the stabilization of E2FC and E2FA (del Pozo et al. 2002; Magyar et al. 2005). Over-expression of the N-terminally deleted ΔE2FC mutant together with DPB led to growth arrest due to compromised cell proliferation and an elevated endocycle (del Pozo et al. 2006), providing further evidence that E2FC represses mitosis but promotes endocycle. Genome wide gene expression data with plants overexpressing E2FA-DPA or E2FC-DPA revealed minimal overlap. Whether the target specificity is also determined by the dimerisation partner, DPA or DPB remains to be established. While E2FC represses CYCB1;1, over-expression of E2FB, or co-over-expression of E2FA with DPA activates it. Whether this opposing function of E2FC and E2FB is through competition of these two factors for the same site or through alternative E2F-binding sites or whether these E2Fs indirectly regulate CYCB1;1 expression is not known. However, it has been established that E2FB and E2FC amounts are oppositely regulated by light conditions. In the dark E2FC amounts are high compared with E2FB, while in the light it is the opposite, and these depend on light-signalling components, DET1 and COP1 (Lopez-Juez et al. 2008; Berckmans et al. 2011a). E2FC was also shown to be destabilised in light through the ubiquitin-SCF pathway (del Pozo et al. 2002). The atypical E2F, *DEL1* is the target for E2FB and E2FC and these two antagonistically regulate *DEL1* levels through competition for a single E2F *cis*-acting binding site (Berckmans et al. 2011a). DEL1 in turn regulates CCS52As and entry into endocycle in response to light conditions (Lammens et al. 2008). The ratio between E2FC and E2FB could also be influenced by auxin. Auxin concentration has a dose-dependent effect on cell division and cell elongation; high auxin promotes cell proliferation, while low auxin leads to exit from the cell cycle and stimulates cell elongation (Scheres and Xu 2006). Auxin-dependent reactivation of E2FA was shown to be an important factor for the asymmetric cell division during lateral root initiation (Berckmans et al. 2011b). Auxin was also shown to stabilise E2FB, and co-over-expression of E2FB with DPA could sustain proliferation in the lack of auxin (Magyar et al. 2005). Cells in the E2FB-DPA plants were extremely small suggesting that E2FB inhibits growth. Over-expression of E2FB alone in *Arabidopsis* plants also led to overproliferation, small cells and these plants had short roots (Sozzani et al. 2006). A possible model that can be derived from these data is that elevated levels of E2FB maintain proliferation in the meristems until the E2FC level rises above E2FB in mature leaves leading to the switch from mitotic cell cycle to endocycle through the repression of mitotic genes (Fig. 6.2).

All these data underline the importance of plant E2Fs in regulating the G2 to M-phase transitions in a similar way to that found for animal E2Fs (Neufeld et al. 1998; Hernando et al. 2004). Elevated E2FB levels in BY-2 tobacco cell lines resulted in a shortened cell cycle which is analogous to what was found for *Drosophila* dE2F1 that in parallel controls two important regulators for the G1 to S and G2 to M-phase transitions (Magyar et al. 2005). E2FB targets the promoter of CDKB1;1 a critical regulator for the G2 to M transition (Magyar and Bogre, unpublished result).

E2FA could be at the top of the hierarchy of the E2F regulatory network. It does not contain an E2F element in its own promoter but has been shown to activate the expression of both E2FB and E2FC (Vandepoele et al. 2005). Plants over-expressing both E2FA and DPA are severely stunted (De Veylder et al. 2002; Kosugi and Ohashi 2003; Magyar et al. 2012); cells in some tissues show over-proliferation while in others over-endoreduplication. E2FA may also repress cell proliferation in mature leaves in a similar way to that shown for E2FC (He et al. 2004). *E2FA* is most abundant in meristems and specifically peaks in S-phase cells, but can also be detected in endoreduplicating cells. Transcriptomics analysis of an E2FA/DPA line showed a strong up-regulation of genes involved in DNA synthesis (e.g. CDC6, ORC1, CDC45, RNRII, MCM3), but genes with

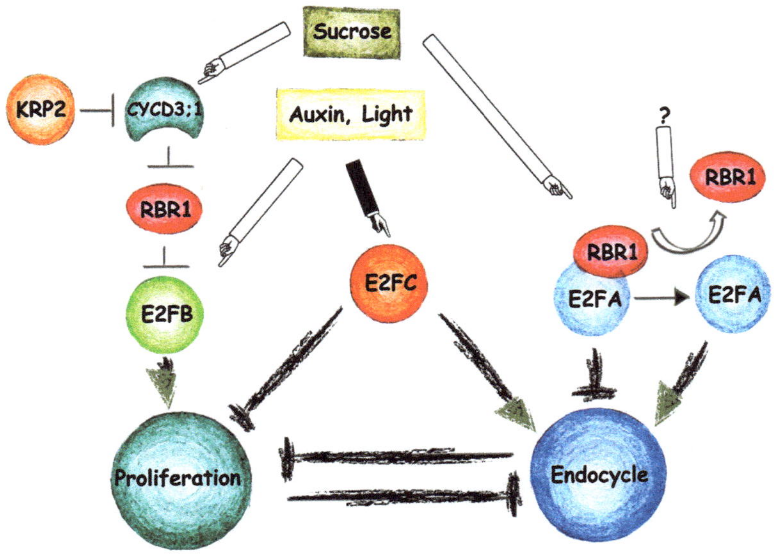

Fig. 6.2 Regulation of cell proliferation versus endocycle and differentiation by the RBR1/E2F transcriptional regulatory switch. RBR1 phosphorylation is regulated by sucrose availability dependent on the CYCD3;1 and KRP2. The released E2FB is driving cells into proliferation. Auxin and light regulate proliferation by oppositely affecting the activator E2FB and repressor E2FC levels. Opposite to E2FB, elevated sucrose levels increase the association of RBR1 with E2FA, which leads to the repression of genes involved in endocycle. In this way the E2FA-RBR1 complex contributes to the maintenance of meristems by the inhibition of cell differentiation. Later on RBR1 dissociates from E2FA by an unknown mechanism, and RBR1-free E2FA stimulates endocycle in differentiated cells. *Arrows* indicate activation, *hammers* repression, *white arms* are stimulatory, *black arm* indicate inhibitory inputs

roles in the G2 to M-phase transition, such as cyclin B1;1, and CDKB1;1 were also upregulated (Vlieghe et al. 2003; Vandepoele et al. 2005). CDKB1;1 was shown experimentally to be a direct target of the E2FA-DPA heterodimer (Boudolf et al. 2004a). The paradox is how E2FA promotes the G2 to M transition when its expression is restricted to the S-phase and how it can regulate the two antagonistic processes, endocycle and G2 to M transition. Recently it was shown that in proliferating cells E2FA forms a stable complex with RBR1 to repress genes involved in endocycle and differentiation, in this way the E2FA-RBR1 complex maintains the competence for cell proliferation in the meristem (Magyar et al. 2012). E2FA over-expression leads to elevated levels of the E2FA-RBR1 complex and increased proliferation. However, outside the meristem E2FA, by some unknown mechanism, dissociates from RBR1 and can stimulate endocycle (Magyar et al. 2012).

In *Arabidopsis*, there is only a single RBR1 protein. How RBR1 functions through the three distinct E2Fs, the timing and regulation of their interaction and the repression of distinct batteries of genes by these different E2F-RBR1 complexes is not well understood. RBR1 protein is most abundant in proliferating cells (Wildwater et al. 2005; Borghi et al. 2010; Umbrasaite et al. 2010).

The first identified *rbr1* mutant allele was found to be non-viable, with defects during both male and female gametogenesis due to over-proliferation, suggesting a repressive function for RBR1 in the mitotic cell cycle (Ebel et al. 2004). RBR1 is also required for cell fate determination in gametophytes. Reduction in RBR1 amounts in somatic tissues also leads to over-proliferation and delayed differentiation (Park et al. 2005; Desvoyes et al. 2006). In addition, RBR1 has been shown to play a role in repressing endocycle and maintaining genome integrity (Henriques et al. 2010; Johnston et al. 2010; Magyar et al. 2012).

Arabidopsis RBR1 has been shown to be important for stem cell maintenance which relies on the canonical CYCD/RBR/E2F pathway; reducing RBR1 amounts by RNAi or elevated RBR1 phosphorylation through the over-expression of CYCD3;1 produced extra stem cell layers while over-expression of RBR1 or hypophosphorylation induced by increased amounts of KRP2 led to the loss of stem cells. In accordance with the canonical model, having more RBR1-free E2Fs in cells by over-expression of both E2FA and DPA also produced more stem cells (Wildwater et al. 2005).

RBR1 over-expression stimulates early differentiation both in shoot and root meristems (Wildwater et al. 2005; Wyrzykowska et al. 2006). Besides CYCD/CDKA, there are also other signalling pathways that can control the transcriptional activity of the E2F-RBR1 complex. It was found that the mutation or silencing of the two ribosomal protein S6 kinase genes in *Arabidopsis* led to proliferation in growth limiting conditions such as lack of sucrose (Henriques et al. 2010). This deregulation in proliferation also resulted in frequent aneuploidisation. S6 kinase 1 was shown to interact with RBR1 and this interaction promotes the nuclear localization of RBR1. How other signalling pathways, such as the mitogen-activated protein kinase pathway, or Aurora kinase can regulate E2F-RBR1 functions remains to be established.

6.4 Transcriptional Regulation at G2 Phase to Mitosis and the Onset of Endoreplication

The most established function for E2F/DP is to control transcription during the G1 to S-phase transition by activating genes required for DNA replication and repair, thereby generating waves of gene expression peaking at this stage. The E2F transcription factors might also regulate the following waves of gene expression at G2, which may be associated with progression through G2 and entry into mitosis. There are a relatively small number of genes that show this pattern of expression, they include plant-specific B1-type CDKs (CDKB1) and A2-type mitotic cyclins (CYCA2); (Menges et al. 2003, 2005). In *Arabidopsis*, E2FA activates the transcription of *CDKB1;1* which has E2F binding sites in its promoter (Boudolf et al. 2004a). Both CDKB1;1 and CYCA2;3 are known to be involved in the onset of endoreduplication during organ development in *Arabidopsis* (Boudolf et al. 2004b; Imai et al. 2006). It has been proposed that these proteins form an active complex that maintains mitotic division and represses endoreduplication onset (Boudolf et al. 2009). There is no evidence that *CYCA2;3* is also directly regulated by E2FA, instead a report shows that it is regulated by INCREASED LEVEL OF POLYPLOIDY1 (ILP1), whose loss-of-function resulted in reduced polyploidy and upregulation of all *CYCA2* members (Yoshizumi et al. 2006). The authors concluded that this putative transcription factor may repress *CYCA2* transcription, thereby promoting exit of mitotic division and onset of endoreplication. Further genome-wide studies are needed to understand if ILP1 acts as a master regulator of G2-phase genes by repressing other genes that are expressed at this stage of the cell cycle.

6.4.1 Genes Expressed During Late G2 and Mitosis

Following the peak of CDKB1 and CYCA2;3 expression at G2, a different group of genes is induced shortly before entry into mitosis. This group contains a much larger number of genes including mitotic cyclins of A1, B1 and B2 types (CYCA1, CYCB1 and CYCB2) and plant-specific B2-type CDKs (CDKB2); (Menges et al. 2005; Kato et al. 2009). Besides these core cell cycle genes with presumable mitotic function, this group also contains many genes acting for spindle checkpoint (e.g., CDC20.1, MAD2, and BUB3), dynamics of mitotic microtubule machinery (many kinesins and other microtubule-associated proteins), and genes involved in cell plate formation (e.g., *KNOLLE*, *HINKEL/AtNACK1*, and *PLEIADE*). We have recently defined 185 genes showing such G2/M-specific expression in *Arabidopsis* (Haga et al. 2011). Most of these genes commonly contain promoter motifs similar to the MSA element, this has previously been identified as a *cis*-acting element important for G2/M-specific transcriptional activation in tobacco cells (Ito et al. 1998). The MSA element may be involved in evolutionarily conserved mechanisms for G2/M-specific promoter activation, because this motif is commonly found in G2/M-specific genes both in dicots, monocots as well as in moss.

6.4.2 Transcriptional Activation of G2/M-Specific Genes

Several lines of evidence suggest that R1R2R3-type MYB transcription factors directly bind to the MSA element, and regulate a suit of G2/M-specific genes (Ito 2005). These types of MYB transcription factors are broadly conserved among eukaryotes and may be prototypes of the diverse R1R2R3-type MYB proteins found in plants. R1R2R3-MYB proteins are structurally subdivided into at least four groups which we call A-, B-, C- and D-type MYBs. Among them, A-type MYBs are conserved in angiosperms and act as transcriptional activators (Fig. 6.3). In transient expression assays in BY-2 protoplasts, tobacco A-type MYB proteins, NtMYBA1 and NtMYBA2, can increase promoter activity of G2/M-specific genes in an MSA-dependent manner (Ito et al. 2001). It has been shown that NtMYBA2 has a negative regulatory domain in its C-terminus whose deletion enhances the activity of NtMYBA2 for transactivation of G2/M-specific genes (Araki et al. 2004). Microarray analysis showed that over-expression of the hyperactive form of NtMYBA2 lacking the negative regulatory domain resulted in up-regulation of many G2/M-specific genes in BY-2 cells (Kato et al. 2009). Conversely, simultaneous mutation of the two A-type MYB genes, *MYB3R1* and *MYB3R4*, resulted in the down-regulation of many G2/M-specific genes in *Arabidopsis* (Haga et al. 2011). The most prominent defect caused by the double *myb3r1 myb3r4* mutation is the frequent occurrence of incomplete cytokinesis during somatic cell division. In agreement with the proposed roles of A-type MYB, this cytokinetic defect is mainly due to reduced expression of *KNOLLE* that is essential for cell plate formation during cytokinesis (Haga et al. 2007).

Cellular activity of A-type MYB proteins is regulated both at the transcriptional and post-transcriptional level

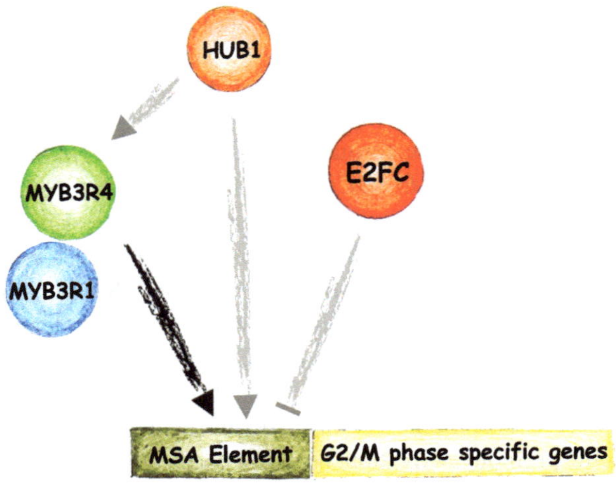

Fig. 6.3 Transcriptional regulation of G2/M-specific genes. R1R2R3-type MYB transcription factors directly bind to the MSA element, and regulate a suit of G2/M-specific genes. A-type MYBs (MYB3R4 and MYB3R1) are transcriptional activators and they positively regulate many G2/M-specific genes. The RING E3 ligase, encoded by *HISTONE MONOUBIQUITINATION1* (*HUB1*) is also involved in regulating the G2/M-specific gene transcription. Whether HUB1 directly regulates promoter activities or acts through the R1R2R3-type MYB transcription factors is not yet known. Opposing these positive regulators is the E2FC transcription factor. Whether E2FC is part of a larger complex together with R1R2R3-type MYBs remains to be established. *Arrows* indicate activation, *hammers* repression, *black lines* are experimentally proven and *grey lines* are hypothetical

during the cell cycle, such that transcription of their target genes is oscillating, with maximum levels at G2/M. In tobacco cells, NtMYBA2 shows cell cycle-regulated expression peaking at G2/M, contains an MSA element in its upstream region, and is up-regulated by the hyperactive form of NtMYBA2 (Ito et al. 2001; Kato et al. 2009). This suggests that transcription of NtMYBA2 is activated by itself in a positive feedback loop. Similarly, NtMYBA2 is directly phosphorylated and activated by a CDK in complex with CYCB1, which is a potential target being transcriptionally activated by NtMYBA2 (Araki et al. 2004). All these data suggest the presence of positive feedback loops in NtMYBA2 activation at transcriptional and post-transcriptional levels, which may enable the rapid increase of the regulatory proteins in the narrow window of the cell cycle.

6.4.3 Transcriptional Repression of G2/M-Specific Genes

In addition to A-type MYB transcriptional activators, repressor-type R1R2R3-MYB proteins also exist and participate in the regulation of G2/M-specific genes. The B-type MYB protein, NtMYBB when transiently over-expressed in tobacco BY-2 cells, was shown to repress the G2/M-specific genes *CYCB1* and *NACK1*, which was dependent on the presence of an MSA-element in their promoters (Ito et al. 2001). Replacement of the C-terminal region of NtMYBB with VP16 activation domain converted this transcriptional repressor into a strong activator. This suggests that the function for transcriptional control is determined by C-terminal regions and not by the N-terminally located MYB domain that is responsible for recognition of the MSA sequence. When NtMYBA2 and NtMYBB were simultaneously over-expressed, NtMYBB could decrease promoter activity that was enhanced by NtMYBA2, suggesting that NtMYBB may act as a competitive repressor (Ito et al. 2001). Therefore, G2/M-specific transcription may be regulated by a dynamic balance between an activator (NtMYBA2) and a repressor (NtMYBB), both of which bind to the same MSA sequence. NtMYBB is expressed constitutively throughout the cell cycle, in contrast to the G2/M-specific expression of NtMYBA2. Based on the findings from R1R2R3-MYB genes, a model was proposed that G2/M-specifc genes are normally repressed by constitutive expression of NtMYBB, whereas their transcription is triggered when activity of NtMYBA2 increases and overcomes NtMYBB-mediated repression. NtMYBB is also expressed in non-dividing cells, but it remains to be determined if this MYB protein is also involved in the maintenance of the repressive state of G2/M-specific genes in quiescent cells.

A recent report suggests that E2F transcription factors may also play a role in repressing G2/M-specific genes. E2FC, one of the three E2F proteins in *Arabidopsis*, is regarded as a transcriptional repressor because it does not have transcriptional activation properties in transient expression assays, and its over-expression causes down-regulation of *CDC6*, a representative E2F target (del Pozo et al. 2002). RNAi-mediated knockdown of E2FC resulted in extra cell divisions during leaf development with reduced levels of endoreplication (del Pozo et al. 2006). In the E2FC knockdown there is a significant up-regulation of S-phase genes in mature leaves but not in young leaves. This suggests that a primary role of E2FC is to establish or maintain the repressive state of S-phase genes during or after cessation of cell divisions in organ development. In addition, a marked increase of *CYCB1;1* transcript was also found in mature but not young leaves of E2FC knockdown plants (del Pozo et al. 2006). Conversely, over-expression of E2FC diminished expression of the CYCB1;1-GUS transgene and the endogenous *KNOLLE* gene. Consistently, E2FC is expressed in mature organs comprising mostly non-dividing cells, such as cotyledons and mature leaves, but as stated above, it is most abundant in meristematic tissues in *Arabidopsis* (del Pozo et al. 2002). One attractive hypothesis is that G2/M-specific genes are regulated positively by MYB3R1/4 and negatively by E2FC, and that their actions are dependent on the stage of organ development (Fig. 6.3). It would be interesting to examine if E2FC knockdown also affects other G2/M-specific genes in microarray analysis, and if such effects are dependent on the MSA element or not.

6.4.4 Other Factors Affecting the G2/M-Specific Transcription

Among G2/M-specific genes, *CYCB1;1* has been extensively studied as a representative gene of this group. Initial genetic approaches identified a mutation which affected GUS reporter expression driven by the *CYCB1;1* promoter in *Arabidopsis*, but that mutation has not yet been mapped (Himanen et al. 2003). Putative regulatory proteins for *CYCB1;1* transcription have been identified by a different strategy. Two putative transcription factors, the ELM2 domain-containing protein and transcription elongation factor TFIIS family protein, were isolated in a protein complex that binds to affinity beads covalently attached to an oligonucleotide of the MYB binding sequence in the *CYCB1;1* promoter (Planchais et al. 2002). While in a more recent study, TCP20, a TCP family transcription factor, has been shown to bind to the GCCCR element found in *CYCB1;1* and some ribosomal protein genes (Li et al. 2005). This element is required for high level transcription from the *CYCB1;1* promoter in the presence of the MSA element. Future reverse genetic studies are needed to confirm if these factors are actually required for transcription of *CYCB1;1* and other G2/M-specific genes.

It has been recently proposed that chromatin modifications are involved in the G2/M-specific transcription in *Arabidopsis*. The RING E3 ligase encoded by *HISTONE MONOUBIQUITINATION1* (*HUB1*) has been identified as a gene responsible for reduced leaf size in *ang4* mutants (Fleury et al. 2007). HUB1 is an ortholog of BRE1 from yeast and human which carries out monoubiquitination of histone H2B, a posttranslational modification important for transcriptionally active chromatin. The reduced leaf size in *ang4/hub1* mutants may be caused by increased cell cycle duration, which is, in turn, caused by the downregulation of many G2/M-specific genes. Microarray analysis of shoot meristems from the *ang4/hub1* mutant showed significant down regulation of 66 out of 82 mitosis-specific genes defined by Menges et al. (2005), most of which contain the MSA motif in their promoter regions. This suggests that HUB1 may positively and selectively affect transcriptional activity, and that such selective action for G2/M-specific genes may be achieved by a mechanism that recognizes the MSA element. One likely explanation is that HUB1 may be recruited to the MSA sites with the aid of R1R2R3-MYB transcription factors that have MSA-binding ability. In such a scenario, transcriptional activation regulated by MYB3R1/4 may be mediated through chromatin modifications, i.e., H2B monoubiquitination, around the MSA sites during mitosis (Fig. 6.3). It remains to be investigated if G2/M-specific genes are actually monoubiquitinated, when such modification occurs in the cell cycle, and if the effects of the *ang4/hub1* mutation are dependent on MYB3R1/4.

6.5 Spatial Organization of Mitosis

Microtubules are required for cellular transport, they serve as retention devices for many signalling molecules, and are used for the construction of the chromosome segregation machinery, the mitotic spindle. Microtubule assembly and orientation is directed from specialized structures such as the spindle pole bodies in fungi and centrosomes in animal cells. Centrosomes are known to organize both the interphase microtubules and the bipolar mitotic spindle in animal cells, where their function is modulated by phosphorylation. Defects of centrosomes result in chromosome instability (Kramer et al. 2011). The major mitotic kinases are the cyclin dependent kinase1/cyclinB complex, Polo-like kinase (Plk) and the Aurora kinases. These mitotic kinases are known to control the spatial and temporal events partially through the phosphorylation of centrosomal proteins (Ma and Poon 2011).

The microtubule network is made up of the conserved tubulin family, commonly through the polymerization of α-tubulin and β-tubulin. γ-Tubulin, in its soluble cytoplasmic state forms small complexes with two additional γ-tubulin complex proteins (GCPs), GCP2 and GCP3. γ-Tubulin is also present as a large ring complex (γTuRC) that is built of small tetrameric γ-tubulin complexes and that is established as a core unit needed for microtubule nucleation from centrosomes. Other specific components of the large γTuRCs, GCP4–6 are not essential for nucleation of microtubules but increase nucleation efficiency (Raynaud-Messina and Merdes 2007). However, non-centrosomal nucleation of microtubules occurs not only in acentrosomal cells such as oocytes but also in cells equipped with centrosomes (Bartolini and Gundersen 2006), and the role of γ-tubulin in the alternative pathway of microtubule nucleation is generally accepted (Janson et al. 2005; Schuh and Ellenberg 2007). Discrete structures for microtubule nucleation, such as centrosomes, are absent in higher plant cells and non-centrosomal microtubules are nucleated from dispersed γ-tubulin positive sites in the cytoplasm, with membranes, and with pre-existing microtubules (Murata et al. 2005; Binarová et al. 2006; Pastuglia et al. 2006). Localization of γ-tubulin around the outer nuclear envelope and its association with pre-kinetochores and kinetochores in early mitosis may depend on phosphorylation by mitotic kinases (Binarová et al. 1998, 2000). Furthermore, γ-tubulin together with the nuclear pore complex Nup107-160, are also localized to kinetochores of mitotic HeLa cells (Mishra et al. 2010), and thus it appears that γ-tubulin has a conserved function in the nucleation of kinetochore microtubules in the early stages of mitosis both in plant and animal cells. Apart from its association with kinetochores, γ-tubulin is also present along plant spindle microtubules where it shows gradual translocation from the kinetochore region to the acentrosomal poles during mitotic progression (Liu et al. 1993; Dryková et al. 2003). Correct localization of γ-tubulin to mitotic spindle microtubules is necessary for the organization of acentrosomal spindles in *Drosophila* oocytes, where γ-tubulin stabilizes the attachment of spindle microtubules at the kinetochores (Hughes et al. 2011). These data suggest that the precise spatial regulation of γ-tubulin localization throughout the cell cycle is required to ensure γ-tubulin functions in both centrosomal and acentrosomal cells. Compared with the extensive data available on the regulation of cell cycle specific centrosomal function, the regulation of dispersed centrosomes in acentrosomal cells is much less understood.

Spin down experiments with taxol-polymerized plant microtubules confirmed the presence of γ-tubulin and Cyclin Dependent Kinase A (CDKA) with microtubules (Weingartner et al. 2001; Dryková et al. 2003). Besides CDKA, other molecules that make up the core of the cell cycle machinery were also shown to be associated with microtubules by mass spectrometry-based proteomics of taxol-stabilised microtubule pellets (unpublished data; Jones, Binarová, Bogre). Centrosome-independent nucleation of spindle microtubules requires augmin (Goshima et al. 2008) and it assembles into a complex with γ-tubulin, dependent on Cyclin Dependent Kinase1 (CDK1), Polo kinase and Aurora signalling (Johmura et al. 2011). Augmin related proteins are important for spindle and phragmoplast organization in plants (Ho et al. 2011). It has been shown that γ-tubulin and GCPs are regulated in a cell cycle dependent manner by phosphorylation. Analysis of cell cycle phosphoproteome of yeast spindle pole bodies revealed dynamic phosphorylation of γ-tubulin and GCPs during cell cycle progression (Keck et al. 2011). Clustering of phosphosites that create charged regions were prominent within yeast centrosomal proteins and suggestive of phosphorylation-mediated protein-protein interactions (Schweiger and Linial 2010). Two G1-phase specific and six M-phase specific phoshosites are present on the γ-tubulin molecule and these are highly conserved in eukaryotes (Keck et al. 2011). It has been shown in two independent studies that the mutation of a single CDK phosphorylation site, Tub4 S360D, causes mitotic delay and aberrant anaphase spindle elongation (Keck et al. 2011; Lin et al. 2011). This phosphosite is conserved on plant γ-tubulin and suggests that CDK phosphorylation may also control plant γ-tubulin function in plants. Our analysis of the co-expression neighbourhood with γ-tubulin uncovered CDKB2.1. It would be interesting to examine experimentally, whether CDKB2.1 is involved in phosphoregulation of γ-tubulin (Fig. 6.4).

γ-Tubulin and GCP3 protein have been identified to be amongst the targets of Aurora signalling in mammalian cells (Sardon et al. 2010). Aurora kinases are central to the coordination of many cellular events of cell division and differentiation. There are three aurora kinases in animal cells, AURORA A–C. AURORA A localizes to centrosomes and along mitotic spindle microtubules, and plays a role in centrosome maturation, spindle pole organization and maintenance in metazoan cells. Depletion of Aurora A delays chromosome condensation and CDK1 activation (Liu and Ruderman 2006). Aurora B, as a chromosomal passenger, is essential for normal chromosome segregation and cytokinesis (Carmena and Earnshaw 2003). As in animal cells, three Aurora kinases have been identified in the *Arabidopsis* genome, named as AtAurora1, 2, and 3. Meristem-defective phenotypes observed in RNAi plants with silenced Aurora kinases suggest they play a role in maintaining meristematic cells in the mitotic cycle (Petrovská et al. 2012). A higher level of endoreduplication occurs in plants with downregulated Aurora kinases, demonstrating that Aurora kinases may interconnect cell cycle signalling with unknown targets involved in the switch from cell division to differentiation. In mammalian cells Aurora B phosphorylates retinoblastoma (RB) protein *in vitro* and *in vivo* at serine 780. Inhibition of Aurora B led to RB hypophosphorylation accompanied by endoreduplication (Nair et al. 2009). Therefore Aurora B kinase directly regulates the RB protein and the postmitotic checkpoint by preventing endoreduplication in cells with aberrant mitosis.

The AtAurora1 kinase localizes with the prophase spindle around the nuclei, with the preprophase band of cortical microtubules, and with mitotic spindle microtubules (Demidov et al. 2005, 2009). AtAurora1 physically interacts with its activator the AtTPX2 protein, and the co-localization of both proteins on preprophase spindle microtubules suggests that active AtAurora1 kinase is involved in γ-tubulin-driven microtubular nucleation during plant mitosis (Petrovská et al. 2012). The AtAurora1 kinase and TPX together with γ-tubulin localize with a specific subset of early phragmoplast microtubules adjacent to chromatin and distal to the cell plate e.g., at the site where phragmoplast microtubules are nucleated. Activity of Aurora A kinase is required for nucleation of microtubules with or without centrosomes (Tsai and Zheng 2005). Presence of AtAurora1 and its activating subunit AtTPX2 with γ-tubulin on microtubules suggests a link between the Aurora kinases and γ-tubulin mediated processes, such as acentrosomal plant microtubule nucleation from existing

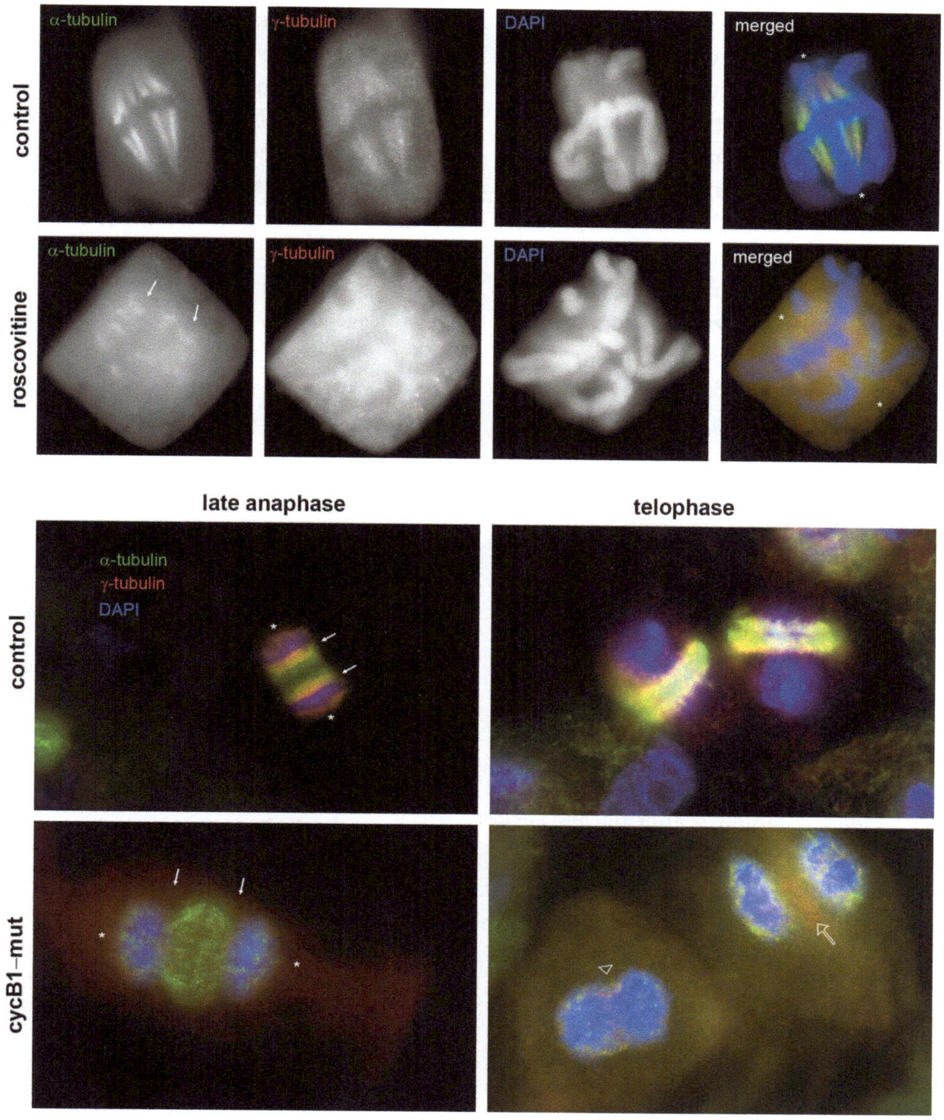

Fig. 6.4 CDK associated with CYCB1 is required for γ-tubulin mediated microtubule nucleation and organization of mitotic and cytokinetic microtubular arrays in acentrosomal plant cells. *Upper panel: Vicia faba* root meristem cells at metaphase stained for α- and γ-tubulin and for DNA with DAPI. γ-Tubulin is co-localized with the kinetochore fibres of the microtubular spindle. γ-Tubulin shows a gradient, being most abundant in the vicinity of spindle poles (*asterisk*) away from the kinetochores. Inhibition of mitotic CDKs by the drug roscovitine results in aberrant metaphase spindles; only remnants of kinetochore microtubular fibres persist (*arrows*). γ-Tubulin is not associated with these remnants of aberrant microtubule fibres, but is dispersed within the cytoplasm, suggesting that CDK might regulate γ-tubulin association with microtubules, or γ-tubulin-mediated microtubule nucleation. *Lower panel:* Triple immunofluorescence labelling for α- and γ-tubulin and for DNA with DAPI in wild type tobacco BY-2 cells, and in cells where a non-destructible mutant CYCB1 form is over-expressed (*cycB1mut*). In late anaphase, γ-tubulin label accumulates in the region with shortening kinetochore fibres on poles (*asterisk*) and forms a gradient in the midzone with enrichment at sites adjacent to chromatin (*arrows*) where phragmoplast microtubules are nucleated. At telophase, γ-tubulin is localized with the phragmoplast microtubules. Cells over-expressing a destruction box mutant form of CYCB1 (*cycB1mut*) maintain an abnormally high CDK activity at anaphase. γ-Tubulin in these cells neither localizes to anaphase poles (*asterisks*) nor accumulates in the midzone adjacent to separating chromatin (*arrows*). We find an abnormal localization pattern for γ-tubulin during the anaphase/telophase transition and a delayed transition from the mitotic spindle organization into phragmoplast. As a consequence, the microtubular cycle lags behind the chromatin cycle. γ-Tubulin becomes completely dispersed in cells with strong *cycB1mut* expression; the phragmoplast and consequently the cell plate are absent and cells become binuclear (*empty arrow*), eventually cells initiate anaphase but lack nuclear division and cytokinesis (*arrowhead*), and the next interphase is restored without regular chromatin

microtubules (Murata et al. 2005). AtAurora1 and the AtTPX2 regulatory module may act with the γ-tubulin nucleation complex and other substrates involved in the spatio-temporal regulation of acentrosomal plant spindle formation.

There is growing evidence to suggest a non-canonical role for γ-tubulin as a multifunctional protein involved in various cellular processes, including cell cycle regulation. Independent of microtubule assembly, γ-tubulin down-regulation in *Arabidopsis* compromises cytokinesis (Binarová et al. 2006).

CyclinB1 degradation and inactivation of mitotic kinases are prerequisites for mitotic exit and for the completition of cytokinesis (Fig. 6.4). In higher plants, microtubules are organized without a well defined microtubule organizing centre (MTOC). At anaphase, the acentrosomal mitotic spindle is formed by shortening microtubular fibres on poles and the midzone microtubules are transformed into the microtubular cytokinetic apparatus, called the phragmoplast. It is still not known whether phragmoplast microtubules are nucleated *de novo* from dispersed sites in the midzone analogous to the equatorial microtubule nucleation sites of *Saccharomyces pombe* or whether phragmoplast is formed by capturing preformed microtubules (Jürgens 2005). Cytokinesis is severely impaired if there is over-expression of nondegradable cyclin B and mitotic activity of CDK in late mitosis (Fig. 6.4) (Weingartner et al. 2004). Coordination between mitosis and cytokinesis is lost in γ-tubulin-conditional mutants of *S. pombe* (Hendrickson et al. 2001) and γ-tubulin mutants of *Aspergillus,* where the function of mitotic checkpoint proteins controlling entry into anaphase is impaired (Prigozhina et al. 2004). Failure to inactivate the anaphase promoting complex/cyclosome in γ-tubulin mutants suggests that regulation of cell division through γ-tubulin may be mediated by its interaction with the APC complex (Nayak et al. 2010).

The accumulating evidence demonstrating the presence of γ-tubulin in both plant and mammalian nuclei (Binárova et al. 2000; Lesca et al. 2005) has led to the intensive search for its nuclear function. γ-Tubulin is phosphorylated by the cell cycle inhibitory WEE1 kinase in *Drosophila* embryo extracts (Stumpff et al. 2005). Also, γ-tubulin was shown to mediate the binding of hypophosphorylated Breast Cancer Associated 1 (BRCA1) protein with the mitotic centrosome (Hsu et al. 2001). During S and G2-phases BRCA1-dependent ubiquitin ligase directs ubiquitination of γ-tubulin resulting in the inhibition of centrosome microtubule nucleation (Parvin 2009). Microtubular and nuclear localization of γ-tubulin requires a functional Nuclear Localization Signal (NLS) sequence; activation by CDK1; association with BRCA1 for loading of γ-tubulin to microtubules; and transportation of the BRCA1 γ-tubulin complex to the nucleus. All these are required for the γ-tubulin function in DNA damage response. (Hubert et al. 2011). Coupling of DNA damage checkpoints with the centrosomal cycle is further supported by data showing that nuclear γ-tubulin interacts with the tumour suppressor C53 protein (Hořejši et al. 2012).

The γ-tubulin forms a complex with the E2F transcription factor during the G1/S transition and the interaction of nuclear γ-tubulin with E2F was found to be important for the G1/S phase checkpoint (Hoog et al. 2011). Moreover, γ-tubulin regulates E2F activity by competing with DP1 in the E2F/DP1 complex. Down-regulation of γ-tubulin or its mutation causes a delay of S-phase but an increase in E2F activity, accompanied by an enrichment of E2F regulated transcripts. The NLS sequence of γ-tubulin is essential for its interaction with the E2F proteins during G1/S regulation. These results provide further evidence that γ-tubulin is essential for all major transitions of the cell cycle, including G1/S, G2/M and the mitotic exit. However, the molecular mechanisms as to how γ-tubulin regulates these checkpoints are not well understood. It could equally be that γ-tubulin acts alone or in complex with GCPs as scaffolding proteins, or it may be part of the checkpoint sensors.

Acknowledgement We thank Dr. Věra Cenklová for providing pictures of *V. faba* for Fig. 6.4.

References

Andersen SU, Buechel S, Zhao Z, Ljung K, Novak O, Busch W, Schuster C, Lohmann JU (2008) Requirement of B2-type cyclin-dependent kinases for meristem integrity in *Arabidopsis thaliana*. Plant Cell 20:88–100

Andriankaja M, Dhondt S, De Bodt S, Vanhaeren H, Coppens F, De Milde L, Muhlenbock P, Skirycz A, Gonzalez N, Beemster GT, Inze D (2012) Exit from proliferation during leaf development in *Arabidopsis thaliana*: a not-so-gradual process. Dev Cell 22:64–78

Araki S, Ito M, Soyano T, Nishihama R, Machida Y (2004) Mitotic cyclins stimulate the activity of c-Myb-like factors for trans-activation of G2/M phase-specific genes in tobacco. J Biol Chem 279:32979–32988

Aslanian A, Iaquinta PJ, Verona R, Lees JA (2004) Repression of the Arf tumor suppressor by E2F3 is required for normal cell cycle kinetics. Genes Dev 18:1413–1422

Aw SJ, Hamamura Y, Chen Z, Schnittger A, Berger F (2010) Sperm entry is sufficient to trigger division of the central cell but the paternal genome is required for endosperm development in *Arabidopsis*. Development 137:2683–2690

Banks JA, Nishiyama T, Hasebe M, Bowman JL, Gribskov M, dePamphilis C, Albert VA, Aono N, Aoyama T, Ambrose BA, Ashton NW, Axtell MJ, Barker E, Barker MS, Bennetzen JL, Bonawitz ND, Chapple C, Cheng C, Correa LG, Dacre M, DeBarry J, Dreyer I, Elias M, Engstrom EM, Estelle M, Feng L, Finet C, Floyd SK, Frommer WB, Fujita T, Gramzow L, Gutensohn M, Harholt J, Hattori M, Heyl A, Hirai T, Hiwatashi Y, Ishikawa M, Iwata M, Karol KG, Koehler B, Kolukisaoglu U, Kubo M, Kurata T, Lalonde S, Li K, Li Y, Litt A, Lyons E, Manning G, Maruyama T, Michael TP, Mikami K, Miyazaki S, Morinaga S, Murata T, Mueller-Roeber B, Nelson DR, Obara M, Oguri Y, Olmstead RG, Onodera N, Petersen BL, Pils B, Prigge M, Rensing SA, Riano-Pachon DM, Roberts AW, Sato Y, Scheller HV, Schulz B, Schulz C, Shakirov EV, Shibagaki N, Shinohara N, Shippen DE, Sorensen I, Sotooka R, Sugimoto N, Sugita M, Sumikawa N, Tanurdzic M, Theissen G, Ulvskov P, Wakazuki S, Weng JK, Willats WW, Wipf D, Wolf PG, Yang L, Zimmer AD, Zhu Q, Mitros T, Hellsten U, Loque D, Otillar R, Salamov A, Schmutz J, Shapiro H, Lindquist E, Lucas S, Rokhsar D, Grigoriev IV (2011) The *Selaginella* genome identifies genetic changes associated with the evolution of vascular plants. Science 332:960–963

Bartolini F, Gundersen GG (2006) Generation of noncentrosomal microtubule arrays. J Cell Sci 119:4155–4163

Berckmans B, Lammens T, Van Den Daele H, Magyar Z, Bogre L, De Veylder L (2011a) Light-dependent regulation of DEL1 is determined by the antagonistic action of E2Fb and E2Fc. Plant Physiol 157:1440–1451

Berckmans B, Vassileva V, Schmid SP, Maes S, Parizot B, Naramoto S, Magyar Z, Kamei CL, Koncz C, Bogre L, Persiau G, De Jaeger G, Friml J, Simon R, Beeckman T, De Veylder L (2011b) Auxin-dependent cell cycle reactivation through transcriptional regulation of *Arabidopsis* E2Fa by lateral organ boundary proteins. Plant Cell 23:3671–3683

Binarová P, Doležel J, Dráber P, Heberle-Bors E, Strnad M, Bogre L (1998) Treatment of *Vicia faba* root tip cells with specific inhibitors to cyclin-dependent kinases leads to abnormal spindle formation. Plant J 16:697–707

Binárova P, Cenklová V, Hause B, Kubatová E, Lysák M, Doležel J, Bogre L, Dráber P (2000) Nuclear gamma-tubulin during acentriolar plant mitosis. Plant Cell 12:433–442

Binarová P, Cenklová V, Procházková J, Doskočilová A, Volc J, Vrlik M, Bogre L (2006) Gamma-tubulin is essential for acentrosomal microtubule nucleation and coordination of late mitotic events in *Arabidopsis*. Plant Cell 18:1199–1212

Bisova K, Krylov DM, Umen JG (2005) Genome-wide annotation and expression profiling of cell cycle regulatory genes in *Chlamydomonas reinhardtii*. Plant Physiol 137:475–491

Bogre L (2011) Sensing microtubule states through the mitogen-activated protein kinase pathway during mitosis and morphogenesis. New Phytol 189:897–900

Bogre L, Magyar Z, Lopez-Juez E (2008) New clues to organ size control in plants. Genome Biol 9:226

Borghi L, Gutzat R, Futterer J, Laizet Y, Hennig L, Gruissem W (2010) *Arabidopsis* RETINOBLASTOMA-RELATED is required for stem cell maintenance, cell differentiation, and lateral organ production. Plant Cell 22:1792–1811

Boruc J, Van den Daele H, Hollunder J, Rombauts S, Mylle E, Hilson P, Inze D, De Veylder L, Russinova E (2010) Functional modules in the *Arabidopsis* core cell cycle binary protein-protein interaction network. Plant Cell 22:1264–1280

Bosco G (2010) Cell cycle: retinoblastoma, a trip organizer. Nature 466:1051–1052

Bosco G, Du W, Orr-Weaver TL (2001) DNA replication control through interaction of E2F-RB and the origin recognition complex. Nat Cell Biol 3:289–295

Boudolf V, Vlieghe K, Beemster GT, Magyar Z, Torres Acosta JA, Maes S, Van Der Schueren E, Inze D, De Veylder L (2004a) The plant-specific cyclin-dependent kinase CDKB1;1 and transcription factor E2Fa-DPa control the balance of mitotically dividing and endoreduplicating cells in *Arabidopsis*. Plant Cell 16:2683–2692

Boudolf V, Barroco R, Engler Jde A, Verkest A, Beeckman T, Naudts M, Inze D, De Veylder L (2004b) B1-type cyclin-dependent kinases are essential for the formation of stomatal complexes in *Arabidopsis thaliana*. Plant Cell 16:945–955

Boudolf V, Lammens T, Boruc J, Van Leene J, Van Den Daele H, Maes S, Van Isterdael G, Russinova E, Kondorosi E, Witters E, De Jaeger G, Inze D, De Veylder L (2009) CDKB1;1 forms a functional complex with CYCA2;3 to suppress endocycle onset. Plant Physiol 150:1482–1493

Buendia-Monreal M, Renteria-Canett I, Guerrero-Andrade O, Bravo-Alberto CE, Martinez-Castilla LP, Garcia E, Vazquez-Ramos JM (2011) The family of maize D-type cyclins: genomic organization, phylogeny and expression patterns. Physiol Plant 143:297–308

Caillaud MC, Paganelli L, Lecomte P, Deslandes L, Quentin M, Pecrix Y, Le Bris M, Marfaing N, Abad P, Favery B (2009) Spindle assembly checkpoint protein dynamics reveal conserved and unsuspected roles in plant cell division. PLoS One 4:e6757

Carmena M, Earnshaw WC (2003) The cellular geography of aurora kinases. Nat Rev Mol Cell Biol 4:842–854

Cayirlioglu P, Bonnette PC, Dickson MR, Duronio RJ (2001) *Drosophila* E2f2 promotes the conversion from genomic DNA replication to gene amplification in ovarian follicle cells. Development 128:5085–5098

Cayirlioglu P, Ward WO, Silver Key SC, Duronio RJ (2003) Transcriptional repressor functions of *Drosophila* E2F1 and E2F2 cooperate to inhibit genomic DNA synthesis in ovarian follicle cells. Mol Cell Biol 23:2123–2134

Chen HZ, Tsai SY, Leone G (2009) Emerging roles of E2Fs in cancer: an exit from cell cycle control. Nat Rev Cancer 9:785–797

Chong JL, Wenzel PL, Saenz-Robles MT, Nair V, Ferrey A, Hagan JP, Gomez YM, Sharma N, Chen HZ, Ouseph M, Wang SH, Trikha P, Culp B, Mezache L, Winton DJ, Sansom OJ, Chen D, Bremner R, Cantalupo PG, Robinson ML, Pipas JM, Leone G (2009) E2f1-3 switch from activators in progenitor cells to repressors in differentiating cells. Nature 462:930–934

Cools T, De Veylder L (2009) DNA stress checkpoint control and plant development. Curr Opin Plant Biol 12:23–28

Costas C, MdlP S, Sequeira-Mendes J, Gutierrez C (2011a) Progress in understanding DNA replication control. Plant Sci 181:203–209

Costas C, Desvoyes B, Gutierrez C (2011b) A chromatin perspective of plant cell cycle progression. Biochim Biophys Acta 1809:379–387

Costas C, Sanchez M De La Paz, Stroud H, Yu Y, Oliveros JC, Feng S, Benguria A, Lopez-Vidriero I, Zhang X, Solano R, Jacobsen SE, Gutierrez C (2011c) Genome-wide mapping of *Arabidopsis thaliana* origins of DNA replication and their associated epigenetic marks. Nat Struct Mol Biol 18:395–400

Coudreuse D, Nurse P (2010) Driving the cell cycle with a minimal CDK control network. Nature 468:1074–1079

Cross FR, Buchler NE, Skotheim JM (2011) Evolution of networks and sequences in eukaryotic cell cycle control. Philos Trans Roy Soc Lond B Bio 366:3532–3544

Danielian PS, Friesenhahn LB, Faust AM, West JC, Caron AM, Bronson RT, Lees JA (2008) E2f3a and E2f3b make overlapping but different contributions to total E2f3 activity. Oncogene 27:6561–6570

De Veylder L, Beeckman T, Beemster GT, de Almeida EJ, Ormenese S, Maes S, Naudts M, Van Der Schueren E, Jacqmard A, Engler G, Inze D (2002) Control of proliferation, endoreduplication and differentiation by the *Arabidopsis* E2Fa-DPa transcription factor. EMBO J 21:1360–1368

De Veylder L, Beeckman T, Inze D (2007) The ins and outs of the plant cell cycle. Nat Rev Mol Cell Biol 8:655–665

del Pozo JC, Boniotti MB, Gutierrez C (2002) *Arabidopsis* E2Fc functions in cell division and is degraded by the ubiquitin-SCF (AtSKP2) pathway in response to light. Plant Cell 14:3057–3071

del Pozo JC, Diaz-Trivino S, Cisneros N, Gutierrez C (2006) The balance between cell division and endoreplication depends on E2FC-DPB, transcription factors regulated by the ubiquitin-SCFSKP2A pathway in *Arabidopsis*. Plant Cell 18:2224–2235

Demidov D, Van Damme D, Geelen D, Blattner FR, Houben A (2005) Identification and dynamics of two classes of aurora-like kinases in *Arabidopsis* and other plants. Plant Cell 17:836–848

Demidov D, Hesse S, Tewes A, Rutten T, Fuchs J, Ashtiyani RK, Lein S, Fischer A, Reuter G, Houben A (2009) Aurora1 phosphorylation activity on histone H3 and its cross-talk with other post-translational histone modifications in *Arabidopsis*. Plant J 59:221–230

Desvoyes B, Ramirez-Parra E, Xie Q, Chua NH, Gutierrez C (2006) Cell type-specific role of the retinoblastoma/E2F pathway during *Arabidopsis* leaf development. Plant Physiol 140:67–80

Dimova DK, Stevaux O, Frolov MV, Dyson NJ (2003) Cell cycle-dependent and cell cycle-independent control of transcription by the *Drosophila* E2F/RB pathway. Genes Dev 17:2308–2320

Dissmeyer N, Weimer AK, Pusch S, De Schutter K, Alvim Kamei CL, Nowack MK, Novak B, Duan GL, Zhu YG, De Veylder L, Schnittger A (2009) Control of cell proliferation, organ growth, and DNA damage response operate independently of dephosphorylation of the *Arabidopsis* Cdk1 homolog CDKA;1. Plant Cell 21:3641–3654

Dissmeyer N, Weimer AK, De Veylder L, Novak B, Schnittger A (2010) The regulatory network of cell-cycle progression is fundamentally different in plants versus yeast or metazoans. Plant Signal Behav 5:1613–1618

Doonan JH, Sablowski R (2010) Walls around tumours—why plants do not develop cancer. Nat Rev Cancer 10:794–802

Dryková D, Cenklová V, Sulimenko V, Volc J, Dráber P, Binarová P (2003) Plant gamma-tubulin interacts with alphabeta-tubulin dimers and forms membrane-associated complexes. Plant Cell 15:465–480

Ebel C, Mariconti L, Gruissem W (2004) Plant retinoblastoma homologues control nuclear proliferation in the female gametophyte. Nature 429:776–780

Edgar BA, Orr-Weaver TL (2001) Endoreplication cell cycles: more for less. Cell 105:297–306

Fleury D, Himanen K, Cnops G, Nelissen H, Boccardi TM, Maere S, Beemster GT, Neyt P, Anami S, Robles P, Micol JL, Inze D, Van Lijsebettens M (2007) The *Arabidopsis thaliana* homolog of yeast BRE1 has a function in cell cycle regulation during early leaf and root growth. Plant Cell 19:417–432

Fulop K, Tarayre S, Kelemen Z, Horvath G, Kevei Z, Nikovics K, Bako L, Brown S, Kondorosi A, Kondorosi E (2005) *Arabidopsis* anaphase-promoting complexes: multiple activators and wide range of substrates might keep APC perpetually busy. Cell Cycle 4:1084–1092

Gaamouche T, Manes CLD, Kwiatkowska D, Berckmans B, Koumproglou R, Maes S, Beeckman T, Vernoux T, Doonan JH, Traas J, Inze D, De Veylder L (2010) Cyclin-dependent kinase activity maintains the shoot apical meristem cells in an undifferentiated state. Plant J 64:26–37

Goshima G, Mayer M, Zhang N, Stuurman N, Vale RD (2008) Augmin: a protein complex required for centrosome-independent microtubule generation within the spindle. J Cell Biol 181:421–429

Guo J, Song J, Wang F, Zhang XS (2007) Genome-wide identification and expression analysis of rice cell cycle genes. Plant Mol Biol 64:349–360

Haga N, Kato K, Murase M, Araki S, Kubo M, Demura T, Suzuki K, Muller I, Voss U, Jürgens G, Ito M (2007) R1R2R3-Myb proteins positively regulate cytokinesis through activation of KNOLLE transcription in *Arabidopsis thaliana*. Development 134:1101–1110

Haga N, Kobayashi K, Suzuki T, Maeo K, Kubo M, Ohtani M, Mitsuda N, Demura T, Nakamura K, Jürgens G, Ito M (2011) Mutations in MYB3R1 and MYB3R4 cause pleiotropic developmental defects and preferential down-regulation of multiple G2/M-specific genes in *Arabidopsis*. Plant Physiol 157:706–707, 717

He SS, Liu J, Xie Z, O'Neill D, Dotson S (2004) *Arabidopsis* E2Fa plays a bimodal role in regulating cell division and cell growth. Plant Mol Biol 56:171–184

Hendrickson TW, Yao J, Bhadury S, Corbett AH, Joshi HC (2001) Conditional mutations in gamma-tubulin reveal its involvement in chromosome segregation and cytokinesis. Mol Biol Cell 12:2469–2481

Henriques R, Magyar Z, Monardes A, Khan S, Zalejski C, Orellana J, Szabados L, de la Torre C, Koncz C, Bogre L (2010) *Arabidopsis* S6 kinase mutants display chromosome instability and altered RBR1-E2F pathway activity. EMBO J 29:2979–2993

Hernando E, Nahle Z, Juan G, Diaz-Rodriguez E, Alaminos M, Hemann M, Michel L, Mittal V, Gerald W, Benezra R, Lowe SW, Cordon-Cardo C (2004) Rb inactivation promotes genomic instability by uncoupling cell cycle progression from mitotic control. Nature 430:797–802

Himanen K, Reuzeau C, Beeckman T, Melzer S, Grandjean O, Corben L, Inze D (2003) The *Arabidopsis* locus RCB mediates upstream regulation of mitotic gene expression. Plant Physiol 133:1862–1872

Hirt H, Pay A, Gyorgyey J, Bako L, Nemeth K, Bogre L, Schweyen RJ, Heberle-Bors E, Dudits D (1991) Complementation of a yeast cell cycle mutant by an alfalfa cDNA encoding a protein kinase homologous to p34cdc2. Proc Natl Acad Sci USA 88:1636–1640

Ho CM, Hotta T, Kong Z, Zeng CJ, Sun J, Lee YR, Liu B (2011) Augmin plays a critical role in organizing the spindle and phragmoplast microtubule arrays in *Arabidopsis*. Plant Cell 23:2606–2618

Hochegger H, Takeda S, Hunt T (2008) Cyclin-dependent kinases and cell-cycle transitions: does one fit all? Nat Rev Mol Cell Biol 9:910–916

Hoog G, Zarrizi R, von Stedingk K, Jonsson K, Alvarado-Kristensson M (2011) Nuclear localization of gamma-tubulin affects E2F transcriptional activity and S-phase progression. FASEB J 25: 3815–3827

Hořejši B, Vinopal S, Sladková V, Dráberová E, Sulimenko V, Sulimenko T, Vosecká V, Philimonenko A, Hozák P, Katsetos CD, Dráber P (2012) Nuclear gamma-tubulin associates with nucleoli and interacts with tumor suppressor protein C53. J Cell Physiol 227:367–382

Hsu LC, Doan TP, White RL (2001) Identification of a gamma-tubulin-binding domain in BRCA1. Cancer Res 61:7713–7718

Hu X, Cheng X, Jiang H, Zhu S, Cheng B, Xiang Y (2010) Genome-wide analysis of cyclins in maize (*Zea mays*). Genet Mol Res 9:1490–1503

Hubert T, Vandekerckhove J, Gettemans J (2011) Cdk1 and BRCA1 target gamma-tubulin to microtubule domains. Biochem Biophys Res Commun 414:240–245

Hughes SE, Beeler JS, Seat A, Slaughter BD, Unruh JR, Bauerly E, Matthies HJ, Hawley RS (2011) Gamma-tubulin is required for bipolar spindle assembly and for proper kinetochore microtubule attachments during prometaphase I in *Drosophila* oocytes. PLoS Genet 7:e1002209

Huysman MJ, Martens C, Vandepoele K, Gillard J, Rayko E, Heijde M, Bowler C, Inze D, Van de Peer Y, De Veylder L, Vyverman W (2010) Genome-wide analysis of the diatom cell cycle unveils a novel type of cyclins involved in environmental signaling. Genome Biol 11:R17

Imai KK, Ohashi Y, Tsuge T, Yoshizumi T, Matsui M, Oka A, Aoyama T (2006) The A-type cyclin CYCA2;3 is a key regulator of ploidy levels in *Arabidopsis* endoreduplication. Plant Cell 18:382–396

Infante A, Laresgoiti U, Fernandez-Rueda J, Fullaondo A, Galan J, Diaz-Uriarte R, Malumbres M, Field SJ, Zubiaga AM (2008) E2F2 represses cell cycle regulators to maintain quiescence. Cell Cycle 7:3915–3927

Inze D, De Veylder L (2006) Cell cycle regulation in plant development. Annu Rev Genet 40:77–105

Ishikawa M, Murata T, Sato Y, Nishiyama T, Hiwatashi Y, Imai A, Kimura M, Sugimoto N, Akita A, Oguri Y, Friedman WE, Hasebe M, Kubo M (2011) *Physcomitrella* cyclin-dependent kinase A links cell cycle reactivation to other cellular changes during reprogramming of leaf cells. Plant Cell 23:2924–2938

Ito M (2005) Conservation and diversification of three-repeat Myb transcription factors in plants. J Plant Res 118:61–69

Ito M, Iwase M, Kodama H, Lavisse P, Komamine A, Nishihama R, Machida Y, Watanabe A (1998) A novel cis-acting element in promoters of plant B-type cyclin genes activates M phase-specific transcription. Plant Cell 10:331–341

Ito M, Araki S, Matsunaga S, Itoh T, Nishihama R, Machida Y, Doonan JH, Watanabe A (2001) G2/M-phase-specific transcription during the plant cell cycle is mediated by c-Myb-like transcription factors. Plant Cell 13:1891–1905

Iwakawa H, Shinmyo A, Sekine M (2006) *Arabidopsis* CDKA;1, a cdc2 homologue, controls proliferation of generative cells in male gametogenesis. Plant J 45:819–831

Janson ME, Setty TG, Paoletti A, Tran PT (2005) Efficient formation of bipolar microtubule bundles requires microtubule-bound gamma-tubulin complexes. J Cell Biol 169:297–308

Johmura Y, Soung NK, Park JE, Yu LR, Zhou M, Bang JK, Kim BY, Veenstra TD, Erikson RL, Lee KS (2011) Regulation of microtubule-based microtubule nucleation by mammalian polo-like kinase 1. Proc Natl Acad Sci USA 108:11446–11451

Johnston AJ, Kirioukhova O, Barrell PJ, Rutten T, Moore JM, Baskar R, Grossniklaus U, Gruissem W (2010) Dosage-sensitive function of retinoblastoma related and convergent epigenetic control are required during the *Arabidopsis* life cycle. PLoS Genet 6:e1000988

Jürgens G (2005) Cytokinesis in higher plants. Annu Rev Plant Biol 56:281–299

Karsenti E, Vernos I (2001) The mitotic spindle: a self-made machine. Science 294:543–547

Kasili R, Walker JD, Simmons LA, Zhou J, De Veylder L, Larkin JC (2010) SIAMESE cooperates with the CDH1-like protein CCS52A1 to establish endoreplication in *Arabidopsis thaliana* trichomes. Genetics 185:257–268

Kato K, Galis I, Suzuki S, Araki S, Demura T, Criqui MC, Potuschak T, Genschik P, Fukuda H, Matsuoka K, Ito M (2009) Preferential up-regulation of G2/M phase-specific genes by overexpression of the hyperactive form of NtmybA2 lacking its negative regulation domain in tobacco BY-2 cells. Plant Physiol 149:1945–1957

Keck JM, Jones MH, Wong CC, Binkley J, Chen D, Jaspersen SL, Holinger EP, Xu T, Niepel M, Rout MP, Vogel J, Sidow A, Yates JR III, Winey M (2011) A cell cycle phosphoproteome of the yeast centrosome. Science 332:1557–1561

Kono A, Umeda-Hara C, Adachi S, Nagata N, Konomi M, Nakagawa T, Uchimiya H, Umeda M (2007) The *Arabidopsis* D-type cyclin CYCD4 controls cell division in the stomatal lineage of the hypocotyl epidermis. Plant Cell 19:1265–1277

Korenjak M, Brehm A (2005) E2F-Rb complexes regulating transcription of genes important for differentiation and development. Curr Opin Genet Dev 15:520–527

Kosugi S, Ohashi Y (2003) Constitutive E2F expression in tobacco plants exhibits altered cell cycle control and morphological change in a cell type-specific manner. Plant Physiol 132:2012–2022

Kramer A, Maier B, Bartek J (2011) Centrosome clustering and chromosomal (in)stability: a matter of life and death. Mol Oncol 5:324–335

Lammens T, Boudolf V, Kheibarshekan L, Zalmas LP, Gaamouche T, Maes S, Vanstraelen M, Kondorosi E, La Thangue NB, Govaerts W, Inze D, De Veylder L (2008) Atypical E2F activity restrains APC/CCCS52A2 function obligatory for endocycle onset. Proc Natl Acad Sci USA 105:14721–14726

Larson-Rabin Z, Li Z, Masson PH, Day CD (2009) FZR2/CCS52A1 expression is a determinant of endoreduplication and cell expansion in *Arabidopsis*. Plant Physiol 149:874–884

Leone G, Nuckolls F, Ishida S, Adams M, Sears R, Jakoi L, Miron A, Nevins JR (2000) Identification of a novel E2F3 product suggests a mechanism for determining specificity of repression by Rb proteins. Mol Cell Biol 20:3626–3632

Lesca C, Germanier M, Raynaud-Messina B, Pichereaux C, Etievant C, Emond S, Burlet-Schiltz O, Monsarrat B, Wright M, Defais M (2005) DNA damage induce gamma-tubulin-RAD51 nuclear complexes in mammalian cells. Oncogene 24:5165–5172

Li C, Potuschak T, Colon-Carmona A, Gutierrez RA, Doerner P (2005) *Arabidopsis* TCP20 links regulation of growth and cell division control pathways. Proc Natl Acad Sci USA 102:12978–12983

Lin TC, Gombos L, Neuner A, Sebastian D, Olsen JV, Hrle A, Benda C, Schiebel E (2011) Phosphorylation of the yeast gamma-tubulin Tub4 regulates microtubule function. PLoS One 6:e19700

Liu Q, Ruderman JV (2006) Aurora A, mitotic entry, and spindle bipolarity. Proc Natl Acad Sci USA 103:5811–5816

Liu B, Marc J, Joshi HC, Palevitz BA (1993) A gamma-tubulin-related protein associated with the microtubule arrays of higher plants in a cell cycle-dependent manner. J Cell Sci 104:1217–1228

Liu DX, Nath N, Chellappan SP, Greene LA (2005) Regulation of neuron survival and death by p130 and associated chromatin modifiers. Genes Dev 19:719–732

Lloyd C (2011) Plant cytokinesis: circles within circles. Curr Biol 21: R926–R927

Lopez-Juez E, Dillon E, Magyar Z, Khan S, Hazeldine S, de Jager SM, Murray JA, Beemster GT, Bogre L, Shanahan H (2008) Distinct light-initiated gene expression and cell cycle programs in the shoot apex and cotyledons of *Arabidopsis*. Plant Cell 20:947–968

Ma HT, Poon RY (2011) Orderly inactivation of the key checkpoint protein mitotic arrest deficient 2 (MAD2) during mitotic progression. J Biol Chem 286:13052–13059

Magyar Z (2008) Keeping the balance between proliferation and differentiation by the E2F transcriptional regulatory network is central to plant growth and development. In: Bogre L, Beemster GT (eds) In plant growth signaling. Springer, Berlin/Heidelberg

Magyar Z, De Veylder L, Atanassova A, Bako L, Inze D, Bogre L (2005) The role of the *Arabidopsis* E2FB transcription factor in regulating auxin-dependent cell division. Plant Cell 17:2527–2541

Magyar Z, Horvath B, Khan S, Mohammed B, Henriques R, De Veylder L, Bako L, Scheres B, Bogre L (2012) *Arabidopsis* E2FA stimulates proliferation and endocycle separately through RBR-bound and RBR-free complexes. EMBO J 31:1480–1493

Maluszynska J, Kolano B, Sas-Nowosielska H (2013) Endopolyploidy in plants. In: Leitch IJ, Greilhuber J, Doležel J, Wendel JF (eds) Plant genome diversity, vol 2, Physical structure, behaviour and evolution of plant genomes. Springer-Verlag, Wien, pp 99–119

Marti A, Wirbelauer C, Scheffner M, Krek W (1999) Interaction between ubiquitin-protein ligase SCFSKP2 and E2F-1 underlies the regulation of E2F-1 degradation. Nat Cell Biol 1:14–19

Menges M, Hennig L, Gruissem W, Murray JA (2003) Genome-wide gene expression in an *Arabidopsis* cell suspension. Plant Mol Biol 53:423–442

Menges M, de Jager SM, Gruissem W, Murray JA (2005) Global analysis of the core cell cycle regulators of *Arabidopsis* identifies novel genes, reveals multiple and highly specific profiles of expression and provides a coherent model for plant cell cycle control. Plant J 41:546–566

Menges M, Pavesi G, Morandini P, Bogre L, Murray JA (2007) Genomic organization and evolutionary conservation of plant D-type cyclins. Plant Physiol 145:1558–1576

Mishra RK, Chakraborty P, Arnaoutov A, Fontoura BM, Dasso M (2010) The Nup107-160 complex and gamma-TuRC regulate microtubule polymerization at kinetochores. Nat Cell Biol 12:164–169

Morgan DO (2007) The cell cycle principles of control. Oxford University Press, Corby

Murata T, Sonobe S, Baskin TI, Hyodo S, Hasezawa S, Nagata T, Horio T, Hasebe M (2005) Microtubule-dependent microtubule nucleation based on recruitment of gamma-tubulin in higher plants. Nat Cell Biol 7:961–968

Nair JS, Ho AL, Tse AN, Coward J, Cheema H, Ambrosini G, Keen N, Schwartz GK (2009) Aurora B kinase regulates the postmitotic endoreduplication checkpoint via phosphorylation of the retinoblastoma protein at serine 780. Mol Biol Cell 20:2218–2228

Nayak T, Edgerton-Morgan H, Horio T, Xiong Y, De Souza CP, Osmani SA, Oakley BR (2010) Gamma-tubulin regulates the anaphase-promoting complex/cyclosome during interphase. J Cell Biol 190:317–330

Neufeld TP, de la Cruz AF, Johnston LA, Edgar BA (1998) Coordination of growth and cell division in the *Drosophila* wing. Cell 93:1183–1193

Nieuwland J, Maughan S, Dewitte W, Scofield S, Sanz L, Murray JA (2009) The D-type cyclin CYCD4;1 modulates lateral root density in *Arabidopsis* by affecting the basal meristem region. Proc Natl Acad Sci U S A 106:22528–22533

Nowack MK, Grini PE, Jakoby MJ, Lafos M, Koncz C, Schnittger A (2006) A positive signal from the fertilization of the egg cell sets off endosperm proliferation in angiosperm embryogenesis. Nat Genet 38:63–67

Nowack MK, Shirzadi R, Dissmeyer N, Dolf A, Endl E, Grini PE, Schnittger A (2007) Bypassing genomic imprinting allows seed development. Nature 447:312–315

Park JA, Ahn JW, Kim YK, Kim SJ, Kim JK, Kim WT, Pai HS (2005) Retinoblastoma protein regulates cell proliferation, differentiation, and endoreduplication in plants. Plant J 42:153–163

Parvin JD (2009) The BRCA1-dependent ubiquitin ligase, gamma-tubulin, and centrosomes. Environ Mol Mutagen 50:649–653

Pastuglia M, Azimzadeh J, Goussot M, Camilleri C, Belcram K, Evrard JL, Schmit AC, Guerche P, Bouchez D (2006) Gamma-tubulin is essential for microtubule organization and development in *Arabidopsis*. Plant Cell 18:1412–1425

Petrovská B, Cenklová V, Pochylová Z, Kourová H, Doskočilová A, Plíhal O, Binarová L, Binarová P (2012) Plant Aurora kinases play a role in maintenance of primary meristems and control of endoreduplication. New Phytol 193:590–604

Planchais S, Perennes C, Glab N, Mironov V, Inze D, Bergounioux C (2002) Characterization of cis-acting element involved in cell cycle phase-independent activation of Arath;CycB1;1 transcription and identification of putative regulatory proteins. Plant Mol Biol 50:111–127

Prigozhina NL, Oakley CE, Lewis AM, Nayak T, Osmani SA, Oakley BR (2004) Gamma-tubulin plays an essential role in the coordination of mitotic events. Mol Biol Cell 15:1374–1386

Rayman JB, Takahashi Y, Indjeian VB, Dannenberg JH, Catchpole S, Watson RJ, te Riele H, Dynlacht BD (2002) E2F mediates cell cycle-dependent transcriptional repression in vivo by recruitment of an HDAC1/mSin3B corepressor complex. Genes Dev 16:933–947

Raynaud-Messina B, Merdes A (2007) Gamma-tubulin complexes and microtubule organization. Curr Opin Cell Biol 19:24–30

Royzman I, Austin RJ, Bosco G, Bell SP, Orr-Weaver TL (1999) ORC localization in *Drosophila* follicle cells and the effects of mutations in dE2F and dDP. Genes Dev 13:827–840

Sage J, Straight AF (2010) RB's original CIN? Genes Dev 24:1329–1333

Sardon T, Pache RA, Stein A, Molina H, Vernos I, Aloy P (2010) Uncovering new substrates for Aurora A kinase. EMBO Rep 11:977–984

Scheres B, Xu J (2006) Polar auxin transport and patterning: grow with the flow. Genes Dev 20:922–926

Schnittger A, Schobinger U, Stierhof YD, Hulskamp M (2002) Ectopic B-type cyclin expression induces mitotic cycles in endoreduplicating *Arabidopsis* trichomes. Curr Biol 12:415–420

Schubert I, Shaw P (2011) Organization and dynamics of plant interphase chromosomes. Trends Plant Sci 16:273–281

Schuh M, Ellenberg J (2007) Self-organization of MTOCs replaces centrosome function during acentrosomal spindle assembly in live mouse oocytes. Cell 130:484–498

Schweiger R, Linial M (2010) Cooperativity within proximal phosphorylation sites is revealed from large-scale proteomics data. Biol Direct 5:6

Sozzani R, Maggio C, Varotto S, Canova S, Bergounioux C, Albani D, Cella R (2006) Interplay between *Arabidopsis* activating factors E2Fb and E2Fa in cell cycle progression and development. Plant Physiol 140:1355–1366

Sterken R, Kiekens R, Coppens E, Vercauteren I, Zabeau M, Inze D, Flowers J, Vuylsteke M (2009) A population genomics study of the *Arabidopsis* core cell cycle genes shows the signature of natural selection. Plant Cell 21:2987–2998

Stumpff J, Kellogg DR, Krohne KA, Su TT (2005) *Drosophila* Wee1 interacts with members of the gamma TURC and is required for proper mitotic-spindle morphogenesis and positioning. Curr Biol 15:1525–1534

Takahashi I, Kojima S, Sakaguchi N, Umeda-Hara C, Umeda M (2010) Two *Arabidopsis* cyclin A3s possess G1 cyclin-like features. Plant Cell Rep 29:307–315

Tsai MY, Zheng Y (2005) Aurora A kinase-coated beads function as microtubule-organizing centers and enhance RanGTP-induced spindle assembly. Curr Biol 15:2156–2163

Tsai SY, Opavsky R, Sharma N, Wu L, Naidu S, Nolan E, Feria-Arias E, Timmers C, Opavska J, de Bruin A, Chong JL, Trikha P, Fernandez SA, Stromberg P, Rosol TJ, Leone G (2008) Mouse development with a single E2F activator. Nature 454:1137–1141

Tyson JJ, Novak B (2008) Temporal organization of the cell cycle. Curr Biol 18:R759–R768

Umbrasaite J, Schweighofer A, Kazanaviciute V, Magyar Z, Ayatollahi Z, Unterwurzacher V, Choopayak C, Boniecka J, Murray JA, Bogre L, Meskiene I (2010) MAPK phosphatase AP2C3 induces ectopic proliferation of epidermal cells leading to stomata development in *Arabidopsis*. PLoS One 5:e15357

van den Heuvel S, Dyson NJ (2008) Conserved functions of the pRB and E2F families. Nat Rev Mol Cell Biol 9:713–724

Van Leene J, Hollunder J, Eeckhout D, Persiau G, Van De Slijke E, Stals H, Van Isterdael G, Verkest A, Neirynck S, Buffel Y, De Bodt S, Maere S, Laukens K, Pharazyn A, Ferreira PC, Eloy N, Renne C, Meyer C, Faure JD, Steinbrenner J, Beynon J, Larkin JC, Van de Peer Y, Hilson P, Kuiper M, De Veylder L, Van Onckelen H, Inze D, Witters E, De Jaeger G (2010) Targeted interactomics reveals a complex core cell cycle machinery in *Arabidopsis thaliana*. Mol Syst Biol 6:397

Van Leene J, Boruc J, De Jaeger G, Russinova E, De Veylder L (2011) A kaleidoscopic view of the *Arabidopsis* core cell cycle interactome. Trends Plant Sci 16:141–150

Vandepoele K, Raes J, De Veylder L, Rouze P, Rombauts S, Inze D (2002) Genome-wide analysis of core cell cycle genes in *Arabidopsis*. Plant Cell 14:903–916

Vandepoele K, Vlieghe K, Florquin K, Hennig L, Beemster GT, Gruissem W, Van de Peer Y, Inze D, De Veylder L (2005) Genome-wide identification of potential plant E2F target genes. Plant Physiol 139:316–328

Vandepoele K, Quimbaya M, Casneuf T, De Veylder L, Van de Peer Y (2009) Unraveling transcriptional control in *Arabidopsis* using cis-regulatory elements and coexpression networks. Plant Physiol 150:535–546

Vanstraelen M, Baloban M, Da Ines O, Cultrone A, Lammens T, Boudolf V, Brown SC, De Veylder L, Mergaert P, Kondorosi E (2009) APC/C-CCS52A complexes control meristem maintenance in the *Arabidopsis* root. Proc Natl Acad Sci USA 106:11806–11811

Vlieghe K, Vuylsteke M, Florquin K, Rombauts S, Maes S, Ormenese S, Van Hummelen P, Van de Peer Y, Inze D, De Veylder L (2003) Microarray analysis of E2Fa-DPa-overexpressing plants uncovers a cross-talking genetic network between DNA replication and nitrogen assimilation. J Cell Sci 116:4249–4259

Weingartner M, Binarova P, Drykova D, Schweighofer A, David JP, Heberle-Bors E, Doonan J, Bogre L (2001) Dynamic recruitment of Cdc2 to specific microtubule structures during mitosis. Plant Cell 13:1929–1943

Weingartner M, Criqui MC, Meszaros T, Binarova P, Schmit AC, Helfer A, Derevier A, Erhardt M, Bogre L, Genschik P (2004) Expression of a nondegradable cyclin B1 affects plant development and leads to endomitosis by inhibiting the formation of a phragmoplast. Plant Cell 16:643–657

Wenzel PL, Chong JL, Saenz-Robles MT, Ferrey A, Hagan JP, Gomez YM, Rajmohan R, Sharma N, Chen HZ, Pipas JM, Robinson ML, Leone G (2011) Cell proliferation in the absence of E2F1-3. Dev Biol 351:35–45

Wildwater M, Campilho A, Perez-Perez JM, Heidstra R, Blilou I, Korthout H, Chatterjee J, Mariconti L, Gruissem W, Scheres B (2005) The RETINOBLASTOMA-RELATED gene regulates stem cell maintenance in *Arabidopsis* roots. Cell 123:1337–1349

Wyrzykowska J, Schorderet M, Pien S, Gruissem W, Fleming AJ (2006) Induction of differentiation in the shoot apical meristem by transient overexpression of a retinoblastoma-related protein. Plant Physiol 141:1338–1348

Yoshizumi T, Tsumoto Y, Takiguchi T, Nagata N, Yamamoto YY, Kawashima M, Ichikawa T, Nakazawa M, Yamamoto N, Matsui M (2006) Increased level of polyploidy1, a conserved repressor of CYCLINA2 transcription, controls endoreduplication in *Arabidopsis*. Plant Cell 18:2452–2468

Yuan L, Yang X, Makaroff CA (2011) Plant cohesins, common themes and unique roles. Curr Protein Pept Sci 12:93–104

Zhu W, Giangrande PH, Nevins JR (2005) Temporal control of cell cycle gene expression mediated by E2F transcription factors. Cell Cycle 4:633–636

Endopolyploidy in Plants

7

Jolanta Maluszynska, Bozena Kolano, and Hanna Sas-Nowosielska

Contents

J. Maluszynska (✉)
Department of Plant Anatomy and Cytology, University of Silesia, Jagiellonska 28, 40-032 Katowice, Poland
e-mail: jolanta.maluszynska@us.edu.pl

I.J. Leitch et al. (eds.), *Plant Genome Diversity Volume 2*,
DOI 10.1007/978-3-7091-1160-4_7, © Springer-Verlag Wien 2013

7.1 Introduction

Plant growth and development is precisely programmed and achieved through three processes: cell division (proliferation), growth and differentiation. These three processes may overlap during plant organ development, when some cells start to differentiate while others continue to divide e.g., leaf epidermal cells (Harashima and Schnittger 2010). Dividing cells, called meristematic cells, increase their number and supply new cells for post-embryonic plant development. Outside the meristems non-dividing cells expand and differentiate. Cell proliferation and expansion result in varied but determined cell sizes specific for the plant, organ and tissue. The next phase in plant development is cell-type specification along with the differentiation processes. The control of all processes and the determination of final cell mass and size are poorly understood but there is increasing knowledge about the molecular mechanisms underpinning the regulatory systems. Cell sizes in plants are usually closely related to their function. There are two strategies to enlarge cell size: one is based on water uptake and vacuolar growth and the other is to increase the nuclear DNA content or the level of polyploidy, this gives rise to endopolyploidy.

7.1.1 Definition and Types of Endopolyploidy

Endopolyploidy is a general term defining the results of the exponential multiplication of nuclear DNA (2^n) in the absence of mitosis. Endopolyploidisation is mainly the result of either the endoreduplication or the endomitosis pathways (D'Amato 1964). Other processes such as the fusion of nuclei or the appearance of multinuclear cells also lead to polyploidisation, but they are rare and not essential for plant development (D'Amato 1984). Exceptions are the acytokinetic mitotic activity in the anther tapetum of most seed plants and in the antipodals and endosperm of some angiosperm taxa.

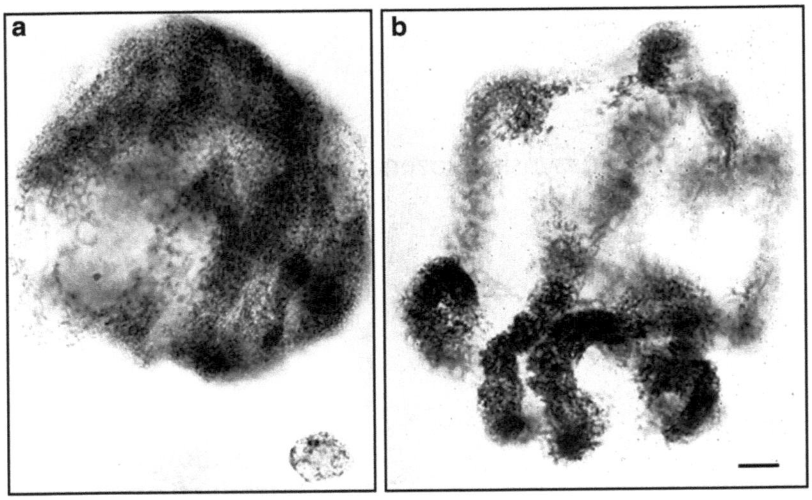

Fig. 7.1 *Aconitum ranunculifolius* ($2n = 16$), highly endopolyploid nuclei from the antipodals in the embryo sac; (**a**) fully developed giant nucleus with diffuse chromatin structure (note a 2C nucleus from the integument at the bottom right hand corner of the figure); (**b**) nucleus with "giant chromosomes" in haploid number ($n = 8$). *Bar*: 5 μm (From Tschermak-Woess (1956b), with permission of Springer)

7.1.1.1 Endomitosis

Endomitosis was first described by Geitler (1939) in the insect *Gerris lateralis* (Heteroptera) as the process of chromosome division without cell division. During the endomitotic cycle chromosomes double and condense (endoprophase, endometaphase) within an intact nuclear envelope. The sister chromatids then separate and decondense in the nucleus leading to endopolyploidy, which is characterised by the multiplication of the visible chromosome number. Endomitotic cycles of this type have been reported in several animal groups but rarely in angiosperms (D'Amato 1989; Weiss and Maluszynska 2001).

7.1.1.2 Endoreduplication

The most common mode of plant cell polyploidisation is endoreduplication. This process was first described by Levan (1939) in the elongation zone of *Allium cepa* roots. It is estimated to occur in over 90% of all angiosperms and is also present in algal and fern cells but absent in gymnosperms (Barow and Meister 2003). Endoreduplication is a process in which chromosomes replicate without a subsequent nucleus and cell division and without any obvious signs of chromatin condensation and decondensation. As a result, chromosomes have 2^n chromatids, which may become visible when such a nucleus is stimulated to enter mitosis again. The chromatids in interphase may fall apart or remain united at certain regions, mostly the centromeres, or, very rarely, along most of their length. In the latter case chromatid bundles, so-called giant chromosomes, are formed. Many successive cycles of DNA replication without the segregation of sister chromatids may result in polytene chromosomes. As proposed by Huskins (1947), the lowest level of polyteny is represented by a diplochromosome (= 4 chromatid chromosome), which is the result of two successive rounds of DNA replication. The best-known examples of true polyteny are the giant salivary gland chromosomes of *Drosophila*. The characteristic of polyteny is the cable-architecture with banded appearance of the chromosome. In plants, giant chromosomes resembling polytene chromosomes of the dipteran type have been observed in only a few species of *Phaseolus* and *Vigna* in ovary or immature seed tissue (embryo suspensor) and the anther tapetum respectively (Carvalheira 2000) or in the antipodal cells of *Aconitum ranunculifolius* (Fig. 7.1) and specialized cells of the embryo, embryo sac and endosperm of a few other plant species (Tschermak-Woess 1956a, b, 1963).

While endomitosis always leads to a visible multiplication of chromosomes, endoreduplication can have different effects, depending on the chromatin characteristics (Fig. 7.2). Various intermediate DNA configurations can occur, differing in the degree of association between duplicated chromatids (Edgar and Orr-Weaver 2001). Both processes, endomitosis and endoreduplication, differ from mitosis by the absence of mitotic spindle formation, and the presence of a nuclear envelope during the entire endocycle.

Although endoreduplication is widespread in eukaryotes, many questions and problems remain. The role and control of the endocycle are especially poorly characterised in plants. Nevertheless, there is a growing interest in this field, particularly focused on (1) the switch between cell proliferation and cell differentiation, and between the mitotic cycle and the endocycle, and (2) the termination of the endoreduplication process. Increasing knowledge about the mechanisms of plant cell cycle regulation should contribute to a better understanding of the regulation of the

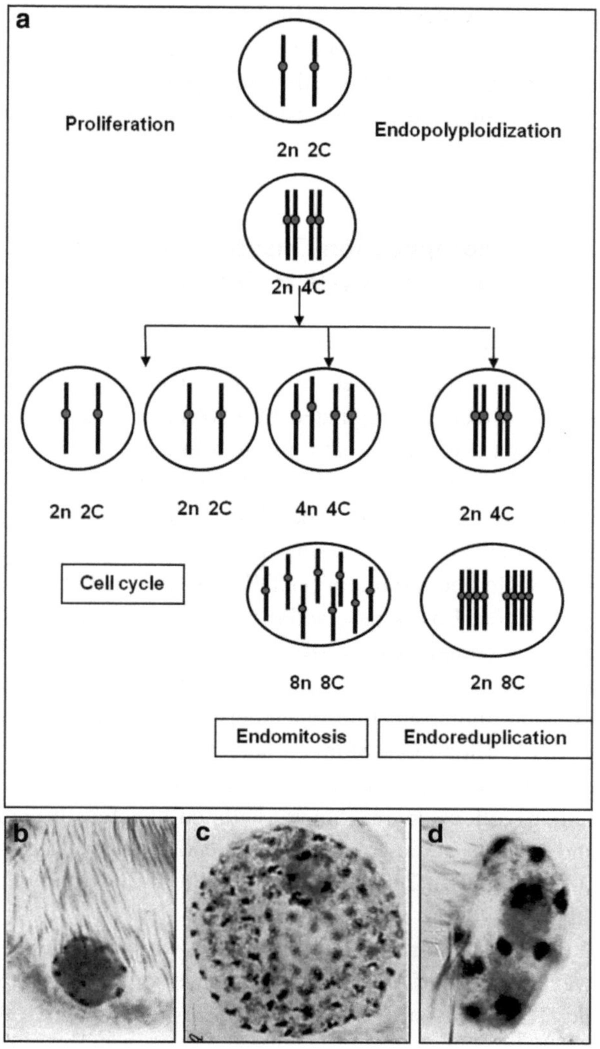

Fig. 7.2 Nuclei during cell cycle and endopolyploidisation. (**a**) Schematic changes in chromosome number and DNA content, (**b–d**) *Gibbaeum heathii* ($2n = 18$), interphase nuclei of raphide cells of secondary leaf; (**b**) 2C nucleus, (**c**) presumably 16C nucleus with single chromocentres in multiplied number, (**d**) presumably 16C nucleus with endochromocentres in diploid number. *Bar*: 5 μm (from Schlichtinger (1956) with permission of Springer)

endocycle and the biological significance of endopolyploidisation in plant development. Recent genetic studies, mostly utilising *Arabidopsis thaliana* mutants, have revealed the key genes involved in plant endopolyploidy regulation, while progress in flow and image cytometry has accelerated investigations and determinations of the endopolyploidisation pattern in new plant species.

7.1.2 Polysomatic and Non-polysomatic Plants

Endopolyploidisation manifests itself at varying ploidy levels (labelled as 2C, 4C, 8C, 16C ….) of different cells in the same individual as a result of several rounds of DNA synthesis without mitosis. The variation of cell ploidy levels is also designated as somatic polyploidy. Endopolyploidisation is closely associated with cell differentiation and is essential for normal development and physiology in many different organs of plants and animals. Plants which undergo endopolyploidisation are called polysomatic plants in contrast to non-polysomatic plants, whose cells remain diploid throughout differentiation (Fig. 7.3).

Endoreduplication is widespread in eukaryotes and is especially common in angiosperms; however, it is not present in all species and what determines which species undergo a polysomatic or non-polysomatic (non-endopolyploid) type of development is unclear. For example, why some cells undergo endopolyploidisation in one species in the early stage of development (e.g., *Chenopodium quinoa*, Kolano et al. 2009) while in others all cells remain diploid and their non-dividing nuclei stay at the 2C DNA level (e.g., *Crepis capillaris*, Brossard 1978; Maluszynska 1990) is poorly understood. Non-polysomatic plants seem to be typical for some plant families. The dominance of non-polysomatic species has been documented in such families as Amaryllidaceae, Asteraceae, Fagaceae, Liliaceae, Ranunculaceae, Rosaceae and others (Barow and Jovtchev 2007).

7.1.3 Pattern of Endoreduplication: Polysomaty

Endopolyploidy in polysomatic plants can be characterised in two ways: (1) the frequency of endopolyploid cells and (2) the degree of ploidy in the particular cells of tissues and organs. These two factors determine the pattern of endopolyploidy, which is characteristic for each species and is known to have an organ-specific character. The degree of endopolyploidy is indicated by the C level where 1C is the DNA content in unreplicated gametophyte nuclei and gametes. 2C and 4C represent the DNA amounts of mitotically active sporophytic cells in the G1 and G2 phase of the cell cycle, respectively. Cells with a DNA content higher than 4C are considered to be endopolyploid. Generally, during endopolyploidisation nuclei undergo several rounds of endoreduplication, reaching levels of 64C DNA. It should be noted that the 4C population may comprise not only of cells in the G2 phase of the cell cycle but also differentiated cells that have already gone through the first round of endoreduplication. However, in mature organs mitotically active cells are not present or their frequency is very low. Higher levels of endoreduplication occur in metabolically highly active cells that have a secretory or nutritive function. Multiple cycles of endoreduplication have been described in many specific cell types like antipodal cells, suspensor cells, endosperm haustoria or tapetum cells (Table 7.1). An

Table 7.1 Examples of endopolyploidy in specific plant tissues (Mauseth 2009)

Species	Tissue	Degree
Carex hirta	Tapetum	8C
Viola declinata	Elaiosome	16C
Cucumis sativus	Anther hairs	64C
Triticum aestivum	Antipodal cells	128C
Geranium phaeum	Integument	512C
Phaseolus vulgaris	Suspensor	2,048C
Scilla bifolia	Elaiosome	4,096C
Phaseolus coccineus	Suspensor	8,192C
Arum maculatum	Endosperm	24,576C

extremely high DNA content, up to 24,576C, which corresponds to 13 endocycles, has been reported for the endosperm cells of *Arum maculatum* (Traas et al. 1998).

7.2 Molecular Mechanisms of Endoreduplication

During the mitotic cell cycle, cells undergo sequential phases of DNA replication (S-phase) and chromosome segregation (mitosis—M-phase), allowing chromosome replication only once per cell cycle. Replication is preceded by the gap phase (G1) during which cells grow and synthesise the proteins needed for S-phase. A second gap phase (G2) also separates the S-phase and mitosis. During the gaps several checkpoints operate which regulate cell cycle progression. The checkpoints act in response to intra- and extracellular regulatory signals (Fig. 7.4a).

A particular class of protein kinases, known as CYCLIN DEPENDENT KINASES (CDK) along with their regulatory proteins, known as CYCLINS (CYCs), are involved in the progression through the cell cycle. Two main families of CDKs act during the cell cycle: (1) continuously active CDKA and (2) temporarily activated, mitotic CDKBs (Fig. 7.4a). These CDKs form complexes with a variety of cyclin family members. Cyclic changes in cyclin levels result in changes in the activity of the CYC-CDK complexes, thus allowing phase succession (Dewitte and Murray 2003; see also Magyar et al. 2013, this volume).

When cells start to differentiate they stop dividing and may enter endoreduplication which operates through a modified cell cycle called an endocycle. A single endocycle consists of just the G1 and S-phase, i.e., it lacks the G2 and M-phases (Edgar and Orr-Weaver 2001). Based on available data, the entire endocycle operates through the action of CDKA1 (Leiva-Neto et al. 2004). Successive rounds of replication separated by G-phases are executed by changes in a CDKA1 cyclin partner. Cyclin determines which of the potential substrates kinase will interact with. Two cyclin families seem to be involved in the endocycle: CYCD and CYCA. CYCD drives the cells into S-phase, while CYCA probably controls the progression of S-phase (Joubes and Chevalier 2000).

7.2.1 Endoreplication Onset: Transition from Mitotic Cell Cycle to Endocycle

The transition from the mitotic cycle to the endocycle is irreversible and controlled by genetic and environmental cues. The process is regulated by several pathways which are thought to lead to the inactivation of the mitosis-promoting factor (MPF) and provide the oscillatory activity of S-phase CYC-CDK complexes. MPF inactivation prevents the cells from passing the G2/M transition checkpoint, while the oscillatory CDK activity enables DNA re-replication. Low kinase activity is thought to allow pre-replication complexes (pre-RC) to set up at the origin of replication (ORI) in the chromatin, while high kinase activity permits the replication machinery to proceed and copy the DNA (Sugimoto-Shirasu and Roberts 2003). This oscillatory activity pattern is achieved mostly by utilising common cell cycle machinery with only a few proteins found to be specific for endoreduplication (Dewitte et al. 2007; Ishida et al. 2008; John and Qi 2008; Chevalier et al. 2011) (Fig. 7.4a).

Little is known about how MPF inactivation is achieved; however, studies of *Arabidopsis* have shed some light on this topic. The family of CDKB1s has been shown to drive the cells through the G2/M transition during the cell cycle (Menges et al. 2006; Gutierrez 2009). Within this family, CDKB1;1 seems to play a special role and its high activity is crucial for keeping the cells in the cell cycle and preventing them from entering the endocycle (Boudolf et al. 2004, 2009). Indeed, analysis of *Arabidopsis* transformants with reduced CDKB1;1 activity has shown that the down-regulation of CDKB1;1 results in the onset of the endocycle (Boudolf et al. 2004). This downregulation in wild type plants is probably achieved by a decrease in the availability of the cyclin CYCA2;3 through its selective degradation, as the stability and activity of CDKB1;1 is dependent on its interactions with CYCA2;3 during the G2/M transition. It has also been suggested that the CYCA2;3-CDKB1 complex is responsible for the temporal downregulation of the CYC-CDKA1 activity, which might result in blockage of the G2/M transition (Verkest et al. 2005a; Menges et al. 2005; Gutierrez 2009).

Precise data about how the cells are able to undergo multiple rounds of DNA replication are lacking; however, it seems that this involves changing the activity of members of the CYCD family. Among CYCD family members,

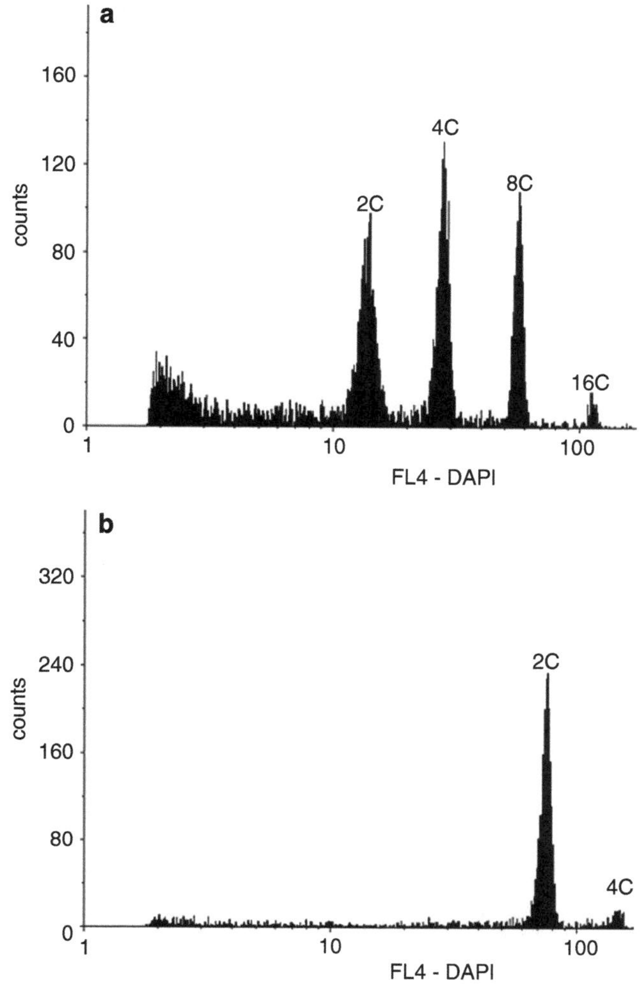

Fig. 7.3 Flow histograms of relative fluorescence of nuclei isolated from hypocotyls from (**a**) polysomatic (*Brassica campestris)* and (**b**) non-polysomatic (*Helianthus annuus*) plants. In (**a**) four peaks are visible showing endopolyploidy up to 16C DNA while in (**b**) the large peak indicates that most nuclei are 2C. The second, much smaller peak shows that there are, however, a few nuclei with a 4C DNA amount. *x-axis* nuclear log DAPI fluorescence intensity, *y-axis* number of nuclei

CYCD3;1 has been shown to be necessary for keeping cells in the cell cycle. Studies of *Arabidopsis* mutants have shown that CYCD3;1 acting in response to a cytokinin signal is involved in the maintenance of the cell cycle while its downregulation seems to be needed for the onset of endoreduplication (Dewitte et al. 2007).

7.2.2 DNA Replication in the Endocycle

The G1/S transition and DNA replication are mainly regulated by the S-phase specific cyclin family CYCD (Menges et al. 2006). During the endocycle a sub-family of CYCD1 seems to play an important role. It has been shown that one of the main targets of the CYCD1-CDKA1 complex is RETINOBLASTOMA-RELATED protein (RBR), which forms an inhibitory complex with the E2FA transcription factor (E2FA TF) when it is unphosphorylated (Joubes and Chevalier 2000; Park et al. 2005; Gutierrez 2009). The phosphorylation of RBR, via the action of the CYCD1-CDKA1 complex, releases the E2FA TF allowing it to form a complex with the DIMERISATION PARTNER A (DPA) protein. The E2FA-DPA complex then induces the expression of the main replication proteins. Consequently, the action of the E2FA-DPA allows the pre-RC to assemble and licences the start of the next S-phase (Joubes and Chevalier 2000; del Castellano et al. 2001, 2004; Park et al. 2005; Berckmans and De Veylder 2009).

DNA replication during the endocycle also involves the specific action of various proteins not found during the mitotic cell cycle replication. Topoisomerase VI (TOPOVI) is a plant-specific topoisomerase involved in chromosome decatenation. Analysis of *Arabidopsis* mutants with non-functional components of TOPOVI complex has shown that they only undergo the first two rounds of

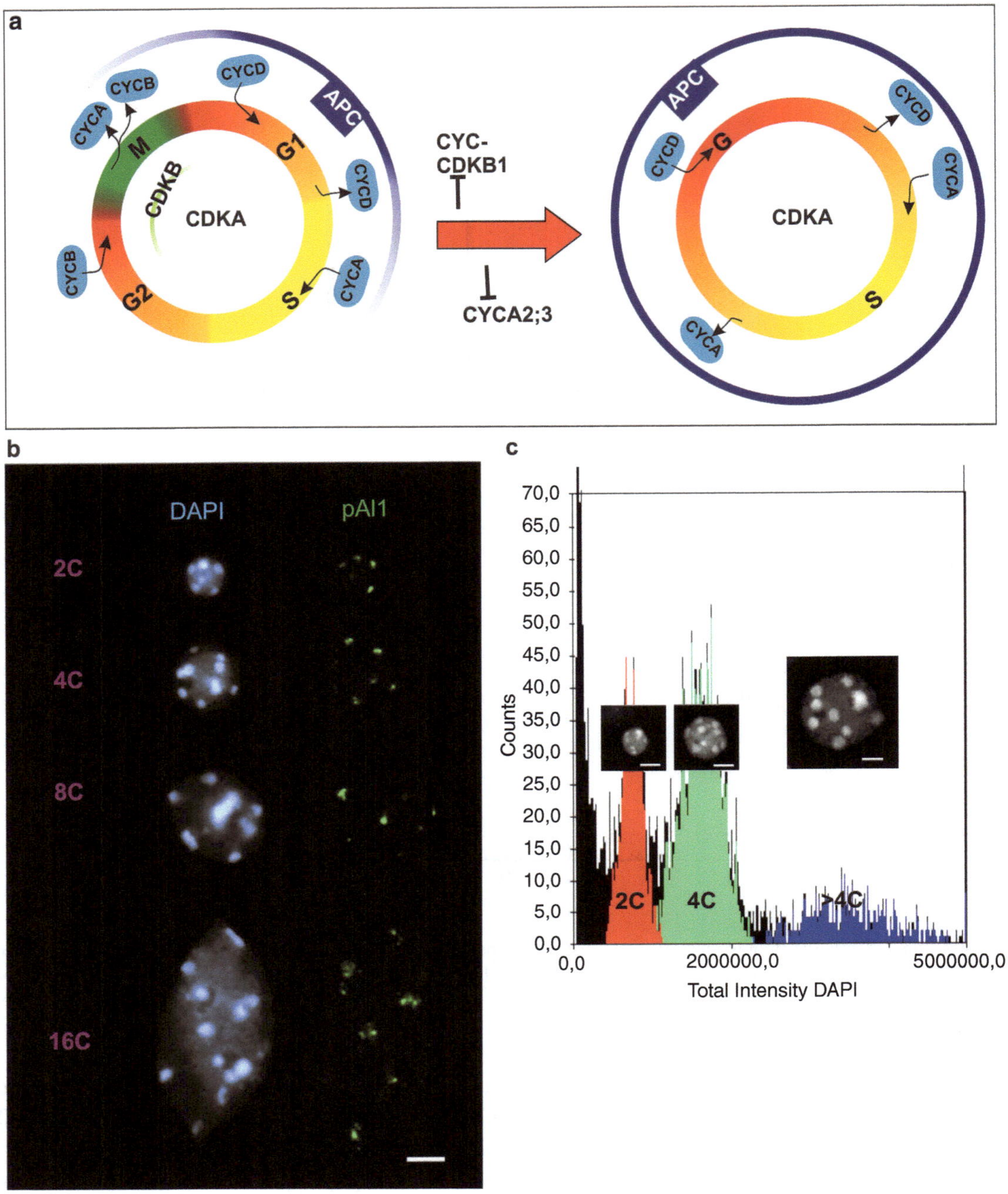

Fig. 7.4 (**a**) Schematic representation of the cell cycle (*left*) and the endocycle (*right*). The cell cycle consists of two main phases: replication (S-phase) and division (M-phase), separated by G1 and G2 phases. Transition from one phase to another is determined by the cyclin-dependent kinases (CDK)—CDKA and CDKB. CDKA is present constantly during the cell cycle, while CDKB is considered to be specific for the G2/M transition. CDK activity and substrate specificity are regulated by cyclin (CYC) proteins, thus the periodic presence of different CYCs, determines cell cycle progression. The anaphase promoting complex (APC) participates in the degradation of mitotic cyclins, thus preventing cells from entering mitosis prematurely. The switch from the cell cycle to the endocycle involves a decrease in the activity of CYC-CDKB complexes and inactivation of CYCA2;3. The endocycle is considered to be the cell cycle without the M-phase, thus it is regulated by similar mechanisms. As for the cell cycle, its progression is regulated by the oscillatory activity of CYCs, however only CYCA and CYCD have been shown to be active during the endocycle. Constant activity of APC ensures that nuclei do not enter mitosis. (**b**) Changes in the nuclear morphology and chromocenter structure during consecutive rounds of endoreplication. *Arabidopsis thaliana* stem leaf nuclei after DAPI staining (*left*) and fluorescence *in situ* hybridization with the centromeric probe, pAl1 (*right*). In nuclei with low ploidy levels centromeric heterochromatin is compacted, indicating sister chromatid association, while in nuclei of higher ploidy levels centromeric heterochromatin is decondesed and appears diffuse. (**c**) Analysis of the DNA content in *A. thaliana* stem leaf nuclei using image cytometry. Slides with stem leaf nuclei were stained with DAPI and examined under the microscope. The DNA contents of the nuclei were evaluated by measuring fluorescence intensity (x-axis) using image cytometry and the results are displayed as a histogram. *Peaks* represent consecutive ploidy levels (2C, 4C and >4C), determined from relative fluorescence intensity. Pictures of representative nuclei for each ploidy level are shown over each peak. *Bar*: 5 μm

endoreduplication and fail to succeed in the following rounds, suggesting a role of TOPOVI in proper DNA replication in nuclei with higher ploidy levels (Sugimoto-Shirasu et al. 2002; Kirik et al. 2007; Sugimoto-Shirasu et al. 2005).

There is also evidence for the involvement of the chromatin assembly factor 1 (CAF-1) in the regulation of the endocycle. CAF-1 is involved in the assembly of nucleosomes onto a newly synthesised DNA strand. In *caf-1* mutants, nuclei were seen to go through one additional endocycle on average. It therefore seems that the replication machinery is also involved in the regulation of the number of endocycles (Exner and Henning 2008; Gutierrez 2009).

7.2.3 Endocycle Regulatory Proteins

7.2.3.1 Transcription Factors in the Endocycle

Control of the endocycle is provided at various levels by many pathways that are interdependent and often form feedback loops (Edgar and Orr-Weaver 2001; Lee et al. 2009). It has been shown that the basic steps needed for DNA replication remain under the transcriptional control of the E2F TF family. These factors can be divided into typical and atypical TFs, depending on whether their action requires dimerisation with DP proteins or whether they work as monomers, respectively. In *Arabidopsis* typically E2Fs act either as transcriptional activators (E2FA and E2FB) or repressors (E2FC) and their main action takes place during the G1/S transition (del Castellano et al. 2001, 2004; Sozzani et al. 2006). E2F TFs may also be involved in the onset of an endocycle, as their target elements have been found in the promoter region of *CDKB1* and *CDKB1;1* (Boudolf et al. 2004; del Pozo et al. 2006). The group of atypical E2F TFs work as transcriptional repressors (Berckmans and de Veylder 2009). Several studies have provided evidence for their role in the regulation of the transition from the cell cycle to the endocycle, e.g., the DP/E2F-like protein DEL1 has been found to control entry into endoreduplication, operating upstream of the proteosomal degradation pathway (Vlieghe et al. 2005; Lammens et al. 2008). Apart from E2F TFs, whose actions are the best studied, other transcription factors have been shown to impact both the onset of an endocycle and their number. Most of them are involved in *CYC* and *CDK* gene transcription; however, their precise mode of action is unknown (Wang et al. 2006; Berckmans and De Vevlder 2009; Tojo et al. 2009).

7.2.3.2 Protein-Protein Interactions in the Regulation of an Endocycle

During endoreduplication an important role is played by the posttranslational control of various regulatory proteins. These proteins can be divided into two groups, depending on whether they execute their action directly, binding to their target, or indirectly, by chemically modifying their target.

The first type of action is performed by the INHIBITOR/INTERACTOR OF THE CDK/KIP-RELATED PROTEINS (ICK/KRP), the plant CYCLIN DEPENDENT KINASE INHIBITORS (CKIs). ICK/KRPs most probably execute their inhibitory action by binding to CYC-CDK complexes and inhibiting their kinase activity (Verkest et al. 2005a). Particular members of ICK/KRP show substrate specificity, which is determined by the cyclin involved in the CYC-CDK complex (Verkest et al. 2005a; Wang et al. 2006). However, even though their main action seems to be focused on CYC-CDK complexes, ICK/KRPs can also interact with cyclins (Coelho et al. 2005; Wang et al. 2006). Several studies have highlighted the role of ICK/KRPs in the onset and progression of an endocycle in a concentration-dependent manner (Schnittger et al. 2003; Coelho et al. 2005; Verkest et al. 2005b; Bisbis et al. 2006). One of the best studied ICK/KRP members is *Arabidopsis* SIAMESE (SIM), a crucial regulator of the onset of an endocycle in trichomes (Churchman et al. 2006). Mutants with truncated SIM protein produce multicellular trichomes with nuclei possessing lower ploidy levels, compared with wild-type plants. Experiments using Fluorescence Resonance Energy Transfer (FRET) methods have shown that SIM interacts with CYCD and CDKA1, blocking the G2/M phase transition and allowing the onset of an endocycle (Churchman et al. 2006; Kasili et al. 2010).

Depending on the effect of their action, proteins introducing chemical modifications can be divided into inhibitors or activators. Among the inhibitors, is WEE1 kinase which has been shown to play an important role during the cell cycle and is also involved in the endocycle. WEE1 is involved in M-phase inhibition, probably acting through CDKA1 phosphorylation. Tomato transformants with an impaired expression of Solly:WEE1 were generally smaller and developed smaller fruits with lower ploidy levels compared with wild-type plants. (Sorrell et al. 2002; Gonzalez et al. 2007). Activating phosphorylation is performed by CYCLIN-DEPENDENT KINASE ACTIVATING KINASE (CAK). It has been suggested that one of the CAKs (CDKF;1), which is necessary for activation of the CYC-CDKA complex during the cell cycle, plays a crucial role as a positive endocycle regulator (Umeda et al. 2000; Shimotohno et al. 2006; Takatsuka et al. 2009).

It has also been shown that SUMO (small ubiquitin-like modifier) ligase E3 may act as an endocycle repressor in different cell types. Seedlings of *Arabidopsis* mutants with inactive SUMO ligase only live up to three weeks and show severe developmental defects. These defects concern the cell cycle, and result in the malformation of the meristem as well as extra rounds of endoreplication in leaf cells (Ishida et al. 2009). The precise mechanism in which SUMOylation affects an endocycle still needs to be evaluated; however, *cyclin*, *CDK* as well as the *E2F* genes seem to be potential targets. A possible SUMO action over the structure maintenance of chromosome (SMC) complex has also been proposed (Ishida et al. 2009).

7.2.3.3 Protein Degradation

The molecular pathways driving the protein degradation necessary for an endocycle remain largely unsolved. It is generally known that endocycle protein activity is controlled through a ubiquitination-dependent proteasomal-degradation pathway. Three enzymes are responsible for performing ubiquitination: ubiquitin activating enzyme (E1), ubiquitin conjugating enzyme (E2), and ubiquitin ligase (E3). The E3 enzyme is a multisubunit complex that performs ubiquitination and is responsible for its substrate specificity (Vierstra 2009). During the endocycle, as in the cell cycle, two E3 ligase types have been shown to be involved: ANAPHASE-PROMOTING COMPLEX (APC) and SKP, CULLIN, F-BOX CONTAINING COMPLEX (SCF) (Kondorosi and Kondorosi 2004; Hershko 2005). Both APC and SCF are nuclear proteins which direct proteins destined for degradation into the 26S proteasome pathway. The specificity of their action is determined by several regulatory subunits which are different for each complex (Vierstra 2009). It has been suggested that APC action focuses on regulating the availability of cyclin, while SCF targets different regulatory proteins as well as the E2F transcription factors (Hershko 2005; Del Pozo et al. 2006; Ren et al. 2008). In different organs the APC substrate specificity is regulated by different factors. For example in *Arabidopsis*, CCS52A and CCS52B proteins determine the substrate specificity of the APC (Kondorosi and Kondorosi 2004; Li et al. 2009). The CCS52A2 regulatory subunit is responsible for driving the ubiquitin-dependent degradation of CYCA2;3 in leaves, whereas a similar role is played by CCS52A1 in roots (Lammens et al. 2008; Boudolf et al. 2009). The role of proteasomal degradation has also been shown for other proteins such as members of pre-RC: CDC6 or CDT1 (del Castellano et al. 2001, 2004). Generally, the constant activity of both APC and SCF, which change their substrate specificity during the endocycle, seems to provide the mechanism by which the oscillatory activity of CDK is achieved. Recently Sako et al. (2010) showed that a mutation in one of the subunits of the 26S proteasome regulatory particle affected the number of endocycles in *Arabidopsis* trichomes, thus indicating the direct involvement of the 26S proteasome pathway in the regulation of an endocycle.

7.2.3.4 Determination of the Number of Endocycles

Each cell type has a determined ploidy level that indicates that the DNA replication cycles need to be stopped at a particular time (Joubes and Chevalier 2000; Jovtchev et al. 2007; Lee et al. 2009). Current data point to several proteins that seem to be specifically involved in the exit from the endocycle. Imai et al. (2006) studied *Arabidopsis* mutants with a truncated action of CYCA2;3. They showed that the loss of CYCA2;3 activity resulted in large phenotypic changes in plants. Compared with wild-type plants the mutants had trichomes with bigger nuclei; their roots, cotyledons, and first leaves had cells with higher ploidy levels; and the first endocycle occurred earlier in development. Molecular studies have shown that the CYCA2;3 activity is negatively correlated with cell ploidy levels and that CYCA2;3 is able to form an active complex with CDKA1 in plants. Thus, it has been suggested that the CYCA2;3-CDKA1 complex is involved in the termination of an endocycle, probably by the down-activation of the pre-RC complexes. Moreover, studies of *Arabidopsis* trichomes have also pointed to the role of GT2 LIKE1 (GTL1) in the termination of an endocycle. GTL1 belongs to the trihelix TF family and acts downstream of TRANSPARENT TESTA GLABRA (TTG1) and GLABROUS 2 (GL2) TFs (Breuer et al. 2009). Unfortunately, the precise targets of GTL1 have not yet been identified.

7.2.4 Factors Involved in Endopolyploidy Modulation

Endoreduplication remains under the developmental control of various phytohormones. The best studied are the actions of gibberellin (GA), auxin and cytokinins (see below); nevertheless, other phytohormones are also known to have an impact on the endocycle. Ethylene has been shown to be involved in regulating the number of endocycles in *Arabidopsis* hypocotyls. Treating wild-type plants with an ethylene precursor resulted in extra rounds of endoreplication (Gendreau et al. 1999). The actions of brassinosteroids seem to be involved in maintaining cells in the endocycle. Studies on the *Arabidopsis* brassinosteroid-insensitive mutant *bin4-1* showed that leaf ploidy levels did not exceed 8C, whereas in wild-type plants they reached 32C. This mutation resulted in the truncation of the TOPOVI complex and was correlated with an increased level of CDKB1;1, an endocycle inhibitor (Breuer et al. 2007). In contrast, abscisic acid (ABA) has been shown to have a negative impact on the progress of an endocycle. Working during the G1/S transition, ABA is probably involved in the regulation of genes involved in replication and CDK regulation (del Castellano et al. 2004; del Pozo et al. 2005; Wang et al. 2006). It has also been shown that environmental factors including light and nutrient availability affect an endocycle and are involved in determining their number.

7.2.4.1 Gibberellic Acid

Gibberellic Acid (GA) signalling is known to be involved in the onset of an endocycle during *Arabidopsis thaliana* trichome differentiation. Using a wide variety of *A. thaliana* trichome mutants with disrupted GA signalling, many proteins participating in the GA signalling pathway have been discovered. It has been shown that during trichome differentiation GA acts in a concentration-dependent manner activating various transcription factors (Perazza et al. 1998; Ishida et al. 2008). Among the genes directly activated by those regulators are the key genes involved in the onset and progress of an endocycle like: *RBR1*, *SIM* and *ICK1/KRP1*. During trichome differentiation *RBR1* is probably involved in the determination of the number of endocycles, whereas *SIM* participates in the induction of endocycles (Churchman et al. 2006; Morohashi and Grotewold 2009). Along with SIM, ICK1/KRP1 also plays a role in the induction of endocycles in trichomes by downregulating the activity of mitotic CYC-CDK complexes (Schnittger et al. 2003).

7.2.4.2 Auxin

In the body of a plant auxin forms concentration gradients which give rise to differences in the local auxin concentration. These differences are responsible for the local character of auxin action, including the regulation of endoreduplication (del Pozo et al. 2005; Gutierrez 2009). A positive correlation between auxin concentration and the onset of an endocycle has been shown in maize endosperm cells (Lur and Setter 1993). On the other hand, in the roots of *Arabidopsis* a low auxin concentration seems to promote endoreduplication, whereas high concentrations determine meristematic cell activity in the proximal meristem (Ishida et al. 2010). Treating *Arabidopsis* seedlings with auxin antagonists results in a reduction of the root meristem size. The same results have been obtained in *Arabidopsis auxin resistant mutant 3* (*axr3*), which is auxin insensitive. Thus, in the *Arabidopsis* root meristem, cells enter an endocycle prematurely when auxin signalling is inhibited. Ishida et al. (2010) suggested that high endogenous levels of auxin, known to be present in the root division zone, have an inhibitory effect on the onset of an endocycle in *Arabidopsis* proximal root cells. Whereas in cells outside the meristematic zone, where auxin concentrations are lower, the repression of endoreduplication might be lifted allowing cells to switch to the endocycle.

In an endocycle, auxin has been shown to act as a KRP2-specific repressor and to modulate the activity of the E2F transcription factors, which are known to regulate the expression of the endocycle genes, including genes necessary for DNA replication (Magyar et al. 2005; Verkest et al. 2005a; Sozzani et al. 2006). Generally, auxin can influence DNA synthesis and epigenetic modifications as well as the degradation of different proteins involved in both the cell cycle and endocycle, and thus execute the developmental program of the cell (Horvath et al. 2006; Ishida et al. 2009, 2010; Perrot-Rechenmann 2010).

7.2.4.3 Cytokinins

Cytokinins have been shown to regulate the transcription of the main cell cycle genes, and thus are considered to be one of the main factors affecting the proliferative activity of plant cells. In most studies their action seems to be antagonistic to auxin (del Pozo et al. 2005). Ishida et al. (2010) studied how the application of exogenous cytokinins influenced the timing of the onset of an endocycle during the differentiation of *Arabidopsis* root cells. After treatment with exogenous cytokinins, an increase in the amount of endogenous cytokinins was noticed. Studied plants also showed a decrease in root meristem size. This led to a suggestion that elevated cytokinin levels are responsible for the earlier onset of an endocycle in *Arabidopsis* proximal root cells. A similar analysis conducted on mutants with disrupted cytokinin synthesis showed a significant delay in the onset of the endocycle. However, this does not seem to be a general rule. During the development of *Arabidopsis* leaf and maize endosperm, high cytokinin levels were positively correlated with a high mitotic activity of the cells (Lur and Setter 1993; Xia et al. 2009). The action of cytokinins is executed by activating the dephosphorylation of mitotic CYC-CDK complexes along with the positive regulation of *CYCD*, particularly *CYCD3*, gene transcription, thus keeping cells in the cell cycle. It is possible that cytokinins might also act through directing specific proteins to a proteasomal pathway (Riou-Khamlichi et al. 1999; Dewitte and Murray 2003).

7.2.4.4 Light

Studies of *phyA* and *phyB* mutants of *A. thaliana* have provided evidence for the involvement of the phytochromes PHYA (far red light receptor) and PHYB (white light and red light receptor), in the regulation of endoreduplication. Both phytochromes act by inhibiting specific rounds of DNA synthesis. In the hypocotyls of *Arabidopsis* seedlings grown in the light, cells went through two rounds of replication reaching 8C ploidy level, while in seedlings grown in the dark an extra round of replication was observed creating nuclei with 16C (Gendreau et al. 1998). The number of endocycles also remains under blue light control, probably executed through the ubiquitination pathway (Tsumoto et al. 2006). In *Arabidopsis* cotyledons, the action of light stimulated the expression of the main cell cycle genes, including those with an impact on endoreduplication (López-Juez et al. 2008).

7.2.4.5 Nutrients

Sucrose is known to regulate G1/S progression through protein phosphatase-dependent *CYCD* gene induction (Riou-Khamlichi et al. 1999; Hirano et al. 2008). Acting down this pathway, sucrose might also have an impact on the progression of an endocycle. Moreover, it has been shown that nutrient starvation affecting plant growth can also influence endoreduplication (Henriques et al. 2010).

7.3 Structure of the Endopolyploid Plant Nucleus

The structure of a plant nucleus and the organization of chromosomes inside, especially endopolyploid nuclei, is still poorly understood. One of the most widely studied are the nuclei of *Arabidopsis thaliana,* a polysomatic plant. The advantage of studying these nuclei is that they possess heterochromatic chromocenters. In meristematic cells, chromocenters are clearly visible after fluorescent staining with DAPI and they correspond to chromosome number. The number of chromocenters also allows the ploidy level of the cell to be estimated. Compared with meristematic nuclei, endoreduplicated nuclei are bigger, in proportion to the increase in DNA ploidy, and usually they have bigger chromocenters (Maluszynska and Heslop-Harrison 1991; Schubert et al. 2006; Henriques et al. 2010). However, the increase in size is not linearly proportional to the DNA content indicating that chromatin reorganization has occurred. There is also evidence for the differential behaviour of particular chromatin domains. For example, there is a greater tendency for NORs (Nucleolar Organizing Regions) to fuse in *A. thaliana* leaf nuclei, compared with the centromeric regions of different chromosomes which stay preferentially separated. However, it should be noted that NOR fusion is most probably connected with their function in the organization of the nucleolus (Pecinka et al. 2004; Schubert et al. 2006; Berr and Schubert 2007).

Very little is known (with exceptions, see Sect. 7.1.1 and Figs. 7.1 and 7.2) about the structure of endoreduplicated chromosomes since they rarely condense and become visible. During endopolyploidisation, consecutive rounds of replication result in the formation of chromosomes built of numerous chromatin strands, sister chromatids, whose number increases exponentially with each endocycle. This implies that only indirect studies of the endoreduplicated chromosome structure and organization are possible. One of the most useful techniques is fluorescence *in situ* hybridization (FISH) using chromosome or chromosome-region specific probes. By analysing the size and localization of the signal insights into the structure of endoreduplicated chromosomes can be gained (Fig. 7.4b). Indeed, using this approach for *A. thaliana* leaf nuclei, it was shown that sister chromatids stayed associated along their entire chromosome length in nuclei with low endopolyploidy levels. However, as ploidy level increased, separation of sister chromatids took place (Schubert et al. 2006). Separation started at the terminal regions, and, as the number of DNA strands increased, this extended into the interstitial regions. With ploidy levels over 16C, dissociation of centromeric regions was observed (Schubert et al. 2006, 2008). It has been suggested that sister chromatid dissociation is probably the result of a decrease in the number of cohesin sites along the chromosome arms (Schubert et al. 2009) and results obtained for endoreduplicated nuclei in the endosperm of *Zea mays* support this thesis. A tighter association of DNA at centromeric and knob regions than in other chromosome regions was also revealed by FISH (Bauer and Birchler 2006).

The chromatin structure remains under epigenetic control. Unfortunately, very little is known about the epigenetic changes associated with endoreduplication. Data from *Triticum durum* have suggested that there is increased methylation in endoreplicated nuclei (Polizzi et al. 1998) and recent studies on tomato fruit indicate that such modifications may be tissue-specific (Teyssier et al. 2008). On the other hand, a comparison of the pattern of epigenetic modifications between *A. thaliana* meristematic and endoreplicated root and leaf nuclei revealed general similarities despite an increased acetylation of the chromatin in the chromocenters (Jasencakova et al. 2003).

7.4 Occurrence of Endopolyploidy

7.4.1 Endopolyploidy in Species

A relationship between taxonomic position and endopolyploidisation was first reported by Tschermak-Woess (1956a) and has since been confirmed by several authors (D'Amato 1964; Nagl 1976; Olszewska and Osiecka 1982; Barow and Meister 2003). The occurrence of endopolyploidy and the degree of endopolyploidisation also seems to be characteristic for plant families. Endopolyploidy has mainly been noted to be common in species with a small genome size (e.g., *Arabidopsis thaliana* (Galbraith et al. 1991), *Solanum lycopersicum* (Smulders et al. 1994), *Lupinus cosentinii* (Hajdera et al. 2003), *Brassica oleracea* (Kudo and Kimura 2001a) and it has been suggested that this may be because certain specialized cells need a minimum amount of nuclear DNA to maintain their specific functional status, something that can be achieved through endopolyploidisation in vegetative cells (Nagl 1976; Galbraith et al. 1991).

To investigate the impact of genome size and taxonomic position on endopolyploidisation, comparative analyses were made between plants belonging to 16 phylogenetically-diverse families covering a wide range of genome sizes in each family (Barow and Meister 2003). Interestingly, a negative correlation between endopolyploidisation and genome size was confirmed for only three of the investigated families (Brassicaceae, Urticaceae and Solanaceae). Some species with small genomes showed low levels of endopolyploidy (e.g., *Haplopappus gracilis, Aquilegia vulgaris*) or were non-polysomatic (e.g., *Oryza sativa*) whereas some species with large genomes in Fabaceae or Alliaceae were polysomatic. The authors concluded that there was only a weak but significant negative correlation between genome size and endopolyploidisation.

Families may differ significantly in the presence and pattern of endopolyploidy. For instance in some families polysomatic species with high level of endopolyploidy predominate (e.g., Amaranthaceae, Cucurbitaceae, Chenopodiaceae and Brassicaceae), while Solanaceae, Alliaceae, and Fabaceae show an intermediate level and Poaceae a low level of endopolyploidy (Barow and Meister 2003; Barow and Jovtchev 2007). Usually, the pattern of endopolyploidy in related species is similar but there are some exceptions (e.g., the polysomatic species *Aquilegia vulgaris* belongs to Ranunculaceae, a family dominated by non-polysomatic species; Barow and Meister 2003).

The endopolyploidy pattern may differ between individuals of the same species belonging to different ecotypes or varieties. For example, the frequency of endopolyploid cells in roots was observed to differ between 18 ecotypes of *Arabidopsis thaliana* (Beemster et al. 2002). Significantly different patterns of endoreduplication have also been described in the endosperm of several inbred lines of *Zea mays*. Flow cytometric analysis showed that the DNA content of nuclei ranged from 3C to 48C in some lines while in others from 3C to 192C (Larkins et al. 2001). Similarly, different endoreduplication patterns and number of endocycles have been reported in mature cotyledons of different genotypes of *Pisum sativum* (Lemontey et al. 2000).

7.4.2 Life Strategy

The level of endopolyploidisation seems to be correlated not only with genome size but also with the type of life cycle. Endopolyploids are more frequent in annual rather than perennial herbs. Among the 34 polysomatic species investigated by Barow and Meister (2003), nine were perennial and 25 annual or biennial. Woody plants were non-polysomatic (Barow and Meister 2003).

Analysis of endoreduplication in root cells during differentiation within different monocotyledonous families revealed a higher level of endopolyploidy in species with a lower DNA content as well as an annual type of life cycle (Olszewska and Osiecka 1982). These observations were confirmed for the root cells of dicotyledonous herbaceous species (Olszewska and Osiecka 1983). In general, a short life cycle is usually correlated with a small genome size and a high level of endopolyploidy. A good example of such a correlation is *Arabidopsis thaliana*, which has a very small genome size (157 Mb, Bennett et al. 2003), a short life cycle and endopolyploidisation in all vegetative organs (Galbraith et al. 1991)

7.4.3 Endopolyploidy in Generatively Polyploid Plants

Considering the weak but significant negative correlation between genome size and endopolyploidy, the question has been raised as to whether polysomaty is affected by a plant's polyploidy level. A comparison of endopolyploidy patterns in diploid and polyploid lines of the same species should answer this question to see whether a certain number of endocycles is typical for a species regardless of polyploidy level (i.e., the number of endocycles is the same between diploid and polyploidy cytotypes) or whether there is a fixed maximum DNA amount for a species in which case the number of endocycles is expected to differ between diploid and polyploid cytotypes. Actually, both patterns have been observed.

Tetraploid plants of *Portulaca grandiflora* induced by colchicine treatment, showed the same number of endocycles as diploid plants following flow cytometric analysis (Mishiba and Mii. 2000) and similar results have also been obtained for artificially-induced tetraploid lines of *Zea mays* (Biradar et al. 1993) and *Solanum lycopersicum* (Smulders et al. 1994).

In contrast, flow cytometry analyses of diploid and induced autotetraploid *Arabidopsis thaliana* plants (ecotype Columbia) revealed different degrees of endoreduplication. A histogram for young leaves of diploid plants showed four peaks corresponding to a 2C, 4C, 8C and 16C DNA content (i.e., three endocycles took place) whereas for the tetraploid plants three peaks were observed (2C, 4C, 8C) indicating just two endocycles had occurred (Fras et al. 2007). The maximum DNA content was similar in diploid and tetraploid plants but a different number of endocycles had occurred. However, the situation for naturally occurring polyploids may be different from artificially induced ones as no differences in the number of endoreduplication rounds were observed in the hypocotyls between diploid ecotypes of *Arabidopsis* and the natural tetraploid Cape Verde islands ecotype. Two endocycles occurred in both cytotypes even though the cells of the tetraploid contained twice as much DNA as the wild-type plant (Gendreau et al. 1998). The results indicate that the endoreduplication pattern for natural and artificial polyploid plants may be different. On the basis of flow cytometric investigations of diploid and tetraploid accessions and species, it was concluded that natural polyploids showed a lower endopolyploidisation level than the corresponding diploid suggesting that the endopolyploidy level needs several generations to become established (Jovtchev et al. 2007). This hypothesis needs more detailed investigations and confirmation for different plant species.

7.4.4 Endopolyploidy in Plant Development

Plant development is the result of genetic predisposition and a range of environmental parameters. It is the effect of two processes—(1) cell division and (2) expansion which can result from endopolyploidisation. In an individual plant different organs can exhibit different patterns of endopolyploidisation. In some species extensive endopolyploidy has been reported in nearly all organs, while in other species it occurs to a low degree or not in all organs and is correlated with plant development.

7.4.4.1 Embryogenesis

Usually, the embryos of the dry fully maturated seeds of various plant species show large amounts of cells at the presynthetic G1 phase of the cell cycle (e.g., Smulders et al. 1994). Depending on the species, in the radicle only 2C nuclei may be observed (*Cichorium endivia*, *Capsicum annuum*), while in others (*Castanea sativa*, *Hordeum vulgare*), besides the 2C nuclei, a small number of 4C DNA nuclei have also been identified (Bino et al. 1993; Gendreau et al. 2008). For example, in the *H. vulgare* embryo, 82% of the nuclei of the radicle tip cells have a 2C content and only 18% have a 4C DNA content indicating that the majority of cells have been arrested in the G1 phase of the cell cycle. Interestingly, the percentages of 4C nuclei were different in the plumule (22%) suggesting that the cell cycle activity in the developing embryo has stopped at different stages depending on the organ (Gendreau et al. 2008). The occurrence of endopolyploidy (8C DNA nuclei) in the embryo has also been reported in some species including *Phaseolus vulgaris*, *Spinacia oleracea*, and *Chenopodium quinoa*. This indicated that endopolyploidisation can take place in the early stages of tissue differentiation during embryo development. Most often the endopolyploid cells have been observed in the radicles, which first appear during seed germination (Bino et al. 1993; Kolano et al. 2009).

7.4.4.2 Post-embryogenic Development

Systemic control of endopolyploidy has been shown in plant species such as *Arabidopsis thaliana*, *Cucumis sativus*, *Solanum lycopericum*, and *Mesembryanthemum crystallinum* (De Rocher et al. 1990; Galbraith et al. 1991; Gilissen et al. 1993; Smulders et al. 1994). Polysomaty in these species may be expressed in roots, leaves, petioles, stem internodes and flowers. Increasing evidence has revealed that a pattern of endopolyploidy is characteristic for tissue type and developmental stage indicating that polysomaty is spatially and temporally regulated. The patterns of endopolyploidisation within a plant of one species are usually different in various organs and correlate with the developmental stage (Galbraith et al. 1991; Gilissen et al. 1993; Barow 2006).

Seedling Development

During seed germination and subsequent seedling development an increasing proportion of endopolyploid nuclei appear. The highest level of endopolyploidy is usually observed in the hypocotyls and primary roots (Fig. 7.5). The extent of endopolyploidisation differs significantly between species, for example hypocotyls of *Solanum lycopersicum*, *Brassica oleracea* and *Chenopodium quinoa* (quinoa) undergo two endocycles (up to 16C DNA) whereas hypocotyls of *Beta vulgaris* (sugar-beet) undergo three endocycles (up to 32C DNA) (Smulders et al. 1994; Kudo and Kimura 2001a; Sliwinska and Lukaszewska 2005; Kolano et al. 2009).

There are also species-specific endopolyploidy patterns in primary roots. The number of endocycles can be as low as one (e.g., in tomato roots) or more numerous as in sugar-beet roots (Smulders et al. 1994; Sliwinska and Lukaszewska 2005). Older organs usually exhibit higher levels of endopolyploidy than younger ones. An increase of endopolyploidy to 32C and 16C DNA has been observed in the hypocotyls and primary roots of cucumber and cabbage seedlings, respectively (Gilissen et al. 1993; Kudo and Kimura 2001a). Another pattern of endoreduplication in seedling development has been observed in quinoa and sugar-beet. Here the endopolyploidy level initially increases and then gradually decreases (Sliwinska and Lukaszewska 2005; Kolano et al. 2009).

The high level of endopolyploidy observed in hypocotyls and primary roots can be explained by the considerable proportion of vascular tissue present in these organs. The increase of endopolyploidy seems to be correlated with their rapid elongation during germination, and this may be connected with vascular tissue development. Endopolyploid cells appear slightly later in the cotyledons during germination and seedling development (e.g., cucumber, tomato, quinoa, sugar-beet). The level of endopolyploidy depends on the species and can reach 4C in *C. quinoa* (Fig. 7.5), 32C in *A. thaliana* or as much as 64C DNA in *Lupinus cosentinii* (Galbraith et al. 1991; Hajdera et al. 2003; Kolano et al. 2009). A very high level of endopolyploidisation in *L. cosentinii* cotyledons seems to be typical for seed storage organs, and could be connected with the high metabolic activity of the tissue (Larkins et al. 2001).

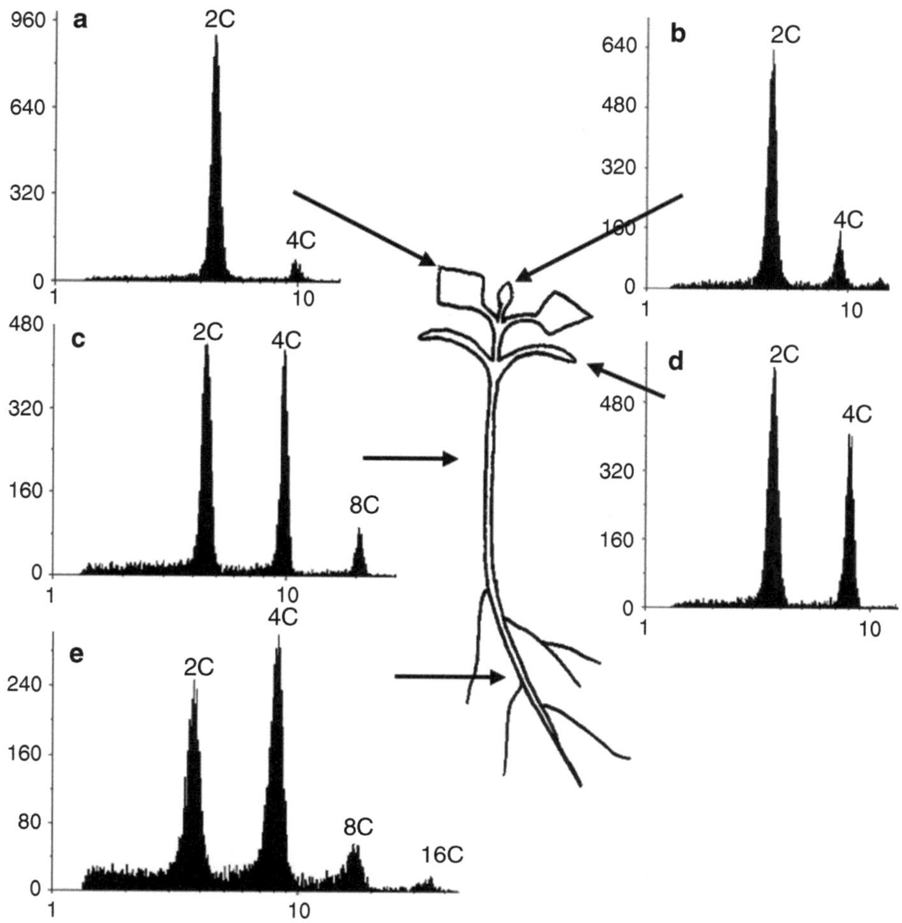

Fig. 7.5 Evaluation of endopolyploidy in seedling of *Chenopodium quinoa* (quinoa) determined by flow cytometry. (**a**) Leaf; (**b**) shoot apex; (**c**) hypocotyl; (**d**) cotyledon; (**e**) root. *x-axis* nuclear log DAPI fluorescence intensity, *y-axis* number of nuclei

Species- and Organ-Specific Endopolyploidisation

Almost all plant organs can become polysomatic during vegetative and reproductive development. However, an absence of endopolyploidy in the shoot and root apical meristems is generally observed (Kudo and Kimura 2001a; Lim and Loh 2003). It has been suggested that repression of endopolyploidisation at the shoot apex might be one of the mechanisms to ensure genetic stability of the germ line (Kudo and Kimura 2001a). Usually the frequency of endopolyploid nuclei is highest in organs rich in vascular tissue (e.g., petioles, leaf main midribs, stem and root) as well as in storage tissue (cotyledons, endosperm) (Barow and Meister 2003; Barow 2006).

The level of endopolyploidy is generally species- and organ-specific but usually older organs exhibit a higher level of endopolyploidy than younger ones. An exception to this may be the leaves of *Cucumis sativus* where the pattern of polysomaty remains unchanged in young and mature leaves (Gilissen et al. 1993). In the same organ, a gradient of endopolyploidy can also be observed. For example, in *Amaranthus caudatus* the number of cells with 2C DNA decreased and the number of cells with a DNA content higher than 4C increased in the basal part of the shoot compared with the youngest most distal part (Fig. 7.6) (Kolano et al. unpublished). In the leaves of the orchid *Vanda* 'Miss Joaquim' the level of endopolyploidy increased from base to tip (acropetally), while in cotyledons of *C. sativus* and leaves of *Spathoglottis plicata*, it increased from tip to base (basipetally) (Gilissen et al. 1993; Yang and Loh 2004).

The level of endopolyploidy in leaf cells differs considerably among species. For example, cells with a DNA content higher than 4C were not observed in the young fully extended lamina of *Chenopodium quinoa* (Kolano et al. 2009) whereas in leaves of *Arabidopsis thaliana* or *Solanum lycopersicum* cells were present that had undergone one or two endocycles. Interestingly, the leaves of succulents such as *Portulaca grandiflora* or *Mesembryanthemum crystallinum* can exhibit high levels of endoreduplication up to 64C. It has been suggested that the high nuclear DNA content in succulent

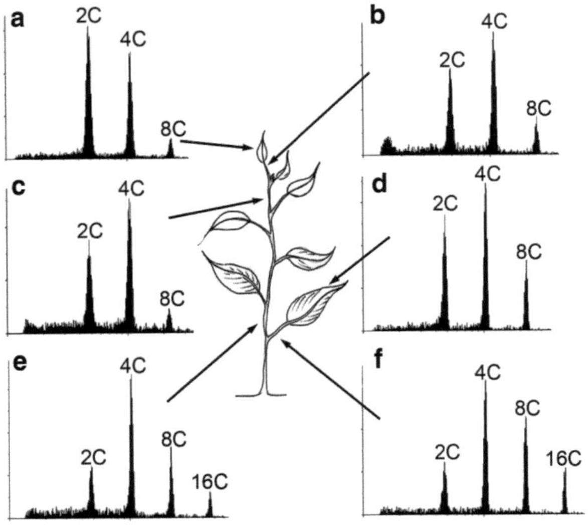

Fig. 7.6 Comparison of endopolyploidy patterns in young and old organs of *Amaranthus caudatus*. (**a**) Young leaf lamina; (**b**) young leaf petiole; (**c**) young internode of the shoot; (**d**) old leaf lamina, (**e**) old internode of the shoot; (**f**) old leaf petiole. *x-axis* nuclear log DAPI fluorescence intensity, *y-axis* number of nuclei

leaves occurs in the water storing mesophyll cells (De Rocher et al. 1990; Mishiba and Mii 2000).

Tissue-Specific Endopolyploidisation

Differences in the pattern of endopolyploidy can occur within an organ. For example, in sugar-beet leaves there was no endopolyploidisation in any part of the leaf lamina, regardless of leaf age. However, endopolyploid nuclei occurred in the main midrib (Lukaszewska and Sliwinska 2007). Even within a single tissue, the cells can differ in endopolyploidy level. An analysis of *Brassica* species showed that in the distal part of a petal nuclei were small and numerous with DNA contents of 2C and 4C, whereas in the proximal part nearly 40% of the epidermal cells were large and extremely elongated, with endopolyploidy up to 32C (Kudo and Kimura 2002). An additional example is found in the epidermis of *Arabidopsis*. Here a regular pattern of endopolyploidy is evident in the epidermal cells, with the nuclei of epidermal pavement cells exhibiting 2C, 4C, and 8C DNA amounts in the stem epidermis and 2C–16C in the leaf epidermis. Leaf trichome nuclei also have elevated ploidy levels of 4C, 8C, 16C, 32C and 64C. In contrast, the guard cell nuclei have a 2C DNA content i.e., typical of cells in the G1 phase of the cell cycle. Interestingly, among epidermal pavement cells a gradual decrease in endopolyploidy was found towards the guard cells (Melaragno et al.1993).

Endopolyploidy in Flowers

Endopolyploidy occurs in most flower organs such as the petals, sepals, carpels and stamens of many species (Mishiba and Mii 2000; Kudo and Kimura 2001b; Barow and Meister 2003; Lukaszewska and Sliwinska 2007). The presence of endopolyploid cells has been shown in different organs of the cabbage flower. Filament tissue contained cells with six ploidy levels (2C–64C) whereas petals had cells with five ploidy levels (2C–32C). A 2C–8C range of ploidy levels in the carpel tissue has been described while a high level of endopolyploidy has been reported for stamens (Kudo and Kimura 2001b).

For many species the highest level of endopolyploidy has been shown to occur in the flower organs (Barow and Meister 2003). For instance, petals of the orchids *Oncidium varicosum* and *Phalaenopsis* spp. and cabbage (*Brassica oleracea*) have been shown to undergo extensive endopolyploidisation with some cells reaching 64C (Kudo and Kimura 2001b; Lee et al. 2004). However, this is not always the case as in some species such as the generally polysomatic plant *A. thaliana*, petal cells remain diploid (Galbraith et al. 1991).

Endopolyploidy in Endosperm

The endosperm is the main source of nutrition for germinating seeds. Endosperm formation starts soon after fertilisation and it is often associated with the switch from a mitotic cell cycle to an endocycle. In maize, up to 90% of the nuclei can undergo endoreduplication but the number of endocycles varies among endosperm cells. The cells in the center of the endosperm enter endoreduplication first and are followed by the most adjacent cells and so forth. The spatial/temporal pattern of the cell cycle/endocycle switch creates a gradient in the DNA content and cell size with the smallest nuclei (3C and 6C) located at the periphery of the endosperm while increasingly larger endopolyploid nuclei are found towards the central region. Most commonly, four to five endocycles occur although individual nuclei with a DNA content of up to 690C have been observed. This endopolyploidisation is correlated with an increase in endosperm cell size and the rapid synthesis of starch, suggesting that endopolyploidy can increase metabolic activity, rRNA synthesis and transcriptional activity. However, a 50% reduction in the mean DNA content of mutant maize endosperm has very little affect on the accumulation of starch and storage proteins. Therefore, it is possible that endopolyploid endosperm may simply provide a mechanism to store nucleotides during the development and germination of the embryo (Larkins et al. 2001; Leiva-Neto et al. 2004; Lee et al. 2009).

Endopolyploidy in Fruit

The input of endopolyploidisation in the development of fruit has been extensively studied in the tomato (*Solanum lycopersicum*—see review by Chevalier et al. 2011). Growth of the tomato fruit starts after the bloom with intensive cell divisions. As development proceeds, the proliferative activity of the cells slows down until it ceases and the population progressively enters the stage of cell enlargement (Bertin et al. 2003). Cessation of cell divisions and an increase in cell size are closely linked to endopolyploidisation (Melaragno et al. 1993; Traas et al. 1998; Joubes and Chevalier 2000). Endopolyploid cells are present in the pericarp, locular gel and central columella at the breaker stage. The three tissues display different ploidy profiles, with the highest C-values (512C) in the pericarp. Sepals also become endopolyploid during fruit growth, but to a lower extent, with C-values up to 32C. Cheniclet et al. (2005) reported a positive correlation between endoreduplication and cell size in the pericarp tissues of the tomato.

7.4.4.3 Endopolyploidy in the Nodulation Process

Endopolyploidisation is involved in the nodulation process in legumes. These species establish a nitrogen-fixing symbiosis with rhizobia bacteria. During this process the root nodule is formed to supply the host plant with nitrogen. (e.g., *Medicago truncatula–Sinorhizobium meliloti* interaction). A longitudinal section of a *M. truncatula* nodule reveals cells at different stages of development, and exhibiting a differentiation gradient from the apical part to the basal tissues attached to the root (Newcomb et al. 1979). The apical meristem (Zone I) is composed of small proliferating, uninfected cells. Endoreduplication is primarily associated with the infection zone (Zone II). Indeed, the symbiotic differentiation program involves the arrest of cell division, followed by several rounds of endoreduplication. Enlarged endopolyploid cells are invaded and host the bacteroids, whereas the small diploid cells remain uninfected. Endoreduplication continues into Zone III where fully differentiated bacteroids develop the capacity to fix nitrogen. The DNA content increases from 2C and 4C levels up to 64C and is accompanied by nuclear and cell enlargement (Kondorosi and Kondorosi 2004; Wildermuth 2010). Endoreduplication and cell enlargement seem to play a central role in nodulation as studies of *Medicago* plants, whose nodules had reduced endopolyploidy levels, exhibited inefficient nitrogen fixation. The repeated endoreduplication cycles during symbiotic cell development might have dual roles, (1) they enable the extreme enlargement of cells to host the bacteroids and (2) they provide an energy and nutrient supply for the bacteroids through the increased transcriptional and metabolic activities in the host cell (Kondorosi and Kondorosi 2004).

7.4.4.4 Relationships Between DNA Content, Nuclear Volume and Cell Size

The contribution and role of endopolyploidy to plant growth and development is not yet fully understood. One problem that needs explaining is the correlation between the nuclear DNA content and cell size (Fig. 7.1). Plant development and cell differentiation are usually accompanied by an increase in cell size and since a correlation between nuclear DNA content and cell size has been reported in *A. thaliana* and other plant species (Melaragno et al. 1993; Bertin 2005; Jovtchev et al. 2006) this has led to the hypothesis of a 'nuclear–cytoplasmic ratio'. This hypothesis states that some control mechanism ensures that the amount of cytoplasm in a cell is proportional to the amount of DNA in its nucleus. Indeed, a simple comparison of genome size and nuclear and cell volume among species supports this theory. Species with larger genomes generally have larger nuclear and cellular volumes and it is well known that polyploid cells are bigger than diploid cells (Cavalier-Smith 2005; Jovtchev et al. 2006; Webster et al. 2009).

The hypothesis that endoreduplication does indeed determine the rate of cell growth has been proposed based on numerous observations. For example, cells that have undergone endopolyploidisation are larger than comparable diploid cells as has been shown in the epidermis of *A. thaliana* and *Brassica* species (Melaragno et al. 1993; Kudo and Kimura 2002; Jovtchev et al. 2006). In addition, a link between cell size and the average C-value has been reported in seeds (Lemontey et al. 2000) and fruits (Bertin 2005). The hypothesis is further supported by the fact that endoreduplication usually precedes cell expansion (Traas et al. 1998). For example, endoreduplication seems to be a prerequisite for cell expansion during the growth of tomato fruit (Bertin 2005). Endoreduplication is also associated with the increase in cell size as observed in *A. thaliana* leaf epidermal cells, etiolated hypocotyls and various mutants that exhibit a truncated size and/or morphology (Melaragno et al. 1993; Gendreau et al. 1997; Sugimoto-Shirasu and Roberts 2003). However, the 'nuclear–cytoplasmic ratio' theory does not explain why cells from different tissues in a given organism with the same amount of DNA differ in nucleus and cell size. For example, different relationships between the endopolyploidy level and cell volume were found in the root cortex of 18 accessions of *A. thaliana* (Beemster et al. 2002). In addition, although endopolyploidisation clearly contributes to increasing cell size, a key feature of this phenomenon seems to be a boost in the cell's capacity for future growth as the determination of the final cell size also depends on the development and function of the cell. The current data therefore actually argue for a physiological role of endopolyploidy as a facilitator of cell growth and as an accelerator for organ growth (Kondorosi et al. 2000; Chevalier et al. 2011).

7.5 Polysomatic and Non-polysomatic Plants in Culture *In Vitro*

Plant organs, or their fragments, are frequently used as a source of explants for tissue and cell culture *in vitro* and for genetic transformation experiments. Explants from polysomatic plants are composed of a heterogeneous population of endopolyploid cells (mixoploid). Such non-dividing, differentiated cells can be induced into mitosis by wounding or phytohormone treatment. In the conditions of an *in vitro* culture endopolyploid cells undergo dedifferentiation and start mitosis thus displaying polyploid chromosome numbers. These polyploid cells are one of the major sources of chromosomal instability in a tissue culture and somaclonal variation in regenerated plants. In contrast, explants of non-polysomatic plants contain only diploid cells and are thus chromosomally stable during *in vitro* culture.

The relationship between the type of primary explant and the occurrence of chromosomal variation during callogenesis, long-term callus culture and capability to regenerate has been investigated in two model plants (1) the non-polysomatic species *Crepis capillaris* ($2n = 2x = 6 + Bs$) and (2) the polysomatic *Arabidopsis thaliana* ($2n = 2x = 10$ and $2n = 4x = 20$). Stability at the diploid level during 1 year of *C. capillaris* callus culture has been reported by several authors, although structural chromosomal aberrations did occur at the diploid level. Polyploidisation occurred in the older callus culture but regenerated plants were only diploid (Maluszynska 1990). Stability at the diploid level has also been reported for callus or cell suspension cultures for other non-polysomatic species such as *Helianthus annuus* (Butcher et al. 1975), *Lilium longiflorum* (Sheridan 1975) and the gymnosperm *Pinus nigra* (Papes et al. 1983).

A cytogenetic analysis of *A. thaliana* leaf explants during callogenesis showed that the first mitotic polyploid cells occurred just after the third day of culture. Flow cytometric analysis revealed additional endoreduplication cycles up to 64C (Fras et al. 2007). In primary callus cultures derived from different explants, callus lines of meristematic origin (i.e., root tips) were characterized by a narrower range of ploidy levels ($2x$–$10x$, i.e., 10–50 chromosomes) than callus lines of non-meristematic origin (leaves, cotyledons, hypocotyls) which had ploidy levels ranging from $2x$ to $15x$ (10–75 chromosomes) although cells with $2x$ and $8x$ were the most common). The frequency of polyploid cells increased with age of culture and was correlated with the type of explants (Fras and Maluszynska 2004). Among *A. thaliana* regenerants a high frequency of polyploids, mostly tri- and tetraploids, have been reported (Sangwan et al. 1992; Altman et al. 1994; Mittelsten Scheid et al. 1996). Correlations between the level of endopolyploidy of the plant organ tissue donor for *in vitro* culture and the frequency of polyploid regenerants were described for *Solanum lycopersicum* (Van den Bulk et al. 1990).

D'Amato (1986) described several processes like endoreduplication, endomitosis and fusion of nuclei as the mechanisms responsible for polyploidisation in tissue cultures derived from polysomatic plants. He also noted that such plants were cytologically unstable and all contained aneuploid cells and cells with a reduced number of chromosomes compared with the donor plants. In contrast, callus derived from non-polysomatic plant explants were observed to be chromosomally stable over long periods of time in *in vitro* culture, remaining diploid and never undergoing endoreduplication. He thus considered that it was the presence of preexisting endopolyploidy that seemed to be crucial for a lack of chromosome stability in a cell or tissue culture and that this may also play a role in contributing to the undesirable somaclonal variation in regenerated plants.

7.6 Methods to Analyse Endopolyploidy

Endopolyploidy can be studied both quantitatively and qualitatively. Quantitative methods allow the ploidy pattern of a tissue, organ or entire organism to be established. For this purpose flow cytometry is the most widely applied technique although image cytometry (ICM) is also generating increased interest (Vilhar et al. 2001). Both techniques enable the fast, high-throughput analysis of nuclear DNA content. Sample preparation for both techniques requires the isolation of a nuclear suspension and this is usually achieved by the mechanical homogenisation of a tissue in an appropriate isolation buffer. Afterwards, the nuclei are stained with a fluorochrome, most often DAPI or propidium iodide, and their DNA content is evaluated on the basis of fluorescence intensity (Doležel and Bartos 2005; Doležel et al. 2007). Data are visualized in the form of a histogram in which the consecutive peaks represent rounds of endoreduplication (e.g., Figs. 7.5 and 7.6). The mean value of endopolyploidisation can be calculated based on the histogram, which enables a comparative analysis of endopolyploidy between different samples.

There are two parameters that determine the endopolyploidy mean value: (1) the mean C-level and (2) the cycle value (Barow and Meister 2003; Barow and Jovtchev 2007). The mean C-level indicates the mean DNA content per nucleus and is calculated from the number of nuclei at each ploidy level multiplied by the corresponding ploidy level. The sum is divided by the total number of nuclei being investigated up to the formula:

$$\text{Mean C-level} = (2 \times n_{2C} + 4 \times n_{4C} + 8 \times n_{8C} + \ldots)/(n_{2C} + n_{4C} + n_{8C} + \ldots)$$

The cycle value represents the mean number of endocycles per nucleus and is calculated from the number of nuclei at each ploidy level multiplied by the number of endoreduplication cycles necessary to reach the corresponding ploidy level. The sum is then divided by the total number of nuclei according to the formula:

$$\text{Cycle value} = (0 \times n_{2C} + 1 \times n_{4C} + 2 \times n_{8C} + 3 \times n_{16C} \ldots)/(n_{2C} + n_{4C} + n_{8C} + n_{16C} \ldots)$$

A cycle value below 0.1 indicates a non-polysomatic type of plant.

Although flow cytometry enables the ploidy levels in large numbers of cells to be measured rapidly, it does not provide any tissue-specific information. The advantage of ICM over flow cytometry is that it combines cytometric data with fluorescent microscopy (Vilhar et al. 2001). Nuclei are investigated on microscopic slides and after (or simultaneously with) the determination of the DNA content, they can be used for studies with other techniques such as fluorescence *in situ* hybridization (FISH) or immunodetection (Fig. 7.4c).

Insights into the organization of chromatin in endoreplicated nuclei can be obtained by FISH which is useful not only to determine the ploidy level in a single nucleus of the tissue being analysed but can also provide information about the mechanisms by which endoreduplication is driven (Weiss and Maluszynska 2001). Coupling FISH with cytometric data allows information about the organization and structure of endoreplicated chromosomes at various ploidy levels to be obtained (Schubert et al. 2006, 2008). *In vivo* methods using transgenes with green fluorescent protein (GFP) linked to cell-type specific promoters are very promising for endopolyploidy studies as they enable the C-value status of specific cell types to be analysed (Zhang et al. 2005; Galbraith 2007). Such approaches have the potential to probe the onset, progression and regulation of endopolyploidy at a much higher resolution than has hitherto been possible. However, so far this approach has only been used for studies of *A. thaliana* (Mathieu et al. 2003; Fang and Spector 2005; Matzke et al. 2005). Nevertheless, its application to other plants which can be transformed seems likely to increase as such techniques become more widely applicable to other systems.

References

Altman T, Damm B, Frommer WB, Martin T, Morris PC, Schweizer D, Wilmitzer L, Schmidt R (1994) Easy determination of ploidy level in *Arabidopsis thaliana* plants by means of pollen size measurement. Plant Cell Rep 13:652–656

Barow M (2006) Endopolyploidy in seed plants. Bioessays 23:271–281

Barow M, Jovtchev G (2007) Endopolyploidy in plants and its analysis by flow cytometry. In: Dolezel J, Greilhuber J, Suda J (eds) Flow cytometry with plant cells. Weinheim, Wiley, pp 349–372

Barow M, Meister A (2003) Endopolyploidy in higher plants is correlated to systematics, life strategy and genome size. Plant Cell Environ 26:571–584

Bauer MJ, Birchler JA (2006) Organization of endoreduplicated chromosomes in the endosperm of *Zea mays* L. Chromosoma 115:383–394

Beemster GTS, De Vusser K, De Tavernier E, Bock D, Inze D (2002) Variation in growth rate between *Arabidopsis* cytotypes is correlated with cell division and A-type cyclin dependent kinase activity. Plant Physiol 129:854–864

Bennett MD, Leitch IJ, Price HJ, Johnston S (2003) Comparisons with Caenorhabditis (100 Mb) and Drosophila (175 Mb) using flow cytometry show genome size in Arabidopsis to be 157 Mb and thus 25% larger than the Arabidopsis genome initiative estimate of 125 Mb. Ann Bot 91:547–557

Berckmans B, De Veylder L (2009) Transcriptional control of the cell cycle. Curr Opin Plant Biol 12:599–605

Berr A, Schubert I (2007) Interphase chromosome arrangement in *Arabidopsis thaliana* is similar in differentiated and meristematic tissues and shows a transient mirror symmetry after nuclear division. Genetics 176:853–856

Bertin N (2005) Analysis of the tomato fruit growth response to temperature and plant fruit load in relation to cell division, cell expansion and DNA endoreduplication. Ann Bot 95:439–447

Bertin N, Borel C, Brunel B, Cheniclet C, Causse M (2003) Do genetic make-up and growth manipulation affect tomato fruit size by cell number, or cell size and DNA endoreduplication? Ann Bot 92:415–424

Bino RJ, Lanteri S, Verhoeven HA, Kraak HL (1993) Flow cytometric determination of nuclear replication stages in seed tissues. Ann Bot 72:181–187

Biradar DP, Rayburn AL, Bullock DG (1993) Endopolyploidy in diploid and tetraploid maize (*Zea mays* L.). Ann Bot 71:417–421

Bisbis B, Delmas F, Joubès J, Sicard A, Hernould M, Inzé D, Mouras A, Chevalier C (2006) Cyclin-dependent kinase (CDK) inhibitors regulate the CDK-cyclin complex activities in endoreduplicating cells of developing tomato fruit. J Biol Chem 281:7374–7383

Boudolf V, Vlieghe K, Beemster GT, Magyar Z, Torres Acosta JA, Maes S, Van Der Schueren E, Inzé D, De Veylder L (2004) The plant-specific cyclin-dependent kinase CDKB1;1 and transcription factor E2Fa-DPa control the balance of mitotically dividing and endoreduplicating cells in *Arabidopsis*. Plant Cell 16:2683–2692

Boudolf V, Lammens T, Boruc J, Van Leene J, Van Den Daele H, Maes S, Van Isterdael G, Russinova E, Kondorosi E, Witters E, De Jaeger G, Inzé D, De Veylder L (2009) CDKB1;1 forms a functional complex with CYCA2;3 to suppress endocycle onset. Plant Physiol 150:1482–1493

Breuer C, Stacey N, Westd C, Zhaoe Y, Choryf J, Tsukaya H, Azumii Y, Maxwell A, Roberts K, Sugimoto-Shirasu K (2007) BIN4, a novel component of the plant DNA topoisomerase VI complex, is required for endoreduplication in *Arabidopsis*. Plant Cell 19:3655–3668

Breuer C, Kawamura A, Ichikawa T, Tominaga-Wada R, Wada T, Kondou Y, Muto S, Matsui M, Sugimoto K (2009) The trihelix transcription factor GTL1 regulates ploidy-dependent cell growth in the *Arabidopsis* trichome. Plant Cell 21:2307–2322

Brossard D (1978) Microspectrophotometric and ultrastructural analyses of a case of cell differentiation without endopolyploidization: the pith of *Crepis capillaris* (L.) Wallr. Protoplasma 93:369–380

Butcher DN, Sogeke AK, Tommerup IC (1975) Factors influencing changes in ploidy and nuclear DNA levels in cells from normal, crown-gall and habituated cultures of *Helianthus annuus* L. Protoplasma 86:295–308

Carvalheira G (2000) Plant polytene chromosomes. Genet Mol Biol 23:1043–1050

Cavalier-Smith T (2005) Economy, speed and size matter: evolutionary forces driving nuclear genome miniaturization and expansion. Ann Bot 95:147–175

Cheniclet C, Rong WY, Causse M, Frangne N, Bolling L, Carde JP, Renaudin JP (2005) Cell expansion and endoreduplication show a large genetic variability in pericarp and contribute strongly to tomato fruit growth. Plant Physiol 139:1984–1994

Chevalier C, Nafati M, Mathieu-Rivet E, Bourdon M, Frangne N, Cheniclet C, Renaudin J-P, Gévaudant F, Hernould M (2011) Elucidating the functional role of endoreduplication in tomato fruit development. Ann Bot 107:1159–1169

Churchman M, Brown M, Kato N, Kirik V, Hülskamp M, Inzé D, De Veylder L, Walker JD, Zheng Z, Oppenheimer DG, Gwin T, Churchman J, Larkin J (2006) SIAMESE, a plant-specific cell cycle regulator, controls endoreplication onset in *Arabidopsis thaliana*. Plant Cell 18:3145–3157

Coelho CM, Dante RA, Sabelli PA, Sun Y, Dilkes BP, Gordon-Kamm WJ, Larkins BA (2005) Cyclin-dependent kinase inhibitors in maize endosperm and their potential role in endoreduplication. Plant Physiol 138:2323–2336

D'Amato F (1964) Endopolyploidy as a factor in plant tissue development. Caryologia 17:41–52

D'Amato F (1984) Role of polyploidy in reproductive organs and tissue. In: Johri BM (ed) Embryology of angiosperms. Springer, Berlin/Heidelberg/New York, pp 519–566

D'Amato F (1986) Cytogenetics of plant cell and tissue cultures and their regenerates. CRC Crit Rev Plant Sci 3:73–112

D'Amato F (1989) Polyploidy in cell differentiation. Caryologia 42:183–211

De Rocher EJ, Harkins KR, Galbraith DW, Bohner HJ (1990) Developmentally regulated systemic endopolyploidy in succulents with small genomes. Science 250:99–100

del Castellano MM, del Pozo JC, Ramirez-Parra E, Brown S, Gutierrez C (2001) Expression and stability of *Arabidopsis* CDC6 are associated with endoreplication. Plant Cell 13:2671–2678

del Castellano M, Boniotti MB, Caro E, Schnittger A, Gutierrez C (2004) DNA replication licensing affects cell proliferation or endoreplication in a cell type-specific manner. Plant Cell 16:2380–2393

Del Pozo JC, Lopes-Matas A, Ramirez-Parra E, Gutierrez C (2005) Hormonal control of the plant cell cycle. Physiol Plant 123:173–183

Del Pozo JC, Diaz-Trivino S, Cisneros N, Gutierrez C (2006) The balance between cell division and endoreplication depends on E2FC-DPB, transcription factors regulated by the ubiquitin-SCFSKP2A pathway in *Arabidopsis*. Plant Cell 18:2224–2235

Dewitte W, Murray JAH (2003) The plant cell cycle. Annu Rev Plant Biol 54:235–264

Dewitte W, Scofield S, Alcasabas A, Maughan S, Menges M, Braun N, Collins C, Nieuwland J, Prinsen E, Sundaresan V, Murray J (2007) *Arabidopsis* CYCD3 D-type cyclins link cell proliferation and endocycles and are rate-limiting for cytokinin responses. Proc Natl Acad Sci USA 104:14537–14542

Doležel J, Bartos J (2005) Plant DNA flow cytometry and estimation of nuclear genome size. Ann Bot 95:99–110

Doležel J, Greilhuber J, Suda J (2007) Estimation of nuclear DNA content in plants using flow cytometry. Nat Protoc 2:2233–2244

Edgar BA, Orr-Weaver TL (2001) Endoreplication cell cycles: more for less. Cell 105:297–306

Exner V, Henning L (2008) Chromatin rearrangements in development. Curr Opin Plant Biol 11:64–69

Fang Y, Spector DL (2005) Centromere positioning and dynamics in living *Arabidopsis* plants. Mol Biol Cell 16:5710–5718

Fras A, Maluszynska J (2004) The correlation between the chromosome variation in callus and genotype of explants of *Arabidopsis thaliana*. Genetica 121:145–154

Fras A, Juchimiuk J, Siwinska D, Maluszynska J (2007) Cytological events in explants of *Arabidopsis thaliana* during early callogenesis. Plant Cell Rep 26:1933–1939

Galbraith DW (2007) Analysis of plant gene expression using flow cytometry and sorting. In: Doležel J, Greilhuber J, Suda J (eds) Flow cytometry with plant cells. Wiley, Weinheim, pp 405–422

Galbraith DW, Harkins KR, Knapp S (1991) Systemic endopolyploidy in *Arabidopsis thaliana*. Plant Physiol 96:985–989

Geitler L (1939) Die Entstehung der polyploiden Somakerne der Heteropteren durch Chromosomenteilung ohne Kernteilung. Chromosoma 1:1–22

Gendreau E, Traas J, Desnos T, Grandjean O, Caboche M, Höfte H (1997) Cellular basis of hypocotyl growth in *Arabidopsis thaliana*. Plant Physiol 114:295–305

Gendreau E, Höfte H, Grandjean O, Brown S, Traas J (1998) Phytochrome controls the number of endoreduplication cycles in the *Arabidopsis thaliana* hypocotyls. Plant J 13:221–230

Gendreau E, Hofte H, Orbovic V, Traas J (1999) Gibberellin and ethylene control endoreduplication levels in the *Arabidopsis thaliana* hypocotyl. Planta 209:513–516

Gendreau E, Romaniello S, Barad S, Leymarie J, Benech-Arnold RL, Corbineau F (2008) Regulation of cell cycle activity in the embryo of barley seeds during germination as related to grain hydration. J Exp Bot 59:203–212

Gilissen LJ, Staveren MJ, Creemers-Molenaar J, Verhoeven HA (1993) Development of polysomaty in seedling and plants of *Cucumis sativus*. Science 91:171–179

Gonzalez N, Gévaudant F, Hernould M, Chevalier C, Mouras A (2007) The cell cycle-associated protein kinase WEE1 regulates cell size in relation to endoreduplication in developing tomato fruit. Plant J 51:642–655

Gutierrez C (2009) The *Arabidopsis* cell division cycle. In: Last R, Chang C, Jander G, Kliebenstein D, McClung R, Millar H (eds) The *Arabidopsis* book. American Society of Plant Biologists, Rockville, pp 1–19

Hajdera I, Siwinska D, Hasterok R, Maluszynska J (2003) Molecular cytogenetic analysis of genome structure in *Lupinus angustifolius* and *Lupinus cosentinii*. Theor Appl Genet 107:988–996

Harashima H, Schnittger A (2010) The integration of cell division, growth and differentiation. Curr Opin Plant Biol 13:66–74

Henriques R, Magyar Z, Monardes A, Khan S, Zalejski C, Orellana J, Szabados L, de la Torre C, Koncz C, Bögre L (2010) *Arabidopsis* S6 kinase mutants display chromosome instability and altered RBR1-E2F pathway activity. EMBO J 29:2979–2993

Hershko A (2005) The ubiquitin system for protein degradation and some of its roles in the control of the cell division cycle. Cell Death Differ 12:1191–1197

Hirano H, Harashima H, Shinmyo A, Sekine M (2008) *Arabidopsis* RETINOBLASTOMA-RELATED PROTEIN 1 is involved in G1 phase cell cycle arrest caused by sucrose starvation. Plant Mol Biol 66:259–275

Horvath BM, Magyar Z, Zhang Y, Hamburger AW, Bako L, Visser RG, Bachem CW, Bogre L (2006) EBP1 regulates organ size through cell growth and proliferation in plants. EMBO J 25:4909–4920

Huskins CL (1947) The subdivision of the chromosomes and their multiplication in nondividing tissues: possible interpretation in term of gene structure and gene action. Am Nat 81:401–434

Imai KK, Ohashi Y, Tsuge T, Yoshizumi T, Matsui M, Oka A, Aoyama T (2006) The A-type cyclin CYCA2;3 is a key regulator of ploidy levels in *Arabidopsis* endoreduplication. Plant Cell 18:382–396

Ishida T, Kurata T, Okada K, Wada T (2008) A genetic regulatory network in the development of trichomes and root hairs. Annu Rev Plant Biol 59:365–386

Ishida T, Fujiwara S, Miura K, Stacey N, Yoshimura M, Schneider K, Adachi S, Minamisawa K, Umeda M, Sugimoto K (2009) SUMO

E3 ligase HIGH PLOIDY2 regulates endocycle onset and meristem maintenance in *Arabidopsis*. Plant Cell 2:2284–2297

Ishida T, Adachi S, Yoshimura M, Shimizu K, Umeda M, Sugimoto K (2010) Auxin modulates the transition from the mitotic cycle to the endocycle in *Arabidopsis*. Development 137:63–71

Jasencakova Z, Soppe WJ, Meister A, Gernand D, Turner BM, Schubert I (2003) Histone modifications in *Arabidopsis*—high methylation of H3 lysine 9 is dispensable for constitutive heterochromatin. Plant J 33:471–480

John PC, Qi R (2008) Cell division and endoreduplication: doubtful engines of vegetative growth. Trends Plant Sci 13:121–127

Joubes J, Chevalier C (2000) Endoreduplication in higher plants. Plant Mol Biol 43:735–745

Jovtchev G, Schubert V, Meister V, Barow M, Schubert I (2006) Nuclear DNA content and cell volume are positively correlated in angiosperms. Cytogenet Genome Res 114:77–82

Jovtchev G, Barow M, Meister A, Schubert I (2007) Impact of environmental and endogenous factors on endopolyploidization in angiosperms. Environ Exp Bot 60:404–411

Kasili R, Walker J, Simmons L, Zhou J, De Veylder L, Larkin JC (2010) SIAMESE cooperates with the CDH1-like protein CCS52A1 to establish endoreplication in *Arabidopsis thaliana* trichomes. Genetics 185:257–268

Kirik V, Schrader A, Uhrig JF, Hulskamp M (2007) MIDGET unravels functions of the *Arabidopsis* topoisomerase VI complex in DNA endoreduplication, chromatin condensation, and transcriptional silencing. Plant Cell 19:3100–3110

Kolano B, Siwinska D, Maluszynska J (2009) Endopolyploidy patterns during development of *Chenopodium quinoa*. Acta Biol Cracov Ser Bot 51:85–92

Kondorosi E, Kondorosi A (2004) Endoreduplication and activation of the anaphase-promoting complex during symbiotic cell development. FEBS Lett 567:152–157

Kondorosi E, Roudier F, Gendreau E (2000) Plant cell-size control: growing by ploidy? Curr Opin Plant Biol 3:488–492

Kudo N, Kimura Y (2001a) Patterns of endopolyploidy during seedling development in cabbage (*Brassica oleracea* L.). Ann Bot 87:275–281

Kudo N, Kimura Y (2001b) Flow cytometric evidence for endopolyploidization in cabbage flowers. Sex Plant Reprod 13:279–283

Kudo N, Kimura Y (2002) Nuclear DNA endoreduplication during petal development in cabbage: relationship between ploidy levels and cell size. J Exp Bot 53:1017–1023

Lammens T, Boudolf V, Kheibarshekan L, Zalmas L, Gaamouche T, Maes S, Vanstraelen M, Kondorosi E, La Thangue N, Govaerts W, Inzé D, de Veylder L (2008) Atypical E2F activity restrains APC/CCCS52A2 function obligatory for endocycle onset. Proc Natl Acad Sci U S A 105:14721–14726

Larkins BA, Dilkes BP, Dante RA, Coelho CM, Woo Y, Liu Y (2001) Investigating the hows and whys of DNA endoreduplication. J Exp Bot 52:183–192

Lee H-C, Chiou D-W, Chen W-H, Markhart AH, Chen Y-H, Lin T-Y (2004) Dynamics of cell growth and endoreduplication during orchid flower development. Plant Sci 166:659–667

Lee HO, Davidson JM, Duronio RJ (2009) Endoreplication: polyploidy with purpose. Genes Dev 23:2461–2477

Leiva-Neto JT, Grafi G, Sabelli PA, Dante RA, Woo YM, Maddock S, Gordon-Kamm WJ, Larkins BA (2004) A dominant negative mutant of cyclin-dependent kinase A reduces endoreduplication but not cell size or gene expression in maize endosperm. Plant Cell 16:1854–1869

Lemontey C, Mousset-Declas C, Munier-Jolain N, Boutin JP (2000) Maternal genotype influences pea seed size by controlling both mitotic activity during early embryogenesis and final endoreduplication level/cotyledon cell size in mature seed. J Exp Bot 51:167–175

Levan A (1939) Cytological phenomena connected with root swelling caused by growth substances. Hereditas 25:87–96

Li Z, Larson-Rabin Z, Masson P, Day C (2009) FZR2/CCS52A1 mediated endoreduplication in *Arabidopsis* development. Plant Signal Behav 4:451–453

Lim WL, Loh CS (2003) Endopolyploidy in *Vanda* 'Miss Joaquim' (Orchidaceae). New Phytol 159:279–287

López-Juez E, Dillon E, Magyar Z, Khan S, Hazeldine S, de Jager SM, Murray JA, Beemster GT, Bögre L, Shanahan H (2008) Distinct light-initiated gene expression and cell cycle programs in the shoot apex and cotyledons of *Arabidopsis*. Plant Cell 20:947–968

Lukaszewska E, Sliwinska E (2007) Most organs of sugar-beet (*Beta vulgaris* L.) plants at the vegetative and reproductive stages of development are polysomatic. Sex Plant Reprod 20:99–107

Lur HS, Setter TL (1993) Role of auxin in maize endosperm development (timing of nuclear DNA endoreduplication, zein expression, and cytokinin). Plant Physiol 10:273–280

Magyar Z, De Veylder L, Atanassova A, Bakó L, Inzé D, Bögre L (2005) The role of the *Arabidopsis* E2FB transcription factor in regulating auxin-dependent cell division. Plant Cell 17: 2527–2541

Magyar Z, Ito M, Binarová P, Mohamed B, Bogre L (2013) Cell cycle modules in plants for entry into proliferation and for mitosis. In: Leitch IJ, Greilhuber J, Doležel J, Wendel JF (eds) Plant genome diversity, vol 2, Physical structure, behaviour and evolution of plant genomes. Springer-Verlag, Wien, pp 77–97

Maluszynska J (1990) B chromosomes of *Crepis capillaris* (L.) Wallr. in vivo and in vitro. University of Silesia, Katowice

Maluszynska J, Heslop-Harrison JS (1991) Localization of tandemly repeated DNA sequences in *Arabidopsis thaliana*. Plant J 1:159–166

Mathieu O, Jasencakova Z, Vaillant I, Gendrel AV, Colot V, Schubert I, Tourmente S (2003) Changes in 5 S rDNA chromatin organization and transcription during heterochromatin establishment in *Arabidopsis*. Plant Cell 15:2929–2939

Matzke AJM, Henttel B, von der Winden J, Matzke M (2005) Use of two-color fluorescence-tagged transgenes to study interphase chromosomes in living plants. Plant Physiol 139:1586–1596

Mauseth JD (2009) Botany: an introduction to plant biology, 4th edn. Jones and Bartlett Publishers, London, p 71

Melaragno JE, Mehrotra B, Coleman AW (1993) Relationship between endopolyploidy and cell size in epidermal tissue of *Arabidopsis*. Plant Cell 5:1661–1668

Menges M, de Jager SM, Gruissem W, Murray JA (2005) Global analysis of the core cell cycle regulators of *Arabidopsis* identifies novel genes, reveals multiple and highly specific profiles of expression and provides a coherent model for plant cell cycle control. Plant J 4:546–566

Menges M, Samland AK, Planchais S, Murray JAH (2006) The D-type cyclin CYCD3;1 is limiting for the G1-to-S phase transition in Arabidopsis. Plant Cell 18:893–906

Mishiba K, Mii M (2000) Polysomaty analysis in diploid and tetraploid *Portulaca grandiflora*. Plant Sci 156:213–219

Mittelsten Scheid O, Jakovleva L, Afsar K, Maluszynska J, Paszkowski J (1996) A change of ploidy can modify epigenetic silencing. Proc Natl Acad Sci USA 93:7114–7119

Morohashi K, Grotewold E (2009) A systems approach reveals regulatory circuitry for *Arabidopsis* trichome initiation by the GL3 and GL1 selectors. PLoS Genet 5:e1000396

Nagl W (1976) DNA endoreduplication and polyteny understood as evolutionary strategies. Nature 261:614–615

Newcomb W, Sippell D, Peterson RL (1979) The early morphogenesis of *Glycine max* and *Pisum sativum* root nodules. Can J Bot 57:2603–2616

Olszewska MJ, Osiecka R (1982) The relationship between 2 C DNA content, life cycle type, systematic position and the level of DNA endoreduplication in nuclei of parenchyma cells during growth and differentiation of roots in some monocotyledonous species. Biochem Physiol Pfl 177:319–336

Olszewska JM, Osiecka R (1983) The relationship between 2 C DNA content, life cycle type, systematic position and the dynamics of DNA replication in parenchyma nuclei during growth and differentiation of roots in some dicotyledonous herbaceous species. Biochem Physiol Pfl 178:581–599

Papes D, Garej-Verhovac V, Jelaska S, Kolevska-Pletikapic B (1983) Chromosome behaviour in cultured cell populations of higher plants. In: Brandham PE, Bennett MD (eds) Kew chromosome conference II. George Allen and Unwin, London, pp 155–163

Park JA, Ahn JW, Kim YK, Kim SJ, Kim JK, Kim WT, Pai HS (2005) Retinoblastoma protein regulates cell proliferation, differentiation, and endoreduplication in plants. Plant J 42:153–163

Pecinka A, Schubert V, Meister A, Kreth G, Klatte M, Lysák MA, Fuchs J, Schubert I (2004) Chromosome territory arrangement and homologous pairing in nuclei of *Arabidopsis thaliana* are predominantly random except for NOR-bearing chromosomes. Chromosoma 113:258–269

Perazza D, Vachon G, Herzog M (1998) Gibberellins promote trichome formation by up-regulating GLABROUS1 in *Arabidopsis*. Plant Physiol 117:375–383

Perrot-Rechenmann C (2010) Cellular responses to auxin: division versus expansion. Cold Spring Harb Perspect Biol 2:a001446

Polizzi E, Natali L, Muscio AM, Giordani T, Cionini G, Cavallini A (1998) Analysis of chromatin and DNA during chromosome endoreduplication in the endosperm of *Triticum durum* Desf. Protoplasma 203:175–185

Ren H, Santner A, del Pozo JC, Murray JA, Estelle M (2008) Degradation of the cyclin-dependent kinase inhibitor KRP1 is regulated by two different ubiquitin E3 ligases. Plant J 53:705–711

Riou-Khamlichi C, Huntley R, Jacqmard A, Murray JA (1999) Cytokinin activation of *Arabidopsis* cell division through a D-type cyclin. Science 283:1541–1544

Sako K, Maki Y, Imai KK, Aoyama T, Goto DB, Yamaguchi J (2010) Control of endoreduplication of trichome by RPT2a, a subunit of the 19 S proteasome in *Arabidopsis*. J Plant Res 123:701–706

Sangwan RS, Bourgeois Y, Dubois F, Sangwan-Norrel BS (1992) In vitro regeneration of *Arabidopsis thaliana* from cultured zygotic embryos and analysis regenerants. J Plant Physiol 140:588–595

Schlichtinger F (1956) Karyologische Untersuchungen an endopolyploiden Chromozentrenkernen von *Gibbaeum Heathii* im Zusammenhang mit der Differenzierung. Österr Bot Z 103:485–528

Schnittger A, Weinl C, Bouyer D, Schöbinger U, Hülskamp M (2003) Misexpression of the cyclin-dependent kinase inhibitor ICK1/KRP1 in single-celled *Arabidopsis* trichomes reduces endoreduplication and cell size and induces cell death. Plant Cell 15:303–315

Schubert V, Klatte M, Pecinka A, Meister A, Jasencakova Z, Schubert I (2006) Sister chromatids are often incompletely aligned in meristematic and endopolyploid interphase nuclei of *Arabidopsis thaliana*. Genetics 172:467–475

Schubert V, Kim YM, Schubert I (2008) *Arabidopsis* sister chromatids often show complete alignment or separation along a 1.2-Mb euchromatic region but no cohesion "hot spots". Chromosoma 117:261–266

Schubert V, Weissleder A, Ali H, Fuchs J, Lermontova I, Meister A, Schubert I (2009) Cohesin gene defects may impair sister chromatid alignment and genome stability in *Arabidopsis thaliana*. Chromosoma 118:591–605

Sheridan WF (1975) Plant regeneration and chromosome stability in tissue cultures. In: Ledoux L (ed) Genetic manipulation with plant material. Plenum Publishing, New York, pp 263–267

Shimotohno A, Ohno R, Bisova K, Sakaguchi N, Huang J, Koncz C, Uchimiya H, Umeda M (2006) Diverse phosphoregulatory mechanisms controlling cyclin-dependent kinase-activating kinases in *Arabidopsis*. Plant J 47:701–710

Sliwinska E, Lukaszewska E (2005) Polysomaty in growing in vitro sugar-beet (Beta vulgaris L.) seedlings of different ploidy level. Plant Sci 168:1067–1074

Smulders MJM, Rus-Kortekaas W, Gilissen LJW (1994) Development of polysomaty during differentiation in diploid and tetraploid tomato (*Lycopersicon esculentum*) plants. Plant Sci 97:53–60

Sorrell DA, Marchbank A, McMaho K, Dickinson JR, Rogers HJ, Francis D (2002) A WEE1 homologue from *Arabidopsis thaliana*. Planta 215:518–522

Sozzani R, Maggio C, Varotto S, Canova S, Bergounioux C, Albani D, Cella R (2006) Interplay between *Arabidopsis* activating factors E2Fb and E2Fa in cell cycle progression and development. Plant Physiol 140:1355–1366

Sugimoto-Shirasu K, Roberts K (2003) "Big it up": endoreduplication and cell-size control in plants. Curr Opin Plant Biol 6:544–553

Sugimoto-Shirasu K, Stacey NJ, Corsar J, Roberts K, McCann MC (2002) DNA topoisomerase VI is essential for endoreduplication in *Arabidopsis*. Curr Biol 12:1782–1786

Sugimoto-Shirasu K, Roberts GR, Stacey NJ, McCann MC, Maxwell A, Roberts K (2005) RHL1 is an essential component of the plant DNA topoisomerase VI complex and is required for ploidy-dependent cell growth. Proc Natl Acad Sci USA 102: 18736–18741

Takatsuka H, Ohno R, Umeda M (2009) The *Arabidopsis* cyclin-dependent kinase-activating kinase CDKF;1 is a major regulator of cell proliferation and cell expansion but is dispensable for CDKA activation. Plant J 59:475–487

Teyssier E, Bernacchia G, Maury S, How Kit A, Stammitti-Bert L, Rolin D, Gallusci P (2008) Tissue dependent variations of DNA methylation and endoreduplication levels during tomato fruit development and ripening. Planta 228:391–399

Tojo T, Tsuda K, Yoshizumi T, Ikeda A, Yamaguchi J, Matsui M, Yamazaki K (2009) *Arabidopsis* MBF1s control leaf cell cycle and its expansion. Plant Cell Physiol 50:254–264

Traas J, Hulskamp M, Gendreau E, Hofte H (1998) Endoreduplication and development: rule without dividing? Curr Opin Plant Biol 1:498–503

Tschermak-Woess E (1956a) Karyologische Pflanzenanatomie. Protoplasma 46:798–834

Tschermak-Woess E (1956b) Notizen über die Riesenkerne und "Riesenchromosomen" von *Aconitum*. Chromosoma 8:114–134

Tschermak-Woess E (1963) Strukturtypen der Ruhekerne von Pflanzen und Tieren. Protoplasmatologia V/1. Springer, Wien

Tsumoto Y, Yoshizumi T, Kuroda H, Kawashima M, Ichikawa T, Nakazawa M, Yamamoto N, Matsui M (2006) Light-dependent polyploidy control by a CUE protein variant in *Arabidopsis*. Plant Mol Biol 61:817–828

Umeda M, Umeda-Hara C, Uchimiya H (2000) A cyclin-dependent kinase-activating kinase regulates differentiation of root initial cells in *Arabidopsis*. Proc Natl Acad Sci USA 97:13396–13400

Van den Bulk RW, Lofller HJM, Lindhout WH, Koornneef M (1990) Somaclonal variation in tomato. Effects of explant source and comparison with chemical mutagenesis. Theor Appl Genet 80:817–825

Verkest A, Manes CL, Vercruysse S, Maes S, Van Der Schueren E, Beeckman T, Genschik P, Kuiper M, Inze D, De Veylder L (2005a) The cyclin-dependent kinase inhibitor KRP2 controls the onset of the endoreduplication cycle during *Arabidopsis* leaf development through inhibition of mitotic CDKA;1 kinase complexes. Plant Cell 17:1723–1736

Verkest A, Weinl C, Inze D, De Veylder L, Schnittger A (2005b) Switching the cell cycle. KIP-related proteins in plant cell cycle control. Plant Physiol 139:1099–1106

Vierstra RD (2009) The ubiquitin-26S proteasome system at the nexus of plant biology. Nat Rev 10:385–397

Vilhar B, Greilhuber J, Dolenc Koce J, Temsch EM, Dermastia M (2001) Plant genome size measurement with DNA image cytometry. Ann Bot 87:719–728

Vlieghe K, Boudolf V, Beemster GTS, Maes S, Magyar Z, Atanassova A, De Almeida EJ, De Groodt R, Inzé D, De Veylder L (2005) The DP-E2F-like gene DEL1 controls the endocycle in *Arabidopsis thaliana*. Curr Biol 15:59–63

Wang H, Zhou Y, Fowke LC (2006) The emerging importance of cyclin-dependent kinase inhibitors in the regulation of the plant cell cycle and related processes. Can J Bot 84:640–650

Webster M, Witkin KL, Cohen-Fix O (2009) Sizing up the nucleus: nuclear shape, size and nuclear-envelope assembly. J Cell Sci 122:1477–1486

Weiss H, Maluszynska J (2001) Molecular cytogenetic analysis of polyploidization in the anther tapetum of diploid and autotetraploid *Arabidopsis thaliana* plants. Ann Bot 87:729–735

Wildermuth MC (2010) Modulation of host nuclear ploidy: a common plant biotroph mechanism. Curr Opin Plant Biol 13:449–458

Xia JC, Zhao H, Liu WZ, Li LG, He YK (2009) Role of cytokinin and salicylic acid in plant growth at low temperatures. Plant Growth Regul 57:211–222

Yang M, Loh CS (2004) Systemic endopolyploidy in *Spathoglottis plicata* (Orchidaceae) development. BMC Cell Biol 5:33

Zhang C, Gong F, Lambert G, Galbraith D (2005) Cell type-specific characterization of nuclear DNA contents within complex tissues and organs. Plant Method 1:7

Meiosis: Recombination and the Control of Cell Division

8

Eric Jenczewski, Raphael Mercier, Nicolas Macaisne, and Christine Mézard

Contents

C. Mézard (✉)
Institut Jean-Pierre Bourgin, Institut National de Recherche Agronomique, route de Saint-Cyr, 78026 Versailles, cedex, France
e-mail: Christine.Mezard@versailles.inra.fr

I.J. Leitch et al. (eds.), *Plant Genome Diversity Volume 2*,
DOI 10.1007/978-3-7091-1160-4_8,

8.1 Introduction

Meiosis is a key step in the reproduction of many species, halving ploidy levels to produce haploid gametes, which are then restored by fertilization. The reduction in ploidy in meiosis results from one cycle of replication followed by two cell divisions (Fig. 8.1). During the first division, which is very specific to meiosis, the homologous chromosomes segregate to opposite poles. The second division during which the sister chromatids separate, is a modified version of mitosis. Thus a diploid cell produces four haploid spores.

However, depending on species-specific mechanisms for sporocyte maturation, it is possible that not all spores become gametes. For example, in *Arabidopsis thaliana* four spores are obtained from female meiosis but only one matures to produce an embryo sac containing an egg cell, while in male meiosis, all four spores give rise to pollen grains.

During the first meiotic division, homologous chromosomes recombine producing mosaic chromosomes that are harmonious patchworks of the original genetic material contained in the meiocytes. The underlying molecular mechanisms for these recombination events are very well conserved among species even if each species has shown some specificity in regulation of the pathways or the molecular actors involved. Recombination is initiated by the formation of programmed DNA double strand breaks (DSBs) (Fig. 8.2a). These breaks are repaired using mainly the homologous chromosome as a template. The repair process leads either to a crossover (CO, i.e., reciprocal exchange of a large fragment of the chromosome) or a gene conversion not associated with a CO (non-reciprocal exchange of small DNA fragments, i.e., a non-crossover, NCO)(see Fig. 8.2).

In this review we summarize and discuss recent findings concerning the main steps of chromosome recombination and segregation during meiosis and the control of cell division in diploid and polyploid plants. Several excellent

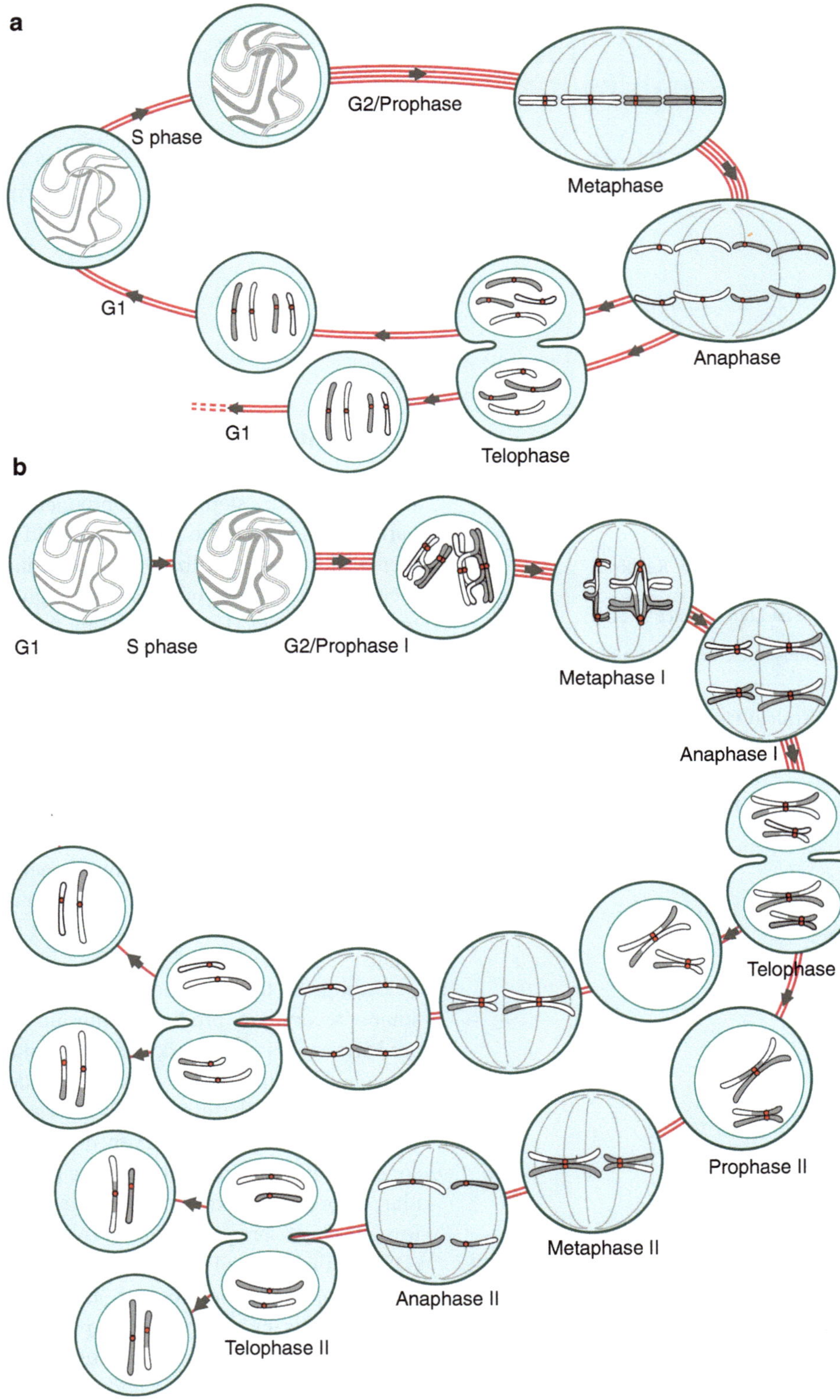

Fig. 8.1 Schematic representation of (**a**) mitosis and (**b**) meiosis. In the mitotic cell cycle: a "mother" cell (2*n*) produces two daughter cells (2*n*) that can re-enter the mitotic cycle. In contrast, in meiosis the "mother" cell (2*n*) produces four haploid (*n*) spores after one cycle of replication followed by two rounds of chromosome segregation. The first division is very specific to meiosis and it segregates homologous chromosomes; the second round is a modified version of the mitotic division, sister chromatids are sent to opposite poles

reviews have been published recently (Hamant et al. 2006; Mercier and Grelon 2008; De Muyt et al. 2009a; Martinez-Perez 2009; Ronceret et al. 2009; Harrison et al. 2010), thus we focus only on new aspects from recent results obtained in plants but also in other kingdoms to better understand the meiotic process as a whole.

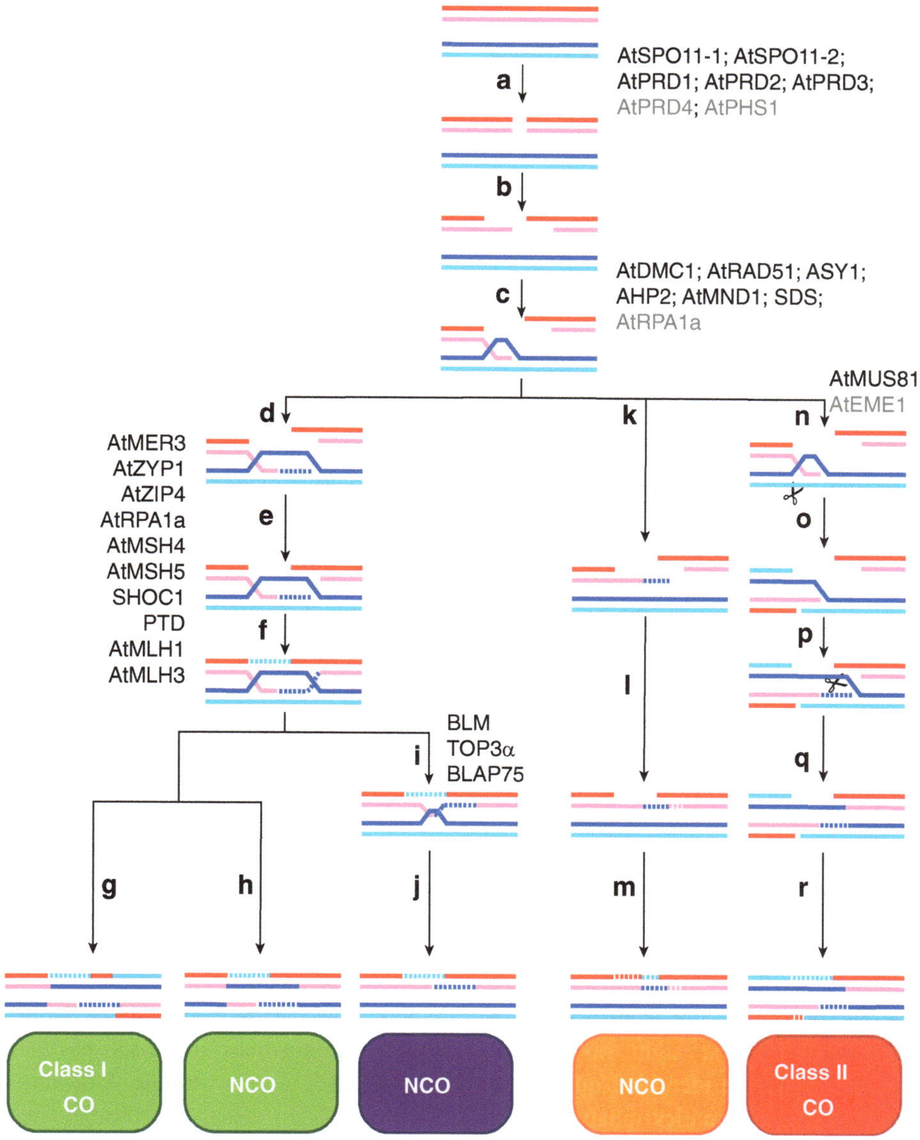

Fig. 8.2 A simplified view of the meiotic recombination mechanisms. (**a**) Recombination is initiated by the programmed formation of DNA double strand breaks (DSBs). (**b**) DSBs are then processed to produce 3′ single stranded DNA ends. (**c**) Next, DNA ends invade allelic sequences on the homologous chromosomes. (**d–f**) A subset of these recombination intermediates form double Holliday junctions (dHJs). Resolution of dHJs leads either to class I crossovers (CO) (**g**) or to non-crossovers (NCOs) (**h**). (**i, j**) On the other hand, decatenation of dHJs produces NCOs. (**k, l, m**) Nascent DSB partner complexes shown in (**c**) may also be processed by Single Strand Annealing (SSA). After DNA invasion and replication, the extended end is displaced from the template (**k**) and anneals with its partner (**l**) producing a NCO (**m**). (**n–q**) Class II COs are produced via a single Holliday Junction. *Arabidopsis* proteins known (*black characters*) or thought to be (*grey characters*) involved are indicated at their presumed step of action

8.2 Meiotic Recombination

8.2.1 Initiation of Meiotic Recombination

In all species, the topoisomerase-like protein Spo11 generates meiotic DSBs (Keeney 2007). Three Spo11 homologs are present in *Arabidopsis thaliana*, *Oryza sativa* (rice) and several other plant genomes (Malik et al. 2007). AtSPO11-1 and AtSPO11-2 are both necessary for DSB formation in *A. thaliana*, whereas AtSPO11-3 is not (Stacey et al. 2006). In rice, functional analysis of OsSPO11-1 demonstrated its role in DSB formation (Yu et al. 2010). In *Saccharomyces cerevisiae* (budding yeast), six other meiotic proteins (Rec102, Rec104, Rec114, Ski8, Mer2 and Mei4) and three proteins, also involved in DSB repair in vegetative cells (Rad50, Mre11, Xrs2), play a role in meiotic DSBs formation. In fission yeast *Schizosaccharomyces pombe*, in addition to Spo11, six genes that have no obvious homologs in *S. cerevisiae* participate in DSB formation (*Mde2, Rec6,*

Rec10, Rec15, Rec24) (Keeney 2007). In plants as in other eukaryotes, several screens (forward or reverse, see Mercier and Grelon 2008) were developed to identify genes that control this DSB formation step. It appears that only a few of these additional DSB-forming proteins are conserved in other kingdoms and even when they show sequence similarities their function in DSB formation may not be entirely conserved. For example, in *A. thaliana*, the DSB processing role of Rad50 and Mre11 is only conserved in vegetative cells and not in meiotic DSB formation (Bleuyard et al. 2004; Puizina et al. 2004). The *Arabidopsis* Ski8 ortholog does not appear to play a role in DSB formation (Jolivet et al. 2006). Nevertheless, two plant DSB formation genes were successfully identified in forward screens: *AtPRD1* in *Arabidopsis* and *PAIR1* in rice (Nonomura et al. 2004; De Muyt et al. 2007).

To improve the identification of new genes involved in DSB formation in *A. thaliana*, a very efficient high throughput analysis forward screen was designed (De Muyt et al. 2009b). A collection of 55,000 mutant lines was successively screened for (1) a reduction in fertility (1,280 lines) then (2) meiotic defects (80 lines) and finally (3) meiotic DSB formation defect-like phenotypes (28 lines). Following fine localization, out of the nine loci identified, three new genes *AtPRD2*, *AtPRD3* (ortholog of *PAIR1* in rice) and *AtPRD4* (Mathilde Grelon, personal communication) were obtained that are involved in DSB formation. Several mutant alleles of each gene were identified, suggesting that screen saturation has been reached and that all genes required for DSB formation have been identified in *Arabidopsis* (unless there is gene duplication, preventing its identification by forward genetics). Although at first sight no obvious orthologs for these proteins could be identified outside the plant kingdom, improved bioinformatics analysis then showed that *AtPRD2* could be the ortholog of Mei4 in mice and yeast (Kumar et al. 2010). For *AtPRD3* and *AtPRD4* as yet no significant homology has been identified outside the plant kingdom.

The *PHS1* gene was identified in maize and appears to be involved in DSB repair but not formation and to prevent synapsis between non-homologous chromosomes (Pawlowski et al. 2004). When the *Arabidopsis PHS1* homolog is inactivated, nuclear import of the DSB repair protein Rad50 is dramatically decreased. However, surprisingly, chromosomal fragmentation was not reported, which is in contrast to the phenotype of the single *rad50* mutant (Bleuyard et al. 2004; Ronceret et al. 2009). A recent report in mice suggested that PHS1 is an ortholog of the Rec114 protein (see above; (Kumar et al. 2010)). In *S. cerevisiae,* Rec114 interacts with Mei4 and both proteins are involved in DSB formation (Keeney 2007). In mice, the role of Mei4 in DSB formation as well as its interaction with Rec114 is conserved suggesting that its function in DSB formation may also be conserved. It would be interesting to reanalyse the *phs1* mutants in maize and *Arabidopsis* with regards to these new data obtained in mice concerning Mei4 and Rec114.

8.2.2 Meiotic DNA Double-Strand Break Repair

DSBs are then processed to produce 3′ single stranded DNA ends (Fig. 8.2b). In yeast, this single stranded DNA interacts with RPA, a heterotrimeric protein that helps the recombinase activity of Rad51 (reviewed in Sung et al. 2003). *Arabidopsis thaliana* has five and *Oryza sativa* (rice) has three homologs of the large subunit RPA1. Analyses of the meiotic proteome of *Brassica oleracea*, a close relative of *Arabidopsis*, showed that *AtRPA1a,* one member of the *AtRPA1* family, is potentially a meiotically-expressed gene (Sanchez-Moran et al. 2005). *OsRPA1a* has the most amino acid sequence similarities with *AtRPA1a* and belongs to the same phylogenetic clade (Chang et al. 2009). However, chromosomal defects in the corresponding mutants are very different. A mutation in *OsRPA1a* leads to severe fragmentation of meiotic chromosomes in male meiocytes consistent with a similar role in rice and yeast. In contrast, no fragmentation is observed in *atrpa1* but crossover (CO) rates are dramatically lower (see below) (Osman et al. 2009) suggesting that this protein functions later in the DNA repair process. Thus, different RPA complexes in rice and *Arabidopsis* could perform the various RPA tasks during meiotic recombination (Chang et al. 2009).

Next, DNA ends invade allelic sequences on the homologous chromosomes and a subset of these recombination intermediates forms double Holliday junctions (dHJs) (Fig. 8.2c–f). In yeast, it is thought that the meiotic dHJs are mainly resolved toward the formation of COs (Hunter 2007). However, in *Arabidopsis*, recent results suggest that some of the meiotic dHJs could mature via a different pathway (Chelysheva et al. 2008; Hartung et al. 2008). A protein complex known as BTB in mammals and RTR in yeast has anti-CO activity through the dissociation of dHJ (or potentially aberrant dHJ) (Raynard et al. 2006). This complex contains a RecQ-like helicase (BLM for Bloom in Human, Sgs1 in yeast), a topoisomerase (Top3α) and a third protein Blap75/Rmi1/Nce4, identified through its interaction with either BLM or Sgs1. However, limited data are available on the role of this complex in meiosis. In *Arabidopsis*, a mutation in BLAP75 or a hypomorphic mutation in Top3α results in extensive fragmentation of meiotic chromosomes, which depends on meiotic DSB formation, and an unusual arrest at the end of telophase I (Chelysheva et al. 2008; Hartung et al. 2008). The activity of BLM and RTR *in vitro* and in vegetative cells suggests that at least in *Arabidopsis,* and potentially in all species,

BLM/RTR could decatenate a subset of normal dHJs or aberrant recombination intermediates containing dHJs toward NCOs.

DNA DSBs are then repaired using either the homologous chromosome or the sister chromatid as a template. In meiosis, it is thought that there is a bias in the choice of template and that repair is mainly directed toward the homolog. In *A. thaliana*, AtDMC1 the meiotic recombinase, is one of the major actors facilitating this bias (Couteau et al. 1999; De Muyt et al. 2009b). Three other proteins also play a role in this process (De Muyt et al. 2009b): ASY1, an axis associated protein related to yeast Hop1 (Sanchez-Moran et al. 2007); AtHP2, a homolog of yeast Hop2 which was suggested to mediate the activity of the yeast Dmc1 protein in a complex with AtMND1 (Chen et al. 2004; Vignard et al. 2007); and SDS, an *Arabidopsis* cyclin that interacts with several Cyclin Dependant Kinases (CDKs) (Azumi et al. 2002). The role of PAIR2, the rice homolog of ASY1, in this process remains to be elucidated (Nonomura et al. 2006).

The *PAIR3* gene was also identified in rice in a large-scale screen of a population of T-DNA insertional mutants for partial or complete sterility (Yuan et al. 2009). The protein has no similarity with any known proteins and no known motifs except a coil-coiled domain. A preliminary characterization of *pair3* mutants found that synapsis between homologous chromosomes at prophase I was absent. Additional studies are needed to determine the role of PAIR3 in meiosis.

The *AtXRI1* gene is highly transcribed in response to gamma rays (Dean et al. 2009). Detailed characterization of *atxri1* mutants showed that its chromosomes displayed extensive fragmentation that did not depend on SPO11-induced DSBs. Thus, *AtXRI1* could play an important role in the S-phase of meiosis together with two other proteins which when depleted cause a similar phenotype, MEI1 and CDC45 (Grelon et al. 2003; Stevens et al. 2004). AtXRI1 could also function in meiotic DSB repair by interacting with MIP1, which in turn interacts with the AtMND1 protein partner of AtHP2 (see above). Thus *AtXRI1* could be involved in various aspects of homologous replication: repair of collapsed replication forks and post-replication repair pathways. Finally, AtXRI1 is also involved in the post-meiotic stages of pollen development.

8.2.3 Crossover Distribution

DSBs can be repaired either as COs or NCOs (Fig. 8.2). In plants, as in many higher eukaryotes, it is thought that the number of NCOs largely exceeds the number of COs. However only indirect evidence supports this supposition, no direct test to measure NCO rates is presently available in plants (reviewed in De Muyt et al. 2009a).

The number of COs per chromosome and per meiosis is tightly controlled. In most species, there is a need for one obligatory CO per pair of homologous chromosomes. Moreover, the distribution of COs along chromosomes is not homogenous. Interference (see below) plays a role in both controlling and constraining the final CO distribution but additional factors are also important in localizing COs. In all species with an available genetic map, the CO rate drops in centromeric regions with estimated decreases between five- to more than 200-fold depending on organisms (Talbert and Henikoff 2010). Nevertheless, intriguing recent results in maize suggest that centromeres are not inert for meiotic recombination. In a mapping population, two gene conversion events not associated with COs were mapped in 53 inbred lines giving a theoretical rate of 10^{-5} conversion events per marker and per generation. This suggests that CO formation in centromeric regions is not prohibited through recombination abolition (e.g., prevention of DSB formation) but by specific control of the DSB repair process. A precise analysis of the frequency and distribution of NCOs is clearly needed to better understand their rate of formation and localization in plant genomes.

GC content is positively correlated with high CO rates in many species such as rat, mice, human, zebra finch, honeybee, maize, even at a broad scale (Jensen-Seaman et al. 2004; Beye et al. 2006; Gore et al. 2009; Backström et al. 2010), although the underlying mechanisms responsible for this correlation are still under discussion (see Duret and Arndt 2008; Marsolier-Kergoat and Yeramian 2009). To add a piece to the puzzle, in *Arabidopsis*, the variation in CO rate of a male–female averaged recombination map was negatively correlated to GC content (Drouaud et al. 2006). Moreover, on the five *Arabidopsis* chromosomes, CO rates negatively correlate with GC content in female meiosis but not in male meiosis (Laurène Giraut, personal communication). Variation in CO rates also correlate with several other genomic features such as transposable elements, the CpG ratio, gene density, nucleotide polymorphisms or chromosome architecture properties like distance to telomeres or centromeres (Petes 2001; Nachman 2002; Jensen-Seaman et al. 2004; Myers et al. 2005; Saintenac et al. 2009). Nevertheless, none of these other characteristics are systematically correlated with CO rate variation across every species. Thus the various features that correlate with non-homogeneity in CO rates may have causal relationships or may be incidentally related.

CO rates and distribution can vary between male and female meiosis in the same species. Haldane (1922) suggested that the heterogametic sex had a lower CO rate as a consequence of selection against recombination between the sex chromosomes. However, this hypothesis, referred to as the Haldane and Huxley rule, has since been called into question. Less recombination in the homogametic sex than

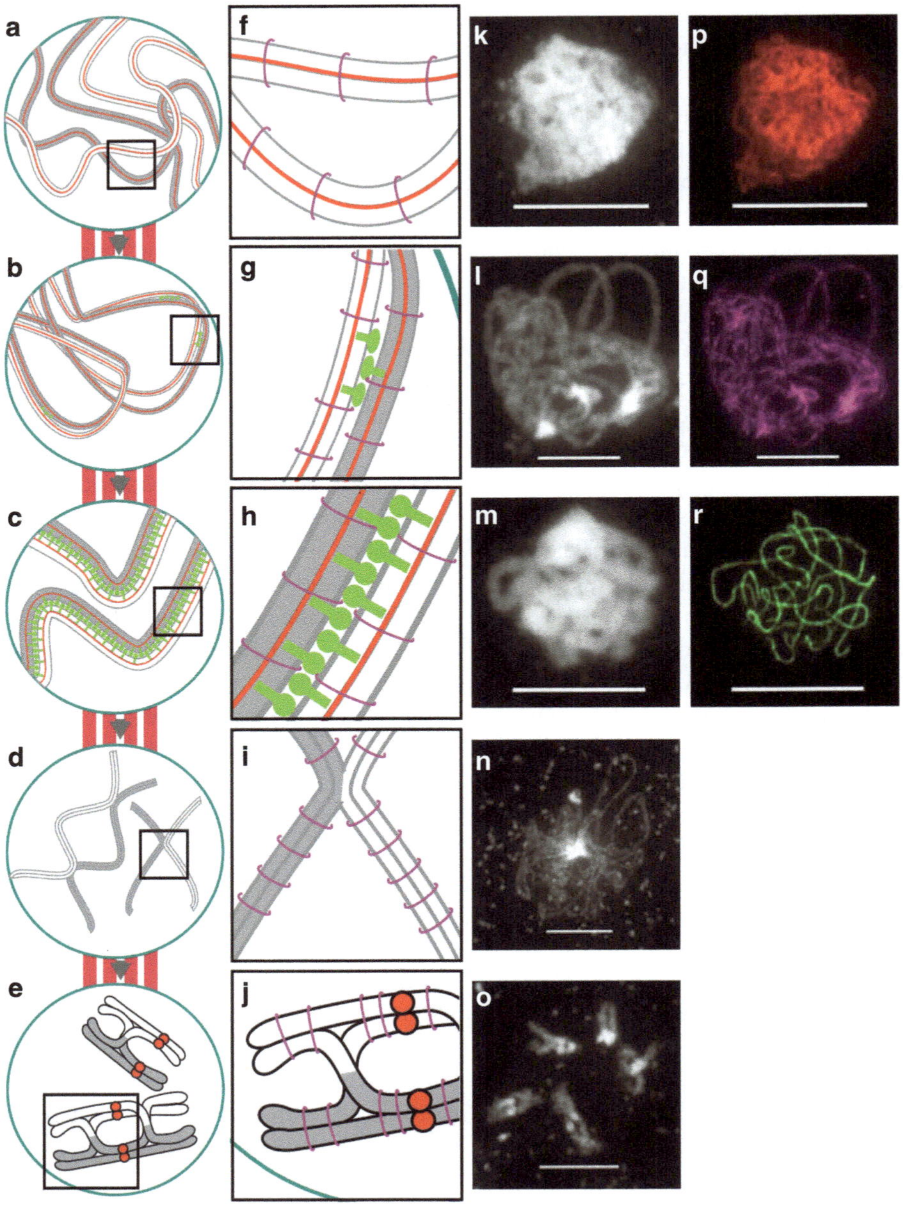

Fig. 8.3 Prophase of the first meiotic division. Leptotene (**a**, **f**, **k**, **p**): chromosomes begin to condense and the axial element (*red line*) along chromosomes is formed (**a**, **f**); *purple structures* = cohesin complexes; *red line*: axial elements. (**k**) DAPI staining of an *Arabidopsis* meiocyte at leptotene. (**p**) Immunolocalization of the ASY1 protein associated with the axial element at leptotene. Zygotene (**b**, **g**, **l**, **q**): homologous chromosomes synapse and the central element between the axial elements, now called lateral elements, is formed to constitute the tripartite structure called the synaptonemal complex (SC) shown in *green* in (**b**) and in greater detail in (**g**). (**l**) DAPI staining of an *Arabidopsis* meiocyte at zygotene. (**q**) Immunolocalization of the cohesin SCC3 protein at zygotene. Pachytene (**c**, **h**, **m**, **r**): synapsis between homologous chromosomes is complete (**c**, **h**). (**m**) DAPI staining of an *Arabidopsis* meiocyte at pachytene. (**r**) Immunolocalization of the Zyp1 protein in the synaptonemal complex at pachytene. Diplotene (**d**, **i**, **n**): chromosomes decondense and the SC is disassembled (**d**, **i**). (**n**) DAPI staining of an *Arabidopsis* meiocyte at diplotene. Diakinesis (**e**, **j**, **o**): chromosomes condense; bivalents individualize; homologous chromosomes remain physically attached through chiasmata and sister chromatid cohesion (**e**, **j**). (**o**) DAPI staining of an *Arabidopsis* meiocyte at diakinesis

the heterogametic one has been observed in some species and heterochiasmy has been found without the presence of sex chromosomes in plants such as *Allium* (Ved Brat 1966), *Brassica oleracea* (Kearsey et al. 1995) and *A. thaliana* (Armstrong and Jones 2001; Drouaud et al. 2007), and in animals like the saltwater crocodile (Miles et al. 2009). Other hypotheses have been proposed (reviewed in Hedrick 2007) but none satisfactorily explain the variations in heterochiasmy in all species.

Strikingly, a correlation was reported between CO number per chromosome and the total length of the synaptonemal complex (SC) (a proteinaceous structure that links homologous chromosomes at the pachytene stage of meiosis I; Fig. 8.3b, c). In *A. thaliana* male meiosis, there is a linear correlation between the mean CO number per chromosome and the SC length (Laurène Giraut and Christine Mézard, personal communication). Moreover, several studies have shown that CO number and SC length vary coordinately, even in situations where DNA length is constant (Kleckner et al. 2003). For example, in human meiosis, males have about half the CO number and total SC length compared with females (Wallace and Hulten 1985). This correlation was also reported in male and female meiocytes in the flatworm *Dendrocoelum lacteum* (Croft and Jones 1989) and zebrafish (Singer et al. 2002; Wallace and Wallace 2003) and in many other species with various individuals of the same population (Fox 1973; Quevedo et al. 1997; Tease and Hulten 2004). The reasons for this correlation are still poorly understood.

8.2.4 Crossover Formation and Interference

In many species, at least two CO pathways coexist (Fig. 8.2). Class I COs are sensitive to interference whereas Class II are not. These two pathways also exist in plants (Copenhaver et al. 2002). The phenomenon of interference was first described at the beginning of the twentieth century as a lower frequency of double COs in disjoint chromosomal intervals than was expected if they occurred independently (Sturtevant 1915). One of the consequences is that the physical distance between adjacent COs increases. When interference is total as in *Caenorhabditis elegans*, only one CO occurs per pair of homologous chromosomes. The mechanism that mediates interference is still poorly understood. Nevertheless, many molecular actors that are involved in this Class I pathway have been identified. In the yeast *S. cerevisiae*, interfering COs depend on the ZMM proteins (Zip1, Zip2, Zip3, Zip4, Msh4, Msh5 and Mer3), Mlh1 and Mlh3 whereas non-interfering COs need the Mus81-Mms4 complex (reviewed in Lynn et al. 2007). In *A. thaliana*, homologs of many of these genes have been characterized: *AtMSH4*, *AtMSH5*, *AtMER3*, two *AtZYP1*, *AtZYP4*, *AtMLH3*, *AtMLH1* and *AtMUS81* (Fig. 8.2). Since a detailed description of these genes was recently reviewed elsewhere (Mercier and Grelon 2008; De Muyt et al. 2009a), here we will only discuss the main conclusions.

In rice, a mutant in the *MER3* homolog displays a similar phenotype to the *Arabidopsis mer3* mutants described (Wang et al. 2009). The *SHOC1* gene is also a likely functional homolog of Zip2 (Macaisne et al. 2008). SHOC1 interacts with PTD, an Ercc1 homolog, suggesting a role for a XPF-Ercc1 complex in Class I CO formation (Nicolas Macaisne, personal communication). Two Mms4 homologs were identified in *Arabidopsis* and the corresponding proteins interact with AtMus81 (Geuting et al. 2009). However, the meiotic phenotype of corresponding mutants was not reported. As yet, no Zip3 homolog has been identified. In *zmm* mutants, CO rates drop down to a minimum of 15% of the wild type level and residual COs do not display interference, suggesting the co-existence of two CO pathways. Even in the absence of both the ZMM and AtMUS81 proteins, and as observed in *S. cerevisiae*, some COs (5–7%) are still detected suggesting either the existence of a third CO formation pathway or the activation of a pathway only active when the two main ones are absent. Unlike observations in yeast, *zmm* mutants in *Arabidopsis* do not show defects in synapsis between homologous chromosomes suggesting that synapsis depends on precocious recombination intermediates rather than Class I COs. The *Arabidopsis* genome contains two *Zip1* orthologs, *AtZYP1a* and *AtZYP1b*. Plants where both ZYP1 genes have been downregulated using interfering RNAs undergo synapsis between non-homologous chromosomes and form aberrant multivalents whereas chiasmata labelled by the MLH1 proteins are only slightly reduced (Higgins et al. 2005). The Zip1 homolog of rice, ZEP1 also shows some specificity compared with other organisms (Wang et al. 2010a). In its absence, chromosomes align along their entire length but the SC is not formed and the number of COs may increase slightly suggesting a role later in meiosis.

In maize, a statistical approach was developed to analyse the position of late recombination nodules, which are electron dense structures thought to designate all Class I and Class II COs (Falque et al. 2009). This analysis demonstrates that (1) both Class I and Class II coexist in maize; (2) the proportion of Class II COs varies from 6% to 23% (with an average of 15%) depending on the chromosomes. A similar proportion of non-interfering COs was identified in two other plants: 30% in tomato (Lhuissier et al. 2007) and 28% in rice (Wang et al. 2009). These values, together with the approximately 15% found in *Arabidopsis*, seem to be in the same range as the values found in *S. cerevisiae* (7–12%) and mice (5–11%) (Broman et al. 2002; Malkova et al. 2004; Guillon et al. 2005).

8.3 Dynamics of Chromosomes, Chromatin Structure and Meiosis

During meiosis, chromosomes undergo a series of dramatic structural changes. They condense, exchange DNA segments, pair, synapse, decondense and move in the nucleus. Probably due to this complexity, meiosis is a relatively long process compared to mitosis. It takes from two (in *Triticum aestivum*) to nine times longer (in *Trillium erectum*) in some species (Bennett 1971). In *Arabidopsis*, the meiotic process is three times longer than the mitotic cell cycle (8.5 h) (Armstrong et al. 2003). Prophase of the first

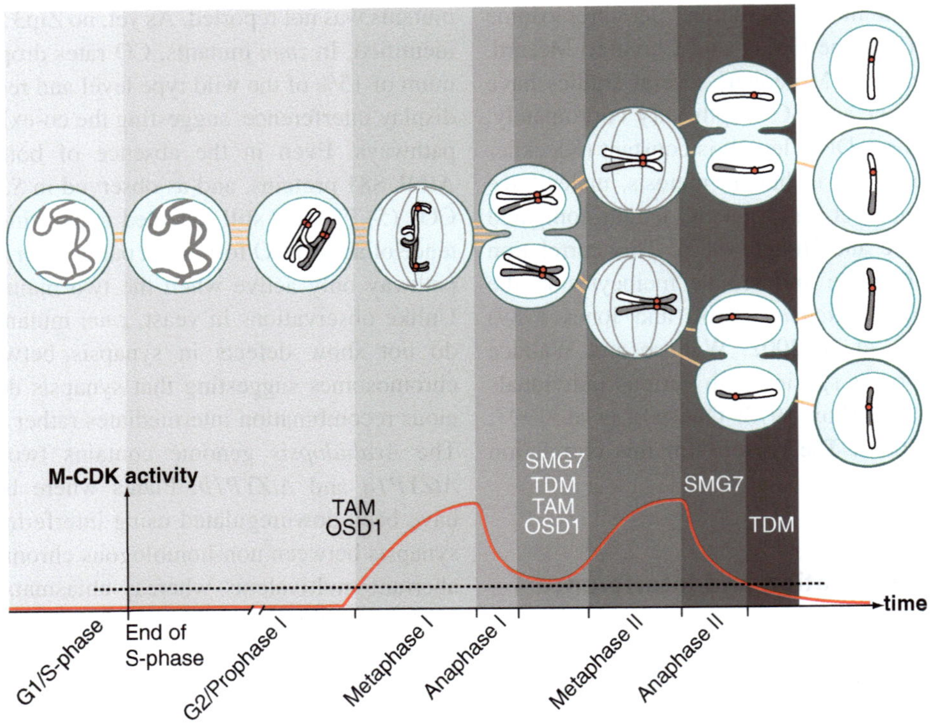

Fig. 8.4 Model of the control of meiosis progression. The *red line* represents a not yet identified cyclin-cyclin dependent kinase (*CDK*) activity that controls meiosis progression. This activity must reach a certain level to allow the G2/prophase to metaphase I transition and reach a peak at metaphase I and metaphase II. The metaphase to anaphase transitions are ensured by a rapid decrease in CDK activity. However, at the end of anaphase I, a certain level of cyclin-CDK activity must be maintained to allow entry into meiosis II and to avoid premature meiosis exit. *Arabidopsis* genes involved in the control of meiosis progression are indicated above the graph at their presumed time(s) of action

meiotic division (Figs. 8.3 and 8.4) is the longest stage reflecting its complexity (21.3 h), the remaining stages occur relatively quickly (2.7 h).

In leptotene (Fig. 8.3a, f, k, p) chromosomes begin to condense and the axial element along the chromosomes is installed. At zygotene (Fig. 8.3b, g, l, q), homologous chromosomes synapse and the central element between the axial elements, now called lateral elements, is formed to constitute the tripartite structure called the SC (see above).

Synapsis is complete at pachytene (Fig. 8.3c, h, m, r). In most species, telomeres cluster during early zygotene at a single site on the nuclear envelope and this structure, called the "bouquet" resumes at the beginning of pachytene. Recombination is initiated by the formation of DSBs in leptotene, thus before synapsis, and is completed during pachytene. At diplotene (Fig. 8.3d, i, n), the SC disappears but homologous chromosomes remain linked to each other by COs and sister chromatid cohesion (a proteinaceous) structure deposited on chromosomes after replication that keeps sister chromatids linked to each other) until metaphase I. At diakinesis (Fig. 8.3e, j, o), chromosomes condense further and move to form the metaphase plate (reviewed in Hamant et al. 2006).

Thus chromosomal movement across the nucleus is essential for recombination, pairing and synapsis. Live observations in *S. pombe* and *S. cerevisiae* have provided evidence of unexpected vigorous chromosome motility during prophase I. In *S. pombe*, before karyogamy, telomeres cluster tightly beneath the spindle pole body (SPB) while centromeres detach from the SPB. During karyogamy, the two telomere clusters fuse and after karyogamy, the diploid nucleus elongates about three times and initiates violent oscillations from one end to the other forming the "horsetail" nucleus that will remain throughout prophase I. This movement is driven by microtubules coming from the SPB and telomeres clustering beneath the SPB located at the leading edge of the moving nucleus (reviewed in Ding and Hiraoka 2007). In *S. cerevisiae*, chromosomal dynamics were visualized using a SC protein tagged with green fluorescent protein (GFP) (Scherthan et al. 2007; Koszul et al. 2008). SCs exhibit dramatic and continuous movements, crossing relatively large distances in the nucleus and changing shape. This motion seems to be driven by actin cables of the cytoskeleton.

Maize is the first higher eukaryote where live imaging of meiosis prophase has been possible (Sheehan and

Pawlowski 2009). Similar to the observations in yeasts, maize chromosomes move extensively during zygotene and pachytene with dynamic deformation of the nuclear envelope. Chromosome motility is dependent upon the actin and tubulin cytoskeletons. It has been proposed that these chromosome movements either promote pairing of homologous chromosomes and/or eliminate unwanted connections (Ding and Hiraoka 2007; Koszul et al. 2008; Sheehan and Pawlowski 2009).

Chromatin structure has an important role to play in the meiotic process. The most striking evidence has come recently from mice and *S. cerevisiae* where a link between trimethylation of the histone H3 at lysine 4 (H3K4) and the formation of DSBs was demonstrated (Borde et al. 2009; Baudat et al. 2010; Myers et al. 2010; Parvanov et al. 2010). In both species, this trimethylation, which is associated with an open chromatin structure, occurs before the formation of DSBs (Borde et al. 2009; Buard et al. 2009). Set1 ensures it in yeast. In mice, Prdm9 does the job: it is a meiosis-specific protein containing a Set1-like domain and a series of Zinc fingers that gives it sequence specificity for H3K4 trimethylation and thus DSB formation. No clear orthologs of these proteins exist in plants and the similarity of mechanisms remains to be demonstrated. However, recently, in maize, based on the parallel between Mu transposon insertion sites and regions with high recombination rates, it was suggested that meiotic recombination occurs in regions marked with open chromatin features in vegetative cells such as a high level of trimethylated H3K4, trimethylated H3K36, acetylated H3K9 and low level of trimethylated H3K27 (Liu et al. 2009). Whether this correlation holds true when DSB forming regions are looked at more precisely remains to be confirmed and whether the mechanisms are similar to those described in budding yeast and mice needs to be elucidated.

In *S. pombe* (fission yeast), a mutation in the Set1 homolog that severely reduces the level of trimethylated H3K4 does not affect sporulation and thus probably not the global level of DSBs (Noma and Grewal 2002). However in fission yeast, histone acetylation plays a role in meiotic recombination (Yamada et al. 2004). Histone acetylation is also correlated with meiotic recombination in *S. cerevisiae* (budding yeast) (Mieczkowski et al. 2007) even if it is now clear that it is not the main player linked to DSB formation. Recent results obtained in *A. thaliana* suggest that histone acetylation plays a role in the meiotic process (Perella et al. 2010). In a mutant line over-expressing the GCN5-related histone N-acetyltransferase, the distribution of meiotic COs appeared to be altered. Overall, the number of COs per meiotic cell was similar to wild type but the number of COs per chromosome was altered with less COs on the longest chromosome 1 and more COs on the shortest chromosome 4. Sister chromatid separation during the second meiotic division was also disrupted suggesting that histone hyperacetylation may have a profound effect on global chromatin structure affecting different aspects of the meiotic process.

8.4 Meiotic COs in Polyploid Plants

It has now been clearly established that *all* flowering plants have experienced at least one, and usually more, rounds of whole genome duplication (i.e., polyploidy events) during their evolution (Soltis et al. 2009; Van de Peer et al. 2009; Fawcett et al. 2013, this volume). Nearly 30% of extant flowering plants, including most of the world's important crops, are polyploids (Wood et al. 2009) while all other angiosperms have been impacted by ancient polyploidy events (they are *palaeo*polyploid species; Jiao et al. 2011).

One immediate consequence of polyploidy is an increase in the number of chromosomes that can compete for pairing, synapsis and recombination at meiosis. In newly-formed synthetic or natural polyploids, the presence of more than two partners usually leads to abnormal meiosis with multiple or illegitimate chiasma associations that result in chromosome missegregation, aneuploidy and partial fertility (Ramsey and Schemske 2002). By contrast, extant wild and cultivated polyploids display a clear tendency towards a diploid-like meiotic behaviour. In autopolyploid species, which are derived from within a single species (see Soltis et al. 2010 for details), this results from an increased number of bivalents forming at meiosis I, to the detriment of multivalents, between any pair of homologs (resulting in polysomic inheritance). In allopolyploid species, which usually arise via hybridization between different species followed by chromosome doubling, regular meiosis requires a non-random assortment of chromosomes into pairs with COs being exclusively formed between homologous rather than homoeologous chromosomes (i.e., inherited from different progenitors). Several reviews have examined diploidization in polyploids from a cytological, genetic, agronomic and evolutionary point of view (Jenczewski and Alix 2004; Able and Langridge 2006; Hamant et al. 2006; Able et al. 2009; Moore and Shaw 2009; Cifuentes et al. 2010a; Yousafzai et al. 2010a). We reiterate here that suppression of COs between homoeologous chromosomes is usually under polygenic control, with one locus having a greater influence than the others and frequent gene-dosage effects.

8.4.1 Changes in CO Distribution

An unexpected result that has emerged from comparisons of CO formation in allopolyploid species and their diploid progenitors is that COs sometimes get a boost in polyploids.

For example, Brubaker et al. (1999) compared the genetic map of allotetraploid cotton with that of its diploid parents and observed that allotetraploidy in *Gossypium* has been accompanied by an increase in CO rates. However we do not yet know if this observation holds true for polyploids in general or if it only illustrates genotypic variation in the propensity to form COs within the diploid species (i.e., it is possible that the true parents of the allopolyploid cotton formed more COs than the diploid plants used to build the diploid genetic maps). Nevertheless, recent analyses in *Brassica* which directly addressed this concern have confirmed the trend observed in cotton. Leflon et al. (2010) analysed the extent of CO variation in a set of related diploid AA, allotriploid AAC and allotetraploid AACC *Brassica* hybrids which were produced using exactly the same genotypes. Combining immunolocalization of MLH1 and genetic analyses, the authors observed that the number of COs in the allotriploid AAC hybrid was higher than in the diploid AA hybrid, with the allotetraploid AACC showing intermediate behaviour (Leflon et al. 2010). Although the precise mechanisms involved are not presently known, these results point towards an unexpected interplay between increased ploidy and enhanced recombination. This in turn may tentatively explain why increased and more variable recombination rates are observed in plants, which have been repeatedly impacted by polyploidy, compared with animals (Gaut et al. 2007; Kejnovsky et al. 2009).

8.4.2 Genetic Control of CO Formation

8.4.2.1 Wheat

At present, most continuing efforts to identify genes involved in the genetic control of recombination in polyploids are focusing on wheat (Crismani et al. 2006; Boden et al. 2007; de Bustos et al. 2007; Lloyd et al. 2007; Khoo et al. 2008; Boden et al. 2009; Bovill et al. 2009; Perez et al. 2010), and most notably on recombination between homoeologous chromosomes (Griffiths et al. 2006; Al-Kaff et al. 2008; Yousafzai et al. 2010b).

A number of these studies have aimed to identify wheat orthologs of key recombination genes including Asy1 (Boden et al. 2007), Msh7 (Lloyd et al. 2007), Mre11 (de Bustos et al. 2007) and Rad50 (Perez et al. 2010). The homoeologous copies of Mre11 and Rad50 in wheat are highly conserved but have slightly different expression levels (de Bustos et al. 2007; Perez et al. 2010); yeast two-hybrid assays revealed that the products of the Rad50 and Mre11 genes show a great affinity for one another that does not depend on their genome of origin (Perez et al. 2010). This example demonstrates that genetic redundancy in recent (<10,000 year-old) polyploids provides a buffer effect for crucial functions.

Another approach has specifically aimed to identify novel candidate meiotic genes. Crismani et al. (2006) produced and analysed a large-scale transcriptomics dataset across a meiotic time series in bread wheat and identified ~1,350 transcripts that were temporally-regulated during the early stages of meiosis. Some of these candidates shared sequence similarity with genes involved in chromatin condensation, SC formation and recombination while most others showed no sequence similarities with genes in the databases. Functional characterization of these candidates has now to be pursued.

CO suppression between homoeologous chromosomes is one expected additional constraint that an allopolyploid species has to contend with during meiosis (compared to a diploid species). In wheat, this control is ensured by a multigenic system (for review: Sears 1976; Jenczewski and Alix 2004), with the *Ph1* locus (Wall et al. 1971) localized on the long arm of chromosome 5B, having the most influence (Riley and Chapman 1958; Sears and Okamoto 1958). This system is efficient in both natural and synthetic wheats (Mestiri et al. 2010) even when suppressors are present as have been found in some genotypes of *Aegilops speltoides* (Dvorak et al. 2006).

A 50-year quest was necessary to characterize the *Ph1* locus at the molecular level (see Yousafzai et al. 2010b for a review of the cloning efforts). *Ph1* has been defined as a region containing a cluster of seven genes related to cyclin-dependant kinases (CDK) (see Sect. 8.5 for details about these proteins) interrupted by a segment of heterochromatin (Griffiths et al. 2006). This makes it distinct from its homoeologs, which contain five (on chromosome 5A) and two (on chromosome 5D) cdk-like genes, respectively (Al-Kaff et al. 2008). Detailed bioinformatics and *in silico* analyses has recently indicated that one of the *Ph1* cdk-like genes shares many functional attributes with human CDK2 (Yousafzai et al. 2010a). Human CDK2 promotes the timely entry into premeiotic DNA replication and the first nuclear division; it also induces transcription of a number of early meiosis-specific genes including *Hop1*, which encodes a component of the lateral element of the SC and plays an important role in pairing of meiotic chromosomes (Szwarcwort-Cohen et al. 2009). Expression profiling has revealed that one 5A cdk-like and five of the *Ph1* cdk-like genes from the 5B locus are transcribed, including two apparent pseudogenes (Al-Kaff et al. 2008). By contrast, transcription of most functional cdk-like genes from 5A and 5D is detected only when the 5B copies are deleted. G. Moore et al. have thus proposed that deletion of *Ph1* could result in a higher level of overall functional cdk-like proteins, which may be responsible for the observed meiotic phenotypes of *ph1* mutants (Al-Kaff et al. 2008). If this hypothesis holds true, then over-expression of cdk2-like target genes would be expected in these mutants. Indeed, this is exactly what Boden et al. (2009) observed; they showed that absence of *Ph1* correlates with increased transcription of *TaASY1*, a gene with significant sequence and functional similarity to *Hop1*. Likewise, Knight et al. (2010) showed that okadaic acid, a potent protein serine/threonine phosphatase inhibitor that activates CDKs, promotes intergenomic recombination in wheat × rye interspecific hybrids, thereby phenocopying the absence of *Ph1*. Even if direct functional evidence is still required to demonstrate that wheat *Ph1* is a functional analogue of human *CDK2*, all current results indicate that it is likely to be so (Yousafzai et al. 2010b). A major challenge of the future will be to reconcile this function with all the meiotic defects that have been observed in *ph1* mutants: non-resolution of incorrect and multiple early associations between chromosomes (Holm and Wang 1988), CO formation between non-homologous chromosome segments (Luo et al. 1996), defects in chromosome condensation and/or scaffold organization (Mikhailova et al. 1998), and asynchronous chromatin remodelling between homologs (Prieto et al. 2004; Colas et al. 2008).

8.4.2.2 *Brassica napus*

Over the last decade, *Brassica napus* has become a second model for deciphering how COs between homoeologous chromosomes are genetically suppressed. In this species, a major locus, named *PrBn* for *Pairing regulator in B. napus*, was shown to regulate CO formation between non-homologous chromosomes in allohaploid plants produced from different varieties (Jenczewski et al. 2003). Comparative analyses between the meiotic phenotype of 363 allohaploids produced from 29 accessions and the maternal origins of the same 29 accessions recently indicated that these different levels of CO suppression between homoeologs in *B. napus* allohaploids were the result of repeated polyploidy in this species (Cifuentes et al. 2010b). Actually *PrBn* shows incomplete penetrance (Cifuentes et al. 2010b) and four to six additive and epistatic QTLs have been detected, which contribute to the meiotic behaviour of *B. napus* haploids (Liu et al. 2006). *PrBn* has also been shown to determine the frequency, but not the distribution of COs between homoeologous chromosomes during meiosis of *B. napus* haploids and also to affect homologous recombination in allotriploid AAC hybrids (Nicolas et al. 2009). By contrast, genotypes with differing *PrBn* activities at the haploid level do not change the rate of CO formation between homologous chromosomes in allotetraploid AACC hybrids (Nicolas et al. 2009) suggesting that the effect of *PrBn* on recombination is dosage-dependent. Experiments are ongoing to characterize the *PrBn* locus at the molecular and phenotypic levels.

8.4.2.3 Cotton

Although it is clear that natural selection has certainly favored the evolution of mechanisms that promote exclusive bivalent formation in allopolyploids, it remains to be understood if the CO suppression between homoeologous chromosomes is paralleled by a similar effect on the other product of meiotic recombination, i.e., NCO or gene conversion events. Recent investigations in allotetraploid cotton have provided very interesting information on this pending question (Salmon et al. 2010). Tetraploid cotton is almost a strictly bivalent-forming species (Kimber 1961) and displays disomic inheritance (Reinisch et al. 1994; Brubaker et al. 1999) indicating that COs are almost exclusively formed between homologous chromosomes. Irrespective of this control, Salmon et al. (2010) used a vast EST dataset to demonstrate that non-reciprocal recombination events have occurred between homoeologous sequences throughout polyploid divergence and speciation in cotton. The authors first identified ESTs in the allopolyploids derived from one parental EST but that displayed at least one homoeoSNP from the other parent. They subsequently sequenced *de novo* genomic DNA corresponding to 20 putative recombination events and validated 14 of these chimeric sequences. Six were sequences from the D genome that have experienced conversion by the A homoeolog, whereas four others exhibited the reciprocal situation. Cloning and sequencing homoeologous copies from a total of five allopolyploid cotton species demonstrated that not a single recombination event was shared among species, which indicated that these events occurred sporadically after allopolyploid speciation.

8.5 The Meiotic Cell Cycle in Plants

Modification of the somatic cell cycle is one of the key features of meiosis (N.B. meiosis is not a cycle, the products of meiosis cannot undergo another round of meiosis, but by analogy with mitosis we use the term cell cycle to describe the control of progression through meiosis) (Figs. 8.1 and 8.4). The most prominent difference between meiosis and the canonical mitotic cell cycle is that two divisions follow a single replication at meiosis whereas replication and division strictly alternate during mitotic proliferation. The meiotic cell cycle follows the same phases as mitosis: G1, which precedes replication; S, the replication phase; G2, between replication and nuclear membrane break-down; and a mitotic phase during which the chromosomes contract and align (prophase to metaphase) and separate (anaphase). But meiosis differs from mitosis in several aspects of the cell cycle. Firstly, the G2 phase is dramatically longer and is the scene of chromosomal pairing and recombination. Secondly, while mitosis follows (G1 – S – G2 – metaphase – anaphase – G1)n cycles, meiosis follows a strikingly different pattern that leads to halving of the chromosomal content, G1 – S – G2 – metaphase I – anaphase I – interphase – metaphase II – anaphase II (Fig. 8.4). Two phases of division follow each other, without an intervening S phase, but just a short interphase without DNA replication.

Precisely how the mitotic machinery is modified for the purpose of meiosis is still poorly understood. Most of the knowledge currently available originates from studies carried out in unicellular fungi, *Xenopus laevis* or mouse oocyte systems (Marston and Amon 2004). Cyclins and CDKs form complexes that are essential for progression through both the mitotic and meiotic cell cycles. High CDK activity is required for entry and progression into metaphase and rapid decrease of CDK activity, mediated by the Anaphase Promoting Complex (APC), is required for induction of anaphase and exit from the division phases. The transition from meiosis I to meiosis II requires a fine balance in cyclin–CDK activity: it must be sufficiently low to exit meiosis I but must nonetheless be maintained at a level sufficiently high to suppress DNA replication and promote entry into meiosis II (Marston and Amon 2004; Pesin and Orr-Weaver 2008).

While only a few years ago nothing was known about the control of meiotic progression at the molecular level in plants this is an exciting time for the field. Recently, CDKA activity was shown to indeed oscillate throughout the course of plant meiosis (Bulankova et al. 2010) and several genes involved in crucial transitions were functionally characterized in *Arabidopsis* (*OSD1*, *TAM*, *SMG7*, *TDM*).

Two of these genes, *TAM* and *OSD1* are required for the transition from meiosis I to meiosis II. Null *osd1* and *tam* mutants fail to enter the second division in both male and female meiosis and exit meiosis after anaphase I, leading to the production of functional 2*n* gametes (d'Erfurth et al. 2009, 2010; Wang et al. 2010b). Interestingly, male meiocytes lacking both *OSD1* and *CYCA1;2/TAM* fail to enter meiotic division I, producing tetraploid spores directly after prophase (d'Erfurth et al. 2010). This shows that in addition to their crucial function in driving the meiosis I to meiosis II transition, these two genes are involved in the prophase to metaphase I transition (Fig. 8.4). This suggests that they both contribute to an activity, most likely mediated by CDK, which promotes entry into metaphase I and into metaphase II. The molecular function of OSD1 is currently unknown, however, by analogy to fission yeast Mes1, it may inhibit APC activity, thus promoting CDK activity (d'Erfurth et al. 2009). The cyclin TAM/CYCA1;2, may directly or indirectly promote the CDK activity controlling cell cycle progression.

The two other genes, *SMG7* and *TDM*, appear to have opposite effects on cell cycle progression from the above genes in the meiotic cycle. In *smg7*, meiosis arrests at late anaphase II and cells do not exit from meiosis (Fig. 8.4). SMG7 is a nonsense-mediated mRNA decay factor suggesting an unappreciated link between the cell cycle machinery and RNA processing (Riehs et al. 2008). Mutants in the *TDM* gene have an even more spectacular phenotype: meiosis progresses normally up to the end, performing the two divisions, but then an aberrant third round of division takes place, without replication (Ross et al. 1997; Glover et al. 1998) (Fig. 8.4). TDM1 has limited similarity with Xe-p9, a regulatory subunit of CDK. Genetic interaction studies revealed complex interactions between these genes. *tdm* is epistatic to *smg7*, suggesting that the anaphase II arrest of *smg7* depends (directly or indirectly) on the presence of TDM (Bulankova et al. 2010). Strikingly, double *tam-2/tdm* and *tam-2/smg7* mutants exhibit the *tdm* and *smg7* phenotype, respectively (Bulankova et al. 2010). Thus, the absence of either TDM or SMG7 removes the requirement for TAM for entry into meiosis II (Fig. 8.4). Remarkably, while *osd1* and *tam* single mutants have very similar phenotypes (d'Erfurth et al. 2010), the double *osd1/tam* mutant shows the *osd1* phenotype with meiotic exit at the end of meiosis I (Raphaël Mercier et al., unpublished data). This suggests that OSD1 and TAM act differently on the progression of meiosis. Altogether, these recent studies highlight the existence of a very complex network (not yet well understood!) to regulate the meiotic process.

Two other genes may also affect the meiotic cell cycle, *AtPS1* and *JASON* (d'Erfurth et al. 2008; Erilova et al. 2009). Both mutants produce diploid spores because of a defect in spindle orientation at the second division in the case of *Atps1* and through an as yet unclear mechanism in the case of *jas*. In both cases it is reasonable to suggest that a defect in cell cycle/phase identity could generate the observed phenotype. Interestingly, AtPS1 possesses a PINc domain that is present in proteins involved in RNA processing, suggesting that, like SMG7, there is a link between RNA decay and the meiotic cell cycle.

The identification of a series of genes involved in meiotic progression and the striking effect of their disruption alone or in combination, provides a unique opportunity to decipher the meiotic cell cycle. Study of the meiotic cell cycle machinery will certainly help to decipher general cell cycle mechanisms, mirroring the massive contribution of meiosis research to the understanding of DNA recombination mechanisms in somatic cells (Szostak et al. 1983; Paques and Haber 1999).

Interestingly cell cycle machinery genes, beyond their function in controlling meiotic progression, appear to have been recruited to control other key features of meiosis such as meiotic recombination. Indeed, the SDS gene, which clearly encodes a cyclin, controls the choice of template (e.g., homologous versus sister chromatid) during recombination (De Muyt et al. 2009b) (see above). Similarly, the wheat *Ph1* locus that controls homoeologous recombination (see above) contains CDK-like genes.

Conclusion

This is an exciting time with an international effort in progress to better understand the molecular processes involved in meiosis in a large variety of higher eukaryotes. In plants, thanks to powerful genetic studies and the development of new efficient molecular tools, these studies have enjoyed a boom in the last 10 years with more than 50 genes identified mainly in *Arabidopsis thaliana* but also in rice, maize, and wheat. Results obtained in plants demonstrate that there is both conservation in the underlying molecular mechanisms but also some specificity compared to other eukaryotic kingdoms. There is now an opportunity when analysing plants to decipher the impact of genome structure on the meiotic process.

References

Able JA, Langridge P (2006) Wild sex in the grasses. Trends Plant Sci 11:261–263

Able JA, Crismani W, Boden SA (2009) Understanding meiosis and the implications for crop improvement. Funct Plant Biol 36:575–588

Al-Kaff N, Knight E, Bertin I, Foote T, Hart N, Griffiths S, Moore G (2008) Detailed dissection of the chromosomal region containing the Ph1 locus in wheat *Triticum aestivum*: with deletion mutants and expression profiling. Ann Bot 101:863–872

Armstrong SJ, Jones GH (2001) Female meiosis in wild-type *Arabidopsis thaliana* and in two meiotic mutants. Sex Plant Reprod 13:177–183

Armstrong SJ, Franklin FCH, Jones GH (2003) A meiotic time-course for *Arabidopsis thaliana*. Sex Plant Reprod 16:141–149

Azumi Y, Liu D, Zhao D, Li W, Wang G, Hu Y, Ma H (2002) Homolog interaction during meiotic prophase I in *Arabidopsis* requires the SOLO DANCERS gene encoding a novel cyclin-like protein. EMBO J 21:3081–3095

Backström N, Forstmeier W, Schielzeth H, Mellenius H, Nam K, Bolund E, Webster MT, Ost T, Schneider M, Kempenaers B, Ellegren H (2010) The recombination landscape of the zebra finch *Taeniopygia guttata* genome. Genome Res 20:485–495

Baudat F, Buard J, Grey C, Fledel-Alon A, Ober C, Przeworski M, Coop G, de Massy B (2010) PRDM9 is a major determinant of meiotic recombination hotspots in humans and mice. Science 327:836–840

Bennett MD (1971) The duration of meiosis. Proc Roy Soc Lond B Bio 178:277–299

Beye M, Gattermeier I, Hasselmann M, Gempe T, Schioett M, Baines JF, Schlipalius D, Mougel F, Emore C, Rueppell O, Sirvio A, Guzman-Novoa E, Hunt G, Solignac M, Page RE (2006) Exceptionally high levels of recombination across the honey bee genome. Genome Res 16:1339–1344

Bleuyard JY, Gallego ME, White CI (2004) Meiotic defects in the *Arabidopsis* rad50 mutant point to conservation of the MRX complex function in early stages of meiotic recombination. Chromosoma 113:197–203

Boden SA, Shadiac N, Tucker EJ, Langridge P, Able JA (2007) Expression and functional analysis of TaASY1 during meiosis of bread wheat (*Triticum aestivum*). BMC Mol Biol 8:65

Boden SA, Langridge P, Spangenberg G, Able JA (2009) TaASY1 promotes homologous chromosome interactions and is affected by deletion of Ph1. Plant J 57:487–497

Borde V, Robine N, Lin W, Bonfils S, Geli V, Nicolas A (2009) Histone H3 lysine 4 trimethylation marks meiotic recombination initiation sites. EMBO J 28:99–111

Bovill WD, Deveshwar P, Kapoor S, Able JA (2009) Whole genome approaches to identify early meiotic gene candidates in cereals. Funct Integr Genomics 9:219–229

Broman KW, Rowe LB, Churchill GA, Paigen K (2002) Crossover interference in the mouse. Genetics 160:1123–1131

Brubaker CL, Paterson AH, Wendel JF (1999) Comparative genetic mapping of allotetraploid cotton and its diploid progenitors. Genome 42:184–203

Buard J, Barthes P, Grey C, de Massy B (2009) Distinct histone modifications define initiation and repair of meiotic recombination in the mouse. EMBO J 28:2616–2624

Bulankova P, Riehs-Kearnan N, Nowack MK, Schnittger A, Riha K (2010) Meiotic progression in *Arabidopsis* is governed by complex regulatory interactions between SMG7, TDM1, and the meiosis I-specific cyclin TAM. Plant Cell 22:3791–3803

Chang Y, Gong L, Yuan W, Li X, Chen G, Zhang Q, Wu C (2009) Replication protein A (RPA1a) is required for meiotic and somatic DNA repair but is dispensable for DNA replication and homologous recombination in rice. Plant Physiol 151:2162–2173

Chelysheva L, Vezon D, Belcram K, Gendrot G, Grelon M (2008) The *Arabidopsis* BLAP75/Rmi1 homologue plays crucial roles in meiotic double-strand break repair. PLoS Genet 4:e1000309

Chen YK, Leng CH, Olivares H, Lee MH, Chang YC, Kung WM, Ti SC, Lo YH, Wang AH, Chang CS, Bishop DK, Hsueh YP, Wang TF (2004) Heterodimeric complexes of Hop2 and Mnd1 function with Dmc1 to promote meiotic homolog juxtaposition and strand assimilation. Proc Natl Acad Sci USA 101:10572–10577

Cifuentes M, Eber F, Lucas MO, Lode M, Chevre AM, Jenczewski E (2010a) Repeated polyploidy drove different levels of crossover suppression between homoeologous chromosomes in *Brassica napus* allohaploids. Plant Cell 22:2265–2276

Cifuentes M, Grandont L, Moore G, Chevre AM, Jenczewski E (2010b) Genetic regulation of meiosis in polyploid species: new insights into an old question. New Phytol 186:29–36

Colas I, Shaw P, Prieto P, Wanous M, Spielmeyer W, Mago R, Moore G (2008) Effective chromosome pairing requires chromatin remodeling at the onset of meiosis. Proc Natl Acad Sci USA 105:6075–6080

Copenhaver GP, Housworth EA, Stahl FW (2002) Crossover interference in *Arabidopsis*. Genetics 160:1631–1639

Couteau F, Belzile F, Horlow C, Grandjean O, Vezon D, Doutriaux MP (1999) Random chromosome segregation without meiotic arrest in both male and female meiocytes of a dmc1 mutant of *Arabidopsis*. Plant Cell 11:1623–1634

Crismani W, Baumann U, Sutton T, Shirley N, Webster T, Spangenberg G, Langridge P, Able JA (2006) Microarray expression analysis of meiosis and microsporogenesis in hexaploid bread wheat. BMC Genomics 7:267

Croft JA, Jones GH (1989) Chromosome pairing and chiasma formation inspermatocytes and oocytes of *Dendrocoelum lacteum* (Turbellaria, Tricladida): a cytogenetical and ultrastructural study. Heredity 63:97–106

d'Erfurth I, Jolivet S, Froger N, Catrice O, Novatchkova M, Simon M, Jenczewski E, Mercier R (2008) Mutations in AtPS1 (*Arabidopsis thaliana* parallel spindle 1) lead to the production of diploid pollen grains. PLoS Genet 4:e1000274

d'Erfurth I, Jolivet S, Froger N, Catrice O, Novatchkova M, Mercier R (2009) Turning meiosis into mitosis. PLoS Biol 7:e1000124

d'Erfurth I, Cromer L, Jolivet S, Girard C, Horlow C, Sun Y, To JP, Berchowitz LE, Copenhaver GP, Mercier R (2010) The cyclin-A CYCA1;2/TAM is required for the meiosis I to meiosis II transition and cooperates with OSD1 for the prophase to first meiotic division transition. PLoS Genet 6:e1000989

de Bustos A, Perez R, Jouve N (2007) Characterization of the gene Mre11 and evidence of silencing after polyploidization in *Triticum*. Theor Appl Genet 114:985–999

De Muyt A, Vezon D, Gendrot G, Gallois JL, Stevens R, Grelon M (2007) AtPRD1 is required for meiotic double strand break formation in *Arabidopsis thaliana*. EMBO J 26:4126–4137

De Muyt AD, Mercier R, Mezard C, Grelon M (2009a) Meiotic recombination and crossovers in plants. Genome Dyn 5:14–25

De Muyt A, Pereira L, Vezon D, Chelysheva L, Gendrot G, Chambon A, Laine-Choinard S, Pelletier G, Mercier R, Nogue F, Grelon M (2009b) A high throughput genetic screen identifies new early meiotic recombination functions in *Arabidopsis thaliana*. PLoS Genet 5:e1000654

Dean PJ, Siwiec T, Waterworth WM, Schlögelhofer P, Armstrong SJ, West CE (2009) A novel ATM dependant X-ray inducible gene is essential for both plant meiosis and gametogenesis. Plant J 58:791–802

Ding D-Q, Hiraoka Y (2007) Nuclear movement enforcing chromosome alignment in fission yeast meiosis without homolog synapsis. In: Egel R, Lankenau DH (eds) Recombination and meiosis. Springer, Berlin/Heidelberg/New York, pp 231–247

Drouaud J, Camilleri C, Bourguignon PY, Canaguier A, Berard A, Vezon D, Giancola S, Brunel D, Colot V, Prum B, Quesneville H, Mezard C (2006) Variation in crossing-over rates across chromosome 4 of *Arabidopsis thaliana* reveals the presence of meiotic recombination "hot spots". Genome Res 16:106–114

Drouaud J, Mercier R, Chelysheva L, Berard A, Falque M, Martin O, Zanni V, Brunel D, Mezard C (2007) Sex-specific crossover distributions and variations in interference level along *Arabidopsis thaliana* chromosome 4. PLoS Genet 3:e106

Duret L, Arndt PF (2008) The impact of recombination on nucleotide substitutions in the human genome. PLoS Genet 4:e1000071

Dvorak J, Deal KR, Luo MC (2006) Discovery and mapping of wheat Ph1 suppressors. Genetics 174:17–27

Erilova A, Brownfield L, Exner V, Rosa M, Twell D, Mittelsten Scheid O, Hennig L, Köhler C (2009) Imprinting of the polycomb group gene MEDEA serves as a ploidy sensor in *Arabidopsis*. PLoS Genet 5:e1000663

Falque M, Anderson LK, Stack SM, Gauthier F, Martin OC (2009) Two types of meiotic crossovers coexist in maize. Plant Cell 21:3915–3925

Fawcett JA, Van de Peer Y, Maere S (2013) Significance and biological consequences of polyploidization in land plants. In: Leitch IJ, Greilhuber J, Doležel J, Wendel JF (eds) Plant genome diversity, vol 2, Physical structure, behaviour and evolution of plant genomes. Springer-Verlag, Wien, pp 277–293

Fox DP (1973) The control of chiasma distribution in the locust, *Schistocerca gregaria* (Forskål). Chromosoma 43:289–328

Gaut BS, Wright SI, Rizzon C, Dvorak J, Anderson LK (2007) Recombination: an underappreciated factor in the evolution of plant genomes. Nat Rev Genet 8:77–84

Geuting V, Kobbe D, Hartung F, Durr J, Focke M, Puchta H (2009) Two distinct MUS81-EME1 complexes from *Arabidopsis* process Holliday junctions. Plant Physiol 150:1062–1071

Glover J, Grelon M, Craig S, Chaudhury A, Dennis E (1998) Cloning and characterization of MS5 from *Arabidopsis*: a gene critical in male meiosis. Plant J 15:345–356

Gore MA, Chia JM, Elshire RJ, Sun Q, Ersoz ES, Hurwitz BL, Peiffer JA, McMullen MD, Grills GS, Ross-Ibarra J, Ware DH, Buckler ES (2009) A first-generation haplotype map of maize. Science 20:1115–1117

Grelon M, Gendrot G, Vezon D, Pelletier G (2003) The *Arabidopsis* MEI1 gene encodes a protein with five BRCT domains that is involved in meiosis-specific DNA repair events independent of SPO11-induced DSBs. Plant J 83:465–475

Griffiths S, Sharp R, Foote TN, Bertin I, Wanous M, Reader S, Colas I, Moore G (2006) Molecular characterization of Ph1 as a major chromosome pairing locus in polyploid wheat. Nature 439:749–752

Guillon H, Baudat F, Grey C, Liskay RM, de Massy B (2005) Crossover and noncrossover pathways in mouse meiosis. Mol Cell 20:563–573

Haldane JBS (1922) Sex ratio and unisexual sterility in hybrid animals. J Genet 12:101–109

Hamant O, Ma H, Cande WZ (2006) Genetics of meiotic prophase I in plants. Annu Rev Plant Biol 57:267–302

Harrison C, Alvey E, Henderson I (2010) Meiosis in flowering plants and other green organisms. J Exp Bot 61:2863–2875

Hartung F, Suer S, Knoll A, Wurz-Wildersinn R, Puchta H (2008) Topoisomerase 3alpha and RMI1 suppress somatic crossovers and are essential for resolution of meiotic recombination intermediates in *Arabidopsis thaliana*. PLoS Genet 4:e1000285

Hedrick PW (2007) Sex: differences in mutation, recombination, selection, gene flow and genetic drift. Evolution 61:2750–2771

Higgins JD, Sanchez-Moran E, Armstrong SJ, Jones GH, Franklin FC (2005) The *Arabidopsis* synaptonemal complex protein ZYP1 is required for chromosome synapsis and normal fidelity of crossing over. Genes Dev 19:2488–2500

Holm PB, Wang X (1988) The effect of chromosome 5B on synapsis and chiasma formation in wheat, *Triticum aestivum* cv. Chinese Spring. Carlsberg Res Commum 5:191–208

Hunter N (2007) Meiotic recombination. In: Aguilera A, Rothstein R (eds) Molecular genetics of recombination. Springer, Berlin/New York, pp 381–442

Jenczewski E, Alix K (2004) From diploids to allopolyploids: the emergence of pairing control genes. Crit Rev Plant Sci 23:21–45

Jenczewski E, Eber F, Grimaud A, Huet S, Lucas MO, Monod H, Chèvre AM (2003) *PrBn*, a major gene controlling homoeologous pairing in oilseed rape (*Brassica napus*) haploids. Genetics 164:645–653

Jensen-Seaman MI, Furey TS, Payseur BA, Lu Y, Roskin KM, Chen CF, Thomas MA, Haussler D, Jacob HJ (2004) Comparative recombination rates in the rat, mouse, and human genomes. Genome Res 14:528–538

Jiao Y, Wickett NJ, Ayyampalayam S, Chanderbali AS, Landherr L et al (2011) Ancestral polyploidy in seed plants and angiosperms. Nature 473:97–100

Jolivet S, Vezon D, Froger N, Mercier R (2006) Non conservation of the meiotic function of the Ski8/Rec103 homolog in *Arabidopsis*. Genes Cells 11:615–622

Kearsey MJ, Ramsay LD, Jennings DE, Lydiate DJ, Bohuon EJR, Marshall DF (1995) Higher recombination frequencies in female compared to male meisoses in *Brassica oleracea*. Theor Appl Genet 92:363–367

Keeney S (2007) Spo11 and the formation of DNA double-strand breaks in meiosis. In: Egel R, Lankenau D-H (eds) Genome dynamics and stability, vol 2, Recombination and meiosis: crossing-over and disjunction. Springer, Berlin/New York, pp 81–123

Kejnovsky E, Leitch IJ, Leitch AR (2009) Contrasting evolutionary dynamics between angiosperm and mammalian genomes. Trends Ecol Evol 24:572–582

Khoo KHP, Jolly HR, Able JA (2008) The RAD51 gene family in bread wheat is highly conserved across eukaryotes, with RAD51A upregulated during early meiosis. Funct Plant Biol 35:1267–1277

Kimber G (1961) Basis of the diploid-like meiotic behaviour of polyploid cotton. Nature 191:98–100

Kleckner N, Storlazzi A, Zickler D (2003) Coordinate variation in meiotic pachytene SC length and total crossover/chiasma frequency under conditions of constant DNA length. Trends Genet 19:623–628

Knight R, Greer E, Draeger T, Thole V, Reader S, Shaw P, Moore G (2010) Inducing chromosome pairing through premature condensation: analysis of wheat interspecific hybrids. Funct Integr Genomics 10:603–608

Koszul R, Kim KP, Prentiss M, Kleckner N, Kameoka S (2008) Meiotic chromosomes move by linkage to dynamic actin cables with transduction of force through the nuclear envelope. Cell 133:1188–1201

Kumar R, Bourbon H, de Massy B (2010) Functional conservation of Mei4 for meiotic DNA double-strand break formation from yeasts to mice. Genes Dev 24:1266–1280

Leflon M, Grandont L, Eber F, Huteau V, Coriton O, Chelysheva L, Jenczewski E, Chevre AM (2010) Crossovers get a boost in *Brassica* allotriploid and allotetraploid hybrids. Plant Cell 22:2253–2264

Lhuissier FG, Offenberg HH, Wittich PE, Vischer NO, Heyting C (2007) The mismatch repair protein MLH1 marks a subset of strongly interfering crossovers in tomato. Plant Cell 19: 862–876

Liu Z, Adamczyk K, Manzanares-Dauleux M, Eber F, Lucas MO, Delourme R, Chevre AM, Jenczewski E (2006) Mapping PrBn and other quantitative trait loci responsible for the control of homeologous chromosome pairing in oilseed rape (*Brassica napus* L.) haploids. Genetics 174:1583–1596

Liu S, Yeh CT, Ji T, Ying K, Wu H, Tang HM, Fu Y, Nettleton D, Schnable PS (2009) Mu transposon insertion sites and meiotic recombination events co-localize with epigenetic marks for open chromatin across the maize genome. PLoS Genet 5:e1000733

Lloyd AH, Milligan AS, Langridge P, Able JA (2007) TaMSH7: a cereal mismatch repair gene that affects fertility in transgenic barley (*Hordeum vulgare* L.). BMC Plant Biol 7:67

Luo MC, Dubcovsky J, Dvorak J (1996) Recognition of homeology by the wheat *Ph1* locus. Genetics 144:1195–1203

Lynn A, Soucek R, Börner GV (2007) ZMM proteins during meiosis: crossover artists at work. Chromosome Res 15:591–605

Macaisne N, Novatchkova M, Peirera L, Vezon D, Jolivet S, Froger N, Chelysheva L, Grelon M, Mercier R (2008) SHOC1, an XPF endonuclease-related protein, is essential for the formation of class I meiotic crossovers. Curr Biol 18:1432–1437

Malik SB, Ramesh MA, Hulstrand AM, Logsdon JM Jr (2007) Protist homologs of the meiotic Spo11 gene and topoisomerase VI reveal an evolutionary history of gene duplication and lineage-specific loss. Mol Biol Evol 24:2827–2841

Malkova A, Swanson J, German M, McCusker JH, Housworth EA, Stahl FW, Haber JE (2004) Gene conversion and crossing over along the 405-kb left arm of *Saccharomyces cerevisiae* chromosome VII. Genetics 168:49–63

Marsolier-Kergoat MC, Yeramian E (2009) GC content and recombination: reassessing the causal effects for the *Saccharomyces cerevisiae* genome. Genetics 183:31–38

Marston AL, Amon A (2004) Meiosis: cell-cycle controls shuffle and deal. Nat Rev Mol Cell Biol 5:983–997

Martinez-Perez E (2009) Meiosis in cereal crops: the grasses are back. Genome Dyn 5:26–42

Mercier R, Grelon M (2008) Meiosis in plants: ten years of gene discovery. Cytogenet Genome Res 120:281–290

Mestiri I, Chague V, Tanguy AM, Huneau C, Huteau V, Belcram H, Coriton O, Chalhoub B, Jahier J (2010) Newly synthesized wheat allohexaploids display progenitor-dependent meiotic stability and aneuploidy but structural genomic additivity. New Phytol 186:86–101

Mieczkowski PA, Dominska M, Buck MJ, Lieb JD, Petes TD (2007) Loss of a histone deacetylase dramatically alters the genomic distribution of Spo11p-catalyzed DNA breaks in *Saccharomyces cerevisiae*. Proc Natl Acad Sci USA 104:3955–3960

Mikhailova EI, Naranjo T, Shepherd K, Wennekes-van Eden J, Heyting C, de Jong JH (1998) The effect of the wheat *Ph1* locus on chromatin organisation and meiotic chromosome pairing analysed by genome painting. Chromosoma 107:33–350

Miles LG, Isberg SR, Glenn TC, Lance SL, Dalzell P, Thomson PC, Moran C (2009) A genetic linkage map for the saltwater crocodile (*Crocodylus porosus*). BMC Genomics 10:339

Moore G, Shaw P (2009) Improving the chances of finding the right partner. Curr Opin Genet Dev 19:99–104

Myers S, Bottolo L, Freeman C, McVean G, Donnelly P (2005) A fine-scale map of recombination rates and hotspots across the human genome. Science 310:321–324

Myers S, Bowden R, Tumian A, Bontrop RE, Freeman C, MacFie TS, McVean G, Donnelly P (2010) Drive against hotspot motifs in primates implicates the PRDM9 gene in meiotic recombination. Science 327:876–879

Nachman MW (2002) Variation in recombination rate across the genome: evidence and implications. Curr Opin Genet Dev 12:657–663

Nicolas SD, Leflon M, Monod H, Eber F, Coriton O, Huteau V, Chevre AM, Jenczewski E (2009) Genetic regulation of meiotic cross-overs between related genomes in *Brassica napus* haploids and hybrids. Plant Cell 21:373–385

Noma K, Grewal SI (2002) Histone H3 lysine 4 methylation is mediated by Set1 and promotes maintenance of active chromatin states in fission yeast. Proc Natl Acad Sci USA 99(Suppl 4):16438–16445

Nonomura K, Nakano M, Fukuda T, Eiguchi M, Miyao A, Hirochika H, Kurata N (2004) The novel gene HOMOLOGOUS PAIRING ABERRATION IN RICE MEIOSIS1 of rice encodes a putative coiled-coil protein required for homologous chromosome pairing in meiosis. Plant Cell 16:1008–1020

Nonomura K, Nakano M, Eiguchi M, Suzuki T, Kurata N (2006) PAIR2 is essential for homologous chromosome synapsis in rice meiosis I. J Cell Sci 119:217–225

Osman K, Sanchez-Moran E, Mann SC, Jones GH, Franklin FC (2009) Replication protein A (AtRPA1a) is required for class I crossover formation but is dispensable for meiotic DNA break repair. EMBO J 28:394–404

Paques F, Haber JE (1999) Multiple pathways of recombination induced by double-strand breaks in *Saccharomyces cerevisiae*. Microbiol Mol Biol Rev 63:349–404

Parvanov ED, Petkov PM, Paigen K (2010) Prdm9 controls activation of mammalian recombination hotspots. Science 327:835

Pawlowski WP, Golubovskaya IN, Timofejeva L, Meeley RB, Sheridan WF, Cande WZ (2004) Coordination of meiotic recombination, pairing, and synapsis by PHS1. Science 303:89–92

Perella G, Consiglio MF, Aiese-Cigliano R, Cremona G, Sanchez-Moran E, Barra L, Errico A, Bressan RA, Franklin FC, Conicella C (2010) Histone hyperacetylation affects meiotic recombination and chromosome segregation in *Arabidopsis*. Plant J 62:796–806

Perez R, Cuadrado A, Chen IP, Puchta H, Jouve N, De Bustos A (2010) The Rad50 genes of diploid and polyploid wheat species. Analysis of homologue and homoeologue expression and interactions with Mre11. Theor Appl Genet 122:251–262

Pesin JA, Orr-Weaver TL (2008) Regulation of APC/C activators in mitosis and meiosis. Annu Rev Cell Dev Biol 24:475–499

Petes TD (2001) Meiotic recombination hot spots and cold spots. Nat Rev Genet 2:360–369

Prieto P, Shaw P, Moore G (2004) Homologue recognition during meiosis is associated with a change in chromatin structure. Nat Cell Biol 6:906–908

Puizina J, Siroky J, Mokros P, Schweizer D, Riha K (2004) Mre11 deficiency in *Arabidopsis* is associated with chromosomal instability in somatic cells and Spo11-dependent genome fragmentation during meiosis. Plant Cell 16:1968–1978

Quevedo C, Del Cerro AL, Santos JL, Jones GH (1997) Correlated variation of chiasma frequency and synaptonemal complex length in *Locusta migratoria*. Heredity 78:515–519

Ramsey J, Schemske DW (2002) Neopolyploidy in flowering plants. Annu Rev Ecol Syst 33:589–639

Raynard S, Bussen W, Sung P (2006) A double Holliday junction dissolvasome comprising BLM, topoisomerase IIIalpha, and BLAP75. J Biol Chem 281:13861–13864

Reinisch AJ, Dong JM, Brubaker CL, Stelly DM, Wendel JF, Paterson AH (1994) A detailed RFLP map of cotton, *Gossypium hirsutum* × *Gossypium barbadense*: chromosome organization and evolution in a disomic polyploid genome. Genetics 138:829–847

Riehs N, Akimcheva S, Puizina J, Bulankova P, Idol RA, Siroky J, Schleiffer A, Schweizer D, Shippen DE, Riha K (2008) *Arabidopsis* SMG7 protein is required for exit from meiosis. J Cell Sci 121:2208–2216

Riley R, Chapman V (1958) Genetic control of the cytologically diploid behaviour of hexaploid wheat. Nature 13:713–715

Ronceret A, Doutriaux MP, Golubovskaya IN, Pawlowski WP (2009) PHS1 regulates meiotic recombination and homologous chromosome pairing by controlling the transport of RAD50 to the nucleus. Proc Natl Acad Sci USA 129:173–183

Ross KJ, Fransz P, Armstrong SJ, Vizir I, Mulligan B, Franklin FC, Jones GH (1997) Cytological characterization of four meiotic mutants of *Arabidopsis* isolated from T-DNA-transformed lines. Chromosome Res 5:551–559

Saintenac C, Falque M, Martin OC, Paux E, Feuillet C, Sourdille P (2009) Detailed recombination studies along chromosome 3B provide new insights on crossover distribution in wheat (*Triticum aestivum* L.). Genetics 181:393–403

Salmon A, Flagel L, Ying B, Udall JA, Wendel JF (2010) Homoeologous nonreciprocal recombination in polyploid cotton. New Phytol 186:123–134

Sanchez-Moran E, Mercier R, Higgins JD, Armstrong SJ, Jones GH, Franklin FC (2005) A strategy to investigate the plant meiotic proteome. Cytogenet Genome Res 109:181–189

Sanchez-Moran E, Santos JL, Jones GH, Franklin FC (2007) ASY1 mediates AtDMC1-dependent interhomolog recombination during meiosis in *Arabidopsis*. Genes Dev 21:2220–2233

Scherthan H, Wang H, Adelfalk C, White EJ, Cowan C, Cande WZ, Kaback DB (2007) Chromosome mobility during meiotic prophase in *Saccharomyces cerevisiae*. Proc Natl Acad Sci USA 104:16934–16939

Sears ER (1976) Genetic control of chromosome pairing in wheat. Annu Rev Genet 10:31–51

Sears ER, Okamoto M (1958) Intergenomic chromosome relationships in hexaploid wheat. In: 10th international congress of genetics. Southam Printing Co., Montreal, pp 258–259

Sheehan MJ, Pawlowski WP (2009) Live imaging of rapid chromosome movements in meiotic prophase I in maize. Proc Natl Acad Sci USA 106:20989–20994

Singer A, Perlman H, Yan Y, Walker C, Corley-Smith G, Brandhorst B, Postlethwait J (2002) Sex-specific recombination rates in zebrafish (*Danio rerio*). Genetics 160:649–657

Soltis DE, Albert VA, Leebens-Mack J, Bell CD, Paterson AH, Zheng C, Sankoff D, dePamphilis CW, Wall PK, Soltis PS (2009) Polyploidy and angiosperm diversification. Am J Bot 96:336–348

Soltis DE, Buggs RJA, Doyle JJ, Soltis PS (2010) What we still don't know about polyploidy. Taxon 59:1387–1403

Stacey NJ, Kuromori T, Azumi Y, Roberts G, Breuer C, Wada T, Maxwell A, Roberts K, Sugimoto-Shirasu K (2006) *Arabidopsis* SPO11-2 functions with SPO11-1 in meiotic recombination. Plant J 48:206–216

Stevens R, Grelon M, Vezon D, Oh J, Meyer P, Perennes C, Domenichini S, Bergounioux C (2004) A CDC45 homolog in *Arabidopsis* is essential for meiosis, as shown by RNA interference-induced gene silencing. Plant Cell 16:99–113

Sturtevant AH (1915) The behavior of the chromosomes as studied through linkage. Mol Gen Genet 13:234–287

Sung P, Krejci L, Van Komen S, Sehorn MG (2003) Rad51 recombinase and recombination mediators. J Biol Chem 278:42729–42732

Szostak JW, Orr-Weaver TL, Rothstein RJ, Stahl FW (1983) The double-strand-break repair model for recombination. Cell 33:25–35

Szwarcwort-Cohen M, Kasulin-Boneh Z, Sagee S, Kassir Y (2009) Human Cdk2 is a functional homolog of budding yeast Ime2, the meiosis-specific CDK-like kinase. Cell Cycle 8:647–654

Talbert PB, Henikoff S (2010) Centromeres convert but don't cross. PLoS Biol 8:e1000326

Tease C, Hulten MA (2004) Inter-sex variation in synaptonemal complex lengths largely determine the different recombination rates in male and female germ cells. Cytogenet Genome Res 107:208–215

Van de Peer Y, Fawcett JA, Proost S, Sterck L, Vandepoele K (2009) The flowering world: a tale of duplications. Trends Plant Sci 14:680–688

Ved Brat S (1966) Genetic systems in *Allium.* Sex differences in meiosis. Chromosomes Today 1:31–40

Vignard J, Siwiec T, Chelysheva L, Vrielynck N, Gonord F, Armstrong SJ, Schlögelhofer P, Mercier R (2007) The interplay of RecA-related proteins and the MND1-HOP2 complex during meiosis in *Arabidopsis thaliana.* PLoS Genet 3:1894–1906

Wall AM, Riley R, Chapman V (1971) Wheat mutants permitting meiotic chromosome pairing. Genet Res 18:311–328

Wallace BM, Hulten MA (1985) Meiotic chromosome pairing in the normal human female. Ann Hum Genet 49:215–226

Wallace BM, Wallace H (2003) Synaptonemal complex karyotype of zebrafish. Heredity 90:136–140

Wang K, Tang D, Wang M, Lu J, Yu H, Liu J, Qian B, Gong Z, Wang X, Chen J, Gu M, Cheng Z (2009) MER3 is required for normal meiotic crossover formation, but not for presynaptic alignment in rice. J Cell Sci 122:2055–2063

Wang M, Wang K, Tang D, Wei C, Li M, Shen Y, Chi Z, Gu M, Cheng Z (2010a) The central element protein ZEP1 of the synaptonemal complex regulates the number of crossovers during meiosis in rice. Plant Cell 22:417–430

Wang Y, Jha AK, Chen R, Doonan JH, Yang M (2010b) Polyploidy-associated genomic instability in *Arabidopsis thaliana.* Genesis 48:254–263

Wood TE, Takebayashi N, Barker MS, Mayrose I, Greenspoon PB, Rieseberg LH (2009) The frequency of polyploid speciation in vascular plants. Proc Natl Acad Sci USA 106:13875–13879

Yamada T, Mizuno K, Hirota K, Kon N, Wahls WP, Hartsuiker E, Murofushi H, Shibata T, Ohta K (2004) Roles of histone acetylation and chromatin remodeling factor in a meiotic recombination hotspot. EMBO J 23:1792–1803

Yousafzai FK, Al-Kaff N, Moore G (2010a) The molecular features of chromosome pairing at meiosis: the polyploid challenge using wheat as a reference. Funct Integr Genomics 10:147–156

Yousafzai FK, Al-Kaff N, Moore G (2010b) Structural and functional relationship between the Ph1 locus protein 5B2 in wheat and CDK2 in mammals. Funct Integr Genomics 10:157–166

Yu H, Wang M, Tang D, Wang K, Chen F, Gong Z, Gu M, Cheng Z (2010) OsSPO11-1 is essential for both homologous chromosome pairing and crossover formation in rice. Chromosoma 119:625–636

Yuan W, Li X, Chang Y, Wen R, Chen G, Zhang Q, Wu C (2009) Mutation of the rice gene PAIR3 results in lack of bivalent formation in meiosis. Plant J 59:303–315

Mechanisms of Chromosome Rearrangements 9

Martin A. Lysák and Ingo Schubert

Contents

M.A. Lysák (✉)
Laboratory of Plant Cytogenomics, CEITEC – Central European Institute of Technology, Masaryk University, Kamenice 5, CZ-625 00 Brno, Czech Republic
e-mail: lysak@sci.muni.cz

I.J. Leitch et al. (eds.), *Plant Genome Diversity Volume 2*,
DOI 10.1007/978-3-7091-1160-4_9,

9.1 Introduction

The striking diversity of land plants is associated with immense genetic variation manifested also by a wide range of genome sizes and chromosome numbers. Nuclear genome size across land plants varies more than 2,300-fold from ~64 Mb (*Genlisea aurea*; Greilhuber et al. 2006) to ~150,000 Mb (*Paris japonica*; Pellicer et al. 2010; see also Leitch and Leitch 2013, this volume). Accordingly, chromosome numbers can vary from $n = 2$ in six angiosperm species (Vanzela et al. 1996; Cremonini 2005) to $n > 320$ in the angiosperm *Sedum suaveolens* (Uhl 1978) and $n =$ c. 720 in the fern *Ophioglossum reticulatum* (Khandelwal 1990). This extensive variation of chromosome numbers among land plants is driven by two main trends in opposite directions: chromosome numbers increase through polyploidy (whole-genome duplications, WGD) and decrease through structural chromosome rearrangements (descending dysploidy). Genomic and cytogenetic analyses indicate that probably all land plants have experienced at least one WGD event (Jaillon et al. 2009; Soltis et al. 2009; Van de Peer et al. 2009; see also Fawcett et al. 2013, this volume) followed by more or less extensive karyotype reshuffling towards diploid-like genomes (e.g., Wolfe 2001; Thomas et al. 2006; Cenci et al. 2010; Mandáková et al. 2010a). Karyotypic changes at a given ploidy level are mediated by chromosome rearrangements such as insertions, duplications, deletions, inversions and translocations altering the size and morphology of chromosomes. Centric fissions and different types of reciprocal translocations combined with meiotic (mis) segregation may lead to a reduction or increase of chromosome number (descending/ascending dysploidy).

Here we discuss mechanisms which alter size, shape and number of chromosomes, reviewed earlier by Darlington (1937), Stebbins (1971), Jones (1998), Levin (2002), Schubert (2007) and Schubert and Lysák (2011). In particular, we focus on chromosome rearrangements that have an impact on the alteration of chromosome number during evolution of land plants and likely play a role in speciation. The emphasis is put on chromosome rearrangements detectable by microscopic techniques, whereas small-range insertion/deletion events (indels), inversions or gene conversion are beyond the scope of this review.

9.2 It Starts With a Break ...

Each chromosome aberration starts with one or more double-strand break(s) (DSB). DSBs arise either directly after the exposure to S-phase-independent mutagens (e.g., ionizing radiation such as X-rays or gamma rays, chemicals such as bleomycin or restriction endonucleases), or indirectly when the replication fork meets a repair-mediated single strand gap. DSBs can be repaired restoring the original DNA structure, whereas a mis-repair of two or more simultaneous DSBs results in a chromosome rearrangement. DSBs are repaired by homologous recombination (HR), using sequences homologous to the broken strands as a template for DNA synthesis and operating mainly in S and G2-phase or by non-homologous end joining (NHEJ, also known as illegitimate recombination), which can join broken ends directly without or with just a few base pairs of homology. HR is a mostly accurate repair mechanism, whereas NHEJ is an error-prone process usually associated with microdeletions and insertions (filler DNA) prior to ligation of broken ends. HR can also occur between ectopic homologous sequences. Non-allelic (ectopic) homologous recombination (NAHR) may occur within one double helix, between sister chromatids, or between chromatids of homologous or non-homologous chromosomes. Recombination using different templates has different outcomes as shown in Fig. 9.1 (see also Gaut et al. 2007; Gu et al. 2008). Intrastrand recombination between direct repeats results in deletion (via "pop-out" of a small circular chromosome segment). If the repeats have the opposite orientation, NAHR yields an inversion of the region between the breakpoints. HR between ectopic homologous sequences of sister chromatids in direct orientation produces chromatids and chromosomes with duplication and deletion, respectively (unequal sister chromatid exchange). This process is analogous to unequal crossover at meiosis. NAHR between repeats on homologous or non-homologous chromosomes results in reciprocal chromosome translocations.

Both repair mechanisms, HR and NHEJ, are active in somatic cells, while allelic meiotic recombination is exclusively based on HR, processing Spo11-induced DSBs. In plants, rearrangements occurring in somatic tissues (shoot meristems) can be transmitted to the next generation.

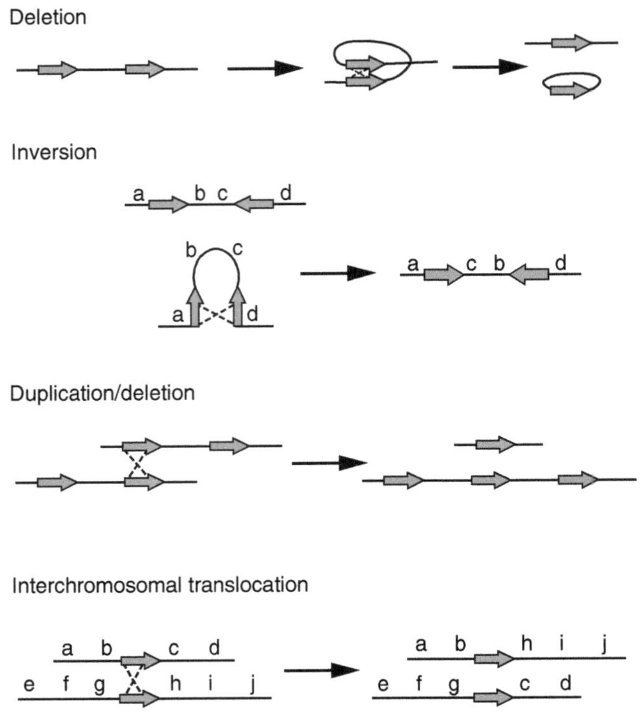

Fig. 9.1 Primary chromosome rearrangements mediated by non-allelic homologous recombination (NAHR). *Grey arrows* represent homologous (repetitive) sequences (redrawn and modified after Griffiths et al. (2007))

9.3 Primary Chromosome Rearrangements

9.3.1 Insertions

DSB repair can be accompanied by insertion of endogenous or alien sequences. A possible mechanism of insertion can be recombination with extrachromosomal circular DNA (Cohen et al. 2008; Navrátilová et al. 2008) (Fig. 9.2a).

9.3.2 Deletions

Deletions can be terminal or interstitial. After loss of a terminal chromosome segment, the broken end has to be healed and stabilized by *de novo* telomere synthesis or by telomere capture (acquiring telomeric sequence from a sister chromatid, a homologous or heterologous chromosome *via* a conversion-like mechanism; Yu and Graf 2010), otherwise the broken chromosome ends enter a fusion-breakage cycle or the harbouring cell dies. Interstitial deletions require two breaks and arise e.g., *via* NHEJ between the most proximal and the most distal break end (Fig. 9.2b). During meiotic pairing in an interstitial-deletion heterozygote, the wild-type chromosome loops out the non-deleted region.

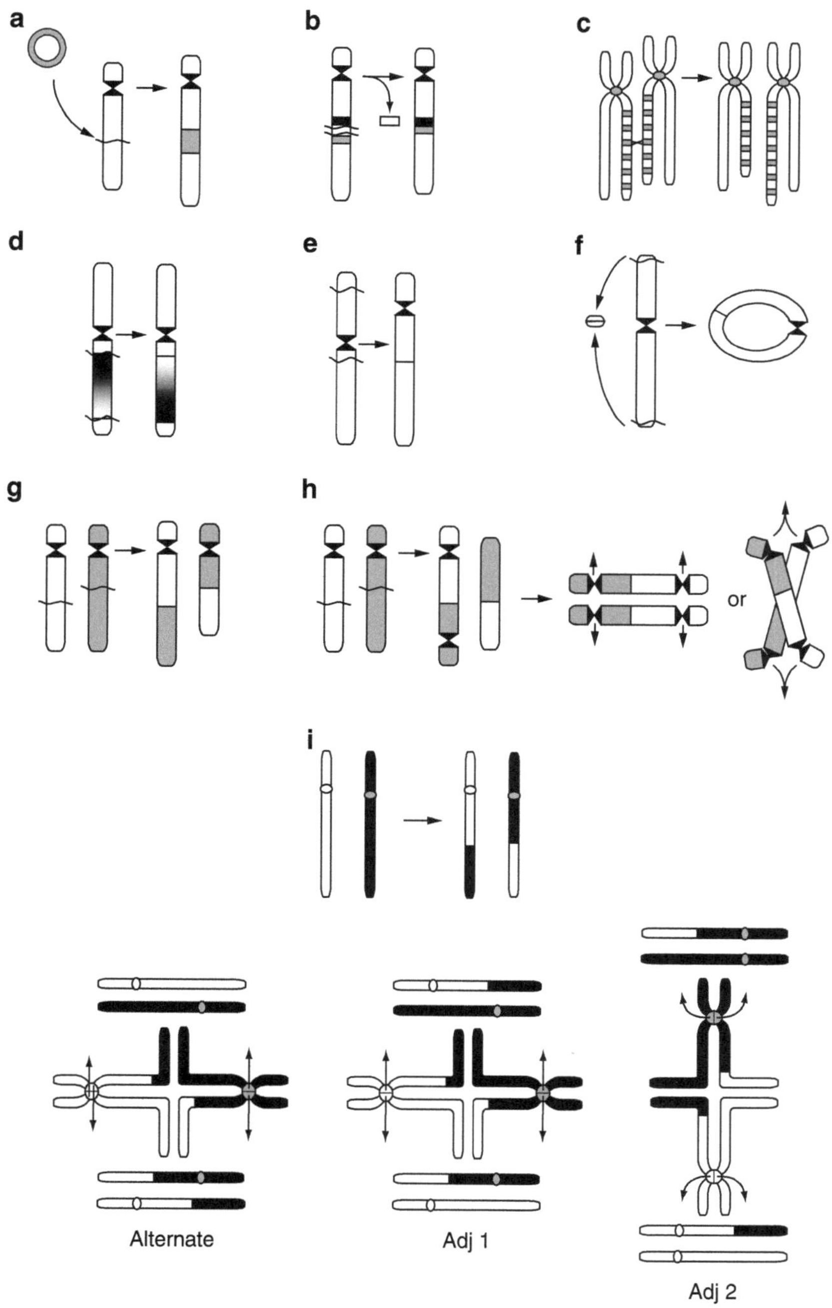

Fig. 9.2 Primary chromosome rearrangements. (**a**) Insertion of circular extrachromosomal DNA. (**b**) Interstitial deletion. (**c**) Duplication and deletion caused by unequal meiotic crossover between homologous chromosomes. (**d**) Paracentric inversion. (**e**) Pericentric inversion. (**f**) Intrachromosomal translocation resulting in a ring chromosome. (**g**) Symmetrical reciprocal translocation. (**h**) Asymmetric reciprocal translocation yielding a dicentric chromosome and an acentric fragment; the dicentric is destroyed by a rupture of the anaphase bridge when there is an intercentromeric sister chromatid twist (far right). (**i**) Meiosis I in a symmetrical translocation heterozygote. The alternate segregation will result in balanced gametes with a "normal" and translocation chromosome complement, respectively. Adjacent (Adj) segregation yields unbalanced gametes with duplications and deletions, *via* adjacent segregation of non-homologous (Adj-1) or homologous centromeres (Adj-2)

9.3.3 Duplications and Deletions

Large-scale (>10 kb) duplications usually coupled with simultaneous deletion events may originate through multiple mechanisms. A duplicated segment is either placed adjacent to the original sequence (tandem duplication) or elsewhere in the complement (insertional duplication). During meiotic prophase of a tandem-duplication heterozygote, the unpaired duplicated region forms a loop. Erroneous repair of two DSBs *via* unequal sister chromatid exchange or *via* unequal meiotic crossover (CO) between two not well aligned homologous chromatids may result in duplication and deletion

(Fig. 9.2c). NHEJ frequently accompanied by exonucleolytic digestion at DSB ends or by insertional filling of sequences into the break may cause deletions and duplications at the submicroscopic level. Tandem duplications generate additional substrates for ectopic homologous recombination (Fig. 9.1); thus increasing the probability of further rearrangements caused by this mechanism.

9.3.4 Inversions

Inversions are intra-chromosomal rearrangements that alter the collinearity due to the opposite orientation of a chromosome segment, without changing the chromosome size. Inversions are induced by two breaks followed by the reunion of the internal fragment with the 'wrong' break ends. Traditionally paracentric and pericentric inversions have been recognized. A paracentric inversion does not include the centromere (Fig. 9.2d) and thus does not change the chromosome morphology, if no breakpoints at different distances on either side of a nucleolus organizing region (NOR) are involved. A pericentric inversion, including the respective centromere, may result in a changed arm ratio, if the breakpoints on either side have a different distance to the centromere (Fig. 9.2e).

Ectopic homologous recombination (see Sect. 9.2), for instance between two transposon copies (Delprat et al. 2009), seems to be a mechanism frequently yielding chromosome inversions. Two DSBs (Runcie and Noor 2009) or the coincidence of two pairs of staggered single-strand breaks eventually repaired by NHEJ (Ranz et al. 2007) may lead to inversions without requiring homologous motifs between the inversion breakpoints.

During meiotic pairing in inversion heterozygotes the inverted region forms an 'inversion loop' with the non-inverted homologous regions (including the centromere in heterozygotes for pericentric inversion). If no CO occurs within the inversion loop, 50% of the gametes will carry the inversion. However a CO within the inversion loop of a heterozygote for a pericentric inversion will result in normal (1), inversion (1) and duplication/deletion (2) chromatid types. In a heterozygote for a paracentric inversion, a CO within the inversion loop connects homologous chromosomes to form a dicentric and an acentric fragment (the latter lost as a micronucleus). At anaphase I, the dicentric bridge is randomly broken yielding two chromatids with terminal deletions. One normal and one inversion chromatid are also produced.

9.3.5 Translocations

Translocations are induced by two DSBs resulting in four DNA ends. An erroneous repair by ligation of the ends of different breaks generates a reciprocal translocation. In this case, DSBs are mis-repaired by NHEJ not requiring sequence homology at the breakpoints, but DSBs can also be joined through HR between ectopic repeats (see Sect. 9.2). NAHR-mediated translocations are facilitated by the abundance of repetitive elements which are common in plants with large genome sizes. An intrachromosomal reciprocal translocation with a break in both arms yields a ring chromosome and a terminal acentric fragment (Fig. 9.2f). The ring chromosome (particularly if very small) can be stably inherited for some time (Murata et al. 2008) but usually both translocation products are unstable and lost. More common are interchromosomal reciprocal translocations (interchanges) which may involve homologous or heterologous chromosomes and can be either symmetric and yield monocentric products (Fig. 9.2g), or asymmetric, resulting in a dicentric and an acentric product (Fig. 9.2h). Dicentric products of asymmetric translocations are destroyed by rupture of anaphase bridges if inter-centromeric sister chromatid twists occur and the two centromeres are pulled to the opposite poles. Dicentrics might be stable if both centromeres are in close vicinity and are pulled to one pole (Fig. 9.2h). Inactivation and/or loss of one centromere can turn the dicentric chromosome into a stable monocentric (see Sect. 9.6).

During the first meiotic division in a translocation heterozygote, normal and translocated chromosomes pair to form a cruciform configuration (Fig. 9.2i). Depending on the number and localization of chiasmata, different chromosome configurations (e.g., a ring or chain of chromosomes) are observed in diplotene and diakinesis. Due to the independent assortment of chromosomes, two types of segregation can occur (Fig. 9.2i). During alternate segregation, the translocation chromosomes are pulled to one pole and "normal" chromosomes to the other pole. This segregation eventually yields balanced gametes with two normal and two translocation chromatids. In so-called adjacent segregation, either both "adjacent" non-homologous or homologous centromeres pass to the same pole. Both types of adjacent segregation result in unbalanced gametes with duplications and deletions. Such gametes are usually not viable, but may be tolerated in polyploids. If so, the progeny genomes may be erroneously considered as bearing a 'non-reciprocal translocation' (in polyploids also called homoeologous non-reciprocal transposition or translocation; Udall et al. 2005; Gaeta et al. 2007; Nicolas et al. 2007) which have not yet been proven experimentally. Translocation events involving unequally sized chromosome regions may also be erroneously interpreted as a "non-reciprocal" translocation. In this case, besides a large translocation product, very small centric or acentric fragments are also generated (see Sect. 9.5.1). As these products may be either undetectable by routine microscopic techniques or eliminated during subsequent mitotic (acentric fragments) or meiotic divisions (centric fragments), the translocation event might appear as non-reciprocal within the progenies.

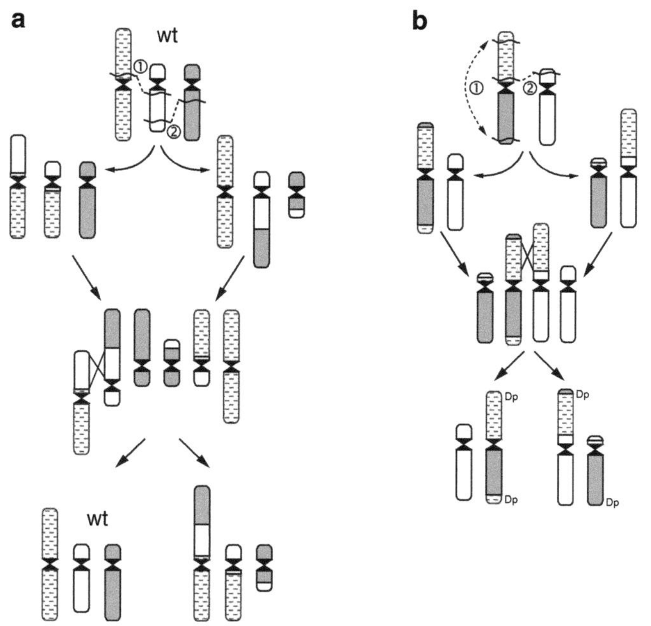

Fig. 9.3 Secondary chromosome rearrangements. (**a**) Two translocations (① and ②) involving three chromosomes of the wild-type (wt) karyotype lead to hexavalents during meiosis in double heterozygous individuals. Cross over between homologous chromosome regions flanked by non-homologous regions generates a novel karyotype and re-establishes the wt complement. (**b**) Origin of new karyotypes through primary chromosome rearrangements and meiotic recombination. One chromosome is involved in an inversion in one individual (①) and in a translocation in another individual (②). In the resulting quadrivalent, crossover within homologous chromosome regions flanked by non-homologous regions generates two new karyotypes with complementary duplication (Dp) and deletions, respectively

Translocated regions might be lost or duplicated due to CO between rearranged chromosomes (Fig. 9.3b), or due to meiotic mis-segregation in translocation heterozygotes (Fig. 9.4g) yielding unbalanced karyotypes.

Short range gene conversion should not to be considered as non-reciprocal translocation.

9.4 Secondary Chromosome Rearrangements

Structural chromosome alterations can additionally arise as secondary chromosome rearrangements (for review see Schubert 2007). Such rearrangements may arise in organisms doubly heterozygous for two primary rearrangements (translocations and/or inversions), if one chromosome is involved in both of them (Fig. 9.3). Meiotic CO between homologous regions of rearranged chromosomes, which differ in the regions distal to the CO, leads to gametes with a new karyotype and to complementary gametes displaying a re-established wild type chromosome complement (Fig. 9.3a). Depending on the type of primary rearrangements involved, unbalanced gametes may also arise, harbouring duplications as well as deletions (Fig. 9.3b; Schubert et al. 1988).

9.5 Changes of Chromosome Number (Dysploidy)

Descending and ascending dysploidy is a karyotypic alteration towards a lower or higher chromosome number. Translocations play a prominent role in decreasing chromosome numbers. The increase of chromosome number, apart from polyploidization, is mediated by chromosome fission (dissociation). The evolutionary success of novel karyotypes depends on the viability of gametes and on whether or not homozygosity is possible under natural selection.

9.5.1 Descending Dysploidy

A Robertsonian (Rb) translocation or centric fusion refers to a translocation between two telocentric or acrocentric chromosomes. This translocation event results in a meta (di)centric 'fusion' chromosome (Sullivan et al. 1996) and a small acentric fragment (Fig. 9.4a), or in two monocentric products of unequal size (Fig. 9.4c). If the minichromosome is eliminated during subsequent cell divisions, the Rb event may lead to a heritable decrease in chromosome number (n−1). Typically, Rb breakpoints are localized at the centric ends of the short arms of telo- or acrocentric chromosomes [i.e., (sub)telomere in telocentrics]. As a consequence, Rb translocations reduce chromosome number but maintain the number of major chromosome arms (the *nombre fondamental*, NF). The ribosomal DNA (rDNA) of terminal NORs was thought to mediate Rb translocations. Often breakpoints were found proximal to NORs on acrocentric chromosomes and the NORs were deleted (Page et al. 1996). Several human Rb metacentrics have originated through variably positioned breakpoints within the short arms and less frequently at the pericentromeres of the two acrocentric chromosomes involved (Page et al. 1996; Sullivan et al. 1996). Hence, 90% of Rb chromosomes in humans are dicentric (Page et al. 1996). Although an exact mechanism of Rb fusion remains to be elucidated, the abundance of satellite DNA arrays on short arms of acrocentrics and at centromeres of Rb chromosomes suggests NAHR based on ectopic homologous sequences as a mechanism (Page et al. 1996; Sullivan et al. 1996). Rb fusions in plant species can also result in dicentric chromosomes as shown for the metacentric chromosome 1 of *Vicia faba* (Schubert 1992). In the crucifers (Brassicaceae) (and likely in other groups), Rb translocations are frequently preceded by a pericentric inversion in one of the chromosomes, rendering a (sub)

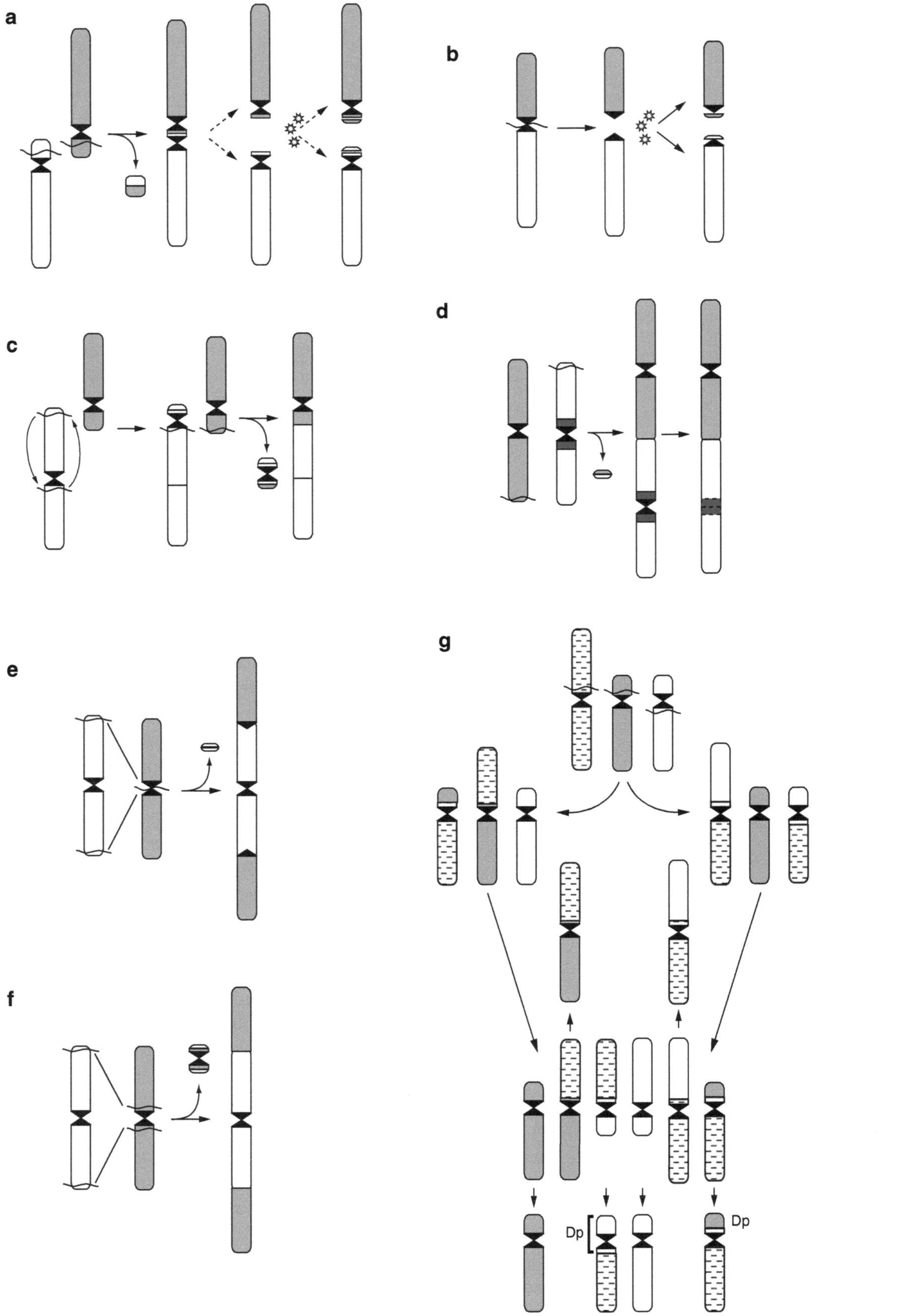

Fig. 9.4 Chromosome rearrangements mediating descending and ascending dysploidy. (**a**) Robertsonian translocation between two telo-/acrocentric chromosomes with breakpoints in short arms results in a meta (di)centric chromosome and an acentric fragment. The dicentric chromosome can be prone to fission which yields two telocentric chromosomes in the presence of telomerase (stars) stabilized by *de novo* added telomeres. (**b**) Centric fission in a monocentric chromosome. Centric ends of fission products are "healed" by telomerase (stars).

metacentric chromosome telo- or acrocentric. This rearrangement allows for a Rb event resulting in a large "fusion" chromosome and a minichromosome prone to loss (Fig. 9.4c; Lysák et al. 2006). For chromosome number reduction in some cruciferous species, centromere inactivation and/or loss has been envisaged to follow asymmetric, end-to-end translocation between two metacentrics (Fig. 9.4d; Mandáková et al. 2010a, b).

Based on genomic data, insertional or nested chromosome fusion (NCF) has been proposed as a prominent mechanism of descending dysploidy in grasses (Luo et al. 2009; Salse et al. 2009; Thiel et al. 2009; Abrouk et al. 2010; International Brachypodium Initiative 2010; Murat et al. 2010), but it is considered to occur less frequently in cruciferous taxa (Mandáková et al. 2010b). In NCF, one chromosome is inserted between the chromosome arms of another (recipient) chromosome. Although not yet proven experimentally, a NCF can most easily be interpreted as the outcome of a multiple breakage and translocation event between two non-homologous chromosomes. Simultaneous breaks at the ends of both arms of the nested chromosome and one around the centromere of the recipient chromosome are followed by a mis-repair of the chromosome breaks (Fig. 9.4e). Disruption of the recipient centromere by a break followed by the inactivation of its parts is a plausible mechanism to avoid an unstable di- or tricentric product. Alternatively, if the recipient chromosome has breaks on either side of its centromere, a symmetrical translocation may yield two monocentric products: the 'fusion' chromosome and a minichromosome comprising the centromere of the recipient chromosome and the telomeres of the inserted chromosome. The small product without essential genes is prone to loss during meiosis (Fig. 9.4f). Theoretically, the two arms of the recipient chromosome could have been translocated to the nested chromosome arm ends via two subsequent events. However, it is hard to explain why the second event in so many cases again involved the same two chromosomes (Luo et al. 2009; Thiel et al. 2009; International Brachypodium Initiative 2010). A simultaneous translocation of broken donor chromosome arms to different recipient chromosomes would bear a higher risk of unbalanced gametes resulting from segregation of meiotic multivalents.

9.5.2 Ascending Dysploidy (Centric Fission)

Centric fission, i.e., dissociation of two arms of usually a metacentric chromosome increasing the chromosome number (n + 1), is considered to counteract Robertsonian fusion (Sect. 9.5.1). Centric fissions are a favoured explanation for the origin of telocentric chromosomes. However, to decide whether acro-/telocentric chromosomes result from fission or from a whole-arm pericentric inversion in a (sub)metacentric has been difficult without recently established chromosome painting or comparative genomics. The mechanism of centric fissions, more often observed in animals than plants, remains controversial. Each of the two telocentric fission products should possess a functional centromere and telomere sequences capping the centric ends. "Chromosome healing" by *de novo* synthesis of telomeres or telomere capture (Fig. 9.4a, b) has been postulated. Moreover, interstitial telomere(−like) repeats identified in centromeric regions of some species may serve as telomere seeds in fission telocentrics. In a fission heterozygote of the plant *Hypochaeris radicata*, both fission products were stabilized at centric ends by transposition of 45S rDNA from a NOR of another chromosome (Hall and Parker 1995). Apparently, rDNA is a potential substrate for telomere synthesis by telomerase at chromosome breakpoints through 2–4-nucleotide target motifs within the rDNA sequence (Tsujimoto et al. 1999). The transposition and amplification of rDNA is concordant with the post-fission extension of constitutive heterochromatin transforming fission telocentrics into acrocentrics as proposed by Imai (1991).

A transversal division of the centromere into two functional parts represents a challenging problem (Fig. 9.4a, b). A stable dicentric Robertsonian translocation chromosome provides functional centromeres when a break occurs between the two centromeric regions (Fig. 9.4a and Schubert et al. 1995). Fission events in monocentrics (Fig. 9.4b) were explained by at least two mechanisms. Perry et al. (2005) carried out detailed molecular analyses of centric fissions in human chromosomes and found evidence for simple fission events as well as for centromere duplication prior to fission. In the latter case duplication of (peri)centromeric sequences is predisposing the centromere to breakage. An alternative

Fig. 9.4 (Continued) (**c**) Translocation between two chromosomes with breakpoints in the pericentromere of the long arm of one chromosome and in the telomeric end of the short arm of another chromosome. This translocation yields a large monocentric "fusion" chromosome and a meiotically unstable minichromosome. Involved chromosomes can become telo-/acrocentric *via* a preceding pericentromeric inversion (far left). (**d**) End-to-end reciprocal translocation between two (sub) metacentric chromosomes followed by inactivation/loss of one centromere. (**e**, **f**) Proposed mechanisms of nested chromosome "fusion" through three (**e**) or four (**f**) simultaneous double strand breaks (DSBs) in two non-homologous chromosomes. In (**e**) the centromere of the recipient chromosome is disrupted by a single DSB, whereas in (**f**) it is lost as a second, meiotically unstable translocation product. (**g**) Simultaneously ascending and descending dysploidy in double heterozygous individuals. Crossing of two karyotypes heterozygous for two different translocations involving three chromosomes may result in mis-segregation from meiotic multivalents. The two metacentric translocation chromosomes segregate to one pole (n−1) and the four acrocentric chromosomes to the other (n + 1). The hypoploid gametes harbour small deletions, whereas hyperploid gametes carry the corresponding duplications (Dp)

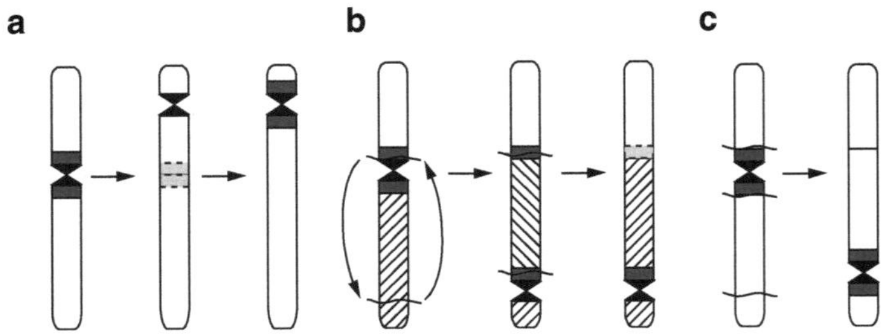

Fig. 9.5 Alternative mechanisms of centromere repositioning (CR). (**a**) CR through the emergence of a neocentromere and inactivation of the original centromere. The old pericentromere (*grey*) decays and new pericentromeric sequences accumulate at the neocentromere. (**b**) CR *via* subsequent peri- and paracentric inversions. The paracentric inversion restores the original collinearity along the chromosome, mimicking the mechanism described under (**a**). Pericentromeric arrays at the original centromere decay and are re-established at the new centromeric position. (**c**) CR mediated by the deletion of a centromere through centromere-flanking DSBs and its re-insertion into another DSB on the same chromosome

mechanism is based on coupling fission events with the appearance of neocentromeres (Nasuda et al. 2005). The origin of human acrocentric chromosomes 14 and 15 from an ancestral metacentric chromosome was interpreted assuming (1) a noncentromeric break within the ancestral metacentric, (2) inactivation of the ancestral centromere and the appearance of a neocentromere on HSA15, and (3) formation of a neocentromere on HSA14 (Fig. 9.1 in Ventura et al. 2003).

9.5.3 Bidirectional Dysploidy

The occurrence of two translocations between three chromosomes (one of the three involved in both translocations) with all breakpoints close to the centromeres may result in simultaneous ascending and descending dysploidy (Fig. 9.4g). In individuals double-heterozygous for both translocations, a hexavalent is formed during meiotic chromosome pairing. At a low frequency, the two translocation metacentrics of the hexavalent segregate to one pole and the four acrocentrics to the other. Consequently, gametes containing the acrocentric translocation chromosomes possess one chromosome more than the parental lines and a duplication of at least one centromere plus two terminal regions, while gametes with the metacentric translocation chromosomes have one chromosome less and the corresponding deletions. When gametes of the same dysploidy fuse, the chromosome number can increase or decrease simultaneously in a homozygous fashion, provided the accompanying duplications and deletions can be tolerated. This dysploidy mechanism has been experimentally proven in *Vicia faba* (Schubert and Rieger 1985). Similar mechanisms have been suggested already by Stebbins (1971). The increase in chromosome number from $n = 10$ to $n = 12$ postulated for a common (tetraploid) ancestor of cereals (Salse et al. 2008) could be explained in the same way. After whole genome duplication (= polyploidy) events, in the course of genome diploidization, a broad range of chromosome rearrangements can be tolerated (Mandáková et al. 2010a, b), and dysploidy, including deletions or duplications, is likely to become fixed. However it can be difficult to interpret these events subsequently as seen, for example, in the descending dysploidy leading to the extant maize genome (Wei et al. 2007; Salse et al. 2008, 2009). The more distant from the centromere the breakpoints in the translocation chromosomes of the double heterozygous individuals are, the larger are the duplicated (and the corresponding deleted) regions in the dysploid progeny karyotypes. The resulting segmental duplications or deletions may therefore give the impression of 'non-reciprocal translocations'.

9.6 Centromere Rearrangements

The classical theory of chromosome rearrangements considered centromeres as evolutionary stable chromosome structures, which may be lost or gained only by translocation or centric fission events. Consequently, until recently (Voullaire et al. 1993) positional changes of centromere position as well as emergence or loss of centromeres have been explained based on primary chromosome rearrangements. Today circumstantial evidence suggests the possibility of so-called centromere repositioning (CR), assuming centromere inactivation at one site and the emergence of a new centromere at another site. These events might occur without distortion of the marker order along the chromosome (Fig. 9.5). Such a centromere shift may change the arm ratio and thus the shape of the corresponding chromosome. CR events are claimed for vertebrates (Ventura et al. 2007; Piras et al. 2010) and have been presumed for homoeologous chromosomes of cucumber and melon (Han Y. et al. 2009). It remains however unclear how during the transition phase of CR (old centromere no longer active and neocentromere just emerging) the chromosome resolves the problem of having either no active centromere or two centromeres.

Although some dicentric human chromosomes were shown to be stable over several cell generations (Stimpson and Sullivan 2010), more experimental data are needed to further elucidate this process. Apart from a presumably epigenetic mechanism of CR (Fig. 9.5a), subsequent peri- and paracentric inversions may alter centromere position and restore collinearity (Fig. 9.5b). Alternatively, CR can be mediated by the deletion of a (peri)centromere and its re-insertion into another region of the same chromosome (Fig. 9.5c).

Centromere inactivation is not only postulated to occur during a shift of centromeres but may also play a role in stabilizing dicentric chromosomes. The inactivation of one centromere on a dicentric maize B chromosome was caused by the loss of epigenetic centromere markers such as CENH3, CENP-C and H3S10P. The inactive centromere could be reactivated when separated from the active centromere by intrachromosomal recombination (Han F. et al. 2009). This is compelling evidence that epigenetically inactivated centromeres can be reactivated. Centromere inactivation has also been observed in dicentric human chromosomes (e.g., Stimpson and Sullivan 2010) and it has been suggested to stabilize primary dicentric chromosomes resulting from telomere-to-telomere translocations in some cruciferous species (Fig. 9.4d; Mandáková et al. 2010a, b). Mechanisms of centromere inactivation are not well understood.

As centromere formation and function seem to be determined epigenetically without specific sequence requirements, a neocentromere could also emerge in acentric fragments and thus increase chromosome number. In plants, the emergence of a new centromere was shown to occur on fragments of barley chromosomes (Nasuda et al. 2005). As shown for plants (Han F. et al. 2009) and vertebrates (Rocchi et al. 2009), some neocentromeres occur at sites of inactivated ancestral centromeres.

9.7 Rearrangements Mediated by Transposable Elements

Barbara McClintock was the first to recognize that the mobility of transposable elements could cause chromosome rearrangements such as translocations, inversions, duplications and deletions (e.g., McClintock 1948). Indeed there is now cytological evidence showing how the origin of translocations and inversions in maize is due to the alternative transposition of *Ac* elements (Pulletikurti et al. 2009). Whereas conventional transposition is just moving a single element to another genomic position, alternative transposition events involving termini of two elements can cause chromosome breakage and major chromosome rearrangements (Weil 2009; Zhang et al. 2009). Models to explain different chromosome rearrangements resulting from alternative transposition of maize *Ac*/*Ds* transposons are available at http://jzhang.public.iastate.edu/transposition.html, and as supplemental materials online in Zhang et al. (2009).

Some transposable elements show the capacity to capture and move other genomic sequences. In maize, for example, *Helitron* transposons replicating through a rolling-circle mechanism frequently capture neighbouring sequences such as gene fragments (Yang and Bennetzen 2009). Transposition of larger chunks of chromosomal DNA has been suggested for repeated arrays of the large ribosomal RNA genes (rDNA) discernible on chromosomes as NORs. Since Schubert and Wobus (1985) showed the mobility of NORs ("jumping NORs"), even in cells of the same individual, intraspecific variation in number, position and extension of rDNA loci has been revealed in a wide range of plant species (e.g., in the tribe Triticeae, Dubcovsky and Dvořák 1995). As the observed variation in number and position of NORs between different individuals or populations is often not associated with collinearity distortion (Datson and Murray 2006), the NOR mobility is explained by transpositional 'hitch hiking' together with flanking transposable elements. Indeed, Raskina et al. (2004, 2008) showed by fibre fluorescence *in situ* hybridization that transposable elements can be interspersed with 5S and 45S rDNA repeats in *Aegilops* and *Triticum* species suggesting that the mobility of transposable elements is responsible for the extensive intraspecific variation in the number and localization of rDNA loci in the analysed grass species.

References

Abrouk M, Murat F, Pont C, Messing J, Jackson S, Faraut T, Tannier E, Plomion C, Cooke R, Feuillet C, Salse J (2010) Palaeogenomics of plants: synteny-based modelling of extinct ancestors. Trends Plant Sci 15:479–487

Cenci A, Combes M-C, Lashermes P (2010) Comparative sequence analyses indicate that *Coffea* (Asterids) and *Vitis* (Rosids) derive from the same paleo-hexaploid ancestral genome. Mol Genet Genomics 283:493–501

Cohen S, Houben A, Segal D (2008) Extrachromosomal circular DNA derived from tandemly repeated genomic sequences in plants. Plant J 53:1027–1034

Cremonini R (2005) Low chromosome number angiosperms. Caryologia 58:403–409

Darlington CD (1937) Recent advances in cytology. J. & A. Churchill, London

Datson PM, Murray BG (2006) Ribosomal DNA locus evolution in *Nemesia*: transposition rather than structural rearrangement as the key mechanism? Chromosome Res 14:845–857

Delprat A, Negre B, Puig M, Ruiz A (2009) The transposon Galileo generates natural chromosomal inversions in *Drosophila* by ectopic recombination. PLoS One 4(11):e7883. doi:10.1371/journal.pone.0007883

Dubcovsky J, Dvorák J (1995) Ribosomal RNA multigene loci: nomads of the Triticeae genomes. Genetics 140:1367–1377

Fawcett JA, Van de Peer Y, Maere S (2013) Significance and biological consequences of polyploidization in land plants. In: Leitch IJ, Greilhuber J, Doležel J, Wendel JF (eds) Plant genome diversity, vol 2, Physical structure, behaviour and evolution of plant genomes. Springer-Verlag, Wien, pp 277–293

Gaeta RT, Pires JC, Iniguez-Luy F, Leon E, Osborn TC (2007) Genomic changes in resynthesized *Brassica napus* and their effect on gene expression and phenotype. Plant Cell 19:3403–3417

Gaut BS, Wright SI, Rizzon C, Dvorák J, Anderson LK (2007) Recombination: an underappreciated factor in the evolution of plant genomes. Nat Rev Genet 8:77–84

Greilhuber J, Borsch T, Müller K, Worberg A, Porembski S, Barthlott W (2006) Smallest angiosperm genomes found in Lentibulariaceae, with chromosomes of bacterial size. Plant Biol 8:770–777

Griffiths AJF, Gelbart WM, Lewontin RC (2007) Introduction to genetic analysis. W. H. Freeman, New York

Gu W, Zhang F, Lupski JR (2008) Mechanisms for human genomic rearrangements. Pathogenetics 1:4

Hall KJ, Parker JS (1995) Stable chromosome fission associated with rDNA mobility. Chromosome Res 3:417–422

Han F, Gao Z, Birchler JA (2009a) Reactivation of an inactive centromere reveals epigenetic and structural components for centromere specification in maize. Plant Cell 21:1929–1939

Han Y, Zhang Z, Liu C, Liu J, Huang S, Jiang J, Jin W (2009b) Centromere repositioning in cucurbit species: implication of the genomic impact from centromere activation and inactivation. Proc Natl Acad Sci USA 106:14937–14941

Imai HT (1991) Mutability of constitutive heterochromatin (C bands) during eukaryotic chromosomal evolution and their cytological meaning. Jpn J Genet 66:635–661

International Brachypodium Initiative (2010) Genome sequencing and analysis of the model grass *Brachypodium distachyon*. Nature 463:763–768

Jaillon O, Aury J-M, Wincker P (2009) "Changing by doubling", the impact of whole genome duplications in the evolution of eukaryotes. C R Biol 332:241–253

Jones K (1998) Robertsonian fusion and centric fission in karyotype evolution of higher plants. Bot Rev 64:273–289

Khandelwal S (1990) Chromosome evolution in the genus *Ophioglossum* L. Bot J Linn Soc 102:205–217

Leitch IJ, Leitch AR (2013) Genome size diversity and evolution in land plants. In: Leitch IJ, Greilhuber J, Doležel J, Wendel JF (eds) Plant genome diversity, vol 2, Physical structure, behaviour and evolution of plant genomes. Springer-Verlag, Wien, pp 307–322

Levin DA (2002) The role of chromosomal change in plant evolution. Oxford University Press, New York

Luo MC, Deal KR, Akhunov ED, Akhunova AR, Anderson OD, Anderson JA, Blake N, Clegg MT, Coleman-Derr D, Conley EJ, Crossman CC, Dubcovsky J, Gill BS, Gu YQ, Hadam J, Heo HY, Huo N, Lazo G, Ma Y, Matthews DE, McGuire PE, Morrell PL, Qualset CO, Renfro J, Tabanao D, Talbert LE, Tian C, Toleno DM, Warburton ML, You FM, Zhang W, Dvorák J (2009) Genome comparisons reveal a dominant mechanism of chromosome number reduction in grasses and accelerated genome evolution in Triticeae. Proc Natl Acad Sci USA 106:15780–15785

Lysák MA, Berr A, Pecinka A, Schmidt R, McBreen K, Schubert I (2006) Mechanisms of chromosome number reduction in *Arabidopsis thaliana* and related Brassicaceae species. Proc Natl Acad Sci USA 103:5224–5229

Mandáková T, Joly S, Krzywinski M, Mummenhoff K, Lysák MA (2010a) Fast diploidization in close mesopolyploid relatives of *Arabidopsis*. Plant Cell 22:2277–2290

Mandáková T, Heenan PB, Lysák MA (2010b) Island species radiation and karyotypic stasis in *Pachycladon* allopolyploids. BMC Evol Biol 10:367

McClintock B (1948) Mutable loci in maize. Carnegie Inst Wash Year Book, vol 47., pp 155–169

Murat F, Xu JH, Tannier E, Abrouk M, Guilhot N, Pont C, Messing J, Salse J (2010) Ancestral grass karyotype reconstruction unravels new mechanisms of genome shuffling as a source of plant evolution. Genome Res 20:1545–1557

Murata M, Yokota E, Shibata F, Kashihara K (2008) Functional analysis of the *Arabidopsis* centromere by T-DNA insertion-induced centromere breakage. Proc Natl Acad Sci USA 105:7511–7516

Nasuda S, Hudakova S, Schubert I, Houben A, Endo TR (2005) Stable barley chromosomes without centromeric repeats. Proc Natl Acad Sci USA 102:9842–9847

Navrátilová A, Koblízková A, Macas J (2008) Survey of extrachromosomal circular DNA derived from plant satellite repeats. BMC Plant Biol 8:90

Nicolas SD, Le Mignon G, Eber F, Coriton O, Monod H, Clouet V, Huteau V, Lostanlen A, Delourme R, Chalhoub B, Ryder CD, Chèvre AM, Jenczewski E (2007) Homeologous recombination plays a major role in chromosome rearrangements that occur during meiosis of *Brassica napus* haploids. Genetics 175:487–503

Page SL, Shin J-C, Han J-Y, Choo KH, Andy SLG (1996) Breakpoint diversity illustrates distinct mechanisms for Robertsonian translocation formation. Hum Mol Genet 5:1279–1288

Pellicer J, Fay MF, Leitch IJ (2010) The largest eukaryotic genome of them all. Bot J Linn Soc 164:10–15

Perry J, Nouri S, La P, Daniel A, Wu Z, Purvis-Smith S, Northrop E, Choo KH, Slater HR (2005) Molecular distinction between true centric fission and pericentric duplication-fission. Hum Genet 116:300–310

Piras FM, Nergadze SG, Magnani E, Bertoni L, Attolini C, Khoriauli L, Raimondi E, Giulotto E (2010) Uncoupling of satellite DNA and centromeric function in the genus *Equus*. PLoS Genet 6: e1000845

Pulletikurti V, Yu C, Zhang J, Peterson T, Weber DF (2009) Cytological evidence that alternative transposition by Ac elements causes reciprocal translocations and inversions in *Zea mays* L. Maydica 54:457–462

Ranz JM, Maurin D, Chan YS, von Grotthuss M, Hillier LW, Roote J, Ashburner M, Bergman CM (2007) Principles of genome evolution in the *Drosophila melanogaster* species group. PLoS Biol 5:e152

Raskina O, Belyayev A, Nevo E (2004) Quantum speciation in *Aegilops*: molecular cytogenetic evidence from rDNA cluster variability in natural populations. Proc Natl Acad Sci USA 101:14818–14823

Raskina O, Barber JC, Nevo E, Belyayev A (2008) Repetitive DNA and chromosomal rearrangements: speciation-related events in plant genomes. Cytogenet Genome Res 120:351–357

Rocchi M, Stanyon R, Archidiacono N (2009) Evolutionary new centromeres in primates. Prog Mol Subcell Biol 48:103–152

Runcie DE, Noor MAF (2009) Sequence signatures of a recent chromosomal rearrangement in *Drosophila mojavensis*. Genetica 136:5–11

Salse J, Bolot S, Throude M, Jouffe V, Piegu B, Quraishi UM, Calcagno T, Cooke R, Delseny M, Feuillet C (2008) Identification and characterization of shared duplications between rice and wheat provide new insight into grass genome evolution. Plant Cell 20:11–24

Salse J, Abrouk M, Bolot S, Guilhot N, Courcelle E, Faraut T, Waugh R, Close TJ, Messing J, Feuillet C (2009) Reconstruction of monocotelydoneous proto-chromosomes reveals faster evolution in plants than in animals. Proc Natl Acad Sci USA 106:14908–14913

Schubert I (1992) Telomeric polymorphism in *Vicia faba*. Biol Zbl 111:164–168

Schubert I (2007) Chromosome evolution. Curr Opin Plant Biol 10:109–115

Schubert I, Lysák MA (2011) Interpretation of karyotype evolution should consider chromosome structural constraints. Trends Genet 27:207–216

Schubert I, Rieger R (1985) A new mechanism for altering chromosome number during karyotype evolution. Theor Appl Genet 70:213–221

Schubert I, Wobus U (1985) In situ hybridization confirms jumping nucleolus organizing regions in *Allium*. Chromosoma 92:143–148

Schubert I, Rieger R, Michaelis A (1988) On the toleration of duplications and deletions by the *Vicia faba* genome. Theor Appl Genet 76:64–70

Schubert I, Rieger R, Fuchs J (1995) Alteration of basic chromosome number by fusion-fission cycles. Genome 38:1289–1292

Soltis DE, Albert VA, Leebens-Mack J, Bell CD, Paterson AH, Zheng CF, Sankoff D, dePamphilis CW, Wall PK, Soltis PS (2009) Polyploidy and angiosperm diversification. Am J Bot 96:336–348

Stebbins GL (1971) Chromosomal evolution in higher plants. Edward Arnold, London

Stimpson KM, Sullivan BA (2010) Epigenomics of centromere assembly and function. Curr Opin Cell Biol 22:1–9

Sullivan BA, Jenkins LS, Karson EM, Leana-Cox J, Schwartz S (1996) Evidence for structural heterogeneity from molecular cytogenetic analysis of dicentric Robertsonian translocations. Am J Hum Genet 59:167–175

Thiel T, Graner A, Waugh R, Grosse I, Close TJ, Stein N (2009) Evidence and evolutionary analysis of ancient whole-genome duplication in barley predating the divergence from rice. BMC Evol Biol 9:209

Thomas BC, Pedersen B, Freeling M (2006) Following tetraploidy in an *Arabidopsis* ancestor, genes were removed preferentially from one homeolog leaving clusters enriched in dose-sensitive genes. Genome Res 16:934–946

Tsujimoto H, Usami N, Hasegawa K, Yamada T, Nagaki K, Sasakuma T (1999) De novo synthesis of telomere sequences at the healed breakpoints of wheat deletion chromosomes. Mol Gen Genet 262:851–856

Udall JA, Quijada PA, Osborn TC (2005) Detection of chromosomal rearrangements derived from homologous recombination in four mapping populations of *Brassica napus* L. Genetics 169:967–979

Uhl CH (1978) Chromosomes of Mexican *Sedum* II. Section Pachysedum. Rhodora 80:491–512

Van de Peer Y, Fawcett JA, Proost S, Sterck L, Vandepoele K (2009) The flowering world: a tale of duplications. Trends Plant Sci 14:680–688

Vanzela ALL, Guerra M, Luceno M (1996) *Rhynchospora tenuis* Link (Cyperaceae), a species with the lowest number of holocentric chromosomes (n = 2). Cytobios 88:219–228

Ventura M, Mudge JM, Palumbo V, Burn S, Blennow E, Pierluigi M, Giorda R, Zuffardi O, Archidiacono N, Jackson MS, Rocchi M (2003) Neocentromeres in 15q24–26 map to duplicons which flanked an ancestral centromere in 15q25. Genome Res 13: 2059–2068

Ventura M, Antonacci F, Cardone MF, Stanyon R, D'Addabbo P, Cellamare A, Sprague LJ, Eichler EE, Archidiacono N, Rocchi M (2007) Evolutionary formation of new centromeres in macaque. Science 316:243–246

Voullaire LE, Slater HR, Petrovic V, Choo KH (1993) A functional marker centromere with no detectable alpha-satellite, satellite III, or CENP-B protein: activation of a latent centromere? Am J Hum Genet 52:1153–1163

Wei F, Coe E, Nelson W, Bharti AK, Engler F, Butler E, Kim H, Goicoechea JL, Chen M, Lee S, Fuks G, Sanchez-Villeda H, Schroeder S, Fang Z, McMullen M, Davis G, Bowers JE, Paterson AH, Schaeffer M, Gardiner J, Cone K, Messing J, Soderlund C, Wing RA (2007) Physical and genetic structure of the maize genome reflects its complex evolutionary history. PLoS Genet 3:e123

Weil CF (2009) Too many ends: aberrant transposition. Genes Dev 23:1032–1036

Wolfe KH (2001) Yesterday's polyploids and the mystery of diploidization. Nat Rev Genet 2:333–341

Yang L, Bennetzen JL (2009) Distribution, diversity, evolution, and survival of Helitrons in the maize genome. Proc Natl Acad Sci USA 106:19922–19927

Yu S, Graf WD (2010) Telomere capture as a frequent mechanism for stabilization of the terminal chromosomal deletion associated with inverted duplication. Cytogenet Genome Res 129:265–274

Zhang J, Yu C, Pulletikurti V, Lamb J, Danilova T, Weber DF, Birchler J, Peterson T (2009) Alternative Ac/Ds transposition induces major chromosomal rearrangements in maize. Genes Dev 23:755–765

Biology and Evolution of B Chromosomes

10

Andreas Houben, Ali Mohammad Banaei-Moghaddam, and Sonja Klemme

Contents

A. Houben (✉)
Leipnitz-Institute of Plant Genetics and Crop Plant Research (IPK), Chromosome Structure and Function Laboratory, Corrensstrasse 3, 06466 Gatersleben, Germany
e-mail: houben@ipk-gatersleben.de

10.1 Introduction

Supernumerary or B chromosomes (Bs) are dispensable components of the genomes of numerous plant, fungi, and animal species. They do not pair with any of the standard A chromosomes (As) at meiosis, and have irregular modes of inheritance. As they are dispensable for normal growth, Bs were considered non-functional and without any essential genes. As a result, B chromosomes follow their own species-specific evolutionary pathways. Because most Bs do not confer any advantages on the organisms that harbour them, they may be thought of as parasitic elements that persist in populations by making use of the cellular machinery required for the inheritance and maintenance of A chromosomes. In low numbers Bs show little or no impact on the hosts. However, increased numbers of Bs cause phenotypic differences and reduced fertility, reviewed in Jones and Rees (1982), Jones (1991, 1995), Bougourd and Jones (1997), Camacho et al. (2000), Camacho (2004, 2005), Jones and Houben (2003), Jenkins and Jones (2004), and Jones et al. (2008a, b).

B chromosomes are a major source of intraspecific variation in nuclear DNA amounts in numerous species of plants (Jones et al. 2008b). The distribution of Bs among different groups of angiosperms is not random. Among flowering plants they are more likely to occur in outcrossing than inbred species, and their presence is also positively correlated with genome size and negatively with chromosome number. They are not found any more frequently in polyploids than in diploids (Palestis et al. 2004; Levin et al. 2005). In many plants different morphological types of Bs exist within a single species. The relationship between genome size and B frequency may be explained on the grounds that species with large genomes can tolerate supernumerary chromosomes more readily (Puertas 2002) or that the greater amount of noncoding DNA, which is what largely constitutes large genomes, is itself a trigger for B formation (Levin et al. 2005). According to a survey of 23,652 angiosperm species, about 8% of monocots and 3% of eudicots have Bs, and of

I.J. Leitch et al. (eds.), *Plant Genome Diversity Volume 2*,
DOI 10.1007/978-3-7091-1160-4_10,

these by far the largest number of species belong to the Poaceae and Asteraceae. However, because these families are also highly speciose, several other families have a higher proportion of species with Bs. Among orders, the two B chromosome 'hot spots' are the Liliales and Commelinales (Palestis et al. 2004; Levin et al. 2005).

10.2 Structure and DNA Composition of B Chromosomes

The size of B chromosomes differs from species to species. Generally Bs are smaller than As, however Bs bigger than As have been detected in some species, e.g., the cyprinid fish *Alburnus alburnus* (Ziegler et al. 2003) and the neotropical fish *Astyanax scabripinnis paranae* (Maistro et al. 1992). In some species even different types of B chromosomes exist, for example in the plant *Brachycome dichromosomatica* large Bs and micro Bs can be observed (Carter and Smith-White 1972). The marsupial frog *Gastrotheca espeletia* possesses three types of Bs which differ among themselves in size and morphology (Schmid et al. 2002).

Generally, B chromosomes have a similar overall DNA composition to the respective As of the host species (Camacho et al. 2000). Early investigations into the DNA composition of Bs were based on gradient density centrifugation and renaturation kinetics. In rye, these experiments showed that the ratio and heterogeneity of repeats in Bs did not differ from As, although a slight increase in cytosine and guanine content was observed in the DNA from B-carrier plants (Rimpau and Flavell 1975; Timmis et al. 1975). Buoyant densities of DNA from maize (*Zea mays*) plants with and without Bs were found to be similar (Chilton and Mccarthy 1973) and the same results were reported in the mealy bug *Pseudococcus obscurus* (Klein and Eckhardt 1976) and the grasshopper *Myrmeleotettix maculatus* (Gibson and Hewitt 1970). Later, Amos and Dover (1981) introduced the use of restriction endonuclease digestion of genomic satellite DNA to characterize Bs. However, only in recent years have techniques like chromosome microdissection (Houben et al. 2001a) and flow-sorting (Kubalakova et al. 2003; Bartos et al. 2008) evolved enough to allow for direct isolation of B chromosome-derived DNA. In the future, efficient and less costly sequencing tools should allow the analysis of megabase-long DNA fragments derived from A and B chromosomes. Comparative sequence analysis will then significantly improve our knowledge on the origin of Bs, and hence of the evolution of genomes.

Many Bs contain large amounts of repetitive DNA, especially mobile elements, although strong amplification of tandem repeats and satellite DNA has also been observed (Langdon et al. 2000; Dhar et al. 2002). One such example is the micro B of *Brachycome dichromosomatica* which is mainly composed of tandem repeats (Houben et al. 2001b). The large amounts of repetitive elements might facilitate the strong heterochromatinization observed in some Bs. Since Bs are cut off from recombination, a greater proportion of mobile elements will accumulate compared with the As, and because Bs are under relaxed selective pressure, they present a kind of safe spot for mobile elements. Nevertheless, as Bs follow a separate path of evolution, the composition of their repetitive fraction is expected to differ from that of the As, depending on the age of the Bs.

Molecular studies have shown that in many species the Bs contain sequences that have originated from one or several A chromosomes (Houben et al. 2001b; Page et al. 2001; Cheng and Lin 2003; Bugrov et al. 2007). Mostly these are non-coding repetitive sequences or mobile elements like in the plant *Crepis capillaris* (Jamilena et al. 1994, 1995), the fish *Prochilodus lineatus* (de Jesus et al. 2003), the fungus *Nectria haematococca* (Enkerli et al. 1997), and the insect *Drosophila subsilvestris* (Gutknecht et al. 1995). In maize, sequence analysis revealed many highly repetitive sequences, including transposons which were present on both A and B chromosomes, but enriched on Bs (Alfenito and Birchler 1993; Stark et al. 1996; Theuri et al. 2005; Lamb et al. 2007a). Only a few sequences are considered B-specific though they are also present on As in minor traces. For example: the Bd49 tandem repeat in *Brachycome dichromosomatica* (Franks et al. 1996) or the B-specific retroelement in *Alburnus alburnus* (Schmid et al. 2006). In rye (*Secale cereale*) two B-specific sequences, D1100 (Sandery et al. 1990) and E3900 (Blunden et al. 1993), have been identified. Both are located at the end of the long arm of this B-type (Wilkes et al. 1995) and reside in the same region that is supposed to contain one or more genes controlling the directed non-disjunction of the rye Bs. More data on characterization of the DNA composition of Bs are given in Jones and Rees (1982) and Camacho (2005).

10.2.1 Centromere Organization of B Chromosomes

An understanding of the structure and regulation of A and B centromeres is a prerequisite for a better understanding of the unique segregation behaviour of Bs. The B-specific repeat ZmBs has been used to characterise extensively the centromeres of maize Bs (Alfenito and Birchler 1993; Kaszas and Birchler 1996, 1998; Kaszas et al. 2002), which are amongst the best-characterized plant centromeres. These studies have shown that the centromeres of maize Bs contain several megabases of a B chromosome-specific repeat (ZmBs), a 156-bp satellite repeat (CentC), and centromere-specific retrotransposons (CRM elements).

However, only a small fraction of the ZmBs repeats interacts with the kinetochore protein CENH3 (the histone H3 variant specific to functional centromeres and also called CENPA). CentC, which marks the CENH3-associated chromatin in maize A centromeres, is restricted to a 700-kb domain within the larger context of the ZmBs repeats (Jin et al. 2005). Clearly, centromere specification must have an epigenetic component as dicentric A-B translocation chromosomes are characterized by the stable inheritance of an inactive state of one of the centromeres over several generations (Han et al. 2006).

A comparison of maize A and B chromosomes seems to show that Bs are enriched with DNA elements that are normally found at or near A centromeres (Lamb et al. 2005). A similar tendency has been described for the rye B, which is characterized by a likely higher copy number of the rye retrotransposon-like centromeric repeat pAWRC.1 (Wilkes et al. 1995). However, in contrast to maize Bs (Lamb et al. 2005; Jin et al. 2008), the rye kinetochore protein CENH3 is present in equal amounts on both As and Bs (Houben unpublished).

The centromeric region of *B. dichromosomatica* standard Bs (of cytodeme A1, A2, and A4) is enriched with a B-specific tandem repeat (Bd49) that is not microscopically detectable on A chromosomes (Franks et al. 1996; Leach et al. 1995). Initially the predominantly centromeric location of the Bd49 repeat suggested a possible role for this sequence in the drive process, but a noncentromeric Bd49 location in *B. dichromosomatica* cytodeme A3 and differences in signal sizes among all the Bs of different cytodemes do not support this assumption (Houben et al. 1999).

10.2.2 B Chromosome-Located Genes

Single or low copy genes are rarely found on Bs. This could be due to a problem of finding them between the massively amplified mobile elements. However, many Bs have been suggested to contain active genes due to visible phenotypic effects on the host (Camacho 2005). In maize, several sequences have been isolated that are homologous to non-coding regions of genes located on the As (Cheng and Lin 2003). Recently, histone genes have been found on the Bs of the migratory locust *Locusta migratoria*, although these genes appear in a repetitive fraction and often carry deleterious mutations (Teruel et al. 2010). Transcriptional activity of gene fragments located on Bs has been shown in rye together with an alteration of the transcription of corresponding gene copies on A chromosomes (Carchilan et al. 2009).

10.2.3 DNA Composition of Rye B Chromosomes

One of the best studied models in B chromosome research is the Bs of rye (*Secale cereale*). Under the microscope the B of rye is easily distinguishable from the As by eye even without a specific marker. In contrast to the metacentric As, the B is acrocentric. Based on Giemsa-banding, the B appears to have both heterochromatic and euchromatin parts, while the B-terminal region is Giemsa-staining positive (Jones and Puertas 1993). The total size of a rye B has been estimated to be c. 400 Mb as 1C content (Jones et al. 2008b). At interphase, Bs show a predisposition for coorientation and a tendency for association, supporting even numbers of Bs. This association was shown in a background of hexaploid wheat (Morais-Cecilio et al. 1996) and in rye (Morais-Cecilio et al. 1997). It has also been suggested that the Bs of rye influence the ribosomal DNA (rDNA) organization at interphase. Studies showed that the presence of Bs increased the condensation state of these regions on the As suggesting the Bs were influencing the transcription of rDNA on the A chromosomes (Delgado et al. 1995). In addition, the presence of B chromosomes was shown to result in the condensation of an A-located satellite repeat. These condensation events were not directly proportional to the number of Bs, suggesting that this effect is caused by the Bs themselves rather than by the increase of total nuclear DNA content (Delgado et al. 2004).

At the DNA level, apart from the terminal region of the long B chromosome arm, a high level of overall similarity exists between As and Bs of rye (Timmis et al. 1975; Tsujimoto and Niwa 1992; Wilkes et al. 1995; Houben et al. 1996). This hints at the intraspecific origin of this type of chromosome. Also rye Bs appear to be monophyletic and very stable, as they are very similar even in other rye species like *Secale segetale*, which is very closely related to *Secale ancestrale* (Niwa and Sakamoto 1995). This is rather unexpected since Bs are expected to have an elevated mutation rate compared with the A genome.

Using genomic DNA from a plant with no B chromosomes (0B) as a probe for genomic *in situ* hybridization (GISH) resulted in signal labelling along the whole of the Bs except the heterochromatic terminal part (Tsujimoto and Niwa 1992; Wilkes et al. 1995; Morais-Cecilio et al. 1996). This region of the B is unstable, prone to rearrangements and favours duplication over deletion (Langdon et al. 2000). It is enriched in B-specific sequences (Houben et al. 1996) such as D1100 (Sandery et al. 1990; Wilkes et al. 1995) and E3900 (Blunden et al. 1993). Sequence comparisons have shown that there are similarities between

D1100 and Miniature Inverted-repeat Transposable Element sequences (MITEs) which appear frequently in temperate cereals (Langdon et al. 2000). E3900 shows similarities to Ty3-gypsy retrotransposons and contains a partial reading frame of a *gag* gene. A possible ancestor of the *gag* fragment that frequently colonizes the centromeres of Poaceae was cloned and described (Langdon et al. 2000). It therefore seems likely that members of the D1100 and E3900 families were formed from parts of the A genome which underwent frequent amplification and simplification steps (Langdon et al. 2000).

Several subtelomeric, high copy sequences common to the As are reported to be missing from the Bs such are pSc74 (Tsujimoto and Niwa 1992; Manzanero and Puertas 2003), pSc119 (Tsujimoto and Niwa 1992; Wilkes et al. 1995; Manzanero and Puertas 2003), pSc200 (Hasterok et al. 2002; Kubalakova et al. 2003; Manzanero and Puertas 2003; Zhou et al. 2010) and pSc250 (Kubalakova et al. 2003). However, other authors have claimed a potential polymorphism on Bs that carry traces of pSc74 and pSc119 in the subtelomeric end of the long arm (Cuadrado and Jouve 1994; Kubalakova et al. 2003). Tsujimoto and Niwa (1994) proposed that the differential presence of high copy tandem repeats (satellite DNA) is caused by the different population genetics of As and Bs. Bs are usually present in only some individuals in a random mating population, therefore, if a spontaneous amplification of a satellite sequence occurs in an individual without the B chromosome, the amplified sequence will not become distributed over the Bs.

The chromosomal ends of Bs do contain typical *Arabidopsis*-type telomere sequences (Manzanero and Puertas 2003). However, whether ribosomal genes are present is still controversial. While Flavell and Rimpau (1975) reported they were present on rye Bs, more recent studies have failed to confirm their findings (Cuadrado and Jouve 1994; Kubalakova et al. 2003). The centromeres of both types of chromosomes are enriched in paWRC.1 sequences, so-called Bilby repeats (Francki 2001), although a more extended distribution of Bilby elements has been reported for Bs (Wilkes et al. 1995). Figure 10.1 summarizes the DNA composition of rye A and B chromosomes.

Several translocations between rye As and Bs have been reported although they mostly arise from artificially-induced events via irradiation (Hasterok et al. 2002) or strains kept under laboratory conditions for very long times (Schlegel and Pohler 1994; Wilkes et al. 1995). Indeed, the contradictory findings of the subtelomeric repeat sequences pSc74 and pSc119 (Tsujimoto and Niwa 1992; Cuadrado and Jouve 1994) have been attributed to such translocations (Wilkes et al. 1995). It has also been found, that these translocations are not stable in their respective populations (Hasterok et al. 2002). One incident was published where a translocation spontaneously arose in a strain sampled from wild populations, which developed a deficient form of the rye B lacking the B-specific heterochromatic distal part (Ribeiro et al. 2004). However, apart from that example, only the analysis of a large number of Bs has enabled sporadic variants to be identified (Kubalakova et al. 2003). It is notable though that many of these translocations appear to involve the A chromosome 1R. Artificially-induced translocations in a wheat rye B addition line have been useful instruments for mapping the Bs (Endo et al. 2008). Other structural variants that arise in experimental strains include iso-chromosomes which contain only the long arm or only the short arm of the standard B (Jones and Puertas 1993).

10.2.4 Chromatin Composition of B Chromosomes

Although chromatin structure is increasingly seen as playing an essential role in different aspects of chromosome function, little information is available on the chromatin composition of Bs, and whether it differs from that of the standard A chromosomes. Based on classical cytological observations (e.g., Giemsa-banding) an early survey suggested that the Bs in about half of the plant species which carry them were heterochromatic (Jones 1975). Indeed, because no genes with specific phenotypic effects necessary for normal development are known for Bs, the finding that some of them are totally euchromatic is surprising.

In some species like the mouse *Apodemus flavicollis* (Tanic et al. 2005), the snail *Helix pomatia* (Evans 1960) and the plants *Scilla vvedenskyi* (Greilhuber and Speta 1976) and *Allium flavum* (Vosa 1973), the Bs are predominantly euchromatic. At interphase, Bs of some plant species, e.g., *Anthoxanthum aristatum* (Östergren 1947), *Puschkinia libanotica* (Barlow and Vosa 1969), *Scilla autumnalis* (Ruiz-Rejon et al. 1980), *Clematis orientalis* and *C. hatherliensis* (Shambulingappa 1965), *Rosa rugosa* (Price et al. 1981), *Tainia laxiflora* (Tanaka and Matsuda 1972) or *Picea glauca* (Teoh and Rees 1977) display so called heterochromatic chromocenters after Giemsa- or Feulgen-staining.

Recent advances in chromatin characterization, in terms of epigenetic marks, has shown the involvement of DNA methylation and post-translational histone modifications in chromatin assembly and maintenance (Richards and Elgin 2002; Craig 2005; Kouzarides 2007). The N-terminal tails of the nucleosomal core histones, extending from the nucleosome surface, are subjected to post-translational modifications such as acetylation, methylation, phosphorylation, ubiquitination, glycosylation, ADP-ribosylation, carbonylation and sumoylation. Acetylation of histones is mainly linked with transcriptional activation, DNA recombination and repair (Grunstein 1997; Struhl 1998; Ikura et al. 2000; Bird et al. 2002). Phosphorylation correlates with

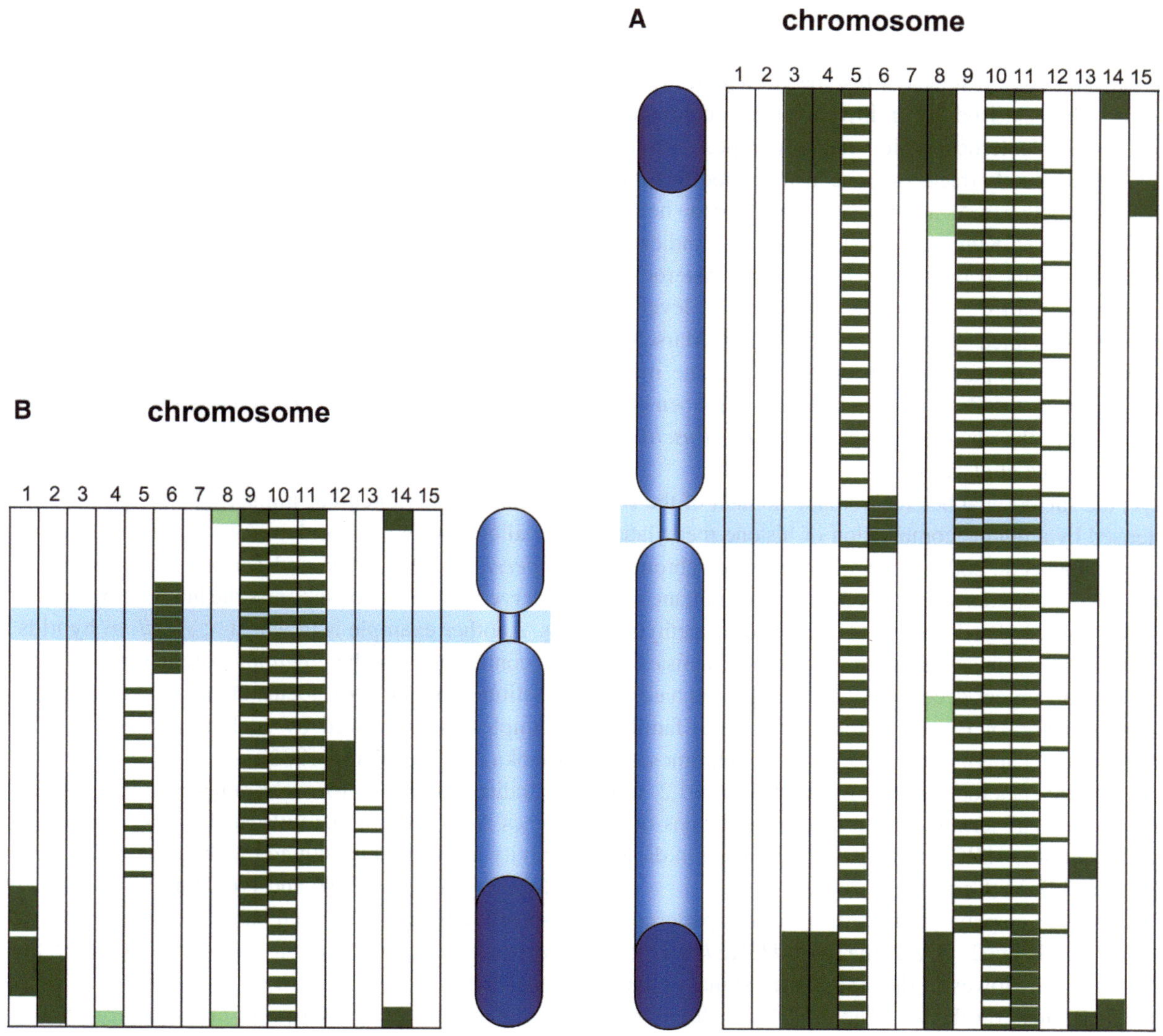

Fig. 10.1 Comparison of the distribution of high-copy sequences between the A and B chromosomes of rye: Chromosomes are schematically shown with their Giemsa-banded regions in *dark blue*. Centromeric regions are highlighted for orientation. *Dark green* bars indicate the location and density of the corresponding repeats as determined by fluorescent *in situ* hybridization. *Light green* indicates contradictory finding or possible polymorphisms. Name of sequences and corresponding references: (*1*) D1100, Sandery et al. (1990); (*2*) E3900, Blunden et al. (1993); (*3*) pSc200, Zhou et al. (2010), Kubalakova et al. (2003), Manzanero and Puertas (2003); (*4*) pSc74, Tsujimoto and Niwa (1992), Cuadrado and Jouve (1994), Manzanero and Puertas (2003); (*5*) R173, Wilkes et al. (1995); (*6*) pAWRC.1, Bilby, Wilkes et al. (1995); (*7*) pSc34, Cuadrado and Jouve (1994), Manzanero and Puertas (2003); (*8*) pSc119.2, Tsujimoto and Niwa (1992), Cuadrado and Jouve (1994), Kubalakova et al. (2003) and Manzanero and Puertas (2003); (*9*) B1334, Carchilan et al. (2009); (*10*) B8149, Carchilan et al. (2009); (*11*) B2465, Carchilan et al. (2009); (*12*) Afa, Kubalakova et al. (2003); (*13*) GAA, Kubalakova et al. (2003); (*14*) *Arabidopsis*-like telomere, Manzanero and Puertas (2003) and (*15*) 5S rDNA Kubalakova et al. (2003)

transcription activation, apoptosis, DNA repair, chromosome condensation, sister chromatid cohesion/segregation and gametogenesis (Kaszas and Cande 2000; Prigent and Dimitrov 2003; Ahn et al. 2005; Krishnamoorthy et al. 2006; Houben et al. 2007). The most characterized histone modification to date is methylation. This modification occurs only on lysine (K) and arginine (R) residues of histone H3 and H4 and is performed by histone methyltransferases (HMTs) (Martin and Zhang 2005). Each arginine can be mono- or dimethylated and each lysine can be either mono-, di-, or trimethylated (Zhang and Reinberg 2001). Several studies have shown that modification of the histone H3 tail by methylation of lysine residues 9 and 27 negatively regulates transcription by mediating a compact chromatin structure. In contrast, euchromatin is marked by methylation of lysine residues 4 and 36 (reviewed by Martin and Zhang 2005). Whereas euchromatin-specific methylation of H3K4 is highly conserved among eukaryotes, heterochromatin indexing by methylation marks at H3K9, 27 and H4K20 is more variable (reviewed by Fuchs et al. 2006).

Although chromosome banding results suggest a similar eu- and heterochromatin composition in the B and A chromosomes of *Brachycome dichromosomatica* (which carries large and micro Bs), the use of highly specific antibodies against various lysine residues of histone H3 has shown that the Bs are characterised by a low level of euchromatic histone marks. Heteropycnotic, tandem repeat-enriched micro Bs revealed only traces of these histone modifications. No differences between As and Bs were found for the heterochromatic marks H3K9me1,2 and H3K27me1,2 indicating that Bs are not marked by an enriched level of heterochromatic histone marks, but rather by a low level of euchromatin-associated histone modifications (Marschner et al. 2007a). A comparable distribution of histone marks was also found for the Bs of *Crepis capillaris* (Houben et al. 2003) and for the heterochromatic Bs of *Puschkinia libanotica* (Kumke et al. 2008).

In rye, the subterminal heterochromatic domain of the B is characterised by a unique combination of histone methylation marks (Carchilan et al. 2007). Contrary to the heterochromatic regions of the A chromosomes, this domain is simultaneously marked by trimethylated histone H3K4 and trimethylated H3K27. In addition, this domain shows a dark Giemsa band at mitosis, but undergoes decondensation during interphase and reveals transcription of B-specific high copy repeat families. The distribution patterns along A and B chromosomes observed for the heterochromatin marks H3K9me1,2 were mainly uniform. The terminal heterochromatic regions of As and Bs showed little H3K27me1 but were enriched in di- and trimethylated H3K27 (Carchilan et al. 2007).

Bs of maize display similar distribution patterns of signals from H3K9me2, H3K27me2 and H3K27me1 compared with the As. However, the H3K27me2 signal on the B was less than those on the As. The H3K4me2 distribution pattern on the B matched the position of cytologically visible euchromatin. Interestingly, only extremely faint H3K12ac signals were observed in the Bs (Jin et al. 2008). A reduced level of histone H3 and H4 acetylation was also shown for the Bs of the grasshopper *Eyprepocnemis plorans* (Cabrero et al. 2007) and of *B. dichromosomatica* (Houben et al. 1997a), respectively.

Not only histones but also cytosine, which is one of the four essential base units of DNA can be extensively methylated. In both animals and plants, cytosine is primarily methylated at the CG dinucleotide position. However, in plants methylation is not restricted to the CG sequence: CHG and the less abundant CHH sequence context are also possible (Gruenbaum et al. 1981). DNA methylation is involved in the silencing of transposable elements and genes (Gehring and Henikoff 2007). The DNA repeats and mobile elements on Bs are also methylated, as reported for the Bd49 tandem repeat in *B. dichromosomatica* (Leach et al. 1995) and E3900 repeat in *S. cereale* (Langdon et al. 2000). However, a higher DNA methylation level of B chromosomes has not been reported for any plant species.

10.3 Effects and Transcripts Associated With B Chromosomes

Although Bs are not essential, some phenotypic effects have been reported, and these are usually cumulative, depending upon the number and not the presence or absence of Bs. In low numbers, Bs have little if any influence on the phenotype, but at high numbers they often have a negative effect on fitness and fertility of the organism (reviewed in Jones and Rees 1982; Jones 1995; Bougourd and Jones 1997; Carlson 2009).

There is evidence that Bs directly or indirectly influence the behaviour of A chromosomes (as noted above for rDNA—see Sect. 10.2.3). One of the most striking of such effects is the potential impact of Bs on diploidization in allopolyploid hybrids, e.g., *Lolium temulentum* × *L. perenne* + B (Evans and Davies 1985), where Bs were observed to prevent or suppress the homoeologous pairing of As. Another example is in wheat × *Aegilops* hybrids where Bs contributed by the *Aegilops* parent seem to be able to substitute for the *Ph1* locus of hexaploid wheat. Further examples are reviewed by Jenkins and Jones (2004) and Tanaka and Kawahara (1982).

Indirect evidence for weak transcriptional activity of Bs comes from the comparative analysis of esterase isozyme activity in two plants with and without Bs-*Scilla autumnalis* (Ruiz-Rejon et al. 1980) and rye (Bang and Choi 1990). In B-positive plants, additional bands were detected by protein electrophoresis, however in both cases whether the additional bands were caused by a B-located gene or whether Bs influenced the transcription behaviour of an A-located gene remained unclear. For grasshoppers, Bs have been demonstrated to alter the expression of A chromosome genes (Teruel et al. 2007).

Except for the B-located 45S rRNA gene of *Crepis capillaris*, in which one of two B-specific members of the rRNA gene family was weakly transcribed (Leach et al. 2005), there was no direct molecular evidence for transcription of B chromosome genes in plants until the transcriptional activity of B-specific repetitive sequences was demonstrated. In maize (Lamb et al. 2007b) and rye (Carchilan et al. 2007) retrotransposon-derived high-copy elements have been shown to be transcriptionally active. In addition in rye two repeat families, E3900 and D1100 clustered at the B chromosome long arm, have now been shown to be transcribed in a tissue-specific manner (Carchilan et al. 2007). The function of these B-transcripts, and the mechanism of transcription of B-repeats, are unknown at present. It has been hypothesized that these transcripts could have a structural function in the organization and regulation of Bs (Carchilan et al. 2007; Lamb et al. 2007b).

Recently, the general transcription activity of rye Bs was analysed by comparative cDNA-AFLP analysis (Carchilan et al. 2009). In addition to weak B chromosome transcription, the data showed that Bs were able to down-regulate A chromosome-localized sequences in a genotype dependent manner. It is likely that Bs may bring about a variety of epigenetic effects, including the differential regulation of A-localized transposable elements through mechanisms such as homology-dependent RNA interference pathways. Since the rye B most likely originated from the A chromosome complement, it seems reasonable to suggest that the transcription alterations of A-located sequences are caused by homology-dependent mechanisms, as has been proposed for the remodelling of gene-activities in newly formed hybrids and allopolyploids (Comai 2005). Another hypothesis for explaining how the Bs exert control over the rest of the genome postulates their effects on the spatial organization of the genome itself. Recent work suggests that spatial positioning of genes and chromosomes can influence gene expression (Misteli 2007). Indeed, Delgado et al. (2004) observed that in rye interphase nuclei, rDNA located on A chromosomes had a more compact distribution in cells with Bs compared with cells without Bs. A more compact distribution of rDNA sites suggests a lower level of rRNA gene activity. A similar effect of an almost gene deficient chromosome has been demonstrated in *Drosophila melanogaster* by Lemos et al. (2008) who showed that the Y chromosome of *D. melanogaster* regulates the activity of hundreds of genes located on other chromosomes.

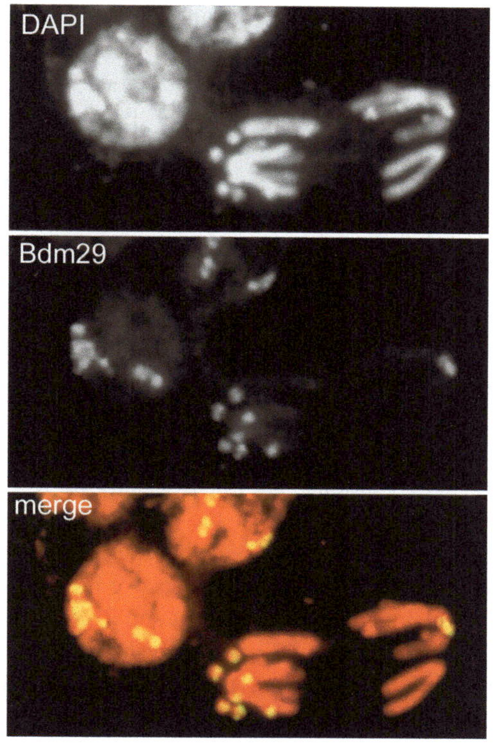

Fig. 10.2 Mitotic nondisjunction of *Brachycome dichromosomatica* ($2n = 4 + Bs$) micro B chromosomes. The B-specific tandem Bdm29 (Houben et al. 1997b) was used as a probe for fluorescence *in situ* hybridization. Note the unequal distribution of micro Bs during anaphase

10.3.1 Mitotic and Meiotic Segregation Behaviour of B Chromosomes

B chromosome inheritance is irregular and non-Mendelian, and therefore polymorphisms exist with respect to the number of Bs within populations or even within different cell lines of an individual carrying Bs. There are several factors which affect the rate of transmission of Bs from one cell to another and from one generation to another. Drive mechanisms play a major role in the equilibrium of B frequency in populations (Jones 1991).

10.3.1.1 Mitotic Behaviour of B Chromosomes

In most species which carry Bs, the mitotic transmission of Bs during growth and development is disjunctional and hence all cells carry the same number of Bs within the individual. However, there are some exceptions in which the Bs show instability during mitosis in somatic tissues and therefore they are absent or present in variable numbers in specific tissues and/or organs. For example, in the grass *Aegilops speltoides* Bs exist in aerial organs but not in roots (Mendelson and Zohary 1972), and in *Crepis capillaris*, while the number of Bs in roots is constant, the number of Bs in aerial organs shows variation (Rutishauser and Röthlisberger 1966). In grasshoppers variation in Bs has been observed to occur specifically among follicles of testis (Nur 1963, 1969). A detailed list of organisms with unstable Bs during mitosis is given by Jones and Rees (1982). The reason of this numerical variation is shown to be nondisjunction of the B chromosome sister chromatids during anaphase of mitosis leading to an absence of Bs in one daughter cell and accumulation in the other (Fig. 10.2). If nondisjunction causes lagging of Bs at anaphase it could also result in a loss of B chromosomes in the daughter cells and consequently exclusion of Bs from those cells.

10.3.1.2 Meiotic Behaviour of B Chromosomes

B chromosomes fail to pair with any member of the A chromosome set during meiosis, although they may pair and form chiasmata among themselves. However, there are several reports indicating that in some species Bs cannot even pair and form chiasmata (Jones and Rees 1982) and the exact reason for this is unclear. Nevertheless, Jones and Rees (1982) have suggested various hypotheses. One possibility is that Bs in these species are not homologs of each other and therefore cannot pair. However, this implies that all analysed individuals in these studies have been carrying structurally different Bs which is unlikely. The other possibility put forward is that Bs in some species such as *Allium cernuum* (Grun 1959) are too short to be able to pair. The large amount of heterochromatin on Bs has also been suggested to be another reason for inhibition of chiasmata formation. However, this is not a general rule as in some species like *Locusta migratoria* (Kayano 1971) Bs are heterochromatic and yet form mainly bivalents.

Depending on the number, Bs can form uni-, bi- or multivalents in metaphase I. The formation of bivalents or multivalents in species with two or more Bs is rarely complete and the formation of univalents among Bs is far higher than the rate of univalent formation from As (Jones and Rees 1982). The fate of these univalents varies in different species under diverse conditions. They will usually be lost during first meiosis or, in the case of the precocious division of sister chromatids, in second anaphase. Like in *Aegilops speltoides* with one B, univalent Bs behave in various ways in different meiotic cells (Mendelson and Zohary 1972). In the majority of pollen mother cells univalent Bs will be lost but in the remaining cells univalent Bs move undivided to one of the daughter nuclei and separate normally in second anaphase. Therefore, only around 10% of tetrads carry one B. In the grasshopper *Myrmeleotettix maculatus* there is no loss of Bs as all univalents move undivided to one of the two telophase nuclei at the first division of meiosis (John and Hewitt 1965). The study of rye plants with two Bs in genotypes with low and high levels of B transmission revealed that the main cause of the difference in the rate of B transmission was their ability to form B bivalents at metaphase I. When the Bs formed bivalents they separated normally in meiosis and were conserved in pollen grains. In contrast, when they formed univalents they were mainly eliminated (Jimenez et al. 1997).

There are several scenarios in which univalent Bs can escape lagging and consequently loss during meiosis. One way to prevent univalent Bs being lost is by the incorporation of nondivided univalent Bs into the first telophase nuclei by association of an univalent B with a non-homologous chromosome. This situation has been observed in the grasshopper *Tetrix ceperoi* in which the B and the X chromosomes form a quasi-bivalent which moves normally to opposite poles in first anaphase (Henderson 1961). In the plant *Parthenium argentatum* Bs form bivalents or multivalents between themselves via chiasmata during prophase. Then, during metaphase the chiasmata are resolved and the majority of undivided re-univalented Bs move to either of the poles and consequently there is no or little loss of Bs at meiosis (Catcheside 1950). In *Plantago serraria* the univalent Bs accumulate near one of the two spindle poles and consequently remain in one of the telophase nuclei after anaphase I, and separate normally in anaphase II (Frost 1959).

10.3.2 Diverse Drive Mechanisms of B Chromosomes

The variety of mechanisms, including segregation failure by which Bs gain advantage in transmission are known as accumulation or drive mechanisms. Drive is the key to understanding Bs and it occurs in a great variety of ways and for none of these is the molecular mechanism known (Burt and Trivers 2006). Since Bs appear to be devoid of essential genes and have no known adaptive advantage, in many cases, their maintenance in natural populations is made possible by their transmission at higher than Mendelian frequencies, and this enables their successful accumulation in populations. The importance of drive is that it retains the B in the population even if its presence has harmful effects on the general viability of its bearers (Kimura and Kayano 1961). Depending on the species, B chromosome drive can be premeiotic, meiotic and post-meiotic although in all cases the outcome is the same i.e., higher number of Bs in the next generation.

In animals, as the gametic nuclei are not replicating, the accumulation mechanism effectively acts either before or during meiosis. Premeiotic drive mechanisms in animals occur in the spermatogonial mitosis in the testes (Jones 1991). Based on the observations of instability in the number of Bs in somatic and germline cells in different species of grasshoppers, it is inferred that a mitotic nondisjunction occurs and the cells containing Bs are preferentially included into germ line cells. Nondisjunction is suggested to happen early in the embryonic divisions as in *Calliptamus palaestinensis* (Nur 1963) or later on like in *Locusta migratoria* (Kayano 1971). Premeiotic drive has also been described in the plant *Crepis capillaris* with the difference that mitotic nondisjunction occurs in the meristematic cells at the time of flower initiation and with the preferential inclusion of Bs in inflorescences during development (Rutishauser and Röthlisberger 1966; Parker et al. 1989). The preferential inclusion of Bs in inflorescences coincides with asymmetrical spindles which occur in shoot meristems during the transition from vegetative to reproductive growth (Rutishauser and Röthlisberger 1966).

Meiotic drive in the megaspore mother cell has been clearly shown in *Lilium callosum* (Kayano 1957), a species with tetrasporic embryo sac development. Here, univalent Bs in egg mother cells tend to be unevenly distributed. Rather than lying on the metaphase plate of metaphase I, univalent Bs were observed to be located on the micropylar side of the spindle in 80% of the analysed cells resulting in their preferential incorporation into the resulting egg cell. A similar situation was observed for the grasshopper *Myrmeleotettix maculatus* in which the Bs were preferentially locating at the egg pole rather than polar bodies pole. Although in the males the Bs were lost during meiosis due to lagging, overall there was a net meiotic drive acting on Bs in *M. maculatus* (Hewitt 1976).

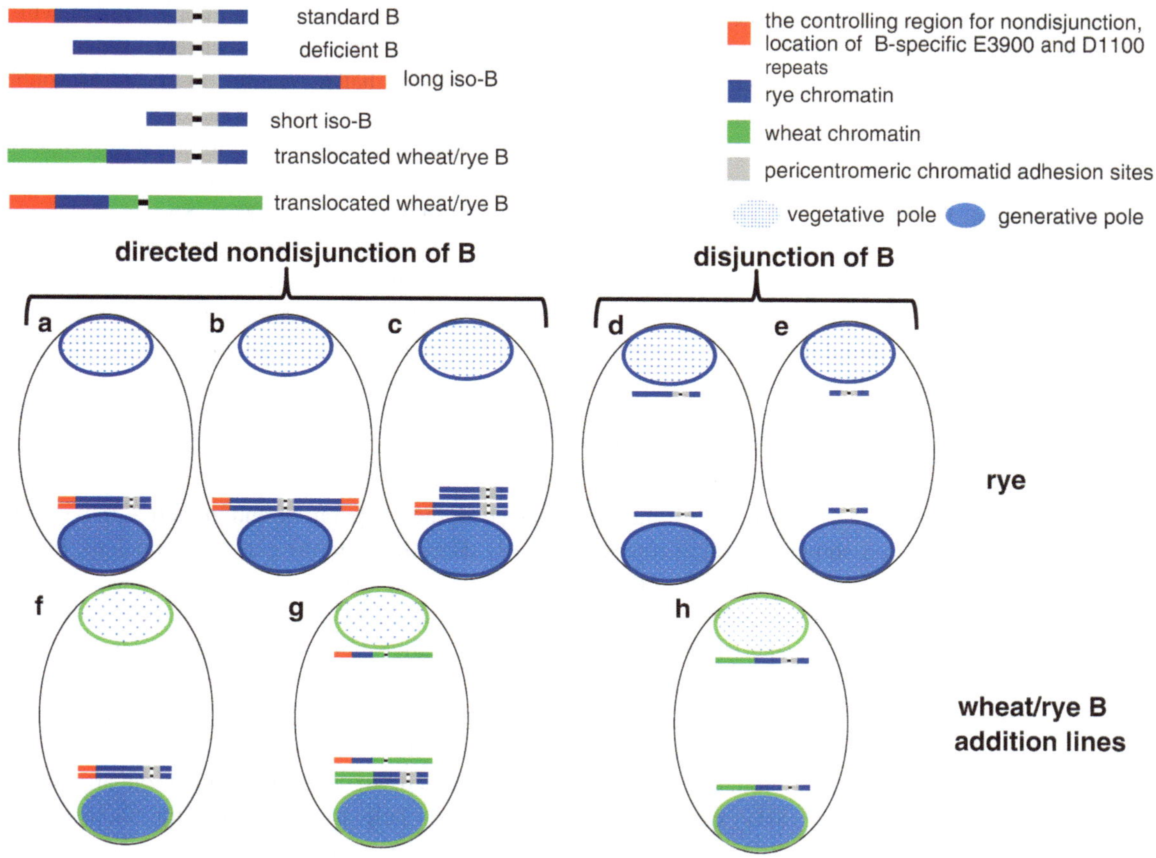

Fig. 10.3 Diagram showing the function of the distal region of the long chromosome arm and the pericentromeric sites in the process of directed nondisjunction of the rye B during first pollen mitosis. The standard B has pericentromeric sticking sites (*grey blocks*) and a heterochromatic distal region on the long arm which contains the controlling region for nondisjunction (*red blocks*). The translocated A/B chromosome is shown in green and blue for wheat and rye chromatin, respectively. (**a**) Sister chromatids of a standard B (Hasegawa 1934) and (**b**) of a B long arm isochromsome (Müntzing 1948). (**c**) If standard and deficient Bs are both present, both B-types show directed nondisjunction (Lima de Faria 1962). (**d**) A chromosome which is deficient for the distal part of the long B arm (Lima de Faria 1962) or (**e**) a short arm isochromosome shows normal disjunction (Müntzing 1948). (**f**) Directed nondisjunction is independent from the background genotype as a standard rye B shows directed nondisjunction if added to wheat (Endo et al. 2008). (**g**) In the case of a reciprocal A/B chromosome translocation, the chromosome possessing the B centromere shows directed nondisjunction. Hence, nondisjunction functions *in trans* (Endo et al. 2008). (**h**) Translocated A/B chromosome without the distal heterochromatic block of the B shows normal disjunction (Endo et al. 2008)

Post-meiotic drive is frequent in flowering plants during male gametophyte maturation. The drive mechanisms of maize and rye Bs are well-studied examples that result in B chromosome accumulation. In rye, directed nondisjunction happens during first pollen grain mitosis resulting in the accumulation of Bs in the generative nucleus and therefore ensuring their transmission at a higher than expected rate to the next generation (Hasegawa 1934). Notably, the B centromere appears to divide normally, but on either side are "sticking sites" which prevent normal anaphase separation of the chromatids and results in nondisjunction at an average frequency of about 86% (Matthews and Jones 1983). In the second pollen grain mitosis the generative nucleus divides to produce two sperm nuclei, each with an unreduced number of Bs. A similar nondisjunction process may occur in the female line as well (Hakansson 1948). Interestingly, nondisjunction works equally well when the rye B is introduced as an addition chromosome into hexaploid wheat (Müntzing 1970; Niwa et al. 1997; Endo et al. 2008) or *Secale vavilovii* (Puertas et al. 1985). Therefore, the segregation behaviour of the B is mainly autonomous and independent of the background genotype. The B itself controls the process of nondisjunction and B-transmission frequency (Romera et al. 1991). The accumulation mechanism of the rye B requires a factor located at the end of its long arm as Bs which lack this terminal region (where two B-specific repeat families E3900 and D1100 reside) undergo normal disjunction (Fig. 10.3a, d) (Müntzing 1948; Endo et al. 2008). This factor may act *in trans* because if a standard B (Fig. 10.3c) (Lima de Faria 1962) or the terminal region of the long arm of the B (Endo et al. 2008) is also present in the same cell containing a B lacking the terminal region then the standard B mediates nondisjunction of both itself and the deficient B (Fig. 10.3c). Interestingly, it has been shown that D1100 and

E3900 are transcriptionally active with higher expression levels in the anthers (Carchilan et al. 2007).

So far no gene, or other DNA sequence on any B has been characterized that plays a role in B chromosome accumulation, and it is therefore tempting to speculate that non-coding RNA is involved in the process of B chromosome nondisjunction. In fission yeast, for example, the repeats flanking the centromere are essential for sister chromatid cohesion and are maintained in a proper heterochromatic state by the RNAi machinery (Volpe et al. 2003). Similarly, pericentromeric heterochromatin is required for proper chromosome cohesion and disjunction in flies and other organisms (Pidoux and Allshire 2005; Vos et al. 2006). It is also noted that the forced accumulation of human centromeric non-coding satellite transcripts leads to defects in the separation of sister chromatids (Bouzinba-Segard et al. 2006) and in plants, RNA molecules have been shown to play a role in establishing centromeric heterochromatin domains (Topp et al. 2004; May et al. 2005). In this context, the transcriptional activity of the D1100 and E3900 repeats (noted above) located in the B chromosome nondisjunction-controlling region of rye is striking (Carchilan et al. 2007). However, no similarity between D1100/E3900 repeats and B-centromere located sequences has been found based on fluorescent *in situ* hybridization (FISH) experiments (Houben et al. unpublished). Nevertheless, it is noted that sequence similarity between non-coding RNA and a target region does not seem to be a functional requirement. For example, dosage compensation in both flies and mammals requires non-coding RNAs which spread *in cis* to coat the X chromosome. The regions of the Xist RNA that are required for localization on the Xi have no obvious sequence homology (Wutz et al. 2002). This also holds true in the case of *Drosophila* roX RNAs, which contain little sequence homology to one another, except for a stretch of 30 nt (Meller and Rattner 2002).

In maize the nondisjunction process differs from that in rye. At least three properties allow the maize B to increase in numbers: nondisjunction at the second pollen grain mitosis, preferential fertilisation of the egg by sperm containing B chromosomes (Roman 1948; Carlson 1969; Rusche et al. 1997), and suppression of meiotic loss when the Bs are unpaired (Carlson and Roseman 1992). The lack of meiotic loss of B univalents is a special feature of maize Bs. In rye, for example, the B univalents are lost in about 80% of 1B × 0B crosses (Jimenez et al. 1997). As in rye, the B-accumulation mechanism in maize requires a factor located on the end of the long arm of the B that may act *in trans* (Roman 1947; Carlson 1978; Lamb et al. 2006). Furthermore, nondisjunction of Bs takes place in the endosperm and in the tapetum. In binucleated tapetal cells, Bs mediate A chromosome instability (Chiavarino et al. 2000; Gonzalez-Sanchez et al. 2004). Sporophytic nondisjunction of Bs occurs if Bs are present at high copy number which could imply that nondisjunction is either repressed at low numbers or so infrequent that it was not observed (Masonbrink and Birchler 2010). One A-located factor seems to codetermine maize B accumulation by preferential fertilization while another factor(s) determines the meiotic loss of Bs (Gonzalez-Sanchez et al. 2003). Sperm nuclei containing deletion derivatives of B-9 (translocation lines involving the B and chromosome 9), which lack the centric heterochromatin and possibly some adjacent euchromatin of the B chromosome, no longer have the capacity for preferential fertilization (Carlson 2007).

Generally there are two models for explaining the mechanism of directed B chromosome nondisjunction in rye (Jones 1991). In the active model, the chromatids of Bs are delayed in their separation and an active movement preferentially takes the Bs to the generative nucleus. In the passive model, the movement of Bs towards the generative nucleus is simply caused by the fact that the equatorial plate is closer to the generative pole and lagging Bs are included in the generative nuclei as the nuclear membrane is formed. In future, the potential interrelationship between unequal cell division and directed nondisjunction should be experimentally addressed. In addition, studies on artificial chromosomes might help to understand the factors affecting B chromosome segregation. As different artificial mammalian chromosomes missegregate over a five-fold range, the data suggest that variable centromeric DNA content, kinetochore composition and/or epigenetic assembly as well as topological positioning in the nucleus can influence the mitotic segregation behaviour of chromosomes (Rudd et al. 2003; Moralli et al. 2009). The presence of pericentromeric heterochromatin, checkpoints, sister chromatid cohesion or other yet unidentified components may be necessary for B chromosome drive, and one or more of these aspects may be different in the Bs.

10.4 Evolution of B Chromosomes

Several scenarios have been proposed for the origin of Bs. Most probably they have arisen in different ways in different organisms, see reviews by Camacho et al. (2000), and Jones and Houben (2003). The most widely accepted view is that they are derived from the A chromosome complement. Some evidence also suggests that Bs can be spontaneously generated in response to the new genomic conditions after interspecific hybridization. The involvement of sex chromosomes has also been argued for their origin in some animals, see Camacho et al. (2000) for examples. Despite the large number of species with Bs, their *de novo* formation is probably a rare event because the occurrence of similar B variants in related species suggests a close relationship between the different variants, e.g., in *Secale* (Jones and Puertas 1993) and *Brachycome* (Houben et al. 1999).

The molecular processes that gave rise to Bs during evolution remain unclear, but the characterization of sequences residing on them has helped to shed some light on their origin and evolution. In maize and *Brachycome dichromosomatica* for example, the Bs contain sequences that originate from several different A chromosomes, so the Bs could represent an amalgamation of these diverse A-derived sequences (Alfenito and Birchler 1993; Houben et al. 2001b; Cheng and Lin 2003; Peng et al. 2005).

The actual process of sequence transfer from As to Bs is not clear, but recent results indicate that transposition of mobile elements may have played an important role (Cheng and Lin 2004; Lamb et al. 2007b; Carchilan et al. 2009). For example, an analysis of large DNA insert clones has shown that maize Bs are composed of B-specific sequences that are intermingled with those in common with the As. The 22 kb-long B-specific *StarkB* element, for example, has been subject to frequent insertions by LTR-type retroelements (Lamb et al. 2007b), in a fashion similar to the nested insertions seen in some intergenic A chromosome regions (SanMiguel et al. 1996). It therefore seems likely that the *StarkB* tandem array formation started with the transposition of a *GrandeB* mobile element into an original *non-GrandeB* sequence, which was then amplified, possibly from sister-strand exchange. Subsequently invasion(s) together with amplification resulted in the loss of the original tandem structure of *StarkB* (Lo et al. 2009). Using the divergence of retroelements interrupting B-specific sequences, Lamb et al. (2007b) have estimated the minimum age of the maize B to be at least 2 million years. The recently established oat-maize B chromosome addition lines (Kynast et al. 2007) are an ideal material to further characterize the sequence composition of the maize B chromosome because of the low level of sequence similarity between oat and maize.

B chromosomes provide an ideal target for transposition of mobile elements (McAllister 1995), and the insertion of such elements may therefore be responsible for generating the structural variability observed in Bs (Camacho et al. 2000). Indeed, a B-specific accumulation of Ty3/gypsy retrotransposons has been reported for the fish *Alburnus alburnus* (Ziegler et al. 2003). In the same context, it is noted that Bs also contain various types of coding and noncoding repeats which are similar to those found in circular extrachromosomal DNA of various organisms (Cohen and Segal 2009). For example, extrachromosomal DNA with similarity to tandem repeat sequences shared by A and B chromosomes of *B. dichromosomatica* has recently been identified (Cohen et al. 2008), and integration of extrachromosomal DNA into Bs may be favoured because transcriptionally less active Bs are subject to reduced negative selection. However, whether an evolutionary link exists between extrachromosomal DNA and the evolution of Bs remains to be determined.

The involvement of rRNA coding repeats in the evolution of B chromosomes does not appear to be accidental, because rDNA loci have been detected on Bs of many species of both plants (e.g., *Crepis capillaris* (Maluszynska and Schweizer 1989)) and animals (e.g., *Rattus rattus* (Stitou et al. 2000), *Trichogramma kaykai* (van Vugt et al. 2005); for review see Green (1990) and Jones (1995)). In the herb *Plantago lagopus* the origin of a B chromosome seems to be associated with the massive amplification of 5S rDNA sequences after chromosome fragmentation of an aneuploid A chromosome (Dhar et al. 2002). Alternatively, but less likely, the location of rDNA sites on B chromosomes could be a consequence of the reported mobile nature of rDNA (Schubert and Wobus 1985) with B chromosomes as the preferred 'landing sites' due to their neutral character. There is also increasing evidence that ribosomal sequences can change position within the genome without corresponding changes in the surrounding sequences (Dubcovsky and Dvorak 1995; Shishido et al. 2000; Datson and Murray 2006). However, it is noted that not all Bs carry rDNA.

Except for the 45S rRNA gene on the B chromosome of *C. capillaris*, there has been no direct molecular evidence demonstrating transcription of B chromosome rDNA in any plant species (Leach et al. 2005). The reason(s) why most rDNA on B chromosomes is not or only weakly transcribed is not clear yet although differences in histone H3 methylation between A and B chromosomes may be responsible (Marschner et al. 2007a). Another possibility is that suppression may occur because of nucleolar dominance, so that the rRNA genes on the As are active at the expense of those on the Bs (Donald et al. 1997).

Available sequence information for B-located rRNA genes has been used to study the likely origin of Bs by determining the relatedness of the internal transcribed spacers (ITS) between the different chromosome types. For example, analysis of ITS sequences from the A and B chromosomes of *C. capillaris* (Leach et al. 2005) and *B. dichromosomatica* (Donald et al. 1997; Marschner et al. 2007b), and comparisons with sequences from related species, indicate that the B chromosome rRNA genes are most likely derived from those of the A chromosomes in the host species.

Some findings imply that B chromosomes arise spontaneously in response to genomic stress following interspecific hybridization, e.g., in *Coix aquatica* and *C. gigantea* (Sapre and Deshpande 1987). After fertilization, the two different parental genomes are combined within a single nucleus, which in most cases is embedded within the maternal cytoplasm. Such a novel genomic constitution may result in conflicts, and as a consequence genomic and epigenetic reorganization of the genomes can occur (Riddle and Birchler 2003; see also Jones and Langdon 2013, this volume). An incomplete loss of one parental genome during hybrid embryogenesis might play a role in the hybrid origin of Bs. Indeed, there is evidence showing that during the uniparental chromosome elimination process, the centromeres of parental chromosomes undergoing elimination are the last to be lost (Gernand et al. 2005). If such a centric fragment is retained, rather than eliminated, a subsequent spontaneous doubling could provide the ideal starting point for the *de novo* formation of a supernumerary chromosome. Indeed, a centric fragment was generated during the introgression of a chromosome region from the wasp *Nasonia giraulti* into *N. vitripennis*. This neo-B showed a lower than normal Mendelian segregation ratio in meiosis, and some mitotic instability; but the transmission rate and mitotic stability then increased over successive generations (Perfectti and Werren 2001).

A novel mechanism for chromosome evolution based on recombination of nonhomologous chromosomes during the DNA double-strand repair process at S-phase has been postulated for the formation of the "zebra" chromosome, which is composed of *Elymus trachycaulus*/*Triticum aestivum* structurally rearranged chromosome fragments (Zhang et al. 2008). Although this restructured chromosome does not represent a B chromosome, it is possible to envisage that a similar mechanism could also result in the formation of a neo-B chromosome.

On the basis of new insights into the mechanisms of chromosome evolution (Hall et al. 2006; Lysák et al. 2006; Schubert 2007; see also Lysák and Schubert 2013, this volume) we are tempted to ask, as did Patton (1977), whether the "by-product" of a Robertsonian translocation between two nonhomologous acrocentric chromosomes with breakpoints close to centromeres could evolve into a B-like chromosome. With the recent development of the comparative chromosome painting technique to reconstruct karyotype evolution, it is becoming clear that chromosome number reductions are often accompanied by pericentromeric inversions and translocations between acrocentric chromosomes (Mandakova and Lysák 2008). Although the minichromosomes formed from Robertsonian translocation events are mainly composed of centromeric sequences, they are frequently lost because of the lack of essential genes and their failure to pair and to segregate properly during meiosis. Centromeric regions are also highly dynamic in sequence composition and display a low recombination frequency (Gaut et al. 2007). Recent findings by Hall et al. (2006) point to (peri)centromeres as genomic regions that may experience selective pressures distinct from those acting on euchromatin. They can tolerate rapid changes in structure and sequence content, such as large insertions of sequences that are typical of B chromosomes, e.g., mobile elements, rDNA arrays, and satellite arrays. When a nonessential centromeric fragment survives, rapid sequence alteration may prevent meiotic pairing with the As, and the gain of a drive (by an unknown mechanism) may put it on the evolutionary pathway to a proto-B.

In addition, tertiary trisomics, which appear in the progenies of translocation heterozygotes, have been hypothesized, under certain circumstances (e.g., suppressed crossing-over, rapid loss of genetic activity to overcome genetic imbalance, and positive selection for plants with an extra chromosome), to be suited for B chromosome formation, e.g., in the garden pea (Berdnikov et al. 2003).

The most widely used approaches to study the evolution and DNA composition of B chromosomes are based on the isolation and characterization of only a single or a small group of mainly repetitive sequences. These approaches have been valuable in tracing the fate of various repeats in a wide range of species. However, they do not allow for the global comparative analysis of sequence profiles required for elucidating evolutionary trends at the whole genome and B chromosome level. To overcome this obstacle in the future, new low-cost sequencing technologies such as 454 pyrosequencing (Margulies et al. 2005) should be used for B chromosome sequence analysis. Recently, 454-sequencing was used to sequence flow sorted plant chromosomes. As few as 10,000 copies of barley chromosome 1H were flow-sorted and used as a template to assess gene content and genomic composition of this chromosome (Mayer et al. 2009). As most Bs are smaller than As, they should be easy to sort by flow cytometry (Kubalakova et al. 2003) and thus seem readily suited for such approaches. For instance, one rye B consists of around 560 Mb, so a single 454-run could identify ~50% of the sequence information of flow-sorted Bs. In the future, the application of next generation sequencing technologies to flow-sorted Bs is likely to generate the amount of sequence data needed for extensive comparative sequence analyses. This will then significantly improve our knowledge of the origin of Bs and hence of the evolution of genomes.

Note added in proof Recently, Martis et al. (2012) reported the results of using 454 pyrosequencing to analyse flow sorted A and B chromosomes from rye. They observed that the rye B chromosome is a mosaic of sequences derived from the A chromosomes as well as the organelles.

Acknowledgments The authors have been supported by the DFG (HO 1779/14-1, HO 1779/10-1).

References

Ahn SH, Cheung WL, Hsu JY, Diaz RL, Smith MM, Allis CD (2005) Sterile 20 kinase phosphorylates histone H2B at serine 10 during hydrogen peroxide-induced apoptosis in *S. cerevisiae*. Cell 120:25–36

Alfenito MR, Birchler JA (1993) Molecular characterization of a maize B chromosome centric sequence. Genetics 135:589–597

Amos A, Dover G (1981) The distribution of repetitive DNAs between regular and supernumerary chromosomes in species of *Glossina* (Tsetse)—a two-step process in the origin of supernumeraries. Chromosoma 81:673–690
Bang JW, Choi HW (1990) Genetic analysis of esterase isozymes in rye (*Secale cereale* L.). Korean J Genet 12:87–94
Barlow PW, Vosa CG (1969) Chromosomes of *Puschkinia libanotica* during mitosis. Chromosoma 27:436–447
Bartos J, Paux E, Kofler R, Havrankova M, Kopecky D, Suchankova P, Safar J, Simkova H, Town CD, Lelley T et al (2008) A first survey of the rye (*Secale cereale*) genome composition through BAC end sequencing of the short arm of chromosome 1R. BMC Plant Biol 8:1–12
Berdnikov VA, Gorel FL, Kosterin OE, Bogdanova VS (2003) Tertiary trisomics in the garden pea as a model of B chromosome evolution in plants. Heredity 91:577–583
Bird AW, Yu DY, Pray-Grant MG, Qiu QF, Harmon KE, Megee PC, Grant PA, Smith MM, Christman MF (2002) Acetylation of histone H4 by Esa1 is required for DNA double-strand break repair. Nature 419:411–415
Blunden R, Wilkes TJ, Forster JW, Jimenez MM, Sandery MJ, Karp A, Jones RN (1993) Identification of the E3900 family, a second family of rye chromosome-B specific repeated sequences. Genome 36:706–711
Bougourd SM, Jones RN (1997) B chromosomes: a physiological enigma. New Phytol 137:43–54
Bouzinba-Segard H, Guais A, Francastel C (2006) Accumulation of small murine minor satellite transcripts leads to impaired centromeric architecture and function. Proc Natl Acad Sci USA 103:8709–8714
Bugrov AG, Karamysheva TV, Perepelov EA, Elisaphenko EA, Rubtsov DN, Warchalowska-Sliwa E, Tatsuta H RNB (2007) DNA content of the B chromosomes in grasshopper *Podisma kanoi* Storozh. (Orthoptera, Acrididae). Chromosome Res 15:315–326
Burt A, Trivers R (2006) Genes in conflict: the biology of selfish genetic elements. Belknap Press of Harvard University Press, Cambridge, Massachusetts, London, England
Cabrero J, Teruel M, Carmona FD, Jimenez R, Camacho JP (2007) Histone H3 lysine 9 acetylation pattern suggests that X and B chromosomes are silenced during entire male meiosis in a grasshopper. Cytogenet Genome Res 119:135–142
Camacho JPM (2004) B chromosomes in the eukaryote genome. Cytogenet Cell Genet 106:147–410
Camacho JPM (2005) B chromosomes. In: Gregory TR (ed) The evolution of the genome. Elsevier, Amsterdam New York, pp 223–285
Camacho JPM, Sharbel TF, Beukeboom LW (2000) B-chromosome evolution. Philos Trans Roy Soc Lond B Bio 355:163–178
Carchilan M, Delgado M, Ribeiro T, Costa-Nunes P, Caperta A, Morais-Cecilio L, Jones RN, Viegas W, Houben A (2007) Transcriptionally active heterochromatin in rye B chromosomes. Plant Cell 19:1738–1749
Carchilan M, Kumke K, Mikolajewski S, Houben A (2009) Rye B chromosomes are weakly transcribed and might alter the transcriptional activity of A chromosome sequences. Chromosoma 118:607–616
Carlson WR (1969) Factors affecting preferential fertilization in maize. Genetics 62:543–554
Carlson WR (1978) B-chromosome of corn. Annu Rev Genet 12:5–23
Carlson WR (2007) Locating a site on the maize B chromosome that controls preferential fertilization. Genome 50:578–587
Carlson W (2009) The B chromosome of maize. In: Bennetzen JL, Hake S (eds) Maize handbook. Volume II: genetics and genomics. Springer, Berlin Heidelberg New York, pp 459–480
Carlson WR, Roseman RR (1992) A new property of the maize B-chromosome. Genetics 131:211–223
Carter CR, Smith-White S (1972) Cytology of *Brachycome lineariloba*. 3. Accessory chromosomes. Chromosoma 39:361–379
Catcheside DG (1950) The B-chromosomes of *Parthenium argentatum*. Genet Iber 2:1–10
Cheng YM, Lin BY (2003) Cloning and characterization of maize B chromosome sequences derived from microdissection. Genetics 164:299–310
Cheng YM, Lin BY (2004) Molecular organization of large fragments in the maize B chromosome: Indication of a novel repeat. Genetics 166:1947–1961
Chiavarino AM, Rosato M, Manzanero S, Jimenez G, Gonzalez-Sanchez M, Puertas MJ (2000) Chromosome nondisjunction and instabilities in tapetal cells are affected by B chromosomes in maize. Genetics 155:889–897
Chilton MD, Mccarthy BJ (1973) DNA from maize with and without B chromosomes: a comparative study. Genetics 74:605–614
Cohen S, Segal D (2009) Extrachromosomal circular DNA in eukaryotes: possible involvement in the plasticity of tandem repeats. Cytogenet Genome Res 124:327–338
Cohen S, Houben A, Segal D (2008) Extrachromosomal circular DNA derived from tandemly repeated genomic sequences in plants. Plant J 53:1027–1034
Comai L (2005) The advantages and disadvantages of being polyploid. Nat Rev Genet 6:836–846
Craig JM (2005) Heterochromatin—many flavours, common themes. Bioessays 27:17–28
Cuadrado A, Jouve N (1994) Highly repetitive sequences in B chromosomes of *Secale cereale* revealed by fluorescence *in situ* hybridization. Genome 37:709–712
Datson PM, Murray BG (2006) Ribosomal DNA locus evolution in *Nemesia*: transposition rather than structural rearrangement as the key mechanism? Chromosome Res 14:845–857
de Jesus CM, Galetti PM, Valentini SR, Moreira O (2003) Molecular characterization and chromosomal localization of two families of satellite DNA in *Prochilodus lineatus* (Pisces, Prochilodontidae), a species with B chromosomes. Genetica 118:25–32
Delgado M, Moraiscecilio L, Neves N, Jones RN, Viegas W (1995) The influence of B-chromosomes on rDNA organization in rye interphase nuclei. Chromosome Res 3:487–491
Delgado M, Caperta A, Ribeiro T, Viegas W, Jones RN, Morais-Cecilio L (2004) Different numbers of rye B chromosomes induce identical compaction changes in distinct A chromosome domains. Cytogenet Genome Res 106:320–324
Dhar MK, Friebe B, Koul AK, Gill BS (2002) Origin of an apparent B chromosome by mutation, chromosome fragmentation and specific DNA sequence amplification. Chromosoma 111:332–340
Donald TM, Houben A, Leach CR, Timmis JN (1997) Ribosomal RNA genes specific to the B chromosomes in *Brachycome dichromosomatica* are not transcribed in leaf tissue. Genome 40:674–681
Dubcovsky J, Dvorak J (1995) Ribosomal RNA multigene loci: nomads of the Triticeae genomes. Genetics 140:1367–1377
Endo TR, Nasuda S, Jones N, Dou Q, Akahori A, Wakimoto M, Tanaka H, Niwa K, Tsujimoto H (2008) Dissection of rye B chromosomes, and nondisjunction properties of the dissected segments in a common wheat background. Genes Genet Syst 83:23–30
Enkerli J, Bhatt G, Covert SF (1997) Nht1, a transposable element cloned from a dispensable chromosome in *Nectria haematococca*. Mol Plant Microbe Interact 10:742–749
Evans GM (1960) Supernumerary chromosomes in wild populations of the snail *Helix pomatia* L. Heredity 15:129–138
Evans GM, Davies EW (1985) The genetics of meiotic chromosome pairing in *Lolium temulentum* x *Lolium perenne* tetraploids. Theor Appl Genet 71:185–192

Flavell RB, Rimpau J (1975) Ribosomal RNA genes and supernumerary B chromosomes of rye. Heredity 35:127–131

Francki MG (2001) Identification of Bilby, a diverged centromeric Ty1-copia retrotransposon family from cereal rye (*Secale cereale* L.). Genome 44:266–274

Franks TK, Houben A, Leach CR, Timmis JN (1996) The molecular organisation of a B chromosome tandem repeat sequence from *Brachycome dichromosomatica*. Chromosoma 105:223–230

Frost S (1959) The cytological behaviour and mode of transmission of accessory chromosomes in *Plantago serraria*. Hereditas 45:191–210

Fuchs J, Demidov D, Houben A, Schubert I (2006) Chromosomal histone modification patterns — from conservation to diversity. Trends Plant Sci 11:199–208

Gaut BS, Wright SI, Rizzon C, Dvorak J, Anderson LK (2007) Recombination: an underappreciated factor in the evolution of plant genomes. Nat Rev Genet 8:77–84

Gehring M, Henikoff S (2007) DNA methylation dynamics in plant genomes. BBA—Gene Struct Expr 1769:276–286

Gernand D, Rutten T, Varshney A, Rubtsova M, Prodanovic S, Bruss C, Kumlehn J, Matzk F, Houben A (2005) Uniparental chromosome elimination at mitosis and interphase in wheat and pearl millet crosses involves micronucleus formation, progressive heterochromatinization, and DNA fragmentation. Plant Cell 17:2431–2438

Gibson I, Hewitt G (1970) Isolation of DNA from B chromosomes in grasshoppers. Nature 225:67–68

Gonzalez-Sanchez M, Gonzalez-Gonzalez E, Molina F, Chiavarino AM, Rosato M, Puertas MJ (2003) One gene determines maize B chromosome accumulation by preferential fertilisation; another gene(s) determines their meiotic loss. Heredity 90:122–129

Gonzalez-Sanchez M, Rosato M, Chiavarino M, Puertas MJ (2004) Chromosome instabilities and programmed cell death in tapetal cells of maize with B chromosomes and effects on pollen viability. Genetics 166:999–1009

Green DM (1990) Muller's ratchet and the evolution of supernumerary chromosomes. Genome 33:818–824

Greilhuber J, Speta F (1976) C-banded karyotypes in the *Scilla hohenackeri* group, *Scilla persica*, and *Puschkinia* (Liliaceae). Plant Syst Evol 126:149–188

Gruenbaum Y, Naveh-Many T, Cedar H, Razin A (1981) Sequence specificity of methylation in higher plant DNA. Nature 292: 860–862

Grun P (1959) Variability of accessory chromosomes in native populations of *Allium cernuum*. Am J Bot 46:218–224

Grunstein M (1997) Histone acetylation in chromatin structure and transcription. Nature 389:349–352

Gutknecht J, Sperlich D, Bachmann L (1995) A species specific satellite DNA family of *Drosophila subsilvestris* appearing predominantly in B chromosomes. Chromosoma 103:539–544

Hakansson A (1948) Behaviour of accessory rye chromosomes in the embryo sac. Hereditas 34:35–59

Hall AE, Kettler GC, Preuss D (2006) Dynamic evolution at pericentromeres. Genome Res 16:355–364

Han F, Lamb JC, Birchler JA (2006) High frequency of centromere inactivation resulting in stable dicentric chromosomes of maize. Proc Natl Acad Sci USA 103:3238–3243

Hasegawa N (1934) A cytological study on 8-chromosome rye. Cytologia 6:68–77

Hasterok R, Jenkins G, Langdon T, Jones RN (2002) The nature and destiny of translocated B-chromosome-specific satellite DNA of rye. Chromosome Res 10:83–86

Henderson S (1961) The chromosomes the of British *Tetrigidae* (Orthoptera). Chromosoma 12:553–572

Hewitt GM (1976) Meiotic drive for B-chromosomes in the primary oocytes of *Myrmeleotettix maculatus* (Orthopera: Acrididae). Chromosoma 56:381–391

Houben A, Kynast RG, Heim U, Hermann H, Jones RN, Forster JW (1996) Molecular cytogenetic characterisation of the terminal heterochromatic segment of the B-chromosome of rye (*Secale cereale*). Chromosoma 105:97–103

Houben A, Belyaev ND, Leach CR, Timmis JN (1997a) Differences of histone H4 acetylation and replication timing between A and B chromosomes of *Brachycome dichromosomatica*. Chromosome Res 5:233–237

Houben A, Leach CR, Verlin D, Rofe R, Timmis JN (1997b) A repetitive DNA sequence common to the different B chromosomes of the genus *Brachycome*. Chromosoma 106:513–519

Houben A, Thompson N, Ahne R, Leach CR, Verlin D, Timmis JN (1999) A monophyletic origin of the B chromosomes *of Brachycome dichromosomatica* (Asteraceae). Plant Syst Evol 219:127–135

Houben A, Field BL, Saunders VA (2001a) Microdissection and chromosome painting of plant B chromosomes. Method Cell Sci 23:115–124

Houben A, Verlin D, Leach CR, Timmis JN (2001b) The genomic complexity of micro B chromosomes of *Brachycome dichromosomatica*. Chromosoma 110:451–459

Houben A, Demidov D, Gernand D, Meister A, Leach CR, Schubert I (2003) Methylation of histone H3 in euchromatin of plant chromosomes depends on basic nuclear DNA content. Plant J 33:967–973

Houben A, Demidov D, Caperta AD, Karimi R, Agueci F, Vlasenko L (2007) Phosphorylation of histone H3 in plants—a dynamic affair. Biochim Biophys Acta 1769:308–315

Ikura T, Ogryzko VV, Grigoriev M, Groisman R, Wang J, Horikoshi M, Scully R, Qin J, Nakatani Y (2000) Involvement of the TIP60 histone acetylase complex in DNA repair and apoptosis. Cell 102:463–473

Jamilena M, Rejon CR, Rejon MR (1994) A molecular analysis of the origin of the *Crepis capillaris* B chromosome. J Cell Sci 107:703–708

Jamilena M, Garridoramos M, Rejon MR, Rejon CR, Parker JS (1995) Characterization of repeated sequences from microdissected B chromosomes of *Crepis capillaris*. Chromosoma 104:113–120

Jenkins G, Jones RN (2004) B chromosomes in hybrids of temperate cereals and grasses. Cytogenet Genome Res 106:314–319

Jimenez MM, Romera F, GonzalezSanchez M, Puertas MJ (1997) Genetic control of the rate of transmission of rye B chromosomes. 3. Male meiosis and gametogenesis. Heredity 78:636–644

Jin WW, Lamb JC, Vega JM, Dawe RK, Birchler JA, Jiang J (2005) Molecular and functional dissection of the maize B chromosome centromere. Plant Cell 17:1412–1423

Jin WW, Lamb JC, Zhang WL, Kolano B, Birchler JA, Jiang JM (2008) Histone modifications associated with both A and B chromosomes of maize. Chromosome Res 16:1203–1214

John B, Hewitt GM (1965) B-chromosome system of *Myrmeleotettix maculatus* (Thunb.). Chromosoma 17:121–138

Jones RN (1975) B-chromosome systems in flowering plants and animal species. Int Rev Cytol 40:1–100

Jones RN (1991) B-chromosome drive. Am Nat 137:430–442

Jones RN (1995) B chromosomes in plants. New Phytol 131: 411–434

Jones N, Houben A (2003) B chromosomes in plants: escapees from the A chromosome genome? Trends Plant Sci 8:417–423

Jones RN, Langdon T (2013) The plant nucleus at war and peace: genome organisation in the interphase nucleus. In: Leitch IJ, Greilhuber J, Doležel J, Wendel JF (eds) Plant genome diversity, vol 2, Physical structure, behaviour and evolution of plant genomes. Springer-Verlag, Wien, pp 13–31

Jones RN, Puertas MJ (1993) The B-chromosomes of rye (*Secale cereale* L.). In: Dhir KK, Sareen TS (eds) Frontiers in plant science research. Bhagwati Enterprises, Delhi, India, pp 81–112

Jones RN, Rees H (1982) B chromosomes, 1st edn. Academic, London, New York

Jones RN, Gonzalez-Sanchez M, Gonzalez-Garcia M, Vega JM, Puertas MJ (2008a) Chromosomes with a life of their own. Cytogenet Genome Res 120:265–280

Jones RN, Viegas W, Houben A (2008b) A century of B chromosomes in plants: so what? Ann Bot 101:767–775

Kaszas E, Birchler JA (1996) Misdivision analysis of centromere structure in maize. EMBO J 15:5246–5255

Kaszas E, Birchler JA (1998) Meiotic transmission rates correlate with physical features of rearranged centromeres in maize. Genetics 150:1683–1692

Kaszas E, Cande WZ (2000) Phosphorylation of histone H3 is correlated with changes in the maintenance of sister chromatid cohesion during meiosis in maize, rather than the condensation of the chromatin. J Cell Sci 113:3217–3226

Kaszas E, Kato A, Birchler JA (2002) Cytological and molecular analysis of centromere misdivision in maize. Genome 45:759–768

Kayano H (1957) Cytogenetic studies in *Lilium callosum*. III. Preferential segregation of a supernumerary chromosome in EMCs. P Jpn Acad B-Phys 33:553–558

Kayano H (1971) Accumulation of B chromosomes in the germ line of *Locusta migratoria*. Heredity 27:119–123

Kimura M, Kayano H (1961) Maintenance of supernumerary chromosomes in wild populations of *Lilium callosum* by preferential segregation. Genetics 46:1699–1712

Klein AS, Eckhardt RA (1976) DNAs of A and B chromosomes of the mealy bug, *Pseudococcus obscurus*. Chromosoma 57:333–340

Kouzarides T (2007) Chromatin modifications and their function. Cell 128:693–705

Krishnamoorthy T, Chen X, Govin J, Cheung WL, Dorsey J, Schindler K, Winter E, Allis CD, Guacci V, Khochbin S et al (2006) Phosphorylation of histone H4 Ser1 regulates sporulation in yeast and is conserved in fly and mouse spermatogenesis. Genes Dev 20:2580–2592

Kubalakova M, Valarik M, Bartos J, Vrana J, Cihalikova J, Molnar-Lang M, Dolezel J (2003) Analysis and sorting of rye (*Secale cereale* L.) chromosomes using flow cytometry. Genome 46:893–905

Kumke K, Jones RN, Houben A (2008) B chromosomes of *Puschkinia libanotica* are characterized by a reduced level of euchromatic histone H3 methylation marks. Cytogenet Genome Res 121:266–270

Kynast RG, Galatowitsch MW, Huettl PA, Phillips RL, Rines HW (2007) Adding B-chromosomes of *Zea mays* L. to the genome of *Avena sativa* L. Maize Genet Coop Newsl 81:17–19

Lamb J, Kato A, Birchler J (2005) Sequences associated with A chromosome centromeres are present throughout the maize B chromosome. Chromosoma 113:337–349

Lamb JC, Han F, Auger DL, Birchler J (2006) A trans-acting factor required for non-disjunction of the B chromosome is located distal to the TB-4Lb breakpoint on the B chromosome. Maize Genet Coop Newsl 80:51–54

Lamb JC, Meyer JM, Corcoran B, Kato A, Han F, Birchler JA (2007a) Distinct chromosomal distributions of highly repetitive sequences in maize. Chromosome Res 15:33–49

Lamb JC, Riddle NC, Cheng YM, Theuri J, Birchler JA (2007b) Localization and transcription of a retrotransposon-derived element on the maize B chromosome. Chromosome Res 15:383–398

Langdon T, Seago C, Jones RN, Ougham H, Thomas H, Forster JW, Jenkins G (2000) De novo evolution of satellite DNA on the rye B chromosome. Genetics 154:869–884

Leach CR, Donald TM, Franks TK, Spiniello SS, Hanrahan CF, Timmis JN (1995) Organization and origin of a B chromosome centromeric sequence from *Brachycome dichromosomatica*. Chromosoma 103:708–714

Leach CR, Houben A, Field B, Pistrick K, Demidov D, Timmis JN (2005) Molecular evidence for transcription of genes on a B chromosome in *Crepis capillaris*. Genetics 171:269–278

Lemos B, Araripe LO, Hartl DL (2008) Polymorphic Y chromosomes harbor cryptic variation with manifold functional consequences. Science 319:91–93

Levin DA, Palestis BG, Jones RN, Trivers R (2005) Phyletic hot spots for B chromosomes in angiosperms. Evolution 59:962–969

Lima de Faria A (1962) Genetic interaction in rye expressed at the chromosome phenotype. Genetics 47:1455–1462

Lo KL, Peng SF, Chen LJ, Lin BY (2009) Tandem organization of StarkB element (22.8 kb) in the maize B chromosome. Mol Genet Genomics 282:131–139

Lysák MA, Schubert I (2013) Mechanisms of chromosome rearrangements. In: Leitch IJ, Greilhuber J, Doležel J, Wendel JF (eds) Plant genome diversity, vol 2, Physical structure, behaviour and evolution of plant genomes. Springer-Verlag, Wien, pp 137–147

Lysák MA, Berr A, Pecinka A, Schmidt R, McBreen K, Schubert I (2006) Mechanisms of chromosome number reduction in *Arabidopsis thaliana* and related Brassicaceae species. Proc Natl Acad Sci USA 103:5224–5229

Maistro EL, Foresti F, Oliveira C, Toledo LFD (1992) Occurrence of macro-B chromosomes in *Astyanax scabripinnis paranae* (Pisces, Characiformes, Characidae). Genetica 87:101–106

Maluszynska J, Schweizer D (1989) Ribosomal RNA genes in B-chromosomes of *Crepis capillaris* detected by non-radioactive in situ hybridization. Heredity 62:59–65

Mandakova T, Lysák MA (2008) Chromosomal phylogeny and karyotype evolution in x = 7 crucifer species (Brassicaceae). Plant Cell 20:2559–2570

Manzanero S, Puertas MJ (2003) Rye terminal neocentromeres: characterisation of the underlying DNA and chromatin structure. Chromosoma 111:408–415

Margulies M, Egholm M, Altman WE, Attiya S, Bader JS, Bemben LA, Berka J, Braverman MS, Chen YJ, Chen ZT et al (2005) Genome sequencing in microfabricated high-density picolitre reactors. Nature 437:376–380

Marschner S, Kumke K, Houben A (2007a) B chromosomes of *B. dichromosomatica* show a reduced level of euchromatic histone H3 methylation marks. Chromosome Res 15:215–222

Marschner S, Meister A, Blattner FR, Houben A (2007b) Evolution and function of B chromosome 45S rDNA sequences in *Brachycome dichromosomatica*. Genome 50:638–644

Martin C, Zhang Y (2005) The diverse functions of histone lysine methylation. Nat Rev Mol Cell Biol 6:838–849

Martis MM, Klemme S, Banaei-Moghaddam AM et al. (2012) Selfish supernumerary chromosome reveals its origin as a mosaic of host genome and organellar sequences. Proc Natl Acad Sci 109:13343–13346

Masonbrink RE, Birchler JA (2010) Sporophytic nondisjunction of the maize B chromosome at high copy numbers. J Genet Genomics 37:79–84

Matthews RB, Jones RN (1983) Dynamics of the B-chromosome polymorphism in rye. 2. Estimates of parameters. Heredity 50:119–137

May BP, Lippman ZB, Fang Y, Spector DL, Martienssen RA (2005) Differential regulation of strand-specific transcripts from *Arabidopsis* centromeric satellite repeats. PLoS Genet 1:e79

Mayer KFX, Taudien S, Martis M, Simkova H, Suchankova P, Gundlach H, Wicker T, Petzold A, Felder M, Steuernagel B et al (2009) Gene content and virtual gene order of barley chromosome 1H. Plant Physiol 151:496–505

McAllister BF (1995) Isolation and characterization of a retroelement from B chromosome (PSR) in the parasitic wasp *Nasonia vitripennis*. Insect Mol Biol 4:253–262

Meller VH, Rattner BP (2002) The roX genes encode redundant male-specific lethal transcripts required for targeting of the MSL complex. EMBO J 21:1084–1091

Mendelson D, Zohary D (1972) Behavior and transmission of supernumerary chromosomes in *Aegilops speltoides*. Heredity 29:329–339

Misteli T (2007) Beyond the sequence: cellular organization of genome function. Cell 128:787–800

Morais-Cecilio L, Delgado M, Jones RN, Viegas W (1996) Painting rye B chromosomes in wheat: Interphase chromatin organization, nuclear disposition and association in plants with two, three or four Bs. Chromosome Res 4:195–200

Morais-Cecilio L, Delgado M, Jones RN, Viegas W (1997) Interphase arrangement of rye B chromosomes in rye and wheat. Chromosome Res 5:177–181

Moralli D, Chan DY, Jefferson A, Volpi EV, Monaco ZL (2009) HAC stability in murine cells is influenced by nuclear localization and chromatin organization. BMC Cell Biol 10:18

Müntzing A (1948) Cytological studies of extra fragment chromosomes in rye. 5. A new fragment type arisen by deletion. Hereditas 34:435–442

Müntzing A (1970) Chromosomal variation in the Lindstroem strain of wheat carrying accessory chromosomes of rye. Hereditas 66:279–285

Niwa K, Sakamoto S (1995) Origin of B-chromosomes in cultivated rye. Genome 38:307–312

Niwa K, Horiuchi G, Hirai Y (1997) Production and characterization of common wheat with B chromosomes of rye from Korea. Hereditas 126:139–146

Nur U (1963) A mitotically unstable supernumerary chromosome with an accumulation mechanism in a grasshopper. Chromosoma 14:407–422

Nur U (1969) Mitotic instability leading to an accumulation of B-chromosomes in grasshoppers. Chromosoma 27:1–19

Östergren G (1947) Heterochromatic B chromosomes in *Anthoxanthum*. Hereditas 33:261–296

Page BT, Wanous MK, Birchler JA (2001) Characterization of a maize chromosome 4 centromeric sequence: evidence for an evolutionary relationship with the B chromosome centromere. Genetics 159:291–302

Palestis BG, Trivers R, Burt A, Jones RN (2004) The distribution of B chromosomes across species. Cytogenet Genome Res 106:151–158

Parker JS, Jones GH, Edgar L, Whitehouse C (1989) The population cytogenetics of *Crepis capillaris*. II. The stability and inheritance of B-chromosomes. Heredity 63:19–27

Patton JL (1977) B chromosome systems in pocket mouse, *Perognathus baileyi*—meiosis and C-band studies. Chromosoma 60:1–14

Peng SF, Lin YP, Lin BY (2005) Characterization of AFLP sequences from regions of maize B chromosome defined by 12 B-10L translocations. Genetics 169:375–388

Perfectti F, Werren JH (2001) The interspecific origin of B chromosomes: experimental evidence. Evolution 55:1069–1073

Pidoux AL, Allshire R (2005) The role of heterochromatin in centromere function. Philos Trans Roy Soc Lond B Bio 360:569–579

Price L, Short KC, Roberts AV (1981) Poor resolution of C-bands and the presence of B-chromosomes in *Rosa rugosa* 'Scabrosa'. Caryologia 34:69–72

Prigent C, Dimitrov S (2003) Phosphorylation of serine 10 in histone H3, what for? J Cell Sci 116:3677–3685

Puertas MJ (2002) Nature and evolution of B chromosomes in plants: a non-coding but information-rich part of plant genomes. Cytogenet Genome Res 96:198–205

Puertas MJ, Romera F, Delapena A (1985) Comparison of B-chromosome effects on *Secale cereale* and *Secale vavilovii*. Heredity 55:229–234

Ribeiro T, Pires B, Delgado M, Viegas W, Jones N, Morais-Cecilio L (2004) Evidence for 'cross-talk' between A and B chromosomes of rye. P Roy Soc Lond B Bio 271:S482–S484

Richards EJ, Elgin SCR (2002) Epigenetic codes for heterochromatin formation and silencing: rounding up the usual suspects. Cell 108:489–500

Riddle NC, Birchler JA (2003) Effects of reunited diverged regulatory hierarchies in allopolyploids and species hybrids. Trends Genet 19:597–600

Rimpau J, Flavell RB (1975) Characterization of rye B chromosome DNA by DNA-DNA hybridization. Chromosoma 52:207–217

Roman H (1947) Mitotic nondisjunction in the case of interchanges involving the B-type chromosome in maize. Genetics 32:391–409

Roman H (1948) Directed fertilization in maize. Proc Natl Acad Sci USA 34:36–42

Romera F, Jimenez MM, Puertas MJ (1991) Genetic control of the rate of transmission of rye B chromosomes. 1. Effects in 2B × 0B crosses. Heredity 66:61–65

Rudd MK, Mays RW, Schwartz S, Willard HF (2003) Human artificial chromosomes with alpha satellite-based de novo centromeres show increased frequency of nondisjunction and anaphase lag. Mol Cell Biol 23:7689–7697

Ruiz-Rejon M, Posse F, Oliver JL (1980) The B-chromosome system of *Scilla autumnalis* (Liliaceae) – effects at the isoenzyme level. Chromosoma 79:341–348

Rusche ML, Mogensen HL, Shi L, Keim P, Rougier M, Chaboud A, Dumas C (1997) B chromosome behavior in maize pollen as determined by a molecular probe. Genetics 147:1915–1921

Rutishauser A, Röthlisberger E (1966) Boosting mechanism of B-chromosomes in *Crepis capillaris*. Chromosomes Today 1:28–30

Sandery MJ, Forster JW, Blunden R, Jones RN (1990) Identification of a family of repeated sequences on the rye B-chromosome. Genome 33:908–913

SanMiguel P, Tikhonov A, Jin YK, Motchoulskaia N, Zakharov D, Melake-Berhan A, Springer PS, Edwards KJ, Lee M, Avramova Z, Bennetzen JL (1996) Nested retrotransposons in the intergenic regions of the maize genome. Science 274:765–768

Sapre B, Deshpande S (1987) Origin of B chromosomes in *Coix* L. through spontaneous interspecific hybridisation. J Hered 78: 191–196

Schlegel R, Pohler W (1994) Identification of an A-B chromosome translocation in diploid rye (*Secale cereale* L). Breed Sci 44:279–283

Schmid M, Ziegler CG, Steinlein C, Nanda I, Haaf T (2002) Chromosome banding in Amphibia – XXIV. The B chromosomes of *Gastrotheca espeletia* (Anura, Hylidae). Cytogenet Genome Res 97:205–218

Schmid M, Ziegler CG, Steinlein C, Nanda I, Schartl M (2006) Cytogenetics of the bleak (*Alburnus alburnus*), with special emphasis on the B chromosomes. Chromosome Res 14:231–242

Schubert I (2007) Chromosome evolution. Curr Opinion Plant Biol 10:109–115

Schubert I, Wobus U (1985) *In situ* hybridisation confirms jumping nucleolus organizing regions in *Allium*. Chromosoma 92:143–148

Shambulingappa KG (1965) Occurrence of B chromosomes in the genus *Clematis*. Curr Sci 34:670–671

Shishido R, Sano Y, Fukui K (2000) Ribosomal DNAs: an exception to the conservation of gene order in rice genomes. Mol Gen Genet 263:586–591

Stark EA, Connerton I, Bennett ST, Barnes SR, Parker JS, Forster JW (1996) Molecular analysis of the structure of the maize B-chromosome. Chromosome Res 4:15–23

Stitou S, Diaz de La Guardia R, Jimenez R, Burgos M (2000) Inactive ribosomal cistrons are spread throughout the B chromosomes of *Rattus rattus* (Rodentia, Muridae). Implications for their origin and evolution. Chromosome Res 8:305–311

Struhl K (1998) Histone acetylation and transcriptional regulatory mechanisms. Genes Dev 12:599–606

Tanaka M, Kawahara T (1982) Cytogenetical effects of B chromosomes in plants – a review. Rep Plant Germplasm Inst Kyoto Univ 5:1–18

Tanaka R, Matsuda T (1972) A high occurrence of accessory chromosomal type in *Tainia laxiflora*, Orchidaceae. Bot Mag Tokyo 85:43–49

Tanic N, Vujosevic M, Dedovi-Tanic N, Dimitrijevic B (2005) Differential gene expression in yellow-necked mice *Apodemus flavicollis* (Rodentia, Mammalia) with and without B chromosomes. Chromosoma 113:418–427

Teoh SB, Rees H (1977) B-chromosomes in white spruce. P Roy Soc Lond B Bio 198:325–344

Teruel M, Cabrero J, Perfectti F, Camacho JP (2007) Nucleolus size variation during meiosis and NOR activity of a B chromosome in the grasshopper *Eyprepocnemis plorans*. Chromosome Res 15:755–765

Teruel M, Cabrero J, Perfectti F, Camacho JPM (2010) B chromosome ancestry revealed by histone genes in the migratory locust. Chromosoma 119:217–225

Theuri J, Phelps-Durr T, Mathews S, Birchler J (2005) A comparative study of retrotransposons in the centromeric regions of A and B chromosomes of maize. Cytogenet Genome Res 110:203–208

Timmis JN, Ingle J, Sinclair J, Jones RN (1975) Genomic quality of rye B chromosomes. J Exp Bot 26:367–368

Topp CN, Zhong CX, Dawe RK (2004) Centromere-encoded RNAs are integral components of the maize kinetochore. Proc Natl Acad Sci USA 101:15986–15991

Tsujimoto H, Niwa K (1992) DNA structure of the B-chromosome of rye revealed by in situ hybridization using repetitive sequences. Jpn J Genet 67:233–241

Tsujimoto H, Niwa K (1994) Evolutionary mechanism of a rye B chromosome viewed from its DNA structure. In: Raupp WJ, Gill BS (eds) Classical and molecular cytogenetic analysis, proceedings of US-Japan symposium. Kansas Agricultural Experiment Station, Kansas State University, Manhattan, pp 90–95

van Vugt J, de Nooijer S, Stouthamer R, de Jong H (2005) NOR activity and repeat sequences of the paternal sex ratio chromosome of the parasitoid wasp *Trichogramma kaykai*. Chromosoma 114:410–419

Volpe T, Schramke V, Hamilton GL, White SA, Teng G, Martienssen RA, Allshire RC (2003) RNA interference is required for normal centromere function in fission yeast. Chromosome Res 11:137–146

Vos LJ, Famulski JK, Chan GK (2006) How to build a centromere: from centromeric and pericentromeric chromatin to kinetochore assembly. Biochem Cell Biol 84:619–639

Vosa CG (1973) The enhanced and reduced quinacrine fluorescence bands and their relationship to the Giemsa patterns in *Allium flavum*. In: Caspersson T, Zech L (eds) Chromosome identification, proceedings of the 23rd Nobel symposium (medical and natural sciences), pp 156–158

Wilkes TM, Francki MG, Langridge P, Karp A, Jones RN, Forster JW (1995) Analysis of rye B-chromosome structure using fluorescence in situ hybridization (FISH). Chromosome Res 3:466–472

Wutz A, Rasmussen TP, Jaenisch R (2002) Chromosomal silencing and localization are mediated by different domains of Xist RNA. Nat Genet 30:167–174

Zhang Y, Reinberg D (2001) Transcription regulation by histone methylation: interplay between different covalent modifications of the core histone tails. Genes Dev 15:2343–2360

Zhang P, Li WL, Friebe B, Gill BS (2008) The origin of a "zebra" chromosome in wheat suggests nonhomologous recombination as a novel mechanism for new chromosome evolution and step changes in chromosome number. Genetics 179:1169–1177

Zhou JP, Yang ZJ, Li GR, Liu C, Tang ZX, Zhang Y, Ren ZL (2010) Diversified chromosomal distribution of tandemly repeated sequences revealed evolutionary trends in *Secale* (Poaceae). Plant Syst Evol 287:49–56

Ziegler CG, Lamatsch DK, Steinlein C, Engel W, Schartl M, Schmid M (2003) The giant B chromosome of the cyprinid fish *Alburnus alburnus* harbours a retrotransposon-derived repetitive DNA sequence. Chromosome Res 11:23–35

Chromosomes and Sex Differentiation

11

Bohuslav Janoušek, Roman Hobza, and Boris Vyskot

Contents

B. Vyskot (✉)
Laboratory of Plant Developmental Genetics, Institute of Biophysics, Academy of Sciences of the Czech Republic, Kralovopolska 135, Brno 61265, Czech Republic
e-mail: vyskot@ibp.cz

I.J. Leitch et al. (eds.), *Plant Genome Diversity Volume 2*,
DOI 10.1007/978-3-7091-1160-4_11, © Springer-Verlag Wien 2013

11.1 Evolution of Plant Sexuality

Plant species can be classified into two major groups: those that permit self-pollination (autogamy) and those that inhibit self-pollination. In mostly self-pollinating species, harmful recessive mutations with a large effect are efficiently eliminated by selection, while slightly deleterious mutations accumulate as a consequence of the reduced effective population size and effective recombination rates (Wright et al. 2008). In contrast, plants that prevent autogamy are able to mask and retain in their genomes harmful recessive mutations with large effects in spite of more efficient selection against slightly deleterious mutations in this group. In cosexual plants, various mechanisms, such as dichogamy, heterostyly or self-incompatibility, prevent self-pollination. Another mechanism is the evolution of unisexual flowers. Populations can be distinguished according to the localization of unisexual flowers: monoecious (male and female on the same plant), gynomonoecious (hermaphrodite and female flowers on the same plant), andromonoecious (male and hermaphrodite flowers on the same plant), dioecious (male and female flowers on different plants), gynodioecious (female and cosexual individuals), androdioecious (male and cosexual individuals), or trioecious (male, female, and cosexual individuals), as reviewed by Dellaporta and Calderon-Urrea (1993). Gymnosperms are mostly monoecious, but also comprise a relatively high percentage of dioecious species. In the gymnosperms, approximately 36 %, namely all *c*. 250 species of cycads, *Ginkgo biloba*, and *c*. 50 Gnetales are dioecious (Ming et al. 2011). In contrast, dioecy has been reported in only about 6 % of angiosperm species (Renner and Ricklefs 1995). Interestingly, dioecy is more widespread in tropical species, and an exceptionally high percentage of the dominant woody species of tropical forests are dioecious (Matallana et al. 2005). New cases of dioecy continue to be found because of the phenomenon of cryptic dioecy (Mayer and Charlesworth 1991).

The taxonomic spread of dioecy indicates that it probably originated many times independently. There are several hypotheses that describe the possible evolutionary history of dioecy. The most popular one is that two basic genetic changes had to occur on the way from cosexuality to dioecy (Charlesworth and Charlesworth 1978). One such change caused male sterility and the other was responsible for female sterility. Gynodioecy was likely the intermediate stage towards complete dioecy, since it occurs much more frequently than androdioecy. According to theoretical studies, the evolution of dioecy through androdioecy is also hindered by males in androdioecious populations having to produce at least twice the pollen of hermaphrodites to be maintained in the population (Charlesworth and Charlesworth 1978). Other possible routes to dioecy include direct evolution of dioecy from monoecy, e.g., species of the genus *Siparuna* (Renner and Won 2001), or evolution of dioecy as a reaction to the loss of autoincompatibility in polyploids (Miller and Venable 2000). The possibility of the evolution of dioecy through heterodichogamy is also theoretically supported (Pannell and Verdú 2006), but there is still no clear example of this route (Renner et al. 2007).

In most dioecious species the male is the heterogametic sex (XY), while females are homogametic (XX). However, there are some exceptions, in which female individuals are heterogametic (ZW), e.g., wild *Fragaria* (Spigler et al. 2008) and *Populus trichocarpa* (Tuskan et al. 2006). In some species, sex is determined by a simple Mendelian genetic system based on the segregation of a few loci such as in *Fragaria virginiana* (Spigler et al. 2008), *Ecballium elaterium* (reviewed in Mather 1949; Gómez-Campo and Casas-Builla 1965), *Mercurialis annua* (Hamdi et al. 1987). In others, such as *Carica papaya*, a short X-Y non-recombining region has recently formed (Yu et al. 2007). These systems are considered to be evolutionary very young. Other dioecious species are evolutionary older, and some of them have evolved heteromorphic sex chromosomes. Heteromorphic sex chromosomes occur in *Silene* (reviewed by Nicolas et al. 2005) and sorrel (*Rumex* spp.; reviewed by Ainsworth et al. 1999), some hop species (*Humulus* spp.; reviewed in Shephard et al. 2000), hemp (*Cannabis sativa*; reviewed by Menzel (1964) and Sakamoto et al. (2000)), and some Cucurbitaceae species (Kumar and Viseveshwaraiah 1952).

11.2 Sex Chromosomes and Sex Determination Systems in Plants

11.2.1 Sex Determination and Sex Chromosomes in Bryophytes

11.2.1.1 Monoecious Bryophytes: Intragametophytic Selfing

Self-fertilization has a different adaptive significance and effects in species with combined sexes than in those with separate sexes. In haploid-dominant species such as liverworts, mosses, and hornworts, self-fertilization in the broad sense (intergametophytic selfing—mating of gametophytes derived from the same spore) is possible in species with combined or separate sexes (monoecious and dioecious, respectively). Monoecious species also possess an additional mode of selfing (intragametophytic selfing) that leads to complete homozygosity in one step. Eppley et al. (2007) found that while there were deficiencies of heterozygotes compared to the null expectation in both monoecious and dioecious mosses, monoecious species had significantly higher levels of heterozygote deficiency than dioecious species. Estimates of selfing rates have suggested that selfing occurs frequently in monoecious populations, but only rarely in dioecious populations. However, significant indications of mixed mating or biparental inbreeding were found in many populations of two dioecious species (*Polytrichadelphus magellanicus* and *Breutelia pendula*) (Eppley et al. 2007).

11.2.1.2 Dioecious Mosses and Liverworts (*Marchantia polymorpha* Main Model)

The sex chromosomes in liverworts and mosses play a direct role in development of gametophyte sex organs. *Marchantia polymorpha* is a model species that has allowed researchers to better understand sex determination in bryophytes and to reveal similarities in the evolution of sex chromosomes between evolutionary distant species. The dominant phase of the *Marchantia polymorpha* life cycle is the haploid gametophyte, which is either male or female. The male possesses a Y chromosome, while the female possesses an X. Asexual propagules (gemmae) are produced by the thallus, so sex-specific cultures can be easily maintained. This experimental system allows researchers to monitor the functions of sex-linked genes without interference from the other sex chromosome.

The X and Y chromosomes are heteromorphic; they differ in size, which has made them good candidates for FISH mapping, permitting direct identification of X- and Y-located sequences. Okada et al. (2000) constructed genomic libraries of male and female plants, and isolated seventy putative male-specific PAC clones based on different intensities of their hybridization with male and female DNAs. Y-specificity of one clone (pMM4G7) was confirmed by Southern blots, PCR analysis and FISH (Okada et al. 2000). Another six male-specific clones were isolated using representational difference analysis (Fujisawa et al. 2001). A detailed analysis of some of the Y-specific clones (pMM4G7 and pMM23-130 F12) revealed many repetitive motifs organised in long stretches. Within these specific repeats, a novel gene family (*ORF162*) was described, which is specifically expressed in male sexual organs and contains a RING motif (Okada et al. 2001). RING finger proteins are known to participate in transcriptional repression and in the ubiquitin-mediated protein turnover processes (Borden 2000). Another five genes amplified on the Y chromosome were described by Ishizaki et al. (2002).

One of the five putative genes shows similarity to a male gamete-specific protein of lily (*Lilium longiflorum*) and is expressed predominantly in male sex organs, suggesting that this gene has a male reproductive function.

In light of this evidence, Ishizaki et al. (2002) have suggested that the Y chromosome evolved by co-amplification of protein-coding genes with unique repeat sequences. Y chromosomes are different from other chromosomes because they do not undergo recombination. Yamato et al. (2007) reported the gene organization of the Y chromosome in *M. polymorpha*. On the 10 Mb Y chromosome, 64 genes were identified, 14 of which were detected only in the male genome. These genes are expressed in reproductive organs but not in the vegetative thalli, suggesting their participation in male reproductive functions. Another 40 genes on the Y chromosome are expressed in the thalli and the male sexual organs. At least six of these genes have diverged from their X-linked counterparts that are expressed in the thalli and sexual organs in female plants, suggesting that these X- and Y-linked genes have essential cellular functions. These findings indicate that the Y and X chromosomes share the same ancestral autosome and support the prediction that in a haploid organism essential genes on the sex chromosomes are more likely to persist than in a diploid organism (Yamato et al. 2007; see also Rensing et al. 2013, this volume).

11.2.2 Sex Determination in Ferns

11.2.2.1 Homosporous Ferns (Main Model *Ceratopteris richardii*)

Environmental sex determination (ESD) is the rule in homosporous ferns (Korpelainen 1998). Sporophytes (the diploid generation) produce free-living gametophytes (the haploid generation) that are potentially bisexual, i.e., they can bear female (archegonia) and male (antheridia) reproductive organs (gametangia) on the same individual. Several environmental factors influence sexual expression. One class of factors, maleness-inducing pheromones, or 'antheridiogens' (Schneller et al. 1990), has received much attention. Antheridiogens are gibberellin-like compounds secreted by large, female or bisexual gametophytes that reduce growth and induce maleness in nearby asexual gametophytes.

Other environmental factors that affect sex in ferns, such as nutrients, have received very little attention in most reviews of sex expression in fern gametophytes (Cousens et al. 1988; Korpelainen 1998; see however Raghavan 1989). Three well known patterns described in gametophyte biology illustrate the plasticity of sex expression in homosporous ferns. Firstly, ontogenetic changes in gametophyte sex undergo sequential hermaphroditism, i.e., the unidirectional change in sex during ontogeny. Klekowski (1969) distinguished four types of sequential hermaphroditism: male to bisexual, male to female to bisexual, female to bisexual, and step-wise change from male through bisexual to female. Secondly, the coexistence of gametophytes of different sexes within a population has been repeatedly reported (Klekowski 1969; Hamilton and Lloyd 1991). Thirdly, harsh growing conditions, such as poor substrate or high density, are known to lead to small and male gametophytes (Miller 1968; Rubin and Paolillo 1983; Rubin et al. 1985; Korpelainen 1995; Huang et al. 2004).

Ceratopteris richardii has been the main fern model for sex determination studies because of its rapid life cycle and easy cultivation (Hickok et al. 1987). Haploid gametophytes of the fern *Ceratopteris* are either male or hermaphroditic. The determinant of sex type is the pheromone antheridiogen, which is secreted by the hermaphrodite and directs male development of young, sexually undetermined gametophytes. Three phenotypic classes of mutations that affect sex-determination have been isolated and include the hermaphroditic (*her*), the transformer (*tra*) and feminization (*fem*) mutations. Eberle and Banks (1996) performed linkage analysis and tests of epistasis among the different mutants to assess the possible interactions among these putative genes. Their results indicate that sex determination in *Ceratopteris* involves at least seven interacting genes, which may interact with antheridiogen, the primary sex-determining signal.

Two models describing how antheridiogen may influence the activity states of these genes and the sex of the gametophyte have been suggested (Talmor-Neiman et al. 2006). To understand how antheridiogen represses the development of female traits at the genetic level, 16 new mutations that feminize the gametophyte in the presence of antheridiogen were characterized (Strain et al. 2001). Seven are tightly linked to the *FEM1* locus, which was previously described by Eberle and Banks (1996). Nine other mutations concern another locus *NOTCHLESS1* (*NOT1*), with mutant plants lacking a meristem notch. Some *not1* mutations also affect sporophyte development when homozygous, indicating that the *not1* mutations are recessive and that *NOT1* is required for normal sporophyte development. The epistatic interactions among *FEM1*, *NOT1*, and other sex-determining genes have been revealed (Strain et al. 2001). According to the current model of sex determination in *Ceratopteris*, the presence of antheridiogen leads to the activation of the *FEM1* gene, which not only promotes the differentiation of male traits, but also represses female development by activating the *NOT1* gene. *NOT1* represses the *TRA* genes necessary for the development of female traits in the gametophyte (Strain et al. 2001).

Kamachi et al. (2007) examined the effect of photomorphogenically active light on antheridiogen-induced male development of gametophytes of *Ceratopteris richardii*.

These workers also revealed the latent antheridiogen-signal transduction pathway. Although blue light did not affect sensitivity to antheridiogen in wild-type gametophytes, it was found that the gametophytes of the *her1* mutant, which are insensitive to antheridiogen, developed into males when grown under blue light in the presence of antheridiogen. The latent antheridiogen-signal transduction pathway is therefore probably activated by blue light. Red light, on the other hand, suppressed male development, and the action of red light seems to dominate that of blue light. The results of experiments with a photomorphogenic mutant also suggest that phytochrome may be involved in the action of red light (Kamachi et al. 2007). Interestingly, while certain *KNOX* genes function similarly in the development of both seed plant and fern sporophyte meristems despite their differences in structure, *KNOX* gene expression is not required for the development of the fern gametophyte. It is therefore supposed that the sporophyte and gametophyte meristems of ferns are not regulated by the same developmental mechanisms at the molecular level (Sano et al. 2005). A systemic gene-silencing method suitable for high throughput, reverse genetic analyses of gene function in fern gametophytes has already been developed (Rutherford et al. 2004). This also opens new possibilities for research into sex determination in ferns.

11.2.2.2 The Heterosporous Ferns and Lycophytes

In contrast to homosporous ferns, heterosporous ferns (e.g., *Marsilea, Azolla*) and some lycophytes (e.g., *Selaginella* and *Isoetes*) form two different types of spores: microspores, which produce male gametophytes, and megaspores which produce female gametophytes. These spores are formed in special structures (microsporangia and megasporangia). In contrast to dioecious bryophytes where the gametophyte type is determined by its genome, the gametophyte type in heterosporous ferns and lycopods is controlled by the type of spore from which it emerged (reviewed in Tanurdzic and Banks 2004). A model for the evolution of heterospory from homospory has been suggested by Haig and Westoby (1988). This model has three phases: (1) a gradual increase in spore size in a homosporous population, (2) the sudden introduction of smaller microspores from sporophytes reproducing predominantly as males, (3) the subsequent divergence in size and specialization of the two spore types. The model proposes that haploid dioecy evolved from pre-existing mechanisms of sex determination, and that endosporic development of megagametophytes arose as a consequence of an increased dependence on spore food reserves for reproduction (Haig and Westoby 1988).

Among the lycophytes, *Selaginella* has perhaps the greatest potential as a useful comparative system for the study of sex determination and the mechanisms leading to the origin of heterospory in plants because many species in this genus have small genome sizes permitting the efficient use of molecular genetic techniques (reviewed by Tanurdzic and Banks 2004).

11.2.3 Gymnosperms

11.2.3.1 Monoecious Gymnosperms

Most gymnosperms are monoecious (Givnish 1980). There is a striking correlation between the breeding system and the mechanism of seed dispersal. Almost all gymnosperms are wind pollinated, except cycads, which are beetle-pollinated. Monoecious gymnosperms also rely on wind to disperse their seeds. In contrast, most dioecious gymnosperms possess fleshy fruits and their seeds are dispersed by animals (Givnish 1980). Exceptions to this rule have been reported; for example, trioecious and dioecious populations occur in *Pinus edulis* (Floyd 1983).

11.2.3.2 Dioecious Gymnosperms

Phylogenetic analyses indicate that dioecy may have been the ancestral condition in gymnosperms (Mathews et al. 2010). In general though, relatively little attention has been paid to the study of sex determination in dioecious gymnosperms. The main reason for this is likely due to long generation times and/or the large size of both plants and their genomes, which complicate genetic studies (see Leitch and Leitch 2013, this volume). However, a few species have attracted the attention of researchers because of their phylogenetic significance (see also Sect. 14.5 in Murray 2013, this volume).

Ginkgo

The staminate and ovulate trees of *Ginkgo biloba* possess the same number of chromosomes ($2n = 24$) and share almost identical chromosome morphology. The only difference is that in the ovulate trees; four chromosomes of the somatic complement have satellites while in the staminate tree only three chromosomes have satellites. In the male plant, the pair of short sub-telocentric chromosomes, only one of which has a satellite, is believed to be the sex chromosomes. An XY type of sex determination is assumed since the male possesses a heteromorphic pair of chromosomes (Lee 1954). To establish the necessary molecular tools to understand the evolution of seeds and pollen, Brenner et al. (2005) created a cDNA library and an EST dataset from the reproductive structures of male (microsporangiate), female (megasporangiate), and vegetative organs (leaves) of *Ginkgo biloba*. The analysis of this EST database from *G. biloba* has revealed genes that are potentially unique to gymnosperms. Many of these genes display a degree of homology with fully sequenced clones from the ESTs of a cycad. Other *Ginkgo* ESTs were found to be similar to developmental regulators in higher plants. This work has set the stage for future studies on *Ginkgo* as a means to better understand seed and pollen evolution, and resolve the ambiguous phylogenetic relationship of *G. biloba* to other gymnosperms.

Cycads

In *Cycas pectinata*, the male and female plants have the same number of chromosomes ($2n = 22$) with almost identical chromosome morphology. The only difference is that in the female plant two chromosomes of the somatic complement (pair III) have satellites, while in the male the same pair is heteromorphic and only one of its members has a satellite. This distinction becomes clearly visible when the two types of haploid complements are observed in pollen mitosis: one type possesses a satellite chromosome, and the other does not (Abraham and Mathew 1962). The construction of a cDNA library and an extensive EST study has already been started in *C. rumphii* (Brenner et al. 2003a).

11.2.4 Angiosperm Plants

The ancestral traits of angiosperm flowers are not yet clear; several hypotheses attempt to explain the origin of the hermaphrodite flower, which is currently the most wide spread flower type. According to the "Out Of Male (OOM)/Out Of Female (OOF) hypotheses," hermaphrodite flowers developed as a modification of strobili containing both male and female flowers (Theissen and Becker 2004). According to the "Mostly Male (MM) theory," the hermaphrodite flower developed from the male flowers of a gymnosperm ancestor by ectopic formation of ovules (Frohlich and Parker 2000). Different hypotheses for the origin of the angiosperm hermaphrodite flower make different predictions concerning the overlap between the genes expressed in the male and female cones of gymnosperms and the genes expressed in the hermaphrodite flowers of angiosperms. The Mostly Male theory predicts that, of genes expressed primarily in male versus female gymnosperm cones, an excess of male orthologs will be expressed in flowers, excluding ovules, while Out Of Male and Out Of Female theories predict no such excess. Data obtained by Tavares et al. (2010) fit better with the Out Of Male and Out of Female theories. However, it is doubtful that female and male ancestral characteristics of the angiosperm flower can be inferred from the number of expressed genes shared by female and male tissues, as differences between sexes are due to only a few genes or are quantitative (Tavares et al. 2010).

11.2.4.1 Monoecious Angiosperm Plants

Cucumis melo* and *C. sativus

Important data concerning the possible mechanisms of sex determination in Cucurbitaceae have been obtained in melons (*Cucumis melo*). In this mostly monoecious species, sex determination is governed by the genes *andromonoecious* (*a*) and *gynoecious* (*g*). The dominant allele of the *a* locus (*CmACS-7* gene; 1-aminocyclopropane-1-carboxylic acid synthase) results in an arrest of stamen development (Boualem et al. 2008), while the dominant allele of the *g* locus causes an arrest of gynoecium development. Monoecious (*A-G-*) plants bear male flowers on the main stem and andromonoecious (*aaG-*) plants bear female or hermaphrodite flowers on the axillary branches. Gynoecious (*AAgg*) and hermaphrodite individuals (*aagg*) bear only female or hermaphrodite flowers respectively. The insertion of the Gyno-hAT transposon in the proximity of the *g* gene (*CmWIP1*) was shown to be a cause of the gynoecious phenotype of several lines (*G* to *g* change by hypermethylation of the promoter of the *g* gene, i.e., *CmWIP1*). The occasional presence of flowers with stamens and reduced ovaries suggests that DNA hypermethylation of *CmWIP1* can be reduced during somatic development of gynoecious plants (Martin et al. 2009). Surprisingly, both *CmACS-7* and its homolog from *C. sativus* are specifically expressed in female buds. The role of 1-aminocyclopropane-1-carboxylic acid synthase in anther arrest seems to be indirect and inter-organ communication is probably responsible for anther arrest (Boualem et al. 2009).

An analysis of the whole genome sequence of *C. sativus* revealed that the evolution of unisexual flowers in cucurbits may have involved the acquisition of two ethylene-responsive elements (AWTTCAAA) and one flower meristem identity gene *LEAFY*-responsive element (CCAATGT) of the *ACS* genes (Huang et al. 2009). Extensive EST analysis in unisexual and bisexual flower buds (using 454 sequencing) showed that six auxin-related genes (auxin can regulate sex expression by stimulating ethylene production) and three short-chain dehydrogenase or reductase genes (homologs to the sex determination gene *ts2* in maize) are more highly expressed in unisexual flowers then in hermaphrodite flowers (Huang et al. 2009).

Zea mays

The formation of unisexual flowers in maize requires the selective elimination and sexual maturation of floral organs in an initially bisexual floral meristem. Elimination of pistil primordia occurs in the primary and secondary florets of tassel spikelets and in the secondary florets of ear spikelets. Ill-fated pistil cells undergo a cell death process associated with nuclear degeneration in a specific spatial-temporal pattern that begins in the subepidermis, eventually aborting the entire organ. The sex determination genes *tasselseed1* (*ts1*) and *tasselseed2* (*ts2*) are required for death of pistil cells, and *ts1* is required for the accumulation of *ts2* mRNA in pistil cells. All pistil primordia express *ts2* RNA but functional pistils found in ear spikelets are protected from cell death by the action of the *silkless1* (*sk1*) gene, which blocks tasselseed-induced cell death in the pistil primordia of primary ear florets. This basic model for the control of pistil fate by the action of the *ts1-ts2-sk1* pathway was proposed by Calderon-Urrea and Dellaporta (1999). *TASSELSEED2* converts steroids with specificities found at positions 3 and 17, and several dicarbonyl and quinone compounds, thus establishing *TASSELSEED2* as a plant 3beta/17beta-hydroxysteroid dehydrogenase and carbonyl/quinone reductase. Taken together, the genetic data and the substrate specificities determined suggest that *TS2* converts specific plant compounds and acts as a pre-receptor control mechanism in a manner similar to that of mammalian hydroxysteroid dehydrogenases (Wu et al. 2007). Later, studies by Parkinson et al. (2007) identified the role of two other genes in maize sex determination. Genes *required to maintain repression (rm6)* and *mediator of paramutation 1* (*mop1; putative RNA-dependent RNA polymerase*) are involved in the suppression of the expression of silkless in the male inflorescence. *Rmr6* maintains maize's monoecious pattern of sex determination by restricting the function of the pistil-protecting factor *SILKLESS1* from the apical inflorescence (Parkinson et al. 2007). The exact mechanism leading from *TASSELSEED2* action to proper tassel formation is still unknown. It is, however, known that further genes are involved in stamen development promotion and gynoecium suppression: indeterminate *spikelet1* (known also as *Tasselseed6*) genes and the group of Squamosa-promoter Binding Protein (SBP) box containing genes. Intron-exon structures as well as phylogenetic data support the division of these family members into six groups. The SBP-box genes upregulated in feminized tassels fall into two groups (out of six groups of SBP-box genes that have been distinguished in *Zea mays* according to phylogenetic analysis) that share common structural motifs and include the presence of a target site for miR156. Small RNA blots showed that miR156 levels are decreased in both *mop1* and *ts1* mutants. While there is a correlation between miR156 levels and SBP-box gene transcript levels, this correlation is not absolute, and thus it is hypothesized that decreased levels of miR156 may provide competency for SBP-box gene upregulation by other common factors yet to be identified. A model that suggests a putative link between *ts1*, *ts2*, *ts4*, *Ts6*, and *mop1* in the sex-determination pathway was put forward by Hultquist and Dorweiler (2008). Progress has also been made in the study of the molecular mechanism of action of the putative master gene of sex determination in maize.

Acosta et al. (2009) positionally cloned and characterized the function of the sex determination gene *ts1*. The TS1 protein encodes a plastid-targeted lipoxygenase with predicted 13-lipoxygenase specificity, which suggests that TS1 may be involved in the biosynthesis of the plant hormone jasmonic acid. In the absence of a functional *ts1* gene, lipoxygenase activity was missing and endogenous jasmonic acid concentrations were reduced in developing inflorescences. Application of jasmonic acid to developing inflorescences rescued stamen development in mutant *ts1* and *ts2* inflorescences, revealing a role for jasmonic acid in male flower development in maize (Acosta et al. 2009; Browse 2009). This suggests that jasmonic acid is likely to be involved in more complex effects than just the control of *ts2*. An updated model of sex determination in *Zea* is summarized in Fig. 11.1, which represents improvements to the original scheme suggested by Hultquist and Dorweiler (2008).

11.2.4.2 Dioecious Angiosperm Plants

Genus *Silene*

Silene latifolia: A Model for the Study of Sex Determination

Silene latifolia (Fig. 11.2a, b) is probably the best-studied plant sexual model. Chromosomes of *S. latifolia* (formerly *Melandrium album*) were first described in 1923 (Blackburn 1923; Winge 1923). Shortly after the species was used by Correns to study sex ratio bias (reviewed in Correns 1928a). The nuclear genomes of *S. latifolia* and *S. dioica* are relatively large and arranged into 12 chromosome pairs (see Fig. 11.3a–d). The pair of sex chromosomes is the largest in the genome: the Y is 1.4 times longer than the X chromosome. Deletion mutants have been an important tool in studies of sex determination in *S. latifolia*. For example, the use of deletion mutants has enabled the physical mapping and functional characterization of regions of interest within the sex chromosomes. Historically, spontaneous aberrant Y chromosomes were first studied by Westergaard (1946). By analysing different types of aberrations, he was able to conclude that in *S. latifolia* three different regions within the Y chromosome were important for correct sex expression in males. The upper region of the q-arm is critical for the suppression of gynoecium formation, the medium region of the Y is necessary for male promotion, and a portion of the q-arm is needed for male fertility (Westergaard 1946).

The default sex in *S. latifolia* is female: when the Y is completely missing, flowers form only pistils. Recently, large scale deletions (disruptions) of the Y chromosome were introduced by means of either X-ray or gamma-irradiation. Lebel-Hardenack et al. (2002) irradiated pollen grains using a Siefert X-ray machine, while Farbos et al. (1999) used ^{60}Co as a source of gamma rays. These studies, in addition to the deletion mutants observed already by Westergaard (1946), revealed hermaphrodites and infertile males. This can be taken as confirmation of Westergaard's model of the Y chromosome in *S. latifolia*. The model of Y chromosome organization was improved by Zluvova et al. (2007), who showed the existence of male fertility genes close to the stamen promoter.

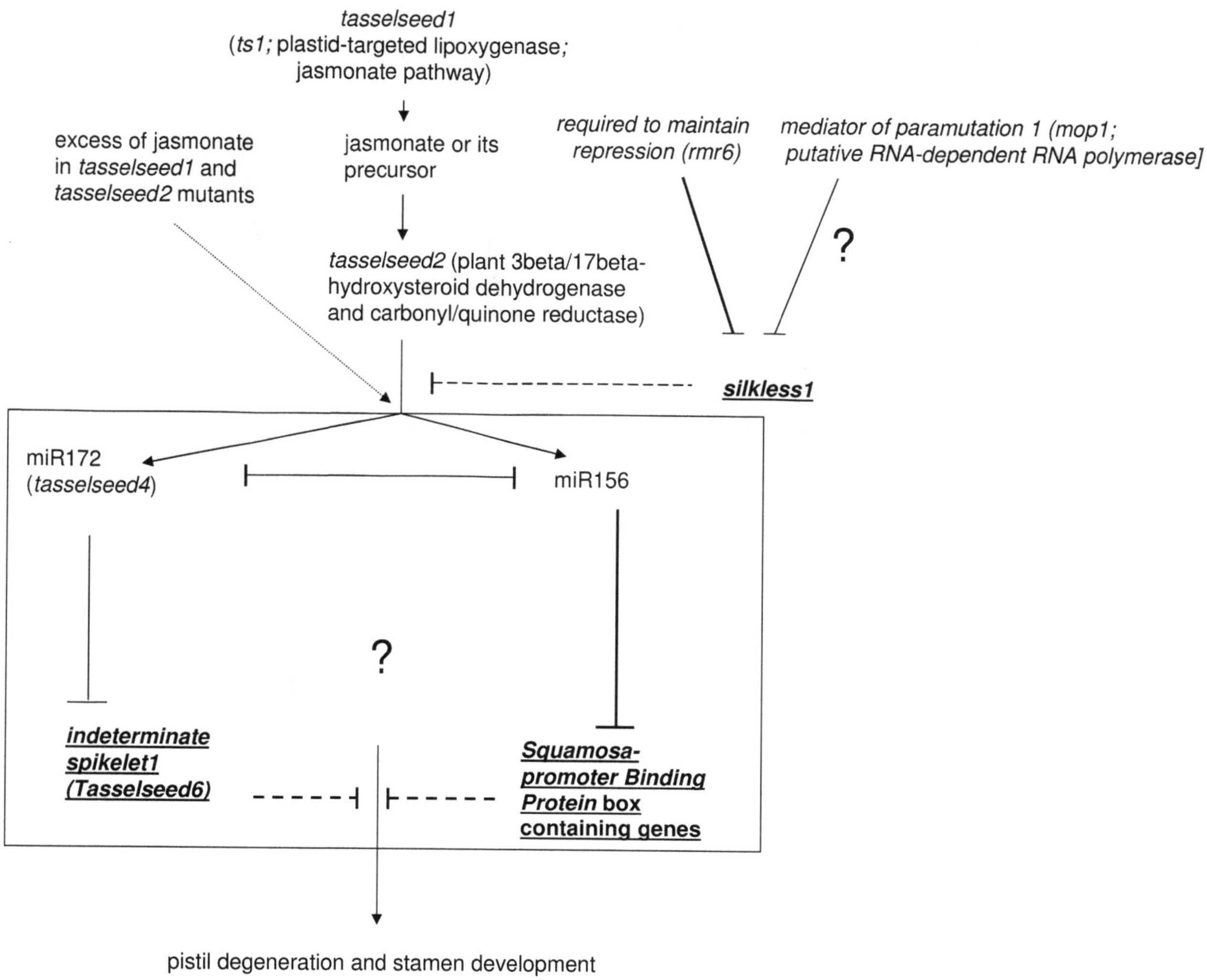

Fig. 11.1 Sex determination in maize (*Zea mays*). This scheme represents an updated "*tasselseed* based" control pathway of sex determination in maize. Only the variant that occurs in tassel is displayed. The genes that push development in the male direction are written in plain text whereas genes pushing development in the female direction are *underlined*. The names of gene products and names of mutants that led to gene identification are in *brackets*. *Broken lines* (e.g., from *silkless1*) denote genes not active in male inflorescences. *The box* symbolizes that the control pathway from *tasselseed2* to pistil abortion and stamen promotion remains mostly unknown, though some genes have been identified. The *dotted line* symbolizes an artificial treatment. As apparent from the scheme, the role of jasmonate is probably more diverse than simply controlling *tasselseed2* expression (modified according to Hultquist and Dorweiler 2008)

At present, few experimental data are available concerning the mechanism of gynoecium suppression in males of *S. latifolia*. Histological studies show a reduction of cell division in the central part of the male flower meristem (Matsunaga et al. 2004) while molecular studies have revealed the role of homologs of *Arabidopsis thaliana SHOOTMERISTEMLESS* (*STM*) and *CUP SHAPED COTYLEDON* (*CUC*) 1 and *CUC2* genes in the arrest of gynoecium development in *S. latifolia* males (Zluvova et al. 2006). The data of Matsunaga et al. (2004) and Zluvova et al. (2006) suggest that the absence of *STM* and the presence of *CUC* 1 and *CUC2* transcripts in the central part of the male flower meristem are the cause of reduced meristematic activity in this region. Independent of this pathway, gynoecium development in *S. latifolia* is also suppressed by the action of the *CLAVATA1* gene, a putative member of the *CLAVATA-WUSCHEL* pathway (Koizumi et al. 2010). The results of Kazama et al. (2009) also indicate a possible role of *SUPERMAN*-like gene in the suppression of anther development in *S. latifolia* females.

Silene latifolia: A Model for the Study of Evolutionary Dynamics of Sex Chromosomes

While the majority of findings concerning sex chromosome evolution have come from research performed in human and animal models, a few plant species appear to be more suitable models for sex chromosome evolution owing to their recently evolved sex chromosomes (Table 11.1). Sex chromosomes, along with B chromosomes, are amongst

Fig. 11.2 Examples of model dioecious flowering plants. (**a**) *Silene latifolia* female flower, (**b**) *S. latifolia* male flower, (**c**) *S. colpophylla* female flower, (**d**) *S. colpophylla* male flower, (**e**) *Rumex acetosa* female flower bud section, (**f**) *R. acetosa* male flower bud section. *Bars* represent 5 mm (a–d) or 5 μm (e–f)

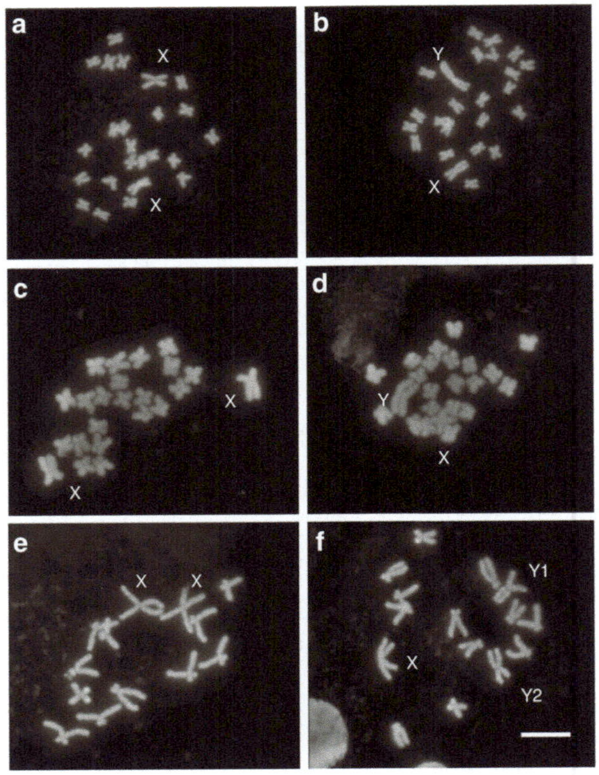

Fig. 11.3 Examples of chromosome preparations of dioecious species with heteromorphic sex chromosomes. (**a**) *Silene latifolia,* female metaphase, (**b**) *S. latifolia,* male metaphase, (**c**) *S. dioica,* female metaphase, (**d**) *S. dioica,* male metaphase, (**e**) *Rumex acetosa,* female metaphase, (**f**) *R. acetosa,* male metaphase. The *bar* represents 5 μm

the few parts of a plant genome that are feasible for chromosome painting techniques. While in animals, ordinary autosomes have specific patterns of DNA organization that allow researchers to distinguish individual chromosomes using complex probes, plant genomes are relatively homogenous and generally have few chromosome-specific DNA structures and sequences. In *S. latifolia*, microdissected sex chromosomes were used by Scutt et al. (1997) to investigate their genomic organization. The experiment was based on a semi-automatic technique of ablation of chromosomes and the mechanical transfer of chromosomes of interest to a membrane (polyester disk). Both X and Y chromosomes were used for DOP-PCR amplification and PCR products were subsequently labelled for FISH experiments (DOP-PCR is the method of choice for amplification of anonymous DNA with partially degenerated primers (Telenius et al. 1992)). Although sex chromosomes in *S. latifolia* are significantly different in their structure, the use of both probes (X- and Y-derived) revealed no sex chromosome-specific signal pattern. Similar data were obtained from experiments by Matsunaga et al. (1999).

A novel approach, which combined microdissection techniques and FISH procedures, was used by Hobza et al. (2004). To avoid any impurities during the dissection process, all manipulations were carried out with a laser beam using the PALM MicroLaser system (Fig. 11.4). The hybridization procedure was significantly shortened (1 h), and the amount of probe was decreased to a low concentration (30 ng/slide). Specific signals were observed using both X and Y probes. The differential labelling patterns of *S. latifolia* sex chromosomes under specific FISH conditions show rapid evolution of repetitive elements in the early stages of sex chromosome divergence (Hobza et al. 2004).

Manual dissection of sex chromosomes was also successfully applied in experiments by Delichere et al. (1999), leading to the discovery of the first active gene to be isolated from a plant Y chromosome. Hernould et al. (1997) also performed microdissection, using an inverted microscope with extended microneedles operated by an electric micromanipulator. The glass tips of microneedles carrying chromosome fragments were broken off and pooled in an Eppendorf tube for subsequent experiments. Delichere et al. (1999) used 10 microdissected Y chromosomes as a template for DOP-PCR. The amplified DNA was used as a probe for screening a premeiotic male flower cDNA library to select Y-linked expressed sequences. To verify Y-linkage, segregation analysis was performed. Individual Y chromosome-derived clones were hybridized with restricted

Table 11.1 Comparison of characteristic properties of the most studied plant dioecious models (chrs = chromosomes, F = female, M = male, My = million years)

	Carica papaya L.	*Silene latifolia* Poiret.	*Rumex acetosa* L.
Common name	Papaya	White campion	Sorrel
Family	*Caricaceae*	*Caryophyllaceae*	*Polygonaceae*
Genome size	1C = 0.372 pg	1C = 2.86 pg	1C = 3.55 pg
Karyotype	$2n$ = 18 sex chromosomes homomorphic	$2n$ = 24 (11pairs of autosomes plus 1 pair of sex chromosomes)	$2n$ = 14(15) (6 pairs of autosomes plus XX or XY_1Y_2)
Heterogametic sex	Male	Male	Male
Sex chrs (F/M)	Not detected	XX/XY (Blackburn 1923)	XX/XY_1Y_2 (Kihara and Ono 1923)
Size of sex chrs	Not detected	The largest Y (9 %), the second X (8 %), smaller autosomes	The largest X (13 %), the second Y1 + Y2 (25 %), small autosomes
Chromatin of sex chrs	Non-recombining region of the Y heterochromatic (Zhang et al. 2008)	Standard as in autosomes (Grant et al. 1994)	The Ys are heterochromatic, histone H4-underacetylated (Lengerova and Vyskot 2001)
Age of sex chrs	2–3 My (Zhang et al. 2008)	Less than 5–10 My (Filatov 2005)	15–20 My (Jamilena et al. 2008)
Type of sex determination	Dominant Y in males (mammalian type)	Dominant Y in males (mammalian type)	X/A ratio (F = 1, M = 0.5) (*Drosophila* type)
Accumulation of Y-repeats	In the non-recombining region of the Y	Plastid DNA, DNA repeats, retrotransposons, microsatellites (Kejnovsky et al. 2009)	Y-specific repeats, microsatellites (Jamilena et al. 2008)
Availability of X-genes	Full genome sequence available (Liu et al. 2004)	At least 10 X (and) Y linked publically available genes	No sex chromosome linked genes available yet

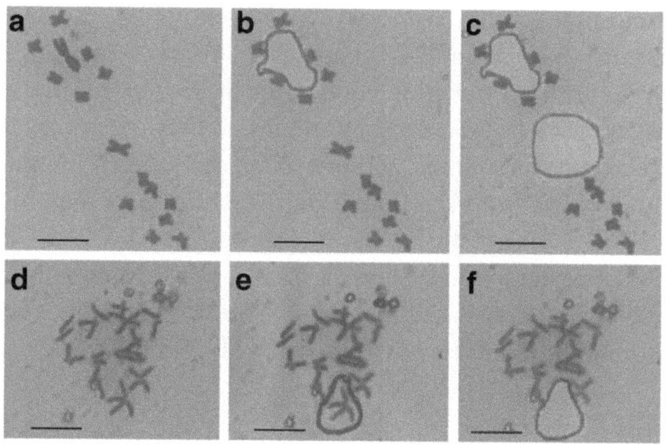

Fig. 11.4 Laser microdissection on *Silene latifolia* and *Rumex acetosa* metaphase chromosomes. The *S. latifolia* Y chromosome is localized under the inverted microscope (**a**). The membrane is cut around the selected chromosome using a laser microbeam, and the chromosome Y (**b**) and X (**c**) is transferred into the cap of a PCR tube. Similarly, the *R. acetosa* X chromosome is selected and separated from the rest of the chromosomes by microdissection (**d–f**)

genomic DNA from male and female parental plants as well as their progeny. Out of the 115 clones selected after hybridization of complex Y probes with the cDNA library, only five clones revealed sex linkage after segregation experiments. A later study reported a level of DNA polymorphism in the Y-linked copy of the gene *SlX1/SlY1* that was twenty times lower than the X-linked copy. These data were the first to suggest that processes involved in Y chromosome degeneration were also acting in the relatively young chromosomes of *S. latifolia* (Filatov et al. 2000).

In spite of their relatively young age it was shown that there is a similar gradient in silent site divergence between the X and Y copies of sex-linked genes on the sex chromosomes *S. latifolia* (Nicolas et al. 2005; Bergero et al. 2007). *Silene latifolia* Y-linked genes tend to evolve faster at the protein level than their X-linked homologs, and they have lower expression levels. Analysis of several Y-linked gene introns suggest they act as sites for transposable-element accumulation, which likely accounts for their increased length (Marais et al. 2008). These signs of degeneration are similar to those observed in animal Y-linked and neo-Y chromosome genes. Cermak et al. (2008) carried out a global survey of all of the major types of transposable elements in *Silene latifolia*. The localization of elements by FISH revealed that most of the Copia elements had accumulated on the Y chromosome. Surprisingly, one type of Gypsy element, which was similar to the Ogre elements known from legumes, was almost absent on the Y chromosome but otherwise uniformly distributed on all chromosomes. Other types of elements were ubiquitous on all chromosomes. Moreover, Cermak et al. (2008) isolated and characterized two new tandem repeats. One of them, STAR-C, was localized at the centromeres of all chromosomes except the Y chromosome, where it was present on the p-arm. Its variant, STAR-Y, which carries a small deletion, was specifically localized on the q-arm of the Y chromosome. FISH analysis of other *Silene* species revealed that some elements (e.g., Ogre-like elements) are confined to the section Elisanthe while others (e.g., Copia or Athila-like elements) are also present in more distantly related species. The unique pattern of repeat distribution found on the Y chromosome, where some elements have accumulated while other elements are conspicuously absent, probably reflects the different forces shaping the evolution of the Y chromosome (Cermak et al. 2008). Kubat et al. (2008) have shown that microsatellites accumulate in the q-arm of the Y chromosome (the arm containing the pseudoautosomal region) and this agrees with the findings of other authors (Morgante et al. 2002) showing that microsatellites are preferentially located in non-repetitive sequences.

However, repetitive sequences are not the only types of DNA to accumulate in the non-recombinant part of plant Y chromosomes. Sequence analysis has revealed that one of the Y chromosome-derived BACs contains part of the plastid genome, indicating that these plastid sequences have been transferred to the Y chromosome and may also contribute to its large size. Kejnovsky et al. (2006) found that plastid sequences located on the Y chromosome had higher rates of divergence in non-genic regions than in genic regions, which showed only very low (max 0.9 %) divergence from their plastid homologs.

The study of the Y chromosome-derived library has also revealed a case of horizontal gene transfer of a DNA fragment from the bacterium—*Ralstonia solanacearum* to the genome of *S. latifolia*. The homologs of this fragment (*MK14*) contain sequences that show similarities to a gene coding bacterial sulphate adenylyltransferase (*CysN*) and to a gene encoding uroporphyrin-III C-methyltransferase (*nirE*) (Talianova 2009). The reaction catalysed by sulphate adenylyltransferase constitutes the first enzymatic step in sulphate utilization following the uptake of sulphate (Leyh et al. 1992). Uroporphyrin-III C-methyltransferase in bacteria is involved in the biosynthesis of corrinoids such as vitamin B12, sirohaem and coenzyme F430 (Vévodová et al. 2004). The homologs of this fragment were subsequently found also in species that do not possess sex chromosomes, suggesting that the Y chromosome is unlikely to be more prone to the accumulation of horizontally acquired genes than regular autosomes.

Other *Silene Species*

Silene is a large genus, where the majority of species possess 12 pairs of chromosomes. In some dioecious species there is one pair of sex chromosomes plus 11 pairs of autosomes. This pattern which is supported by molecular data (Nicolas et al. 2005; Bergero et al. 2007), suggests that one pair of autosomes evolved into the pair of sex chromosomes in subgenus *Behenantha*, section *Melandrium* (classification according to Rautenberg et al. 2010). In contrast, the origin of sex chromosomes in subgenus *Silene* (classification according to Eggens et al. 2007, and Rautenberg et al. 2010) is probably different as indicated by the phylogenetic and genetic mapping data in *S. colpophylla*. Though many species share the same number of chromosomes, there are big differences in genome size among *Silene* species. For example, *S. vulgaris* and *S. pendula* belong to a group of small-genomed species (about 2 pg/C), while *S. latifolia* and *S. chalcedonica* possess relatively large genomes (3–5 pg/C, respectively). These differences raise questions about the mechanisms that have contributed to genome size evolution. Recent data indicate that there are large blocks of subtelomeric heterochromatin missing in some of the smaller genomes, which may account for much of these differences (Cermak et al. 2008). Subgenus *Behenantha* contains only five dioecious species: *S. latifolia*, *S. dioica*, *S. diclinis*, *S. marizii*, and *S. heuffelii*. Subgenus *Silene* contains several dioecious and subdioecious species, the best studied being *S. otites* and *S. colpophylla* (Fig. 11.2c, d) of the former section *Otites*, Wrigley (1986), now subsection *Otites* of the section *Siphonomorpha*, according to Oxelman (2010). Dioecy in section *Melandrium* is of monophyletic origin: the same pair of autosomes evolved into sex chromosomes in all species. In section *Otites*, genetic and phylogenetic analyses have been conducted to understand the sex-determining system and the origin of sex chromosomes in *S. colpophylla* (Mrackova et al. 2008). Here, it was shown that genes that are sex-linked in *S. latifolia* are also linked to each other in *S. colpophylla*, but they are not sex-linked. This finding demonstrates that the sex chromosomes in *S. colpophylla* (which are homomorphic) evolved from a different pair of autosomes than in *S. latifolia*. Phylogenetic analyses also support the view that the sex determination system of *S. colpophylla*, although it is XX/XY (similarly to section *Melandrium*), evolved independently in section *Otites* (Mrackova et al. 2008).

Genus *Rumex*

Various types of reproductive systems occur in *Rumex*: hermaphroditism, polygamy, gynodioecy, monoecy and dioecy (reviewed in Navajas-Pérez et al. 2005). In dioecious *Rumex* species, two different sex-chromosomal systems and sex-determining mechanisms have been described: XX/XY with an active Y chromosome (e.g., *Rumex acetosella*) and XX/XY_1Y_2 with sex determination based on the X/A ratio (e.g., *Rumex acetosa*, Figs. 11.2e, f and 11.3e, f). There is one exceptional species, *Rumex hastatulus*, which has two chromosomal "races": the Texas race possessing XX/XY system and the North Carolina race with an $XX/X\ Y_1Y_2$ system. In this species, the X/A ratio controls sex determination, but the presence of the Y chromosome is necessary for male fertility (Smith 1963). In *R. acetosa* repetitive sequence similarity between both Y chromosomes suggests that they probably originated from one Y chromosome that underwent centromere fission and gave rise to a pair of metacentric chromosomes possessing identical chromosomal arms (isochromosomes).

These isochromosomes were subsequently modified by deletions (Rejon et al. 1994). A recent phylogenetic study (Navajas-Pérez et al. 2005) indicates that all dioecious *Rumex* species evolved from a common hermaphroditic ancestor. The switch from a sex-determining mechanism based on the active role of the Y chromosome to a mechanism based on the X/A ratio occurred at least twice (Navajas-Pérez et al. 2005). The role of the X/A ratio in the sex determination of *R. acetosa* (reviewed by Parker and Clark 1991) resembles the sex-determining system of *Drosophila*, where the primary genetic sex-determining signal is provided by the ratio of X-linked genes to autosomal genes (Pomiankowski et al. 2004).

Hemp (*Cannabis sativa*)

Cannabis sativa or hemp is one of the few dioecious plant species possessing heteromorphic sex chromosomes, albeit the size difference between the X and Y chromosome is small (reviewed in Peil et al. 2003). Males are heterogametic (XY) and females homogametic (XX). The sex-determining system is the *Drosophila* type: the ratio of Xs to autosomes determines the sex rather than the active Y chromosome (revieved by Parker and Clark 1991). In contrast to other dioecious models (e.g., *Silene latifolia* and *Rumex acetosa*) the Y chromosome of *C. sativa* probably does not contain the genes necessary for male fertility. After stamen induction by ethylene synthesis inhibiting drugs (silver nitrate and silver thiosulphate anionic complex), fertile stamens develop in genetically female plants of hemp (Mohan Ram and Sett 1982). Sex determination in hemp is therefore probably under control of plant hormones, as changes in their levels often lead to sex changes (Chailakhyan and Khryanin 1978, 1979). Some features of Y chromosome degeneration have been found in hemp, for example, an accumulation of LINE-like retrotransposons at the terminal region of the longer arm of the Y chromosome has been reported (Sakamoto et al. 2000). Several genes that are involved in sex determination and/or sexual dimorphism have been identified using cDNA AFLP (Moliterni et al. 2004).

Hop (*Humulus lupulus*)

In spite of its agronomical importance, relatively little is known about the sex determination mechanism in hops. It is known that hops possess heteromorphic sex chromosomes and sex determination of XX/XY_1Y_2 i.e., sex determination is based on the ratio of the dosage of X chromosomes and autosomes (Parker and Clark 1991). Shephard et al. (2000) showed that obvious sex-related differences were already present in the plant at the stage of initiation of inflorescences suggesting that the genes involved in sex determination are probably already at work in advance of floral organogenesis (Shephard et al. 2000). Numerous sex-specific markers have been isolated in *H. lupulus* (Jakse et al. 2008) and a linkage map is available together with cytogenetic methods to distinguish between hop chromosomes (Karlov et al. 2003).

Papaya (*Carica papaya*)

Papaya is a favorite dioecious model because of its small nuclear genome size and importance as a crop. Papaya is polygamic and forms three basic sexual types: males, females, and hermaphrodites. Sex in papaya is determined by a single locus, which may have three different alleles (*M1*—male, *M2*—hermaphrodite, and *m*—female) (Storey 1969). Males must have the *M1* dominant allele, while hermaphrodites possess the *M2* allele and *mm* forms females. The combination of any two dominant alleles leads to seed abortion. Since males are heterogametic and females are homogametic, the sex determination can be classified as the XY type with a dominant Y chromosome. A large amount of DNA polymorphisms near the sex locus led to the recent discovery of the Y chromosome in papaya (Ma et al. 2004). There is a small 4–5 Mb region, called the MSY (male specific region Y), which harbors the primary sex-determining genes. Yu et al. (2007) proposed the hypothesis that recombination suppression in the sex-determining region of the Y chromosome is the result of its close proximity to the centromere.

Meadow-rue (*Thalictrum dioicum*)

Meadow-rue is a common weed in the Ranunculaceae family. The unisexual flowers lack any rudiments of the opposite sex since developmental arrest occurs very early during floral ontogeny. *Thalictrum dioicum* has homomorphic sex chromosomes and sex is genetically determined by a few loci. Males are heterogametic, and the genetic system of sex determination looks like the dominant Y system that is present in *Silene latifolia* and humans. The Y chromosomes in dioecious species of *Thalictrum* are probably not largely degenerate since YY males are viable even when no X copy is present in *T. fendleri* (Kuhn 1939). In dioecious plants like *T. dioicum* and *Spinacia oleracea* (see below), where vestiges of the opposite sex organs are missing, homeotic MADS-box genes are good candidates for sex determination genes. In *T. dioicum*, Di Stilio et al. (2005) isolated the organ identity genes (*Agamous, Apetala3*, and *Pistillata*) and showed that they had been duplicated and had diverged in expression. Efficient virus-induced gene silencing (VIGS), using tobacco rattle virus (TRV) vectors, has been achieved, permitting deep analysis of sex determination mechanisms (Di Stilio et al. 2010).

Spinach (*Spinacia oleracea*)

Spinach is predominantly a dioecious species. However, genetically-determined monoecious lines also exist (Janick and Stevenson 1954), and sex expression also depends on environmental influences. For example, high temperatures promote maleness in monoecious lines (reviewed in Iizuka and Janick 1962). The sex-determining system is XX/XY (reviewed in Haga 1934). X and Y chromosomes are of similar size and morphology, but some Y chromosomes lack a 45S rDNA locus (Lan et al. 2006). Sex determination in *S. oleracea* is based on a single locus with three alleles: X, X^m and Y (Janick and Stevenson 1954). The allele X^m is incompletely dominant because X^mX^m individuals are monoecious, but individuals with X^mX produce a higher proportion of pistillate flowers. The Y allele is completely dominant, and thus XY and X^mY plants show the male phenotype.

The exact molecular basis of the sex determination system in *S. oleracea* is unknown. However, a study of the B class floral identity genes *SpAPETALA3* and *SpPISTILLATA* in *S. oleracea* showed their effect on sexual dimorphism and suggested a role in sex determination (Pfent et al. 2005). The gene *SpAgamous* is involved in sex-specific gynoecium control in spinach (Sather et al. 2005). Interestingly, sexual dimorphism in this species can vary in intensity depending on the presence of a special locus located on the Y chromosome that causes a "bracted male" phenotype showing a habit different from female plants (reviewed in Iizuka and Janick 1962). Indeed, there are many indications that the sex chromosomes of *S. oleracea* are young, but this does not necessarily imply that the sex determination system is also young and it is possible that the sex determination system is derived from monoecy. In this case, a single gene controlling the number and/or position of male and female flowers could have been directly transformed into a sex determining gene by sexually antagonistic selection. Given the small size of the spinach genome, it is suitable for basic genomic studies (Khattak et al. 2006) and recent RNAi technology applying a VIGS approach has permitted a functional genomic strategy to be adopted. Such studies showed how the suppression of B class genes led to sex reversal via a homeotic transformation of stamens into carpels (Sather et al. 2010) confirming the view that sex determination in spinach is based on the regulation of B class gene expression.

Mercurialis annua

Mercurialis annua has a simple genetically-based sex determining system. There seem to be three unlinked loci—*A*/*a*, *B1*/*b1*, and *B2*/*b2*—which act complementarily. The male plants usually need the dominant *A* allele and at least one dominant *B* allele. Moreover, in *Mercurialis* sex can be modulated by exogenously applied growth regulators. In females, the presence of trans-zeatin was shown. An application of auxin leads to masculinisation of female flowers (Hamdi et al. 1987). Khadka et al. (2002) reported the characterization of a previously identified male-specific DNA marker, OPB01-1562, from the diploid dioecious *M. annua*. RNA blot hybridization with OPB01-1562 and MARL-1 detected a transcript that was expressed strongly in the stems and flowers of females but not males. The *M. annua* female-expressed (Mafex) transcript may thus play a role in sex determination (Khadka et al. 2005).

Hybridization and polyploidy are widely believed to be important sources of evolutionary novelty in plant evolution (see for review, Soltis et al. 2010; Fawcett et al. 2013, this volume). Both can lead to novel gene combinations and/or novel patterns of gene expression, which in turn provide the variation on which natural selection can act. Obbard et al. (2006) used nuclear and plastid gene trees, in conjunction with morphological data and genome size measurements, to show that both processes have been important in shaping the evolution of *Mercurialis*. Their results indicate that hexaploid populations of *M. annua*, in which the otherwise rare sexual system androdioecy is common (the occurrence of males and hermaphrodites), is of allopolyploid origin involving hybridization between a monoecious autotetraploid lineage of *M. annua* and the related dioecious diploid species *M. huetii*—an event that brought together the genes for specialist males with those for hermaphrodites (Obbard et al. 2006). Dorken and Pannell (2008) found that the progeny sex ratio is strongly dependent on density, with fewer males produced when plants are grown at low density. This occurred in part because of a flexible adjustment in pollen production by hermaphrodites, which produced more pollen when grown at low density than at high density. These results provide support for the prediction that environmental conditions govern sex ratios through their effects on the relative fertility of unisexual versus hermaphrodite individuals (Dorken and Pannell 2008).

Bryonia dioica: The First Dioecious Model in Which the Genetic Basis of Sex Determination was Elucidated

Most species of the Cucurbitaceae family comprising c. 800 species have unisexual flowers, 460 are monoecious, and 340 are dioecious. Some species produce a mixture of bisexual, female, and male flowers in various intra- and inter-individual patterns, and populations can be andromonoecious, androdioecious, gynomonoecious or gynodioecious (Kocyan et al. 2007). Phylogenetic studies suggest that dioecy is the ancestral state in this family (Zhang et al. 2006). However, as various switches to monoecy or to other types of reproductive systems, including androdioecy, have occurred during the evolution of Cucurbitaceae, it is therefore difficult to precisely ascertain how old the sex chromosomes in a given species are (Renner et al. 2007). The best-studied genus containing dioecious species is *Bryonia*. Genetic crosses between the dioecious *B. dioica* and the monoecious *B. alba* in 1903 provided the first clear evidence for Mendelian inheritance of sexual phenotypes (dioecy) and made *B. dioica* the first organism for which the XY sex-determination was experimentally proven (Correns 1928b). Applying molecular tools to this system, Oyama et al. (2009) showed that the size of the non-recombining region may differ between the north-European and southern-European populations. Important data concerning the possible mechanisms of flower sex determination in monoecious Cucurbitaceae were recently obtained in melons (*C. melo*) and in the cucumber (*C. sativus*, see Sect. 11.2.4.1.1 in this chapter).

Date Palm (*Phoenix dactylifera*)

The date palm is a dioecious species displaying strong dimorphism between pistillate and staminate flowers. Siljak-Yakovlev et al. (1996) described a cytological method using chromomycin staining that demonstrates the occurrence of sex chromosomes based on distinctive nucleolar heterochromatin. This offers the possibility of identifying male and female individuals by a simple analysis of root meristems. This observation has been extended by *in situ* rDNA hybridization, confocal microscopy and dual-label flow cytometry of nuclei (Siljak-Yakovlev et al. 1996). The mechanism of arrest of gynoecium development in the date palm appears to be similar to that in *Silene latifolia*. The earliest sex-related difference in flower buds is observed when the number of cells in the gynoecium of pistillate flowers is higher than that in their staminate counterparts. In the pistillate flower, staminodes (sterile stamens) display precocious arrest of development followed by cell differentiation. In the staminate flower, pistillodes (sterile gynoecium) undergo some degree of differentiation, but their development ceases shortly after the ovule has been initiated. Staminode and pistillode cells exhibit nuclear integrity although they do not show any accumulation of histone H4 gene transcripts. These results suggest that the developmental arrest of sterile sex organs and the subsequent unisexuality of date palm flowers result from a cessation of cell division and precocious cell differentiation rather than from cell death (Daher et al. 2010).

11.3 Sexual Dimorphism

Sexual dimorphism, the sex-specific expression of some genes not involved directly in sex determination, presumably results from different modes of selection operating in males and females: males are limited in their reproductive success by access to mates, whereas females are more limited by resources (Bateman 1948; Jones et al. 2002). In animals, the evolution of sexual dimorphism is primarily driven by competition between males and selection for traits recognized by females as indicators of male fitness (reviewed by Hall et al. 2000). Similar principles may be at work in animal-pollinated plants. In *S. latifolia*, odor-compounds involved in pollinator attraction differ significantly between sexes, suggesting that selection for higher attractiveness among competing males is mediated by the sensory ecology of the pollinator (Waelti et al. 2009). In addition, males produce on average up to 16 times more flowers than pollinated females (Laporte and Delph 1996). This difference in flower number may be driven by a combination of male competition and, at least partly, by a higher consumption of resources by developing seeds in pollinated female flowers, which results in a trade-off between seed size and flower number. The difference in flower number is less pronounced in non-pollinated females, which produce on average only four times fewer flowers than males (Laporte and Delph 1996; Steven et al. 2007). It also seems reasonable to expect that differences in the vegetative parts of plants evolved in concert with the flower types or architecture of inflorescences carried by the plant. Many sexually-dimorphic traits could evolve as a consequence of their correlation with other sexually dimorphic traits and so they need not be of adaptive value (Dawson and Geber 1999). For instance, correlations between flower size and the size of the stem leaves have been reported in many species (reviewed by Dawson and Geber 1999). Variation in sex-limited genes with pleiotropic effects and/or linkage between sex-limited loci also occurs in *S. latifolia* (Steven et al. 2007). These authors statistically predicted that selection for increased flower number in males along with weak selection for increased flower size in females could lead to dimorphic evolution in several other traits including leaf mass.

Almost all of the sexually dimorphic traits in *S. latifolia* described so far become apparent only after the initiation of flowering. Notable exceptions to this include: sex-dimorphism in the long-term survival of buried seeds and burial induced dormancy in *S. latifolia* (Purrington and Schmitt 1995), and sex-dimorphism in emergence time and time to flowering (Doust et al. 1987; Purrington and Schmitt 1998). Zluvova et al. (2010) provided molecular evidence that sexually dimorphic gene expression is present in *S. latifolia* at the rosette stage, a long time before the initiation of flowering.

Recently, Janoušek and Mrackova (2010) have taken up the so far neglected question of how a system based on two separate sex-controlling pathways regulated by genes present on the Y chromosome (Charlesworth and Charlesworth 1978) is transformed into a system controlled by the X/A ratio in some species (as in some species of *Rumex*). They suggest the following scenario. In the first step (Fig. 11.5a), dioecy evolves from gynodioecy via a gynoecium-suppressing gene in the proximity of the male fertility controlling gene. This process is promoted by sexually antagonistic selection. The original theory (Charlesworth and Charlesworth 1978) supposes just one

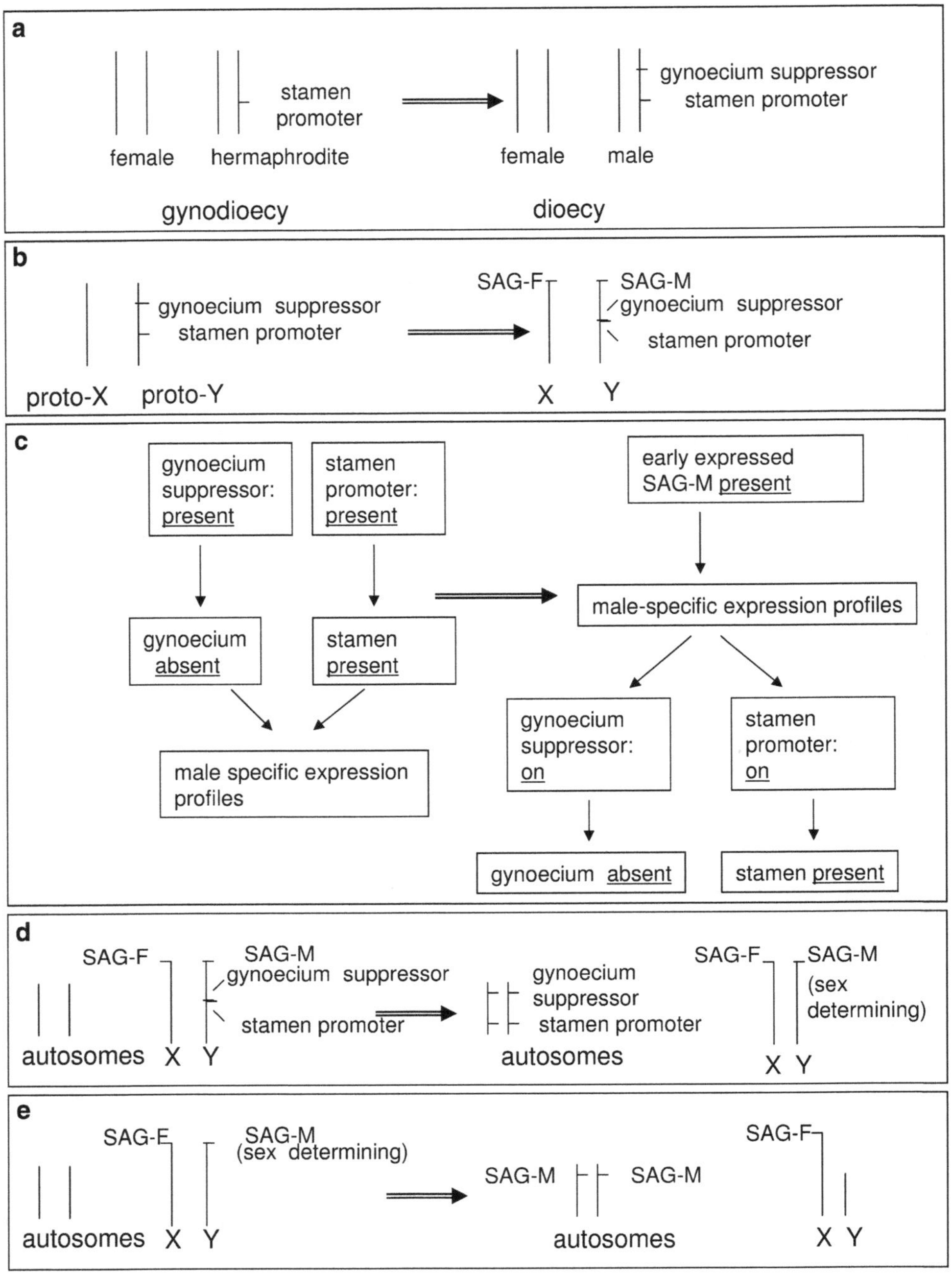

Fig. 11.5 The evolution of the sex-determining pathways in plants. (**a**) The theory of the origin of dioecy via male sterility as suggested by Charlesworth and Charlesworth (1978). Both the gynoecium suppressor and the stamen (anther) promoter act independently, but their coordination is achieved by their proximity on the Y chromosome or by their location in the non-recombining region of the Y chromosome. (**b**) Formation of sex chromosomes. Accumulation of sexually antagonistic genes (SAG) and the reduction of recombination frequency between gynoecium suppressing and male fertility controlling genes creates sex chromosomes. For simplicity, only one sexually antagonistic gene is presented. SAG-F means a sexually antagonistic allele advantageous for females, and SAG-M means a male advantageous sexually antagonistic allele of the same gene. (**c**) A sexually antagonistic gene(s)-based switch in the sex-determining pathway. Sexual dimorphism (controlled by SAG-M) is improved step-by-step and starts to act before the Y-linked genes involved in female and stamen development. At a certain stage, the expression of both the gynoecium suppressor and the stamen promoter becomes sex limited as a consequence of their adaptation to sex specific expression profiles of other genes. (**d**) The restructuring of the sex chromosomes. The gynoecium suppressor and stamen promoting gene (s) are lost from the Y chromosome and transferred to the autosome(s). (**e**) The origin of an X/A based sex-determining system. SAG-M is lost from the Y chromosome and transferred to an autosome. The X/A ratio becomes crucial for sex determination as SAG-M pushes development toward the male direction in contrast to SAG-F that pushes development toward the female direction

male sterility mutation, implying that the male fertility locus of the gynodioecious ancestor is identical with the anther-promoting gene of the resulting dioecious species. The possibility of a step-wise shift in stamen arrest was discussed by Zluvova et al. (2005), but does not essentially change the predictions of the model. In a second step (Fig. 11.5b), sexually antagonistic selection continues and improves the linkage of the sex-determining loci. Sex chromosomes that are created by this process can continue to accumulate sexually antagonistic alleles (Rice 1984). Even in species that are at this stage of sex determination, it is possible to find early expressed sexual dimorphism (e.g., *Silene latifolia*; Zluvova et al. 2010).

In the third step (Fig. 11.5c), sex-determining genes start to accommodate to the sex-specific gene expression patterns controlled by the sexually antagonistic gene(s) (SAG) and their expression starts to be controlled by these genes. Eventually, the gene(s) that previously controlled sexual dimorphism become(s) sex-determining gene(s). It is known that sexually antagonistic genes evolve fast (reviewed in Qvarnström and Bailey 2009) and a good example of this in plants is the Y-linked genes from *S. latifolia.* The lack of a Y chromosome in *S. latifolia* cannot be completely compensated for by the presence of the genome of the related species *S. viscosa,* and anther defects in hybrids between *S. latifolia* and *S. viscosa* resemble two different mutants lacking part of the Y chromosome (Zluvova et al. 2005). The active role of the Y chromosome in sex determination is still preserved because the new sex-determining locus is still located on the Y chromosome. An important difference from the previous stages is that connection between the control of stamen promotion and anther suppression is established. This creates new possibilities for the evolution of the sex-determining system. In a plant species possessing this kind of sex determination, a single master gene mutation should be able to cause a male to female transformation.

In the fourth step (Fig. 11.5d), two things could have happened. The original sex-determining loci could be lost from the Y chromosome by chance or the translocation of the original sex-determining region to the autosomes could be supported by Y chromosome degeneration due to the absence of recombination. Since stamen promotion and gynoecium suppression are already controlled by the single controlling pathway, the genotypes possessing a translocation of these genes to autosomes can be selected for because they can escape from the process of degeneration. The change of the position of these genes does not influence the sexual phenotype, as they have sex-limited expression and are controlled by the gene derived from the gene that initially controlled only sexual dimorphism.

In the fifth step (Fig. 11.5e), the secondary sex-determining locus loses its controlling role as well, and control of sex determination is co-opted by alternative mechanisms, such as an X/A ratio mechanism. The switch could be similar to the mechanisms described in fishes and insects (reviewed by Schütt and Nöthiger 2000; Volff et al. 2007). One of the genes, "a slave" in the sex-determining pathway located on the X chromosome, can become "a master". Simultaneously, the function of this new master gene is influenced by genes located on autosomes. An important prerequisite of this kind of transformation of the sex-determining system is that stamen promotion and gynoecium suppression are controlled by the same pathway. This fifth step in the evolution of the sex-determining pathway may have been reached in hemp (*Cannabis*), hop (*Humulus*), and sorrel (*Rumex*).

Veltsos et al. (2008) performed computer modelling experiments in order to explain spread of neo-sex chromosomes in the grasshopper *Podisma pedestris*. In this species, neo-X chromosomes (autosomes fused to original X) and neo-Y chromosomes (unfused autosome) have spread and become fixed in a large geographical territory. Computer modelling experiments showed that the continuous invasion of the neo-Y chromosome (carrying a sexually antagonistic allele) in the hybrid zone can cause selection in favour of the neo-X chromosome. Afterwards, also the neo-Y chromosome that is, in the initial phase, continuously removed by selection can be selected for as a consequence of the neo-X accumulation. Interestingly, the same mechanism may be involved in the spread and fixation of the degenerated Y chromosomes. It is apparent that, in many plant (e.g., *Carica papaya*, *Silene latifolia*, and *Asparagus officinalis*, reviewed by Ming et al. 2007) and animal species (e.g., most mammals, reviewed by Wilhelm et al. 2007) this evolution is not complete because many species still rely on an active role of the Y chromosome in sex determination.

Instability of sex expression and/or cytologically homomorphic sex chromosomes are sometimes taken as a sign of the primitive status of the evolution of sex-determining systems (e.g., Vyskot and Hobza 2004). Data obtained in the animal models, however, suggest that even very advanced sex-determining systems (as in mammals) can show a considerable plasticity of sex expression (e.g., Bianchi 2002). Only comparative sequencing and phylogenetic analyses can answer the question of the age of sex-determining pathways.

11.4 The Current Status of Research and Prospects for the Future

An overview of currently used approaches for studying sex determination in plants and their inter-connection is provided in Fig. 11.6. Separation of the chromosome(s) by microdissection and chromosome sorting is the most

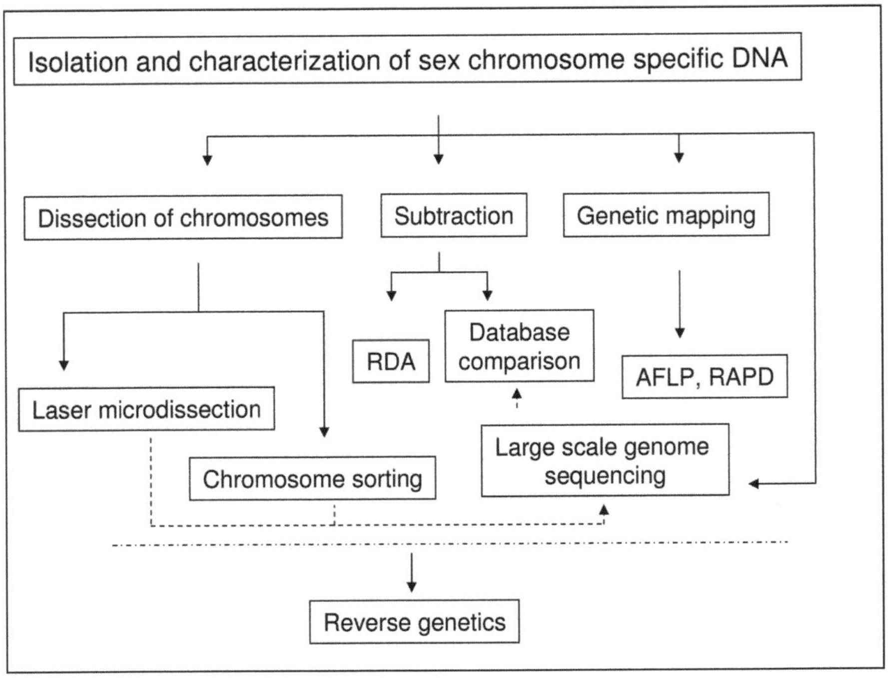

Fig. 11.6 A diagrammatic sketch of the methods frequently used for the isolation and characterization of sex-linked DNA. Dissection of chromosomes is the method of choice for generating complex probes for FISH (fluorescence *in situ* hybridization), constructing chromosome-specific libraries and as a tool for mapping DNA markers using specific chromosomes as a template. A subtractive approach (e.g., RDA—representational difference analysis) is used for the isolation of sample-specific DNA (e.g., in the comparison of male and female genomic DNA or the comparison of sex-specific expression patterns). Nowadays, traditional molecular subtractive methods are replaced by comparisons of different DNA databases. Genetic mapping is still the basic method for analysis of DNA markers and their linkage to phenotypically interesting mutations (AFLP—amplified fragment length polymorphisms; RAPD—Random Amplification of Polymorphic DNA). A combination of genetic maps with information on complete genome sequences is a key approach for the identification and isolation of novel genes and the deduction of their function

straightforward technique for isolating chromosome-specific DNA. The use of these techniques enabled researchers to isolate the first active plant Y-linked gene (Delichere et al. 1999). Moreover, the construction and utilization of -chromosome-specific libraries and the use of sorted chromosomes as complex probes has improved our knowledge of sex chromosomes, mainly in *Rumex acetosa* and *Silene latifolia*. However, only the few species with morphologically different chromosomes are suitable for dissection techniques.

High-throughput sequencing methods accelerate whole-genome sequencing (Tuskan et al. 2006; Ming et al. 2008) and permit the construction of huge EST and genomic databases. Comparisons of different pools of sequenced samples (males versus females, generative tissue versus vegetative tissue) may help identify sex-determining genes and to characterize sex expression pathways (Blavet et al. 2011).

Important questions regarding the basic biological processes affecting the evolution of sex chromosomes that could be answered in the next decade are: How do sex chromosomes degenerate? Is degeneration of genes in non-recombining regions really a one-way ticket as it is in many animal species? Is dosage compensation a general phenomenon for regulating the level of expression in organisms with degenerating sex chromosomes?

Acknowledgment This research was supported by the Grant Agency of the Czech Republic (grants P501/10/0102, 522/09/0083, and 521/08/0932).

References

Abraham A, Mathew PM (1962) Cytological studies in the cycads: sex chromosomes in *Cycas*. Ann Bot 26:261–266

Acosta IF, Laparra H, Romero SP, Schmelz E, Hamberg M, Mottinger JP, Moreno MA, Dellaporta SL (2009) *Tasselseed1* is a lipoxygenase affecting jasmonic acid signaling in sex determination of maize. Science 323:262–265

Ainsworth CC, Lu J, Winfield M, Parker JS (1999) Sex determination by X:autosome dosage: *Rumex acetosa* (sorrel). In: Ainsworth CC (ed) Sex determination in plants. Bios Scientific, Oxford, pp 121–136

Bateman AJ (1948) Intra-sexual selection in *Drosophila*. Heredity 2:34–39, 368

Bergero R, Forrest A, Kamau E, Charlesworth D (2007) Evolutionary strata on the X chromosomes of the dioecious plant *Silene latifolia*: evidence from new sex-linked genes. Genetics 175:1945–1954

Bianchi NO (2002) *Akodon* sex reversed females: the never ending story. Cytogenet Genome Res 96:60–65

Blackburn KB (1923) Sex chromosomes in plants. Nature 112:687–688
Blavet N, Charif D, Oger-Desfeux C, Marais GAB, Widmer A (2011) Comparative high-throughput transcriptome sequencing and development of SiESTa, the *Silene* EST annotation database. BMC Genomics 12:376
Borden KL (2000) RING domains: master builders of molecular scaffolds? J Mol Biol 295:1103–1112
Boualem A, Fergany M, Fernandez R, Troadec C, Martin A, Morin H, Sari M, Collin F, Flowers JM, Pitrat M et al (2008) A conserved mutation in an ethylene biosynthesis enzyme leads to andromonoecy in melons. Science 321:836–838
Boualem A, Troadec C, Kovalski I, Sari M, Perl-Treves R, Bendahmane A (2009) A conserved ethylene biosynthesis enzyme leads to andromonoecy in two *Cucumis* species. PLoS One 4: e6144
Brenner ED, Stevenson DW, McCombie RW, Katari MS, Rudd SA, Mayer KF, Palenchar PM, Runko SJ, Twigg RW, Dai G et al (2003a) Expressed sequence tag analysis in *Cycas*, the most primitive living seed plant. Genome Biol 4:78–78
Brenner ED, Katari MS, Stevenson DW, Rudd SA, Douglas AW, Moss WN, Twigg RW, Runko SJ, Stellari GM, McCombie WR et al (2005) EST analysis in *Ginkgo biloba*: an assessment of conserved developmental regulators and gymnosperm specific genes. BMC Genomics 6:143
Browse J (2009) Jasmonate: preventing the maize tassel from getting in touch with his feminine side. Sci Signal 2:pe9
Calderon-Urrea A, Dellaporta SL (1999) Cell death and cell protection genes determine the fate of pistils in maize. Development 126:435–441
Cermak T, Kubat Z, Hobza R, Koblizkova A, Widmer A, Macas J, Vyskot B, Kejnovsky E (2008) Survey of repetitive sequences in *Silene latifolia* with respect to their distribution on sex chromosomes. Chromosome Res 16:961–976
Chailakhyan MK, Khryanin VN (1978) The influence of growth regulators absorbed by the root on sex expression in hemp plants. Planta 138:181–184
Chailakhyan MK, Khryanin VN (1979) The role of leaves in sex expression in hemp and spinach. Planta 144:205–207
Charlesworth B, Charlesworth D (1978) A model for the evolution of dioecy and gynodioecy. Am Natur 112:975–997
Correns C (1928a) Bestimmung, Vererbung und Verteilung des Geschlechtes bei den höheren Pflanzen. In: Baur E, Hartmann M (eds) Handbuch der Vererbungswissenschaft, vol 2. Borntraeger, Berlin, pp 56–58
Correns C (1928b) Bestimmung, Vererbung und Verteilung des Geschlechtes bei den höheren Pflanzen. In: Baur E, Hartmann M (eds) Handbuch der Vererbungswissenschaft, vol 2. Borntraeger, Berlin, pp 48–50
Cousens MI, Lacey DG, Scheller JM (1988) Safe sites and the ecological life history of *Lorinseria areolata*. Am J Bot 75:797–807
Daher A, Adam H, Chabrillange N, Collin M, Mohamed N, Tregear JW, Aberlenc-Bertossi F (2010) Cell cycle arrest characterizes the transition from a bisexual floral bud to a unisexual flower in *Phoenix dactylifera*. Ann Bot 106:255–266
Dawson TE, Geber MA (1999) Dimorphism in physiology and morphology. In: Geber MA, Dawson TE, Delph LF (eds) Gender and sexual dimorphism in flowering plants. Springer, Berlin, pp 175–215
Delichere C, Veuskens J, Hernould M, Barbacar N, Mouras A, Negrutiu I, Monéger F (1999) *SlY1*, the first active gene cloned from a plant Y chromosome, encodes a WD-repeat protein. EMBO J 18:4169–4179
Dellaporta SL, Calderon-Urrea A (1993) Sex determination in flowering plants. Plant Cell 5:1241–1251
Di Stilio VS, Kramer EM, Baum DA (2005) Floral MADS box genes and homeotic gender dimorphism in *Thalictrum dioicum* (Ranunculaceae)—a new model for the study of dioecy. Plant J 41:755–766
Di Stilio VS, Kumar RA, Oddone AM, Tolkin TR, Salles P, McCarty K (2010) Virus-induced gene silencing as a tool for comparative functional studies in *Thalictrum*. PLoS One 5:e12064
Dorken ME, Pannell JR (2008) Density-dependent regulation of the sex ratio in an annual plant. Am Nat 171:824–830
Doust JL, O'Brien G, Doust LL (1987) Effect of density on secondary sex characteristics and sex ratio in *Silene alba* (Caryophyllaceae). Am J Bot 74:40–46
Eberle JR, Banks JA (1996) Genetic interactions among sex-determining genes in the fern *Ceratopteris richardii*. Genetics 142:973–985
Eggens F, Popp M, Nepokroeff M, Wagner WL, Oxelman B (2007) The origin and number of introductions of the Hawaiian endemic *Silene* species (Caryophyllaceae). Am J Bot 94:210–218
Eppley SM, Taylor PJ, Jesson LK (2007) Self-fertilization in mosses: a comparison of heterozygote deficiency between species with combined versus separate sexes. Heredity 98:38–44
Farbos I, Veuskens J, Vyskot B, Oliveira M, Hinnisdaels S, Aghmir A, Mouras A, Negrutiu I (1999) Sexual dimorphism in white campion: deletion on the Y chromosome results in a floral asexual phenotype. Genetics 151:1187–1196
Fawcett JA, Van de Peer Y, Maere S (2013) Significance and biological consequences of polyploidization in land plants. In: Leitch IJ, Greilhuber J, Doležel J, Wendel JF (eds) Plant genome diversity, vol 2, Physical structure, behaviour and evolution of plant genomes. Springer-Verlag, Wien, pp 277–293
Filatov DA (2005) Evolutionary history of *Silene latifolia* sex chromosomes revealed by genetic mapping of four genes. Genetics 170:975–979
Filatov DA, Monéger F, Negrutiu I, Charlesworth D (2000) Low variability in a Y-linked plant gene and its implications for Y-chromosome evolution. Nature 404:388–390
Floyd ME (1983) Dioecy in five *Pinus edulis* populations in the south-western United States. Am Midl Nat 110:405–411
Frohlich MW, Parker DS (2000) The mostly male theory of flower evolutionary origins: from genes to fossils. Syst Bot 25:155–170
Fujisawa M, Hayashi K, Nishio T, Bando T, Okada S, Yamato KT, Fukuzawa H, Ohyama K (2001) Isolation of X and Y chromosome-specific DNA markers from a liverwort, *Marchantia polymorpha*, by representational difference analysis. Genetics 159:981–985
Givnish TJ (1980) Ecological constraints on the evolution of breeding systems in seed plants: dioecy and dispersal in gymnosperms. Evolution 34:959–972
Gómez-Campo C, Casas-Builla M (1965) Radiation-induced sex reversal in *Ecballium elaterium*. Nature 208:805–805
Grant S, Houben A, Vyskot B, Siroky J, Pan WH, Macas J, Saedler H (1994) Genetics of sex determination in flowering plants. Dev Genet 15:214–230
Haga T (1934) Sex and chromosomes in *Spinacia oleracea* L. Jpn J Genet 10:218–222
Haig D, Westoby M (1988) A model for the origin of heterospory. J Theor Biol 134:257–272
Hall DW, Kirkpatrick M, West B (2000) Runaway sexual selection when female preferences are directly selected. Evolution 54:1862–1869
Hamdi S, Teller G, Louis J (1987) Master regulatory genes, auxin levels, and sexual organogeneses in the dioecious plant *Mercurialis annua*. Plant Physiol 85:393–399
Hamilton RG, Lloyd RM (1991) Antheridiogen in the wild: the development of fern gametophyte communities. Funct Ecol 5:804–809
Hernould M, Glimelius K, Veuskens J, Bergman P, Mouras A (1997) Microdissection and amplification of coding sequences from a chromosome fragment restoring male fertility in alloplasmic male-sterile tobacco. Plant J 12:703–709

Hickok LG, Warne TR, Slocum MK (1987) *Ceratopteris richardii*: applications for experimental plant biology. Am J Bot 74:1304–1316
Hobza R, Lengerova M, Cernohorska H, Rubes J, Vyskot B (2004) FAST-FISH with laser beam microdissected DOP-PCR probe distinguishes the sex chromosomes of *Silene latifolia*. Chromosome Res 12:245–250
Huang Y, Chou H, Chiou W (2004) Density affects gametophyte growth and sexual expression of *Osmunda cinnamomea* (Osmundaceae: pteridophyta). Ann Bot 94:229–232
Huang S, Li R, Zhang Z, Li L, Gu X, Fan W, Lucas WJ, Wang X, Xie B, Ni P et al (2009) The genome of the cucumber, *Cucumis sativus* L. Nat Genet 41:1275–1281
Hultquist JF, Dorweiler JE (2008) Feminized tassels of maize mop1 and ts1 mutants exhibit altered levels of miR156 and specific SBP-box genes. Planta 229:99–113
Iizuka M, Janick J (1962) Cytogenetic analysis of sex determination in *Spinacia oleracea*. Genetics 47:1225–1241
Ishizaki K, Shimizu-Ueda Y, Okada S, Yamamoto M, Fujisawa M, Yamato KT, Fukuzawa H, Ohyama K (2002) Multicopy genes uniquely amplified in the Y chromosome-specific repeats of the liverwort *Marchantia polymorpha*. Nucleic Acids Res 30:4675–4681
Jakse J, Stajner N, Kozjak P, Cerenak A, Javornik B (2008) Trinucleotide microsatellite repeat is tightly linked to male sex in hop (*Humulus lupulus* L.). Mol Breed 21:139–148
Jamilena M, Mariotti B, Manzano S (2008) Plant sex chromosomes: molecular structure and function. Cytogenet Genome Res 120:255–264
Janick J, Stevenson E (1954) A genetic study of the heterogametic nature of the staminate plant in spinach (*Spinacia oleracea* L.). P Am Soc Hort Sci 63:444–446
Janoušek B, Mrackova M (2010) Sex chromosomes and sex determination pathway dynamics in plant and animal models. Biol J Linn Soc 100:737–752
Jones AG, Arguello JR, Arnold SJ (2002) Validation of Bateman's principles: a genetic study of sexual selection and mating patterns in the rough-skinned newt. P Biol Sci 269:2533–2539
Kamachi H, Iwasawa O, Hickok LG, Nakayama M, Noguchi M, Inoue H (2007) The effects of light on sex determination in gametophytes of the fern *Ceratopteris richardii*. J Plant Res 120:629–634
Karlov GI, Danilova TV, Horlemann C, Weber G (2003) Molecular cytogenetics in hop (*Humulus lupulus* L.) and identification of sex chromosomes by DAPI-banding. Euphytica 132:185–190
Kazama Y, Fujiwara MT, Koizumi A, Nishihara K, Nishiyama R, Kifune E, Abe T, Kawano S (2009) A *SUPERMAN*-like gene is exclusively expressed in female flowers of the dioecious plant *Silene latifolia*. Plant Cell Physiol 50:1127–1141
Kejnovsky E, Kubat Z, Hobza R, Lengerova M, Sato S, Tabata S, Fukui K, Matsunaga S, Vyskot B (2006) Accumulation of chloroplast DNA sequences on the Y chromosome of *Silene latifolia*. Genetica 128:167–175
Kejnovsky E, Hobza R, Cermak T, Kubat Z, Vyskot B (2009) The role of repetitive DNA in structure and evolution of sex chromosomes in plants. Heredity 102:533–541
Khadka DK, Nejidat A, Tal M, Golan-Goldhirsh A (2002) DNA markers for sex: molecular evidence for gender dimorphism in dioecious *Mercurialis annua* L. Mol Breed 9:251–257
Khadka DK, Nejidat A, Tal M, Golan-Goldhirsh A (2005) Molecular characterization of a gender-linked DNA marker and a related gene in *Mercurialis annua* L. Planta 222:1063–1070
Khattak JZK, Torp AM, Andersen SB (2006) A genetic linkage map of *Spinacia oleracea* and localization of a sex determination locus. Euphytica 148:311–318
Kihara H, Ono T (1923) Cytological studies on *Rumex* L. I. Chromosomes of *Rumex acetosa* L. Bot Mag Tokyo 37:84–90
Klekowski EJ Jr (1969) Reproductive biology of the pteridophyta. II. Theoretical considerations. Bot J Linn Soc 62:347–359
Kocyan A, Zhang L, Schaefer H, Renner SS (2007) A multi-locus chloroplast phylogeny for the *Cucurbitaceae* and its implications for character evolution and classification. Mol Phylogenet Evol 44:553–577
Koizumi A, Yamanaka K, Nishihara K, Kazama Y, Abe T, Kawano S (2010) Two separate pathways including *SlCLV1*, *SlSTM* and *SlCUC* that control carpel development in a bisexual mutant of *Silene latifolia*. Plant Cell Physiol 51:282–293
Korpelainen H (1995) Growth and reproductive characteristics in artificially formed clonal gametophytes of *Dryopteris filix-mas* (Dryopteridaceae). Plant Syst Evol 196:195–206
Korpelainen H (1998) Labile sex expression in plants. Biol Rev 73:157–180
Kubat Z, Hobza R, Vyskot B, Kejnovsky E (2008) Microsatellite accumulation on the Y chromosome in *Silene latifolia*. Genome 51:350–356
Kuhn E (1939) Selbstbestäubungen subdiözischer Blütenpflanzen, ein neuer Beweis für die genetische Theorie der Geschlechtsbestimmung. Planta 30:457–470
Kumar LSS, Viseveshwaraiah S (1952) Sex mechanism in *Coccinia indica* Wight and Arn. Nature 170:330–331
Lan T, Zhang S, Liu B, Li X, Chen R, Song W (2006) Differentiating sex chromosomes of the dioecious *Spinacia oleracea* L. (spinach) by FISH of 45S rDNA. Cytogenet Genome Res 114:175–177
Laporte MM, Delph LF (1996) Sex-specific physiology and source-sink relations in the dioecious plant *Silene latifolia*. Oecologia 106:63–72
Lebel-Hardenack S, Hauser E, Law TF, Schmid J, Grant SR (2002) Mapping of sex determination loci on the white campion (*Silene latifolia*) Y chromosome using amplified fragment length polymorphism. Genetics 160:717–725
Lee CL (1954) Sex chromosomes in *Ginkgo biloba*. Am J Bot 41:545–549
Leitch IJ, Leitch AR (2013) Genome size diversity and evolution in land plants. In: Leitch IJ, Greilhuber J, Doležel J, Wendel JF (eds) Plant genome diversity, vol 2, Physical structure, behaviour and evolution of plant genomes. Springer-Verlag, Wien, pp 307–322
Lengerova M, Vyskot B (2001) Sex chromatin and nucleolar analyses in *Rumex acetosa* L. Protoplasma 217:147–153
Leyh TS, Vogt TF, Suo Y (1992) The DNA sequence of the sulfate activation locus from *Escherichia coli* K-12. J Biol Chem 267:10405–10410
Liu Z, Moore PH, Ma H, Ackerman CM, Ragiba M, Yu Q, Pearl HM, Kim MS, Charlton JW, Stiles JI et al (2004) A primitive Y chromosome in papaya marks incipient sex chromosome evolution. Nature 427:348–352
Ma H, Moore PH, Liu Z, Kim MS, Yu Q, Fitch MMM, Sekioka T, Paterson AH, Ming R (2004) High-density linkage mapping revealed suppression of recombination at the sex determination locus in papaya. Genetics 166:419–436
Marais GAB, Nicolas M, Bergero R, Chambrier P, Kejnovsky E, Monéger F, Hobza R, Widmer A, Charlesworth D (2008) Evidence for degeneration of the Y chromosome in the dioecious plant *Silene latifolia*. Curr Biol 18:545–549
Martin A, Troadec C, Boualem A, Rajab M, Fernandez R, Morin H, Pitrat M, Dogimont C, Bendahmane A (2009) A transposon-induced epigenetic change leads to sex determination in melon. Nature 461:1135–1138
Matallana G, Wendt T, Araujo DSD, Scarano FR (2005) High abundance of dioecious plants in a tropical coastal vegetation. Am J Bot 92:1513–1519
Mather K (1949) Genetics of dioecy and monoecy in *Ecballium*. Nature 163:926
Mathews S, Clements MD, Beilstein MA (2010) A duplicate gene rooting of seed plants and the phylogenetic position of flowering plants. Philos Trans Roy Soc Lond B Bio 365:383–395

Matsunaga S, Kawano S, Michimoto T, Higashiyama T, Nakao S, Sakai A, Kuroiwa T (1999) Semi-automatic laser beam microdissection of the Y chromosome and analysis of Y chromosome DNA in a dioecious plant, *Silene latifolia*. Plant Cell Physiol 40:60–68

Matsunaga S, Uchida W, Kawano S (2004) Sex-specific cell division during development of unisexual flowers in the dioecious plant *Silene latifolia*. Plant Cell Physiol 45:795–802

Mayer SS, Charlesworth D (1991) Cryptic dioecy in flowering plants. Trends Ecol Evol 6:320–325

Menzel MY (1964) Meiotic chromosomes of monoecious Kentucky hemp (*Cannabis sativa*). Bull Torrey Bot Club 91:193–205

Miller JH (1968) Fern gametophytes as experimental material. Bot Rev 34:361–440

Miller JS, Venable DL (2000) Polyploidy and the evolution of gender dimorphism in plants. Science 289:2335–2338

Ming R, Wang J, Moore PH, Paterson AH (2007) Sex chromosomes in flowering plants. Am J Bot 94:141–150

Ming R, Hou S, Feng Y, Yu Q, Dionne-Laporte A, Saw JH, Senin P, Wang W, Ly BV, Lewis KLT et al (2008) The draft genome of the transgenic tropical fruit tree papaya (*Carica papaya* Linnaeus). Nature 452:991–996

Ming RA, Bendahmane A, Renner SS (2011) Sex chromosomes in land plants. Ann Rev Plant Biol 62:485–514

Mohan Ram HY, Sett R (1982) Induction of fertile male flowers in genetically female *Cannabis sativa* plants by silver nitrate and silver thiosulphate anionic complex. Theor Appl Genet 62:369–375

Moliterni VMC, Cattivelli L, Ranalli P, Mandolino G (2004) The sexual differentiation of *Cannabis sativa* L.: a morphological and molecular study. Euphytica 140:95–1, 06

Morgante M, Hanafey M, Powell W (2002) Microsatellites are preferentially associated with nonrepetitive DNA in plant genomes. Nat Genet 30:194–200

Mrackova M, Nicolas M, Hobza R, Negrutiu I, Monéger F, Widmer A, Vyskot B, Janoušek B (2008) Independent origin of sex chromosomes in two species of the genus *Silene*. Genetics 179:1129–1133

Murray B (2013) Karyotype variation and evolution in gymnosperms. In: Leitch IJ, Greilhuber J, Doležel J, Wendel JF (eds) Plant genome diversity, vol 2, Physical structure, behaviour and evolution of plant genomes. Springer-Verlag, Wien, pp 231–243

Navajas-Pérez R, de la Herrán R, López González G, Jamilena M, Lozano R, Ruiz Rejón C, Ruiz Rejón M, Garrido-Ramos MA (2005) The evolution of reproductive systems and sex-determining mechanisms within *Rumex* (Polygonaceae) inferred from nuclear and chloroplastidial sequence data. Mol Biol Evol 22:1929–1939

Nicolas M, Marais G, Hykelova V, Janoušek B, Laporte V, Vyskot B, Mouchiroud D, Negrutiu I, Charlesworth D, Monéger F (2005) A gradual process of recombination restriction in the evolutionary history of the sex chromosomes in dioecious plants. PLoS Biol 3:e4

Obbard DJ, Harris SA, Buggs RJA, Pannell JR (2006) Hybridization, polyploidy, and the evolution of sexual systems in *Mercurialis* (Euphorbiaceae). Evolution 60:1801–1815

Okada S, Fujisawa M, Sone T, Nakayama S, Nishiyama R, Takenaka M, Yamaoka S, Sakaida M, Kono K, Takahama M et al. (2000) Construction of male and female PAC genomic libraries suitable for identification of Y-chromosome-specific clones from the liverwort, *Marchantia polymorpha*. Plant J 24:421–428

Okada S, Sone T, Fujisawa M, Nakayama S, Takenaka M, Ishizaki K, Kono K, Shimizu-Ueda Y, Hanajiri T, Yamato KT et al. (2001) The Y chromosome in the liverwort *Marchantia polymorpha* has accumulated unique repeat sequences harboring a male-specific gene. Proc Natl Acad Sci USA 98:9454–9459

Oxelman B (2010) Sileneae biodiversity informatics project. Department of Plant and Environmental Sciences. Göteborg University. (http://www.sileneae.info)

Oyama RK, Volz SM, Renner SS (2009) A sex-linked SCAR marker in *Bryonia dioica* (Cucurbitaceae), a dioecious species with XY sex-determination and homomorphic sex chromosomes. J Evol Biol 22:214–224

Pannell JR, Verdú M (2006) The evolution of gender specialization from dimorphic hermaphroditism: paths from heterodichogamy to gynodioecy and androdioecy. Evolution 60:660–673

Parker J, Clark M (1991) Dosage sex-chromosome systems in plants. Plant Sci 80:79–92

Parkinson SE, Gross SM, Hollick JB (2007) Maize sex determination and abaxial leaf fates are canalized by a factor that maintains repressed epigenetic states. Dev Biol 308:462–473

Peil A, Flachowsky H, Schumann E, Weber WE (2003) Sex-linked AFLP markers indicate a pseudoautosomal region in hemp (*Cannabis sativa* L.). Theor Appl Genet 107:102–109

Pfent C, Pobursky KJ, Sather DN, Golenberg EM (2005) Characterization of *SpAPETALA3* and *SpPISTILLATA*, B class floral identity genes in *Spinacia oleracea*, and their relationship to sexual dimorphism. Dev Genes Evol 215:132–142

Pomiankowski A, Nöthiger R, Wilkins A (2004) The evolution of the *Drosophila* sex-determination pathway. Genetics 166:1761–1773

Purrington CB, Schmitt J (1995) Sexual dimorphism of dormancy and survivorship in buried seeds of *Silene latifolia*. J Ecol 83:795–800

Purrington CB, Schmitt J (1998) Consequences of sexually dimorphic timing of emergence and flowering in *Silene latifolia*. J Ecol 86:397–404

Qvarnström A, Bailey RI (2009) Speciation through evolution of sex-linked genes. Heredity 102:4–15

Raghavan V (1989) Developmental biology of fern gametophytes. Cambridge University Press, Cambridge, New York

Rautenberg A, Hathaway L, Oxelman B, Prentice HC (2010) Geographic and phylogenetic patterns in *Silene* section Melandrium (Caryophyllaceae) as inferred from chloroplast and nuclear DNA sequences. Mol Phylogenet Evol 57:978–991

Rejon RC, Jamilena M, Garrido Ramos M, Parker J, Ruiz Rejon M (1994) Cytogenetic and molecular analysis of the multiple sex chromosome system of *Rumex acetosa*. Heredity 72:209–215

Renner SS, Ricklefs RE (1995) Dioecy and its correlates in the flowering plants. Am J Bot 82:596–606

Renner SS, Won H (2001) Repeated evolution of dioecy from monoecy in Siparunaceae (Laurales). Syst Biol 50:700–712

Renner SS, Beenken L, Grimm GW, Kocyan A, Ricklefs RE (2007) The evolution of dioecy, heterodichogamy, and labile sex expression in *Acer*. Evolution 61:2701–2719

Rensing SA, Beike AK, Lang D (2013) Evolutionary importance of generative polyploidy for genome evolution of haploid-dominant land plants. In: Leitch IJ, Greilhuber J, Doležel J, Wendel JF (eds) Plant genome diversity, vol 2, Physical structure, behaviour and evolution of plant genomes. Springer-Verlag, Wien, pp 295–305

Rice WR (1984) Sex chromosomes and the evolution of sexual dimorphism. Evolution 38:735–742

Rubin G, Paolillo DJ (1983) Sexual development of *Onoclea sensibilis* on agar and soil media without the addition of antheridiogen. Am J Bot 70:811–815

Rubin G, Robson DS, Paolillo DJ (1985) Effects of population density on sex expression in *Onoclea sensibilis* L. on agar and ashed soil. Ann Bot 55:205–215

Rutherford G, Tanurdzic M, Hasebe M, Banks JA (2004) A systemic gene silencing method suitable for high throughput, reverse genetic analyses of gene function in fern gametophytes. BMC Plant Biol 4:e6

Sakamoto K, Ohmido N, Fukui K, Kamada H, Satoh S (2000) Site-specific accumulation of a LINE-like retrotransposon in a sex chromosome of the dioecious plant *Cannabis sativa*. Plant Mol Biol 44:723–732

Sano R, Juárez CM, Hass B, Sakakibara K, Ito M, Banks JA, Hasebe M (2005) *KNOX* homeobox genes potentially have similar function in both diploid unicellular and multicellular meristems, but not in haploid meristems. Evol Dev 7:69–78

Sather DN, York A, Pobursky KJ, Golenberg EM (2005) Sequence evolution and sex-specific expression patterns of the C class floral identity gene, *SpAGAMOUS*, in dioecious *Spinacia oleracea* L. Planta 222:284–292

Sather DN, Jovanovic M, Golenberg EM (2010) Functional analysis of B and C class floral organ genes in spinach demonstrates their role in sexual dimorphism. BMC Plant Biol 10:e46

Schneller JJ, Haufler CH, Ranker TA (1990) Antheridiogen and natural gametophyte populations. Am Fern J 80:143–152

Schütt C, Nöthiger R (2000) Structure, function and evolution of sex-determining systems in Dipteran insects. Development 127:667–677

Scutt CP, Kamisugi Y, Sakai F, Gilmartin PM (1997) Laser isolation of plant sex chromosomes: studies on the DNA composition of the X and Y sex chromosomes of *Silene latifolia*. Genome 40:705–715

Shephard HL, Parker JS, Darby P, Ainsworth CC (2000) Sexual development and sex chromosomes in hop. New Phytol 148:397–411

Siljak-Yakovlev S, Cerbah M, Sarr A, Benmalek S, Bounaga N, Coba de la Pena T, Brown SC (1996) Chromosomal sex determination and heterochromatin structure in date palm. Sex Plant Reprod 9:127–132

Smith BW (1963) The mechanism of sex determination in *Rumex hastatulus*. Genetics 48:1265–1288

Soltis DE, Buggs RJA, Doyle JJ, Soltis PS (2010) What we still do not know about polyploidy. Taxon 59:1387–1403

Spigler RB, Lewers KS, Main DS, Ashman T (2008) Genetic mapping of sex determination in a wild strawberry, *Fragaria virginiana*, reveals earliest form of sex chromosome. Heredity 101:507–517

Steven JC, Delph LF, Brodie ED (2007) Sexual dimorphism in the quantitative-genetic architecture of floral, leaf, and allocation traits in *Silene latifolia*. Evolution 61:42–57

Storey WB (1969) Pistillate papaya flower: a morphological anomaly. Science 163:401–405

Strain E, Hass B, Banks JA (2001) Characterization of mutations that feminize gametophytes of the fern *Ceratopteris*. Genetics 159:1271–1281

Talianova M (2009) Horizontal gene transfer—new cases in plants. In: Ninth international congress on plant molecular biology, St. Louis, Missouri, USA, 25–30 Oct 2009. Poster presentation

Talmor-Neiman M, Stav R, Klipcan L, Buxdorf K, Baulcombe DC, Arazi T (2006) Identification of trans-acting siRNAs in moss and an RNA-dependent RNA polymerase required for their biogenesis. Plant J 48:511–521

Tanurdzic M, Banks JA (2004) Sex-determining mechanisms in land plants. Plant Cell 16:S61–S71

Tavares R, Cagnon M, Negrutiu I, Mouchiroud D (2010) Testing the recent theories for the origin of the hermaphrodite flower by comparison of the transcriptomes of gymnosperms and angiosperms. BMC Evol Biol 10:e240

Telenius H, Carter NP, Bebb CE, Nordenskjodl M, Ponder BAJ, Tunnacliffe A (1992) Degenerate oligonucleotide-primed PCR: general amplification of target DNA by a single degenerate primer. Genomics 13:718–725

Theissen G, Becker A (2004) Gymnosperm orthologues of class B floral homeotic genes and their impact on understanding flower origin. Crit Rev Plant Sci 23:129–148

Tuskan GA, Difazio S, Jansson S, Bohlmann J, Grigoriev I, Hellsten U, Putnam N, Ralph S, Rombauts S, Salamov A et al (2006) The genome of black cottonwood, *Populus trichocarpa* (Torr. & Gray). Science 313:1596–1604

Veltsos P, Keller I, Nichols RA (2008) The inexorable spread of a newly arisen neo-Y chromosome. PLoS Genet 4:e1000082

Vévodová J, Graham RM, Raux E, Schubert HL, Roper DI, Brindley AA, Ian Scott A, Roessner CA, Stamford NPJ, Elizabeth Stroupe M et al (2004) Structure/function studies on a S-adenosyl-l-methionine-dependent uroporphyrinogen III C methyltransferase (SUMT), a key regulatory enzyme of tetrapyrrole biosynthesis. J Mol Biol 344:419–433

Volff J, Nanda I, Schmid M, Schartl M (2007) Governing sex determination in fish: regulatory putsches and ephemeral dictators. Sex Dev 1:85–99

Vyskot B, Hobza R (2004) Gender in plants: sex chromosomes are emerging from the fog. Trends Genet 20:432–438

Waelti MO, Page PA, Widmer A, Schiestl FP (2009) How to be an attractive male: floral dimorphism and attractiveness to pollinators in a dioecious plant. BMC Evol Biol 9:e190

Westergaard M (1946) Aberrant Y-chromosomes and sex expression in *Melandrium album*. Hereditas 32:419–443

Wilhelm D, Palmer S, Koopman P (2007) Sex determination and gonadal development in mammals. Physiol Rev 87:1–28

Winge O (1923) On sex chromosomes, sex determination and preponderance of females in some dioecious plants. C R Trav Lab Carlsberg 15:1–26

Wright SI, Ness RW, Foxe JP, Barrett SCH (2008) Genomic consequences of outcrossing and selfing in plants. Int J Plant Sci 169:105–118

Wrigley F (1986) Taxonomy and chorology of *Silene* section Otites (Caryophyllaceae). Ann Bot Fenn 23:69–81

Wu X, Knapp S, Stamp A, Stammers DK, Jörnvall H, Dellaporta SL, Oppermann U (2007) Biochemical characterization of *TASSELSEED 2*, an essential plant short-chain dehydrogenase/reductase with broad spectrum activities. FEBS J 274:1172–1182

Yamato KT, Ishizaki K, Fujisawa M, Okada S, Nakayama S, Fujishita M, Bando H, Yodoya K, Hayashi K, Bando T et al (2007) Gene organization of the liverwort Y chromosome reveals distinct sex chromosome evolution in a haploid system. Proc Natl Acad Sci USA 104:6472–6477

Yu Q, Hou S, Hobza R, Feltus FA, Wang X, Jin W, Skelton RL, Blas A, Lemke C, Saw JH et al (2007) Chromosomal location and gene paucity of the male specific region on papaya Y chromosome. Mol Genet Genomics 278:177–185

Zhang L, Simmons MP, Kocyan A, Renner SS (2006) Phylogeny of the *Cucurbitales* based on DNA sequences of nine loci from three genomes: implications for morphological and sexual system evolution. Mol Phylogenet Evol 39:305–322

Zhang W, Wang X, Yu Q, Ming R, Jiang J (2008) DNA methylation and heterochromatinization in the male-specific region of the primitive Y chromosome of papaya. Genome Res 18:1938–1943

Zluvova J, Lengerova M, Markova M, Hobza R, Nicolas M, Vyskot B, Charlesworth D, Negrutiu I, Janoušek B (2005) The inter-specific hybrid *Silene latifolia* x *S. viscosa* reveals early events of sex chromosome evolution. Evol Dev 7:327–336

Zluvova J, Nicolas M, Berger A, Negrutiu I, Monéger F (2006) Premature arrest of the male flower meristem precedes sexual dimorphism in the dioecious plant *Silene latifolia*. Proc Natl Acad Sci USA 103:18854–18859

Zluvova J, Georgiev S, Janoušek B, Charlesworth D, Vyskot B, Negrutiu I (2007) Early events in the evolution of the *Silene latifolia* Y chromosome: male specialization and recombination arrest. Genetics 177:375–386

Zluvova J, Zak J, Janoušek B, Vyskot B (2010) Dioeciouos *Silene latifolia* plants show sexual dimorphism in the vegetative stage. BMC Plant Biol 10:e208

Holocentric Chromosomes

12

Petr Bureš, František Zedek, and Michaela Marková

Contents

12.1 Introduction

In contrast to the "normal type" of monocentric mitotic chromosomes, where spindle attachment is restricted to a single kinetochore, holocentric[1] chromosomes are those in which spindle microtubules attach along the whole length through kinetochores that cover a substantial part of their poleward surfaces during mitosis. In addition, holocentric sister chromatids are interconnected along their whole length before anaphase disjunction, unlike monocentric chromatids, which cohere only in the pericentromeric area. The morphological distinctions between monocentric and holocentric chromosomes are associated with differences in chromatin structure and modified mitosis or meiosis, as well as karyotype evolution of the holocentrics themselves. In this chapter, we will survey these aspects of holocentrism and also discuss some hypotheses on the origin of holocentric chromosomes, their patterns of occurrence and methods of verification, particularly among plants.

12.2 Recognition and Verification of Holocentrism

Multiple spindle attachment along chromosomes was first unwittingly observed by pioneers of the study of chromosomes, such as the Belgian Édouard van Beneden and the Germans Richard Hertwig and Theodor Boveri, at the end of the nineteenth century. Using light microscopy and sophisticated staining techniques, they depicted a distinct and novel pattern of cell division in their cytogenetic model, the horse roundworm *Ascaris megalocephala* (= *Parascaris univalens*; Nematoda, Ascaridida). The first to

[1] To avoid confusion with the more narrowly defined term "holokinetic", which describes meiotic chromosome behaviour as the opposite of telokinetic behaviour, we prefer to use the term "holocentric" to indicate the opposite of the monocentric chromosomal nature.

P. Bureš (✉)
Department of Botany & Zoology, Masaryk University, Kotlářská 2, Brno 611 37, Brno 621 00, Czech Republic
e-mail: bures@sci.muni.cz

I.J. Leitch et al. (eds.), *Plant Genome Diversity Volume 2*,
DOI 10.1007/978-3-7091-1160-4_12, © Springer-Verlag Wien 2013

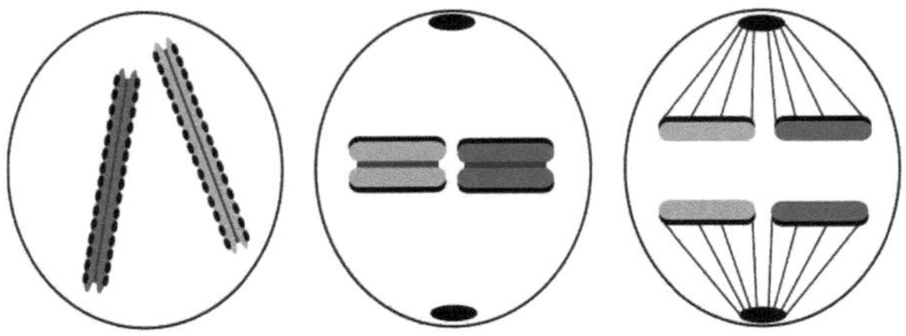

Fig. 12.1 Mitosis in holocentric organisms. Sister chromatids are represented by the same shading, either *dark* or *light grey* and both the kinetochore and microtubules are *black*. (**a**) Prophase: chromosomes after replication. The chromatids are connected lengthwise by cohesins, and kinetochore formation begins (CENH3-bound domains in *black*). (**b**) Metaphase: condensed metaphase chromosomes congress at the equatorial plate, while kinetochores occupy the whole poleward sides of the chromosomes. (**c**) Anaphase: migration of sister chromatids parallel to the equatorial plate

recognize the peculiarity of holocentric chromosomes was the American zoologist and cytologist Franz Schrader (1935) in mealy bugs (Hemiptera). The holocentric chromosomal structure was confirmed experimentally by Schrader's wife Sally and Hans Ris using X-radiation to generate chromosome fragments that behaved like individual chromosomes and were consistently transmitted to the next cell generation (Hughes-Schrader and Ris 1941). To prove the existence of holocentric chromosomes in angiosperms, this reliable experimental method was first applied to *Luzula* (Juncaceae) by the Portuguese cytogeneticists de Castro et al. (1949).

Besides X-rays (Roentgen rays), gamma rays can also be applied to seeds, flower (inflorescence) buds, rhizome buds or young root tips to study putative holocentric plants. Using this method, the presence of chromosome fragments in subsequent cell generations and the absence of potential micronuclei are considered proof of holocentrism (La Cour 1953; Godward 1954; Håkansson 1954; Nordenskiöld 1963, 1964; Flach 1966; Bokhari 1976; Tanaka and Tanaka 1979; Sheikh et al. 1995; Vanzela and Colaço 2002). The presence of holocentric chromosomes can be further confirmed by immunostaining methods using a labelled antibody designed to bind one of the proteins in the centromere/kinetochore complex (usually CENH3). In the case of holocentric chromosomes, the signal from the labelled antibody is present along the whole chromosome, whereas the signal is localized only at the centromere in monocentric chromosomes (Nagaki et al. 2005; see Sect. 12.4). Using light microscopy, the holocentric nature of chromosomes can also be observed at mitotic metaphase, when the absence of a primary constriction is more apparent, and the sister chromatids are oriented parallel to the equatorial plate (Fig. 12.1b). The positioning of the chromatids during mitotic anaphase, when they are pulled broadside towards opposite poles due to the attachment of the spindle fibres along the entire length of the chromosome is also typical for holocentric chromosomes (Fig. 12.1c). In contrast, during anaphase of monocentric mitosis, a typical "V-shape" of chromosomes can be observed due to the specific binding of microtubules to the centromere. The parallel segregation of holocentric chromosomes results in the formation of duplicate anaphase plates, suggesting that the metaphase plate has been split equatorially into two mirrored halves.

12.3 Occurrence of Holocentric Chromosomes in Plants and Other Organisms

Holocentrism is a stable, species-specific feature.[2] Although this phenomenon is much rarer than monocentrism, its occurrence is not randomly distributed, but phylogenetically clustered in several eukaryotic lineages. It is a synapomorphic trait for some families, genera, and subgenera in land plants and for some classes, orders and families in animals.

In early diverging angiosperms, holocentrism has been identified in *Myristica fragrans* (Myristicaceae,[3] Magnoliales, Flach 1966). In the monocot clade, it has been found in the sister families of Juncaceae[4] and Cyperaceae[5] (Poales, de Castro et al. 1949; La Cour 1953; Håkansson 1958; Nordenskiöld

[2] A similar state, sometimes considered to be transitional between monocentrism and holocentrism, is the formation of a neocentromere, a chromosomal region that differs in its sequence and structure but can function like a centromere in meiotic chromosomes. However, the neocentromere in plants is more likely just an aberrant structure that occurs occasionally in some individuals or populations (Dawe and Hiatt 2004). The majority of plant neocentromeres are the outcome of meiotic drive (Dawe and Hiatt 2004), which is one of the possible mechanisms that has been suggested to explain the origin of holocentric chromosomes (see Sect. 12.8).

[3] Holocentric chromosomes have been studied in detail only in the economically important nutmeg tree *Myristica fragrans*. Although chromosome counts have been reported for seven additional species in Myristicaceae, accounting for approximately 2% of the species richness of the family (Bolkhovskikh et al. 1969; Goldblatt and Johnson 2010), the lack of detailed cytological analysis means that the presence of holocentric chromosomes is unclear.

[4] Although holocentrism has been experimentally confirmed several times in the genus *Luzula*, it has been proposed to exist but not satisfactorily proven in the genus *Juncus* (Godward 1985) and other monotypic genera of Juncaceae (but see Sect. 12.7.3). In addition, in the closely related family Thurniaceae (comprising four species of *Thurnia* and *Prionium*), there is currently no karyological research so it is unknown whether holocentric chromosomes are present here (Bolkhovskikh et al. 1969; Goldblatt and Johnson 2010).

[5] This represents the third most species-rich monocot family after orchids and grasses. Numerous studies from the latter half of the twentieth century reported localized centromeres in Cyperaceae from S Asia, but these studies were rightfully doubted by Greilhuber (1995), who considered them to be based on chromosomal constrictions observed during mitotic prophase or metaphase, when chromosomes are not entirely contracted. Holocentrism has since been definitively confirmed experimentally in several taxa and should thus be considered as a family synapomorphy (Greilhuber 1995).

1963, 1964) and separately in the genus *Chionographis* (Melanthiaceae,[6] Liliales,[7] Tanaka and Tanaka 1979). In the eudicot clade of angiosperms, it is present in the genera *Cuscuta*[8] (Convolvulaceae, Solanales, Pazy and Plitmann 1994) and *Drosera* (Droseraceae,[9] Caryophyllales, Sheikh et al. 1995). In total, holocentric chromosomes are estimated to occur in almost 5,500 angiosperm species (approximately 1.5–2% of flowering plants), covering a wide variety of life strategies including trees (*Myristica*), perennial graminoids or annual herbs (Juncaceae, Cyperaceae), parasites (*Cuscuta*), bulbous geophytes (*Chionographis*), and carnivorous hemicryptophytes (*Drosera*). Holocentric plants exhibit variation in breeding strategies because although many are hermaphrodites, they also include some dioecious taxa (*Myristica*, some species of *Carex*, *Kobresia*, *Didymiandrum* and *Scirpus*) and even gynodioecious taxa (*Chionographis*, *Eriophorum vaginatum*, *Juncus kraussii*, and *J. roemerianus*). Agamospermy alone has never been satisfactorily confirmed (when tested) in holocentric plants. There is no apparent geographical preference among holocentric taxa. Although the most species-rich holocentric genus *Carex*, which accounts for almost 2,000 species, is distributed primarily in the temperate to the arctic climatic zones of the Northern hemisphere, many other holocentric genera occur preferentially in tropical areas.

Outside the angiosperms, holocentric chromosomes have been reported in some green algae (Desmidiales and Zygnematales; Godward 1954; King 1960) but so far they have never been found in gymnosperms, ferns or bryophytes.[10] Among animals, holocentric chromosomes are widespread throughout several invertebrate groups, including Nematoda, Onychophora, Dermaptera, Heteroptera, Sternorrhyncha, Auchenorrhyncha, Lepidoptera, Odonata, Phthiraptera, Psocoptera, Trichoptera, and Zoraptera, and are occasionally found in some families or orders in Arachnida or Chilopoda. So far they have not been reported in vertebrates (reviewed in more detail by Mola and Papeschi 2006).

Because the majority of karyological studies in plants have been limited to counting chromosomes, monocentrism seems to be the predominant state among angiosperms. In part this is because it has automatically been assumed to be present in all taxa in which holocentric chromosomes have not been proven using other cytogenetic techniques. In fact, in addition to the group of taxa in which holocentrism has been conclusively identified, it is possible that it may also be present within genera which have been assumed to be all monocentric but not studied in detail. For instance, in the largely monocentric genus *Cuscuta*, it was only through detailed karyotype analysis that holocentric chromosomes were identified in subgenus *Cuscuta*.

12.4 Chromatin Structure of Holocentric Chromosomes

An initial study by Ray and Venketeswaran (1978) analysed the distribution of constitutive heterochromatin and determined that C-bands were dispersed along the entire length of the holocentric chromosomes of *Luzula elegans* (= *L. purpurea*) and *L. multiflora*, in contrast with the centromeric location of C-bands in *Vicia faba*, the representative monocentric organism used by these authors. However, the dispersed pattern of heterochromatin was not confirmed in subsequent investigations of holocentric taxa using C- or Q-banding (*Drosera*: Sheikh and Kondo 1995; *Rhynchospora*: Vanzela and Guerra 2000; *Cuscuta*: Guerra and García 2004). In the chromosomes of these holocentric taxa, heterochromatin was observed to be localized in the terminal or, less frequently, interstitial or median regions. This is therefore not so different from species with monocentric chromosomes where, in addition to the centromeric location that prevails in small chromosomes (<3 μm), terminal or interstitial heterochromatin regions are also frequently observed (Guerra 2000).

[6] The subfamily Chionographioideae comprises two monotypic genera—*Chionographis* and *Chamaelirium*. Given the presence of holocentric chromosomes in *Chionographis*, it has been suggested that *Chamaelirium luteum* may also possess such chromosomes. However, cytological studies have so far failed to show whether the 18 chromosomes, which are very small compared with other species in Melanthiaceae, are holocentric or monocentric (Thomas Meagher, personal communication).

[7] The putative holocentrism reported by Chakravorti (1948a, b) in the families Zingiberaceae and Musaceae from the monocots was not confirmed in later studies.

[8] Holocentrism is present in the *Cuscuta* subgenus *Cuscuta*, whereas monocentrism has been confirmed in the subgenus *Monogyna*. Recently, holocentric chromosomes were also found in some taxa of the subgenus *Grammica* (Guerra et al. 2010), in which monocentrism was expected. In the subgenus *Grammica*, the sizes of the genomes and chromosomes are highly divergent (McNeal et al. 2007), which could indirectly suggest the presence of holocentrism (see Sect. 12.7.1).

[9] In *Aldrovanda*, a related monotypic genus of Droseraceae, chromosomes are very small and isodiametric without any apparent primary constriction, but their holocentric nature has never been satisfactorily confirmed. The monotypic genus *Dionaea* from the same family, however, possesses monocentric chromosomes, as does *Drosophyllum*, a monotypic genus recently separated into a different monotypic family.

[10] The unclear centromeric status reported by Vaarama (1954) in the moss *Pleurozium schreberi* which is sometimes considered to be holocentric, is actually neocentromeric activity (Dawe and Hiatt 2004).

No clear differences between holocentric and monocentric chromosomes were found in either the location of ribosomal DNA (rDNA) arrays or the portion of rDNA arrays associated with the nucleolar organizer (Greilhuber 1995; Vanzela et al. 1998, 2003; Guerra and García 2004; da Silva et al. 2010). Nevertheless, substantial structural differences between holocentric and monocentric plant chromosomes have been recognized in (1) the amount and distribution of the kinetochore/centromeric proteins, (2) the number and location of centromere-like satellites and (3) the phosphorylation pattern of the canonical histone H3, which is temporally and spatially associated with the cohesion of sister chromatids.

The proteins of the kinetochore/centromere complex are conserved across all eukaryotes. The centromere-specific variant of histone H3 (CENH3) plays a key role in this complex. In monocentric chromosomes, CENH3 is interspersed with canonical H3 only at the centromeres, deliminating the boundaries of the centromere (Jiang et al. 2003). To see whether a similar pattern was also found in species with holocentric chromosomes Nagaki et al. (2005) studied the distribution of LnCENH3 (= CENH3 in *Luzula nivea*). As expected, immunostaining with an antibody against LnCENH3 confirmed previous ultramicroscopic observations which showed that the functional kinetochore extends along the entire length of the poleward faces of the mitotic metaphase chromosomes, with the exception of the telomeric regions (Bokhari and Godward 1980). In cross-section, in which the chromosome is viewed axially (end-on from the telomere), the kinetochore was observed to be embedded in the outside groove, that runs along the chromosomal axis (Bokhari and Godward 1980; Nagaki et al. 2005). This groove-embedding of the kinetochore seems to be common, at least in the genus *Luzula*, given that it was also observed in the transverse sections of mitotic chromosomes in *Luzula elegans* (= *L. purpurea*; Braselton 1971).

Unlike monocentric organisms, in which the amount of CENH3 is constant throughout the cell cycle, the amount of LnCENH3 varies throughout the cell cycle in *Luzula nivea*. During interphase, LnCENH3 is clustered in small, dispersed areas, and its levels increase significantly from prophase to metaphase and then decrease from anaphase to interphase (Nagaki et al. 2005). A similar pattern has also been observed in the holocentric nematode model organism *Caenorhabditis elegans* (Buchwitz et al. 1999), suggesting that varying amounts of CENH3 during the cell cycle may provide a way to distinguish holocentric chromosomes from monocentric ones.

In contrast to the conservation of centromere/kinetochore proteins across eukaryotes, centromeric DNA is highly variable in (1) the arrangement and number of satellite repeats and (2) the length and nucleotide sequence of the respective monomers. The amount of centromeric DNA also varies widely between and within species, from tens of kilobases to several megabases. Large arrays of centromeric satellite repeats may be intermingled with centromeric retrotransposons (Jiang et al. 2003; Ma et al. 2007), as observed in grasses with specialized centromeric retrotransposons of the Ty3-*gypsy* superfamily. In monocentric chromosomes, the length and motif of the centromeric satellite DNA monomer may be species-specific. Examples of this species specificity are the 180 bp pAL1 repeat in *Arabidopsis thaliana* (Nagaki et al. 2003), the 156 bp CentC repeat in maize (Ananiev et al. 1998) and the 155 bp CentO repeat in rice (Cheng et al. 2002).

Centromeric satellites have also been found in holocentric chromosomes. Based on the homology to RCS2 (rice centromeric sequence 2, another name for CentO), Haizel et al. (2005) isolated and characterized a *Luzula* centromeric satellite (LCS1) from *Luzula nivea* (Juncaceae). In this species and nine other congeners, the LCS1 monomer was shown to range in length from 173 to 178 bp and was arranged in tandem arrays within the genome. The total size of these tandem arrays was about 50 kb, which is less than the shortest known centromeric satellite array in flowering plants (65 kb in centromere 8 of *Oryza sativa*: Cheng et al. 2002). It is possible that the uncommonly short LCS1 arrays may be connected with the origin of holocentric chromosomes (see Sect. 12.8).

LCS1 arrays in chromosomes of *Luzula nivea* are unique not only in size, but also in location. They occur along all 12 chromosomes of *L. nivea*, with at least five clusters per chromosome, whereas the centromeric satellite arrays of monocentric organisms are restricted to the core of the localized centromere. The arrays of LCS1 are associated with heterochromatin and co-localize with the kinetochore, suggesting that they are incorporated into regions that function as centromeres in holocentric chromosomes (Haizel et al. 2005). During interphase, the centromeric histone homolog LnCENH3 may be associated with LCS1, and together, this complex may determine the chromosomal regions from which the formation of the linear-shaped kinetochore begins. However, the assembly of the kinetochore along the whole chromosome raises a question as to how the formation of the kinetochore can span euchromatic, expressed regions. In *Caenorhabditis elegans*, this problem seems to be solved by a mechanism related to RNA interference. Initially, the transcripts from protein-coding regions become substrates for RNA-dependent RNA polymerase, which creates new antisense transcripts. Subsequently, ribonuclease cleaves the newly formed transcripts to generate small RNAs that are antisense to the original transcripts of the protein-coding genes. These small RNAs interact with specific proteins, and together, they stabilize chromatin and kinetochores, similar to the pericentromeres of monocentric chromosomes (Claycomb et al. 2009). It is likely that a similar mechanism based on the RNA interference-related pathway also works in holocentric plants.

The cell cycle-dependent phosphorylation of canonical histone H3, which seems to be required for the cohesion of sister chromatids (Gernand et al. 2003; Houben et al. 2007), is another feature that distinguishes the structure of holocentric and monocentric chromosomes. In monocentric plants, the phosphorylation of H3 is restricted to the pericentromeric regions during mitosis and meiosis II, whereas H3 is phosphorylated along the entire chromosome during prophase I, when the sister chromatids are held together along their whole length (Manzanero et al. 2000). In contrast, in holocentric plants such as *Luzula nivea* and *L. luzuloides* (Juncaceae) or *Rhynchospora tenuis* (Cyperaceae), the phosphorylation of H3 occurs along entire chromosomes during both mitosis and meiosis (Gernand et al. 2003; Nagaki et al. 2005; Guerra et al. 2006).

12.5 Mitosis in Holocentrics

Upon entry into mitosis, both monocentric and holocentric chromosomes become more compact in structure (reviewed by Stack and Anderson 2001), and sister chromatids must be held together to prevent aneuploidy caused by premature segregation. In this process, condensin and cohesin proteins have a key role. Condensin I facilitates the general condensation of chromatin, while condensin II creates a rigid chromosome structure (Maddox et al. 2004). The adhesion of chromatids is ensured by cohesins (Ono et al. 2003).

The stronger tension of holocentric chromatids due to the large target area for microtubule attachment creates a special requirement for more compact condensation and more extensive cohesion compared with monocentric chromatids. The more rigid structure of holocentric chromatids is ensured by the dominant activity of the condensin II complex along the entire length of the chromatid. In monocentric chromosomes, condensin II is enriched only in the centromere/kinetochore region, so their arms are not as rigid as those in holocentric chromosomes. Extensive cohesion along the whole length of sister chromatids ensures their close parallel orientation before spindle attachment and that their distance does not dramatically increase when bipolar attachments form (Maddox et al. 2004). Theoretically, the large surface area of the kinetochore in holocentric chromosomes increases the likelihood of attachments to the microtubules emanating from the opposite spindle poles (merotelic attachment), which could result in improper chromosome segregation and, thus, aneuploidy (Knowlton et al. 2006). However, the lengthwise cohesion of rigid chromatids leads to a fixed (back-to-back), bipolar chromatid orientation and thus restricts the attachment of a particular kinetochore to a single spindle pole, ensuring that each of the sister kinetochores attaches to the opposite spindle pole (Stear and Roth 2002).

The structure of the diffuse kinetochore is very similar to the organization of the classical trilaminar kinetochore identified in monocentric chromosomes using electron microscopy (Rieder 1982). The gradual formation of the kinetochore can be visualized using the centromere/kinetochore protein CENH3. While CENH3 occupies discrete domains during interphase (Fig. 12.1a), these regions move closer to form a continuous lengthwise kinetochore at the poleward side of each chromatid during mitosis (Fig. 12.1b). Holocentric and monocentric chromosome behaviour is roughly similar during prometaphase, in which kinetochores attach end-on to dynamic microtubules to congress chromosomes to a very tight metaphase plate. During metaphase, the holocentric sister chromatids are oriented parallel to the equatorial plate, just as monocentric chromatids are. During anaphase (Fig. 12.1c), sister chromatid cohesion is resolved, and chromatids are pulled along their entire length to the spindle poles. As a result, their parallel orientation persists during the anaphase movement, and the holocentrics lack the "V-shape" typical of monocentrics. Because the sister chromatids are pulled towards the poles by the spindle microtubules which act along the entire length of the chromosomes neither the preferential orientation nor the spatial localization of the telomeres (Rabl-orientation) form in interphase nuclei.

12.6 Meiosis in Holocentrics

The main scenario for meiotic prophase I is the same in holocentrics as in monocentrics. In both, sister[11] chromatids are interconnected over their whole length by cohesion and homologous chromosomes align to enable the formation of the synaptonemal complex and recombination via crossover.

[11] In mitosis, the sister chromatids are clearly defined both in holocentrics and monocentrics as two identical (daughter) chromatids derived from duplication (replication) of one and the same chromosome during of the cell cycle [see e.g., King RC, Stansfield WD, Mulligan PK (2006) A dictionary of genetics. 7th Ed. Oxford Univ. Press, Oxford]. In meiosis, the term "sister" regarding to patchwork chromatids recombined via crossover is not so easy to define in holocentrics as in monocentrics where connection of centromeres defines which two chromatids are sister while those which are not joined by a common centromere are non-sister (= homolog) chromatids [see e.g., Brooker RJ (2009) Genetics: analysis & principles. 3rd Ed. McGraw-Hill Irwin, New York, p. 49, 821]. In holocentrics, the "majority rule" defines sister and non-sister chromatids meaning that sister chromatids share longer identical segments originated via replication, while non-sister chromatids share shorter absolutely identical segments. From a practical viewpoint, the terms "sister" and "non-sister" well fitted to mitotic chromosomes are universally conserved by such additional criteria (common centromere or "majority rule") for any type (mitosis or meiosis, holocentric or monocentric) or phase (pre- or postrecombinational) of cell division to overcome the difficulty that no original sister chromatids are present anymore after recombination.

Each round of meiotic division in both chromosomal types is associated with one-step loss of cohesion between chromatids. The main differences between holocentric and monocentric chromosomes are (1) the way in which sister chromatid cohesion is lost, (2) the number of crossover sites per bivalent and (3) the orientation of the chiasma within the bivalent to manage the segregation.

In monocentrics, sister chromatid cohesion in the arms is released, and only the centromeric regions remain connected during meiosis I. The connection between non-sister chromatids is retained at the chiasmata while the remaining internal cohesion of the sister chromatids is protected at the centromeres by members of the MEI-S332/Shugoshin protein family (Sakuno et al. 2009). Moreover, the sister kinetochores are covered by the kinetochore protein MIS12, which causes both kinetochores to present a shared face to the microtubule-binding proteins and allows the "co-orientation" of the sister chromatids during meiosis I (Li and Dawe 2009). This "co-orientation" is later changed to a "bi-orientation", and the remaining cohesion between the centromeres is resolved during meiosis II, similar to mitotic division.

Holocentric chromosomes lack a predefined site for the protection of cohesion in meiosis I, so the chiasma serves as an alternative point of reference. In contrast to monocentrics, the number of crossover sites appears to be limited in holocentrics, occurring only once or twice per bivalent (White 1973). Indeed, a deleterious effect caused by the formation of more than two chiasmata was observed by Nokkala et al. (2004). Each chiasma persists to the end of metaphase I, holding the four chromatids in a cross-shaped bivalent (Fig. 12.2b), and following further condensation this gives rise to the so-called "box configuration" (Fig. 12.2c, d, g; Nordenskiöld 1962). The orientation of the meiotic bivalents on the metaphase plate mediates the stepwise loss of sister chromatid cohesion. The cohesion is released in those parts of the bivalent lying on the equatorial plate (shown in yellow in Fig. 12.2d, g), allowing segregation toward the opposite poles during meiosis I, whereas those chromatid parts lying above and below the equatorial plate (oriented axially) remain connected (Fig. 12.2e, h).

Because there is an absence of detailed knowledge about meiosis in plant holocentrics, the detailed studies in *Caenorhabditis elegans* can serve as a model to show how the crossover position can direct the holocentric chromosome orientation. It should be noted that *C. elegans* is an animal model and that molecular mechanisms described below might differ from those of holocentric plants, especially in the proteins involved in meiotic division, although it is likely that general principles would be similar. Moreover, *C. elegans* exhibits telokinetic meiosis whereas most holocentric plants show holokinetic behaviour (see below and Sects. 12.6.1 and 12.6.2). During meiosis in *C. elegans*, the number of crossover sites is limited to a single chiasma for each pair of homologs (Meneely et al. 2002). During late prophase, the homologs desynapse and are remodeled into a cross-shaped bivalent (Fig. 12.2a, b). The position of the chiasma determines the shape of the bivalent where the longer axis separates sister chromatids while the shorter axis separates nonsister chromatids regarding majority rule (see footnote 11). Cohesins and the HORMA proteins (HTP-3 and HIM-3) remain homogeneously distributed, connecting chromatids on both the long and short arms of the bivalent (Nabeshima et al. 2005). In contrast, the distribution of other proteins is associated with particular chromosomal axes and the anaphase-specific release or retention of these other proteins specifies the direction of the bisection of the bivalent. The meiosis-specific proteins HTP-1 and HTP-2 (of the HORMA domain family) and LAB-1 (a nematode-specific protein) are depleted on the short axis and are retained on the long axis, where cohesion is retained until meiosis II (de Carvalho et al. 2008; Martinez-Perez et al. 2008). In contrast, the SYP proteins associate with the short arms, (where cohesion is lost during meiosis I) (Pasierbek et al. 2001; shown in yellow in Fig. 12.2g and hypothetically expected and shown also in yellow in postreductional type in Fig. 12.2b, c, d), and occupy domains where HTP-1/2 and LAB-1 are depleted (de Carvalho et al. 2008; Martinez-Perez et al. 2008). The persistence of SYP-1/2 on the short axis correlates also with the recruitment of the Aurora-like kinase AIR-2, which directs the loss of sister chromatid cohesion through the phosphorylation of the cohesin subunit REC-8 and its subsequent cleavage by separase (Rogers et al. 2002). This spatiotemporal protein diversification allows sister chromatids to switch between the "co-orientation" and the "bi-orientation" during meiotic anaphases.

Whereas the attachment of microtubules in mitosis always occurs along the whole poleward face of each chromatid (holokinetic), two types of microtubule attachments exist in meiosis. Depending on whether the orientation of the meiotic bivalent is equatorial or axial, microtubule attachment is holokinetic (Fig. 12.2e, f) or telokinetic (Fig. 12.2h, i), respectively.[12] Both types differ from conventional monocentric meiosis, which is pre-reductional (i.e the reductional division is followed by the equational division) (Fig. 12.2j–l). Assuming two alternative successions of the

[12] As a consequence of the fission or fusion of chromosomes, trivalents are sometimes formed during meiosis: two fragments are homologous to one non-fragmented chromosome, or one fused chromosome is homologous to the two non-fused ones. Two types of segregation are possible in this case. In the first case (telokinetic meiosis), the larger chromosome migrates to one pole, whereas the two shorter chromosomes migrate to the other pole during meiosis I. In the second case (holokinetic meiosis), two triads of individual chromatids segregate. This specific situation, if observed, could be also used to identify the holokinetic or telokinetic type of meiosis.

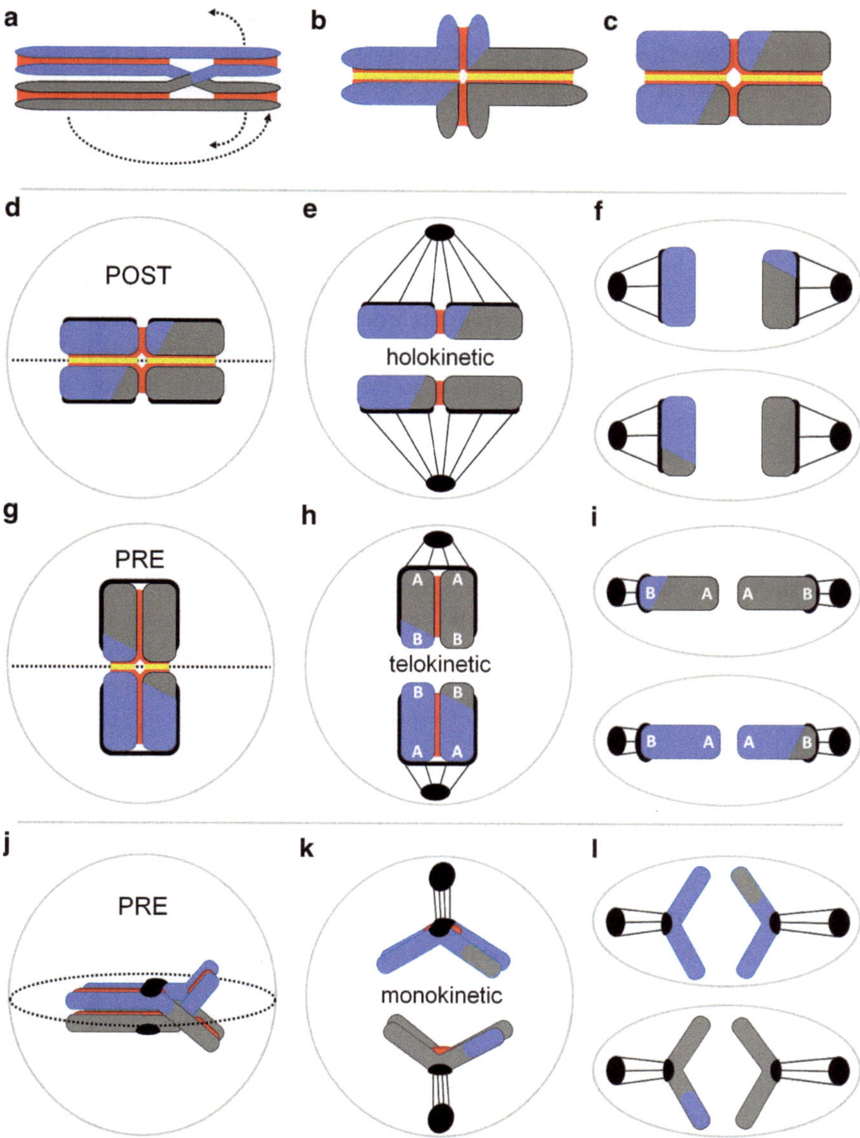

Fig. 12.2 Comparison of meiosis in holocentric (**a–i**) and monocentric (**j–l**) organisms. Sister segments of chromatids are represented by the same color, either *grey* or *blue*, kinetochores and microtubules are *black*, cohesion between chromatids is indicated in *red*, proteins associate specifically with the axis laying in the equatorial plate are in *yellow* (in **d** [*Luzula*] hypothetically expected based on analogy with *Caenorhabditis*), and the equatorial plate is indicated by the *dotted line*. Holocentric chromosome behaviour is shown in prophase I (**a–c**; here postreductional type), holokinetic meiosis (**d–f**; *Luzula*), telokinetic meiosis (**g–i**; *Caenorhabditis*), monocentric meiosis (**j–l**), metaphase I (**d**, **g**, **j**), anaphase I (**e**, **h**, **k**), and anaphase II (**f**, **i**, **l**). (**a**) Removal of the synaptonemal complex and the remodelling of chromosomes. (**b**) Cross-shaped bivalent. (**c**) The "box configuration" of the bivalent resulting from the further condensation of the chromosome and the chiasma resolution. (**d**) Equatorial orientation of bivalents. (**e**) "Sister" chromatids (corresponding to larger sister segments) segregate (equational step), while half-bivalents (short axis) remain connected. (**f**) Chromatids are dragged broadside (reductional step). (**g**) Axial orientation of a bivalent. (**h**) "Sister" chromatids (corresponding to larger sister segments) move together (reductional step) pulled by the chromosome ends. (**i**) Chromatids segregate, pulled by their ends, which were inactive during meiosis I (equational division). (**j**) Equatorial orientation of the bivalent; sister parts of chromatids are connected. (**k**) Chromosome homologs segregate, retaining the junction only in the centromeres (reductional step). (**l**) Sister chromatids segregate, pulled at the centromeres (equational division)

divisions, the post-reductional type was identified in holocentric species, in contrast to the pre-reductional in monocentrics (Wahl 1940). Actually, regarding the position of the crossover site in the telokinetic (pre-reductional) meiosis, the cohesion proteins are released from the short axis so that the whole homologous chromosomes move to opposite poles. In contrast, holokinetic meiosis begins with the separation of the long axis (the sister chromatids), meaning that the reductional step occurs during the second division. Thus, the terms holokinetic and telokinetic seem to be more appropriate for the description of chromosomal meiotic movement in holocentrics. Regardless of the type of chromosomal movement (holokinetic, telokinetic, or monokinetic), the final product of meiosis (i.e., four gametes), is the same.

12.6.1 Holokinetic Meiosis

Most data on holokinetic meiosis have been obtained from Juncaceae (Nordenskiöld 1962), Cyperaceae (Da Silva et al. 2005) and *Cuscuta* (Convolvulaceae; Pazy and Plitmann 1987, 1994). Although this type of meiotic behaviour seems to be typical for plants with holocentric chromosomes, it has also been observed in some animals, for example, during meiosis in the female mealybug *Planococcus citri* (Bongiorni et al. 2004). The segregation of sister chromatids has also been documented for univalent chromosomes, such as the X and Y chromosomes of some holocentric insects (Nokkala et al. 2004) and the B chromosome of *Cuscuta babylonica* (Pazy 1997).

During metaphase I, the long axes of the cross-shaped meiotic bivalents lie in the equatorial plate so that most of the sister chromatid cohesin is released upon anaphase I (Fig. 12.2d). This type of release allows the "bi-orientation" of sister chromatids and their segregation to opposite spindle poles (equational division; Fig. 12.2e). Connections along the non-sister chromatids (short axis) are retained at the end of the chromosome in a small region connected with chiasma (shown in red in Fig. 12.2e). This chromosome structure is also known as a half-bivalent or demi-bivalent (Pazy and Plitmann 1987). The kinetochore plate covers the entire chromatid in a manner similar to that seen during the holocentric mitotic anaphase and the chromatids are pulled along their entire length (Nordenskiöld 1962, 1963). In *Luzula* species, homologous chromatids re-associate during the short meiotic interphase and remain paired until the end of metaphase II (Nordenskiöld 1962), enabling them to become correctly oriented to opposite poles during metaphase II and to subsequently segregate in a holokinetic manner during anaphase II (Fig. 12.2f). In contrast, only telomeric associations between homologous chromatids remain until the second meiosis in *Cuscuta*, enabling a reductional division (Pazy and Plitmann 2002). Both of these types of chromosome associations (i.e., re-associated and only terminally associated) were observed by Strandhede (1965a) in *Eleocharis palustris* subsp. *palustris*.

12.6.2 Telokinetic Meiosis

The appropriate model of telokinetic meiosis is the nematode *Caenorhabditis elegans* (Maddox et al. 2004; Schwarzstein et al. 2010). In *C. elegans*, as noted above, only one crossover forms between homologs (Meneely et al. 2002), and the cross-shaped meiotic bivalent is oriented axially so that the short axis lies in the equatorial plate (Fig. 12.2g; Albertson and Thomson 1993). In telokinetic meiosis, the kinetic activity of the holocentric chromosomes changes from dispersed in mitosis to localized in meiosis, and the kinetochore forms a cup-like shape (Dumont et al. 2010). This conformation ensures the encasement of sister chromatids together to promote their movement as a single functional unit during meiosis I (Monen et al. 2005), similar to the connection of the monocentric sister kinetochores (Li and Dawe 2009). The ends of the chromosomal axes represent two potential "localized meiotic centromeres" that can mediate chromosome spindle interactions. Depending on the crossover site, one end becomes active. Thus, the active chromosome end is oriented to the spindle pole during meiosis I (Fig. 12.2h: end A), and the other end (Fig. 12.2i: end B) is then active during meiosis II.

In *C. elegans*, the kinetic activity is localized at the chromosome end that is opposite to the chiasma, whereas in the insect *Triatoma infestans* (Heteroptera), the kinetic activity may alternatate between the chromosome ends, which are differentiated by the presence or absence of a terminal C-band, irrespective of the chiasma position (Pérez et al. 1997). Similarly, in different somatic cell types of the nematode *Parascaris univalens*, the kinetic activity can be associated with either heterochromatin or euchromatin (Goday and Pimpinelli 1989), indicating that the kinetic activity in holocentric chromosomes is not dependent on a specific nucleotide sequence but rather is epigenetically controlled (Sullivan et al. 2001). A clear example of a telokinetic plant species is from *Chionographis* (Melanthiaceae).[13] Tanaka and Tanaka (1979) determined that there were four Japanese *Chionographis* taxa with holocentric chromosomes. The meiotic behaviour of two of these—*C. japonica* and *C. japonica* var. *hisauchiana*—was later (Tanaka and Tanaka 1980) studied, revealing that *Chionographis* bivalents orient axially during meiosis and undergo reductional separation in the first meiotic division and equational division in the second.

[13] In plants, telokinetic meiosis was reported in *Luzula elegans* by Malheiros et al. (1947), who suggested that during anaphase I and II, the kinetic activity is restricted to both ends of each chromosome. This species was later analysed more precisely by Östergren (1949), who claimed that *L. elegans* chromosomes undergo inverted meiosis (i.e., holokinetic behaviour). The holokinetic chromosomal behaviour has also been reported in other species of *Luzula* (Nordenskiöld 1962, 1963; Braselton 1971, 1981).

12.7 Holocentric Karyotypes and Their Evolution

The range of chromosome numbers encountered in holocentric plants is almost as wide as that in monocentric plants. Despite the rarity of holocentrism (see Sect. 12.3), we find that within the very small group of six angiosperm species which possess the lowest chromosome number so far reported in angiosperms ($n = 2$),[14] the holocentric species *Rhynchospora tenuis* (Vanzela et al. 1996). Indeed, very low chromosome numbers are not rare among other holocentric species, as $n = 3$ has been recorded in *Eleocharis subarticulata*, *Fimbristylis narayanii*, *F. umbellaris*, *Luzula elegans*, and *Drosera roseana* (Malheiros-Garde and de Castro 1947; Rath and Patnaik 1978; Nijalingappa and Tejavathi 1984; Kondo et al. 1994; da Silva et al. 2005), and $n = 4$ has been found in *Cuscuta babylonica* (Pazy and Plitmann 1987, 1994). At the other end of the scale, we find some holocentric plants with extremely high chromosome numbers, such as some taxa of the family Cyperaceae including *Eleocharis kuroguwai* ($n =$ c. 98; Hoshino 1987), *Cyperus pangorei* ($n = 104$; Nijalingappa and Tejavathi 1984), and *C. cyperoides* ($n = 110$; Tejavathi 1988). Similarly, a wide range of chromosome numbers seems to be common in holocentric animals. For example, the extremely low chromosome numbers of $n = 1$ and $n = 2$ have been reported in the germline cells of the nematode *Parascaris univalens* (Ascaridae; Goday et al. 1988) and in the insect *Lethocerus* sp. (Belostomatidae, Heteroptera; Papeschi and Bressa 2006), respectively, whereas exceptionally high numbers of chromosomes have been found in some butterfly species: $n = 147-150$ in *Plebicula dorylas* (Lepidoptera, Lycaenidae; Kandul et al. 2004), $n =$ c. 150 in *Abananote erinome* (Lepidoptera, Ithomiidae; Brown et al. 2007), and $n = 223$ in *Polyommatus atlantica* (= *Lysandra atlantica*; Lepidoptera, Lycaenidae; de Lesse 1970). The latter is the highest chromosome number so far reported for any animal (Marec et al. 2009). Holocentric desmid algae also exhibit chromosome numbers that vary extensively, ranging from $n = 2$ in *Spirogyra cylindrica* to $n =$ c. 200 in *Micrasterias rotata*, and $n = 592$ in *Netrium digitus* (King 1960). As in monocentric plants, both polyploidy and chromosomal rearrangements are likely to be the main forces that increase the range of chromosomal variation. In animals, however, where polyploidy is less frequent, chromosomal rearrangements are likely to generate the wide range of karyotype variation more effectively in holocentric taxa than in monocentric ones. Indeed, a much higher chromosomal rearrangement rate was detected in holocentric organisms, such as the nematodes and butterflies, compared with monocentric organisms (Brassicaceae, Poaceae, Solanaceae, Mammalia; Table 12.1) using the number of chromosomal disruptions in the available genome sequences of the taxa in relation to their genome size (Mb) and divergence time (millions of years).

Table 12.1 Rate of chromosomal rearrangement in holocentric and monocentric taxa

Chromosome type/taxon	Rearrangement rate[a]
Holocentrics	
Lepidoptera	1.370–3.300[b]
Nematoda	0.500–0.700[b]
Monocentrics	
Diptera	0.05253–0.08485[c]
Brassicaceae	0.00357–0.01071[c]
Poaceae	0.00013–0.00041[c]
Solanaceae	0.00010–0.00029[c]
Mammalia	0.00006–0.00030[c]

[a]Number of chromosomal breakages (Mb/millions of years); not calibrated by generation time
[b]Data taken from d'Alençon et al. (2010)
[c]Data taken from Ranz et al. (2001)

Within groups of related monocentric plants, a pure arithmetic relationship among chromosome numbers is usually sufficient to detect recent polyploidy events (or true aneuploidy[15]). However, in holocentric taxa, this empirical pattern is often obscured by frequent chromosomal rearrangements, particularly chromosomal fissions (agmatoploidy) and fusions (symploidy). Moreover, in some holocentric organisms, the relatively clear presence of polyploidy could be illusory due to complete agmatoploidy (concerted fission), which occurs, for example, in *Luzula* (see Sect. 12.7.3.). Therefore, chromosome counting must be combined at least with the measurement of DNA content to reliably evaluate the role of polyploidy in holocentric taxa. A combination of both parameters, such as average chromosome size ($C/n = 2C$ DNA/$2n$; Šmarda et al. 2008) could serve as a criterion in which stability (conditioned by a linear relationship between 2C and $2n$) indicates the presence of polyploidy or true aneuploidy if chromosome number is variable. Conversely, an increase or decrease in C/n could suggest fusion or fission events, respectively (see Nishikawa et al. 1984). Unfortunately, as has been observed in some monocentric organisms, the proliferation or removal of retrotransposons can also be very rapid over short evolutionary scales in holocentric organisms, and can thus also be responsible for increases or decreases in chromosome size within

[14] Such a low chromosome number has been found in only five other angiosperms: *Zingeria biebersteiniana*, *Colpodium versicolor*, *Brachycome dichromosomatica*, *Haplopappus gracilis*, and *Ornithogalum tenuifolium* (Cremonini 2005).

[15] The gain or loss of a particular chromosome or chromosomes due to non-disjunction during irregular meiotic chromosomal segregation.

genera, subgenera or other groups of related taxa independent of chromosome number changes. The estimation of the retrotransposon density in the genome of a particular species is a further parameter that should be taken into consideration in the study of holocentric karyotype evolution (see Zedek et al. 2010). Additionally, the study of multivalent formation during the first meiotic metaphase or a comparative analysis of the FISH patterns in particular species could be helpful for detecting individual chromosomal fission/fusion events or other chromosomal rearrangements, particularly at the intra-specific level (see da Silva et al. 2005, 2008a).

12.7.1 Symploidy and Agmatoploidy

Although in monocentrics, chromosomal fission usually results in the deletion of acentric fragments (see Lysák and Schubert 2013, this volume), such usually deleterious effects are largely prevented in holocentric species by the nature of their chromosomes. Moreover, the extended kinetochore reduces the risk of forming dicentric chromosomes, which after fusion, are often lethal in monocentric organisms. Given these observations and predictions it is perhaps not surprising that agamatoploidy and symploidy can play prominent roles in karyotype evolution of holocentric species compared with monocentric ones. (Heilborn 1924; Luceño and Castroviejo 1991; Hipp 2007; Hipp et al. 2007; Escudero et al. 2008; Hipp et al. 2009). In agreement with the higher rearrangement rate detected in holocentric organisms (Table 12.1), the probabilistic models of karyotype evolution in *Carex* (Cyperaceae) favor the gain or loss of an individual chromosome as the most probable mode of karyotype evolution in sedges, suggesting that fission or fusion has a more important role than polyploidy in this process (Hipp et al. 2009; Mayrose et al. 2010).

Recurrent chromosomal fission and/or fusion during karyotype evolution can lead not only to a wide range of chromosome numbers (see Sect. 12.7), but also to an almost continuous (dysploid) series of chromosome numbers within particular holocentric families, genera, sections or other infrageneric taxa.[16] Examples of karyotype variation in some holocentric plant genera are given in Table 12.2. The extremely broad and continuous karyotype variation on a small evolutionary scales can also be found in animals with holocentric chromosomes, such as the butterflies of Nymphalidae (the most specious family of Lepidoptera) where the chromosome number ranges nearly continuously from $n = 5$ to $n = 134$ (Brown et al. 2004, 2007; Marec et al. 2009). Furthermore, out of all the animal genera which have been examined cytologically, the 48-fold range in chromosome numbers in the insect genus *Apiomorpha* ($n = 2-96$; Hemiptera, Eriococcidae; Cook 2000) and the 13.4-fold range encountered in the butterfly genus *Agrodiaetus* ($n = 10-134$; Lepidoptera, Lycaenidae; Lukhtanov and Dantchenko 2002; Lukhtanov et al. 2006) are considered to be the most extensive (Cook 2000) and both these genera possess holocentric chromosomes.

Indirect evidence that agmatoploidy and symploidy are ongoing processes in plant species with holocentric chromosomes is suggested by the occurrence of univalents or heteromorphic trivalents, which are frequently observed in meiosis I in several taxa of Cyperaceae, particularly *Carex* (Wahl 1940; Tanaka 1949; Faulkner 1972; Hoshino 1981; Luceño and Castroviejo 1991; Luceño 1994; Hoshino and Waterway 1994; Escudero et al. 2008) and *Eleocharis palustris* subsp. *palustris* (Strandhede 1965a). Furthermore, intraplant variations in chromosome number suggest a high frequency of fission or fusion, such as observed in *Carex disticha* (Luceño 1992), *Eleocharis palustris* subsp. *waltersii* (Strandhede 1965b, 1966; Bureš et al. 2004) and in *Luzula multiflora*, *L. nivea* and *L. luzuloides* (Kuta et al. 2004), where the stability of DNA content contrasts strongly with the extensive variability in chromosome number.

A detailed analysis of particular chromosomal fusions in local populations of *Eleocharis maculosa* was performed by da Silva et al. (2008a). These authors found three cytotypes with $2n = 6$, 7, and 8, compared with the common cytotype of $2n = 10$, and explained their origins by fusions of particular non-homologous chromosomes using a sophisticated combination of conventional mitotic methods, FISH (using 45S rDNA, 5S rDNA and telomere probes) and an analysis of meiotic chromosome pairing. The structural heterozygosity of all three *Eleocharis maculosa* cytotypes suggests that they could not be assumed to be stabilized, creating chromosomal races (see da Silva et al. 2008a). Although meiosis could be irregular in holocentric structural heterozygotes (possessing an unbroken chromosome together with its homolog fragments or a fused chromosome with its non-fused homologs), the newly fused (or fragmented) karyotype could easily be homologized via selfing or backcrossing and subsequently fixed, for example, by mechanisms of genetic drift resulting in local chromosomal races. Indeed, this outcome was observed in *Carex conica* by Hoshino and Waterway (1994), *C. sociata* by Ohkawa et al. (2000), and *C. scoparia* var. *tessellata* by Hipp et al. (2010).[17] Some

[16] In some holocentric taxa, the wide range of intraspecific variation may also be associated with true aneuploidy, for example, in *Eleocharis uniglumis* s. l. ($2n = 44-51$, $53-56$, $59-89$, 92; Strandhede 1965b, 1966; Bureš 1998), where the chromosome number and DNA content are linearly correlated (Bureš and Kneřová unpubl.). Other examples of true aneuploidy in *Carex* are reviewed by Hipp et al. (2009).

[17] The presence of more than one cytotype in the species was documented in more than 100 species of *Carex* and many other taxa of Cyperaceae (Roalson 2008; Hipp et al. 2009).

Table 12.2 Variation in chromosome number in some holocentric plant genera

Family/genus	Range	Haploid chromosome number (n)
Droseraceae		
Drosera[a, b, c]	13.3-fold	3–7, 9–17, 20, 23, 30, 32, 36, 40
Melanthiaceae		
Chionographis[d]	1.8-fold	12, 21, 22
Juncaceae		
Luzula[a, b, e]	14.0-fold	3, 6–16, 18, 21, 23, 24, 26, 31, 33, 35, 36, 42
Juncus[a, b]	10.6-fold	9, 10, 13, 15–24, 30, 32, 34, 35, 40, 42, 45, 50, 53, 54, 60, 66, 67, 85
Cyperaceae		
Carex[f, g]	11.6-fold	5–47, 50, 52–58
Eleocharis[f, h]	36.0-fold	3–16, 18–30, 36–44, 47–50, 68, 86–92, 100, 108
Cyperus[f]	22.4-fold	5, 8, 9, 12, 13, 15–32, 34–45, 47–50, 52–60, 62–64, 66–69, 76, 80, 93, 98, 104, 110, 112
Rhynchospora[f, i]	12.5-fold	2, 4–13, 15, 18–22, 24, 25

[a]Bolkhovskikh et al. (1969)
[b]Goldblatt and Johnson (2010)
[c]Rivadavia et al. (2003)
[d]Tanaka and Tanaka (1979)
[e]Nordenskiöld (1951); Kirschner (1992)
[f]Roalson (2008)
[g]Rotreklová et al. (2011)
[h]Strandhede (1965b, 1966); Bureš (1998)
[i]Luceño et al. (1998); Vanzela et al. (2000, 2003)

stable, species-specific bimodal karyotypes, such as those in *Eleocharis palustris* subsp. *palustris* and *E. mamillata* ($2n = 12M + 4L$; Strandhede 1965b; Bureš et al. 2004), *Fimbristylis complanata* ($2n = 12M + 4L$; Yano and Hoshino 2006), or *Luzula alpina* ($2n = 12AL + 24BL$; Kirschner 1992; Bačič et al. 2007), could be assumed to have originated via symploidy or agmatoploidy then fixed through this mechanism. On the other hand, in holocentric organisms, the degree of complexity of such karyotype rearrangements should probably be higher than that in monocentric organisms in order to create a sufficiently effective reproductive barrier necessary for speciation. This point was argued by Hipp et al. (2010), who analysed the rate of gene flow among populations within the sedge species *Carex scoparia* and *C. pachystachya*, which both display extreme inter-population divergence in chromosome number.

A rather intriguing question is whether these symploid or agmatoploid "small steps" in the evolution of holocentric karyotypes have resulted from a stochastic or a deterministic process. The geographic trend in karyotype orthoselection[18] can be inferred from the chromosome variation in *Carex laevigata*, in which $2n$ increases from 69 in the Pyrenees Mountains, continues to increase along the northern and western shores, and reaches 84 at the southern tip of the Iberian Peninsula (Luceño and Castroviejo 1991; Escudero et al. 2008). Hipp et al. (2009), who also reviewed other geographic patterns in chromosome variation in *Carex*, considered these intraspecific chromosomal geographic clines in sedges to most likely have resulted from the stochastic effects of chromosome divergence associated with species range expansion rather than by any environmentally directed adaptive selection in chromosome number.

A phylogenetic trend of decreasing chromosome number associated with an increase in chromosomal size has been suggested for *Eleocharis*. In this genus species with the largest chromosomes are found in the evolutionary youngest subgenus (*Eleocharis* subgen. *Eleocharis*), while those species with numerous and very small chromosomes are found in the early diverging phylogenetic lineages comprising subgenera *Limnochloa* and *Zinserlingia* (Yano et al. 2004; da Silva et al. 2008b; Zedek et al. 2010). A similar pattern might also be present in *Schoenus* (Bhatti et al. 2007). In contrast, the *Luzula* species with the largest chromosomes, *L. elegans*, occupies a position that is sister to all other *Luzula*s based on the latest phylogenetic tree of this genus (Záveská-Drábková and Vlček 2010). In *Carex*, an evolutionary trend from lower to higher chromosome numbers was assumed by the early studies of Heilborn (1924) and in light of recent phylogenetic data the Heilborn hypothesis seems to hold generally over the broad evolutionary scale. *Carex* species with the lowest numbers and largest chromosomes (*C. siderosticta*, *C. pachygyna*, and *C. ciliatomarginata*) belong to section *Siderosticae* which forms a clade sister to the rest of the genus (Waterway et al. 2009). In this genus, however, the opposite trend was also found at a finer scale by Hipp (2007), who tested various probabilistic models on the North American species of *Carex* sect. *Ovales* and demonstrated that the processes of increasing and

[18] Karyotypic orthoselection occurs when a particular type of chromosomal rearrangement is present recurrently in a phylogenetic lineage (White 1973). In this case, the presence of fission or fusion is assumed to have occurred.

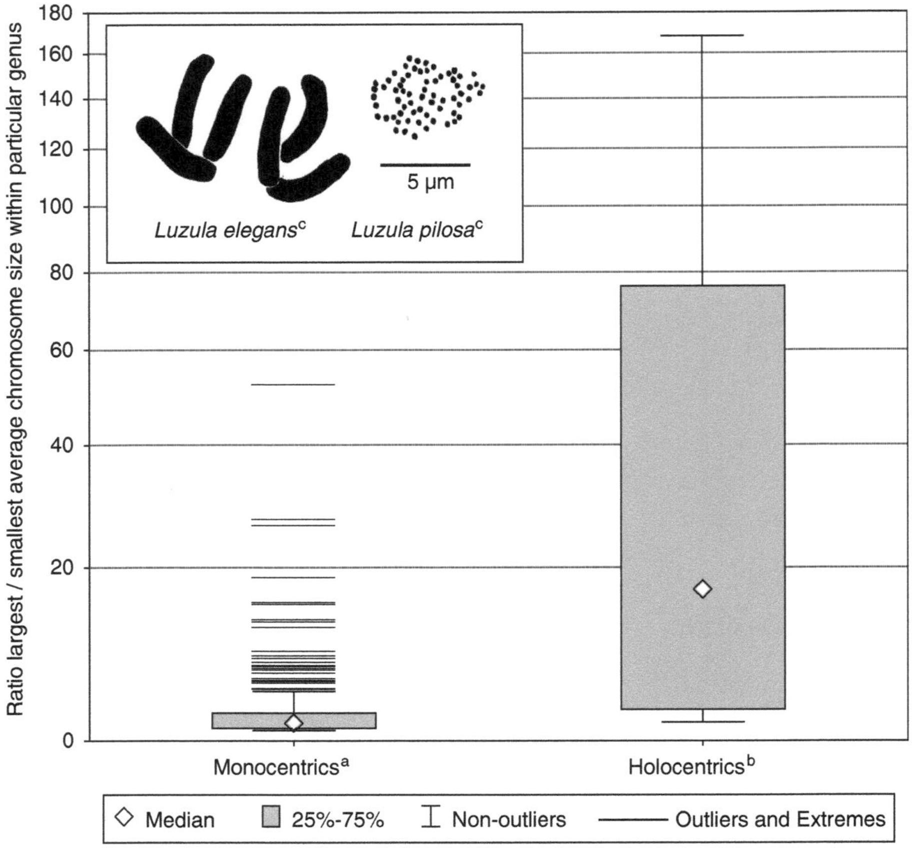

Fig. 12.3 Divergence in chromosome size within monocentric and holocentric plant genera in which the chromosome number and the amount of nuclear DNA are known for at least two species. The *inner box* illustrates the 168-fold difference in the size of chromosomes in the genus *Luzula*. (**a**) 2C and 2*n* data for the 333 genera taken from the Plant DNA C-values Database (Bennett and Leitch 2005). For each species in which the somatic DNA amount and chromosome number are both known, the average chromosome size was calculated. Subsequently, for all genera containing at least two species with a known average chromosome size, the ratio between the largest and the smallest chromosome was calculated. (**b**) Eight genera are represented using data for *Drosera*, *Scirpus*, *Cyperus*, *Carex*, and *Juncus* from the Plant DNA C-values Database (Bennett and Leitch 2005) and data for *Eleocharis* (Zedek et al. 2010), *Cuscuta* (subgenera *Cuscuta* and *Grammica*; McNeal et al. 2007) and *Luzula* (Barlow and Nevin 1976). (**c**) Taken from Nordenskiöld (1951); *Luzula elegans*: $2n = 6$, $2C = 8.55$ pg; *L. pilosa*: $2n = 66$, $2C = 0.55$ pg (Barlow and Nevin 1976)

decreasing chromosome numbers were non-random and were present in various phylogenetic lineages. This observation suggested that different kinds of orthoselection may have modulated rapid karyotype shifts during the early divergence of this section, resulting in recently different karyotypic equilibria within particular clades.

If such divergent orthoselection is also common in other species with holocentric chromosomes, it should result in the rapid divergence of chromosome size over a relatively short evolutionary time scale, for example, within genera. In fact, such extensive divergence in chromosome size is clearly apparent more often in holocentric plant genera compared with monocentric genera (Mann-Whitney, $p = 0.0003$, Fig. 12.3).

If fission and fusion are the main causes of rapid chromosomal size divergence in holocentrics, then one might expect that a plot of chromosome number (x-axis) versus genome size (y-axis) would give a straight line parallel to the x-axis due to the extreme variations in chromosome number (x-axis). Surprisingly, this was not found, instead a plot of data for 64 holocentric species showed that the relationship was negative; the more chromosomes, the smaller the genome size (Nishikawa et al. 1984; Roalson et al. 2007; Fig. 12.4).

This negative 2C/2*n* correlation could be explained by chromosomal fission sometimes resulting in minute fragments that are lost because they lack kinetochoric properties. However, the loss of artificially-induced small fragments has rarely been observed in holocentrics (de Castro et al. 1949). Instead, although no clear explanation for the negative 2C/2*n* relationship has so far been found, it is possible that agmatoploidy is accompanied by an unrecognized mechanism of DNA diminution in holocentric organisms. Indeed, if the genome sizes of particular phylogenetic lineages of holocentrics are compared with their sister monocentric relatives (Fig. 12.5), in almost all cases, holocentrism is associated with the smaller genome size.

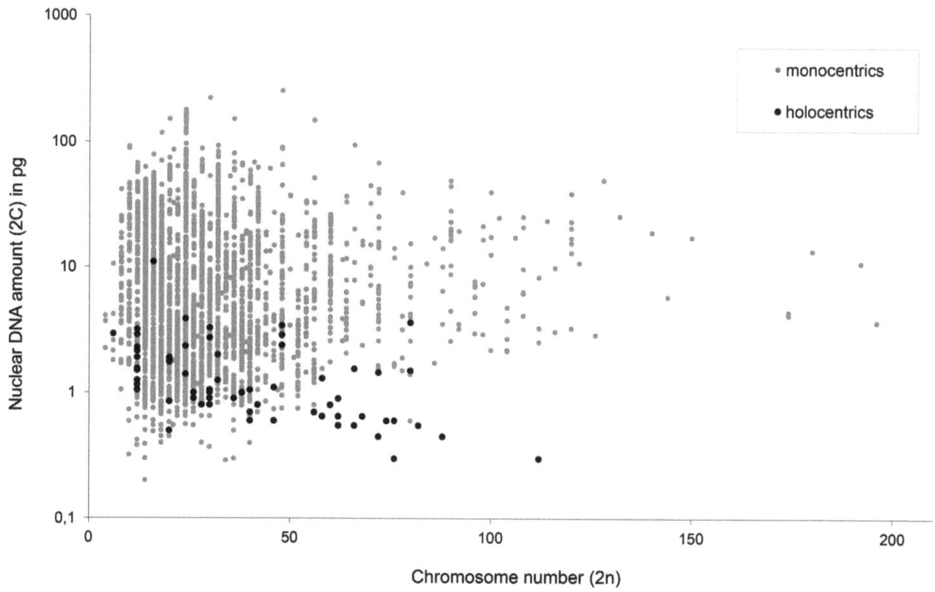

Fig. 12.4 Genome size versus chromosome number in holocentric (*black*; N = 64) and monocentric (*grey*; N = 3624) angiosperm species for which both the genome size and chromosome number are known (Data taken from the Plant DNA C-values Database, Bennett and Leitch 2005). For holocentrics, Spearman's Rho = −0.55; for monocentrics, Spearman's Rho = −0.09 ($p < 10^{-5}$ for both). The monocentric palm *Voanioala gerardii* (Arecaceae; $2n = 596$; $2C = 60$ pg) is outside of the scatter plot scale but was included in the test

An alternative explanation for the observed negative 2C/2*n* relationship is that chromosomal fusion (i.e., symploidy) has been accompanied by retrotransposon proliferation. Indeed, in the only study of retrotransposons in plants with holocentric chromosomes Zedek et al. (2010) found that massive retrotransposon proliferation was responsible for the increased chromosome sizes in species of genus *Eleocharis*.[19] In fact, the 2C/2*n* relationship is significantly positive in this genus, at least in the recently diverged *Eleocharis* subser. *Eleocharis*, where both the largest chromosomes and the highest density of retrotransposons in the genus are found (Zedek et al. 2010).

12.7.2 Polyploidy

The positive 2C/2*n* linear correlation detected in *Eleocharis* (Zedek et al. 2010) suggests that in addition to agmatoploidy and symploidy, polyploidy is as important mechanism of holocentric karyotype evolution. The periodicities in the chromosome numbers indicate that polyploidy is likely to be present in several holocentric genera, such as *Drosera* (Droseraceae; Rivadavia et al. 2003), *Luzula* (Juncaceae; Kirschner 1992, 1995: also confirmed by allozyme analysis), and *Juncus* (Juncaceae; Cope and Stace 1985; Rooks 2008: also confirmed by flow cytometry).

In Cyperaceae, polyploidy (as described above) occurs in *Eleocharis* (da Silva et al. 2008b, 2010: autopolyploidy in *E. sellowiana* confirmed by FISH by a doubled number of rDNA sites) and in *Rhynchospora* (Vanzela et al. 2000, 2003: autopolyploidy in *R. tenuis* confirmed by rDNA FISH). Based on the periodicity of prevailing chromosome numbers, polyploidy is also assumed to occur in *Fimbristylis* (Yano and Hoshino 2006), *Bulbostylis* and *Abildgaardia* (Luceño et al. 1998). In *Carex*, the most specious genus of Cyperaceae with more than 2,000 species, polyploidy appears to be surprisingly rare (Hipp 2007; Hipp et al. 2007, 2009; Roalson 2008). The best-documented but rare example of intraspecific ploidy variation is in *Carex siderosticta*, where autotetraploidy was confirmed based on the formation of tetravalents during meiosis (Tanaka 1940, 1949). The other examples of proven polyploidy in *Carex dolichostachya*, *C. jackiana*, and *C. roraimensis* are reviewed in detail by Hipp et al. (2009); while some other cases are also discussed by Rotreklová et al. (2011). To explain the lower frequency of polyploidy in this genus, Heilborn (1934) hypothesized that the formation of unreduced pollen dyads during pseudomonad pollen development may be prevented. During ovular gametogenesis, three of the four megaspores are aborted, and only one develops into the ovule sac, whereas during regular pollen development in most angiosperms, all four microspore nuclei develop into pollen grains, similar to spermatogenesis

[19] In those species with the largest chromosomes, Ty1-*copia* LTR retrotransposons (assuming that all were full-length) might form up to 70% of the genomic DNA (Zedek et al. 2010)

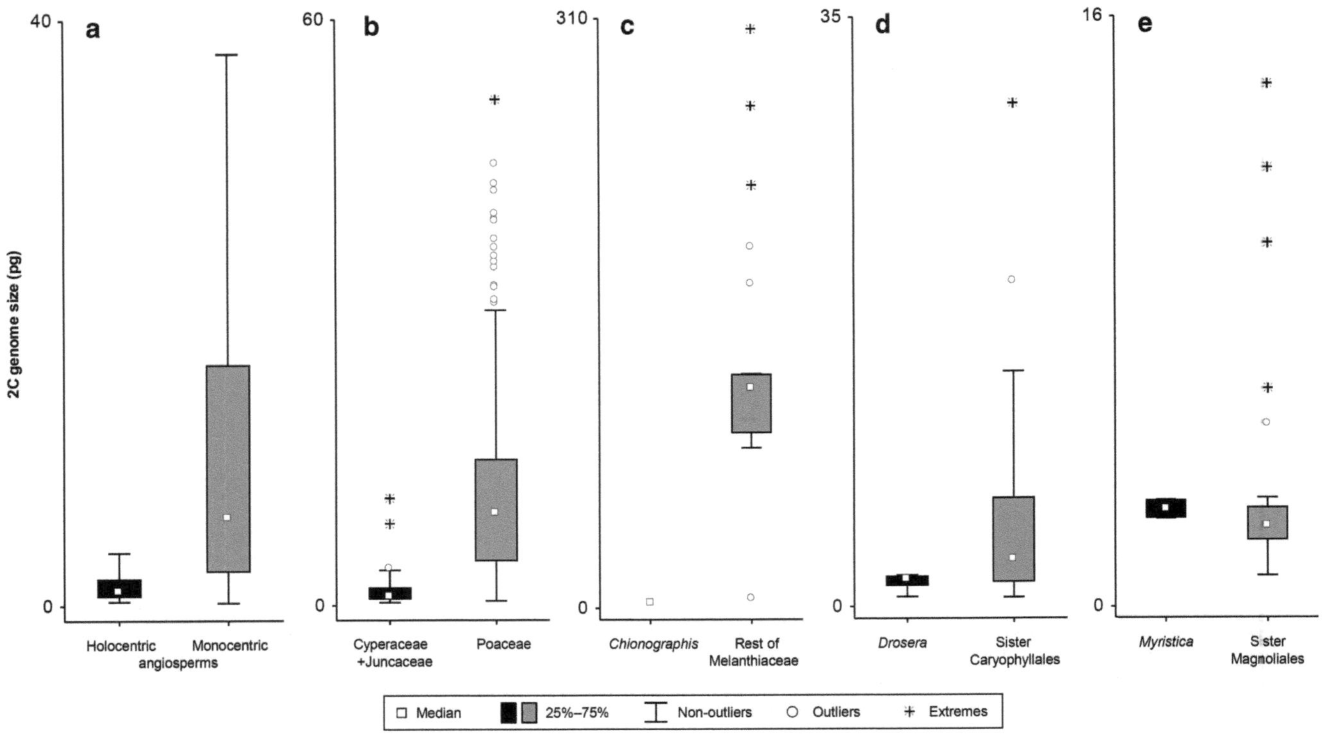

Fig. 12.5 Genome size in holocentric plant species (*black*) compared with related monocentric plants (*grey*) (Data taken from the Plant DNA C-values Database, Bennett and Leitch 2005). (**a**) Holocentric angiosperm species in which the somatic DNA amount is known (N = 79) have significantly smaller genomes than the remaining putatively monocentric angiosperm species (N = 4348; Mann Whitney, p = 0.000; outliers and extremes are not shown in the boxplot but are included in the test). (**b**) Holocentrics in the clades of Cyperaceae and Juncaceae (N = 71) have significantly smaller genomes than the monocentrics in the sister family Poaceae (N = 470; Mann-Whitney, p = 0.000). (**c**) Holocentric *Chionographis japonica* has a smaller genome than any other species of the family Melanthiaceae, to which it belongs (completed with the data from Pellicer et al. 2010; not tested). (**d**) Holocentric species in the genus *Drosera* (N = 6; Droseraceae) have significantly smaller genomes than the putative monocentric species in the sister families of Drosophyllaceae, Nepenthaceae, Tamaricaceae, Plumbaginaceae, and Polygonaceae (N = 26; Mann-Whitney, p = 0.045). (**e**) The genome size of the holokinetic *Myristica* (N = 2) does not differ from that of the putatively non-holocentric species in the sister families of Magnoliaceae, Eupomatiaceae, and Annonaceae (N = 26; Mann-Whitney, not significant)

in animals. In contrast, during pseudomonad pollen meiosis in Cyperaceae, the degeneration of three of the four nuclei occurs, similar to megasporogenesis. Among other angiosperms, this type of pollen development is rare, and currently only reported in the family Ericaceae (in some genera of the subfamily Styphelioideae; Furness and Rudall 2010). The remaining question is whether the opportunity for polyploidy during "regular" pollen and egg development depends more or exclusively on unreduced pollen production, as was implicitly assumed by Heilborn. Recently, the unreduced egg, not the unreduced pollen, was shown to be the most likely source for the origin of polyploids (Krahulcová et al. 2004; Ramsey 2007), even if the natural frequency of producing non-reduced gametes is similar during microsporogenesis and megasporogenesis (Ramsey and Schemske 1998). Moreover, the rare occurrence of unreduced pollen in *Carex podogyna* (Tanaka 1941), *C. recta* (Cayouette and Morisset 1986a), *C. aquatilis* (Cayouette and Morisset 1986b), and *Eleocharis mamillata* (Strandhede 1965c) and in the hybrids *Carex canescens* × *C. lachenalii* (Toivonen 1981) and *C. remota* × *C. paniculata* (Luceño 1994) was clearly shown during pseudomonad development. The presence of polyploidy in other Cyperaceae[20] (see above) suggests the absence of a causal relationship between pseudomonad microsporogenesis and the low frequency of polyploidy in *Carex*.

12.7.3 The Enigmatic Case of *Luzula*: Concerted Fission and True Polyploidy

In the karyotypes of the holocentric genus *Luzula* (Juncaceae), at least three distinct size categories of chromosomes are recognized: large (AL), medium (BL), and small (CL), where the relation between their sizes is

[20] The pseudomonad pollen formation was shown to be synapomorphic for the more derived clades of Cyperaceae, but absent in the early diverging clade of the subfamily Mapanioideae (Simpson et al. 2003).

1AL = 2BL = 4CL.[21] The somatic chromosome numbers of particular species are usually regular "euploids" with a basic chromosome number of x = 6. In the larger size category (AL), four regular ploidy levels are present: diploids ($2n = 12$ AL; e.g., *L. campestris* subsp. *campestris*, *L. pallidula*, and *L. taurica*), tetraploids ($2n = 24$ AL; e.g., *L. divulgata*), hexaploids ($2n = 36$ AL; e.g., *L. multiflora* subsp. *frigida*) and octoploids ($2n = 48$ AL; e.g., *L. congesta*). In the medium size category, only one unimodal karyotype, $2n = 24$ BL, is known, and it is found in *L. fallax* and *L. calabra*. The same is true in the smallest category, where $2n = 48$ CL is present in *L. sudetica*. Based on this pattern, agmatoploidy supposedly operates at the diploid level, and agmatoploid chromosome sets are then carried to higher ploidy levels via allopolyploidy (Nordenskiöld 1961; Kirschner 1992). Support for this so-called complete agmatoploidy (Luceño and Guerra 1996) or concerted fission has been obtained from observations of the meiotic behaviour of chromosomes in crosses between *L. pallescens* ($2n = 12$ AL) and *L. sudetica* ($2n = 48$ CL), in which each of the six large chromosomes (AL) was predominantly associated with four smaller ones (CL; Nordenskiöld 1951, 1964). The estimation of DNA content in particular species also supports the presence of concerted fission (or concerted fusion) combined with polyploidy (Barlow and Nevin 1976; Bačič et al. 2007).[22]

In a few rare cases, Nordenskiöld (1951) found two cytotypes (fragmented and non-fragmented) within particular species, such as *Luzula spicata*, where both the regular diploid and complete agmatoploid patterns of $2n = 12$ AL and $2n = 24$ BL, respectively, were observed. A putatively similar pattern was also documented in *Juncus biglumis* which belongs to the same family as *Luzula*. Here Schönswetter et al. (2007) found three cytotypes, of which two, $2n = 60$ and $2n = 120$, differed an average of just 1.06-fold in their DNA content, suggesting the possibility of complete agmatoploidy in generating the $2n = 120$ karyotype. A descending multiplicative decrease in the number of chromosomes has also been reported in some phylogenetic lineages of butterflies in the family Nymphalidae which have holocentric chromosomes (Brown et al. 2007), suggesting an inverted process of concerted fusion.

Although this enigmatic pattern of concerted fission has been known for almost 60 years, no clear mechanism has been found to be responsible for the half-breaking of the chromosomes in *Luzula* (Greilhuber 1995; Luceño and Guerra 1996; Madej and Kuta 2001; Guerra et al. 2010). Roalson et al. (2007) hypothesized that the possible chromosomal break points could be the remainders of telomeres within previously fused chromosomes.[23] The typical chromosomal "fragile sites" have been shown to be comprised of expanded microsatellites or AT-rich minisatellite arrays (Sutherland et al. 1998) and it is possible that the break points on *Luzula* chromosomes are also comprised of satellite repeats possibly related to the LCS1 repeat identified in *Luzula nivea* ($2n = 12$AL) by Haizel et al. (2005) and shown to be present in at least five sites in each chromosome.

Talbert et al. (2009) speculated that the chromosome "halves" or "quarters" that result from symmetric chromosome breaks during karyotype evolution in *Luzula* could be analogous to the developmentally programmed DNA rearrangements that occur during the ontogenetic cycles of the round worm *Parascaris univalens* (Nematoda). In this holocentric nematode, two chromosomes are present in the germline cell lineage, whereas somatic cells contain about 60 small chromosomes that originate via fission of the two large germline chromosomes. Moreover, the amount of DNA is much lower in the somatic nuclei than in the germline nuclei because these small fragmented chromosomes originate from just the central part of the germline chromosomes. The distal ends of the germline chromosomes do not form daughter nuclei and degenerate in the cytoplasm of somatic cells in a process known as chromatin diminution; a process that has also been detected in several parasitic nematodes (Pimpinelli and Goday 1989; Tobler and Müller 2001). The vast majority of this eliminated DNA is non-coding, but in rare cases, it contains a large number of genes (Nemetschke et al. 2010). Interestingly, chromosomal breakpoints in *Parascaris* do not contain the expected repetitive sequence motifs (Tobler and Müller 2001) that are found in the "fragile sites" of chromosomes (see above). The process of concerted fusion or fission could be facilitated by some kind of karyotype orthoselection as proposed by Greilhuber (1995), for example, by meiotic drive (see Sect. 12.8).

[21] In addition to these three main size categories, giant (A_0) and tiny chromosomes are present in this genus in *Luzula elegans* (= *L. purpurea*) and *L. pilosa*, respectively. See inner box in Fig. 12.3.

[22] An extremely high intraspecific and intraplant variation in chromosome number was found by Kuta et al. (2004) in the cultivated seedlings of *Luzula multiflora*, in which the somatic chromosome number varied 6.83-fold (from 12 to 84), whereas the DNA content varied just 1.04-fold (from 1.796 to 1.864 pg). This observation suggests a high frequency of agmatoploidy even during ontogenesis; however, other studies have not documented such chromosome variation in this species (Nordenskiöld 1951; Kirschner 1992; Bačič et al. 2007). Because samples were based on seeds obtained from the Botanical Garden of Stuttgart (Kuta et al. 2004), testing whether that observed pattern is also present in natural populations or among progeny cultivated from the seeds collected from wild plants of this species would be appropriate.

[23] The probable presence of such fragile points on the chromosomes of *Carex* has also been speculated by Luceño (1994).

12.8 How and Why Holocentric Chromosomes Originate

The scattered phylogenetic distribution of holocentrism suggests that it has evolved independently several times during the course of evolution (Dernburg 2001). However, some authors theorize that the diffuse centromere is the ancestral state and that localized centromeres represent a specialized, derived structure (Mola and Papeschi 2006 and references therein). This might be true in animals, where large holocentric clades alternate with monocentric lineages. In land plants, however, holocentrics are only found among angiosperms, where they are found in the terminal positions of phylogenetic trees, they should therefore be considered as a derived character (Greilhuber 1995). The absence of transient forms suggests that the change from monocentrism to holocentrism represents a significant jump similar to "major evolutionary transitions" (*sensu* Maynard Smith and Szathmáry 1995), in which the complexity of chromosomes is abruptly increased. The exact mechanisms that give rise to holocentrics are still unknown although it is tempting to speculate that the origin of holocentric chromosomes is a very simple process, given that they have arisen independently several times during the evolutionary history of plants.

Even though several hypotheses have been proposed, none of them have yet been confirmed or disproven. First, the holocentric state is proposed to arise through the spread of centromeric sequences from one location to multiple sites along the chromosomal arms. Such a spread could occur by the proliferation of transposable elements that carry a centromeric sequence or by centric fusions and subsequent pericentromeric inversions (Greilhuber 1995; Diaz et al. 2010). The transposition of retroelements in subtelomeric regions combined with the modification of the tubulin-based cytoskeleton was proposed by Villasante et al. (2007) to be responsible for the origin of primitive centromeres from telomeres during "major evolutionary transitions" from prokaryotic to eukaryotic chromosomes. Such initially unstable "ditelocentric" chromosomes could subsequently evolve into chromosomes with a holocentric organization[24] by the continuous spreading of end sequences throughout the chromosome[25] or to metacentric (monocentric) chromosomes, as these authors suggested.

Another elegant and simple mechanism for monocentric/holocentric transition could be that the orthogonal direction of kinetochore formation against the chromosomal axes has turned by 90°, allowing the extension of kinetochores along the chromosome axis (Nagaki et al. 2005). This new kinetochore formation would become embedded in the longitudinal groove along the poleward chromatid surface, as is observed in holocentrics (see Sect. 12.4), rather than develop into the primary constriction seen in monocentrics. This putative change in the direction of kinetochore formation is sometimes considered to be causally related to a lack of condensin I in holocentric chromosomes and co-localization of condensin II with CENH3 along the entire chromosomal length (Ono et al. 2003, 2004; Hauf and Watanabe 2004), but is more likely to be a consequence, rather than a cause, of holocentrism (see Sect. 12.5).

Research efforts should now focus on finding common features among the isolated cases of holocentrism that could help to elucidate to what extent the transition from monocentric chromosomes to holocentric chromosomes (or vice versa) is a general predisposition for all eukaryotes. However, even if there is a common (epi-)genetic predisposition to holocentrism, an open question remains: What are holocentric chromosomes good for?

One explanation for the origin of holocentric chromosomes is based on the mechanics of mitosis and meiosis (Diaz et al. 2010). During anaphase, monocentric chromosomes are pulled toward opposite spindle poles with the centromere leading and the chromosomal arms lagging, resulting in "V-shaped" chromosomes. Pulling chromosomes by one point (centromere) may impose a mechanical penalty because the chromosomes must be pulled far enough to ensure that they are not in the way of the structures that divide the cell during cytokinesis. Indeed, such a mechanical penalty would be proportional to the length of the chromosome. In contrast, holocentric chromosomes attach spindle microtubules along their entire length and move to the poles in an orientation that is parallel to the equatorial plate. Thus, holocentric chromosomes could make cell division easier, regardless of the length of the chromosomes (Diaz et al. 2010).

Prevention of deleterious effects of chromosomal fission or fusion due to the function of the kinetochore along the full chromosome length seems to be a rather anthropocentric

[24] Considering that terminal blocks of heterochromatin exhibiting "centromeric activity" in telokinetic (holocentric) germline chromosomes of the nematode *Parascaris univalens* originated by the segmental amplification of repeats derived from telomeric repeats (Niedermaier and Moritz 2000).

[25] Tandem repeats (including the telomeric repeat TTAGGC) are indeed dispersed at multiple internal sites of the holocentric chromosomes of the nematode *Caenorhabditis elegans*, although they are more abundant at the ends (C. elegans Sequencing Consortium 1998).

explanation because death of an individual due to chromosomal breakage usually has no impact on the survival of a species itself. A theoretical exception might exist if some agent (e.g., irradiation) generates frequent chromosomal breakages and thus confers selective advantage to the species in which survival is not diminished, i.e., holocentric relative to monocentric species. Theoretically, fragmentation would increase if selection favored recombination, whereas fusion would be the result of selection for reduced recombination (Stephan 2007). The nature of the pressure that would select for a certain recombination rate, however, remains a mystery. In some holocentrics (*Caenorhabditis*, *Luzula*), the crossover rate has been documented to be only one or two per homologous pair of chromosomes (Monen et al. 2005; Nordenskiöld 1962). Agmatoploidy might compensate for such lower recombination rates in holocentric organisms because it leads to an increase in chromosome numbers and thus also in the number of possible chromosomal combinations in zygotes (see Sect. 12.7).

An interesting recent hypothesis suggests that holocentric chromosomes might have evolved as a defense against the accumulation of centromeric satellites via "centromere drive", a special type of meiotic drive (Talbert et al. 2009). The centromere drive model, which has only been described in monocentrics to date, postulates that some homologous centromere variants can be preferentially transmitted to the next generation through non-random segregation in asymmetric meiosis (Henikoff et al. 2001), after which only one of four meiotic products remains viable and becomes a megaspore in plants or an egg in animals. The success of a certain centromere variant depends on its orientation during the first meiotic division (Malik and Henikoff 2009). The question remains as to how a centromere can achieve a preferential orientation toward one of the poles. To explain this phenomenon, the meiotic spindle has been proposed to have functional asymmetry (de Villena and Sapienza 2001a), meaning that the egg pole and the polar body pole differ in the number of microtubules emanating from each and consequently, in their ability to capture centromeres. Thus, the egg pole captures more or "stronger" centromeres than the polar body pole or vice versa. Chromosomal fusion, for example, Robertsonian translocation (the fusion of two acrocentric chromosomes into one large metacentric chromosome), reduces the number of centromeres. In human female meiosis, Robertsonian chromosomes are preferentially transmitted to the egg over the two non-fused, acrocentric homologs, whereas the opposite behaviour is observed in mouse females. In other words, the human egg pole preferentially captures fewer centromeres, whereas the mouse egg pole captures more centromeres (de Villena and Sapienza 2001b; Underkoffler et al. 2005). Similarly, the functional asymmetry of the meiotic spindle may maintain B chromosomes in a population despite their potentially negative effect on the fitness of their carriers (Talbert et al. 2009).

The expansion of centromeric satellite repeats can create a larger (i.e., "stronger") centromere that recruits more kinetochore proteins and better attracts microtubules (Malik and Henikoff 2009; Talbert et al. 2009), which may provide an advantage or disadvantage, depending on the preferences of the poles, for this "stronger" centromere in the competition for the megaspore pole. A centromere with such an advantage would quickly spread throughout a population and might even eventually become fixed. However, the spreading of the "stronger" centromere within a population might be accompanied by a number of negative effects. Any deleterious mutations carried on the respective chromosome would also spread within the population. In addition, the heterozygosity of the centromere variants could cause imbalances and non-disjunctions during male meiosis, which often results in aneuploidy and/or increased male sterility (Henikoff et al. 2001). In plants, the deleterious effects of centromere drive caused by the expansion of centromeric satellites on male fertility were documented in monkeyflowers (*Mimulus*; Fishman and Saunders 2008). Centromere drive may also lead to skewed sex ratios in a population, as has been shown in birds (Rutkowska and Badyaev 2008).

Any process or mechanism that suppresses centromere drive would obviously be favored (Malik and Henikoff 2009).[26] A simple switch in the preference of the spindle poles, for example, from capturing more/"stronger" centromeres to capturing fewer/"weaker" centromeres, would function against centromere drive. These switches have occurred many times during the evolution of mammals (de Villena and Sapienza 2001c). Mutations in kinetochore proteins, which interact with centromeric satellite repeats, can also suppress centromere drive (Henikoff et al. 2001). The specialized and fundamental function of centromeric DNA binding proteins, such as CENH3 or CENP-C (centromeric protein C), in kinetochore assembly is well-conserved across eukaryotes. Hence, they would be expected to evolve under the strong pressure of negative (purifying) selection. Surprisingly, the reverse is true. The DNA binding domains of either CENH3 or CENP-C have been reported to have evolved rapidly in an adaptive manner in every investigated eukaryotic organism displaying asymmetric meiosis,

[26] Actually, this observation would be true in animals, which are usually gonochoric. In plants, the male sterility of structural heterozygotes (individuals possessing different types of homologous centromeres) should not decrease population fitness because these male-sterile individuals have to produce sexual progeny exclusively via outcrossing, which would thus counterbalance the eventual inbreeding depression that originates due to the selfing of hermaphrodites. In other words, a pattern similar to gynodioecy would result.

including plants (Talbert et al. 2004; Malik and Henikoff 2009). The adaptive mutations of CENH3 or CENP-C could weaken the "stronger" centromeres or reinforce the "weaker" centromeres, thus restoring balance and suppressing centromere drive (Dawe and Henikoff 2006).

In holocentric chromosomes, the expansion of satellite repeats seems to have no function because the kinetochores assemble along the entire chromosome, meaning that there should not be any "weak" or "strong" centromeres. The hypothesis proposing that holocentric chromosomes are an adaptation against the accumulation of centromeric satellites via "centromere drive" might be supported by the findings in *Luzula nivea*. LCS1 was found in arrays that were about 50 kb in size, which is smaller than any other reported centromeric satellite array in plants (Haizel et al. 2005). Similarly, *Caenorhabditis elegans* almost completely lacks satellite repeats (Emmons 1988; Talbert et al. 2009). Although the accumulation of centromeric satellites might be suppressed in holocentrics, centromere drive could still occur and thus drive variations in chromosome size or number. For example, in a structural heterozygote that possesses a larger fused chromosome together with its two non-fused homologs, the larger chromosome should have a higher or lower chance to attract spindle microtubules emanating from the pole that anchors more or fewer microtubules, respectively, in the competition with its non-fused homologs. Indeed, the ovular pole could emanate more or fewer microtubules, which would then increase or decrease, respectively, the chance of the larger chromosome being transmitted to the next generation. Alternations in such "holocentric drive" could be the reason for the divergent karyotype orthoselection in chromosome numbers observed in *Carex* sect. *Ovales* (see Sect. 12.7.1.). Here, as in most other Cyperaceae, asymmetric meiosis also occurs during pollen development (see Sect. 12.7.2.). Switching between the "weak" or "strong" egg poles during "holocentric drive" could also result in a large variation in chromosomal size, which is found more often in holocentric genera compared to monocentric genera (see Sect. 12.7.1, Fig. 12.3). "Holocentric drive" combined with species range expansion could also explain the geographic patterns in the intraspecific variation of chromosome number observed in some sedges (see Sect. 12.7.1.). Finally, the massive retrotransposon proliferation that resulted in the formation of large chromosomes during the evolution of the holocentric genus *Eleocharis* (Zedek et al. 2010) could also have been facilitated by the same mechanism.

Acknowledgments Our study was supported by the Czech Science Foundation (Grant no. GACR206/09/1405) and by the Ministry of Education, Youth and Sports of the Czech Republic (Grants no. MSM0021622416, and LC06073).

References

Albertson DG, Thomson NJ (1993) Segregation of holocentric chromosomes at meiosis in the nematode, *Caenorhabditis elegans*. Chromosome Res 1:15–26

Ananiev EV, Phillips RL, Rines HW (1998) Chromosome-specific molecular organization of maize (*Zea mays* L.) centromeric regions. Proc Natl Acad Sci USA 95:13073–13078

Bačič T, Jogan N, Dolenc Koce J (2007) *Luzula* sect. *Luzula* in the south-eastern Alps—karyology and genome size. Taxon 56:129–136

Barlow PW, Nevin D (1976) Quantitative karyology of some species of *Luzula*. Plant Syst Evol 125:77–86

Bennett MD, Leitch IJ (2005) Angiosperm DNA C-values database (release 6.0, Oct 2005) http://www.kew.org/cvalues/

Bhatti N, Datson P, Murray B (2007) Chromosome number, genome size and phylogeny in the genus *Schoenus* (Cyperaceae). Chromosome Res 15(suppl 2):33–34

Bokhari FS (1976) Meiosis in untreated and irradiated *Cyperus eragrostis*. Cytologia 41:607–614

Bokhari FS, Godward MBE (1980) The ultrastructure of the diffuse kinetochore in *Luzula nivea*. Chromosoma (Berl) 79:125–136

Bolkhovskikh Z, Grif V, Matvejeva T, Zakharyeva O (1969) Chromosome numbers of flowering plants. Nauka, Leningrad

Bongiorni S, Fiorenzo P, Pippoletti D, Prantera G (2004) Inverted meiosis and meiotic drive in mealybugs. Chromosoma 112:331–341

Braselton JP (1971) The ultrastructure of the non-localized kinetochores of *Luzula* and *Cyperus*. Chromosoma 36:89–99

Braselton JP (1981) The ultrastructure of meiotic kinetochores of *Luzula*. Chromosoma (Berl) 82:143–151

Brown KS Jr, von Schoultz B, Suomalainen E (2004) Chromosome evolution in Neotropical Danainae and Ithomiinae (Lepidoptera). Hereditas 141:216–236

Brown KS Jr, Freitas AVL, Wahlberg N, von Schoultz B, Saura AO, Saura A (2007) Chromosomal evolution in the South American Nymphalidae. Hereditas 144:137–148

Buchwitz BJ, Ahmad K, Moore LL, Roth MB, Henikoff S (1999) Cell division: a histone-H3-like protein in *C. elegans*. Nature 401:547–548

Bureš P (1998) A high polyploid *Eleocharis uniglumis* s.l. (Cyperaceae) from Central and Southeastern Europe. Folia Geobot 33:429–439

Bureš P, Rotreklová O, Stoneberg Holt SD, Pikner R (2004) Cytogeographical survey of *Eleocharis* subser. *Eleocharis* in Europe 1: *Eleocharis palustris*. Folia Geobot 39:235–257

C. elegans Sequencing Consortium (1998) Genome sequence of the nematode *C. elegans*: a platform for investigating biology. Science 282:2012–2018

Cayouette J, Morisset P (1986a) Chromosome studies on the *Carex salina* complex (Cyperaceae, section *Cryptocarpae*) in northeastern North America. Cytologia 51:817–856

Cayouette J, Morisset P (1986b) Chromosome studies on *Carex paleacea* Wahl., *C. nigra* (L.) Reichard and *C. aquatilis* Wahl. in northeastern North America. Cytologia 51:857–883

Chakravorti AK (1948a) Multiplication of chromosome numbers in relation to speciation in Zingiberaceae. Sci Cult 14:137–140

Chakravorti AK (1948b) Theory of fragmentation of chromosomes and evolution of species. Sci Cult 13:309–312

Cheng ZK, Dong FG, Langdon T, Shu OY, Buell CR, Gu MH, Blattner FR, Jiang JM (2002) Functional rice centromeres are marked by a satellite repeat and a centromere-specific retrotransposon. Plant Cell 14:1691–1704

Claycomb JM, Batista PJ, Pang KM, Gu WF, Vasale JJ, van Wolfswinkel JC, Chaves DA, Shirayama M, Mitani S, Ketting RF et al (2009) The argonaute CSR-1 and its 22G-RNA cofactors are required for holocentric chromosome segregation. Cell 139:123–134

Cook LG (2000) Extraordinary and extensive karyotypic variation: a 48-fold range in chromosome number in the gall-inducing scale insect Apiomorpha (Hemiptera: Coccoidea: Eriococcidae). Genome 43:255–263

Cope TA, Stace CA (1985) Cytology and hybridization in the *Juncus bufonius* L. aggregate in Western Europe. Watsonia 15:309–320

Cremonini R (2005) Low chromosome number angiosperms. Caryologia 58:403–409

d'Alençon E, Sezutsu H, Legeai F, Permal E, Bernard-Samain S, Gimenez S, Gagneur C, Cousserans F, Shimomura M, Brun-Barale A, Flutre T, Couloux A, East P, Gordon K, Mita K, Quesneville H, Fournier P, Feyereisen R (2010) Extensive synteny conservation of holocentric chromosomes in Lepidoptera despite high rates of local genome rearrangements. Proc Natl Acad Sci USA 107:7680–7685

da Silva CRM, González-Elizondo MS, Vanzela ALL (2005) Reduction in chromosome number in *Eleocharis subarticulata* (Cyperaceae) by multiple translocation. Bot J Linn Soc 149:457–464

da Silva CRM, González-Elizondo MS, LdNadA R, Torezan JMD, Vanzela ALL (2008a) Cytogenetical and cytotaxonomical analysis of some Brazilian species of *Eleocharis*. Aust J Bot 56:82–90

da Silva CRM, González-Elizondo MS, Vanzela ALL (2008b) Chromosome reduction in *Eleocharis maculosa* (Cyperaceae). Cytogenet Genome Res 122:175–180

da Silva CRM, Quintas CC, Vanzela ALL (2010) Distribution of 45S and 5S rDNA sites in 23 species of *Eleocharis* (Cyperaceae). Genetica 138:951–957

Dawe RK, Henikoff S (2006) Centromeres put epigenetics in the driver's seat. Trends Biochem Sci 31:662–669

Dawe RK, Hiatt EN (2004) Plant neocentromeres: fast, focused, and driven. Chromosome Res 12:655–669

de Carvalho CE, Zaaijer S, Smolikov S, Gu Y, Schumacher JM, Colaiacovo MP (2008) LAB-1 antagonizes the Aurora B kinase in *C. elegans*. Genes Dev 22:2869–2885

de Castro D, Camara A, Malheiros N (1949) X-rays in the centromere problem of *Luzula purpurea* LINK. Genet Iber 1:49–54

de Lesse H (1970) Les nombres de chromosomes dans le groupe de *Lysandra argester* et leur incidence sur la taxonomie. Bull Soc Entomol Fr 75:64–68

de Villena FPM, Sapienza C (2001a) Female meiosis drives karyotypic evolution in mammals. Genetics 159:1179–1189

de Villena FPM, Sapienza C (2001b) Nonrandom segregation during meiosis: the unfairness of females. Mamm Genome 12:331–339

de Villena FPM, Sapienza C (2001c) Transmission ratio distortion in offspring of heterozygous female carriers of Robertsonian translocations. Hum Genet 108:31–36

Dernburg AF (2001) Here, there, and everywhere: kinetochore function on holocentric chromosomes. Cell Biol 6:F33–F38

Diaz MO, Maynard R, Brum-Zorrilla N (2010) Diffuse centromere and chromosome polymorphism in haplogyne spiders of the families Dysderidae and Segestriidae. Cytogenet Genome Res 128:131–138

Dumont J, Oegema K, Desai A (2010) A kinetochore-independent mechanism drives anaphase chromosome separation during acentrosomal meiosis. Nat Cell Biol 12:894–901

Emmons SW (1988) The genome. In: Wood WB (ed) The nematode *Caenorhabditis elegans*. Cold Spring Harbor Laboratory Press, New York, pp 47–79

Escudero M, Valcárcel V, Vargas P, Luceño M (2008) Evolution in *Carex* L. sect. *Spirostachyae* (Cyperaceae): a molecular and cytogenetic approach. Organ Divers Evol 7:271–291

Faulkner JS (1972) Chromosome studies on *Carex* section *Acutae* in northwest Europe. Bot J Linn Soc 65:271–301

Fishman L, Saunders A (2008) Centromere-associated female meiotic drive entails male fitness costs in monkeyflowers. Science 322:1559–1562

Flach M (1966) Diffuse centromeres in a dicotyledonous plant. Nature 209:1369–1370

Furness CA, Rudall PJ (2010) Selective microspore abortion correlated with aneuploidy: an indication of meiotic drive. Sex Plant Reprod. doi:10.1007/s00497-010-0150-z

Gernand D, Demidov D, Houben A (2003) The temporal and spatial pattern of histone H3 phosphorylation at serine 28 and serine 10 is similar in plants but differs between mono- and polycentric chromosomes. Cytogenet Genome Res 101:172–176

Goday C, Pimpinelli S (1989) Centromere organization in meiotic chromosomes of *Parascaris univalens*. Chromosoma 98:160–166

Goday C, Ciofi-Luzzatto A, Pimpinelli S (1988) Centromere ultrastructure in germ-line chromosomes of *Parascaris*. Chromosoma 91:121–125

Godward MDE (1954) The 'diffuse' centromere or polycentric chromosomes in *Spirogyra*. Ann Bot 70:143–156

Godward MBE (1985) The kinetochore. Int Rev Cytol 94:77–106

Goldblatt P, Johnson DE (eds) (2010) Index to plant chromosome numbers [1973–2003]. Missouri Botanical Garden, St. Louis, http://mobot.mobot.org/W3T/Search/ipcn.html

Greilhuber J (1995) Chromosomes of the monocotyledons (general aspects). In: Rudall PJ, Cribb PJ, Cutler DF, Humphries CJ (eds) Monocotyledons: systematics and evolution. Kew Royal Botanic Gardens, Surrey, pp 379–414

Guerra M (2000) Patterns of heterochromatin distribution in plant chromosomes. Genet Mol Biol 23:1029–1041

Guerra M, García MA (2004) Heterochromatin and rDNA sites distribution in the holocentric chromosomes of *Cuscuta approximata* Bab. (Convolvulaceae). Genome 47:134–140

Guerra M, Brasileiro-Vidal AC, Arana P, Puertas MJ (2006) Mitotic microtubule development and histone H3 phosphorylation in the holocentric chromosomes of *Rhynchospora tenuis* (Cyperaceae). Genetica 126:33–41

Guerra M, Cabral G, Cuacos M, González-García M, González-Sánchez M, Vega J, Puertas MJ (2010) Neocentrics and holokinetics (holocentrics): chromosomes out of the centromeric rules. Cytogenet Genome Res 129:82–96

Haizel T, Lim YK, Leitch AR, Moore G (2005) Molecular analysis of holocentric centromeres of *Luzula* species. Cytogenet Genome Res 109:134–143

Håkansson A (1954) Meiosis and pollen mitosis in x-rayed and untreated spikelets of *Eleocharis palustris*. Hereditas 15:325–345

Håkansson A (1958) Holocentric chromosomes in *Eleocharis*. Hereditas 44:531–540

Hauf S, Watanabe Y (2004) Kinetochore orientation in mitosis and meiosis. Cell 119:317–327

Heilborn O (1924) Chromosome numbers and dimensions, species-formation and phylogeny in the genus *Carex*. Hereditas 5:129–216

Heilborn O (1934) On the origin and preservation of polyploidy. Hereditas 19:233–242

Henikoff S, Ahmad K, Malik HS (2001) The centromere paradox: stable inheritance with rapidly evolving DNA. Science 293:1098–1102

Hipp AL (2007) Nonuniform processes of chromosome evolution in sedges (*Carex*: Cyperaceae). Evolution 61:2175–2194

Hipp AL, Rothrock PE, Reznicek AA, Berry PE (2007) Chromosome number changes associated with speciation in sedges: a phylogenetic study in *Carex* section *Ovales* (Cyperaceae) using AFLP data. Aliso 23:193–203

Hipp AL, Rothrock PE, Roalson EH (2009) The evolution of chromosome arrangements in *Carex* (Cyperaceae). Bot Rev 75:96–109

Hipp AL, Rothrock PE, Whitkus R, Weber JA (2010) Chromosomes tell half of the story: the correlation between karyotype

rearrangements and genetic diversity in sedges, a group with holocentric chromosomes. Mol Ecol 19:3124–3138
Hoshino T (1981) Karyomorphological and cytological studies on aneuploidy in *Carex*. J Sci Hiroshima Univ B 2(217):155–238
Hoshino T (1987) Karyomorphological studies on 6 taxa of *Eleocharis* in Japan. Bull Okayama Univ Sci 22A:305–312
Hoshino T, Waterway MJ (1994) Cytogeography and meiotic chromosome configurations of six intraspecific aneuploids of *Carex conica* Boott (Cyperaceae) in Japan. J Plant Res 107:131–138
Houben A, Demidov D, Caperta AD, Karimi R, Agueci F, Vlasenko L (2007) Phosphorylation of histone H3 in plants: a dynamic affair. Biochim Biophys Acta 1769:308–315
Hughes-Schrader S, Ris H (1941) The diffuse spindle attachment of Coccids, verified by the mitotic behaviour of induced chromosome fragments. J Exp Zool 87:429–456
Jiang JM, Birchler JA, Parrott WA, Dawe RK (2003) A molecular view of plant centromeres. Trends Plant Sci 8:570–575
Kandul NP, Lukhtanov VA, Dantchenko AV, Coleman JW, Sekercioglu CH, Haig D, Pierce NE (2004) Phylogeny of *Agrodiaetus* Hübner 1822 (Lepidoptera: Lycaenidae) inferred from mtDNA sequences of COI and COII and nuclear sequences of EF1-a: karyotype diversification and species radiation. Syst Biol 53:278–298
King GC (1960) The cytology of the desmids: the chromosomes. New Phytol 59:65–72
Kirschner J (1992) Karyological differenciation of *Luzula* sect. *Luzula* in Europe. Thaiszia 2:11–39
Kirschner J (1995) Allozyme analysis of *Luzula* sect. *Luzula* (Juncaceae) in Ireland: evidence of the origin of tetraploids. Folia Geobot 30:283–290
Knowlton AL, Lan W, Stukenberg PT (2006) Aurora B is enriched at merotelic attachment sites, where it regulates MCAK. Current Biol 16:1705–1710
Kondo K, Sheikh SA, Hoshi Y (1994) New finding of another $2n = 6$ species in the angiosperm, *Drosera roseana* Marchant. Chrom Inf Service 57:3–4
Krahulcová A, Papoušková S, Krahulec F (2004) Reproduction mode in the allopolyploid facultatively apomictic hawkweed *Hieracium rubrum* (Asteraceae, *H.* subgen. *Pilosella*). Hereditas 141:19–30
Kuta E, Bohanec B, Dubas E, Vižintin L, Przywara L (2004) Chromosome and nuclear DNA study on *Luzula*: a genus with holokinetic chromosomes. Genome 47:246–256
La Cour LF (1953) The *Luzula* system analysed by x-rays. Heredity Suppl 6:77–81
Li X, Dawe RK (2009) Fused sister kinetochores initiate the reductional division in meiosis I. Nat Cell Biol 11:1103–1108
Luceño M (1992) Cytotaxonomic studies in Iberian and Macaronesian species of *Carex* (Cyperaceae). Willdenowia 22:149–165
Luceño M (1994) Cytotaxonomic studies in Iberian, Balearic, North African, and Macaronesian species of *Carex* (Cyperaceae). II. Can J Bot 72:587–596
Luceño M, Castroviejo S (1991) Agmatoploidy in *Carex laevigata* (Cyperaceae). Fusion and fission of chromosomes as the mechanism of cytogenetic evolution in Iberian populations. Plant Syst Evol 177:149–159
Luceño M, Guerra M (1996) Numerical variation in species exhibiting holocentric chromosomes: a nomenclatural proposal. Caryologia 49:301–309
Luceño M, Vanzela A, Guerra M (1998) Cytotaxonomic studies in Brazilian *Rhynchospora* (Cyperaceae), a genus exhibiting holocentric chromosomes. Can J Bot 76:440–449
Lukhtanov VA, Dantchenko AD (2002) Principles of the highly ordered arrangement of metaphase I bivalents in spermatocytes of *Agrodiaetus* (Insecta, Lepidoptera). Chromosome Res 10:5–20
Lukhtanov VA, Vila R, Kandul NP (2006) Rearrangement of the *Agrodiaetus dolus* species group (Lepidoptera, Lycaenidae) using a new cytological approach and molecular data. Insect Syst Evol 37:325–334
Lysák MA, Schubert I (2013) Mechanisms of chromosome rearrangements. In: Leitch IJ, Greilhuber J, Doležel J, Wendel JF (eds) Plant genome diversity, vol 2, Physical structure, behaviour and evolution of plant genomes. Springer-Verlag, Wien, pp 137–147
Ma JX, Wing RA, Bennetzen JL, Jackson SA (2007) Plant centromere organization: a dynamic structure with conserved functions. Trends Genet 23:134–139
Maddox PS, Oegema K, Desai A, Cheeseman IM (2004) "Holo"er than thou: chromosome segregation and kinetochore function in *C. elegans*. Chromosome Res 12:641–653
Madej A, Kuta E (2001) Holokinetic chromosomes of *Luzula luzuloides* (Juncaceae) in callus culture. Acta Biol Cracov Ser Bot 43:33–43
Malheiros N, De Castro D, Camara A (1947) Cromosomas sem centrómero localizado. O caso da *Luzula purpurea* Link. Agron Lusitana 9:51–71
Malheiros-Garde N, de Castro D (1947) Chromosome number and behavior in *Luzula purpurea* Link. Nature 160:156
Malik HS, Henikoff S (2009) Major evolutionary transitions in centromere complexity. Cell 138:1067–1082
Manzanero S, Arana P, Puertas MJ, Houben A (2000) The chromosomal distribution of phosphorylated histone H3 differs between plants and animals at meiosis. Chromosoma 109:308–317
Marec F, Sahara K, Traut W (2009) Rise and fall of the W chromosome in Lepidoptera. In: Goldsmith MR, Marec F (eds) Molecular biology and genetics of the Lepidoptera. CRC Press, Boca Raton/London/New York, pp 49–63
Martinez-Perez E, Schvarzstein M, Barroso C, Lightfoot J, Dernburg AF, Villeneuve AM (2008) Crossovers trigger a remodeling of meiotic chromosome axis composition that is linked to two-step loss of sister chromatid cohesion. Genes Dev 22:2886–2901
Maynard Smith J, Szathmáry E (1995) The major transitions in evolution. Oxford University Press, Oxford, UK
Mayrose I, Barker MS, Otto SP (2010) Probabilistic models of chromosome number evolution and the inference of polyploidy. Syst Biol 59:132–144
McNeal JR, Arumugunathan K, Kuehl JV, Boore JL, dePamphilis CW (2007) Systematics and plastid genome evolution of the cryptically photosynthetic parasitic plant genus *Cuscuta* (Convolvulaceae). BMC Plant Biol 5:55
Meneely PM, Farago AF, Kauffman TM (2002) Crossover distribution and high interference for both the X chromosome and an autosome during oogenesis and spermatogenesis in *Caenorhabditis elegans*. Genetics 162:1169–1177
Mola LM, Papeschi AG (2006) Holokinetic chromosomes at a glance. J Basic Appl Genet 17:17–33
Monen J, Maddox PS, Hyndman F, Oegema K, Desai A (2005) Differential role of CENP-A in the segregation of holocentric *C. elegans* chromosomes during meiosis and mitosis. Nat Cell Biol 7:1248–1255
Nabeshima K, Villeneuve AM, Colaiacovo MP (2005) Crossing over is coupled to late meiotic prophase bivalent differentiation through asymmetric disassembly of the SC. J Cell Biol 168:683–689
Nagaki K, Talbert PB, Zhong CX, Dawe RK, Henikoff S, Jiang JM (2003) Chromatin immunoprecipitation reveals that the 180-bp satellite repeat is the key functional DNA element of *Arabidopsis thaliana* centromeres. Genetics 163:1221–1225
Nagaki K, Kashihara K, Murata M (2005) Visualization of diffuse centromeres with centromere-specific histone H3 in the holocentric plant *Luzula nivea*. Plant Cell 17:1886–1893
Nemetschke L, Eberhardt AG, Hertzberg H, Streit A (2010) Genetics, chromatin diminution, and sex chromosome evolution in the parasitic nematode genus *Strongyloides*. Curr Biol 20:1687–1696

Niedermaier J, Moritz KB (2000) Organization and dynamics of satellite and telomere DNAs in *Ascaris*: implications for formation and breakdown of compound chromosomes. Chromosoma 109:439–452
Nijalingappa BH, Tejavathi DH (1984) Reports. In: Löve Á (ed) Chromosome number reports LXXXIII. Taxon vol 33, pp 352–353
Nishikawa K, Furuta Y, Ishitobi K (1984) Chromosomal evolution in genus *Carex* as viewed from nuclear DNA content, with special reference to its aneuploidy. Jpn J Genet 59:465–472
Nokkala S, Kuznetsova VZ, Maryanska-Nadachowska A, Nokkala C (2004) Holocentric chromosomes in meiosis. I. Restriction of the number of chiasmata in bivalents. Chromosome Res 12:733–739
Nordenskiöld H (1951) Cytotaxonomical studies in the genus *Luzula*. I. Somatic chromosome and chromosome numbers. Hereditas 37:325–355
Nordenskiöld H (1961) Tetrad analysis and the course of meiosis in three hybrids of *Luzula campestris*. Hereditas 47:203–238
Nordenskiöld H (1962) Studies of meiosis in *Luzula purpurea*. Hereditas 48:503–519
Nordenskiöld H (1963) A study of meiosis in the progeny of X-irradiated *Luzula purpurea*. Hereditas 49:33–47
Nordenskiöld H (1964) The effect of irradiation on diploid and polyploid *Luzula*. Hereditas 51:344–374
Ohkawa T, Yokota M, Hoshino T (2000) Aneuploidal population differentiation in *Carex sociata* Boott (Cyperaceae) of the Ryukyu Islands, Japan. Bot J Linn Soc 132:337–358
Ono T, Losada A, Hirano M, Myers MP, Neuwald AF, Hirano T (2003) Differential contributions of condensin I and condensin II to mitotic chromosome architecture in vertebrate cells. Cell 115:109–121
Ono T, Fang Y, Spector DL, Hirano T (2004) Spatial and temporal regulation of condensins I and II in mitotic chromosome assembly in human cells. Mol Biol Cell 15:3296–3308
Östergren G (1949) *Luzula* and the mechanism of chromosome movements. Hereditas 4:445–468
Papeschi AG, Bressa MJ (2006) Evolutionary cytogenetics in Heteroptera. J Biol Res 5:3–21
Pasierbek P, Jantsch M, Melcher M, Schleiffer A, Schweizer D, Loidl J (2001) A *Caenorhabditis elegans* cohesion protein with functions in meiotic chromosome pairing and disjunction. Genes Dev 15:1349–1360
Pazy B (1997) Supernumerary chromosomes and their behaviour in meiosis of the holocentric *Cuscuta babylonica* Ghoisy. Bot J Linn Soc 123:173–176
Pazy B, Plitmann U (1987) Persisting demibivalents: a unique meiotic behavior in *Cuscuta babylonica* Choisy. Genome 29:63–66
Pazy B, Plitmann U (1994) Holocentric chromosome behavior in *Cuscuta*. Plant Syst Evol 191:105–109
Pazy B, Plitmann U (2002) New perspectives on the mechanisms of chromosome evolution in parasitic flowering plants. Bot J Linn Soc 138:117–122
Pellicer J, Fay MF, Leitch IJ (2010) The largest eukaryotic genome of them all? Bot J Linn Soc 164:10–15
Pérez R, Panzera F, Page J, Suja JA, Rufas JS (1997) Meiotic behaviour of holocentric chromosomes: orientation and segregation of autosomes in *Triatoma infestans* (Heteroptera). Chromosome Res 5:47–56
Pimpinelli S, Goday C (1989) Unusual kinetochores and chromatin diminution in *Parascaris*. Trends Genet 5:310–315
Ramsey J (2007) Unreduced gametes and neopolyploids in natural populations of *Achillea borealis* (Asteraceae). Heredity 98:143–150
Ramsey J, Schemske DW (1998) Pathways, mechanisms, and rates of polyploid formation in flowering plants. Annu Rev Ecol Syst 29:467–501
Ranz JM, Casals F, Ruiz A (2001) How malleable is the eukaryotic genome? Extreme rate of chromosomal rearrangement in the genus *Drosophila*. Genome Res 11:230–239
Rath SP, Patnaik SN (1978) Cytological studies in Cyperaceae. Proc Indian Sci Congr Assoc 65:107–108
Ray JH, Venketeswaran S (1978) Constitutive heterochromatin distribution in monocentric and polycentric chromosomes. Chromosoma 66:341–350
Rieder CL (1982) The formation, structure, and composition of the mammalian kinetochore and kinetochore fiber. Int Rev Cytol 79:1–58
Rivadavia F, Kondo K, Kato M, Hasebe M (2003) Phylogeny of the sundews, *Drosera* (Droseraceae), based on chloroplast rbcL and nuclear 18S ribosomal DNA sequences. Am J Bot 90:123–130
Roalson EH (2008) A synopsis of chromosome number variation in the Cyperaceae. Bot Rev 74:209–393
Roalson E, McCubbin AG, Whitkus R (2007) Chromosome evolution in Cyperales. In: Columbus JT, Friar EA, Porter JM, Prince LM, Simpson MG (eds) Monocots: comparative biology and evolution of Poales. Allen, Claremont, pp 62–71
Rogers E, Bishop JD, Waddle JA, Schumacher JM, Lin R (2002) The aurora kinase AIR-2 functions in the release of chromosome cohesion in *Caenorhabditis elegans* meiosis. J Cell Biol 157:219–229
Rooks F (2008) The *Juncus bufonius* polyploid complex in central Europe. Ph.D. thesis, Charles University, Prague
Rotreklová O, Bureš P, Řepka R, Grulich V, Šmarda P, Hralová I, Zedek F, Koutecký T (2011) Chromosome numbers in *Carex*. Preslia 83:25–58
Rutkowska J, Badyaev AV (2008) Meiotic drive and sex determination: molecular and cytological mechanisms of sex ratio adjustment in birds. Philos Trans Roy Soc Lond B Bio 363:1675–1686
Sakuno T, Tada K, Watanabe Y (2009) Kinetochore geometry defined by cohesion within the centromere. Nature 458:852–858
Schönswetter P, Suda J, Popp M, Weiss-Schneeweiss H, Brochmann C (2007) Circumpolar phylogeography of *Juncus biglumis* (Juncaceae) inferred from AFLP fingerprints, cpDNA sequences, nuclear DNA content and chromosome numbers. Mol Phylogenet Evol 42:92–103
Schrader F (1935) Notes on the mitotic behaviour of long chromosomes. Cytologia 6:422–430
Schwarzstein M, Wignall SM, Villeneueve AM (2010) Coordinating cohesion, co-orientation, and congression during meiosis: lessons from holocentric chromosomes. Genes Dev 24:219–228
Sheikh SA, Kondo K (1995) Differential staining with orcein, Giemsa, CMA and DAPI for comparative chromosome study of 12 species of Australian *Drosera* (Droseraceae). Am J Bot 82:1278–1286
Sheikh SA, Kondo K, Hoshi Y (1995) Study of diffused centromeric nature of *Drosera* chromosomes. Cytologia 60:43–47
Simpson DA, Furness CA, Hodkinson TR, Muasya AM, Chase MW (2003) Phylogenetic relationships in Cyperaceae subfamily Mapanioideae inferred from pollen and plastid DNA sequence data. Am J Bot 90:1071–1086
Šmarda P, Bureš P, Horová L, Foggi B, Rossi G (2008) Genome size and GC content evolution of *Festuca*: ancestral expansion and subsequent reduction. Ann Bot 101:421–433
Stack MS, Anderson LK (2001) A model for chromosome structure during the mitotic and meiotic cell cycle. Chromosome Res 9:175–198
Stear JH, Roth MB (2002) Characterization of HCP-6, a *C. elegans* protein required to prevent chromosome twisting and merotelic attachment. Genes Dev 16:1498–1508
Stephan W (2007) Evolution of genome organization. In: Encyclopedia of life sciences. Wiley, Chichester. doi:10.1002/9780470015902.a0001699.pub2
Strandhede S-O (1965a) Chromosome studies in *Eleocharis*, subser. *Palustres*. I. Meiosis in some forms with 15 chromosomes. Hereditas 53:47–62

Strandhede S-O (1965b) Chromosome studies in *Eleocharis*, subser. *Palustres*. III. Observations on Western European taxa. Opera Bot 9(2):1–86

Strandhede S-O (1965c) Chromosome studies in *Eleocharis*, subser. *Palustres*. IV. A possible case of an extra, reductional division giving rise to hemi-haploid pollen nuclei. Bot Not 118:243–253

Strandhede S-O (1966) Morphologic variation and taxonomy in European *Eleocharis*, subser. *Palustres*. Opera Bot 10(2):1–187

Sullivan BA, Blower MD, Karpen GH (2001) Determining centromere identity: cyclical stories and forking paths. Nat Rev Genet 2:584–596

Sutherland GR, Baker E, Richards RI (1998) Fragile sites still breaking. Trends Genet 14:501–506

Talbert PB, Bryson TD, Henikoff S (2004) Adaptive evolution of centromere proteins in plants and animals. BMC Biol 3:18

Talbert PB, Bayes JJ, Henikoff S (2009) Evolution of centromeres and kinetochores: a two-part fugue. In: De Wulf P, Earnshaw WC (eds) The kinetochore: from molecular discoveries to cancer therapy. Springer, New York, pp 193–229

Tanaka N (1940) Chromosome studies in Cyperaceae. VIII. Meiosis in diploid and tetraploid forms of *Carex siderosticta* Hance. Bot Mag (Tokyo) 10:282–310

Tanaka N (1941) Chromosome studies in Cyperaceae. XIII. Aneuploid plants of *Carex podogyna* Franch et Sav. with special reference to an abnormal pollen. Bot Mag (Tokyo) 55:181–186

Tanaka N (1949) Chromosome studies in the genus *Carex* with special reference to aneuploidy and polyploidy. Cytologia 15:15–29

Tanaka N, Tanaka N (1979) Chromosome studies in *Chionographis* (Liliaceae). II. Morphological characteristics of the somatic chromosomes of four Japanese members. Cytologia 44:935–949

Tanaka N, Tanaka N (1980) Chromosome studies in *Chionographis* (Liliaceae). III. The mode of meiosis. Cytologia 45:809–817

Tejavathi DH (1988) Somatic instability in the populations of *Cyperus cyperoides* (L.) O. Kuntze (Cyperaceae). Current Sci 57:724–728

Tobler H, Müller F (2001) Chromatin diminution. In: Encyclopedia of life sciences. Wiley, Chichester. doi:10.1038/npg.els.0001181

Toivonen H (1981) Spontaneous *Carex* hybrids of *Heleonastes* and related sections in Fennoscandia. Acta Bot Fenn 116:1–51

Underkoffler LA, Mitchell LE, Abdulali ZS, Collins JN, Oakey RJ (2005) Transmission ratio distortion in offspring of mouse heterozygous carriers of a (7.18) Robertsonian translocation. Genetics 169:843–848

Vaarama A (1954) Cytological observations on *Pleurozium schreberi*, with special reference to centromere evolution. Ann Bot Soc Vanamo 28:1–59

Vanzela ALL, Colaço W (2002) Mitotic and meiotic behavior of γ irradiated holocentric chromosomes of *Rhynchospora pubera* (Cyperaceae). Acta Sci Maringá 24:611–614

Vanzela ALL, Guerra M (2000) Heterochromatin differenciation in holocentric chromosomes of *Rhynchospora* (Cyperaceae). Genet Mol Biol 23:453–546

Vanzela ALL, Guerra M, Luceño M (1996) *Rhynchospora tenuis* Link (Cyperaceae), a species with the lowest number of holocentric chromosomes. Cytobios 88:219–228

Vanzela ALL, Cuadrado A, Jouve N, Luceño M, Guerra M (1998) Multiple locations of the rDNA sites in holocentric chromosomes of *Rhynchospora* (Cyperaceae). Chromosome Res 6:345–349

Vanzela ALL, Luceño M, Guerra M (2000) Karyotype evolution and cytotaxonomy in Brazilian species of *Rhynchospora* Vahl (Cyperaceae). Bot J Linn Soc 134:557–566

Vanzela ALL, Cuadrado A, Guerra M (2003) Localization of 45 S rDNA and telomeric sites on holocentric chromosomes of *Rhynchospora tenuis* Link (Cyperaceae). Genet Mol Biol 26:199–201

Villasante A, Abad JP, Méndez-Lago M (2007) Centromeres were derived from telomeres during the evolution of the eukaryotic chromosome. Proc Natl Acad Sci USA 104:10542–10547

Wahl HA (1940) Chromosome numbers and meiosis in the genus *Carex*. Am J Bot 27:458–470

Waterway MJ, Hoshino T, Masaki T (2009) Phylogeny, species richness, and ecological specialization in Cyperaceae tribe Cariceae. Bot Rev 75:138–159

White MJD (1973) Animal cytology and evolution. Cambridge University Press, Cambridge, UK

Yano O, Hoshino T (2006) Phylogenetic relationships and chromosomal evolution of Japanese *Fimbristylis* (Cyperaceae) using nrDNA ITS and ETS 1f sequence data. Acta Phytotax Geobot 57:205–217

Yano O, Katsuyama T, Tsubota H, Hoshino T (2004) Molecular phylogeny of Japanese *Eleocharis* (Cyperaceae) based on ITS sequence data, and chromosomal evolution. J Plant Res 117:409–419

Záveská-Drábková L, Vlček Č (2010) Molecular phylogeny of the genus *Luzula* DC. (Juncaceae, Monocotyledones) based on plastome and nuclear ribosomal regions: a case of incongruence, incomplete lineage sorting and hybridisation. Mol Phylogenet Evol 57:536–551

Zedek F, Smerda J, Smarda P, Bureš P (2010) Correlated evolution of LTR retrotransposons and genome size in the genus *Eleocharis*. BMC Plant Biol 10:265

Karyotype Diversity and Evolutionary Trends in Angiosperms 13

Hanna Weiss-Schneeweiss and Gerald M. Schneeweiss

Contents

13.1 Introduction

Each species has a characteristic chromosome complement, its karyotype, which represents the "phenotypic appearance of the somatic chromosomes in contrast to their genic contents" (Levitzky 1931). In other words, the karyotype is the highest level of structural and functional organization of the nuclear genome. By comparing the chromosomes of different taxa much can be learned about patterns and mechanisms of karyotype evolution and its significance for diversification and speciation.

Karyotype constancy, both in chromosome number and structure, results from the necessary fidelity of chromosome replication and ensures the faithful transfer of genetic material to the next generation. On the other hand, karyotype variation confers evolutionary change. There are two major categories of karyotype variation: numerical (resulting usually from meiotic or mitotic spindle malfunction) and structural (usually arising from errors in crossing over). These two major classes are, however, interrelated and the actual evolutionary change of a karyotype is usually complex, involving various mechanisms.

Karyotypic features most commonly recorded for comparative evolutionary analyses are chromosome number, size and symmetry, as well as position and type of chromosomal landmarks such as centromeres, secondary constrictions, or repetitive DNA types. These parameters may be complemented with genome size estimations and data on chromosomal behavior, especially during meiosis. This chapter will describe the most relevant features and discuss their variation and evolutionary trends within angiosperms, including their role in diversification and speciation.

H. Weiss-Schneeweiss (✉)
Department of Systematic and Evolutionary Botany, University of Vienna, Rennweg 14, Vienna A-1030, Austria
e-mail: hanna.schneeweiss@univie.ac.at

I.J. Leitch et al. (eds.), *Plant Genome Diversity Volume 2*,
DOI 10.1007/978-3-7091-1160-4_13,

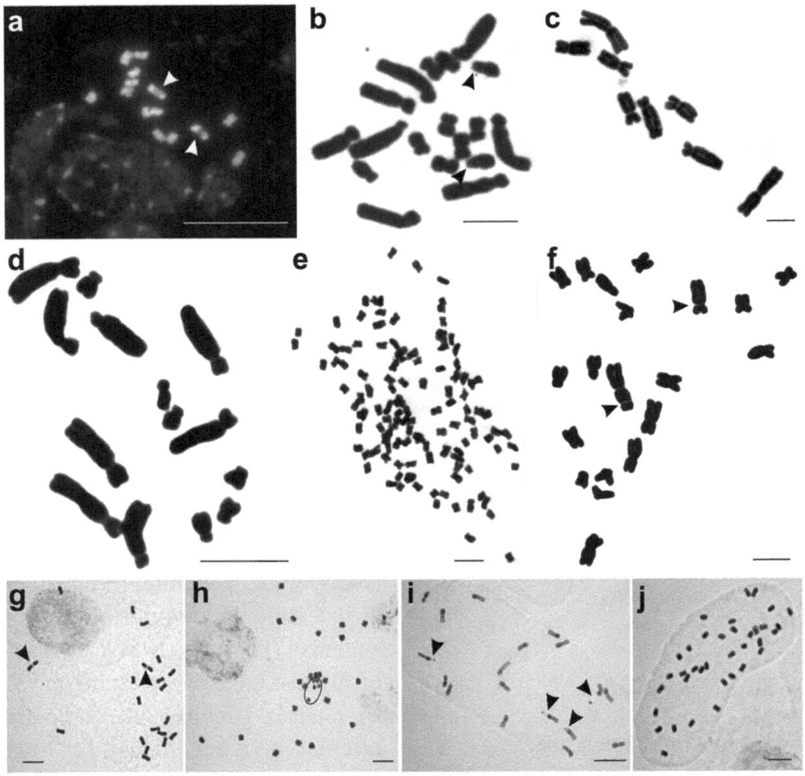

Fig. 13.1 Chromosome number and structure diversity in angiosperms. (**a**) *Oxalis namaquana*/Oxalidaceae: $2n = 2x = 14$ (*arrowheads* indicate secondary constrictions; DAPI staining); (**b**) *Barnardia numidica*/Hyacinthaceae (formerly in genus *Scilla*): $2n = 2x = 18$ (*arrowheads* indicate satellited chromosomes); (**c**) *Bellevalia paradoxa*/Hyacinthaceae: $2n = 2x = 8$; (**d**) bimodal karyotype of *Aloe paralellifolia*/Asphodelaceae: $2n = 2x = 14$; (**e**) *Senecio carniolicus*/Asteraceae: $2n = 6x = 120$; (**f**) *Nectaroscilla hyacinthoides*/Hyacinthaceae: $2n = 2x = 20$ (*arrowheads* indicate polymorphic homologous chromosome pair); (**g**) *Melampodium longipilum*/Asteraceae: $2n = 2x = 20$ (*arrowheads* indicate secondary constrictions); (**h**) *Melampodium montanum*/Asteraceae: $2n = 2x = 22 + 5B$ chromosomes (B chromosomes *encircled*); (**i**) *Melampodium mimulifolium*/Asteraceae: $2n = 2x = 18$ (*arrowheads* indicate satellited chromosomes); (**j**) tetraploid *Melampodium paniculatum*/Asteraceae $2n = 4x = 36$. [All photos H. Weiss-Schneeweiss (**a**, with J. Suda, Charles University, Prague)]. Scale bar represents 5 μm

13.2 Trends in Chromosome Number Evolution

13.2.1 Chromosome Numbers

Chromosome number is still the most consistently and frequently recorded cytological character. In part, this is due to the relative ease of its observation and its nature of being a discrete character, whereas other karyotypic features, such as chromosome structure or composition, are technically more challenging to collect and often more difficult to quantify.

Chromosome numbers have been determined for about 25–30% of angiosperms (Bennett 1998; Guerra 2008; see Fig. 13.1 for examples). The taxonomic coverage is, however, very uneven. In some genera, chromosome numbers are characterized in detail in many or even all species and from numerous populations (e.g., *Brachypodium*/Poaceae: Watanabe et al. 1999; *Crepis*/Asteraceae: Babcock 1947; Dimitrova and Greilhuber 2001; *Hypochaeris*/Asteraceae: Stebbins 1971; Cerbah et al. 1998; Weiss-Schneeweiss et al. 2003, 2007a, 2008; *Melampodium*/Asteraceae: Weiss-Schneeweiss et al. 2009; *Nicotiana*/Solanaceae: Goodspeed 1933; Narayan 1987; Lim et al. 2000), whereas other groups remain poorly characterized chromosomally (e.g., Lentibulariaceae and many tropical plant families). To ease the navigation through continuously published, numerous and highly dispersed chromosome number data several comprehensive chromosome number indices have been compiled over the years. These include, for example, Darlington and Wylie (1955), Fedorov (1969), Moore (e.g., 1974, 1977), Goldblatt and Johnson (1979—; accompanied by an online version hosted by Missouri Botanical Garden: http://www.tropicos.org/Project/IPCN). Compilations may also be available for specific taxa, such as the online Index to Chromosome Numbers in Asteraceae (http://www.lib.kobe-u.ac.jp/infolib/meta_pub/G0000003asteraceae_e), or for certain regions, usually specific countries (e.g., Löve and Löve 1974; Dobeš and Vitek 2000; Góralski et al. 2009).

Known angiosperm chromosome numbers are commonly reported as sporophytic diploid numbers $2n$, but also as gametic haploid numbers n. For comparative analyses the (usually inferred) haploid chromosome number of a clade's (mostly hypothetical) ancestor is referred to as base chromosome number x for the group or clade. This x is not to be confused with the monoploid number x: the lowest haploid chromosome number of a polyploid series (Guerra 2008).

Known angiosperm chromosome numbers vary 160-fold. They range from $2n = 4$ (*Zingeria biebersteiniana* and *Colpodium versicolor*/Poaceae, *Ornithogalum tenuifolium*/Hyacinthaceae, *Brachycome dichromosomatica* and *Haplopappus gracilis*/Asteraceae: Cremonini 2005; *Rhynchospora tenuis*/Cyperaceae: Vanzela et al. 1996) to $2n =$ c. 600 in the palm *Voanioala gerardii* (Arecaceae: Johnson et al. 1989; Röser 1994) and $2n = 640$ in the stonecrop *Sedum suaveolens* (Crassulaceae: Uhl 1978). Chromosome numbers of the majority of angiosperms range from $n = 7$ to $n = 20$ (Grant 1982; Masterson 1994). Chromosome number diversification does not correlate with morphological diversification, as is evident from morphologically diverse but chromosomally uniform island radiations (Stuessy and Crawford 1998; Mandáková et al. 2010a) and morphologically uniform, but chromosomally diverse monocot lineages (*Crocus vernus*/Iridaceae: Frello and Heslop-Harrison 2000; *Prospero*/Hyacinthaceae: Ainsworth et al. 1983, Fig. 13.2).

The wide range of chromosome numbers is not uniformly distributed among orders, families and genera of angiosperms. Some groups show no or very limited chromosome number variation (*Lilium*/Liliaceae: Stewart 1947; Smyth et al. 1989; *Hepatica*/Ranunculaceae: Weiss-Schneeweiss et al. 2007b; *Galtonia*/Hyacinthaceae: Forrest and Jong 2004; many woody angiosperm lineages: Ehrendorfer 1970; Grant 1982), while others are very variable ($2n = 6$, 8, 10, 12, 16 in *Hypochaeris*/Asteraceae: Stebbins 1971; Cerbah et al. 1998; Weiss-Schneeweiss et al. 2003, 2008; $2n = 18$, 20, 22, 24, 28, 36, 40, 46, 48, 54, 56, 60, 66 in *Melampodium*/Asteraceae: Stuessy 1971; Weiss-Schneeweiss et al. 2009; $2n = 4$, 6, 8, 10, 12, 14, 16, 18, 22, 24, 26, 27, 28, 30, 36 in *Brachycome*/Asteraceae: Watanabe et al. 1999). In plant groups where more than one base chromosome number is present ($x = 5$, 6, 7 in *Lotus*/Fabaceae: Grant 1991; $x = 9$, 10, 11, 12, 14 in *Melampodium*/Asteraceae: Blöch et al. 2009; $x = 3$, 4, 5, 6 in *Crepis*/Asteraceae: Babcock and Jenkins 1943), the ancestral base chromosome number might be identified either as the dominant one and/or the one found in the most closely related taxa. Nowadays, however, base chromosome numbers are usually inferred within an explicit (most often molecular) phylogenetic context (Torrell et al. 1999; Mast et al. 2001; Ellison et al. 2006; Enke and Gemeinholzer 2008; Hipp et al. 2009) using, for instance,

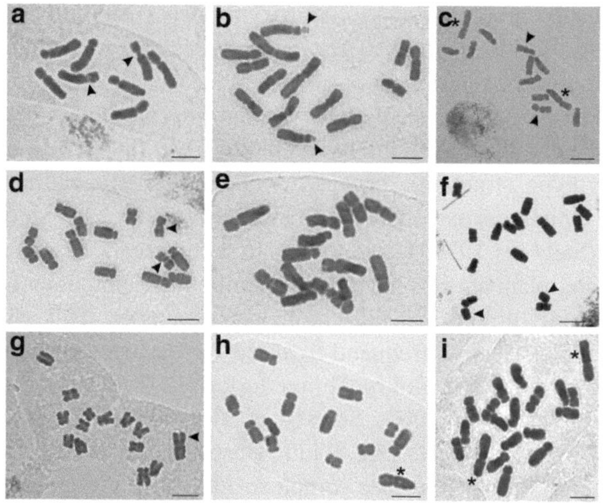

Fig. 13.2 Chromosome number and structure variation in a group of closely related species and cytotypes in *Prospero*/Hyacinthaceae (formerly in genus *Scilla*; asterisks indicate large metacentric "fusion" chromosome, where present). (**a**) *P. obtusifolium*: $2n = 2x = 8$ (*arrowheads* indicate secondary constrictions); (**b**) *P. hanburyi*: $2n = 2x = 14$ (*arrowheads* indicate satellited chromosomes); (**c**) *P. autumnale* s. l. cytotype B6B6: $2n = 2x = 12$ (*arrowheads* indicate secondary constrictions); (**d**) *P. autumnale* s. l. cytotype B7B7: $2n = 2x = 14$ (*arrowheads* indicate secondary constrictions); (**e**) *P. autumnale* s. l. cytotype AA: $2n = 2x = 14$; (**f**) *P. autumnale* s. l. cytotype B7B7: $2n = 2x = 14 + 2B$ chromosomes (*arrowheads* indicate B chromosomes); (**g**) *P. autumnale* s. l. cytotype B7B7: $2n = 2x = 14$ (*arrowhead* indicates supernumerary segment in one of the homologs of chromosome 1); (**h**) *P. autumnale* s. l., diploid homoploid hybrid B6B7: $2n = 2x = 13$; (**i**) *P. autumnale* s. l. triploid hybrid B6B6B7: $2n = 3x = 19$. (All photos T.-S. Jang and H. Weiss-Schneeweiss). *Scale bar* represents 5 μm

maximum parsimony or likelihood based models of chromosome number reconstruction as implemented in software such as Mesquite (Maddison and Maddison 2009) or chromEvol (Mayrose et al. 2010).

Inferences of chromosome base numbers are becoming increasingly difficult (or even impossible) with increasing phylogenetic depth as identical chromosome base numbers will occur in unrelated lineages, already within a single genus (e.g., Cerbah et al. 1998; Enke and Gemeinholzer 2008). Additionally, karyotype rearrangements (with or without polyploidy: see Sect. 13.2.3) as major drivers of stepwise chromosome number changes (see Sect. 13.2.2) might occur frequently and often rapidly on an evolutionary time scale (Brassicaceae: Lysák et al. 2006; *Prospero*/Hyacinthaceae: Ainsworth et al. 1983; Vaughan et al. 1997). This also affects efforts to determine the ancestral basic chromosome number of angiosperms. Consequently, both early estimates by Raven (1975) of $x = 7$ as well as more recent reconstructions of $x = 6$ by Soltis et al. (2005) need to be viewed with caution. Evidently, more in-depth karyotypic information for identification of chromosomal "building blocks" (Schranz et al. 2006) as well as more realistic chromosomal evolutionary models (Dobigny et al. 2004; White et al. 2010) will be necessary for reliable chromosome number reconstructions.

For a long time the dogma was that a species possesses a single chromosome number and ploidy level (as well as one karyotype structure), because deviating somatic chromosome numbers usually confer meiotic irregularities or sterility (Stebbins 1971; Levin 2002). In recent years, however, several plant species have been identified that possess more than one diploid somatic chromosome number. This phenomenon is most frequently caused by the occurrence of intraspecific polyploidy (Ehrendorfer 1958; Frello and Heslop-Harrison 2000; Stuessy et al. 2004; Suda et al. 2007; see also Sect. 13.2.3 and Husband et al. 2013, this volume). In other species complexes, it entails the presence of a dysploid series at the diploid level, which may later be accompanied by hybridization and polyploidy creating complex patterns of chromosome numbers. Examples include *Scilla* (*Barnardia*) *scilloides*/Hyacinthaceae ($2n = 16$, 18, 26, 27, 34, 35, 36, 43: Haga and Noda 1976; Choi et al. 2008) or *Prospero autumnale*/Hyacinthaceae ($2n = 10$, 12, 14, 25, 26, 27, 28, 42 and so on: Ainsworth et al. 1983; Ebert et al. 1996). Such intraspecific numerical and/or structural (see Sect. 13.3) karyotypic variants are called cytotypes or chromosomal races/cytoraces (Stebbins 1971; Stuessy 2009; Fig. 13.2) and may in fact constitute cryptic species.

13.2.2 Dysploidy and Aneuploidy

13.2.2.1 Dysploidy

Dysploidy refers to the stepwise change of the haploid chromosome number among related species (Stebbins 1971; Guerra 2008) and such a series of haploid chromosome numbers constitutes a dysploid series (Ehrendorfer 1964). Although chromosome number change may involve the loss or gain of entire chromosomes (aneuploidy; Guerra 2008), much more frequently it is achieved via genome restructuring resulting in centromere/chromosome number reduction or increase with or without significant loss of genetic material (unbalanced and balanced rearrangements, respectively; see Lysák and Schubert 2013, this volume). A relatively simple example of balanced rearrangement is centric fusion-fission ("Robertsonian translocation" in the broad sense: Guerra 2008). Here, the karyotype involved in a centric fusion or fission has one chromosome fewer or more, respectively, while retaining an intact number of chromosomal arms. Robertsonian fission/fusion events occur regularly in animals (Imai 1978; Rowell et al. 2002) including apes and humans (Yunis and Prakash 1982), and are also documented in some angiosperm groups (reviewed in Jones 1977, 1998; e.g., *Nothoscordum*/Alliaceae: Souza et al. 2009; *Gibasis*/Commelinaceae: Kenton 1981; *Tradescantia*/Commelinaceae: Jones 1998; *Christensonella*/Orchidaceae: Koehler et al. 2008; Moreas et al. 2012). Recent data, however, suggest that in the majority of angiosperms dysploid changes are more often achieved via complex mechanisms involving (usually) pericentric inversions and reciprocal (unequal) translocations acting in concert (Schubert 2007; for details on mechanisms see Lysák and Schubert 2013, this volume). This has been elegantly shown in Brassicaceae (Lysák et al. 2006) and has been suggested for Fabaceae (Choi et al. 2004) and grasses (Devos 2005).

The majority of earlier studies considered descending dysploidy (i.e., chromosome number reduction) as the main trend of dysploid chromosome number change in angiosperms (Jones 1977), often accompanied by the evolution of an annual life form (Stebbins 1971). Examples include $x = 6 \rightarrow 3$ in *Crepis*/Asteraceae (Babcock 1947), $x = 9 \rightarrow 2$ in *Haplopappus*/Asteraceae (Jackson 1962; Anderson et al. 1974) and $x = 15 \rightarrow 12$ in *Crocus*/Iridaceae (Brighton 1978). For some plant groups, ascending dysploidy (i.e., chromosome number increase) has also been inferred ($x = 7 \rightarrow 9$ in *Allium*/Alliaceae: Levan 1932). These largely intuitive inferences might, however, turn out to be oversimplified. For instance, recent molecular phylogenetic analyses of *Crepis* show that chromosome numbers are largely uncorrelated with phylogeny, and that there is no simple descending dysploidy (Enke and Gemeinholzer 2008) as suggested previously. Therefore, the direction of dysploid chromosomal changes not only is largely group specific, but often bidirectional with both types of dysploidy (ascending and descending) co-occurring within a given group (mixed dysploidy; Bakker et al. 2000; Barber et al. 2000; Samuel et al. 2003; Enke and Gemeinholzer 2008; Blöch et al. 2009). Indeed, mechanisms which can lead to simultaneous bidirectional dysploid changes have been previously proposed (Schubert and Rieger 1985). Irrespective of the direction, dysploid chromosome number changes clearly significantly contribute to angiosperm diversification and speciation (Stebbins 1971; Grant 1981; Levin 2002).

13.2.2.2 Aneuploidy

In contrast to dysploidy, aneuploidy refers to the loss or gain of entire chromosomes (Guerra 2008). As aneuploid variation significantly changes the genetic balance of an organism, it is usually disadvantageous or even lethal for its carrier and thus of little relevance on an evolutionary timescale (in contrast to dysploidy; Ehrendorfer 1964, 1980). Exceptions do, however, exist. The most impressive one is *Claytonia virginica* (Portulacaceae) that shows a nearly continuous series from $2n = 12$ up to at least $2n = 100$ (occasionally up to $2n = 191$) with $x = 6$, 7, 8 (Rothwell and Kump 1965; Lewis et al. 1967). Intrapopulational chromosomal variation was shown to manifest itself not only in chromosome number variation, but also in plants with the same chromosome numbers but occasionally with varied karyotype structures. In addition, different organs of the same plants were shown to possess different chromosome numbers (aneusomaty), and climatically driven seasonal variation was also reported (Lewis et al. 1967). This tremendous chromosome variation in *C. virginica* is considered to be the result of hybridization and polyploidization between partly isolated diploid populations with different haploid chromosome numbers and subsequent meiotic irregularities, and/or of ecological factors (Lewis et al. 1967).

In general, the extent of genetic imbalance imposed by aneuploidy is lower in polyploids. Consequently, aneuploidy might be expected to be more frequently fixed in polyploids, resulting in its higher evolutionary potential in a polyploid background (Lewis et al. 1967; Ainsworth et al. 1983; Weiss-Schneeweiss et al. 2009). It is however, not always easy to distinguish dysploidy from aneuploidy in polyploids.

13.2.3 Polyploidy

Polyploidy refers to the instantaneous multiplication of entire chromosome sets (Stebbins 1971; Wendel 2000; Leitch and Leitch 2008). The most widely used classification concerns the distinction of auto- and allopolyploids (Kihara and Ono 1927). Whereas in autopolyploids the (nearly) identical chromosomes within a single (sub)species are multiplied (Ramsey and Schemske 1998, 2002), allopolyploids are the result of hybridization between two (sub)species accompanied by polyploidization (Stebbins 1971). An alternative to this taxonomic definition has been the use of polysomic vs. disomic segregation (accompanied by multivalent vs. bivalent formation in meiosis) suggesting autopolyploid and allopolyploid origin, respectively (Rieseberg and Doyle 1989; Doyle et al. 2008; but see also Stebbins 1971). From the standpoint of chromosome behavior, genetic allopolyploids should behave like diploids with more sets of chromosomes, showing bivalent formation and disomic segregation. Both types of definitions are useful and usually lead to a congruent conclusion, but exceptions do exist. These include, for instance, allopolyploids, that though disomic, may exhibit multivalent formation in early stages after formation (Ownbey 1950; Ramsey and Schemske 2002) or autopolyploids that undergo meiosis with no chromosomal irregularities (Weiss and Małuszyńska 2000). The difficulties might in fact be more pronounced since the polysomic condition is usually considered to be an initial and transient stage leading to chromosomal diploidization and disomy. Thus, old polyploids with disomic behaviour might also have originated as autopolyploids (Doyle et al. 2008). Yet another type of polyploid classification is based on the polyploid's age and point of origin in the evolutionary history, recognizing (in order of decreasing age) paleo-, meso- and neopolyploids (Ehrendorfer 1980; Ramsey and Schemske 1998, 2002; Comai 2005).

The majority of well-analysed polyploids are allopolyploids. Allopolyploidy might result in a novel chromosome base number, if taxa with different chromosome base numbers are involved as parents (dibasic polyploidy; e.g., *Brassica*/Brassicaceae: U 1935). Contrary to earlier views (Stebbins 1950, but see Darlington 1956), where autopolyploids were conventionally treated as intraspecific cytotypes of a lower ploidy taxon, they are now increasingly recognized as being equally important for speciation as allopolyploids (Soltis et al. 2007).

The propensity for polyploidization appears to be unequally distributed among angiosperms (46% of infrageneric polyploidy in higher monocots vs. 32% in rosids: Wood et al. 2009; scarce polyploidy in woody angiosperms vs. frequent, likely ancient, polyploidy in tropical humid forests: Ehrendorfer 1970). The underlying causes remain, however, unknown mostly due to the lack of adequate, in-depth genetic and genomic information on the extent of polyploid ancestry.

In contrast to dysploidy, polyploidization is an essentially unidirectional process towards higher ploidy levels (often resulting in polyploid series or polyploid complexes; Stebbins 1940, 1971), a feature termed "polyploidy ratchet" (Meyers and Levin 2006). Whereas early estimates indicated that 30% of flowering plant species underwent at least one round of polyploidization (Stebbins 1950), current estimates suggest that all angiosperms are ancient polyploids (paleopolyploids, Jiao et al. 2011) being affected by at least two ancestral whole genome duplications (WGD; Bowers et al. 2003; Jaillon et al. 2007; Fawcett et al. 2009; Jiao et al. 2011). These WGDs have been suggested to play causative roles for the evolutionary success of angiosperms (Van de Peer et al. 2009; Jiao et al. 2011). Additional WGDs have been inferred for specific angiosperm lineages (Schranz and Mitchell-Olds 2006; Jaillon et al. 2007; Proost et al. 2011). Recent years have provided the wealth of data demonstrating that polyploidy not only is frequent, but is one of the major forces in angiosperm diversification and speciation. Estimates of the frequency of polyploid speciation (speciation events accompanied by ploidy level increase) range from 3.8% (Levin and Wilson 1976), 2–4% (Otto and Whitton 2000) to 15% (Wood et al. 2009) (see Fawcett et al. 2013, this volume).

Polyploidization is a recurrent phenomenon in many angiosperm lineages (Soltis and Soltis 1999), with numerous examples of both autopolyploids (Suda et al. 2007; Trávníček et al. 2011) and allopolyploids (Dixon et al. 2009; Symonds et al. 2010). Consequently, Soltis et al. (2009) state: The question is no longer "What proportion of angiosperms are polyploid?", but "How many episodes of polyploidy characterize any given lineage?" As independently originated lineages might apparently show different evolutionary trajectories (Doyle et al. 2004) and constitute separate species, the contribution of polyploidy to speciation might well be underestimated by current taxonomic practice (Soltis et al. 2010).

The most obvious effect of polyploidization on the karyotype is the increase in chromosome number. Consequently, reproductive isolation conferred by meiotic problems in hybrids between polyploids and their lower-ploidy ancestors has long been identified as an important mechanism leading to polyploid speciation. Polyploidization, often coupled with hybridization does, however, also trigger a cascade of processes operating at the genomic, epigenomic and proteomic level (Leitch and Leitch 2008; Fawcett et al. 2013, this volume). These include activation of transposable elements, epigenetic phenomena leading to heterochromatin formation, and eventually accumulation of structural chromosomal mutations (e.g., via interhomeologous translocations; Clarkson et al. 2005; Shoemaker et al. 2006; Doyle et al. 2008), which ultimately lead to complete genome diploidization (Wendel 2000; Doyle et al. 2008; Mandáková et al. 2010b).

Genetic and cytological genome diploidization of polyploids is often accompanied by a reduction in genome size (Leitch and Bennett 2004; Pellicer et al. 2010) and over evolutionary time, chromosome numbers of polyploids might also be considerably reduced, even below the original level of the ancestral diploid taxa (Lysák et al. 2006). Therefore, assessment of the incidence of polyploidy from known chromosome numbers alone will be biased downwards. Such cycles of revolutionary (polyploidization) and evolutionary (cytological diploidization) phases have so far only been well documented for a few angiosperm groups, for which enough genomic data are available, including Brassicaceae (Lysák et al. 2006; Mandáková and Lysák 2008), *Oryza*/Poaceae (Wang et al. 2007) or *Gossypium*/Malvaceae (Shoemaker et al. 2006).

13.3 Trends in Karyotype Structure Evolution

Whereas data on chromosome numbers accumulate relatively quickly, detailed karyotype information has been obtained at a much slower pace. Karyotype features detectable in classically stained material (e.g., by Feulgen method or by acetocarmine; Fukui and Nakayama 1996) include, in addition to chromosome number, individual chromosome sizes (absolute and relative), haploid (or diploid) karyotype length, karyotype symmetry, and the positions of morphologically discernible chromosome landmarks, most notably primary and secondary constrictions. The application of more advanced staining techniques, including different banding techniques (C-banding or fluorescent banding: Deumling and Greilhuber 1982; Greilhuber 1982, 1995; Lukaszewski and Gustafson 1983) and, recently and most prominently, fluorescence *in situ* hybridization (Findley et al. 2010), provides many more chromosomal landmarks. These may enable identification of individual (homologous) chromosomes/chromosome arms in a given karyotype or even the 'painting' of all chromosomes (Lysák et al. 2006). However, in contrast to the majority of animals, the ability to label whole chromosomes using specific chromosome 'paints' in angiosperms is difficult, mostly due to the high content of dispersed transposable elements across the genome and their relatively quick turnover (Lim et al. 2007a). The karyotype and its features can be schematically represented by an idiogram or karyogram.

13.3.1 Morphological Features

13.3.1.1 Chromosome Size

Individual chromosome sizes in angiosperms (given as *mean* chromosome length per species) have been estimated to vary from 0.6 to more than 14.6 μm, and the total chromosome length per diploid genome from 14.6 to 250 μm (Levin and Funderburg 1979). Even within a single karyotype of a taxon a relatively wide range of chromosome sizes can occur, with individual chromosome pairs ranging from 1 to 2 μm to more than 10 μm (Schubert 2007; Hamouche et al. 2010). Despite such variation, most plant species possess relatively uniform small to medium-sized chromosomes (c. 1–3 μm; Stace 2000). Possibly the smallest angiosperm chromosomes known so far are those in *Genlisea*/Lentibulariaceae ("bacterial size chromosomes", Greilhuber et al. 2006), which also holds the record for the lowest genome size. The largest chromosomes are found in plants with very large genomes, but not many chromosomes (*Lilium*/Liliaceae, $2n = 24$, 20 μm: Peruzzi et al. 2009; *Paris*/Melanthiaceae, $2n = 40$, some chromosomes larger than 30 μm: Pellicer et al. 2010).

Chromosome size is limited in both directions. The upper limit is determined by the length of the longest chromosome arm. This cannot exceed half of the average spindle axis length at telophase so as not to impair chromatid separation during mitosis (Schubert and Oud 1997; Schubert 2001; Hudakova et al. 2002). The lower limit is more difficult to define. As chromosomes smaller than 1% of the genome fail to segregate correctly during meiosis, even if supposedly fully intact with an original centromere, telomeres and replication origins (Schubert 2001; Murata et al. 2006), the lower limit appears to be determined by bivalent stability and chiasma formation allowing stable transmission during meiosis. Changes in chromosome size can be correlated with changes in genome size (achieved via polyploidy and/or repetitive DNA amplification; see below) or might result from interchromosomal rearrangements, such as centric fusions/fissions or unequal translocations (see Lysák and Schubert 2013, this volume).

13.3.1.2 Karyotype Length and Genome Size

Genome size variation rarely reflects chromosome number variation because the major mechanisms responsible for changes in genome size are different from those leading to chromosome number change. This is best illustrated by the fact that the species with the smallest genome size, found in *Genlisea* (Lentibulariaceae; Greilhuber et al. 2006), have similar chromosome numbers (i.e., $2n$ = c. 40) to the species with the largest genome size, *Paris japonica* (Melanthiaceae; $2n = 40$: Pellicer et al. 2010). Total karyotype length is correlated with genome size because both are shaped by the same processes. These include polyploidization and accumulation of repetitive DNA, most importantly retroelements (Bennetzen 2000; Tenaillon et al. 2011), or DNA deletions via unequal and illegitimate recombination (Bennetzen 2002; Hawkins et al. 2008). Changes in genome size are best reflected in changes in the size (or more accurately volume and composition) of chromosomes. Genome size changes can affect the entire karyotype uniformly (Ainsworth et al. 1983; Raina and Rees 1983) or they may be restricted to a subset of chromosomes usually via amplification/deletion of blocks of certain families of tandem repeats (de la Herrán et al. 2001; Koo and Jiang 2008). For more information on genome size evolution in angiosperms see Leitch and Leitch (2013, this volume).

13.3.1.3 Centromere Position, Holocentric Chromosomes

In mitotic chromosomes, the centromere appears as a cytologically visible primary constriction and represents the site for kinetochore assembly and spindle attachment thereby ensuring faithful segregation of chromosomes during cell division. Most angiosperms possess chromosomes with localized centromeres (monocentric chromosomes), which divide a chromosome into two arms.

Centromere position and consequently arm length ratios (long arm length/short arm length) have been used as criteria in the two most commonly used systems of chromosome nomenclature (Levan et al. 1964; Stebbins 1971; Guerra 1986). Levan et al. (1964) distinguished six chromosome types (arm ratio in square brackets): M (median point; [1]), m (median region; [>1–1.7]), sm (submedian region; [>1.7–3.0]), st (subterminal region; [>3.0–7.0]), t (terminal region; [>7.0–infinity]), and T (terminal point; [infinity]). In contrast, Stebbins (1971), modified by Guerra (1986), distinguished metacentric [1–1.49], submetacentric [1.50–2.99], acrocentric [3.00–infinity] and telocentric [infinity] chromosomes.

Mechanisms leading to shifts in centromere position include structural rearrangements (pericentric inversions and translocations; see Lysák and Schubert 2013, this volume) as well as centromere deactivation and *de novo* centromere formation (Han et al. 2006). Indeed, studies have shown that the presence of centromere-specific tandem repeats are neither necessary nor sufficient to form a functional centromere (Nasuda et al. 2005). A fully functional centromere might be deactivated despite retaining the centromere-specific repeats (for example in dicentric chromosomes; Han et al. 2006) and a functional centromere can be formed *de novo* in an acentric chromosome fragment in the absence of specific centromeric repeats (Topp et al. 2009; Guerra et al. 2010).

In contrast to this, in a few genera and species, centromeres are not localized at specific centromeric constrictions on the chromosomes. Instead they are dispersed over the entire chromosome length. Such holocentric or holokinetic chromosomes (Guerra 2008; Guerra et al. 2010) are found in a few angiosperm lineages, including Cyperaceae (*Carex*, *Rhynchospora*, *Eleocharis*; Luceño et al. 1998; Vanzela et al. 2000; da Silva et al. 2005; Hipp 2007; Zedek et al. 2010) or Convolvulaceae (a single dodder/*Cuscuta* species: Pazy and Plitmann 1995; Guerra and García 2004), where they are often associated with a specialized form of inverted meiosis (Greilhuber 1995). The lack of a localized centromere not only reduces the number of chromosomal landmarks, but may also affect karyotype dynamics. Specifically it has been suggested that chromosome number change (i.e., dysploidy) can occur rapidly in plants with holocentrics, since chromosome number change due to fragmentation (agmatoploidy/agmatopolyploidy) or fusion (symploidy) is not expected to have deleterious effects. Such changes have been documented in natural populations (in *Luzula*/Juncaceae: Kuta et al. 2004; *Carex*/Cyperaceae: Chung et al. 2011), and callus tissue culture (Madej and Kuta 2001). The dynamics of holocentric chromosome number changes can also involve polyploidy (*Luzula*/Juncaceae: Kuta et al. 2004; *Juncus bigumis*/Juncaceae: Schönswetter et al. 2007; note that the presence of holocentric species in *Juncus* is still contentious: Bailey cited in Stace 2000) and "standard" dysploidy (Guerra 2008; Chung et al. 2011). Furthermore, their chromatin organization and genome size dynamics appear to be little influenced by centromeric organization (Guerra and García 2004; Haizel et al. 2005; Zedek et al. 2010). For a detailed account of holocentric chromosomes see Bureš et al. (2013, this volume).

13.3.1.4 Karyotype Symmetry, Bimodal Karyotypes

Karyotype symmetry is determined by the size of the chromosomes and/or centromere position. Most angiosperms have similarly sized mostly meta- and submetacentric chromosomes, and consequently possess uniform, symmetric karyotypes (Levitzky 1931; Stebbins 1971;

Fig. 13.1). However, some lineages have karyotypes comprising chromosomes with a range of sizes and centromere positions giving rise to asymmetric karyotypes. The majority of these are found among monocots with larger chromosomes and genomes (Stedje 1989; Choi et al. 2008; Hamouche et al. 2010), but they also occur in other groups with otherwise relatively uniform karyotypes (Babcock 1947; Cerbah et al. 1998). Changes from symmetric to asymmetric karyotypes most often involve changes in chromosome morphology (from metacetric to telocentric chromosomes), sometimes accompanied by dysploidy and/or non-uniform chromosome size changes (Stebbins 1971; de la Herrán et al. 2001).

Various indices have been suggested to describe karyotype asymmetry (reviewed and extended in Paszko 2006). Interest in these indices, which can serve as concise descriptors of several karyotype features, also results from attempts to identify evolutionary trends in karyotype evolution. Specifically, Stebbins (1971), following Levitzky's principle of karyotype symmetry vs. asymmetry (Levitzky 1931), postulated that asymmetric karyotypes are derived and evolutionarily younger than symmetric ones, because asymmetric karyotypes have often been found in plants with specialized morphological features or in evolutionary younger derivatives. A potential explanation is provided by the minimum interaction hypothesis (Imai et al. 1986), which states that karyotype evolution has been mostly shaped by selection to reduce the occurrence of fitness-reducing chromosomal mutations, such as reciprocal translocations. In species with big genome size to nuclear volume ratios, this can be achieved by an increasing chromosome number, e.g., by centric fission, leading to acrocentric chromosomes (Brighton 1978). To our knowledge, the pattern expected under this hypothesis, i.e., a positive correlation between karyotype asymmetry and the genome size to nuclear volume ratio, has never been rigorously tested in plants. The generality of the derived nature of an asymmetric karyotype has, however, been questioned (Jones 1978; Dimitrova and Greilhuber 2000). The main criticism is that it is difficult to assess the direction of such karyotype evolution as many reversals of the main trend might have occurred (Stace 2000). Recent analyses in Brassicaceae actually suggest that karyotype asymmetry might be a transitory state rather than simply a derived evolutionary endpoint (Lysák et al. 2006; Mandáková and Lysák 2008). More data of this type (i.e., detailed information on chromosomal restructuring studied within an explicit phylogenetic context) will be necessary to test general hypotheses on evolutionary trends in karyotype asymmetry.

A special case of asymmetry is the bimodal karyotype, which is characterized by the presence of two sharply distinct size classes of chromosomes, without any gradual transition. Bimodal karyotypes are well known from monocots, especially in Asphodelaceae (*Aloe*: Fig. 13.1d, *Haworthia*, *Gasteria*: Brandham 1971; Brandham and Doherty 1998; Vosa 2005) and Agavaceae (*Agave*, *Yucca*: Vosa 2005; Robert et al. 2008). Various hypotheses have been proposed to explain the origin of bimodal karyotypes including the suggestion that they are derived from the symmetrical karyotypes of polyploids with the small chromosomes representing the products of differential loss of chromosome segments (Darlington 1963). Alternative hypotheses suggest that bimodal karyotypes are the result of periodical unequal translocations (Stebbins 1971) or the unequal selective amplification of certain types of heterochromatin within a subset of chromosomes (de la Herrán et al. 2001).

13.3.1.5 Secondary Constrictions

In addition to primary constrictions, one or more chromosomes may also possess secondary constrictions. These can be located interstitially or, more often, subterminally, whereby a more or less spherical terminal chromosome piece called the satellite is segmented off from the chromosome arm (Figs. 13.1 and 13.2). Secondary constrictions are the sites of origin of nucleoli (NORs, nucleolar organizing regions) harboring up to thousands of tandemly arranged arrays of 35S (18S-5.8S-25/28S) ribosomal RNA (rRNA) genes (Volkov et al. 2004) and their spacers. The rRNA genes are found in all living organisms ("house-keeping genes") and code for a subset of rRNA for building the large and small ribosome units (Schwarzacher and Wachtler 1986). Active NORs can be detected either by the presence of secondary constrictions and/or satellites or by silver-staining, which detects proteins associated with active NOR regions (Schubert et al. 1979). Detection of all (both active and inactive) NORs in the genome requires fluorescence *in situ* hybridization techniques, and will be discussed in Sect. 13.3.2.2.

The number of morphologically detectable NORs (active NORs) may vary greatly among plant groups (reviewed in Małuszyńska et al. 1998) and to a lesser extent between closely related species (Kamstra et al. 1997; Cerbah et al. 1998), but rarely within species (*Phaseolus vulgaris*/Fabaceae: Pedrosa-Harand et al. 2006; *Alstroemeria pelegrina*/Alstroemeriaceae: Baeza et al. 2007). One prominent trend in angiosperms is reduction of the number of NORs after polyploidization as a result of deactivation or loss of loci (see also Sect. 13.3.2.2).

13.3.1.6 Supernumerary Chromosomal Material

Karyotypes of many species include supernumerary genetic material that by definition is not part of a taxon's regular karyotype. If physically integrated into the regular karyotype (i.e., the A-chromosomes) often appearing as a polymorphic block(s), this material is referred to as a supernumerary segment (SS; John and Miklos 1979). Supernumerary segments can be identified in homologous chromosomes in the heterozygous condition due to chromosome length differences. Although SSs are particularly frequent in some insect groups (locusts and grasshoppers; John and Miklos 1979), so far they have only been identified in a few plant groups, particularly in Hyacinthaceae (Greilhuber and Speta 1978; Ruiz Rejón and Oliver 1981; Garrido-Ramos et al. 1998; Weiss-Schneeweiss et al. 2004; Fig. 13.2g) or the chromosomally variable *Rumex acetosa* group (Polygonaceae; Wilby and Parker 1988), although it is likely that they are much more common (J. Parker, pers. comm.). In most taxa, SSs are heterochromatic (SSH; Ruiz Rejón and Oliver 1981; Weiss-Schneeweiss et al. 2004), and it is only in a few cases that they have been shown to be euchromatic (SSE; Ainsworth et al. 1983; Ebert et al. 1996; Fig. 13.2g). Regardless of the type, SSs can be located either interstitially or terminally within the chromosomes. They appear to be fixed and of ancient origin, often transgressing cytotype, population, or even species boundaries (Ainsworth et al. 1983; J. Parker, pers. comm.). One of the hypotheses concerning the origin of SSs suggests that these segments are relics of chromosome material lost from many, but not all populations (tolerated deletions; Camacho and Cabrero 1987). Alternatively, SSs might represent novel, selectively amplified and maintained segments of the genome containing, for instance, satellite DNA (Shibata et al. 2000) or heterochromatic segments resulting from unequal crossing-over (Smith 1976). The role of SSs in karyotype evolution remains elusive (Wilby and Parker 1988).

If supernumerous genetic material is present as separate chromosomes these are referred to as B chromosomes (accessory or supernumerary chromosomes; Joneset al. 2008; Figs. 13.1h and 13.2f). Characteristics of B chromosomes are (1) lack of recombination with A chromosomes, (2) non-Mendelian and irregular inheritance resulting in fluctuating numbers (including nil), and (3) evolutionary trajectories independent of those of the A chromosomes. B chromosomes are usually smaller than A chromosomes (at least than the largest A chromosomes; Lewis 1951; Żuk 1969; Ainsworth et al. 1983; Parker et al. 1991) and in some cases they are so small that it is impossible to indentify chromosome arms via light microscopy (microchromosomes; e.g., in *Hypochaeris maculata*/Asteraceae: Parker 1976). In overall structure, B chromosomes are similar to A chromosomes (being acrocentric to metacentric in type, sometimes variable within the same species: Ruiz Rejón et al. 1980; Guillén and Ruiz Rejón 1984). B chromosomes are typically thought to possess no or very few genes (such as rRNA genes: Małuszyńska and Schweizer 1989; Dhar et al. 2002; genes coding genetic information for their own transmission: Jones et al. 2008). It is, however, not clear whether any genes that are present are transcriptionally active (Jones et al. 2008) (see also chapter by Houben et al. 2013, this volume).

It is now widely accepted that B chromosomes have multiple origins. Most commonly, B chromosomes are considered to have originated from A chromosomes following chromosome number reductions, especially via unequal translocations (*Crepis capillaris*/Asteraceae: Jones and Rees 1982); from genomic rearrangements of trisomics (*Plantago lagopus*/Plantaginaceae: Dhar et al. 2002); from spontaneous generation following genomic rearrangements after interspecific hybridization (*Coix*/Poaceae: Sapre and Deshpande 1987) or following spontaneous amplification of tandemly repeated sequences and *de novo* recruitment of centromeres and telomeres (*Brachycome dichromosomatica*/Asteraceae: Houben et al. 2001; *Zea mays*/Poaceae: Cheng and Lin 2003).

B chromosomes are widespread in many plant and animal groups. Unlike the corresponding m-chromosomes in bryophytes, B chromosomes in angiosperms are of no taxonomic or diagnostic significance (Guerra 2008). In angiosperms, B chromosomes have so far been reported in c. 1,500 species, but their distribution is non-random with clear hot-spots (27.2% of species in Commelinales and 41.8% in Melanthiaceae in Liliales: Levin et al. 2005) and cold-spots of occurrence (basal dicots, inbreeders: Jones et al. 2008). Generally, monocots seem to be richer in B chromosomes than dicots (8 vs. 3%, respectively) and two families are particularly rich in B chromosomes: Poaceae (152 spp.; 8.6%) and Asteraceae (171 spp.; 5.6%: Levin et al. 2005). Although there is no notable difference between the frequency of B chromosomes in diploids and polyploids, there is a trend that at the family level the frequency of B chromosomes is positively correlated with mean 1C-genome size values (Levin et al. 2005).

B chromosomes cause intraspecific DNA content variation and may, at least partly, be responsible for reports of chromosome number variation, especially in taxa with relatively small A chromosomes and structurally variable B chromosomes, where it is difficult to distinguish these two types (Jones et al. 2008). B chromosomes might also influence the evolution of the A chromosomes, for instance, by acting as diploidizing agents for chromosome pairing in certain allopolyploid hybrids (Evans and Macefield 1972; Jenkins and Jones 2004) or changing recombination patterns by affecting chiasma frequency and distribution in the A chromosomes (Jones and Rees 1967; Jones et al. 2008). For more details on B chromosomes see Houben et al. (2013, this volume).

13.3.2 Chromosome Banding and Molecular Landmarks

With the advent of molecular cytogenetic techniques, particularly *in situ* hybridization (fluorescence and genomic *in situ*

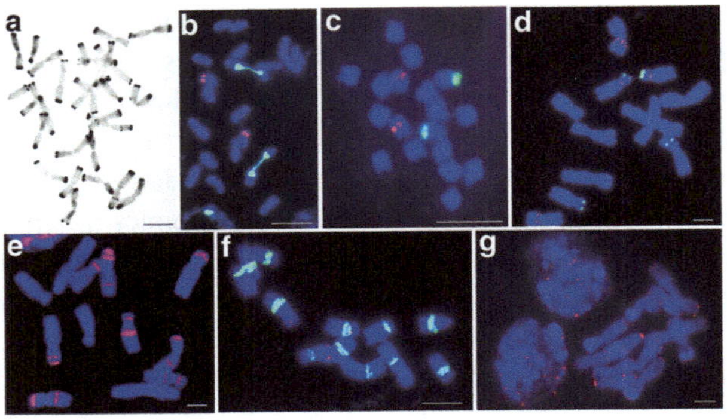

Fig. 13.3 Diversity of molecular chromosomal landmarks in angiosperms. (**a**) DAPI C-banding in *Hepatica nobilis* var. *japonica*/Ranunculaceae: $2n = 4x = 28$; (**b**) rDNA in *Melampodium longipilum*/Asteraceae: $2n = 2x = 20$, interstitial NOR (5S rDNA in *red*, 35S rDNA in *green*); (**c**) rDNA in *Melampodium gracile*/Asteraceae: $2n = 2x = 18$ (5S rDNA in *red*, 35S rDNA in *green*); (**d**) *Othocallis mischtschenkoana*/Hyacinthaceae (formerly in genus *Scilla*): $2n = 2x = 12$ with some of 5S and 35S rDNA loci in the same chromosome pairs (5S rDNA in *red*, 35S rDNA in *green*); (**e**) GC-rich satellite DNA (Deumling 1981) in *Othocallis amoena*/Hyacinthaceae (formerly in genus *Scilla*): $2n = 2x = 12$; (**f**) pericentric loci of satellite DNA (H Weiss-Schneeweiss et al. unpubl.) in *Prospero autumnale*/Hyacinthaceae (formerly in genus *Scilla*): $2n = 2x = 12$ (satellite DNA in *green*, 5S rDNA in *red*); (**g**) human type telomeric loci in *Othocallis siberica*/Hyacinthaceae (formerly in genus *Scilla*): $2n = 2x = 12$. [All photos H. Weiss-Schneeweiss (**e–g**, with D. Schweizer; **f**, with T.-S. Jang; both University of Vienna)]. *Scale bar* represents 5 μm

hybridization [FISH and GISH, respectively]: Schwarzacher and Heslop-Harrison 2000; Fig. 13.3), chromosomal landmarks are no longer restricted to structural features detectable from conventionally-stained chromosomes. Whereas detection of some of these (NORs, telomeric sequences) by FISH requires no *de novo* sequence information (primers are conserved for larger angiosperm groups), others (e.g., transposable elements, tandem satellite DNA repeats) are more demanding and, therefore, their detection has been more restricted to cultivated taxa and/or model plants and their relatives (e.g., *Nicotiana*/Solanaceae: Lim et al. 2000, 2007a; *Beta*/Chenopodiaceae: Dechyeva et al. 2003; *Alstroemeria*/Alstroemeriaceae: Kuipers et al. 2002; *Secale*/Poaceae: Cuadrado and Jouve 2002). Since repetitive DNA is the most dynamic fraction of the plant genome and may encompass taxon-specific sequence types, it is well suited for in depth study to characterize karyotype structure. Being a major force in karyotype differentiation, repetitive DNA may also indirectly contribute to race differentiation and eventually speciation.

13.3.2.1 Heterochromatin and C-Banding

Heitz (1928) was the first to distinguish heterochromatin, which is defined as chromosomal regions that, unlike euchromatin, do not decondense in telophase to the same degree as euchromatin. Heterochromatin, in contrast to diffuse euchromatin, is densely packed (and thus strongly stained), typically replicates late in S-phase and shows reduced rates of crossing over (Bennetzen 2000). Heterochromatin usually contains only a few or no genes (e.g., gene-rich NOR regions) and is often enriched in repetitive DNA (Lamb et al. 2007).

Heterochromatin rich in repetitive DNA usually remains condensed everywhere in an organism and is thus called constitutive heterochromatin, whereas facultative heterochromatin contains less repetitive DNA and is condensed only in some cells and/or ontogenetic stages (Brown 1966). Recent genomic analyses have revealed that heterochromatin is a term that describes many different types of condensed chromatin, possibly with many different features and roles and of different origin (Bennetzen 2000; Lamb et al. 2007). Specifically, heterochromatin includes two major classes of repetitive DNA (see Sect. 13.3.2): tandem repeats (rRNA genes, telomeric sequences, satellite DNA) and dispersed repeats (mostly transposable elements; Schmidt and Heslop-Harrison 1998; Schwarzacher 2003; Gaut and Ross-Ibarra 2008). However, the function of heterochromatin remains largely unknown, even in well understood functional regions such as centromeres (see also Sect. 13.3.2.6).

Euchromatin and heterochromatin distribution was first visualized in chromosomes and interphase nuclei by classical banding techniques (mostly Giemsa C-banding: Sumner 1972; Fig. 13.3a). Such studies showed that the number and size of bands may differ among closely related species (Greilhuber 1982, 1995; Vosa 1985; Sumner 1990; Guerra 2000) or even within a single species, as in *Scilla (Othocallis) siberica*/Hyacinthaceae (Greilhuber and Speta 1978), and in *Alstroemeria aurea* and *A. ligtu*/Alstroemeriaceae (Buitendijk et al. 1998). Given sufficient variation in C-band distribution, each chromosome can be identifed (Greilhuber 1995). Despite this variability, a comparison of the distribution of C-bands between species has shown that there is a general tendency for heterochromatin to be preferentially located in subtelomeric and pericentric regions, NORs and knobs (Bennetzen 2000; Guerra 2000). If interstitial bands are present, these are less conserved in their positions (Marks and Schweizer 1974; Greilhuber 1995; Buitendijk et al. 1998). The precise position of C-bands seems to be at least partly dependent on chromosome size, since proximal bands prevail in karyotypes with small-sized chromosomes (Guerra 2000).

It is noted that C-banding does not discriminate between different types of heterochromatin but this can be accomplished by using fluorescent banding with base-specific fluorochromes (Schweizer 1976, 1981). These approaches enable most of the GC-rich and AT-rich heterochromatin fractions to be visualized. The majority of heterochromatic bands have been found to be rather AT-rich, and they are more frequently found in the interstitial regions in species with medium and large chromosomes (Guerra 2000). 35S rRNA genes have most often been found to coincide with GC-rich bands (Guerra 2000). Although advanced techniques allow detailed characterization of specific types of repetitive DNA (see the following sections), the indiscriminate localization of overall heterochromatin via C-banding remains a valuable tool to enable a first and coarse characterization of the genome, especially in poorly studied groups.

13.3.2.2 rDNA

Genes encoding 35S (18S-5.8S-25/28S) and 5S rRNAs are ubiquitous in eukaryotes (Richard et al. 2008). The 35S rDNA cistron contains 18S, 5.8S and 25/28S rRNA genes separated by the Internal Transcribed Spacers (ITS) 1 and 2 and preceded by an External Transcribed Spacer (ETS). These cistrons are separated by the Non-Transcribed Spacer (NTS) and are arranged tandemly (Volkov et al. 2004). 5S rDNA repeats consist of the 121 bp 5S rRNA gene separated by Non-Transcribed Spacers of variable length (Volkov et al. 2004). In angiosperms, both the number and localization of 35S rDNA and 5S rDNA loci are largely independent from one another (Małuszyńska et al. 1998; Fig. 13.3b–d). An exception are some clades of Asteraceae, where these loci are physically linked (Garcia et al. 2010), and also in some early land plants (Sone et al. 1999). The phylogenetic distribution of such linked arrangements suggests its recurrent origin and/or reversal (Garcia et al. 2010). A survey of 35S rDNA loci number and distribution published for 749 species (175 genera) indicated that the average number of 35S rDNA loci per genome was 4, and that they most often occurred at subterminal chromosomal positions (45%), usually within the short arms (Roa and Guerra 2010).

Several factors render 5S and 35S rDNA loci an excellent choice for markers for karyotype characterization: (1) they are ubiquitous and their coding regions are conserved over long evolutionary distances, thus they can be localized in various plant groups without detailed genomic information; (2) they are highly variable with respect to localization and number (between one and ten or even more: Małuszyńska et al. 1998; Baeza et al. 2007) at the interspecific (Ali et al. 2005; Hasterok et al. 2006; Weiss-Schneeweiss et al. 2008), intraspecific (Pedrosa-Harand et al. 2006) or even intrapopulational level (Frello and Heslop-Harrison 2000; Baeza et al. 2007), and this variability is independent from chromosome number; (3) they are known hotpots of intra- and intergenomic mobility and thus might be involved in species differentiation (Schubert and Wobus 1985; Hall and Parker 1995; Raskina et al. 2008); (4) they contain widely applicable molecular markers (ITS and ETS in 35S rDNA, the non-transcribed spacer of the 5S rDNA array), enabling direct correlation of chromosomal loci evolution with phylogenetic relationships. Although there is some redundancy with morphological landmarks (morphologically discernible NORs represent active 35S rDNA loci), the detection of 5S and 35S rDNA loci using FISH allows identification of all loci, including small and/or inactive ones (the number of active loci does not correlate with the overall number of loci). Examples of successful use of one or both types of rDNA loci in evolutionary questions are aplenty (e.g., Lim et al. 2000; Vanzela et al. 2003; Hasterok et al. 2006; Weiss-Schneeweiss et al. 2007a, b, 2008).

The rates and mechanisms of evolutionary changes differ between the 5S and 35S rDNA loci within the genome, and might be group-specific. 35S rDNA loci are generally more prone to rapid homogenization, silencing, and loss of loci, particularly in polyploids (Clarkson et al. 2005; Kovařík et al. 2005; Weiss-Schneeweiss et al. 2008; Kotseruba et al. 2010). Both rDNA types are affected (but often to different extents) by genome diploidization in polyploids and this usually involves a gradual reduction of loci number, roughly correlating with the polyploid's age (Clarkson et al. 2005). Several mechanisms have been hypothesized to be responsible for the high evolutionary rate and mobility, particularly of 35S rDNA. These include, among others unequal recombination, transposition, conversion/homogenization of repeats among loci (changing the type of repeats but not the locus position), minor loci amplification/major loci reduction, often reversible and in hybrids usually parent-specific loci inactivation (nucleolar dominance: Pikaard 2000; Volkov et al. 2004) via epigenetic modifications affecting the number of satellites/secondary constrictions, and loci loss (Schubert and Wobus 1985; Dubcovsky and Dvorak 1995; Raskina et al. 2008).

13.3.2.3 Telomeric Sequences

Telomeres are nucleoprotein structures at the ends of linear chromosomes and are vital for protecting chromosome ends from shortening and from being recognized and processed as DNA breaks (Blackburn 2001; Zellinger and Riha 2007). Telomeres in most plants are maintained by telomerase and consist of tandem arrays of telomeric repeats. The $(TTTAGGG)_n$ repeat characterized originally in *Arabidopsis thaliana* (Richards and Ausubel 1988) is found in the majority of analysed plant species (Fuchs et al. 1995) and is considered to be ancestral for plants (Fajkus et al. 2005). Not surprisingly, exceptions do exist, most notably in a large clade of the monocot order Asparagales. Here, chromosome termini contain the vertebrate-type telomeric repeat $(TTAGGG)_n$ maintained by telomerase (Weiss and Scherthan 2002; Weiss-Schneeweiss et al. 2004; Fajkus et al. 2005). However, neither of these canonical telomeric repeats is found in *Allium* (which is within the clade of Asparagales with vertebrate-type telomeric sequences—Pich and Schubert 1998), and at present the nature of the sequences present at the telomeres of *Allium* chromosomes remains unknown (Sýkorová et al. 2006).

Telomeric sequences are of relatively little use as chromosomal landmarks for plant chromosomes, as they are invariably found at chromosome ends and are conserved (Fuchs et al. 1995; Fig. 13.3g). Unlike in animals or some gymnosperms, telomeric-like sequences in angiosperms are rarely found at interstitial chromosomal positions being probably rarely retained in places of chromosomal rearrangements (Fuchs et al. 1998; Uchida et al. 2002; Hanmoto et al. 2007). However, *de novo* synthesis of telomeric repeats in places of chromosomal breakage stabilizes resulting chromosomal fragments ("chromosome healing": Tsujimoto et al. 1999) and ultimately contributes to the evolutionary success of new karyotypic variants.

13.3.2.4 Transposable Elements

Transposable elements are a major fraction of repetitive DNA in all eukaryotes and are one of the main drivers of genome and chromosome size differentiation (Bennetzen et al. 2005; Hawkins et al. 2006). In angiosperms, the most abundant type are class I transposable elements, which transpose and amplify via copy and paste mechanisms using reverse transcriptase machinery (Kumar and Bennetzen 1999). Angiosperm genomes are particularly rich in LTR (long terminal repeat)-retroelements of the Ty1-*copia* and the Ty3-*gypsy* type (Kumar and Bennetzen 1999; Bennetzen et al. 2005). Generally, retrotransposons are rather uniformly dispersed over the entire karyotype, but they may be overrepresented in some regions (centromeres: Neumann et al. 2011; sex chromosomes: Kejnovský et al. 2009) and might be underrepresented in others (NORs: Kumar and Bennetzen 1999; Ruas et al. 2008; but see Chester et al. 2008).

Retroelements are, in addition to polyploidy, major players in genome size variation. There is, however, no clear trend with respect to the type of retroelements being involved in genome size increase. Whereas Ty3-*gypsy* elements dominate in, for example, *Helianthus*/Asteraceae hybrids (Ungerer et al. 2006), *Fritillaria*/Liliaceae (Ambrožová et al. 2010), or Fabaceae (*Vicia pannonica*: Neumann et al. 2006; *Pisum sativum*: Macas et al. 2007), Ty1-*copia* elements dominate in *Musa*/Musaceae (Hřibová et al. 2010) and *Prospero autumnale*/Hyacinthaceae (H. Weiss-Schneeweiss et al. unpublished). Sometimes, a single retrotransposon family may occupy nearly 40% of the genome (Neumann et al. 2006).

Activation of transposable elements, often in connection with species diversification without (*Vicia*/Fabaceae) or with hybridization and/or polyploidy (diploid hybrids of *Helianthus*/Asteraceae: Ungerer et al. 2006; allopolyploid *Gossypium*/Malvaceae: Hawkins et al. 2006), usually leads to a uniform DNA amount increase over all chromosomes and hence does not significantly change the overall karyotype structure (Bennetzen et al. 2005). However, newly amplified retroelements also have the potential to increase the frequency of unequal and illegitimate (ectopic) recombination. This can lead to large structural genome rearrangements and changes in chromosome number and results in an altered karyotype structure (Kumar and Bennetzen 1999). In addition, some families of retroelements may be preferentially amplified in certain genomic regions or excluded from others (Kumar and Bennetzen 1999; Ruas et al. 2008; Karlov et al. 2010) leading to further changes in karyotype symmetry.

13.3.2.5 Tandemly Repeated DNA (Satellite DNA)

Satellite DNA is composed of arrays of head-to-tail arranged basic repeat units (monomers), each usually tens to hundreds of nucleotides long, that can exist in millions of copies in the genome (Macas et al. 2002; Plohl et al. 2008). Satellite repeats are among the most dynamic components of eukaryotic genomes and are, therefore, good markers for studying karyotype evolution within closely related plants (Lim et al. 2000; Cuadrado and Jouve 2002; Navrátilová et al. 2003; Pires et al. 2004). Because of their rapid changes in sequence and abundance (Macas et al. 2000, 2006), individual families of satellite DNA are often group-specific and are maintained as highly homogenized repeat types due to concerted evolution (Elder and Turner 1995). Most satellite DNA repeat types localize in the constitutive heterochromatin (see Sect. 13.3.2.1; Schmidt and Heslop-Harrison 1998; Macas et al. 2000, 2007; Fig. 13.3e–f). Specific and novel satellite DNA types are particularly known from sex and B-chromosomes (Sandery et al. 1990; Kejnovský et al. 2009) or from allopolyploids, where these novel types often replace the most abundant types inherited from either or both parental species (Skalicka et al. 2005; Koukalova et al. 2010).

The genomic localization, abundance and diversity of tandem repeats varies considerably between different plant groups (Schmidt and Heslop-Harrison 1998; Chester et al. 2010). Some genera are characterized by containing many different types of satellite DNA families but each family is only weakly amplified (e.g., *Pisum*/Fabaceae: Macas et al. 2007; *Beta*/Chenopodiaceae: Menzel et al. 2008). In contrast, in other genera only a few satellite DNA types are present but these may be amplified moderately (*Musa*/Musaceae: Hřibová et al. 2010) or massively (*Fritillaria*/Liliaceae: Ambrožová et al. 2010; *Scilla* (*Othocallis*) *siberica*/Hyacinthaceae: Deumling and Greilhuber 1982;

Prospero/Hyacinthaceae: H. Weiss-Schneeweiss et al. unpubl., Fig. 13.3f) affecting karyotype structure and symmetry (de la Herrán et al. 2001). Several monocot lineages are known for their propensity to amplify and tolerate large amounts of satellite DNA. Due to different positions of tandem repeat blocks in homologous chromosomes they may be heteromorphic (Liliaceae: Peruzzi et al. 2009; Ambrožová et al. 2010; Hyacinthaceae: Greilhuber 1995; *Crocus*/Iridaceae: Frello and Heslop-Harrison 2000; *Alstroemeria*/Alstroemeriaceae: Kamstra et al. 1997; Kuipers et al. 2002; see Sect. 13.3.1.6).

It has been suggested that different satellite sequences can coexist in genomes of related species, where they form a library of satellite DNAs (library hypothesis: Ugarković and Plohl 2002; Plohl et al. 2008). Initially present in different copy numbers constituting major and minor satellite DNAs, differential amplification after species divergence as well as gradual and differential sequence evolution of repeats (including loss of old and/or emergence of new satellite types) without obvious quantitative change contribute to the diversity of satellites DNA types (Ugarković and Plohl 2002).

13.3.2.6 Centromere-Specific Tandem Repeats and Retroelements

Typical functional centromeres are characterized by the presence of, among other components (see Hirsch and Jiang 2012), specific centromeric repetitive DNA consisting of both tandem repeats and retrotransposons (Houben and Schubert 2003; Ma et al. 2007; Neumann et al. 2011). Centromeric repetitive DNA is not conserved among related plant taxa (Heslop-Harrison et al. 2003; Lee et al. 2005; Koo and Jiang 2008; Koo et al. 2010), thus specific centromeric tandem repeats can be used as chromosomal landmarks once they are isolated from the given taxon/taxa group (facilitated by the availability of ChipSeq: Johnson et al. 2007). Recently, variants of centromeric repeats used as components of a probe cocktail for fluorescence *in situ* hybridization (FISH) permitted simultaneous identification of all chromosome pairs in soybean (Findley et al. 2010).

13.3.2.7 GISH and Chromosome Painting

Genomic *in situ* hybridization (GISH), first demonstrated in synthetic cereal hybrids (Schwarzacher et al. 1989), allows the identification of parental subgenomes in intraspecific hybrids (Marasek et al. 2006; Choi et al. 2008) and allopolyploids (Lim et al. 2007b; Chester et al. 2010) as well as demonstrating their genome and karyotype evolution (reviewed in Chester et al. 2010). The success of this technique depends, among others, on the age and genetic closeness of the parental taxa (Chester et al. 2010). The most comprehensive information on chromosomal evolution (including evolution of the repetitive DNA fraction) can however be achieved via individual chromosome painting, i.e., the identification of individual chromosomes using chromosome-specific probes (Lysák et al. 2010). Whereas this is state of the art in some animal groups (e.g., mammals: Ferguson-Smith and Trifonov 2007), in angiosperms only taxa with relatively small genome sizes (due to the low amount of transposable elements, which by their ubiquity limit the technique's resolution) and extensive genomic resources (BAC or other types of libraries, screened and mapped to chromosomes) are amenable to chromosome painting. So far, *Arabidopsis* and other Brassicaceae are the only examples where whole-genome chromosome painting has been successfully applied using pooled BAC clones representative of an ancestral karyotype (Mandáková and Lysák 2008).

13.4 Specialized Chromosomes and Chromosome Systems

13.4.1 Sex Chromosomes

In angiosperms, dioecy is found in 5–6% of genera distributed in c. 75% of families (Renner and Ricklefs 1995; Vyskot and Hobza 2004). Of these, a few have developed differentiated sex chromosomes, which have evolved from regular autosomal chromosome pairs (Vyskot and Hobza 2004) via repression of recombination between proto-sex chromosomes and accumulation of sex-specific genes (for details see Janoušek et al. 2013, this volume). Initially, sex chromosomes will be morphologically indistinguishable from the autosomes (homomorphic sex chromosomes; *Carica papaya*/Caricaceae: Liu et al. 2004). However, in later stages of sex chromosome evolution they become morphologically differentiated (heteromorphic sex chromosomes; chromosome X and Y in *Cannabis* and *Humulus*/Cannabaceae, *Silene*/Caryophyllaceae, or *Coccinia*/Cucurbitaceae; reviewed in Vyskot and Hobza 2004) usually by the accumulation of repetitive DNA (transposable elements, satellite DNA) and subsequent changes in size and structure of one or both sex chromosomes (Navajas-Pérez et al. 2006; Jamilena et al. 2008; Kejnovský et al. 2009). Sex determination either employs the active Y system (*Silene latifolia*/Caryophyllaceae) or an X to autosome dosage system, whereby sex is determined by the ratio of the number of X chromosomes to the number of autosome sets (e.g., 2:2 in females vs. 1:2 in males in *Rumex acetosa*/Polygonaceae and *Humulus*/Cannabaceae: Vyskot and Hobza 2004; see also Janoušek et al. 2013, this volume).

13.4.2 Permanent Translocation Heterozygotes

Translocation and inversion heterozygosity is quite frequent within populations of many species (Levin 2002). A few plant groups, however, have developed permanent heterozygosity, where none of the chromosomes in the complement has a fully homologous partner, although the sum of chromosomal arms is constant. It is known in at least 57 species (Levin 2002), mostly in Onagraceae (50 species of *Oenothera*: Cleland 1972), but also Commelinaceae (*Rhoeo discolor*: Golczyk et al. 2005) or Paeoniaceae (*Paeonia brownii* and *P. californica*: Zhang and Sang 1998). It is considered to originate recurrently with all transitional forms present (Levin 2002). Heterozygosity is maintained via the existence of two multichromosomal superlinkage groups (Renner complexes: Cleland 1972), and results in extreme inbreeding (homozygotes are lethal: Stebbins 1950). Meiotic pairing is also peculiar in these plants as the chromosomes form a ring when pairing, followed by anaphase during which the centromeres are distributed into the daughter cells in a strictly alternating way. This results in the chromosomes of the original parental sets staying together in either of the daughter cells (Hejnowicz and Feldman 2000).

13.5 Karyotype Evolution in the Genomic Era

Comparative genetic and physical maps have enabled detailed insights into the structure and evolution of the coding part of the genome, as well as into rates and timing of gross chromosomal changes in several genetically well characterized angiosperm groups (Solanaceae: Wu and Tanksley 2010; *Cucumis*/Cucurbitaceae: Huang et al. 2009; *Brachypodium*/Poaceae: Vogel et al. 2010). At the chromosomal level, mapping available bacterial artificial chromosomes (BACs) via multicolor FISH on homoeologous chromosomes of related species (e.g., cross-species multicolor BAC–FISH) offers a unique tool for comparative genomic analyses bridging genetic and physical maps. Additionally, it allows detailed analyses of the size, position, and evolution of repeat-rich regions of the genome (Tang et al. 2008; Lysák et al. 2010; Szinay et al. 2010). For the majority of plants, however, comprehensive analyses of the repetitive fraction of the genome have remained difficult, mostly because the development of new specific chromosomal markers has, until now, been very laborious.

The recent introduction of next generation sequencing (NGS) and its bioinformatic analyses has brought genomic approaches within reach for any plant biologist, providing a rapid way to identify new repeats. These techniques are applicable to any plant group, even if no prior genomic information is available (although basic information on genome size and chromosome number is clearly helpful), and at modest costs. Although one may consider traditional karyotype analysis obsolete because a thorough genome characterization or even whole genome sequencing can be achieved via NGS, it is in fact quite the opposite for two reasons: (1) basic karyotypic information, such as chromosome number and size and organization, genome size, or position of landmarks, will remain important, as it cannot easily be retrieved from NGS data irrespective of their amount, but is necessary for correct data interpretation; (2) the composition of plant genomes containing large amounts of repetitive DNA is no less an obstacle for whole genome assembly from NGS data than it was for genome assembly from traditional Sanger sequencing data. NGS is, thus, a perfect approach with which to rapidly extend the available toolbox for in-depth karyotype analyses. The main challenge for the near future will be to overcome the NGS-inherent bottleneck of bioinformatic data analyses, especially with respect to plant repetitive DNA, where bioinformatic tools are notoriously scarce (but see Novák et al. 2010).

Most importantly, due to the tremendous amount of data, NGS allows detailed insights into plant genome composition and dynamics of different types of repetitive DNA. Apart from a general characterization of the repeat types, including identification of new retroelement families (Macas et al. 2007; Swaminathan et al. 2007; Wicker et al. 2009), NGS also enables rapid isolation and characterization of novel tandem repeat families (Macas et al. 2007; Hřibová et al. 2010), which until now required highly laborious and not always overwhelmingly successful approaches. Importantly, NGS for the first time provides us with the means to conduct comprehensive comparative analyses on the genomic repeat composition of any plant species (Macas et al. 2007; Hřibová et al. 2010) and their contribution to genome size change (Tenaillon et al. 2011). It also allows one to test broader hypotheses concerning group-specific patterns of repetitive DNA evolution, for instance, large-scale amplification of one/few families of tandem repeats versus small-scale amplifications of numerous small repeat families (Macas et al. 2007; Swaminathan et al. 2007; Ambrožová et al. 2010; Hřibová et al. 2010; H. Weiss-Schneeweiss et al. unpubl.), or insights into retroelement-mediated faster genome rearrangement rates suggested for monocots (Leitch et al. 2010).

Acknowledgments The authors acknowledge the financial support of the Austrian Science Fund (FWF), especially projects T218 (Hertha-Firnberg Fellowship) and P21440 to HWS.

References

Ainsworth CC, Parker JS, Horton DM (1983) Chromosome variation and evolution in *Scilla autumnalis*. In: Brandham PE, Bennett MD (eds) Kew chromosome conference II. Georg Allen and Unwin, London, pp 261–268

Ali HBM, Lysák MA, Schubert I (2005) Chromosomal localization of rDNA in the Brassicaceae. Genome 48:341–346

Ambrožová K, Mandáková T, Bureš P, Neumann P, Leitch IJ, Koblízková A, Macas J, Lysák MA (2010) Diverse retrotransposon families and an AT-rich satellite DNA revealed in giant genomes of *Fritillaria* lilies. Ann Bot 107:255–268

Anderson LC, Kyhos DW, Mosquin T, Powell AM, Raven PH (1974) Chromosome numbers in compositae. IX. *Haplopappus* and other Astereae. Am J Bot 61:665–671

Babcock EB (1947) The genus *Crepis* I. The taxonomy, phylogeny, distribution and evolution of *Crepis*. University of California Press, Berkeley/Los Angeles

Babcock EB, Jenkins JA (1943) Chromosomes and phylogeny in *Crepis* III: the relationships of one hundred and thirteen species. Univ Calif Publ Agric Sci 18:241–292

Baeza CM, Schrader O, Budahn H (2007) Characterization of geographically isolated accessions in five *Alstroemeria* L. species (Chile) using FISH of tandemly repeated DNA sequences and RAPD analysis. Plant Syst Evol 269:1–14

Bakker FT, Culham A, Pankhurst CV, Gibby M (2000) Mitochondrial and chloroplast DNA-based phylogeny of *Pelargonium* (Geraniaceae). Am J Bot 87:727–734

Barber JC, Ortega JF, Santos-Guerra A, Marrero A, Jansen RK (2000) Evolution of endemic *Sideritis* (Lamiaceae) in Macaronesia: insights from a chloroplast DNA restriction site analysis. Syst Bot 25:633–647

Bennett MD (1998) Plant genome values: how much do we know? Proc Natl Acad Sci USA 95:2011–2016

Bennetzen JL (2000) The many hues of plant heterochromatin. Genome Biol 1:107.1–107.4

Bennetzen JL (2002) Mechanisms and rates of genome expansion and contraction in flowering plants. Genetica 115:29–36

Bennetzen JL, Ma J, Devos KM (2005) Mechanisms of recent genome size variation in flowering plants. Ann Bot 95:127–132

Blackburn EH (2001) Switching and signaling at the telomere. Cell 106:661–673

Blöch C, Weiss-Schneeweiss H, Schneeweiss GM, Barfuss MHJ, Rebernig CA, Villaseñor JL, Stuessy TF (2009) Molecular phylogenetic analyses of nuclear and plastid DNA sequences support dysploid and polyploid chromosome number changes and reticulate evolution in the diversification of *Melampodium* (Millerieae, Asteraceae). Mol Phylogenet Evol 53:220–233

Bowers JE, Chapman BA, Rong J, Paterson AH (2003) Unravelling angiosperm genome evolution by phylogenetic analysis of chromosomal duplication events. Nature 422:433–438

Brandham PE (1971) The chromosomes of the Liliaceae: II. Polyploidy and karyotype variation in the Aloineae. Kew Bull 25:381–389

Brandham PE, Doherty MJ (1998) Genome size variation in the Aloaceae, an angiosperm family displaying karyotypic orthoselection. Ann Bot 82(Suppl A):67–73

Brighton CA (1978) Telocentric chromosomes in Corsican *Crocus* L. (Iridaceae). Plant Syst Evol 129:299–314

Brown SW (1966) Heterochromatin. Science 151:417–425

Buitendijk J, Peters A, Quene R, Ramanna M (1998) Genome size variation and C-band polymorphism in *Alstroemeria aurea*, *A. ligtu* and *A. magnifica* (Alstroemeriaceae). Plant Syst Evol 212:87–106

Bureš P, Zedek F, Marková M (2013) Holocentric chromosomes. In: Leitch IJ, Greilhuber J, Doležel J, Wendel JF (eds) Plant genome diversity, vol 2, Physical structure, behaviour and evolution of plant genomes. Springer-Verlag, Wien, pp 187–208

Camacho JPM, Cabrero J (1987) New hypotheses about the origin of supernumerary chromosome segments in grasshoppers. Heredity 58:341–343

Cerbah M, Coulaud J, Siljak-Yakovlev S (1998) rDNA organization and evolutionary relationships in the genus *Hypochaeris* (Asteraceae). J Hered 89:312–318

Cheng YM, Lin BY (2003) Cloning and characterization of maize B chromosome sequences derived from microdissection. Genetics 164:299–310

Chester M, Sýkorová E, Fajkus J, Leitch AR (2008) Single integration and spread of a *copia*-like sequence nested in rDNA intergenic spacers of *Allium cernuum* (Alliaceae). Cytogenet Genome Res 129:35–46

Chester M, Leitch AR, Soltis PS, Soltis DE (2010) Review of the application of modern cytogenetic methods (FISH/GISH) to the study of reticulation (polyploidy/hybridisation). Genes 1:166–192

Choi H-K, Mun J-H, Kim D-J, Zhu H, Baek J-M, Mudge J, Roe B, Ellis N, Doyle J, Kiss GB, Young ND, Cook DR (2004) Estimating genome conservation between crop and model legume species. Proc Natl Acad Sci USA 101:15289–15294

Choi H-W, Kim J-S, Lee S-H, Bang J-W (2008) Physical mapping by FISH and GISH of rDNA loci and discrimination of genomes A and B in *Scilla scilloides* complex distributed in Korea. J Plant Biol 51:408–412

Chung K-S, Weber JA, Hipp AL (2011) Dynamics of chromosome number and genome size variation in a cytogenetically variable sedge (*Carex scoparia* var. *scoparia*, Cyperaceae). Am J Bot 98:122–129

Clarkson JJ, Lim KY, Kovařík A, Chase MW, Knapp S, Leitch AR (2005) Long-term genome diploidization in allopolyploid *Nicotiana* section Repandae (Solanaceae). New Phytol 168:241–252

Cleland RE (1972) *Oenothera*: cytogenetics and evolution. Academic, New York

Comai L (2005) The advantages and disadvantages of being polyploid. Nat Rev Genet 6:836–846

Cremonini R (2005) Low chromosome number angiosperms. Caryologia 58:403–409

Cuadrado A, Jouve N (2002) Evolutionary trends of different repetitive DNA sequences during speciation in the genus *Secale*. J Hered 93:339–345

da Silva CRM, González-Elizondo MS, Vanzela ALL (2005) Reduction of chromosome number in *Eleocharis subarticulata* (Cyperaceae) by multiple translocations. Bot J Linn Soc 149:457–464

Darlington CD (1956) Chromosome botany. George Allen and Unwin, London

Darlington CD (1963) Chromosome botany and the origins of cultivated plants. Hafner, New York

Darlington CD, Wylie AP (1955) Chromosome atlas of flowering plants. Georg Allen and Unwin, London

de la Herrán R, Robles F, Cuñado N, Santos JL, Ruiz Rejón M, Garrido-Ramos MA, Ruiz Rejón C (2001) A heterochromatic satellite DNA is highly amplified in a single chromosome of *Muscari* (Hyacinthaceae). Chromosoma 110:197–202

Dechyeva D, Gindullis F, Schmidt T (2003) Divergence of satellite DNA and interspersion of dispersed repeats in the genome of the wild beet *Beta procumbens*. Chromosome Res 11:3–21

Deumling B (1981) Sequence arrangement of a highly methylated satellite DNA of a plant, *Scilla*: a tandemly repeated inverted repeat. Proc Natl Acad Sci USA 78:338–342

Deumling B, Greilhuber J (1982) Characterization of heterochromatin in different species of the *Scilla siberica* group (Liliaceae) by *in situ*

hybridization of satellite DNAs and fluorochrome banding. Chromosoma 84:535–555
Devos K (2005) Updating the 'crop circle'. Curr Opin Plant Biol 8:155–162
Dhar MK, Friebe B, Koul AK, Gill BS (2002) Origin of an apparent B chromosome by mutation, chromosome fragmentation and specific DNA sequence amplification. Chromosoma 111:332–340
Dimitrova D, Greilhuber J (2000) Karyotype and DNA content in ten species of *Crepis* (Asteraceae) distributed in Bulgaria. Bot J Linn Soc 132:281–297
Dimitrova D, Greilhuber J (2001) C-banding patterns and quantitative karyotype characteristics of Bulgarian species of *Crepis* (Asteraceae). Plant Biol 3:88–97
Dixon CJ, Schönswetter P, Suda J, Wiedermann MM, Schneeweiss GM (2009) Reciprocal Pleistocene origin and postglacial range formation of an allopolyploid and its sympatric ancestors (*Androsace adfinis* group, Primulaceae). Mol Phylogenet Evol 50:74–83
Dobeš C, Vitek E (2000) Documented chromosome number checklist of Austrian vascular plants. Naturhistorisches Museum Wien, Wien
Dobigny G, Ducroz J-F, Robinson TJ, Volobouev V (2004) Cytogenetics and cladistics. Syst Biol 53:470–484
Doyle JJ, Doyle JL, Rauscher JT, Brown AHD (2004) Evolution of the perennial soybean polyploid complex (*Glycine* subgenus Glycine): a study of contrasts. Biol J Linn Soc 82:583–597
Doyle JJ, Flagel LE, Paterson AH, Rapp RA, Soltis DE, Soltis PS, Wendel JF (2008) Evolutionary genetics of genome merger and doubling in plants. Annu Rev Genet 42:443–461
Dubcovsky J, Dvorak J (1995) Ribosomal DNA multigene loci are nomads in the Triticeae genomes. Genetics 140:1367–1377
Ebert I, Greilhuber J, Speta F (1996) Chromosome banding and genome size differentiation in *Prospero* (Hyacinthaceae): diploids. Plant Syst Evol 203:143–177
Ehrendorfer F (1958) Ein Variabilitätszentrum als "fossiler" Hybrid-Komplex: Der ost mediterrane *Galium graecum* L.-G. *canum* Req.-Formenkreis. Eine Monographie. (Zur Phylogenie der Gattung *Galium*, VI.). Plant Syst Evol 105:229–279
Ehrendorfer F (1964) Cytologie, Taxonomie und Evolution bei Samenpflanzen. Vistas Bot 4:99–186
Ehrendorfer F (1970) Evolutionary patterns and strategies in seed plants. Taxon 19:185–195
Ehrendorfer F (1980) Polyploidy and distribution. In: Lewis WH (ed) Polyploidy: biological relevance. Plenum Press, New York/London, pp 45–59
Elder JF Jr, Turner BJ (1995) Concerted evolution of repetitive DNA sequences in eukaryotes. Quart Rev Biol 70:297–320
Ellison NW, Liston A, Steiner JJ, Williams WM, Taylor NL (2006) Molecular phylogenetics of the clover genus (*Trifolium*–Leguminosae). Mol Phylogenet Evol 39:688–705
Enke N, Gemeinholzer B (2008) Babcock revisited: new insights into generic delimitation and character evolution in *Crepis* L. (Compositae: Cichorieae) from ITS and *mat*K sequence data. Taxon 57:756–768
Evans GM, Macefield AJ (1972) The suppression of homoeologous pairing by B-chromosomes in a *Lolium* species hybrid. Nature 236:110–111
Fajkus J, Sýkorová E, Leitch AR (2005) Telomeres in evolution and evolution of telomeres. Chromosome Res 13:469–479
Fawcett JA, Maere S, Van de Peer Y (2009) Plants with double genomes might have a better chance to survive the Cretaceous-Tertiary extinction event. Proc Natl Acad Sci USA 106:5735–5742
Fawcett JA, Van de Peer Y, Maere S (2013) Significance and biological consequences of polyploidization in land plants. In: Leitch IJ, Greilhuber J, Doležel J, Wendel JF (eds) Plant genome diversity, vol 2, Physical structure, behaviour and evolution of plant genomes. Springer-Verlag, Wien, pp 277–293
Fedorov AN (ed) (1969) Chromosome numbers of flowering plants. Acad Sci USSR, Leningrad
Ferguson-Smith MA, Trifonov V (2007) Mammalian karyotype evolution. Nat Rev Genet 8:950–962
Findley SD, Cannon S, Varala K, Du J, Ma J, Hudson ME, Birchler JA, Stacey G (2010) A fluorescence *in situ* hybridization system for karyotyping soybean. Genetics 185:727–744
Forrest LL, Jong K (2004) Karyotype asymmetry in *Galtonia* and *Pseudogaltonia* (Hyacinthaceae). Edinburgh J Bot 60:569–579
Frello S, Heslop-Harrison JS (2000) Chromosomal variation in *Crocus vernus* Hill (Iridaceae) investigated by *in situ* hybridization of rDNA and a tandemly repeated sequence. Ann Bot 86:317–322
Fuchs J, Brandes A, Schubert I (1995) Telomere sequence localization and karyotype evolution in higher plants. Plant Syst Evol 196:227–241
Fuchs J, Strehl S, Brandes A, Schweizer D, Schubert I (1998) Molecular-cytogenetic characterization of the *Vicia faba* genome: heterochromatin differentiation, replication patterns and sequence localization. Chromosome Res 6:219–230
Fukui K, Nakayama S (1996) Plant chromosomes. Laboratory methods. CRC Press, Boca Raton
Garcia S, Panero JL, Siroky J, Kovařík A (2010) Repeated reunions and splits feature the highly dynamic evolution of 5S and 35S ribosomal RNA genes (rDNA) in the Asteraceae family. BMC Plant Biol 10:176
Garrido-Ramos MA, Jamilena M, de la Herrán R, Ruiz Rejón C, Camacho JPM, Ruiz-Rejón M (1998) Inheritance and fitness effects of a pericentric inversion and a supernumerary chromosome segment in *Muscari comosum* (Liliaceae). Heredity 80:724–731
Gaut BS, Ross-Ibarra J (2008) Selection on major components of angiosperm genomes. Science 320:484–486
Golczyk H, Hasterok R, Joachimiak AJ (2005) FISH-aimed karyotyping and characterization of Renner complexes in permanent heterozygote *Rhoeo spathacea*. Genome 48:145–153
Goldblatt P, Johnson DE (eds) (1979) Index to plant chromosome numbers. Missouri Botanical Garden, St. Louis
Goodspeed TH (1933) Chromosome number and morphology in *Nicotiana* VI. Chromosome numbers of forty species. Proc Natl Acad Sci USA 19:649–653
Góralski G, Lubczyńska P, Joachimiak AJ (2009) Chromosome number database. http://www.binoz.uj.edu.pl:8080/chromosomes/ accessed 30.03.2011
Grant V (1981) Plant speciation. Columbia University Press, New York
Grant V (1982) Chromosome number patterns in primitive angiosperms. Bot Gaz 143:390–394
Grant WF (1991) Chromosomal evolution and aneuploidy in *Lotus*. In: Tsuchiya T, Gupta PK (eds) Chromosome engineering in plant genetics: genetics, breeding, evolution. Part B. Elsevier, Amsterdam, pp 429–447
Greilhuber J (1982) Trends in der Chromosomenevolution von *Scilla* (Liliaceae). Stapfia 10:11–51
Greilhuber J (1995) Chromosomes of the monocotyledons (general aspects). In: Rudall PJ, Cribb PJ, Cutler DF, Humphries CJ (eds) Monocotyledons: systematics and evolution. Royal Botanic Gardens, Kew, pp 379–414
Greilhuber J, Speta F (1978) Quantitative analysis of the C-banded karyotypes and systematics in the cultivated species of the *Scilla siberica* group (Liliaceae). Plant Syst Evol 129:63–109
Greilhuber J, Borsch T, Müller K, Worberg A, Porembski S, Barthlott W (2006) Smallest angiosperm genomes found in Lentibulariaceae with chromosomes of bacterial size. Plant Biol 8:770–777
Guerra M (1986) Reviewing chromosome nomenclature of Levan et al. Rev Brasil Genet 4:741–743
Guerra M (2000) Patterns of heterochromatin distribution in plant chromosomes. Genet Mol Biol 23:1029–1041
Guerra M (2008) Chromosome numbers in plant cytotaxonomy: concepts and implications. Cytogenet Genome Res 120:339–350

Guerra M, García MA (2004) Heterochromatin and rDNA sites distribution in the holocentric chromosomes of *Cuscuta approximata* Bab. (Convolvulaceae). Genome 47:134–140

Guerra M, Cabral G, Cuacos M, González-García M, González-Sánchez M, Vega J, Puertas MJ (2010) Neocentrics and holokinetics (holocentrics): chromosomes out of the centromeric rules. Cytogenet Genome Res 129:82–96

Guillén A, Ruiz Rejón M (1984) Structural variability and chromosome numbers variation in natural populations of *Scilla autumnalis* (Liliaceae). Plant Syst Evol 144:201–207

Haga T, Noda S (1976) Cytogenetics of the *Scilla scilloides* complex. Genetica 46:161–176

Haizel T, Lim YK, Leitch AR, Moore G (2005) Molecular analysis of holocentric centromeres of *Luzula* species. Cytogenet Genome Res 109:134–143

Hall KJ, Parker JS (1995) Stable chromosome fission associated with rDNA mobility. Chromosome Res 3:417–422

Hamouche Y, Amirouche N, Misset M-T, Amirouche R (2010) Cytotaxonomy of autumn flowering species of Hyacinthaceae from Algeria. Plant Syst Evol 285:177–187

Han F, Lamb JC, Birchler JA (2006) High frequency of centromere inactivation resulting in stable dicentric chromosomes of maize. Proc Natl Acad Sci USA 103:3238–3243

Hanmoto H, Kataoka R, Ohmido N, Yonezawa Y (2007) Interstitial telomere-like repeats in the *Haplopappus gracilis* (Asteraceae) genome revealed by fluorescence *in situ* hybridization. Cytologia 72:483–488

Hasterok R, Wolny E, Hosiawa M, Kowalczyk M, Kulak-Ksiazczyk S, Ksiazczyk T, Heneen W, Małuszyńska J (2006) Comparative analysis of rDNA distribution in chromosomes of various species of Brassicaceae. Ann Bot 97:205–216

Hawkins JS, Kim HR, Nason JD, Wing RA, Wendel JF (2006) Differential lineage-specific amplification of transposable elements is responsible for genome size variation in *Gossypium*. Genome Res 16:1252–1261

Hawkins JS, Grover CE, Wendel JF (2008) Repeated big bangs and the expanding universe: directionality in plant genome size evolution. Plant Sci 174:557–562

Heitz E (1928) Das Heterochromatin der Moose. Jahrb Wiss Botanik 69:762–818

Hejnowicz Z, Feldman LJ (2000) The consequences of a non-uniform tension across kinetochores: lessons from segregation of chromosomes in the permanent translocation heterozygote *Oenothera*. Chromosome Res 8:165–172

Heslop-Harrison JS, Brandes A, Schwarzacher T (2003) Tandemly repeated DNA sequences and centromeric chromosomal regions of *Arabidopsis* species. Chromosome Res 11:241–253

Hipp AL (2007) Non-uniform processes of chromosome evolution in sedges (*Carex*: Cyperaceae). Evolution 61:2175–2194

Hipp AL, Rothrock PE, Roalson EH (2009) The evolution of chromosome arrangements in *Carex* (Cyperaceae). Bot Rev 75:96–109

Hirsch CD, Jiang J (2012) Centromeres: sequences, structure, and biology. In: Wendel JF, Greilhuber J, Doležel J, Leitch IJ (eds) Plant genome diversity, vol 1, Plant genomes, their residents, and their evolutionary dynamics. Springer-Verlag, Wien, pp 59–70

Houben A, Schubert I (2003) DNA and proteins of plant centromeres. Curr Opin Plant Biol 6:554–560

Houben A, Verlin D, Leach CR, Timmis JN (2001) The genomic complexity of micro B chromosomes of *Brachycome dichromosomatica*. Chromosoma 110:451–459

Houben A, Moghaddam AMB, Klemme S (2013) Biology and evolution of B chromosomes. In: Leitch IJ, Greilhuber J, Doležel J, Wendel JF (eds) Plant genome diversity, vol 2, Physical structure, behaviour and evolution of plant genomes. Wien, pp 149–165

Hřibová E, Neumann P, Matsumoto T, Roux N, Macas J, Doležel J (2010) Repetitive part of the banana (*Musa acuminata*) genome investigated by low-depth 454 sequencing. BMC Plant Biol 10:204

Huang S, Li R, Zhang Z, Li L, Gu X, Fan W, Lucas WJ, Wang X, Xie B, Ni P, Ren Y, Zhu H, Li J, Lin K, Jin W, Fei Z, Li G, Staub J, Kilian A, van der Vossen EAG, Wu Y, Guo J, He J, Jia Z, Ren Y, Tian G, Lu Y, Ruan J, Qian W, Wang M, Huang Q, Li B, Xuan Z, Cao J, Asan WuZ, Zhang J, Cai Q, Bai Y, Zhao B, Han Y, Li Y, Li X, Wang S, Shi Q, Liu S, Cho WK, Kim J-Y, Xu Y, Heller-Uszynska K, Miao H, Cheng Z, Zhang S, Wu J, Yang Y, Kang H, Li M, Liang H, Ren X, Shi Z, Wen M, Jian M, Yang H, Zhang G, Yang Z, Chen R, Liu S, Li J, Ma L, Liu H, Zhou Y, Zhao J, Fang X, Li G, Fang L, Li Y, Liu D, Zheng H, Zhang Y, Qin N, Li Z, Yang G, Yang S, Bolund L, Kristiansen K, Zheng H, Li S, Zhang X, Yang H, Wang J, Sun R, Zhang B, Jiang S, Wang J, Du Y, Li S (2009) The genome of the cucumber, *Cucumis sativus* L. Nat Genet 41:1275–1281

Hudakova S, Künzel G, Endo TR, Schubert I (2002) Barley chromosome arms longer than half of the spindle axis interfere with nuclear divisions. Cytogenet Genome Res 98:101–107

Husband BC, Baldwin SJ, Suda J (2013) The incidence of polyploidy in natural plant populations: major patterns and evolutionary processes. In: Leitch IJ, Greilhuber J, Doležel J, Wendel JF (eds) Plant genome diversity, vol 2, Physical structure, behaviour and evolution of plant genomes. Springer-Verlag, Wien, pp 255–276

Imai HT (1978) On the origin of telocentric chromosomes in mammals. J Theor Biol 78:619–637

Imai HT, Maruyama T, Gojobori T, Inoue Y, Crozier RH (1986) Theoretical bases for karyotype evolution. 1. The minimum interaction hypothesis. Am Nat 128:900–920

Jackson RC (1962) Interspecific hybridization in *Haplopappus* and its bearing on chromosome evolution in the Blepharodon section. Am J Bot 49:119–132

Jaillon O et al, The French–Italian Public Consortium for Grapevine Genome Characterization (2007) The grapevine genome sequence suggests ancestral hexaploidization in major angiosperm phyla. Nature 449:463–467

Jamilena M, Mariotti B, Manzano S (2008) Plant sex chromosomes: molecular structure and function. Cytogenet Genome Res 120:255–264

Janoušek B, Hobza R, Vyskot B (2013) Chromosomes and sex diffrentiation. In: Leitch IJ, Greilhuber J, Doležel J, Wendel JF (eds) Plant genome diversity, vol 2, Physical structure, behaviour and evolution of plant genomes. Springer-Verlag, Wien, pp 167–186

Jenkins GM, Jones RN (2004) B chromosomes in hybrids of temperate cereals and grasses. Cytogenet Genome Res 106:314–319

Jiao Y, Wickett NJ, Ayyampalayam S, Chanderbali AS, Landherr L, Ralph PE, Tomsho LP, Hu Y, Liang H, Soltis PS, Soltis DE, Clifton SW, Schlarbaum SE, Schuster SC, Ma H, Leebens-Mack J, dePamphilis CW (2011) Ancestral polyploidy in seed plants and angiosperms. Nature 473:97–100

John B, Miklos GLG (1979) Functional aspects of heterochromatin and satellite DNA. Int Rev Cytol 58:1–114

Johnson MAT, Kenton AY, Bennett MD, Brandham PE (1989) *Voanioala gerardii* has the highest known chromosome number in the monocotyledons. Genome 32:328–333

Johnson DS, Mortazavi A, Myers RM, Wold B (2007) Genome-wide mapping of *in vivo* protein-DNA interactions. Science 316:1497–1502

Jones K (1977) The role of Robertsonian change in karyotype evolution in higher plants. In: de la Chapelle A, Sorea M (eds) Chromosomes today. Elsevier, Amsterdam, pp 121–129

Jones K (1978) Aspects of chromosome evolution in higher plants. Adv Bot Res 6:119–194

Jones K (1998) Robertsonian fusion and centric fission in karyotype evolution of higher plants. Bot Rev 64:273–289

Jones RN, Rees H (1967) Genotypic control of chromosome behaviour in rye. XI. The influence of B-chromosomes on meiosis. Heredity 22:333–347

Jones RN, Rees H (1982) B chromosomes. Academic, New York

Jones RN, Viegas W, Houben A (2008) A century of B chromosomes in plants: so what? Ann Bot 101:767–775

Kamstra SA, Kuipers AG, De Jeu MJ, Ramanna MS, Jacobsen E (1997) Physical localisation of repetitive DNA sequences in *Alstroemeria*: karyotyping of two species with species-specific and ribosomal DNA. Genome 40:652–658

Karlov GI, Fesenko IA, Andreeva GN, Khrustaleva LI (2010) Chromosome organization of Ty1-*copia*-like retrotransposons in the tomato genome. Russ J Genet 46:677–681

Kejnovský E, Hobza R, Čermák T, Kubat Z, Vyskot B (2009) The role of repetitive DNA in structure and evolution of sex chromosomes in plants. Heredity 102:533–541

Kenton A (1981) Chromosome evolution in the *Gibasis linearis* alliance (Commelinaceae). Chromosoma 84:291–304

Kihara H, Ono T (1927) Chromosomenzahlen und systematische Gruppierung der *Rumex*-Arten. Zeitschr wiss Biol B 4:475–481

Koehler S, Cabral JS, Whitten WM, Williams NH, Singer RB, Neubig KM, Guerra M, Souza AP, Amaral Mdo C (2008) Molecular phylogeny of the neotropical genus *Christensonella* (Orchidaceae, Maxillariinae): species delimitation and insights into chromosome evolution. Ann Bot 102:491–507

Koo D-H, Jiang J (2008) Extraordinary tertiary constrictions of *Tripsacum dactyloides* chromosomes: implications for karyotype evolution of polyploids driven by segmental chromosome losses. Genetics 179:1119–1123

Koo D-H, Nam Y-W, Choi D, Bang J-W, de Jong H, Hur Y (2010) Molecular cytogenetic mapping of *Cucumis sativus* and *C. melo* using highly repetitive DNA sequences. Chromosome Res 18:325–336

Kotseruba V, Pistrick K, Blattner FR, Kumke K, Weiss O, Rutten T, Fuchs J, Endo T, Nasuda S, Ghukasyan A, Houben A (2010) The evolution of the hexaploid grass *Zingeria kochii* (Mez) Tzvel. ($2n = 12$) was accompanied by complex hybridization and uniparental loss of ribosomal DNA. Mol Phylogenet Evol 56:146–155

Koukalova B, Moraes AP, Renny-Byfield S, Matyášek R, Leitch AR, Kovařík A (2010) Fall and rise of satellite repeats in allopolyploids of *Nicotiana* over c. 5 million years. New Phytol 186:148–160

Kovařík A, Pires JC, Leitch AR, Lim KY, Sherwood A, Matyášek R, Rocca J, Soltis DE, Soltis PS (2005) Rapid concerted evolution of nuclear ribosomal DNA in two allopolyploids of recent and recurrent origin. Genetics 169:931–944

Kuipers AG, Kamstra SA, de Jeu MJ, Visser RG (2002) Molecular characterization and physical localization of highly repetitive DNA sequences from Brazilian *Alstroemeria* species. Chromosome Res 10:389–398

Kumar A, Bennetzen JL (1999) Plant retrotransposons. Annu Rev Genet 33:479–532

Kuta E, Bohanec B, Dubas E, Vizintin L, Przywara L (2004) Chromosome and nuclear DNA study on *Luzula*—a genus with holokinetic chromosomes. Genome 47:246–256

Lamb JC, Yu W, Han F, Birchler JA (2007) Plant chromosomes from end to end: telomeres, heterochromatin and centromeres. Curr Opin Plant Biol 10:116–122

Lee H-R, Zhang W, Langdon T, Jin W, Yan H, Cheng Z, Jiang J (2005) Chromatin immunoprecipitation cloning reveals rapid evolutionary patterns of centromeric DNA in *Oryza* species. Proc Natl Acad Sci USA 102:11793–11798

Leitch IJ, Bennett MD (2004) Genome downsizing in polyploid plants. Biol J Linn Soc 82:651–663

Leitch AR, Leitch IJ (2008) Genomic plasticity and the diversity of polyploid plants. Science 320:481–483

Leitch IJ, Leitch AR (2013) Genome size diversity and evolution in land plants. In: Leitch IJ, Greilhuber J, Doležel J, Wendel JF (eds) Plant genome diversity, vol 2, Physical structure, behaviour and evolution of plant genomes. Springer-Verlag, Wien, pp 307–322

Leitch IJ, Beaulieu JM, Chase MW, Leitch AR, Fay MF (2010) Genome size dynamics and evolution in monocots. J Bot 2010:1–18, Article ID 862516

Levan A (1932) Cytological studies in *Allium*. I. Chromosome morphological contributions. Hereditas 16:257–264

Levan A, Fredga K, Sandberg AA (1964) Nomenclature for centromeric position on chromosomes. Hereditas 52:201–220

Levin DA (2002) The role of chromosomal change in plant evolution. Oxford University Press, Oxford

Levin DA, Funderburg SW (1979) Genome size in angiosperms: temperate versus tropical species. Am Nat 114:784–795

Levin DA, Wilson AC (1976) Rates of evolution in seed plants: net increase in diversity of chromosome numbers and species numbers through time. Proc Natl Acad Sci USA 73:2086–2090

Levin DA, Palestis BG, Jones RN, Trivers R (2005) Phyletic hot spots for B chromosomes in angiosperms. Evolution 59:962–969

Levitzky GA (1931) The karyotype in systematics. Bull Appl Bot Genet Plant Breed 27:220–240

Lewis H (1951) The origin of supernumerary chromosomes in natural populations of *Clarkia elegans*. Evolution 5:142–157

Lewis WH, Oliver RL, Suda Y (1967) Cytogeography of *Claytonia virginica* and its allies. Ann Miss Bot Garden 54:153–171

Lim KY, Matyášek R, Lichtenstein CP, Leitch AR (2000) Molecular cytogenetic analyses and phylogenetic studies in the *Nicotiana* section Tomentosae. Chromosoma 109:245–258

Lim KY, Kovařík A, Matyášek R, Chase MW, Clarkson JJ, Grandbastien MA, Leitch AR (2007a) Sequence of events leading to near-complete genome turnover in allopolyploid *Nicotiana* within five million years. New Phytol 175:756–763

Lim KY, Matyášek R, Kovařík A, Leitch AR (2007b) Parental origin and genome evolution in the allopolyploid *Iris versicolor*. Ann Bot 100:219–224

Liu Z, Moore PH, MA H, Ackerman CM, Ragiba M, Yu Q, Pearl HM, Kim MS, Charlton JW, Stiles JI, Zee FT, Paterson AH, Ming R (2004) A primitive Y chromosome in papaya marks incipient sex chromosome evolution. Nature 427:348–352

Löve Á, Löve D (1974) Cytotyxonomical atlas of the Slovenian flora. Cramer, Lehre

Luceño M, Vanzela ALL, Guerra M (1998) Cytotaxonomic studies in Brazilian *Rhynchospora* (Cyperaceae), a genus exhibiting holocentric chromosomes. Can J Bot 76:440–449

Lukaszewski A, Gustafson JP (1983) Translocations and modifications of chromosomes in triticale wheat hybrids. Theor Appl Genet 64:239–248

Lysák MA, Schubert I (2013) Mechanisms of chromosome rearrangements. In: Leitch IJ, Greilhuber J, Doležel J, Wendel JF (eds) Plant genome diversity, vol 2, Physical structure, behaviour and evolution of plant genomes. Springer-Verlag, Wien, pp 137–147

Lysák MA, Berr A, Pecinka A, Schmidt R, McBreen K, Schubert I (2006) Mechanisms of chromosome number reduction in *Arabidopsis thaliana* and related Brassicaceae species. Proc Natl Acad Sci USA 103:5224–5229

Lysák M, Mandáková T, Lacombe E (2010) Reciprocal and multi-species chromosome BAC painting in crucifers (Brassicaceae). Cytogenet Genome Res 129:184–189

Ma J, Wing RA, Bennetzen JL, Jackson SA (2007) Plant centromere organization: a dynamic structure with conserved functions. Trends Genet 23:134–139

Macas J, Pozárková D, Navrátilová A, Nouzová M, Neumann P (2000) Two new families of tandem repeats isolated from genus *Vicia* using genomic self-priming PCR. Mol Gen Genet 263:741–751

Macas J, Meszaros T, Nouzová M (2002) PlantSat: a specialized database for plant satellite repeats. Bioinformatics 18:28–35

Macas J, Navrátilová A, Koblízková A (2006) Sequence homogenization and chromosomal localization of VicTR-B satellites differ between closely related *Vicia* species. Chromosoma 115:437–447

Macas J, Neumann P, Navrátilová A (2007) Repetitive DNA in the pea (*Pisum sativum* L.) genome: comprehensive characterization using 454 sequencing and comparison to soybean and *Medicago truncatula*. BMC Genomics 8:427

Maddison WP, Maddison DR (2009) Mesquite: a modular system for evolutionary analysis. http://mesquiteproject.org, accessed 30.03.2011

Madej A, Kuta E (2001) Holokinetic chromosomes of *Luzula luzuloides* (Juncaceae) in callus culture. Acta Biol Cracov Bot 43:33–43

Małuszyńska J, Schweizer D (1989) Ribosomal RNA genes in B chromosomes of *Crepis capillaris* detected by non-radioactive *in situ* hybridization. Heredity 62:59–65

Małuszyńska J, Hasterok R, Weiss H (1998) rRNA genes: their distribution and activity in plants. In: Małuszyńska J (ed) Plant cytogenetics. Silesian University Press, Katowice, pp 75–95

Mandáková T, Lysák MA (2008) Chromosomal phylogeny and karyotype evolution in $x = 7$ crucifer species (Brassicaceae). Plant Cell 20:2559–2570

Mandáková T, Heenan PB, Lysák MA (2010a) Island species radiation and karyotypic stasis in *Pachycladon* allopolyploids. BMC Evol Biol 10:367

Mandáková T, Joly S, Krzywinski M, Mummenhof K, Lysák M (2010b) Fast diploidization in close mesopolyploid relatives of *Arabidopsis*. Plant Cell 22:2277–2290

Marasek A, Hitoshi M, Keiichi O (2006) The origin of Darwin hybrid tulips analyzed by flow cytometry, karyotype analyses and genomic *in situ* hybridization. Euphytica 151:279–290

Marks GE, Schweizer D (1974) Giemsa banding: karyotype differences in some species of *Anemone* and in *Hepatica nobilis*. Chromosoma 44:405–416

Mast AR, Kelso S, Richards AJ, Lang DJ, Feller DMS, Conti E (2001) Phylogenetic relationships in *Primula* L. and related genera (Primulaceae) based on noncoding chloroplast DNA. Int J Plant Sci 162:1381–1400

Masterson J (1994) Stomatal size in fossil plants—evidence for polyploidy in majority of angiosperms. Science 264:1759–1763

Mayrose I, Barker MS, Otto SP (2010) Probabilistic models of chromosome number evolution and the inference of polyploidy. Syst Biol 59:132–144

Menzel G, Dechyeva D, Wenke T, Holtgräwe D, Weisshaar B, Schmidt T (2008) Diversity of a complex centromeric satellite and molecular characterization of dispersed sequence families in sugar beet (*Beta vulgaris*). Ann Bot 102:521–530

Meyers LA, Levin DA (2006) On the abundance of polyploids in flowering plants. Evolution 60:1198–1206

Moraes AP, Leitch IJ, Leitch AR (2012) Chromosome studies in Orchidaceae: Karyotype divergence in Neotropical genera in subtribe Maxillariinae. Bot J Linn Soc 170:29–39

Moore RJ (1974) Index to plant chromosome numbers: 1972. Regnum Vegetabile 91:1–108

Moore RJ (1977) Index to plant chromosome numbers: 1973–1974. Regnum Vegetabile 96:1–256

Murata M, Shibata F, Yokota E (2006) The origin, meiotic behavior, and transmission of a novel minichromosome in *Arabidopsis thaliana*. Chromosoma 115:311–319

Narayan RKJ (1987) Nuclear DNA changes, genome differentiation and evolution in *Nicotiana* (Solanaceae). Plant Syst Evol 157:161–180

Nasuda S, Hudakova S, Schubert I, Houben A, Endo TR (2005) Stable barley chromosomes without centromeric repeats. Proc Natl Acad Sci USA 102:9842–9847

Navajas-Pérez R, Schwarzacher T, de la Herrán R, Ruiz Rejón C, Ruiz Rejón M, Garrido-Ramos MA (2006) The origin and evolution of the variability in a Y-specific satellite-DNA of *Rumex acetosa* and its relatives. Gene 368:61–71

Navrátilová A, Neumann P, Macas J (2003) Karyotype analysis of four *Vicia* species using *in situ* hybridization with repetitive sequences. Ann Bot 91:921–926

Neumann P, Koblížková A, Navrátilová A, Macas J (2006) Significant expansion of *Vicia pannonica* genome size mediated by amplification of a single type of giant retroelement. Genetics 173:1047–1056

Neumann P, Navrátilová A, Koblížková A, Kejnovský E, Hřibová E, Hobza R, Widmer A, Doležel J, Macas J (2011) Plant centromeric retrotransposons: a structural and cytogenetic perspective. Mobile DNA 2:4

Novák P, Neumann P, Macas J (2010) Graph-based clustering and characterization of repetitive sequences in next-generation sequencing data. BMC Bioinformatics 11:378

Otto SP, Whitton J (2000) Polyploid incidence and evolution. Annu Rev Genet 34:401–437

Ownbey M (1950) Natural hybridization and amphiploidy in the genus *Tragopogon*. Am J Bot 37:487–499

Parker JS (1976) The B-chromosome system of *Hypochoeris maculata*. I. B-distribution, meiotic behaviour and inheritance. Chromosoma 59:167–177

Parker JS, Lozano R, Taylor S, Ruiz Rejón M (1991) Chromosomal structure of populations of *Scilla autumnalis* in the Iberian Peninsula. Heredity 67:287–297

Paszko B (2006) A critical review and a new proposal of karyotype asymmetry indices. Plant Syst Evol 258:39–48

Pazy B, Plitmann U (1995) Chromosome divergence in the genus *Cuscuta* and its systematic implications. Caryologia 48:173–180

Pedrosa-Harand A, de Almeida CC, Mosiolek M, Blair MW, Schweizer D, Guerra M (2006) Extensive ribosomal DNA amplification during Andean common bean (*Phaseolus vulgaris* L.) evolution. Theor Appl Genet 112:924–933

Pellicer J, Fay MF, Leitch IJ (2010) The largest eukaryotic genome of them all? Bot J Linn Soc 164:10–15

Peruzzi L, Leitch IJ, Caparelli KF (2009) Chromosome diversity and evolution in Liliaceae. Ann Bot 103:459–475

Pich U, Schubert I (1998) Terminal heterochromatin and alternative telomeric sequences in *Allium cepa*. Chromosome Res 6:315–321

Pikaard CS (2000) Nucleolar dominance: uniparental gene silencing on multi-megabase scale in genetic hybrids. Plant Mol Biol 43:163–177

Pires JC, Lim KY, Kovařík A, Matyásek R, Boyd A, Leitch AR, Leitch IJ, Bennett MD, Soltis PS, Soltis DE (2004) Molecular cytogenetic analysis of recently evolved *Tragopogon* (Asteraceae) allopolyploids reveal a karyotype that is additive of the diploid progenitors. Am J Bot 91:1022–1035

Plohl M, Luchetti A, Mestrović N, Mantovani B (2008) Satellite DNAs between selfishness and functionality: structure, genomics and evolution of tandem repeats in centromeric (hetero)chromatin. Gene 409:72–82

Proost S, Pattyn P, Gerats T, Van de Peer Y (2011) Journey through the past: 150 million years of plant genome evolution. Plant J 66:58–65

Raina SN, Rees H (1983) DNA variation between and within chromosome complements in *Vicia* species. Heredity 51:335–346

Ramsey J, Schemske DW (1998) Pathways, mechanisms and rates of polyploid formation in flowering plants. Annu Rev Ecol Syst 29:467–501

Ramsey J, Schemske DW (2002) Neopolyploidy in flowering plants. Annu Rev Ecol Syst 33:589–639

Raskina O, Barber JC, Nevo E, Belyayev A (2008) Repetitive DNA and chromosomal rearrangements: speciation-related events in plant genomes. Cytogenet Genome Res 120:351–357

Raven PH (1975) The bases of angiosperm phylogeny: cytology. Ann Miss Bot Garden 62:724–764

Renner SS, Ricklefs RE (1995) Dioecy and its correlates in the flowering plants. Am J Bot 82:596–606

Richard GF, Kerrest A, Dujon B (2008) Comparative genomics and molecular dynamics of DNA repeats in eukaryotes. Microbiol Mol Biol Rev 72:686–727

Richards EJ, Ausubel FM (1988) Isolation of a higher eukaryotic telomere from *Arabidopsis thaliana*. Cell 53:127–136

Rieseberg LH, Doyle MF (1989) Tetrasomic segregation in the naturally occurring autotetraploid *Allium nevii* (Alliaceae). Hereditas 11:3–36

Roa F, Guerra M (2010) Trends on the distribution of the 45S rDNA ribosomal DNA in plants. In: Annual meeting of the Society for Experimental Biology, Abstract book. Prague, p 262

Robert ML, Lim KY, Hanson L, Sanchez-Teyer F, Bennett MD, Leitch AR, Leitch IJ (2008) Wild and agronomically important *Agave* species (Asparagaceae) show proportional increases in chromosome number, genome size, and genetic markers with increasing ploidy. Bot J Linn Soc 158:215–222

Röser M (1994) Pathways of karyological differentiation in palms (Arecaceae). Plant Syst Evol 189:89–118

Rothwell NV, Kump JG (1965) Chromosome numbers in populations of *Claytonia virginica* from the New York metropolitan area. Am J Bot 52:403–407

Rowell DM, Rockman MV, Tait NN (2002) Extensive Robertsonian rearrangement: implications for the radiation and biogeography of *Planipapillus* Reid (Onychophora: Peripatopsidae). J Zool 257:171–179

Ruas CF, Weiss-Schneeweiss H, Stuessy TF, Samuel MR, Pedrosa-Harand A, Tremetsberger K, Ruas PM, Schlüter PM, Ortiz Herrera MA, König C, Matzenbacher NI (2008) Characterization, genomic organization and chromosomal distribution of Ty1-*copia* retrotransposons in species of *Hypochaeris* (Asteraceae). Gene 412:39–49

Ruiz Rejón M, Oliver JL (1981) Genetic variability in *Muscari comosum* (Liliaceae). I. A comparative analysis of chromosome polymorphisms in Spanish and Aegean populations. Heredity 47:403–407

Ruiz Rejón M, Posse F, Oliver JL (1980) The B chromosome system of *Scilla autumnalis* (Liliaceae): effects at the isozyme level. Chromosoma 79:341–348

Samuel R, Stuessy TF, Tremetsberger K, Baeza CM, Siljak-Yakovlev S (2003) Phylogenetic relationships among species of *Hypochaeris* (Asteraceae, Cichorieae) based on ITS, plastid *trn*L intron, *trn*L-F spacer, and *mat*K sequences. Am J Bot 90:496–507

Sandery MJ, Forster JW, Blunden R, Jones N (1990) Identification of a family of repeated sequences on the rye B chromosome. Genome 33:908–913

Sapre AB, Deshpande D (1987) Origin of B chromosomes in *Coix* L. through spontaneous interspecific hybridisation. J Hered 78:191–196

Schmidt T, Heslop-Harrison JS (1998) Genomes, genes and junk: the large scale organization of plant chromosomes. Trends Plant Sci 3:195–199

Schönswetter P, Suda J, Popp M, Weiss-Schneeweiss H, Brochmann C (2007) Circumpolar phylogeography of *Juncus biglumis* (Juncaceae) inferred from AFLP fingerprints, cpDNA sequences, nuclear DNA content and chromosome numbers. Mol Phylogenet Evol 42:92–103

Schranz ME, Mitchell-Olds T (2006) Independent ancient polyploidy events in the sister families Brassicaceae and Cleomaceae. Plant Cell 18:1152–1165

Schranz ME, Lysák MA, Mitchell-Olds T (2006) The ABC's of comparative genomics in the Brassicaceae: building blocks of crucifer genomes. Trends Plant Sci 11:535–542

Schubert I (2001) Alteration of chromosome numbers by generation of minichromosomes: is there a lower limit of chromosome size for stable segregation? Cytogenet Cell Genet 93:175–181

Schubert I (2007) Chromosome evolution. Curr Opin Plant Biol 10:109–115

Schubert I, Oud JL (1997) There is an upper limit of chromosome size for normal development of an organism. Cell 88:515–520

Schubert I, Rieger R (1985) A new mechanism for altering chromosome number during karyotype evolution. Theor Appl Genet 70:213–221

Schubert I, Wobus U (1985) *In situ* hybridization confirms jumping nucleolus organizing regions in *Allium*. Chromosoma 92:143–148

Schubert I, Anastassova-Kristeva M, Rieger R (1979) Specificity of NOR staining in *Vicia faba*. Exp Cell Res 120:433–435

Schwarzacher T (2003) DNA, chromosomes, and *in situ* hybridization. Genome 46:953–962

Schwarzacher T, Heslop-Harrison P (2000) Practical *in situ* hybridization, 2nd edn. BIOS, Oxford, UK

Schwarzacher HG, Wachtler F (1986) Nucleolus organizer regions and nucleoli: cytological findings. Chromosomes Today 9:252–260

Schwarzacher T, Leitch AR, Bennett MD, Heslop-Harrison JS (1989) *In situ* localization of parental genomes in a wide hybrid. Ann Bot 64:315–324

Schweizer D (1976) Reverse fluorescent banding with chromomycin and DAPI. Chromosoma 58:307–324

Schweizer D (1981) Counterstain-enhanced chromosome banding. Hum Genet 57:1–14

Shibata F, Hizume M, Kuroki Y (2000) Molecular cytogenetic analysis of supernumerary heterochromatic segments in *Rumex acetosa*. Genome 43:391–397

Shoemaker RC, Schlueter J, Doyle JJ (2006) Paleopolyploidy and gene duplication in soybean and other legumes. Curr Opin Plant Biol 9:104–109

Skalicka K, Lim KY, Matyášek R, Matzke M, Leitch AR, Kovařík A (2005) Preferential elimination of repeated DNA sequences from the paternal, *Nicotiana tomentosiformis* genome donor of a synthetic, allotetraploid tobacco. New Phytol 166:291–303

Smith GP (1976) Evolution of repeated DNA sequences by unequal crossover. Science 191:528–535

Smyth DR, Kongsuwan K, Wisudharomn S (1989) A survey of C-band patterns in chromosomes of *Lilium* (Liliaceae). Plant Syst Evol 163:53–69

Soltis DE, Soltis PS (1999) Polyploidy: recurrent formation and genome evolution. Trends Ecol Evol 14:348–352

Soltis DE, Soltis PS, Endress PK, Chase MW (2005) Evolution of genome size and base chromosome number. In: Soltis DE, Soltis PS, Endress PK, Chase MW (eds) Phylogeny and evolution of angiosperms. Sinauer Associates, Sunderland, pp 287–302

Soltis DE, Soltis PS, Schemske DW, Hancock JF, Thompson JN, Husband BC, Judd WS (2007) Autopolyploidy in angiosperms: have we grossly underestimated the number of species? Taxon 56:13–30

Soltis DE, Albert VA, Leebens-Mack J, Bell CD, Paterson AH, Zheng C, Sankoff D, dePamphilis CD, Wall PK, Soltis PS (2009) Polyploidy and angiosperm diversification. Am J Bot 96:336–348

Soltis DE, Buggs RJA, Doyle JJ, Soltis PS (2010) What we still don't know about polyploidy. Taxon 59:1387–1403

Sone T, Fujisawa M, Takenaka M, Nakagawa S, Yamaoka S, Sakaida M, Nishiyama R, Yamato KT, Ohmido N, Fukui K, Fukuzawa H, Ohyama K (1999) Bryophyte 5S rDNA was inserted into 45S rDNA repeat units after the divergence from higher land plants. Plant Mol Biol 41:679–685

Souza LGR, Crosa O, Winge H, Guerra M (2009) The karyotype of *Nothoscordum arenarium* Herter (Gilliesioideae, Alliaceae): a populational and cytomolecular analysis. Genet Mol Biol 32:111–116

Stace CA (2000) Cytology and cytogenetics as a fundamental taxonomic resource for the 20th and 21st centuries. Taxon 49:451–477

Stebbins GL (1940) The significance of polyploidy in plant evolution. Am Nat 74:54–66

Stebbins GL (1950) Variation and evolution in plants. Columbia University Press, New York

Stebbins GL (1971) Chromosomal evolution in higher plants. Edward Arnold, London

Stedje B (1989) Chromosome evolution within the *Ornithogalum tenuifolium* complex (Hyacinthaceae), with special emphasis on the evolution of bimodal karyotypes. Plant Syst Evol 166:79–89

Stewart RN (1947) The morphology of somatic chromosomes in *Lilium*. Am J Bot 34:9–26

Stuessy TF (1971) Chromosome numbers and phylogeny in *Melampodium* (Compositae). Am J Bot 58:732–736

Stuessy TF (2009) Plant taxonomy: the systematic evaluation of comparative data. Columbia University Press, New York

Stuessy TF, Crawford DJ (1998) Chromosomal stasis during speciation in angiosperms of oceanic islands. In: Stuessy TF, Ono M (eds) Evolution and speciation of island plants. Cambridge University Press, Cambridge, pp 307–324

Stuessy TF, Weiss-Schneeweiss H, Keil DJ (2004) Diploid and polyploid cytotype distribution in *Melampodium cinereum* and *M. leucanthum* (Asteraceae, Heliantheae). Am J Bot 91:889–898

Suda J, Weiss-Schneeweiss H, Tribsch A, Schneeweiss GM, Trávnícek P, Schönswetter P (2007) Complex distribution patterns of di-, tetra- and hexaploid cytotypes in the European high mountain plant *Senecio carniolicus* (Asteraceae). Am J Bot 94:1391–1401

Sumner AT (1972) A simple technique for demonstration of centromeric heterochromatin. Exp Cell Res 75:304–306

Sumner AT (1990) Chromosome banding. Unwin and Hyman, London

Swaminathan K, Varala K, Hudson ME (2007) Global repeat discovery and estimation of genomic copy number in a large, complex genome using a high-throughput 454 sequence survey. BMC Genomics 8:132

Sýkorová E, Fajkus J, Meznikova M, Lim KY, Neplechova K, Blattner FR, Chase MW, Leitch AR (2006) Minisatellite telomeres occur in the family Alliaceae but are lost in *Allium*. Am J Bot 93:814–823

Symonds VV, Soltis PS, Soltis DE (2010) Dynamics of polyploid formation in *Tragopogon* (Asteraceae): recurrent formation, gene flow, and population structure. Evolution 64:1984–2003

Szinay D, Bai Y, Visser R, de Jong H (2010) FISH applications for genomics and plant breeding strategies in tomato and other solanaceous crops. Cytogenet Genome Res 129:199–210

Tang X, Szinay D, Lang C, Ramanna MS, van der Vossen EAG, Datema E, Lankhorst RK, de Boer J, Peters SA, Bachem C, Stiekema W, Visser RGF, de Jong H, Bai Y (2008) Cross-species bacterial artificial chromosome–fluorescence *in situ* hybridization painting of the tomato and potato chromosome 6 reveals undescribed chromosomal rearrangements. Genetics 180:1319–1328

Tenaillon MI, Hufford MB, Gaut BS, Ross-Ibarra J (2011) Genome size and transposable element content as determined by high-throughput sequencing in maize and Zea *luxurians*. Genome Biol Evol 3:219–229

Topp CN, Okagaki RJ, Melo JR, Kynast RG, Phillips RL, Dawe RK (2009) Identification of a maize neocentromere in an oat-maize addition line. Cytogenet Genome Res 124:228–238

Torrell M, Garcia-Jacas N, Susanna A, Valles J (1999) Infrageneric phylogeny of the genus *Artemisia* L. (Asteraceae, Anthemidae) based on nucleotide sequences of nuclear ribosomal DNA internal transcribed spacers (ITS). Taxon 48:721–736

Trávníček P, Kubátová B, Čurn V, Rauchová J, Krajníková E, Jersáková J, Suda J (2011) Remarkable coexistence of multiple cytotypes of the *Gymnadenia conopsea* aggregate (the fragrant orchid): evidence from flow cytometry. Ann Bot 107:77–87

Tsujimoto H, Usami N, Hasegawa K, Yamada T, Nagaki K, Sasakuma T (1999) *De novo* synthesis of telomere sequences at the healed breakpoints of wheat deletion chromosomes. Mol Gen Genet 262:851–856

U N (1935) Genome analysis in *Brassica* with special reference to the experimental formation of *B. napus* and peculiar mode of fertilization. Jpn J Bot 7:389–452

Uchida W, Matsunaga S, Sugiyama R, Kawano S (2002) Interstitial telomere-like repeats in the *Arabidopsis thaliana* genome. Genes Genet Syst 77:63–67

Ugarković D, Plohl M (2002) Variation in satellite DNA profiles: causes and effects. EMBO J 21:5955–5959

Uhl CH (1978) Chromosomes of Mexican *Sedum* II. Section Pachysedum. Rhodora 80:491–512

Ungerer MC, Strakosh SC, Zhen Y (2006) Genome expansion in three hybrid sunflower species is associated with retrotransposon proliferation. Curr Biol 16:R872–R873

Van de Peer Y, Maere S, Meyer A (2009) The evolutionary significance of ancient genome duplications. Nat Rev Genet 10:725–732

Vanzela ALL, Guerra M, Luceno M (1996) *Rhynchospora tenuis* Link (Cyperaceae), a species with the lowest number of holocentric chromosomes. Cytobios 88:219–228

Vanzela ALL, Luceño M, Guerra M (2000) Karyotype evolution and cytotaxonomy in Brazilian species of *Rhynchospora* Vahl (Cyperaceae). Bot J Linn Soc 134:557–566

Vanzela ALL, Cuadrado A, Guerra M (2003) Localization of 45S rDNA and telomeric sites on holocentric chromosomes of *Rhynchospora tenuis* Link (Cyperaceae). Genet Mol Biol 26:199–201

Vaughan HE, Taylor S, Parker JS (1997) The ten cytological races of the *Scilla autumnalis* species complex. Heredity 79:371–379

Vogel J et al, The International Brachypodium Initiative (2010). Genome sequencing and analysis of the model grass *Brachypodium distachyon*. Nature 463:763–768

Volkov RA, Medina FJ, Zentgraf U, Hemleben V (2004) Organization and molecular evolution of rDNA, nucleolar dominance, and nucleolus structure. Prog Bot 65:106–146

Vosa CG (1985) Chromosome banding in plants. In: Sharma AK, Sharma A (eds) Advances in chromosome and cell genetics. Oxford/IBH Publishing, New Delhi, pp 79–104

Vosa CG (2005) On chromosome uniformity, bimodality and evolution in the tribe Aloineae (Asphodelaceae). Caryologia 581:83–85

Vyskot B, Hobza R (2004) Gender in plants: sex chromosomes are emerging from the fog. Trends Genet 20:432–438

Wang X, Tang H, Bowers JE, Feltus FA, Paterson AH (2007) Extensive concerted evolution of rice paralogs and the road to regaining independence. Genetics 77:1753–1763

Watanabe K, Yahara T, Denda T, Kosuge K (1999) Chromosomal evolution in the genus *Brachyscome* (Asteraceae, Astereae): statistical tests regarding correlation between changes in karyotype and habit using phylogenetic information. J Plant Res 112:145–161

Weiss H, Małuszyńska J (2000) Chromosomal rearrangement in autotetraploid plants of *Arabidopsis thaliana*. Hereditas 133:255–261

Weiss H, Scherthan H (2002) *Aloe* spp. – plants with vertebrate-like telomeric sequences. Chromosome Res 10:155–164

Weiss-Schneeweiss H, Stuessy TF, Siljak-Yakovlev S, Baeza CM, Parker J (2003) Karyotype evolution in South American species of *Hypochaeris* (Asteraceae, Lactuceae). Plant Syst Evol 241:171–184

Weiss-Schneeweiss H, Riha K, Jang CG, Puizina J, Scherthan H, Schweizer D (2004) Chromosome termini of the monocot plant *Othocallis siberica* are maintained by telomerase, which specifically synthesizes vertebrate-type telomere sequences. Plant J 37:484–493

Weiss-Schneeweiss H, Stuessy TF, Tremetsberger K, Urtubey E, Valdebenito HA, Beck SG, Baeza CM (2007a) Chromosome numbers and karyotypes of South American species and populations of *Hypochaeris* (Asteraceae). Bot J Linn Soc 153:49–60

Weiss-Schneeweiss H, Schneeweiss GM, Stuessy TF, Mabuchi T, Park J-M, Jang C-G, Sun B-Y (2007b) Chromosomal stasis in diploids contrasts with genome restructuring in auto- and allopolyploid taxa of *Hepatica* (Ranunculaceae). New Phytol 174:669–682

Weiss-Schneeweiss H, Tremetsberger K, Schneeweiss GM, Parker JS, Stuessy TF (2008) Karyotype diversification and evolution in diploid and polyploid South American *Hypochaeris* (Asteraceae) inferred from rDNA localization and genetic fingerprint data. Ann Bot 101:909–918

Weiss-Schneeweiss H, Stuessy TF, Villaseñor JL (2009) Chromosome numbers, karyotypes, and evolution in *Melampodium* (Asteraceae). Int J Plant Sci 170:1168–1182

Wendel JF (2000) Genome evolution in polyploids. Plant Mol Biol 42:225–249

White TA, Bordewich M, Searle JB (2010) A network approach to study karyotypic evolution: the chromosomal races of the common shrew (*Sorex araneus*) and house mouse (*Mus musculus*) as model systems. Syst Biol 59:262–276

Wicker T, Taudien S, Houben A, Keller B, Graner A, Platzer M, Stein N (2009) A wholegenome snapshot of 454 sequences exposes the composition of the barley genome and provides evidence for parallel evolution of genome size in wheat and barley. Plant J 59:712–722

Wilby AS, Parker JS (1988) Mendelian and non-Mendelian inheritance of newly-arisen chromosome rearrangements. Heredity 60:263–268

Wood TE, Takebayashi N, Barker MS, Mayrose I, Greenspoon PB, Rieseberg LH (2009) The frequency of polyploid speciation in vascular plants. Proc Natl Acad Sci USA 106:13875–13879

Wu F, Tanksley SD (2010) Chromosomal evolution in the plant family Solanaceae. BMC Genomics 11:182

Yunis JJ, Prakash O (1982) The origin of man: a chromosomal pictorial legacy. Science 215:1525–1530

Zedek F, Šmerda J, Šmarda P, Bureš P (2010) Correlated evolution of LTR retrotransposons and genome size in the genus *Eleocharis*. BMC Plant Biol 10:265

Zellinger B, Riha K (2007) Composition of plant telomeres. Biochim Biophys Acta 1769:399–409

Zhang D, Sang T (1998) Chromosomal structural rearrangement of *Paeonia brownii* and *P. californica* revealed by fluorescence *in situ* hybridization. Genome 41:848–853

Żuk J (1969) The additional heterochromatic chromosome and its influence on sex chromosome pairing in *Rumex*. Heredity 24:69–74

Karyotype Variation and Evolution in Gymnosperms

14

Brian G. Murray

Contents

B.G. Murray (✉)
School of Biological Sciences, The University of Auckland, 3 Symonds Street, Private Bag 92019, Auckland Mail Centre, Auckland 1142, New Zealand
e-mail: b.murray@auckland.ac.nz

I.J. Leitch et al. (eds.), *Plant Genome Diversity Volume 2*,
DOI 10.1007/978-3-7091-1160-4_14, © Springer-Verlag Wien 2013

14.1 Introduction

The extant gymnosperms are the modern representatives of the most ancient group of seed-bearing plants that first appeared in the Carboniferous, approximately 300 million years before the present. The gymnosperms have a worldwide distribution and form the dominant component of many temperate forests in both the Northern and Southern Hemispheres. They are also widely grown in plantation forests and provide the majority of timber for construction and wood pulp for paper making as well as drugs and other phytochemicals. They are grouped into four sub-classes, Ginkgooidae (one monotypic genus *Ginkgo*), Gnetidae (*Gnetum*, *Ephedra* and *Welwitschia*), Cycadidae (the cycads) and the Pinidae (the pines and other conifers). The first three of these sub-classes are relatively small with just one species in the Ginkgooidae, approximately 50 in the Gnetidae and about 250 in the Cycadidae; the majority of gymnosperms, some 550 species, belong to the Pinidae. Although most gymnosperms are substantial trees, with the redwood of California (*Sequoia sempervirens*) and kauris (*Agathis*) of South East Asia and Oceania as prominent examples, there are some like the pigmy pine, *Lepidothamnus laxifolius*, that are small shrubs that seldom grow to more than 1 m in height and some *Gnetum* species that are lianes. The evolutionary relationships and taxonomic positions of these four groups of gymnosperms remain to be fully resolved (Rai et al. 2008; Zgurski et al. 2008; Chase and Reveal 2009; Eckenwalder 2009). Some studies have placed Gnetales either as the sister group to all conifers ('gnetifer' hypothesis, Chaw et al. 2000) or within conifers as sister to cupressophytes (= 'gnecup' hypothesis; e.g., Chumley et al. 2008) or Pinaceae (= 'gnepine' hypothesis, e.g., Werner et al. 2009). In contrast, others have found little or no support for a sister-group relationship between the conifers and Gnetidae (Palmer et al. 2004). Within both the conifers and cycads there has also been disagreement on the number and relationships of their component families with, for example, the maintenance of Cephalotaxaceae and Phyllocladaceae as separate families (Page 1990) or their merger into Taxaceae and Podocarpaceae, respectively (Rai et al. 2008; Eckenwalder 2009). In the cycads Page (1990) recognizes four families whereas Chaw et al. (2005) recognize just the Cycadaceae and Zamiaceae as separate families. In this chapter the conifer classification of Eckenwalder (2009) and the cycad one of Chaw et al. (2005) have been adopted and the four groups of gymnosperms are considered as sub-classes as proposed by Chase and Reveal (2009).

14.2 Survey of Chromosome Number and Size Variation

The gymnosperms are probably the best studied group of land plants with regard to chromosome number. Table 14.1 lists the 83 genera in the four subclasses of gymnosperms and shows that chromosome numbers are known for some species in all but three of the genera, *Austrotaxus*, *Neocallitropsis* and *Papuacedrus*, all monotypic genera found either on New Caledonia or New Guinea.

One remarkable feature of this survey is the narrow range in overall chromosome numbers, from $2n = 14$ to 66, a c. five-fold variation; this contrasts with a range from $2n = 4$ to c. 640, a 160-fold variation in angiosperms (Weiss-Schneeweiss and Schneeweiss 2013). A closer examination of individual families, where there is more than one genus, (Table 14.1) shows an even greater uniformity. In Cupressaceae, 26 of the 27 genera that have been counted have the same basic number of $x = 11$, in Pinaceae nine of the 11 genera have $x = 12$, one, *Pseudolarix*, has $x = 11$ and the other, *Pseudotsuga* has basic numbers of both $x = 12$ and $x = 13$. Even in Podocarpaceae, the family with the widest range of basic numbers, these range only from $x = 9$ to $x = 14$. In cycads, Cycadaceae has eight species of *Cycas* with $2n = 22$ while Zamiaceae has a wider, though still not great, range of numbers from $2n = 16$ to $2n = 28$. The three families of Gnetidae, Ephedraceae, Gnetaceae and Welwitschiaceae are all monogeneric. *Ephedra* species are generally $2n = 14$ or $2n = 28$ although a few hexaploid and one octoploid cytotypes have been reported (see Leitch and Leitch 2013), *Gnetum* species are $2n = 22$ or $2n = 44$ while the single species of *Welwitschia* has $2n = 42$. The single species in Ginkgooidea, *Ginkgo biloba*, has $2n = 24$.

In the majority of gymnosperm genera, all species have the same basic number, the exceptions include *Halocarpus*, *Lepidothamnus*, *Nageia*, *Prumnopitys* and *Podocarpus* in the Podocarpaceae, *Fokenia* in Cupressaceae and *Zamia* in Zamiaceae. *Amentotaxus argotaenia* in Taxaceae is the only species in the Pinidae to show possible intraspecific variation in chromosome number, Chuang and Hu (1963) reported a count of $2n = 14$, Guan et al. (1993) reported $2n = 40$ and Zhou et al. (2000) found $2n = 36$. All three papers have clear photographic illustrations of the chromosomes so this variation is not easily explained. Zhou et al. (2000) suggested that the material examined by Chuang and Hu was *A. formosana*, however, Eckenwalder (2009) does not recognise this as a distinct species. Zhou et al. (2000) point to a similar number of chromosome arms, the *nombre fondamental* (NF; Matthey 1945), to explain the difference in their count from that of Guan et al. (1993). It would appear that more work is needed on the chromosomes of this species or species complex. In the Cycadidae, *Zamia loddigesii* provides a striking example of chromosome number and karyotype variation, with five different chromosome numbers and varying numbers of metacentrics and telocentrics in populations from the Yucatan peninsula (Vovides and Olivares 1996). This contrasts with the uniformity within a species found in other cycad genera.

Gymnosperm chromosomes are characteristically large (Fig. 14.1). A sample of measurements from three genera shows that the maximum and minimum metaphase chromosome lengths within each genus were 6.4–16.2 μm for 36 species of *Pinus*, 5.4–14.5 μm for 16 species of *Picea* and 4.4–11.6 μm for six species of *Larix* (Hizume 1988). A survey of genome size (C-value) variation (Murray 1998) found a relatively small range of values, 14.4-fold compared to nearly 2,400-fold in angiosperms (Pellicer et al. 2010; Leitch and Leitch 2013, this volume) and, with the exception of *Gnetum,* all species in that survey had 2C DNA amounts greater than 13 pg. Since then the amount of data has nearly doubled, with values now available for 204 species in the Gymnosperm DNA C-values database (Leitch et al. 2001; Murray et al. 2010). Yet despite this additional data, genome sizes still only range 16-fold from $2C = 5.5-72.0$ pg. These relatively large C-values together with the low chromosome numbers typical of most gymnosperms explain the large size of individual chromosomes in gymnosperm karyotypes.

14.3 Incidence of Polyploidy

Polyploidy is exceedingly rare in most gymnosperm subclasses and this has been commented upon by many authors over the years (Khoshoo 1959; Williams 2009; Fawcett et al. 2013, this volume). It is most widespread in Cupressaceae, with examples in *Cryptomeria*, *Fitzroya*, *Juniperus*, *Sequoia* and *Taiwania*. *Fitzroya* and *Sequoia* are both monotypic with $2n = 4x = 44$ and $2n = 6x = 66$ respectively, high numbers for gymnosperms. In contrast, *Cryptomeria*, *Juniperus* and *Taiwania* contain both diploid and polyploid species. *Pseudolarix* in Pinaceae, with $2n = 44$, is also probably a derived polyploid as all other species in the remaining 10 genera of the family have $2n = 24$. The chromosome numbers reported for

Table 14.1 Summary of chromosome numbers in gymnosperms. – = genera where no data are available

Subclass	Family	Genus	2*n* (No. of species counted)
Ginkgoidae	Ginkgoaceae	*Ginkgo*	24 (1)
Pinidae	Araucariaceae	*Agathis*	26 (2)
		Araucaria	26 (13)
		Wollemia	26 (1)
	Cupressaceae	*Actinostrobus*	22 (1)
		Athrotaxis	22 (3)
		Austrocedrus	22 (1)
		Callitris	22 (11)
		Calocedrus	22 (6)
		Chamaecyparis	22 (6)
		Cryptomeria	22, 33, 44 (2)
		Cunninghamia	22 (3)
		Cupressus	22 (12) 2 with Bs
		Diselma	22 (1)
		Fitzroya	44 (1)
		Fokienia	22, 24
		Glyptostrobus	22 (1)
		Juniperus	22 (15) 44 (2)
		Libocedrus	22 (3)
		Macrobiota	22 (2)
		Metasequoia	22 (1)
		Microbiota	22 (1)
		Neocallitropsis	–
		Papuacedrus	–
		Platycladus	22 (1)
		Sequoia	66 (1) 1 with Bs
		Sequoiadendron	22 (1)
		Taiwania	22, 33 (2)
		Taxodium	22 (3) 1 with Bs
		Tetraclinis	22 (1)
		Thuja	22 (4)
		Thujopsis	22 (1)
		Widdringtonia	22 (2)
	Podocarpaceae	*Acmopyle*	20 (1)
		Afrocarpus	24 (1)
		Dacrycarpus	20 (3)
		Dacrydium	20 (6)
		Falcatifolium	20 (1)
		Halocarpus	18 (1) 22 (1) 24 (1)
		Lagarostrobos	30 (1)
		Lepidothamnus	28 (1) 30 (2)
		Manoao	20 (1)
		Microcachrys	30 (1)
		Microstrobos	26 (2)
		Nageia	20 (1) 26 (1)
		Parasitaxus	36 (1)
		Phyllocladus	18 (3)
		Prumnopitys	36 (1) 38 (1)
		Podocarpus	34 (2) 38 (2)
		Retrophyllum	20 (3)
		Saxegothaea	24 (1)
	Sciadopityaceae	*Sciadopitys*	20 (1)
	Taxaceae	*Austrotaxus*	–

(continued)

Table 14.1 (continued)

Subclass	Family	Genus	2*n* (No. of species counted)
		Amentotaxus	14, 36, 40 (2)
		Cephalotaxus	24 (8)
		Pseudotaxus	24 (1)
		Taxus	24 (8) 1 with Bs
		Torreya	22 (3)
	Pinaceae	*Abies*	24 (23)
		Cathaya	24 (1)
		Cedrus	24 (2)
		Keteleeria	24 (2)
		Larix	24 (18)
		Nothotsuga	24 (1)
		Picea	24 (28) 6 with Bs
		Pinus	24 (68) 1 with Bs
		Pseudolarix	44 (2)
		Pseudotsuga	24 (3), 26 (1)
		Tsuga	24 (8)
Cycadidae	Cycadaceae	*Cycas*	22 (12)
	Zamiaceae	*Bowenia*	18 (2)
		Ceratozamia	16 (6)
		Chigua	18 (1)
		Dioon	18 (3)
		Encephalartos	18 (8)
		Lepidozamia	18 (2)
		Macrozamia	18 (9)
		Microcycas	26 (1)
		Stangeria	16 (1)
		Zamia	16–28 (35)
Gnetidae	Ephedraceae	*Ephedra*	14 (10) 28 (9) 14 & 28 (2)
			56 (2), 1 with Bs
	Gnetaceae	*Gnetum*	22 (1) 44 (1)
	Welwitschiaceae	*Welwitschia*	42 (1)

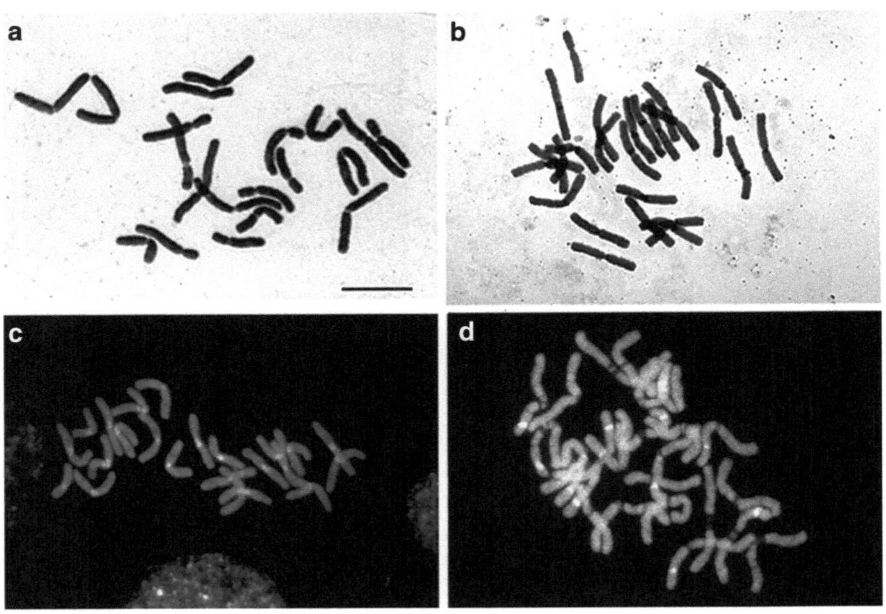

Fig. 14.1 (**a**) Mitotic chromosomes of *Manoao colensoi* showing a 'typical' gymnosperm complement comprising a low number of large metacentric and acrocentric chromosomes; (**b**) C-banded chromosomes of *Pinus radiata*; (**c**) CMA-banded chromosomes of *Pinus radiata*; (**d**) DAPI-banded chromosomes of *Pinus radiata*. Scale bar = 10 μm

Amentotaxus (Taxaceae) are $2n = 14$, 36 and 40 (Chuang and Hu 1963; Guan et al. 1993; Zhou et al. 2000) which is suggestive of polyploidy in one of the species. In Gnetidae there is one reported case of polyploidy in *Gnetum* while in *Ephedra* it appears more common with several species comprising both diploid and polyploid subspecies (Table 14.1). *Welwitschia* with $2n = 42$ would also appear to have undergone polyploidy in its ancestry (Khoshoo and Ahuja 1963) although available molecular data is equivocal (Cui et al. 2006).

14.4 B Chromosomes

Only 13 species spread across seven genera have been reported to have B chromosomes (Table 14.1). These are typically large, approximately 1/3–1/4 the length of the largest chromosome of the normal complement (Teoh and Rees 1977; Hizume et al. 1989, 1991). The relative rarity of B chromosomes in gymnosperms is interesting, given that Trivers et al. (2004) reported a positive correlation between the presence of B chromosomes and increasing genome size in angiosperms. Consequently, we might then expect gymnosperms, with characteristically large genome sizes, to have a large number of species with Bs. It is possible that B chromosomes have a more widespread systematic distribution because in many instances only a few individuals per species have been analysed. A more detailed discussion of B chromosomes in plant genomes is given by Houben et al. (2013, this volume).

14.5 Sex Chromosomes

Although the majority of gymnosperm species are hermaphrodite and monoecious, many are dioecious including *Ginkgo biloba*, all Cycadidae and Gnetidae as well as species scattered in many families of the Pinidae (Givnish 1980). In *Ginkgo*, claims have been made that the number of chromosomes with satellites (= secondary constrictions) differed between males and females and some of these findings have been summarized by Hizume (1997). However, he analysed 10 male and 15 female plants and found that while the number of satellites did vary, this variation was not related to the sex of the plant. Similar claims have been made about sex chromosomes in cycads (for example, Sangduen et al. 2009). A heteromorphic pair associated with sex differences was described in *Stangeria* by Marchant (1968) while a clear difference in one pair of chromosomes of *Cycas revoluta* (homomorphic in females and heteromorphic in males) was described by Segawa et al. (1971). In both examples the male is the heterogametic sex. However, some doubt surrounds these observations as subsequent studies in *Cycas revoluta* have not reported a heteromorphic pair in males (Hizume et al. 1992b; Hizume 1995), while in several *Zamia* species where both male and female plants have been analysed, no obvious chromosome differences between the sexes have been found (Tagashira and Kondo 1999). Nevertheless, Kokubugata and Kondo (1994) have shown that chromosome condensation in *Cycas revoluta*, which like most cycads has very large chromosomes, is highly variable and consequently the identification of homologous pairs in the absence of differential chromosome banding and FISH may be unreliable. More recently Hizume et al. (1998a) applied molecular cytogenetic techniques to this problem in *C. revoluta* and were able to distinguish a heteromorphic pair, one member of which lacked a large terminal segment of AT-rich, telomere-related DNA, in the karyotype and a heteromorphic bivalent at metaphase I, both present only in male plants. These results suggest an XYmale/XXfemale sex chromosome system may be present in this species. Other gymnosperms have different sex chromosome systems. For example, *Podocarpus macrophyllus* (native to southern Japan and China) is interesting in that it is the first example of an XXY sex chromosome system in gymnosperms (Hizume et al. 1988b). Males have $2n = 37$ and females have $2n = 38$, with a unique large, submetacentric chromosome present in all males but absent in females. In contrast, Davies et al. (1997) were unable to identify sex chromosomes in any of the New Zealand Podocarpaceae, several genera of which are dioecious. A more extensive discussion of plant sex chromosomes is given by Janoušek et al. (2013, this volume).

14.6 Karyotype Diversity and Chromosome Banding Patterns

Initial chromosome studies of many gymnosperm families using solid (Feulgen, orcein or carmine) staining suggested a high degree of uniformity in overall karyotype structure, However, more detailed studies, such as those of Hizume (1988), have shown that within this apparent uniformity there is significant, often genus-specific, variation. Hizume (1988) classified the karyotypes of Pinaceae into seven groups that differed in the degree of karyotype symmetry. For example, *Pinus* is characterized by species having large metacentric chromosomes with one or two pairs of smaller, often sub-metacentric ones. In contrast, other genera such as *Larix* and *Pseudolarix* show karyotypes with varying degrees of bimodality.

Despite their overall structural similarity, studies using differential chromosome banding and fluorescence *in situ* hybridization (FISH) have shown that the chromosomes of species in genera like *Pinus* are far from uniform (Figs. 14.1 and 14.2).

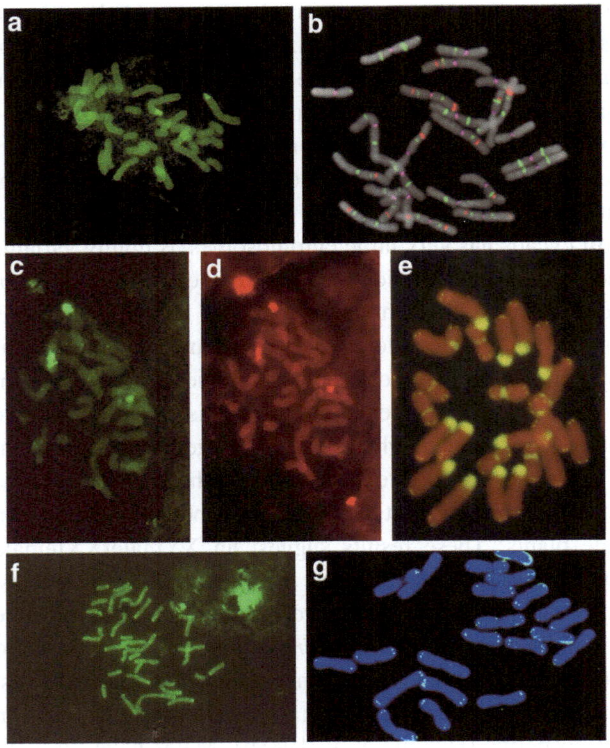

Fig. 14.2 *In situ* hybridization of repetitive DNA sequences to gymnosperm chromosomes. (**a**) 45S rDNA in *Podocarpus totara*; (**b**) 45S and 5S rDNA (*red*), '*Arabidopsis* type' telomere repeat (*green*) and a centromeric repeat, *PCSR*, (*magenta*) on chromosomes of *Pinus densiflora*; (**c** and **d**) colocalization of the 45S (**c**) and 5S (**d**) rDNA hybridization sites in the F_1 hybrid *Podocarpus lawrencei* × *P. nivalis*; (**e**, **f** and **g**) the '*Arabidopsis* type' telomere repeat in *Cycas revoluta;* (**e**) *Podocarpus totara* (**f**) and *Cryptomeria japonica* (**g**) [(a, c, d and f) from Murray et al. 2002 by permission of Oxford University Press; (b, e and g) with permission from Dr M Hizume and the Editors of Chromosome Science]

This was first demonstrated following studies which showed that the Giemsa C-banding technique could be used to locate heterochromatin in gymnosperms such as *Cycas revoluta*, *Ginkgo biloba* and *Pinus densiflora* (Tanaka and Hizume 1980). For example, MacPherson and Filion (1981) examined five species of *Pinus*, two in sub-genus *Strobus* and three in sub-genus *Pinus,* and found that pericentromeric C-bands were limited to species in sub-genus *Pinus*. Further differences between the five species were seen in the number and position of intercalary bands. In other *Pinus* species differences between C-banding patterns have also been reported, including examples where no consistent C-bands were seen (Kupila-Ahvenniemi and Hohtola 1977; Jacobs et al. 2000).

Many more species of gymnosperm have been investigated with the base-specific fluorochromes chromomycin A_3 (CMA) and 4′,6-diamidino-2-phenylindole (DAPI) that stain preferentially GC or AT rich regions respectively of heterochromatin in the chromosomes (Fig. 14.1c, d). These staining techniques have revealed that the amount and distribution of these two types of heterochromatin can vary greatly. Only a limited number of bands are seen in some species such as *Ginkgo biloba*, which shows just four CMA-positive bands that are also C-bands (Hizume 1997). Similarly, in eight species of Taxodiaceae, Hizume et al. (1988a) found no DAPI bands and six species had only two or four prominent CMA bands. An absence of DAPI bands was also reported in all the New Zealand endemic gymnosperms belonging to three families, Araucariaceae, Cupressaceae and Podocarpaceae, and each species had just a single, prominent CMA-positive band (Davies et al. 1997). In contrast, two other Cupressaceae species *Cunninghamia lanceolata* and *Metasequoia glyptostroboides* showed small bands at all their centromeres in addition to two terminal or six proximal large, prominent bands, respectively. Many species of *Pinus* and *Larix* also have large numbers of either or both CMA and DAPI bands. For example, recent studies of *Pinus heldreichii* and *P. taeda* have demonstrated multiple DAPI bands distributed on all chromosomes (Bogunic et al. 2006; Islam-Faridi et al. 2007). Similarly, Hizume et al. (1993) studied six European *Larix* species and found they all had numerous CMA and DAPI bands although their number and position differed between species.

In the Cycadidae, as in Pinidae, there are contrasting patterns of fluorescent chromosome banding between species and genera. Kokubugata and Kondo (1996) found that four species of *Cycas* with essentially similar karyotypes showed a common pattern of DAPI and CMA bands, with DAPI bands at the centromeres of all chromosomes and CMA bands widely distributed on most chromosomes at one or both telomeres. A few interstitial CMA bands were also seen on different chromosomes of the various species. In subsequent studies of Zamiaceae (Kokubugata and Kondo 1998; Tagashira and Kondo 1999), no DAPI bands were observed in *Microcycas calocoma* but they were seen as small dots at the centromeres of all chromosomes of *Ceratozamia mexicana* and at four to 12 centromeres of nine different species of *Zamia*. Only two CMA bands were seen in a centromeric location on the single metacentric pair of chromosomes in *M. calocoma* but in *C. mexicana* the CMA bands were located terminally on seven chromosomes. CMA bands in *Zamia* are very variable; in some species such as *Z. muricata* 20 CMA bands have been observed whereas in *Z. angustifolia* only six are reported (Tagashira and Kondo 1999).

Intraspecific variation in the number and size of CMA bands has been reported in several species of Cupressaceae (Hizume et al. 1988a), Pinaceae (Hizume and Tanaka 1990; Hizume and Akiyama 1992) and Podocarpaceae (Davies et al. 1997). For example, in *Pseudotsuga menziesii* (Pinaceae) and *Dacrydium cupressinum* (Podocarpaceae) bands in comparable positions on homologous chromosomes show clear size differences (Hizume and Akiyama 1992; Davies et al. 1997) and in *Picea brachytyla* var. *complanata* (Pinaceae) a survey of 35 seedlings showed that though all shared some common bands others had additional unique bands (Hizume et al. 1991).

14.7 *In situ* Hybridization

Fluorescence *in situ* hybridization (FISH) has been used extensively to locate a variety of repetitive DNA sequences on chromosomes of *Ginkgo biloba* and many species in Pinidae and Cycadidae.

14.7.1 Ribosomal DNA (rDNA)

Both the 18S-5.8S-26S (denoted hereafter as 45S) rDNA and 5S rDNA repeats have been mapped in a huge variety of gymnosperms from all the subclasses except Gnetidae (Table 14.2).

In many cases this has enabled the identification of individual chromosome pairs in the often uniform gymnosperm karyotype. The number and location of 45S sites varies greatly (Fig. 14.2) from a single pair in some *Podocarpus* (Murray et al. 2002), cycad (Kokubugata et al. 2000) and *Larix* (Hizume et al. 1995) species to a massive 38 sites in *Pinus taeda* (Islam-Faridi et al. 2007). The size of the hybridization sites also varies greatly both between chromosomes within a species and also between species. Many are large and are usually at an interstitial position, though in some cycads and *Podocarpus* species they are adjacent to the telomeres (Kokubugata and Kondo 1998; Kokubugata et al. 2000; Murray et al. 2002) (Fig. 14.2). Small, often variable sites, in addition to the major ones, have been reported at the centromeres of some *Pinus* species (Fig. 14.2) (Liu et al. 2003; Islam-Faridi et al. 2007) and this location has been confirmed by Hizume et al. (2001) who used microdissection followed by degenerate oligonucleotide-primed PCR and cloning to identify these sequences and use them as probes in FISH experiments on the chromosomes of *Pinus densiflora*.

There is considerably less variation in the number of 5S sites with two or four being the predominant number (Table 14.2). The 5S and 45S rDNA sites often occur on the same chromosome and Puizina et al. (2008) have summarized the linkage patterns of these sites in several genera of the Pinaceae. They found that *Pinus* shows the greatest diversity of patterns and that all the patterns seen in other genera, such as *Picea*, *Abies*, *Pseudotsuga* and *Larix*, are also found in *Pinus*. In several species of *Podocarpus*, the 45S and 5S signals colocalize to very similar, if not the same site in an interspersed fashion on one pair of chromosomes (Fig. 14.2c, d) (Murray et al. 2002). This pattern of linkage has also been described in liverworts (Sone et al. 1999) and angiosperms (Garcia et al. 2007). The diversity of rDNA locations is probably a consequence of their movement by transposition (Datson and Murray 2006) as meiotic analyses (see Sect. 14.8 below) suggest little structural rearrangement in the evolution of gymnosperm genomes.

14.7.2 Telomere Sequences

The presence of *Arabidopsis*-type telomere sequences ($TTTAGGG_n$) in gymnosperms was first demonstrated using FISH in *Pinus sylvestris* and *Zamia furfuracea* (Fuchs et al. 1995). They found the expected hybridization sites at the ends of all the chromosomes of both species but in *P. sylvestris* large interstitial sites were also seen. Since then other species have been investigated and a variety of patterns has emerged. In *Cycas revoluta* all chromosomes have terminal sites that are conspicuously large on the telocentric chromosomes of the complement. In addition there are sites at the centromeres of the acrocentric chromosomes (Hizume et al. 1998a; Fig. 14.2e). Large interstitial signals, up to four on some chromosomes, have since been reported in all *Pinus* species examined (Lubaretz et al. 1996; Hizume et al. 2000; Schmidt et al. 2000; Hizume et al. 2002a; Islam-Faridi et al. 2007). In four species of *Podocarpus*, Murray et al. (2002) found terminal sites and a dispersed pattern of small sites scattered along the length of all chromosomes (Fig. 14.2f). Sites at the ends of telocentric chromosomes were not conspicuously different from those at the ends of non-telocentric chromosomes in these species. Chromosomes with only terminal hybridization sites have been reported in *Abies alba*, *Cryptomeria japonica*, *Ginkgo biloba*, *Larix decidua* and *Picea abies* (Fig. 14.2g) (Lubaretz et al. 1996; Hizume et al. 2000; Puizina et al. 2008). Albeit with a limited sample of species, it would appear as though *Pinus* amongst the gymnosperms is, to date, the only genus with exceptionally large and conspicuous interstitial telomere hybridization sites (e.g., Fig. 14.2b).

14.7.3 Other Repeats

The highly repetitive nature of DNA comprising the genomes of gymnosperms first became apparent following studies on the reassociation kinetics of DNA from *Cycas revoluta* and several species in Cupressaceae and Pinaceae (Miksche and Hotta 1973; Rake et al. 1980; Kurdi-Haidar et al. 1983; Kriebel 1985). A surprising finding from these studies was how large the percentage of 'unique' sequences was of the total genome, (i.e., in the region of 20–30%). Nevertheless

Table 14.2 The total number of 45S and 5S rDNA hybridization sites detected by FISH in different gymnosperm genomes

Species	45S	5S	Reference
Abies alba	10	4	Shibata et al. (2004), Puizina et al.(2008)
Abies nordmania	10	4	Shibata et al. (2004)
Abies pinsapo	10	4	Shibata et al. (2004)
Abies sachalinensis	8	4	Shibata et al. (2004)
Bowenia serrulata	4	–	Kokubugata et al. (2000)
Bowenia spectabilis	2	–	Kokubugata et al. (2000)
Bowenia sp. 'Tinaroo'	3	–	Kokubugata et al. (2000)
Ceratozamia hildae	–	2	Kokubugata et al. (2004)
Ceratozamia kuesteriana	–	2	Kokubugata et al. (2004)
Ceratozamia mexicana	7/8	2	Kokubugata and Kondo (1998), Tagashira and Kondo (2001), Kokubugata et al. (2004)
Ceratozamia norstogii	–	2	Kokubugata et al. (2004)
Cryptomeria japonica	2, 4	–	Hizume et al. (1998b)
Cycas revoluta	16	2	Hizume et al. (1992b), Hizume (1995)
Ginkgo biloba	4	–	Hizume (1997)
Larix decidua	6	2	Lubaretz et al. (1996)
Larix potaninii var. *macrocarpa*	2	2	Hizume et al. (1995)
Microcycas calocoma	2	–	Kokubugata and Kondo (1998)
Picea abies	12	2	Lubaretz et al. (1996), Siljak-Yakovlev et al. (2002)
Picea crassifolia	10[a]	–	Hizume et al. (1999)
Picea jezoenisi var. *hondoensis*	10	2	Hizume and Kuzukawa (1995)
Picea glauca	14	4	Brown et al. (1993), Brown and Carlson (1997)
Picea koraiensis	10[a]	4	Hizume et al. (1999)
Picea omorika	16	2	Siljak-Yakovlev et al. (2002)
Picea sitchensis	10	4	Brown and Carlson (1997)
Pinus densata	18	3	Liu et al. (2003)
Pinus densiflora	14	4	Hizume et al. (1992a, 2002a)
Pinus elliottii	16	6	Doudrick et al. (1995)
Pinus massoniana	20	2	Liu et al. (2003)
Pinus merkusii	16	4	Liu et al. (2003)
Pinus nigra	18	4	Hizume et al. (2002a)
Pinus radiata	20	4	Jacobs et al. (2000)
Pinus sylvestris	14	4	Lubaretz et al. (1996), Hizume et al. (2002a)
Pinus tabuliformis	24	4	Liu et al. (2003)
Pinus taeda	38	4	Jacobs et al. (2000), Islam-Faridi et al. (2007)
Pinus thunbergii	12, 14	4	Hizume et al. (1992a, 2002a)
Pinus yunanensis	20	4	Liu et al. (2003)
Podocarpus acutifolius	2	2	Murray et al. (2002)
Podocarpus lawrencei	4	2	Murray et al. (2002)
Podocarpus nivalis	2	2	Murray et al. (2002)
Podocarpus totara	2	2	Murray et al. (2002)
Pseudotsuga menziesii	6	2	Hizume et al. (1996)
Stangeria eriopus	16	2	Kokubugata et al. (2002, 2004)
Zamia angustifolia	6	2	Tagashira and Kondo (2001)
Zamia integrifolia	6	2	Tagashira and Kondo (2001)
Zamia pumila	7	2	Tagashira and Kondo (2001)
Zamia pygmaea	6	2	Tagashira and Kondo (2001)
Zamia furfuracea	14	2	Tagashira and Kondo (2001)
Zamia loddigesii	14	2	Tagashira and Kondo (2001)
Zamia skinneri	6	2	Tagashira and Kondo (2001)
Zamia vazquezii	10	2	Tagashira and Kondo (2001)
Zamia muricata	20	2	Tagashira and Kondo (2001)

[a] Additional weak signals observed in some plants

the repetitive fraction still comprised a very significant component of the genome. Elsik and Williams (2000) suggested that the high frequency of 'unique' sequences was due to the presence of divergent retrotransposons which contributed to the excess low-copy DNA and indeed, recent sequence data support this. In a large scale transcriptome sequencing project of *Pinus contorta* Parchman et al. (2010) identified a surprisingly large number of transcriptionally-active retrotransposon sequences in their sample. While in *Pinus taeda* Morse et al. (2009) uncovered large numbers of *gypsy* and *copia* retrotransposons in the low copy fraction of a Cot-based fractionation of genomic DNA. Further verification of Elsik and Williams' proposal and a clearer idea of the contribution and composition of 'unique' sequences in gymnosperm genomes will no doubt come as high throughput sequencing technologies are increasingly applied to gymnosperms.

Cafasso et al. (2003) isolated an unusual GC-rich satellite repeat from *Zamia paucijuga* that was shown to be located in the sub-terminal regions of most chromosomes using FISH. The repeat also existed as a dispersed repeat that comprised two to four satellite repeats flanked by a 0.6 kb AT-rich repeat. These dispersed elements were found in 30 other *Zamia* species but not in *Ceratozamia* and *Microcycas* and the amount and dispersal pattern of the repeats differed between *Zamia* species (Cafasso et al. 2009). In some species the repeats were at low or almost undetectable levels whereas other showed very strong signals on Southern blots and they often showed evidence of secondary amplification events.

In contrast to these cycad results, Brown et al. (1998) identified an AT-rich satellite sequence from *Picea glauca* that had a centromeric distribution on a sub-set of chromosomes of *P. glauca* and *P. sitchensis*. Southern hybridization showed that the sequence was present in 18 *Picea* species but absent from *Pinus*, *Pseudotsuga* and *Thuja* and that amongst *Picea* species there was little evidence of structural alteration of the sequence. Subsequently, Hizume et al. (1999) used the sequence as a probe onto chromosomes of two Chinese *Picea* species and found clear hybridization signals at the centromeres of three pairs of chromosomes, fewer than the four or five pairs found by Brown et al. (1998). Hizume et al. (2002b) have since found a similar distribution pattern for two AT-rich sequences, one a 170 bp repeat and the other a 220 bp repeat that was a partial duplication of it, both located near to the centromeres of 22 chromosomes in *Larix leptolepis* but only 14 chromosomes of *L. chinensis*.

Various microsatellite or simple sequence repeats (SSRs) have been identified and in some cases mapped to chromosomes of Podocarpaceae and Pinaceae. Many sequences represent a large fraction of the repetitive DNA in the genome and have dispersed FISH hybridization patterns showing that the sequences are widely distributed across the chromosomes (Schmidt et al. 2000; Murray et al. 2002). In several species, standard telomere repeats have given rise to mutated or degenerate copies that may have non-terminal chromosomal locations and some are present at very high repeat frequency and constitute significant chromosome landmarks (Schmidt et al. 2000; Shibata et al. 2005).

The distribution and evolution of retrotransposons has also been studied in gymnosperms and, as noted above, both *gypsy* and *copia* elements have been found (Kamm et al. 1996; Brandes et al. 1997; Friesen et al. 2001). When hybridized to *Pinus elliottii* chromosomes, a cloned *Ty1-copia*-like element showed a dispersed pattern of hybridization over all the chromosomes, with the exception of centromeres and nucleolar organizing regions (NORs) (Kamm et al. 1996). Friesen et al. (2001) looked at the distribution of various *gypsy* and *copia* elements in *Picea abies* and *Pinus pinaster*. While some *gypsy* elements showed stronger signals at the chromosome ends (e.g., *Pagy11* in *Picea abies*) or associated with 18S rDNA and centromeric regions (e.g., *Ppgy1* in *Pinus pinaster*) most showed a dispersed pattern of hybridization, similar to that reported for another *gypsy* element (*Gymny*) in *Pinus taeda* (Morse et al. 2009). Indeed, such a dispersed distribution pattern which has been reported for many other pine *gypsy* elements contrasts with observations in angiosperms where most *gypsy* elements appear localized in the centromeric and pericentromeric regions (Grover and Wendel 2010). Such observations hint at fundamental differences in the organization of DNA between angiosperms and gymnosperms and this has recently been discussed by Leitch and Leitch (2013).

14.8 Karyotype Homology Across Species and Genera

The uniformity of Pinidae karyotypes begs the question of whether the similarity is superficial, with overall chromosome numbers and morphology being conserved, or whether variation is widespread but karyotype orthoselection (i.e., the maintenance of karyotype uniformity through the occurrence of characteristic structural mutations; White 1973) masks extensive DNA sequence rearrangements. As discussed above, studies using chromosome banding and FISH have revealed extensive variation in the location of a variety of repetitive DNA sequences, but to assess whether chromosomal conservation at a large syntenic scale has occurred requires evidence either from comparative linkage analysis using orthologous loci or the analysis of chromosome pairing at meiotic prophase I/metaphase I in interspecific hybrids. Available data from both approaches support the general idea of karyotype conservation across specific and generic divides. For example, fertile hybrids between the Japanese *Larix kaempferi* (syn. *L. leptolepis*) and *L. decidua*, from Northern and Central Europe, showed regular meiotic pairing with a similar chiasmata and univalent frequency to the parental species (Sax 1932). *L. kaempferi* has also been crossed successfully with *L. sibirica* and *L. gmelenii* (Eckenwalder 2009). Many more hybrid combinations have been studied in *Pinus* (Sax 1960; Saylor and Smith 1966; Williams et al. 2002) and these also show that there have been few major structural changes in the evolution of *Pinus* karyotypes. In Podocarpaceae, introgressive hybridization has been reported between *Podocarpus totara* and *P. acutifolius*. This implies that the hybrids are fertile with presumably regular meiosis and hence karyotype uniformity (Wardle 1972). Quinn and Rattenbury (1972) studied F_1 hybrids between *Dacrydium laxifolium* and *D. intermedium* and observed mostly bivalent formation with two to four univalents in 75% of cells and one or two dicentric bridges and associated fragments in a similar percentage of anaphase I cells. They concluded that these two species were basically similar in karyotype structure but differentiated by two paracentric inversions. None of these gymnosperm hybrids showed any evidence of chromosome translocations.

Genetic maps have been produced for several species of *Pinus* and these show a remarkable conservation of gene distribution and order between species. These data have enabled comparative analyses to be made between various *Pinus* genomes and that of *Pseudotsuga menziesii* (Krutovsky et al. 2004). *Pseudotsuga menziesii* differs from all other Pinaceae in having 26 as opposed to 24 chromosomes, it has 11 pairs of metacentrics/sub-metacentrics and two pairs of telocentrics compared with 12 pairs of metacentrics/sub-metacentrics found in all other species. Comparative mapping has shown that overall there is conservation of syntenic loci between the species of *Pinus* analysed as well as between *P. taeda* and *P. menziesi*, thus providing further support for the proposal that Pinaceae genomes have been remarkably stable over c. 45 million years, the time when *Pseudotsuga* first appeared in the fossil record. It is possible that there has been even greater stability as fossils corresponding to the subsections of *Pinus*, which share common karyotypes, have been found in Cretaceous floras over 65 million years old (Stockey and Nishida 1986; Miller 1993).

The Cycadidae provide contrasting patterns of karyotype stability in different genera. *Lepidozamia*, *Macrozamia*, *Encephalartos* and *Dioon*, for example, all have the same chromosome number, $2n = 18$, and essentially similar karyotypes, though *Encephalartos* lacks a pair of telocentric chromosomes in its complement which are characteristic of the other three genera (Marchant 1968; Kokubugata et al. 1999; Sangduen et al. 2009). *Cycas* species also show essentially common patterns of fluorochrome banding (Kokubugata and Kondo 1996). In contrast, *Zamia* is chromosomally variable with numbers ranging from $2n = 16$ to $2n = 28$. Much of this variation has been attributed to centric fission or fusion as the species with lower numbers of chromosomes have more metacentrics and fewer telocentrics compared with species with higher chromosome numbers which show the reverse pattern. However the direction of change remains controversial (Marchant 1968; Moretti 1990; Caputo et al. 1996). Whether there is large-scale conservation of gene order in cycad chromosomes is not known as no large-scale genomic mapping exercises have been carried out.

Conclusions

The overall uniformity of chromosome number and karyotype in the major groups of gymnosperms is in striking contrast to the situation in angiosperms (Weiss-Schneeweiss and Schneeweiss 2013, this volume). The observation of conserved biological form and function is strongly susceptible to an adaptive interpretation but an obvious selective explanation is lacking for gymnosperm karyotypes. White (1973) differentiated between karyotype orthoselection, the maintenance of karyotype uniformity through the occurrence of characteristic structural mutations, and karyotype conservation, the maintenance of similar karyotype morphology in different taxa through a lack of structural mutations. Karyotype orthoselection has been put forward as an explanation (Williams 2009), however it would appear that it is karyotype conservation that is important here as there is little evidence for widespread structural change in most gymnosperms (with the exception of some cycads). What the selective mechanism underlying gymnosperm karyotype conservation is remains a significant goal for understanding the evolution of gymnosperm karyotypes.

Nevertheless, the application of molecular cytogenetic techniques to gymnosperm genomes has revealed considerable variation in the location and composition of a wide variety of repetitive DNA sequences and this variation has proved to be invaluable for karyotyping and chromosome identification. This is amply illustrated by two examples from *Pinus*. In *P. densiflora* Hizume et al. (2002a) and Shibata et al. (2005) used FISH to map a variety of different repeats to the chromosomes (Fig. 14.2b) and this has allowed the identification of all members of the complement. Similarly, Islam-Faridi et al. (2007) have produced a reference karyotype for *P. taeda* and provided a key for the identification of each member of the complement. Although both these studies have used many comparable sequences Islam-Faridi et al. (2007) point out that it is still very difficult to identify homologous chromosomes between *P. taeda* and *P. densiflora*. The development of a wider range of probes and their application to a range of species should better facilitate the comparative physical mapping of gymnosperm chromosomes. Mapping studies suggest the conservation of syntenic groups of genes (though *P. densiflora* has not been included in such studies) across many millions of years (Krutovsky et al. 2004) but repeats such as the rDNA sequences do not appear to be under the same constraints.

The increasing speed and falling costs of high-throughput genome sequencing approaches are expected to ultimately lead to the complete sequencing of genomes such as those of *Pinus* and *Picea*, despite their large size. Nevertheless, already transcriptome sequencing data are available for some gymnosperm species (e.g., *Pinus contorta*—Parchman et al. 2010, *Ginkgo biloba*—Brenner et al. 2005) and these are providing a wealth of novel data on the functioning and evolution of gymnosperm genomes (Barker et al. 2010). Such data together with improved understanding of the phylogenetic relationships between different gymnosperm lineages will provide enhanced insights into the evolutionary processes which have shaped the karyotypes of these economically very important plants.

Acknowledgments I would like to thank Dr Masahiro Hizume for allowing me to use some of his images to illustrate this chapter.

References

Barker MS, Graham SW, Rieseberg LH (2010) Comparative gymnosperm transcriptomics. Abstract for Botany 2010 meeting, USA. http://2010.botanyconference.org/engine/search/index.php?func=detail&aid=437, Accessed on 18 June 2011

Bogunic F, Muratovic E, Siljak-Yakovlev S (2006) Chromosomal differentiation between *Pinus heldreichii* and *Pinus nigra*. Ann For Sci 63:267–274

Brandes A, Heslop-Harrison JS, Kamm A, Kubis S, Doudrick RL, Schmidt T (1997) Comparative analysis of the chromosomal and genomic organization of *Ty1-copia*-like retrotransposons in pteridophytes, gymnosperms and angiosperms. Plant Mol Biol 33:11–21

Brenner ED, Katari MS, Stevenson DW, Rudd SA, Douglas AW, Moss WN, Twigg RW, Runko SJ, Stellari GM, McCombie WR, Coruzzi GM (2005) EST analysis in *Ginkgo biloba*: an assessment of conserved developmental regulators and gymnosperm specific genes. BMC Genomics 6:143

Brown GR, Carlson JE (1997) Molecular cytogenetics of the genes encoding 18S-5.8S-26S rRNA and 5S rRNA in two species of spruce (*Picea*). Theor Appl Genet 95:1–9

Brown GR, Amarasinghe V, Kiss G, Carlson JE (1993) Preliminary karyotype and chromosomal location of ribosomal DNA in white spruce using fluorescence in situ hybridization. Genome 36:310–316

Brown GR, Newton CH, Carlson JE (1998) Organization and distribution of a *Sau*3A tandem repeated DNA sequence in *Picea* (Pinaceae) species. Genome 41:560–565

Cafasso D, Cozzolino S, Chinali G, De Luca P (2003) An unusual satellite DNA from *Zamia paucijuga* (Cycadales) characterized by two different organizations of the repetitive unit in the plant genome. Gene 311:69–77

Cafasso D, Cozzolino S, Vereecken NJ, De Luca P, Chinali G (2009) Organization of a dispersed repeated DNA element in the *Zamia* genome. Biol Plant 53:28–36

Caputo P, Cozzolino S, Gaudio L, Moretti A, Stevenson DW (1996) Karyology and phylogeny of some Mesoamerican species of *Zamia* (Zamiaceae). Am J Bot 83:1513–1520

Chase MW, Reveal JL (2009) A phylogenetic classification to accompany APG III. Bot J Linn Soc 161:122–127

Chaw SM, Parkinson CL, Cheng YC, Vincent TM, Palmer JD (2000) Seed plant phylogeny inferred from all three plant genomes: monophyly of extant gymnosperms and origin of Gnetales from conifers. Proc Natl Acad Sci USA 97:4086–4091

Chaw S-M, Walters TW, Chang C-C, Hu S-H, Chen S-H (2005) A phylogeny of cycads (Cycadales) inferred from chloroplast *matK* gene, *trnK* intron, and nuclear rDNA ITS region. Mol Phylogenet Evol 37:214–234

Chuang T-I, Hu WWL (1963) Study of *Amentotaxus argotaenia* (Hance) Pilger. Bot Bull Acad Sin 4:10–14

Chumley TW, McCoy SKR, Raubeson LA (2008) Gne-deep: exploring Gnetalean affinities in seed plant phylogeny with 83 plastid genes. Abstract for Botany 2008 meeting, USA.http://2008.botanyconference.org/engine/search/index.php?func=detail&aid=770

Cui L, Wall PK, Leebens-Mack JH, Lindsay BG, Soltis DE et al. (2006) Widespread genome duplications throughout the history of flowering plants. Genome Res 16:738–749

Datson PM, Murray BG (2006) Ribosomal DNA locus evolution in *Nemesia*: transposition rather than structural rearrangement as the key mechanism? Chromosome Res 14:845–867

Davies BJ, O'Brien IEW, Murray BG (1997) Karyotypes, chromosome bands and genome size variation in New Zealand endemic gymnosperms. Plant Syst Evol 208:169–185

Doudrick RL, Heslop-Harrison JS, Nelson CD, Schmidt T, Nance WL, Schwarzacher T (1995) Karyotype of slash pine (*Pinus elliottii* var. *elliottii*) using patterns of fluorescence in situ hybridization and fluorochrome banding. J Hered 86:289–296

Eckenwalder JE (2009) Conifers of the world. Timber Press, Portland

Elsik CG, Williams CG (2000) Retroelements contribute to the excess low-copy-number DNA in pine. Mol Gen Genet 264:47–55

Fawcett JA, Van de Peer Y, Maere S (2013) Significance and biological consequences of polyploidization in land plants. In: Leitch IJ, Greilhuber J, Doležel J, Wendel JF (eds) Plant genome diversity, vol 2, Physical structure, behaviour and evolution of plant genomes. Springer-Verlag, Wien, pp 277–293

Friesen N, Brandes A, Heslop-Harrison JS (2001) Diversity, origin and distribution of retrotransposons (*gypsy* and *copia*) in conifers. Mol Biol Evol 18:1176–1188

Fuchs J, Brandes A, Schubert I (1995) Telomere sequence localization and karyotype evolution in higher plants. Plant Syst Evol 196:227–241

Garcia S, Garnatje T, Hidalgo O, McArthur ED, Siljak-Yakovlev S, Vallès J (2007) Extensive ribosomal DNA (18S-5.8S-26S and 5S) colocalization in the North American endemic sagebushes (subgenus *tridentatae*, *Artimisia*, Asteraceae) revealed by FISH. Plant Syst Evol 267:79–92

Givnish TJ (1980) Ecological constraints on the evolution of breeding systems in seed plants: dioecy and dispersal in gymnosperms. Evolution 34:959–972

Grover C, Wendel JF (2010) Recent insights into mechanisms of genome size change in plants. J Bot 2010:1–8, Article ID 382732

Guan Q-L, Yu Z-L, Feng Y-L (1993) Studies of chromosomes in *Amentotaxus argotaenia*. Acta Bot Yunnan 15:385–391

Hizume M (1988) Karyomorphological studies in the family Pinaceae. Mem Fac Educ Ehime Univ Nat Sci 8:1–108

Hizume M (1995) Physical mapping of 5S rRNA genes in *Cycas revoluta*. Cytologia 60:389–393
Hizume M (1997) Chromosomes of *Ginkgo biloba*. In: Hori T et al. (eds) *Ginkgo biloba*—a global treasure. Springer, Tokyo, pp 109–118
Hizume M, Akiyama M (1992) Size variation of chromomycin A_3-band in chromosomes of Douglas fir, *Pseudotsuga menziesii*. Jpn J Genet 67:425–435
Hizume M, Kuzukawa Y (1995) Fluorescent chromosome banding in *Picea* II. Relationships between rDNA loci and chromomycin A3-bands in somatic chromosomes of *Picea jezoensis* var. *hondoensis*. La Kromosomo II 79–80:2754–2759
Hizume M, Tanaka A (1990) Fluorescent chromosome bandings in two American larches, *Larix occidentalis* and *L. laricina*. La Kromosomo II 58:1979–1987
Hizume M, Abe KK, Tanaka A (1988a) Fluorescent chromosome banding in the Taxodiaceae. La Kromosomo II 50:1609–1619
Hizume M, Shiraishi H, Tanaka A (1988b) A cytological study of *Podocarpus macrophyllus* with special reference to sex chromosomes. Jpn J Genet 63:413–423
Hizume M, Kishimoto K, Kubo Y, Tanaka A (1989) Fluorescent chromosome banding in *Picea* I. Differences in chromomycin A_3 band pattern between *P. jesoensis* var. *jesoensis* and *P. jesoensis* var. *hondoensis*. La Kromosomo II 53:1736–1744
Hizume M, Kitazawa N, Gu Z, Kondo K (1991) Variation of fluorescent chromosome band in *Picea brachytyla* var. *complanata* collected in Yunnan, China. La Kromosomo II 63–64:2149–2158
Hizume M, Ishida F, Murata M (1992a) Multiple locations of the rRNA genes in chromosomes of pines, *Pinus densiflora* and *P. thunbergii*. Jpn J Genet 67:389–396
Hizume M, Ishida F, Kondo K (1992b) Differential staining and in situ hybridization of nucleolar organizers and centromeres in *Cycas revoluta* chromosomes. Jpn J Genet 67:381–387
Hizume M, Tominaga HH, Kondo K, Gu Z, Yue Z (1993) Fluorescent chromosome banding in six taxa of Eurasian *Larix*, Pinaceae. La Kromosomo II 69:2342–2354
Hizume M, Kuzukawa Y, Kondo T, Yang Q, Hong D, Tanaka R (1995) Localization of rDNA and fluorescent bandings in chromosomes of *Larix potaninii* var. *macrocarpa* collected in Sichuan, China. La Kromosomo II 78:2689–2694
Hizume M, Kuzukawa Y, Kondo T (1996) Physical mapping of 5S rDNA locus on chromosomes in *Pseudotsuga menziesii*, Pinaceae. La Kromosomo II 83–84:2901–2908
Hizume M, Kurose N, Shibata F, Kondo K (1998a) Molecular cytogenetic studies on sex chromosomes and proximal heterochromatin containing telomere-like sequences in *Cycas revoluta*. Chromosome Sci 2:63–72
Hizume M, Shibata F, Kondo T (1998b) Fluorescence in situ hybridization of ribosomal RNA gene in *Cryptomeria japonica*, Taxodiaceae. Chromosome Sci 2:99–102
Hizume M, Shibata F, Kondo K, Hoshi Y, Kondo T, Ge S, Yang Q, Hong D (1999) Identification of chromosomes in two Chinese spruce species by multicolour fluorescence in situ hybridization. Chromosome Sci 3:37–41
Hizume M, Shibata F, Matsusaki Y, Kondo K (2000) Chromosomal localization of telomere sequence repeats in five gymnosperm species. Chromosome Sci 4:39–42
Hizume M, Shibata F, Maruyama Y, Kondo T (2001) Cloning of DNA sequences localized on proximal fluorescent chromosome bands by microdissection in *Pinus densiflora* Sieb. Zucc Chromosoma 110:345–351
Hizume M, Shibata F, Matsusaki Y, Garajova Z (2002a) Chromosome identification and comparative karyotypic analyses of *Pinus* species. Theor Appl Genet 105:491–497
Hizume M, Shibata F, Matsumoto A, Maruyama Y, Hayashi E, Kondo T, Kondo K, Zhang S, Hong D (2002b) Tandem repeat DNA localizing on the proximal DAPI bands of chromosomes in *Larix*, Pinaceae. Genome 45:777–783
Houben A, Moghaddam AMB, Klemme S (2013) Biology and evolution of B chromosomes. In: Leitch IJ, Greilhuber J, Doležel J, Wendel JF (eds) Plant genome diversity, vol 2, Physical structure, behaviour and evolution of plant genomes. Wien, pp 149–165
Ickert-Bond SM (2003) Systematics of New World *Ephedra L.* (Ephedraceae): integrating morphological and molecular data. Thesis, Arizona State University, pp 363
Islam-Faridi MN, Nelson CD, Kubisiak TL (2007) Reference karyotype and cytomolecular map for loblolly pine (*Pinus taeda* L.). Genome 50:241–251
Jacobs MD, Gardner RC, Murray BG (2000) Cytological characterization of heterochromatin and rDNA in *Pinus radiata* and *P. taeda*. Plant Syst Evol 223:71–79
Janoušek B, Hobza R, Vyskot B (2013) Chromosomes and sex diffrentiation. In: Leitch IJ, Greilhuber J, Doležel J, Wendel JF (eds) Plant genome diversity, vol 2, Physical structure, behaviour and evolution of plant genomes. Springer-Verlag, Wien, pp 167–186
Kamm A, Doudrick RL, Heslop-Harrison JS, Schmidt T (1996) The genomic and physical organization of *Ty1-copia*-like sequences as a component of large genomes in *Pinus elliottii* var. *elliottii* and other gymnosperms. Proc Natl Acad Sci USA 93:2708–2713
Khoshoo TN (1959) Polyploidy in gymnosperms. Evolution 13:24–39
Khoshoo TN, Ahuja MR (1963) The chromosomes and relationships of *Welwitschia mirabilis*. Chromosoma 14:522–533
Kokubugata G, Kondo K (1994) Quantitative variability in karyotype of *Cycas revoluta*. La Kromosomo II 75–76:2613–2618
Kokubugata G, Kondo K (1996) Differential fluorescent-banding patterns in chromosomes of four species of *Cycas* (Cycadaceae). Bot J Linn Soc 120:51–55
Kokubugata G, Kondo K (1998) Comparative karyotype analysis of *Ceratozamia mexicana* and *Microcycas calocoma* (Zamiaceae) using fluorochrome banding (CMA/DAPI) and fluorescence in situ hybridization of ribosomal DNA. Plant Syst Evol 210:41–50
Kokubugata G, Kondo K, Randall LM (1999) Comparison of chromosome number and karyotype of two *Lepidozamia* species (Zamiaceae, Cycadales). Ann Tsukuba Bot Gard 18:65–69
Kokubugata G, Kondo K, Wilson GW, Randall LM, van der Schans A, Morris DK (2000) Comparison of karyotype and rDNA-distribution in somatic chromosomes of *Bowenia* species (Stangeriaceae, Cycadales). Austral Syst Bot 13:15–20
Kokubugata G, Hill KD, Kondo K (2002) Ribosomal DNA distribution in somatic chromosomes of *Stangeria eriopus* (Stangeriaceae, Cycadales) and molecular-cytotaxonomic relationships to some other cycad genera. Brittonia 54:1–5
Kokubugata G, Vovides AP, Kondo K (2004) Mapping 5S ribosomal DNA on somatic chromosomes of four species of *Ceratozamia* and *Stangeria eriopus* (Cycadales). Bot J Linn Soc 145:499–504
Kriebel HB (1985) DNA sequence components of the *Pinus strobus* genome. Can J For Res 15:1–4
Krutovsky KV, Troggio M, Brown GR, Jermstad KD, Neale DB (2004) Comparative mapping in the Pinaceae. Genetics 168:447–461
Kupila-Ahvenniemi S, Hohtola A (1977) Structure of the chromosomes of Scotch pine. Hereditas 87:185–188
Kurdi-Haidar B, Shalhoub V, Dib-Haji S, Deeb S (1983) DNA sequence organization in the genome of *Cycas revoluta*. Chromosoma 88:319–327
Leitch AR, Leitch IJ (2012) Ecological and genetic factors linked to contrasting genome dynamics in seed plants. New Phytol 194:629–646
Leitch IJ, Leitch AR (2013) Genome size diversity and evolution in land plants. In: Leitch IJ, Greilhuber J, Doležel J, Wendel JF (eds) Plant genome diversity, vol 2, Physical structure, behaviour and evolution of plant genomes. Springer-Verlag, Wien, pp 307–322
Leitch IJ, Hanson L, Winfield M, Parker J, Bennett MD (2001) Nuclear DNA C-values complete familial representation in gymnosperms. Ann Bot 88:843–849

Liu Z-L, Zhang D, Hong D-Y, Wang X-R (2003) Chromosomal localization of 5S and 18S-5.8S-25S ribosomal DNA sites in five Asian pines using fluorescence in situ hybridization. Theor Appl Genet 106:198–204

Lubaretz O, Fuchs J, Ahne R, Meister A (1996) Karyotyping of three Pinaceae species via fluorescent in situ hybridization and computer-aided chromosome analysis. Theor Appl Genet 92:411–416

MacPherson P, Filion WG (1981) Karyotype analysis and the distribution of constitutive heterochromatin in five species of *Pinus*. J Hered 72:193–198

Marchant CJ (1968) Chromosome patterns and nuclear phenomena in the cycad families Stangeriaceae and Zamiaceae. Chromosoma 24:100–134

Matthey R (1945) L'évolution de la formule chromosomiale chez les Vertébrés. Experientia 1(50–56):78–86

Miksche JP, Hotta Y (1973) DNA base composition and repetitious DNA in several conifers. Chromosoma 41:29–36

Miller CI (1993) Impact of the Eocene on the evolution of *Pinus* L. Ann Missouri Bot Gard 80:471–498

Moretti A (1990) Cytotaxonomy of cycads. Mem NY Bot Gard 57:114–122

Morse AM, Peterson DG, Islam-Faridi MN, Smith KE, Magbanua Z, Garcia SA, Kubisiak TL, Amerson HV, Carlson JE, Nelson CD, Davis JM (2009) Evolution of genome size and complexity in *Pinus*. PLoS One 4:e4332

Murray BG (1998) Nuclear DNA amounts in gymnosperms. Ann Bot 82(Suppl A):3–15

Murray BG, Friesen N, Heslop-Harrison JS (2002) Molecular cytogenetic analysis of *Podocarpus* and comparison with other gymnosperm species. Ann Bot 89:483–489

Murray BG, Leitch IJ, Bennett MD (2010) Gymnosperm DNA *C*-values database (release 4.0, Dec 2010). http://data.kew.org/cvalues/

Page CN (1990) Gymnosperms. In: Kramer KU, Green PS (eds) The families and genera of vascular plants. Springer, Berlin, pp 280–361

Palmer JD, Soltis DE, Chase MW (2004) The plant tree of life: an overview and some points of view. Am J Bot 91:1437–1445

Parchman T, Geist K, Grahnen J, Benkman C, Buerkle CA (2010) Transcriptome sequencing in an ecologically important tree species: assembly, annotation, and marker discovery. BMC Genomics 11:180

Pellicer J, Fay MF, Leitch IJ (2010) The largest eukaryotic genome of them all? Bot J Linn Soc 164:10–15

Puizina J, Sviben T, Krajacic-Sokol I, Zoldos-Pecnik V, Siljak-Yakovlev S, Papes D (2008) Cytogenetic and molecular characterization of the *Abies alba* genome and its relationship with other members of the Pinaceae. Plant Biol 10:256–267

Quinn CJ, Rattenbury JA (1972) Structural hybridity in New Zealand *Dacrydium*. New Zealand J Bot 10:427–436

Rai HS, Reeves PA, Peakall R, Olmstead RG, Graham SW (2008) Inference of higher-order conifer relationships from a multi-locus plastid data set. Botany 86:658–669

Rake AV, Miksche JP, Hall RB, Hansen KM (1980) DNA reassociation kinetics of four conifers. Can J Genet Cytol 22:69–79

Sangduen N, Toahsakul M, Hongtrakul V (2009) Comparative karyomorphological study between male and female plants of some *Cycas* and *Zamia* species. Kasetsart J (Nat Sci) 43:476–485

Sax HJ (1932) Chromosome pairing in *Larix* species. J Arnold Arb 13:368–373

Sax K (1960) Meiosis in interspecific pine hybrids. Forest Sci 6:135–138

Saylor LC, Smith BW (1966) Meiotic irregularity in species and interspecific hybrids of *Pinus*. Am J Bot 53:453–468

Schmidt A, Doudrick RL, Heslop-Harrison JS, Schmidt T (2000) The contribution of short repeats of low sequence complexity to large conifer genomes. Theor Appl Genet 101:7–14

Segawa M, Kishi S, Tatsuno S (1971) Sex chromosomes of *Cycas revoluta*. Jpn J Genet 46:33–39

Shibata F, Hizume M, Garajova Z (2004) Conserved FISH karyotypes in four species of *Abies* (Pinaceae). Chromosome Sci 8:95–98

Shibata F, Matsusaki Y, Hizume M (2005) AT-rich sequences containing *Arabidopsis*-type telomere sequence and their chromosomal distribution in *Pinus densiflora*. Theor Appl Genet 110:1253–1258

Siljak-Yakovlev S, Cerbah M, Coulaud J, Stoian V, Brown SC, Zoldos V, Jelenic S, Papes D (2002) Nuclear DNA content, base composition, heterochromatin and rDNA in *Picea omorika* and *Picea abies*. Theor Appl Genet 104:505–512

Sone T, Fujisawa M, Takenaka M, Nakagawa S, Yamaoka S, Sakaida M, Nishiyama R, Yamato KT, Ohimodo N, Fukui K, Fukuzawa H, Ohyama K (1999) Bryophyte 5S rDNA was inserted into 45S rDNA repeat units after the divergence from higher land plants. Plant Mol Biol 41:679–685

Stockey RA, Nishida M (1986) *Pinus harborensis sp. nov*. and affinities of permineralized leaves from the Upper Cretaceous of Japan. Can J Bot 64:1856–1866

Tagashira N, Kondo K (1999) A karyotype comparison of nine species of aneuploid *Zamia* by using the conventional orcein staining and the fluorochrome CMA-DAPI differential staining methods. Cytologia 64:449–458

Tagashira N, Kondo K (2001) Chromosome phylogeny of *Zamia* and *Ceratozamia* by means of Robertsonian changes detected by fluorescence in situ hybridization (FISH) technique of rDNA. Plant Syst Evol 227:145–155

Tanaka R, Hizume M (1980) C banding treatment for the chromosomes of some gymnosperms. Bot Mag Tokyo 93:167–170

Teoh SB, Rees H (1977) B chromosomes in white spruce. P Roy Soc Lond B Bio 198:325–344

Trivers R, Burt A, Palestris BG (2004) B chromosomes and genome size in flowering plants. Genome 47:1–8

Vovides AP, Olivares M (1996) Karyotype polymorphism in the cycad *Zamia loddigesii* (Zamiaceae) of the Yucatan Peninsula, Mexico. Bot J Linn Soc 120:77–83

Wardle P (1972) *Podocarpus totara* var. *waihoensis* var. nov.: the result of introgressive hybridization between *P. totara* and *P. acutifolius*. New Zealand J Bot 10:195–201

Weiss-Schneeweiss H, Schneeweiss GM (2013) Karyotype diversity and evolutionary trends in angiosperms. In: Leitch IJ, Greilhuber J, Doležel J, Wendel JF (eds) Plant genome diversity, vol 2, Physical structure, behaviour and evolution of plant genomes. Springer-Verlag, Wien, pp 209–230

Werner T, Braukmann TWA, Kuzmina M, Stefanovic S (2009) Loss of all plastid *ndh* genes in Gnetales and conifers: extent and evolutionary significance for the seed plant phylogeny. Curr Genet 55:323–337

White MJD (1973) Animal cytology and evolution. Cambridge University Press, Cambridge

Williams CG (2009) Conifer reproductive biology. Springer, Dordrecht

Williams CG, Joyner KL, Auckland LD, Johnston S, Price HJ (2002) Genomic consequences of *Pinus* spp. hybridization. Biol J Linn Soc 75:503–508

Zgurski JM, Rai HS, Fai QM, Bogler DJ, Francisco-Ortega J, Graham SW (2008) How well do we understand the overall backbone of cycad phylogeny? New insights from a large, multigene plastid data set. Mol Phylogenet Evol 47:1232–1237

Zhou Q-x, Z-j Gu, Yue Z-S (2000) Karyomorphology and relationships of *Amentotaxus* Pilg. Acta Phytotax Sin 36:522–527

15 Karyotype and Genome Evolution in Pteridophytes

Michael S. Barker

Contents

M.S. Barker (✉)
Department of Ecology & Evolutionary Biology, University of Arizona, Tucson, USA
e-mail: msbarker@email.arizona.edu

I.J. Leitch et al. (eds.), *Plant Genome Diversity Volume 2*,
DOI 10.1007/978-3-7091-1160-4_15, © Springer-Verlag Wien 2013

15.1 Introduction

Although genomics is a field often discussed as a recent development brought upon by the advent of next-generation sequencing technologies, the roots of genomics extend to at least the early twentieth century. Rather than running a gel or assembling sequence reads on a server, early cytologists squashed and stained actively dividing cells and viewed them under a microscope to reveal a variety of chromosome features including numbers, sizes, and pairing behaviour. These data revolutionized our perspective of plant species, and provoked numerous questions about genome evolution, some of which endure today. Among these long-standing questions is how the high chromosome numbers of homosporous ferns and lycophytes evolved and are maintained.

By the 1950s, it was clear that fern nuclear genomes, particularly those of homosporous species, possessed exceptionally high chromosome numbers (Manton 1950). Comparison of chromosome counts among related species of many genera clearly showed that series of currently recognizable polyploids (neopolyploids) such as tetraploids and hexaploids are frequent among ferns and lycophytes, eventually reaching the highest known chromosome count among extant eukaryotes in *Ophioglossum reticulatum* ($n > 600$; Khandelwal 1990). These observations indicated that whole genome duplication is an ongoing process and a significant feature of fern and lycophyte evolution. In addition to the abundant occurrence of neopolyploids, it became clear that, with rare exceptions, even the lowest chromosome numbers in each genus were much higher than those of other plant groups. Cytologists studying angiosperms and other heterosporous species rarely required more than fingers and toes to count chromosomes; flowering plants have an average of $n = 15.99$ chromosomes (Klekowski and Baker 1966). However, homosporous ferns and lycophytes required more digits and patience to count their chromosomes. On average, homosporous fern and lycophyte genomes contain $n = 57.05$ chromosomes, over three-fold more than the average flowering plant (Klekowski and Baker 1966). This striking difference between homosporous and heterosporous plants spawned a number of hypotheses.

15.2 High Chromosome Numbers and Hypotheses

Although Manton's (1950) cytological studies focused attention on homosporous plant chromosome numbers, Klekowski and Baker (1966) provided the first synthetic hypothesis for their origin and maintenance. Historically, haploid chromosome numbers higher than 14 were generally considered to be polyploids in angiosperms (Grant 1981). If this rule were applied to homosporous ferns, with their average $n = 57.05$, more than 95% would be diagnosed as polyploids. A significant exception to the generally high chromosome numbers in ferns are the heterosporous water ferns; the haploid chromosome number for heterosporous ferns averages 13.6, and 90% of these species have chromosome numbers less than 28. A similar situation is encountered in the lycophytes, where the homosporous Lycopodiaceae have significantly higher chromosome numbers than their sister groups (Pryer et al. 2004), the heterosporous Selaginellaceae and Isoetaceae (Löve et al. 1977). Thus, homosporous pteridophytes have significantly higher chromosome numbers compared to their close heterosporous relatives and seed plants.

Based on the large number of chromosomes, Klekowski and Baker (1966) posited that most homosporous ferns were polyploids or species with more than two complete sets of chromosomes in their somatic cells. They proposed that homosporous vascular plants evolved high numbers of chromosomes and ploidy levels to provide an extra source of genetic variation to compensate for their putative primary mode of reproduction, intragametophytic self-fertilization, an extreme form of inbreeding that results in 100% homozygosity in a single generation (Klekowski 1973). Under this perspective, homosporous ferns potentially suffer severe losses of heterozygosity more frequently than heterosporous plants. According to Klekowski and Baker (1966), additional non-Mendelian genetic variation could be generated by abnormal pairing during meiosis of different versions of chromosomes, or homoeologs, rather than the normal homologous pairing. Homoeologous pairing would be unaffected by this extreme self-fertilization and release genetic variation in these putatively polyploid genomes that would otherwise be "fixed." Putatively, heterosporous ferns should have lower chromosome numbers than homosporous ferns because they are obligately outcrossing and therefore do not experience sharp reductions in heterozygosity through intragametophytic selfing. A variety of subsequent studies supported Klekowski and Baker's hypothesis. For example, Hickok and Klekowski (1974) and Hickok (1978) demonstrated homoeologous pairing in a small percentage (<1–3%) of self-fertilized offspring of the tetraploid *Ceratopteris thalictroides*. These studies demonstrated that homoeologous pairing can occur and may operate to provide heterozygosity in otherwise largely homozygous vascular plants.

Additional data to address the Klekowski and Baker (1966) hypothesis came from isozymes rather than additional cytological research. By providing direct insight into patterns of genetic variation and inheritance, isozymes were instrumental for characterizing fern and lycophyte genomes. An early isozyme study allegedly supported the Klekowski and Baker hypothesis by finding non-Mendelian inheritance of multiple enzyme bands, consistent with homoeologous pairing in a neopolyploid (Chapman et al. 1979). However, a subsequent investigation demonstrated that a homosporous fern with the lowest chromosome number for its genus possessed a diploid, not polyploid, gene expression profile with Mendelian inheritance (Gastony and Gottlieb 1982). Furthermore, it was demonstrated that the complex isozyme banding patterns of Chapman et al. (1979) were mostly attributable to subcellularly compartmentalized isozymes (Gastony and Darrow 1983; Wolf et al. 1987). Continued isozyme investigations supported Gastony and Gottlieb's observation that homosporous ferns with base chromosome numbers for their genus had diploid expression profiles despite possessing relatively high chromosome numbers (Gastony and Gottlieb 1985; Haufler and Soltis 1986; Soltis 1986; Haufler 1987). These results differed from those in angiosperms, which showed that angiosperms with high chromosome numbers did indeed have duplicated sets of isozymes (reviewed in Gottlieb 1982; Soltis and Soltis 1990).

Analyses of fern breeding systems also cast serious doubt on the Klekowski and Baker (1966) hypothesis. Rather than finding evidence of extensive intragametophytic self-fertilization, analyses of natural fern populations also found that they were predominantly outcrossing. Haufler and Soltis (1984) found that 83% of the examined populations of diploid *Bommeria hispida* had heterozygous enzyme banding patterns attributable to outcrossing between genetically different gametophytes. In a more extensive analysis of diploid populations of *Pellaea andromedifolia*, Gastony and Gottlieb (1985) determined that at least 81.3% of 145 sporophytes examined from natural populations arose through outcrossing between gametophytes carrying different alleles. Because of the unique breeding system of homosporous vascular plants, Holsinger (1987) developed a specialized statistical technique for estimating rates of self-fertilization and applied it to the data of Gastony and Gottlieb (1985). He found no evidence of intragametophytic self-fertilization in two of Gastony and Gottlieb's populations with sufficient sample sizes for his statistical analyses and concluded that intragametophytic self-fertilization occurs only a small fraction of the time, if at all. Thus Klekowski and Baker's (1966) original rationale for the polyploid homoeologous heterozygosity hypothesis—high levels of intragametophytic self-fertilization—was rejected.

By the end of the isozyme era, it was clear that the previous explanation for high numbers of chromosomes in homosporous genomes—high levels of neopolyploidy—was no longer supported. Alternative explanations that do not involve polyploidy for the high chromosome numbers, including ancestral high chromosome numbers and high rates of chromosomal fission, have never gained much traction because counts within genera and clades often follow a euploid pattern and rates of neopolyploidy are high among ferns (Wood et al. 2009). To explain the paradox of high chromosome numbers and diploid isozyme expressions in homosporous ferns, Haufler (1987) hypothesized that homosporous ferns may have acquired high chromosome numbers with diploid gene expression through repeated cycles of polyploidization and subsequent gene silencing without the loss of chromosomes. If such a mechanism is common, the genomes of extant homosporous ferns should contain multiple, silenced copies of each nuclear gene. In support of Haufler's hypothesis, a few studies have identified silenced nuclear genes in homosporous fern genomes. Gastony (1991) captured the process of gene silencing by documenting that the duplicated expression of a phosphoglucoisomerase isozyme in the neotetraploid homosporous fern *Pellaea rufa* progressively diminished to a diploid level in several populations. However, isozyme analyses can assess only the expression of enzyme-coding loci and not the presence of these genes. Therefore, it cannot be determined whether homosporous ferns with the lowest chromosome numbers in their genus have diploid expression of isozyme loci because they have never experienced polyploidization or because they have gone through the cycle of polyploidization and gene silencing, unless sequences of gene copies are examined (reviewed in Soltis and Soltis 1987). Using DNA sequence based probes for chlorophyll a/b binding protein (CAB) genes in the homosporous fern *Polystichum munitum*, Pichersky et al. (1990) identified several copies of this high copy number gene and found that several gene copies had been pseudogenized, consistent with Haufler's hypothesis.

Subsequent studies evaluating gene copy number in homosporous ferns utilized Restriction Fragment Length Polymorphisms (RFLPs). RFLPs may be used to estimate the number of genes present in a genome by using DNA probes to detect length differences in restriction enzyme-digested genomic fragments. Mutations at paralogous loci will introduce changes in the length of fragments by either directly introducing length differences (e.g., indels) or by altering restriction sites. Using RFLPs with several cDNA probes, McGrath et al. (1994) found multiple copies of most genes tested in the genome of the diploid homosporous fern species *Ceratopteris richardii*, consistent with Haufler's hypothesis. Therefore, even diploid species of homosporous ferns seem to have multiple gene copies, although many appear to be silenced pseudogenes. McGrath et al. also compared the RFLP profiles of diploid *C. richardii* and tetraploid *C. thalictroides* and expected that twice as many gene copies should be present in *C. thalictroides*. In contrast to their expectation, the tetraploid species had on average only 30% (marginally significant) more gene copies and some loci were single-copy. Although some paralogs may have been undetected if they co-migrate on the gel or if their sequences are too diverged from the probe sequences, it seems reasonable to assume that at least in some cases, genes duplicated through polyploidization have been not only silenced but also completely deleted from the genome. These results are consistent with the spectrum of observations in angiosperm genomes and polyploids (Wendel 2000).

15.3 Recent Advances

In the last decade, the study of plant genomes has advanced with the advent of new sequencing technologies. Fern and lycophyte genomics are beginning to benefit from these advancements and progress is being made on the long-standing questions surrounding the evolution of high chromosome numbers (Barker and Wolf 2010). Traditional marker based approaches to the study of genomics often have limited power, especially if mapping populations are not employed to evaluate inheritance patterns. For example, RFLP studies cannot distinguish genome-wide polyploidization events from small scale duplications because the number of detected RFLP fragments tends to overestimate the number of gene copies if restriction sites exist within the probe hybridizing sites. Critically, the relative positions of gene copies in the genome are unknown in RFLP studies without analysis of segregating populations. Some of these limitations can be overcome by fluorescent *in situ* hybridization (FISH) or "chromosome painting," which allows one to visualize the physical position of genes corresponding to the fluorescently labelled hybridizing probes on the fixed chromosomes. Using a ribosomal DNA (rDNA) probe, FISH analyses in *C. richardii* showed that multiple rDNA gene copies exist on different chromosomes (McGrath and Hickok 1999), supporting the hypothesis of paleopolyploidy in homosporous ferns. However, it is difficult to determine relative positions among different marker loci with FISH, and the approach is technically challenging and labour intense, especially when hybridizing many low copy markers. Other techniques have provided more information about fern and lycophyte genomes with more reliability.

Genetic linkage mapping has provided the most comprehensive view of synteny, or gene order colinearity among duplicated regions. Linkage maps describe the relative locations of polymorphic markers in genomes inferred from the segregation patterns of parental alleles in mapping populations. Inference of marker positions is based on a simple assumption: the frequency of recombination between two markers increases with the physical distance between these loci, although physical and genetic map distances are often not proportional because recombination rates are unevenly distributed across the genome. Linkage mapping is a powerful tool for investigating paleopolyploidy not only because gene copy number and relative marker positions can be estimated fairly accurately, but also because mapping a large number of duplicated genes allows us to infer large scale duplication (e.g., polyploidy) if sets of duplicated loci occur in similar orders in different parts of the genome. Nakazato et al. (2006) constructed a genetic linkage map of the homosporous fern *C. richardii* based on 729 markers (368 RFLPs, 358 Amplified Fragment Length Polymorphisms (AFLPs), and 3 isozymes) using a mapping population consisting of 488 Doubled Haploid Lines (DHLs). Most genes (85%) were found in multiple copies based on screens of 1,037 RFLPs hybridized to fluorescently labelled cDNA probes, consistent with the results of previous RFLP studies. After correcting for potential multiple restriction sites within hybridizing fragments, 24% of genes were estimated to be single copy. This is one of the lowest proportions of single copy genes among plant species, at least as measured by RFLP linkage mapping. Local duplications such as tandem duplications are common, yet a large number of duplications were detected between rather than within chromosomes. Although these data favor the paleopolyploidy hypothesis as in previous studies, Nakazato et al. (2006) were unable to find conclusive evidence of paleopolyploidy. First, the detected gene duplicates appear to be scattered more or less haphazardly across the genome rather than occurring in recognizable homoeologous chromosomal segments (Fig. 15.1). If there was an ancient polyploidization, syntenic chromosomal blocks would be expected to have been recognizable even after millions of years. For example, the genomes of *Brassica* species currently recognized as diploid are mainly composed of three sets of gene copies with short syntenic regions, suggesting that they are ancient hexaploids (Lagercrantz and Lydiate 1996).

The absence of significant homoeologous synteny in Nakazato et al.'s mapping study suggests that polyploidization has not occurred in at least the last several million years in the *C. richardii* lineage and perhaps not in diploid homosporous ferns in general. Statistical analyses of the distribution of gene copies, however, indicate that there is significant clustering in the *C. richardii* genome; that is, the copies of a set of markers in a given linkage group tend to co-occur on a different linkage group, although the pattern is not strong enough to be visually apparent. This result suggests that there may have been ancient whole genome duplications in the history of *C. richardii*, and likely other related polypod ferns, that have been masked by subsequent chromosomal and gene copy number evolution.

To gain further insight into the origin of the high chromosome numbers of homosporous ferns, modern genomic tools have recently been applied to the question. In large and putatively complex genomes such as those of ferns, whole genome sequencing has not yet been economically feasible. An alternative to sequencing a whole nuclear genome is to sequence a large portion of the transcriptome, i.e., that part of the genome that is transcribed into RNA. Such whole-scale sequencing of cDNA libraries—either normalized Expressed Sequence Tags (ESTs) where highly expressed transcripts are reduced in abundance or non-normalized RNA-Seq approaches—has been particularly useful in ferns. Using traditional Sanger long reads or many more but shorter next-generation technologies such as Illumina or 454 sequencing, entire transcriptomes and the "gene catalog" may be economically sequenced. Computational tools specifically developed for large scale sequence analyses (e.g., EvoPipes.net; Barker et al. 2010) are used to efficiently analyse and summarize these large data sets. Transcriptome approaches are particularly well suited for studying the genomics of non-model organisms because genome complexity is reduced by focusing on transcribed regions of the genome. Considering the low cost and analytic flexibility of ESTs, they will likely play a significant role in furthering our understanding of fern genomes.

How can transcriptome data be used to evaluate the paleopolyploid hypothesis and the origin of high chromosome numbers in ferns? Phylogenies of each gene family may be constructed from a transcriptome library for a species, and common gene duplication events identified. By plotting the ages of all gene duplications, typically in terms of their number of synonymous substitutions (K_s or dS), ancient genome duplications may be inferred as peaks in the age distribution (Fig. 15.2). These peaks reflect the very large number of duplications of similar age that one expects to result from ancient polyploidy, against a backdrop of ongoing small scale gene duplication birth and death. By using relative rate corrections and multiple species gene family phylogenies to account for variation in substitution rates across lineages, it is possible to discern whether paleopolyploidizations observed in two or more taxa are shared (Barker et al. 2008, 2009). Applied to the transcriptomes of two polypod ferns (*Ceratopteris richardii* and *Adiantum capillus-veneris*) similar analyses have

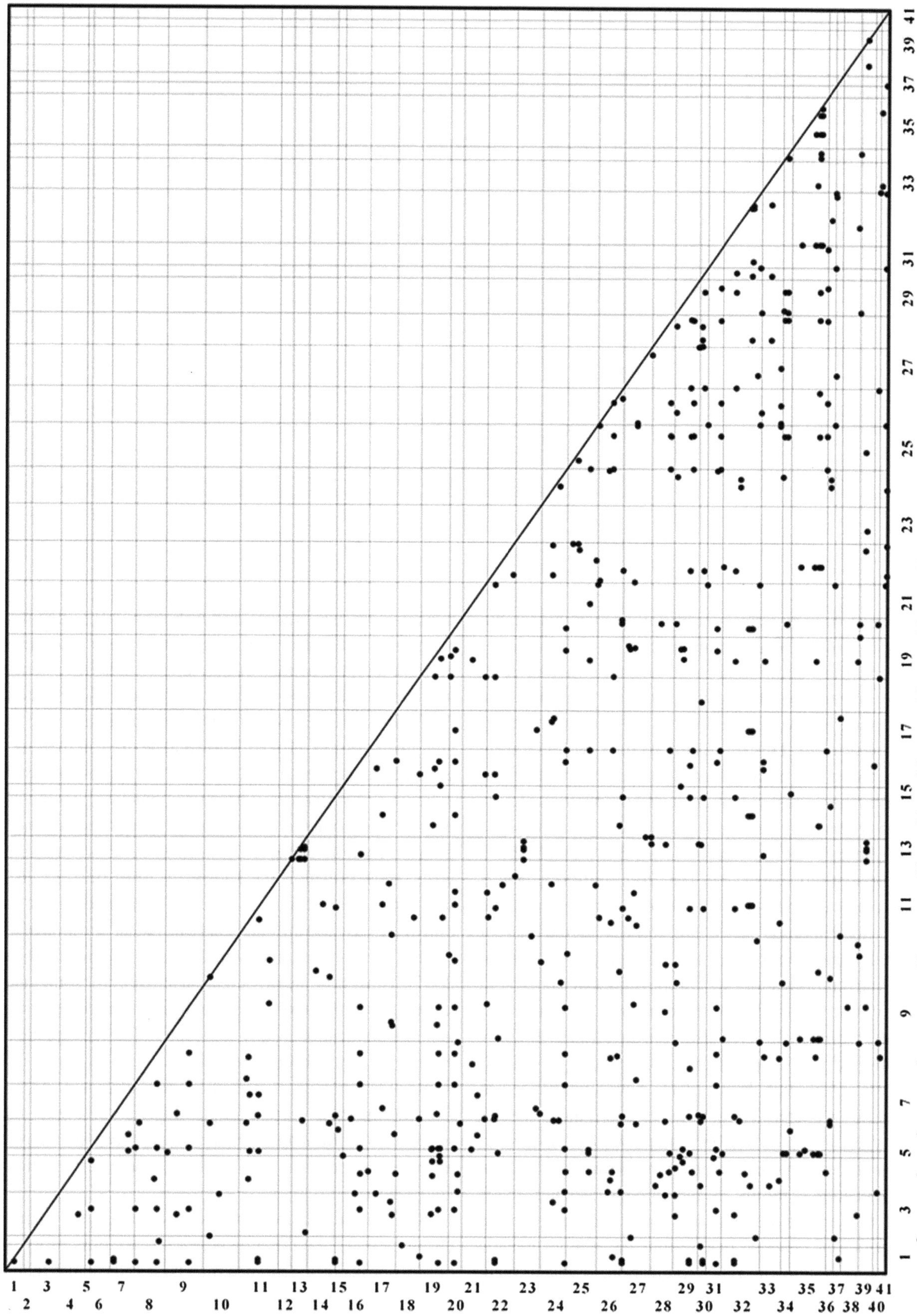

Fig. 15.1 Plot of duplicated gene copies shared by pairs of linkage groups for the homosporous fern *Ceratopteris richardii* (indicated by the *numbers*). The width and the height of each block are proportional to the cM map length of the corresponding linkage group (After Nakazato et al. 2006)

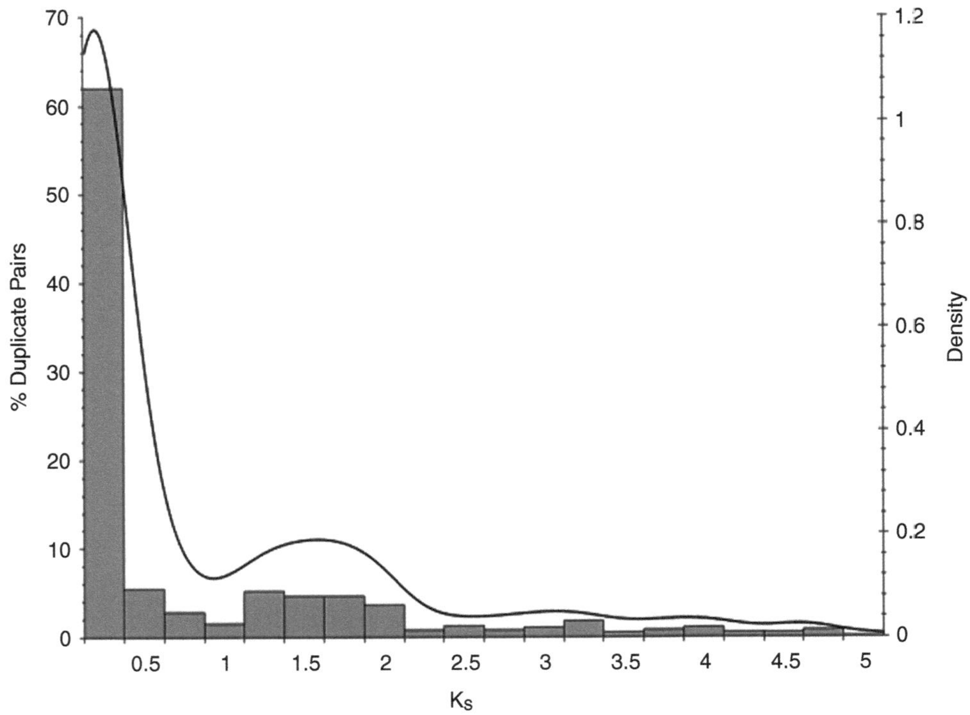

Fig. 15.2 Histogram of *Ceratopteris richardii* duplicate K_s values demonstrating a peak in duplications from K_s 0.96 to 1.84 (Barker 2009). The *solid line* is a significant SiZer (Chaudhuri and Marron 1999) kernel density estimate, with the peak significant at $p < 0.05$

revealed each lineage shares only a single detectable genome duplication with a median peak of Ks ~ 1.6 (Barker 2009; Fig. 15.2). By combining fossil dates (Schneider et al. 2004) with nuclear gene phylogenies from the transcriptome data, this duplication event is thought to have occurred nearly 180 million years ago. This places the paleopolyploidization somewhere along the branch leading to all or most of the polypod ferns, the largest extant clade of homosporous ferns.

Although we have evidence for one ancient genome duplication in the ancestry of most extant ferns, it is unlikely that a single paleopolyploidy over the last 200 million years is sufficient to create and maintain the extraordinary chromosome numbers of homosporous ferns. Differences in the rate of chromosomal change and loss may also play a significant role. Following multiplication, the chromosomes of a polyploid genome may form complexes of three or more chromosomes (multivalents) during meiosis. A distinguishing trait of diploidy is the formation of pairs (*di*-ploidy) of chromosomes (bivalents) during meioses. Although many plants have experienced at least one round of ancient genome duplication, these species all behave as genetic diploids rather than polyploids. So how do most plant genomes, which have experienced rounds of ancient genome duplications, return to this diploid genetic state? Subsequent to genome duplication, plant nuclear genomes undergo a series of changes that restore the diploid genetic system, a process known as diploidization. Typical plant genomes experience, at varying rates, rearrangements and often a net loss of genetic material, via mechanisms such as chromosomal fusion, illegitimate recombination, and transposition (see Lysák and Schubert 2013, this volume). Combined with silencing of duplicated genes via mutation or outright loss, the actions of this suite of genomic changes lead to diploidization (Doyle et al. 2008; see also Fawcett et al. 2013, this volume). Significant variation in the rate of these genomic changes is well known from the numerous whole genome sequences available for the angiosperms. At one extreme is *Arabidopsis*, whose small nuclear genome contains only five chromosomes but has experienced at least three rounds of whole genome duplication in less than 200 million years (Bowers et al. 2003; Tang et al. 2008; Barker et al. 2009). In contrast, *Vitis* appears to have experienced only one duplication in the past 200 million years (the oldest one in *Arabidopsis*) yet has a larger genome with 17 chromosomes (Jaillon et al. 2007). Thus, genome organization and chromosome number is a product of variation in diploidization processes as well as genome duplication. Many angiosperm lineages appear to have experienced two to three rounds of paleopolyploidy in a time frame of $K_s < 2$ (e.g., Cui et al. 2006; Barker et al. 2008, 2009). Considering that only a single duplication event has been observed in current analyses of fern genomes they may experience a lower rate of whole genome duplication than angiosperms.

So, why do homosporous ferns have so many more chromosomes than angiosperms? One possibility is that homosporous fern genomes may be less dynamic than angiosperm genomes, and may experience chromosomal loss at a much slower rate. Consistent with this perspective is the exceptionally low density of genes observed in a fern

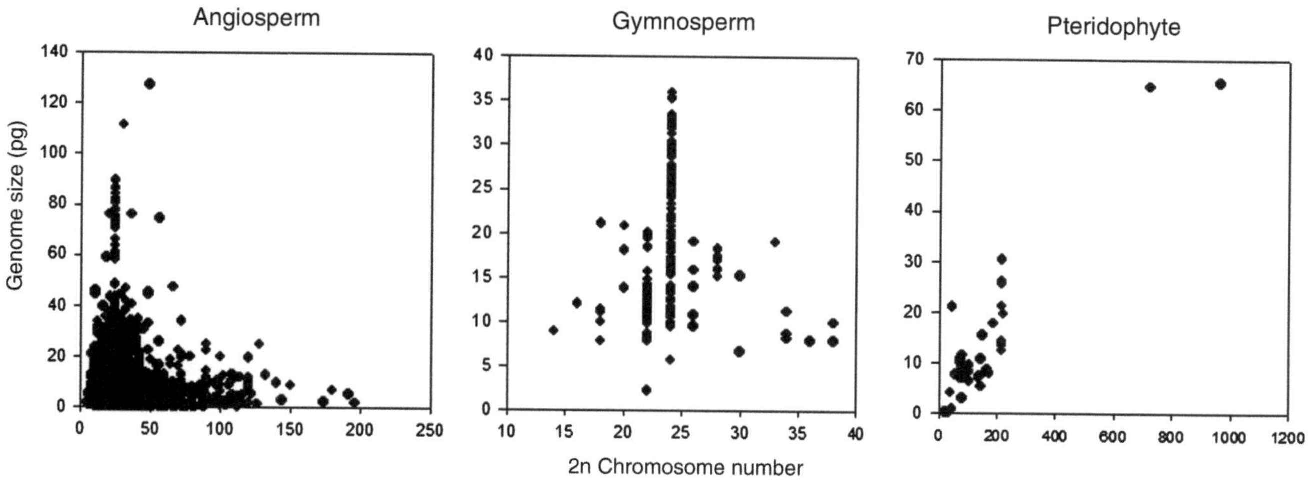

Fig. 15.3 Correlation between chromosome numbers and genome size (1C-value in picograms) for angiosperms, gymnosperms, and pteridophytes. (After Nakazato et al. 2008)

genome (Rabinowicz et al. 2005) and the observation that genome size and chromosome number are strongly correlated in ferns (Fig. 15.3; Nakazato et al. 2008). Similarly, earlier observations from cytological studies also found that fern chromosomes are generally highly uniform in size and structure across genera (reviewed in Wagner and Wagner 1980). These observations imply that although fern genomes appear to lose duplicate genes through gene silencing, physical genetic material may be lost at a slow rate compared to many angiosperms and that chromosomal evolution is less dynamic among ferns.

One explanation that may reconcile these observations is that repetitive elements are less active or abundant in fern genomes. Repetitive element expansion and contraction has been long established as a primary determinant of genome size variation in angiosperms (Leitch et al. 2005), with a well-established positive correlation of genome size and numbers of retrotransposons, particularly Long Terminal Repeat (LTR) retrotransposons (Flavell 1980). Recent genome analyses in angiosperms, especially grasses, indicate an increase in genome size in the last 10 million years largely due to transposable element (TE) expansion (SanMiguel and Bennetzen 1998; Vicient et al. 1999; Shirasu et al. 2000; Hawkins et al. 2006). These elements comprise over 60% of large genomes like maize, wheat, and barley, but less than 50% in rice and about 10% in *Arabidopsis* (reviewed in Bennetzen 2002). Although little is known about the numbers and kinds of TEs in ferns, reduced copy number of *Ty1-copia*-like retrotransposons were recently reported in the homosporous fern *Pteris cretica* compared to numbers in other plant species, despite the fact that it possesses a large genome size and appears to be a neopolyploid. Thus one possible explanation for the evolution of stable genome sizes and chromosome numbers in ferns is suppression of TEs, but this is highly speculative and needs further evaluation. It is not entirely clear why this would occur, but additional information, including transcriptome data from more fern lineages, as well as fern whole genome sequences are needed to determine how, and ultimately why, fern nuclear genomes are different from seed plants, and whether such differences are caused by different genomic compositions or related to homospory and heterospory.

15.4 Future Directions for Fern Nuclear Genomics

Furthering our understanding of the evolution of genome structure and organization of ferns—and vascular plants in general—will require the application of new technologies as well as analytical tools. Several new directions have been initiated in recent years which will be important for the next generation of advancements. The complete sequence for a lycophyte, *Selaginella moellendorffii,* was recently completed (Banks et al. 2011). Analyses revealed that unlike angiosperms with low numbers of chromosomes, *S. moellendorffii* does not contain evidence of any ancient polyploidy. This result is consistent with the lower frequency of ancient polyploidy observed in homosporous ferns relative to angiosperms (Barker and Wolf 2010). Additional nuclear genomes, as well as phylogenetically diverse transcriptomes, will undoubtedly be sequenced for ferns and lycophytes over the next few years and insights into vascular plant genome evolution will likely come quickly relative to the past. Combined with novel bioinformatic tools for studying genome evolution, such as chromEvol (Mayrose et al. 2010) and EvoPipes.net (Barker et al. 2010), many of the twentieth century's cytological mysteries may finally be resolved.

References

Banks JA et al. (2011) The *Selaginella* genome identifies genetic changes associated with the evolution of vascular plants. Science 332:960–963

Barker MS (2009) Evolutionary genomic analyses of ferns reveal that high chromosome numbers are a product of high retention and fewer rounds of polyploidy relative to angiosperms. Am Fern J 99:136–141

Barker MS, Wolf PG (2010) Unfurling fern biology in the genomics age. Bioscience 60:177–185

Barker MS, Kane NC, Matvienko M, Kozik A, Michelmore RW, Knapp SJ, Rieseberg LH (2008) Multiple paleopolyploidizations during the evolution of the Compositae reveal parallel patterns of duplicate gene retention after millions of years. Mol Biol Evol 25:2445–2455

Barker MS, Vogel H, Schranz ME (2009) Paleopolyploidy in the Brassicales: analyses of the *Cleome* transcriptome elucidate the history of genome duplications in *Arabidopsis* and other Brassicales. Genome Biol Evol 1:391–399

Barker MS, Dlugosch KM, Dinh L, Challa RS, Kane NC, King MG, Rieseberg LH (2010) EvoPipes.net: bioinformatic tools for ecological and evolutionary genomics. Evol Bioinform 6:143–149

Bennetzen JL (2002) Mechanisms and rates of genome expansion and contraction in flowering plants. Genetica 115:29–36

Bowers JE, Chapman BA, Rong JK, Paterson AH (2003) Unravelling angiosperm genome evolution by phylogenetic analysis of chromosomal duplication events. Nature 422:433–438

Chaudhuri P, Marron JS (1999) SiZer for exploration of structures in curves. J Amer Stat Assoc 94:807–823.

Chapman RH, Klekowski EJ, Selander RK (1979) Homoeologous heterozygosity and recombination in the fern *Pteridium aquilinum*. Science 204:1207–1209

Cui LY, Wall PK, Leebens-Mack JH, Lindsay BG, Soltis DE, Doyle JJ, Soltis PS, Carlson JE, Arumuganathan K, Barakat A, Albert VA, Ma H, dePamphilis CW (2006) Widespread genome duplications throughout the history of flowering plants. Genome Res 16:738–749

Doyle JJ, Flagel LE, Paterson AH, Rapp RA, Soltis DE, Soltis PS, Wendel JF (2008) Evolutionary genetics of genome merger and doubling in plants. Annu Rev Genet 42:443–461

Fawcett JA, Van de Peer Y, Maere S (2013) Significance and biological consequences of polyploidization in land plants. In: Leitch IJ, Greilhuber J, Doležel J, Wendel JF (eds) Plant genome diversity, vol 2, Physical structure, behaviour and evolution of plant genomes. Springer-Verlag, Wien, pp 277–293

Flavell R (1980) The molecular characterization and organization of plant chromosomal DNA sequences. Annu Rev Plant Phys 31:569–596

Gastony GJ (1991) Gene silencing in a polyploid homosporous fern: paleopolyploidy revisited. Proc Natl Acad Sci USA 88:1602–1605

Gastony GJ, Darrow DC (1983) Chloroplastic and cytosolic isozymes of the homosporous fern *Athyrium filix-femina* L. Am J Bot 70:1409–1415

Gastony GJ, Gottlieb LD (1982) Evidence for genetic heterozygosity in a homosporous fern. Am J Bot 69:634–637

Gastony GJ, Gottlieb LD (1985) Genetic variation in the homosporous fern *Pellaea andromedifolia*. Am J Bot 72:257–267

Gottlieb LD (1982) Conservation and duplication of isozymes in plants. Science 216:373–380

Grant V (1981) Plant speciation. Columbia University Press, New York

Haufler CH (1987) Electrophoresis is modifying our concepts of evolution in homosporous pteridophytes. Am J Bot 74:953–966

Haufler CH, Soltis DE (1984) Obligate outcrossing in a homosporous fern: field confirmation of a laboratory prediction. Am J Bot 71:878–881

Haufler CH, Soltis DE (1986) Genetic evidence suggests that homosporous ferns with high chromosome numbers are diploid. Proc Natl Acad Sci USA 83:4389–4393

Hawkins JS, Kim H, Nason JD, Wing RA, Wendel JF (2006) Differential lineage-specific amplification of transposable elements is responsible for genome size variation in *Gossypium*. Genome Res 16:1252–1261

Hickok LG (1978) Homoeologous chromosome pairing and restricted segregation in the fern *Ceratopteris*. Am J Bot 65:516–521

Hickok LG, Klekowski EJ (1974) Inchoate speciation in *Ceratopteris*: an analysis of the synthesized hybrid *C. richardii* × *C. pteridoides*. Evolution 28:439–446

Holsinger KE (1987) Gametophytic self-fertilization in homosporous plants: development, evaluation, and application of a statistical method for evaluating its importance. Am J Bot 74:1173–1183

Jaillon O et al (2007) The grapevine genome sequence suggests ancestral hexaploidization in major angiosperm phyla. Nature 449:463–467

Khandelwal S (1990) Chromosome evolution in the genus *Ophioglossum* L. Bot J Linn Soc 102:205–217

Klekowski EJ (1973) Sexual and subsexual systems in homosporous pteridophytes: a new hypothesis. Am J Bot 60:535–544

Klekowski EJ, Baker HG (1966) Evolutionary significance of polyploidy in the Pteridophyta. Science 153:305–307

Lagercrantz U, Lydiate DJ (1996) Comparative genome mapping in *Brassica*. Genetics 144:1903–1910

Leitch IJ, Soltis DE, Soltis PS, Bennett MD (2005) Evolution of DNA amounts across land plants (Embryophyta). Ann Bot 95:207–217

Löve Á, Löve D, Pichi-Sermolli REG (1977) Cytotaxonomical atlas of the Pteridophyta. J. Cramer, Vaduz

Lysák MA, Schubert I (2013) Mechanisms of chromosome rearrangements. In: Leitch IJ, Greilhuber J, Doležel J, Wendel JF (eds) Plant genome diversity, vol 2, Physical structure, behaviour and evolution of plant genomes. Springer-Verlag, Wien, pp 137–147

Manton I (1950) Problems of cytology and evolution in the Pteridophyta. Cambridge University Press, Cambridge

Mayrose I, Barker MS, Otto SP (2010) Probabilistic models of chromosome number evolution and the inference of polyploidy. Syst Biol 59:132–144

McGrath JM, Hickok LG (1999) Multiple ribosomal RNA gene loci in the genome of the homosporous fern *Ceratopteris richardii*. Can J Bot 77:1199–1202

McGrath JM, Hickok LG, Pichersky E (1994) Assessment of gene copy number in the homosporous ferns *Ceratopteris thalictroides* and *C. richardii* (Parkeriaceae) by restriction fragment length polymorphisms. Plant Syst Evol 189:203–210

Nakazato T, Jung MK, Housworth EA, Rieseberg LH, Gastony GJ (2006) Genetic map-based analysis of genome structure in the homosporous fern *Ceratopteris richardii*. Genetics 173:1585–1597

Nakazato T, Barker MS, Rieseberg LH, Gastony GJ (2008) Evolution of the nuclear genome of ferns and lycophytes. In: Ranker TA, Haufler CH (eds) Biology and evolution of ferns and lycophytes. Cambridge University Press, Cambridge, pp 175–198

Pichersky E, Soltis DE, Soltis PS (1990) Defective chlorophyll a/b-binding protein genes in the genome of a homosporous fern. Proc Natl Acad Sci USA 87:195–199

Pryer KM, Schuettpelz E, Wolf PG, Schneider H, Smith AR, Cranfill R (2004) Phylogeny and evolution of ferns (monilophytes) with a focus on the early leptosporangiate divergences. Am J Bot 91:1582–1598

Rabinowicz PD, Citek R, Budiman MA, Nunberg A, Bedell JA, Lakey N, O'Shaughnessy AL, Nascimento LU, McCombie WR, Martienssen RA (2005) Differential methylation of genes and repeats in land plants. Genome Res 15:1431–1440

SanMiguel P, Bennetzen JL (1998) Evidence that a recent increase in maize genome size was caused by the massive amplification of intergene retrotransposons. Ann Bot 82:37–44

Schneider H, Schuettpelz E, Pryer KM, Cranfill R, Magallón S, Lupia R (2004) Ferns diversified in the shadow of angiosperms. Nature 428:553–557

Shirasu K, Schulman AH, Lahaye T, Schulze-Lefert P (2000) A contiguous 66-kb barley DNA sequence provides evidence for reversible genome expansion. Genome Res 10:908–915

Soltis DE (1986) Genetic evidence for diploidy in *Equisetum*. Am J Bot 73:908–913

Soltis DE, Soltis PS (1987) Polyploidy and breeding systems in homosporous Pteridophyta—a reevaluation. Am Nat 130:14219–14232

Soltis DE, Soltis PS (1990) Isozyme evidence for ancient polyploidy in primitive angiosperms. Syst Bot 15:328–337

Tang HB, Bowers JE, Wang XY, Ming R, Alam M, Paterson AH (2008) Synteny and collinearity in plant genomes. Science 320: 486–488

Vicient CM, Suoniemi A, Anamthawat-Jonsson K, Tanskanen J, Beharav A, Nevo E, Schulman AH (1999) Retrotransposon *BARE*-1 and its role in genome evolution in the genus *Hordeum*. Plant Cell 11:1769–1784

Wagner WH, Wagner FS (1980) Polyploidy in pteridophytes. In: Lewis WH (ed) Polyploidy: biological relevance. Plenum Press, New York, pp 199–214

Wendel JF (2000) Genome evolution in polyploids. Plant Mol Biol 42:225–249

Wolf PG, Haufler CH, Sheffield E (1987) Electrophoretic evidence for genetic diploidy in the bracken fern (*Pteridium aquilinum*). Science 236:947–949

Wood TE, Takebayashi N, Barker MS, Mayrose I, Greenspoon PB, Rieseberg LH (2009) The frequency of polyploid speciation in vascular plants. Proc Natl Acad Sci USA 106:13875–13879

The Incidence of Polyploidy in Natural Plant Populations: Major Patterns and Evolutionary Processes

16

Brian C. Husband, Sarah J. Baldwin, and Jan Suda

Contents

16.1 Introduction

Polyploidy, the multiplication of complete chromosome sets per nucleus ($n = x$ in haploid phase, $2n = 2x$ in diploid phase, where n is the haploid phase chromosome number and x is the monoploid chromosome number; Winge 1917; Grant 1981) is a prominent and recurring transition in the evolution of eukaryotic genomes (Otto and Whitton 2000). It has been documented within every major lineage of eukaryotes (protists, plants, fungi, animals). Among these groups, polyploidy is arguably most prevalent in land plants (i.e., angiosperms, gymnosperms, ferns and allies, lycopods, and bryophytes). As in other groups, there is evidence for widespread taxonomic heterogeneity in the occurrence of polyploidy within plants, and polymorphism at all taxonomic levels including species and populations (Stebbins 1950; Otto and Whitton 2000). Heterogeneity in the incidence of polyploidy has also been associated with geographic location, environment as well as life history, morphology and functional traits (Otto and Whitton 2000; Levin 2002). Collectively, this heterogeneity has fuelled the long-standing question in evolutionary biology: what accounts for the evolutionary success of polyploids or, more specifically, what ecological, phenotypic and genetic factors promote the formation and establishment of polyploid taxa (Thompson and Lumaret 1992)?

Our understanding of what polyploidy is has broadened dramatically since its discovery in 1907 (Lutz 1907), a transformation that has occurred in parallel with advances in genetic technologies and a better understanding of genome structure and diversity. Conventionally, polyploidy is defined as an increase in the number of complete chromosome sets relative to the ancestral state. For nearly 100 years, the primary source of evidence for polyploidy was, by definition, an increase in chromosome number in multiples of the base (monoploid) number. The formation of multivalents at meiosis and multisomic segregation of alleles in progeny are also symptomatic of having additional copies of

B.C. Husband (✉)
Department of Integrative Biology, University of Guelph, Guelph, Ontario N1G 2W1, Canada
e-mail: bhusband@uoguelph.ca

I.J. Leitch et al. (eds.), *Plant Genome Diversity Volume 2*,
DOI 10.1007/978-3-7091-1160-4_16, © Springer-Verlag Wien 2013

homologous chromosomes and have been important criteria for identifying certain kinds of polyploids, specifically autopolyploids (Müntzing 1936; Jackson and Casey 1982). Autopolyploids are generally defined as those whose duplicated chromosome sets arise from the same species, and have a high degree of homology among genome copies, whereas allopolyploids contain chromosome sets derived from different species and of weaker or incomplete homology (Kihara and Ono 1926; Ramsey and Schemske 1998) (see Sect. 16.4).

In the era of genomics, the phrase 'whole genome duplication (WGD)' has also been closely associated and used synonymously with the term polyploidy. This reflects the addition of a more cryptic form of evidence for polyploidy. Specifically, phylogenetic, genomic sequence and comparative genomic approaches continue to reveal polyploidy in the form of multiple copies of whole genome sequences or genome segments (Soltis and Soltis 1993; see also Fawcett et al. 2013, this volume). This evidence can occur without a concomitant duplication in chromosome number, suggesting that either genomic restructuring and duplication are occurring on a large scale within an existing set of chromosomes (not polyploidy) or polyploidization has occurred in the distant evolutionary past and was then followed by a reduction of chromosomes and/or genes through selection toward a diploid state (diploidization; Wolfe 2001; Tate et al. 2005). This reinforces the view that polyploids may fall anywhere along an evolutionary continuum from initial chromosome doubling to varying degrees of chromosomal and genomic restructuring that approaches a more diploid cytogenetic state. Not all polyploids will follow the same paths. While many ancient polyploids exhibit multiple gene/genome copies distributed throughout large portions of their genome, they will not necessarily exhibit the same consequences as having multiple chromosome sets (e.g., sterility; Stebbins 1950; Ramsey and Schemske 2002) or double the nuclear DNA content (Bennett and Leitch 2005), characteristic of young polyploids. Here, we mostly focus on patterns of polyploidy as conventionally defined, using chromosome number as the primary criterion.

After a century of research, our understanding of the incidence and evolution of polyploids in plants is still in flux. The early 1900s saw significant advances through the application of cytogenetic techniques to a wide range of taxa and consideration of its taxonomic implications. The mid and late-1900s saw large advances in the spatial scale of research due to methods such as flow cytometry and growing interest in the population processes giving rise to polyploids (Levin 1975; Felber 1991; Thompson and Lumaret 1992). More recently, comparisons of nucleotide sequences for large portions of whole genomes have offered a more complete glimpse into the evolution and organization of genome sequences and their function. Each advance has contributed and altered our inferences about the evolution and evolutionary significance of polyploids. Many scholarly reviews on the incidence of polyploidy have been published in recent decades with the purpose of exploring specific aspects such as their taxonomic distribution (Stebbins 1950), modes and rates of origin (de Wet 1980; Soltis and Soltis 1993; Ramsey and Schemske 1998; Soltis and Soltis 1999) and impact on genetic and evolutionary divergence (Otto and Whitton 2000; Soltis and Soltis 2000; Tate et al. 2005).

In this chapter, we examine some of the dominant patterns of polyploid incidence that have emerged in the literature but for which detailed explanations are still unresolved. These patterns are: (1) variation among taxonomic groups at or above species, (2) intraspecific variation, (3) variation in mechanisms of formation (auto vs. allo), (4) geographic/ecological patterns of polyploid incidence, and (5) associations between ploidy and life history (reproduction). Beginning with a summary of the incidence of polyploidy within major groups of plants, we review the state of evidence with respect to major patterns of incidence and the progress and challenges associated with understanding their evolutionary causes.

16.2 Taxonomic Variation

16.2.1 Variation Among Major Plant Groups

One of the most widely recognized patterns in the incidence of polyploidy in plants is the variation in frequency among different taxonomic groups (Table 16.1). With increased scope of sampling and higher throughput analyses, our estimates have become more complete if nothing else. From this evidence, it is apparent that polyploidy is widely distributed but that major plant lineages vary dramatically in both the frequency of polyploid species and the maximum ploidy level observed (Table 16.1). In this section, we provide an up-to-date account of the presence of polyploidy in major plant groups within the Archaeplastida (*sensu* Adl et al. 2005), i.e., the monophyletic clade of primary photosynthetic eukaryotes. The Archaeplastida (or plants *sensu lato*) comprises three major evolutionary lineages: (1) Glaucophytes, (2) Rhodophytes (red algae), and (3) Viridiplantae (green algae plus land plants).

16.2.1.1 Glaucophyta

Glaucophytes are a small group (3 genera, 13 species) of microscopic algae occurring in freshwater habitats (Keeling 2004). Despite their pivotal role in our understanding of the evolution of photosynthesis in eukaryotes, little is known about their chromosome number, ploidy level or genome size, perhaps at least partly due to the fact that none of the species is particularly common. Detailed karyological

Table 16.1 Frequency of polyploidy, the highest known ploidy level and the record-holder species in different plant groups. The groups do not represent comparable phylogenetic/taxonomic categories (Data compiled from different sources)

Plant group	Number of extant species	Incidence of polyploidy	Maximum ploidy level	Record-holder
Glaucophytes (Glaucophyta)	13	Unknown	Unknown	–
Red algae (Rhodophyta)	~6,000	Frequent	Moderate (~16-ploid)	*Polyides rotundus*, $n = 68–72$ (Cole 1990)
Chlorophyta (Green algae s.s.)	~3,800	Frequent	Moderate (~20-ploid)	*Eraemosphaera viridis*, n = 80 (Mainx 1927)
Charophyta (Green algae, streptophyte algae)	~5,000	(Very) frequent	(Very) high	*Netrium digitus*, $n = 592$ (King 1960)
Liverworts (Marchantiophyta)	~5,000	Rare (~8%)	Rather low (~12-ploid)	*Riccia macrocarpa*, $n = 48$ (Przywara and Kuta 1995)
Mosses (Bryophyta s.s.)	~12,000	Ambiguous (c. 20–80%)	Moderate (~16-ploid)	*Leptodictyum riparium*, *Physcomitrium pyriforme*, $n = 72$ (Przywara and Kuta 1995)
Hornworts (Anthocerotophyta)	~150	Absent	–	*Anthoceros sampalocensis*, $n = 10$ (Przywara and Kuta 1995)
Lycopods (Lycopodiophyta)	~900	(Very) frequent	Very high (~50-ploid)	*Huperzia prolifera*, $2n = \sim556$ (Tindale and Roy 2002)
Ferns and allies (Monilophyta)	~11,000	Very frequent (up to ~95%)	Very high (~96-ploid)	*Ophioglossum reticulatum*, $2n = \sim1440$ (Khandelwal 1990)
Cycads (Cycadophyta)	~250	Absent	–	*Zamia paucijuga*, *Z. prasina*, $2n = 28$ (Moretti and Sabato 1984)
Ginkgo (Ginkgophyta)	1	Absent	–	$2n = 24$
Conifers (Pinophyta)	~550	Very rare (<2%)	Low (hexaploid)	*Sequoia sempervirens*, $2n = 66$ (Hirayoshi and Nakamura 1943)
Gnetophytes (Gnetophyta)	~50	Moderate (~30%)	Low (octoploid)	*Ephedra funerea*, *E. gerardiana*, $2n = 56$ (Ickert-Bond 2003)
Monocots	~60,000	Very frequent (~75%)	Very high (~50-ploid)	*Voanioala gerardii*, $2n = \sim596$ (Johnson et al. 1989)
Dicots s.l.	~200,000	Very frequent (~75%)	Very high (~80-ploid)	*Sedum suaveolens*, $2n = \sim640$ (Uhl 1978)

investigations of glaucophytes are needed to ascertain the role of genome duplication in the early evolution of photosynthetic eukaryotes and the recently sequenced genome of the glaucophyte of *Cyanophora paradoxa* (Price et al. 2012) should shed light on this.

16.2.1.2 Rhodophyta (Red Algae)

The red algae (Rhodophyta) are a large and diverse group of microscopic algae and macroalgae (c. 6,000 species in 700 genera; Kapraun 2005) that mostly occur in marine environments. Some chromosome counts (Cole 1990) and estimates of nuclear DNA contents (Bennett and Leitch 2010) are available although still quite fragmentary. Haploid numbers range from $n = 2$ to 68–72 (Cole 1990; Kapraun 2005); some orders (e.g., Bangiales) have low chromosome numbers whereas others have predominantly high numbers (e.g., Ceramiales, Gigartinales or Gracilariales). Difficulties establishing the base chromosome number have precluded precise estimates of the ploidy level, but the maximum value in the haplophasic gametophytic generation is ~$16x$ (analogous to $32x$ in groups with predominant sporophytes such as seed plants). Allopolyploidy has been suggested as a major evolutionary force in rhodophytes (Nichols 1980) and more recent investigations show that multiple polyploidy events must have occurred independently in several lineages (Kapraun 2005). In addition, intraspecific ploidy heterogeneity recorded in several species (Cole 1990) suggests that genome duplications have occurred in the recent evolutionary history of red algae.

16.2.1.3 Viridiplantae (Green Plants)

Following Lewis and McCourt (2004) and Smith et al. (2009), the green plants (Viridiplantae) contain two major clades: (1) Chlorophyta (or green algae *sensu stricto*), and (2) Streptophyta. The latter is further divided into Charophyta (or strephophyte green algae) and Embryophyta (higher plants), which includes the non-vascular land plants (bryophytes) and vascular land plants (lycopods, ferns and allies, gymnosperms, angiosperms), both of which are karyologically relatively well known.

Green Algae

Green algae are a paraphyletic group of uni- and multicellular alga with plastids containing chlorophylls a and b. Recent evidence supports dividing this group into two major clades (Chlorophyta and Charophyta), which have distinct ecological preferences (predominance in marine and freshwater habitats, respectively; Becker and Marin 2009), plus a residuum of related organisms (Prasinophyta; Lewis and McCourt 2004; Kapraun 2007). The two major lineages have differences in genomic properties, particularly in the extent of polyploidy (e.g., Sarma 1982). A number of technical challenges to studying their chromosomes (e.g., small chromosome size, difficulties with sample preparation and in obtaining suitable material) preclude many general conclusions.

The Prasinophyta, a paraphyletic and morphologically diverse group, comprises tiny free-floating microorganisms with low cellular complexity (Becker and Marin 2009). Most published genome sizes of prasinophytes are small enough to suggest that the incidence of polyploidy is unlikely (range from 0.005 to 0.35 pg/1C, with a modal value around 0.01 pg/1C; Kapraun 2007). One exception is *Ostreococcus tauri*, the smallest free-living eukaryotic organism. Although the genome is as small as the chloroplast genome of other green algae, cytogenetic studies show that it is fragmented into a comparatively high number of chromosomes from 14 (Courties et al. 1998) to ~20 (Grimsley et al. 2010), raising the possibility of polyploidy in prasinophycean green algae.

Although polyploidy has been identified in many groups in each major lineage, the overall proportion of polyploid taxa and the maximum ploidy level in the Chlorophyta may be lower than that of the Charophyta (Kapraun 2005, 2007). The orders Desmidiales and Zygnematales, which are within Charophyta, are particularly rich in polyploid species, and are the most speciose groups of charophytes with approximately 4,000 described species. The incidence of polyploidy in these two groups has taxonomic implications in that cell dimensions are often used as diagnostic criterion for species delineation. Although difficult to determine, the rate of interspecific and intraspecific ploidy variation in the Desmidiales may be quite high (Sarma 1982; Hoshaw and McCourt 1988), and may be comparable to that of the angiosperms (see below). One species in this group, *Netrium digitus*, is the green alga with the highest known number of chromosomes. Different strains of this common species were found to possess from ~30 to ~592 chromosomes in the haplophase (King 1960; Table 16.1). The Charales consist of six extant genera with ~400 species (Kapraun 2007), many of which exhibit polyploidy and have chromosome numbers up to $n = 70$ (Sarma 1982). In addition, Charales possess the largest genome sizes of all algae (up to 19.6 pg/1C in *Chara contraria*; Bennett and Leitch 2010), with values being higher than the upper limit for older lineages of land plants (e.g., bryophytes). This is particularly important for evolutionary interpretations because several multigene phylogenies have placed Charales as a sister taxon to land plants (Becker and Marin 2009) although other molecular studies suggest Zygnematales holds this position (Wodniok et al. 2011).

Bryophytes

Roughly 19,000 bryophytes (i.e., non-vascular land plants) are classified into three monophyletic lineages, which differ markedly in the incidence of genome duplication: (1) liverworts (Marchantiophyta), (2) mosses (Bryophyta), and (3) hornworts (Anthocerotophyta). Chromosomal data have been compiled for more than 2,250 species (Fritsch 1991; Przywara and Kuta 1995), making bryophytes a moderately well explored plant group from a cytogenetic point of view.

There is a general consensus that genome duplication is rare in liverworts (~8% of species, plus ~6% that have both diploid and polyploid cytotypes) and absent in hornworts (using $n > 10$ as a threshold for polyploidy; Przywara and Kuta 1995); the proportion of polyploid mosses is a matter of debate. In contrast to other bryophytes, mosses show striking variation in base chromosome numbers, from $x = 4$ to $x = 19$. Assuming that polyploids have gametophytic numbers $n > 9$, Kuta and Przywara (1997) concluded that 84% of extant mosses have undergone polyploidy and 28% are considered high polyploids ($n > 15$). Other authors have challenged these estimates and suggest that genome duplication likely plays only a minor role in mosses (i.e., as low as 20% of species). Support for the paucity of polyploids has come largely from studies of genome size, which reveal a relatively small overall variation (~12-fold) and relatively constant values within genera and even families (Voglmayer 2000). Interestingly, the range of genome sizes in liverworts is much greater, and species with comparatively large genomes have recently been recorded (Temsch et al. 2010; see also Leitch and Leitch 2013, this volume). The evolutionary constraints on polyploidy are not clear but may be linked to strong selection for motile biflagellate male gametes. The majority of known polyploids in mosses seem to be of autopolyploid origin although there is growing evidence that allopolyploids are more frequent than originally supposed (e.g., Wyatt et al. 1988; Orzechowska et al. 2010) and may even dominate in some groups such as peat mosses (e.g., Ricca et al. 2008; Karlin et al. 2010). Regardless of the frequency of polyploids and mode of origin, the maximum ploidy in bryophytes is rather low, reaching only about 16*x* in haplophase. One of the record-holders is *Physcomitrium pyriforme*, a polyploid complex that encompasses several different cytotypes with up to $n = 72$ (Table 16.1).

Ferns and Allies (Monilophytes)

The highest incidence of polyploidy among all plants is found in monilophytes, a monophyletic group comprising five groups of spore-bearing vascular plants (Psilotales—whisk ferns; Ophioglossales—ophioglossoid ferns; Equisetales—horsetails; Marattiales—marattioid ferns; and Leptosporangiatae—leptosporangiate ferns), with around 95% of species believed to have experienced one or more rounds of genome duplication in their evolutionary history (assuming that plants with chromosome numbers above $n = 14$ are polyploid; Grant 1981). The average gametic chromosome number in monilophytes (e.g., ~57 in homosporous leptosporangiate ferns) is higher than in any other group of vascular plants except lycopods (for example it is just ~16 in angiosperms; Klekowski and Baker 1966). However, this theory was later questioned by the results of protein electrophoretic analyses, which used the number of isozymes per enzyme system as a proxy for ploidy level (Haufler 1987; Soltis and Soltis 1987). Despite often very high chromosome numbers, many leptosporangiate homosporous ferns and horsetails were found to be genetically diploid. These results led some authors to conclude that the level of polyploidy in ferns and allies may in fact be overestimated and supported much more conservative estimates (e.g., Vida (1976) suggested that just 43.5% of species were polyploid) (see also Barker 2013, this volume).

The observed incongruence between karyological and protein electrophoresis data can be explained either by the origin of monilophytes with high chromosome numbers in the diploid state or, more probably, by the evolution of this lineage through several cycles of polyploidy followed by chromosomal diploidization, gene silencing, and extinction of diploid progenitors (Haufler 1987). In general, the incidence of polyploidy is higher in groups with subterranean gametophytes (whisk ferns, ophioglossoid ferns) than those with photosynthetic and aboveground gametophytes, and higher in homosporous leptosporangiate ferns than in heterosporous ones (Klekowski and Baker 1966; Haufler 1987). Some authors have suggested that polyploidy provides a mechanism for maintaining genetic diversity in groups with subterranean and spatially isolated monoecious gametophytes, which have a higher chance of self-fertilization and increased homozygosity. Perhaps not surprisingly, the maximum ploidy level ever recorded in any plant is from the pantropical adder's-tongue species *Ophioglossum reticulatum* which may reach up to 96*x* in individuals with 1,440 somatic chromosomes (Khandelwal 1990; Table 16.1). Gametic chromosome numbers above 200 (and correspondingly high levels of polyploidy) are also common in other ophioglossoid ferns and some related whisk ferns (Brownsey and Lovis 1987). On the contrary, polyploidy is of lower evolutionary significance (and less frequent) in heterosporous ferns, which have separate male and female gametophytes that facilitate out-crossing.

Lycopods

Polyploidy is extensive in the Lycopodiophyta, the oldest extant lineage of spore-bearing vascular plants (tracheophytes), which is sister to all other tracheophytes. Usually, three families are recognized within this lineage, the homosporous Lycopodiaceae (clubmosses and firmosses) and the heterosporous Isoetaceae (quillworts) and Selaginellaceae (spikemosses). The distribution of polyploidy in lycopods generally mirrors the pattern described above for monilophytes (e.g., higher frequency in homosporous species). Considering $x = 11$ to be the base chromosome number of the genus *Huperzia* (Löve et al. 1977), the level of polyploidy in lycopods may reach ~50*x* (Table 16.1). However, as in ferns, several lycopod species have been found to possess isozyme patterns typical of diploid plants despite their high number of chromosomes (Soltis and Soltis 1988). Interestingly, aquatic heterosporous quillworts contain more than 60% polyploid species, forming polyploid series ranging from 3*x* to 12*x*, and allopolyploidy is considered to be one of the major modes of speciation in this group (Troìa 2001).

Gymnosperms

The extant gymnosperms (cycads, *Ginkgo*, conifers, and gnetophytes) form a diverse but relatively small group of woody seed-bearing plants with exposed ovules and seeds. Due to their limited number of species (<1,000), showy appearance, dominance in many terrestrial ecosystems and high value in forestry, no other group of plants can compete with gymnosperms in the wealth of karyological data.

There is compelling evidence that polyploidy is absent in cycads and *Ginkgo* and extremely rare in conifers (Table 16.1). Despite intensive research, only four polyploid coniferous species, all belonging to family Cupressaceae, have been found (Khoshoo 1959; Ahuja 2005): three tetraploids—*Fitzroya cupressoides*, *Juniperus chinensis* 'Pfitzeriana' and *Juniperus squamata* 'Meyeri', and one hexaploid—*Sequoia sempervirens*. While *Fitzroya* is likely an autotetraploid, *Juniperus chinensis* likely originated by allopolyploidy (the origin of the other two taxa is uncertain). The lack of polyploidy in gymnosperms may be, at least partly, explained by their large monoploid genome sizes, which impose constraints on plant development. Indeed, experimentally-induced conifer polyploids showed reduced growth, dwarfism and premature mortality (Ahuja 2005). In addition, concomitant changes associated with polyploidy such as increased cell size might not be compatible with the proper formation of woody fibres, thus constraining the rise of ploidy level in woody plants in general.

Gnetophytes are unique among gymnosperms in having a much higher proportion of polyploids (Table 16.1) although duplicated genomes (based on $x = 7$) have been observed with certainty only in the genus *Ephedra* (Ahuja 2005). About half of *Ephedra* species are polyploid but generally only at the tetraploid level with the exception of octoploid counts in *E. funerea* and *E. gerardiana* (Ickert-Bond 2003). No obvious geographic pattern exists in the distribution of diploid and polyploid species, since polyploidy occurs worldwide. Despite high chromosome numbers in *Gnetum* ($2n = 44$) and *Welwitschia* ($2n = 42$) relative to other gymnosperms, whether these genera are polyploid is unclear. A detailed karyological investigation of the latter species (Khoshoo and Ahuja 1963) revealed an unusual chromosome complement consisting of only telocentric chromosomes, with one pair bearing a satellite, quite dissimilar from that of *Ephedra*. The high chromosome number in *Welwitschia* does not necessarily imply that the species is hexaploid. In fact, genome sizes of both *Gnetum* and *Welwitschia* are at the lower end of published values for gymnosperms (Bennett and Leitch 2010; see also Leitch and Leitch 2013, this volume).

Angiosperms

Chromosome numbers have been determined in about 25% of angiosperm species (Stace 2000), which provides a solid basis for spatial comparisons of the incidence of polyploidy and assessing relationships between genome duplication and other plant traits (described below). Based on chromosome number, genome duplication is frequent in both dicots and monocots (Table 16.1) and can attain high levels, such as ~80*x* in the dicotyledonous Mexican stonecrop *Sedum suaveolens* (Uhl 1978) and ~50*x* in the monocotyledonous Madagascan palm *Voanioala gerardii* (Johnson et al. 1989).

Estimates of the standing frequency of polyploidy vary from 30% to 80% (Masterson 1994) depending on how they are calculated. Stebbins (1938) estimated that 20–40% of angiosperms were polyploid based on the mean proportion of species per genus that have chromosome numbers that are more than double the base number of the genus. This estimate is similar to Wood et al. (2009) for vascular plants (35%), based on current phylogenetic data and similar criteria. Undoubtedly these values underestimate polyploidy as they do not consider chromosome multiplication that occurs in association with the origin of different genera. In contrast, Grant (1963) assumed that the chromosome base number for angiosperms was $n = 7-9$ and that a species with $n \geq 14$ was by definition polyploid. Based on this premise, he estimated that 57% of species were polyploid. In turn, Goldblatt (1980) and Masterson (1994) argued that any organism with $n \geq 11$ constituted a polyploid, which amounted to 70% of flowering plants. Otto and Whitton (2000) estimated the incidence of polyploidy (polyploid index) based on abundance of haploid chromosomes that are even-numbered compared to those that are odd-numbered. From this, they estimated moderate values for both monocots (31.7%) and dicots (17.7%).

16.2.2 Evolutionary Implications of Taxonomic Variation

16.2.2.1 Preadaptations for Polyploidy

Heterogeneity in the standing frequency of polyploidy is a well-established feature of plants, regardless of taxonomic level. Such variation suggests that taxonomic groups have either different predispositions to polyploid evolution or different amounts of time to accumulate polyploids (Meyers and Levin 2006), or both. Heterogeneity among sister clades, for which divergence time is controlled, suggests that at least some lineages are preadapted to polyploidization. For example, within the genus *Chamerion* (formerly *Epilobium*), sister species *C. angustifolium* and *C. latifolium* have both evolved autotetraploid forms in parallel (Mosquin and Small 1971), whereas the sister clade, consisting of *C. dodonaei* and *C. fleischeri*, exhibit no polyploidy. Many such examples exist and raise the question: what attributes of these species pairs contribute to the difference in polyploidy? The answer may rest with a range of attributes that influence either the formation of polyploids (e.g., unreduced gamete production) or their establishment (clonality, self-fertilization, ecological differentiation). The former are largely unexamined in natural populations (although see Bretagnolle and Thompson 1995; Ramsey and Schemske 2002) and the latter are invoked primarily through comparative approaches (Levin 2002), and so the specific traits involved are rarely unambiguous.

In addition to these specific phenotypic attributes, the incidence of polyploidy likely varies as a function of intrinsic genetic and genomic attributes. For example, Wood et al. (2009) show clearly that the incidence of polyploidy is highest in genera with a low base chromosome number (Fig. 16.1). Similarly, Grif (2000) found a negative association between polyploid incidence and genome size. This pattern suggests that polyploid formation/establishment is subject to intrinsic constraints determined by the size of the genome itself. The biological basis for this constraint is not fully understood but could reflect the increasing negative effects of high gene dosage, large cell size and increased meiotic irregularities associated with high genome copy number. Interestingly, recent genetic analyses of polyploid yeast indicate that genes controlling mitotic spindle formation, chromosome cohesion and homologous recombination, and DNA repair may be particularly important in determining the stability of polyploids (Štorchová et al. 2006; Thorpe et al. 2007). Whatever the cause, the negative relation between genome size and frequency of polyploidy within genera helps to explain the progressive decline in the incidence of higher ploidy levels on the landscape. This pattern is clearly depicted in a recent survey of chromosome number and ploidy in the Californian flora (Fig. 16.2; J. Ramsey and B. C. Husband, unpublished).

16.2.2.2 Polyploidy and Evolutionary Diversification

The prevalence of polyploidy within plants has led to debate about the influence of genome duplication on net species diversification (Otto and Whitton 2000). Researchers have

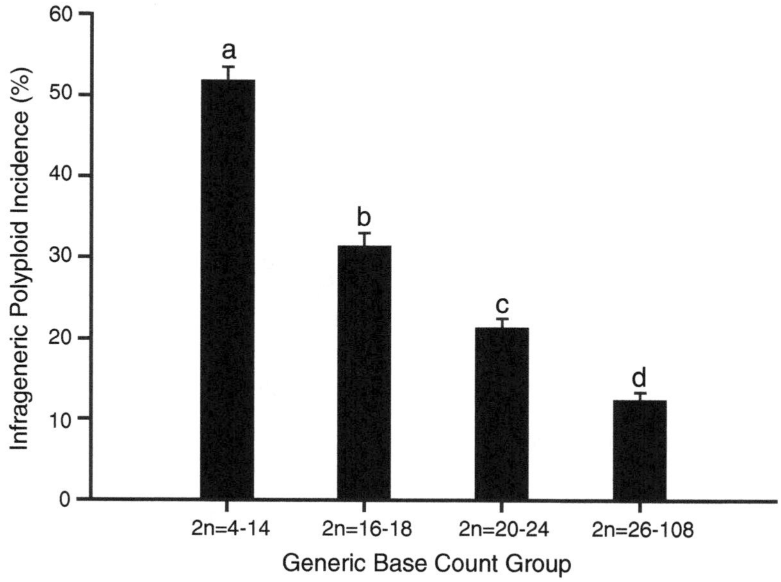

Fig. 16.1 Relationship between the base chromosome number per genus and the incidence of polyploidy (relative to base chromosome number) in angiosperms. (From Wood et al. 2009 with permission)

argued, on one hand, that polyploidy may simply accumulate as a unidirectional mutation with little influence on the rate or trajectory of evolutionary divergence (Stebbins 1971; Meyers and Levin 2006). This idea is based on the premise that ploidy is more likely to increase than decrease in time. Although reversals do occur, the resulting haploids are relatively rare and often of reduced fitness and high sterility (Stebbins 1980; Ramsey and Schemske 2002). Alternatively, polyploidy itself may enhance evolutionary diversification leading to increased divergence and species richness (Otto and Whitton 2000). Elevated diversification may occur if polyploidy accelerates adaptive divergence and evolution of reproductive isolation (Otto and Whitton 2000) or reduces species extinction. This divergence may be enhanced because of the ability to differentiate extra gene copies (subfunctionalization, neofunctionalization) or the effects of genomic restructuring on standing genetic and phenotypic variation (Schranz and Osborn 2004; Tate et al. 2005; see also Fawcett et al. 2013, this volume).

Several studies have tested for elevated divergence rates in polyploids by looking for a positive relationship between the incidence of polyploidy in specific taxonomic groups (e.g., genera) and species richness. Using this approach, Petit and Thompson (1999), Otto and Whitton (2000), Vamosi and Dickinson (2006), and Wood et al. (2009) found a weak increase in species richness per genus with the incidence of polyploidy. However, this method confounds species richness with age of lineage (Meyers and Levin 2006). Vamosi and Dickinson (2006) controlled for time by comparing diploid and polyploidy sister taxa in the Rosaceae and, in that case, found no difference with respect to species number. Similarly, Wood et al. (2009) compared the species richness of 59 sister genera, separated by a ploidy increase, from 34 family-level phylogenies. They, too, found no tendency for polyploid genera to have increased species diversification. Taken together, the evidence suggests that although polyploidy may contribute significantly to species diversification, there is no indication that polyploid lineages have a higher rate of diversification than non-polyploid lineages. This result is corroborated by experimental selection studies, which find that higher ploidy lineages most often have reduced responses to selection compared to their lower ploidy progenitors (Zeyl et al. 2003; Anderson et al. 2004).

Quantitative estimates of the contribution of polyploidy to speciation have been lacking from the literature until recently. Otto and Whitton (2000) developed a model for estimating the incidence of polyploidy per speciation event using the proportion of species with even versus odd chromosome numbers, expressed as a proportion of the minimum number of speciation events associated with a chromosome number change. They estimated that the percentage of speciations associated with genome duplication is ~7% in ferns and 2–4% in angiosperms. Most recently, Wood et al. (2009) used 120 phylogenetic trees for angiosperms and 20 trees for ferns to estimate more directly the rate of polyploid speciation; they estimated that 15% of angiosperm speciation events and 31% of ferns are associated with a ploidy increase.

16.2.2.3 Plants versus Animals

Although the incidence of polyploidy in animals is better understood and recognized as more common than once thought (Gregory and Mable 2005), it is still clear that the phenomenon is more widespread in plants, particularly among sexually reproducing organisms. Why polyploidy is more prevalent in plants than animals has perplexed biologists and is still unresolved. The dominant explanations relate to differences in the likelihood of formation of polyploids or the potential consequences for the function and survival of

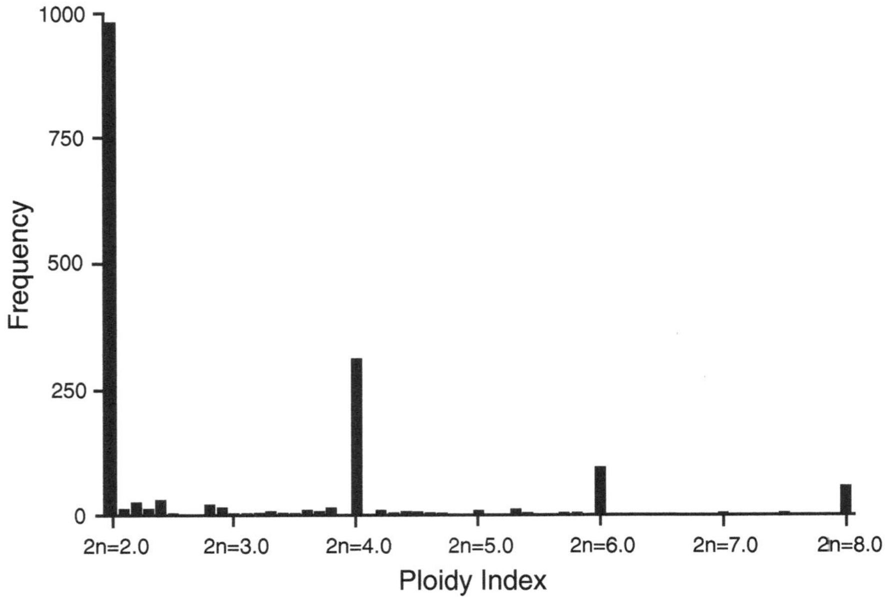

Fig. 16.2 Distribution of somatic ploidy level (2*n*) among species of the Californian flora (N = 3,163 biological species; 346 genera). Ploidy index for each species is expressed as a multiple of the base chromosome number for the respective genus. Values range from $2n = 2x$ to $2n = 20x$ in the data set but are only shown to $2n = 8x$. The modal peaks illustrate the prevalence of orthoploid-euploids compared to anorthoploids and dysploids or aneuploids among taxa (J. Ramsey and B. C. Husband unpublished)

polyploid individuals (Gregory and Mable 2005). Jackson (1976) postulated that triploid infertility posed a barrier to tetraploid formation in animals. However, triploids are also prevalent in plants and, in fact, it has been argued they are the very pathway by which polyploids are produced (de Wet 1980; Husband 2004). Muller (1925) suggested that polyploidy interferes with the sex-determining mechanisms, specifically the sex chromosome : autosome ratios in species where the X chromosome is dominant. Unfortunately this mode of sex-determination is not prevalent among animals, and Muller's explanation cannot account for the absence of polyploidy in species with a Y-dominant sex-determining mechanism. Moreover, Muller's hypothesis that polyploidy disrupts sex expression is not corroborated by the abundance of dioecious plant species that contain polyploid series or can be duplicated experimentally (e.g., Li et al. 2010). Orr (1990) suggested that in species with a degenerate sex chromosome, polyploidy disrupts the X : autosomal balance and hence dosage compensation mechanisms with fatal effects. Plants are less vulnerable to this because they rarely have degenerate sex chromosomes. Thus, while these specific mechanisms may apply in organisms with dominant X chromosomes and with degenerate sex chromosomes, it cannot account for the general scarcity of polyploidy in animals (Gregory and Mable 2005). Other considerations, such as the effect of duplication on development, reproductive isolation, the incidence of hybridization and mobility may also be important (Gregory and Mable 2005).

16.3 Intraspecific Variation

Although polyploidization is often associated with species diversification (beginning with Hagerup 1927) due to the barriers to gene flow that result from chromosome multiplication, ploidy variation is also observed within taxonomic species. The presence of intraspecific ploidy diversity has three main causes. First, genome duplication is an ongoing genomic mutation within many natural populations and ploidy polymorphism is therefore expected when polyploid formation is common. In fact, Ramsey and Schemske (1998) estimated that polyploids form at the rate of 10^{-5} per individual per generation, well above the genic mutation rate, and in some cases it may be even higher than that (Husband 2004). Second, depending on the frequency and pathways of polyploid formation, diploids and polyploids may be maintained as genetically homogeneous populations, which are justifiably treated as single taxonomic species. Finally, ploidy polymorphism may arise as an artefact of our imperfect taxonomic approach. Since morphological divergence is often used as a criterion for species delineation, polyploid lineages may not be recognized at the species level, especially in autopolyploids, which often have a close resemblance to their diploid progenitors (Soltis et al. 2007). In many cases, the patterns of variation within mixed-ploidy species can provide insights into the early stages of polyploid evolution.

16.3.1 Occurrence in Natural Populations

Intraspecific ploidy variation has been documented in many plant groups. Based on a broad survey of species, Wood et al. (2009) reported that 12–13% of angiosperm species and 17% of fern species are variable for ploidy. Similarly, a survey of the Californian flora indicated that nearly 13% of the 2,647 species were polymorphic for ploidy (J. Ramsey and B. C. Husband, unpublished; Soltis et al. 2007). Intraspecific ploidy variation has also been observed in most other plant groups such as red algae (Cole 1990), green algae (Kapraun 2007), liverworts (Przywara and Kuta 1995), and mosses (Kuta and Przywara 1997). Surprisingly, ploidy heterogeneity has even been detected in horsetails (*Equisetum*; Bennert et al. 2005), which have long been a textbook example of a chromosomally-uniform group. However, to date, no intraspecific ploidy variation has been detected in most gymnosperms, with the possible exception of a few *Ephedra* species (e.g., *E. breana*; Khoshoo 1959).

The global frequency of ploidy variation within plant species is difficult to estimate at this time because the vast majority of species have not been investigated in enough detail to confidently describe their ploidy composition. Historically, the frequency of intraspecific ploidy variation was unknown because large-scale karyotypic surveys were time and cost intensive and not deemed essential for cytotaxonomic treatments. The consequence is illustrated well in a review of ploidy estimates in 185 species of wild potato (*Solanum* section *Petota*). Hijmans et al. (2007) estimated that, due to inadequate sampling, only 28% of the mixed-ploidy species in this group have likely been discovered. Fortunately, new high-throughput tools, such as flow cytometry, are enabling more species to be analysed with much higher sample sizes (Kron et al. 2007). Since the application of flow cytometry for ploidy determination, there have been many new observations of additional ploidy levels. For example, nonaploids have now been detected in the mostly hexaploid *Elytrigia repens* and *E. intermedia* ($9x = 3.8\%$, $6x = 96.2\%$, $N = 238$; Mahelka et al. 2005) and penta- and heptaploids in the mostly hexaploid *Senecio carniolicus* ($5x = 1.1\%$, $7x = 0.3\%$, $6x = 55.5\%$, $N = 380$; Suda et al. 2007). In the latter study the sample size has now been increased from 380 to 5,033, and three additional ploidy levels ($3x$, $8x$, $9x$) have been detected, making *S. carniolicus* one of the most ploidy-diverse plant species ($2x$–$9x$ inclusive; Sonnleitner et al. 2010; Table 16.2). Overall it is likely that ploidy heterogeneity in species is underestimated and may continue to increase with more intensive sampling. Furthermore, because of the variation in sampling intensity, caution must be taken when comparing rates of intraspecific polymorphism between plant groups.

In species with intraspecific ploidy variation, the number of ploidy states varies from at least two to eight. The distribution of ploidy states across species depends on how the data were collected. In a survey of chromosome numbers among polymorphic species of the Californian flora ($N = 334$), which is based on moderate sampling per species, most species had two ploidy states and the frequency declined sharply with progressively higher numbers of ploidy states (Fig. 16.3; J. Ramsey and B. C. Husband, unpublished). In contrast, using studies of 26 species with more intensive sampling (average sample size of 1,744), species with three cytotypes were most common and the frequency decreased less consistently as ploidy number increased (Fig. 16.3, Table 16.2). The most common two-cytotype combination was $2x/4x$ in both surveys while the most common three-cytotype combination was $2x/4x/6x$ in the floristic analysis and $2x/3x/4x$ in the intensive studies. In many of the intensive studies, triploids occurred at low frequencies as in *Chamerion angustifolium* (3.3%; Sabara 2008), *Galax urceolata* (11%; Burton and Husband 1999), *Heuchera grossulariifolia* (1.4%; Thompson et al. 1997), and *Melampodium leucanthum* (0.2%; Stuessy et al. 2004). One other notable difference between the two datasets was the high incidence of odd ploidies ($3x$, $5x$, etc.) in the intensively studied species (84.6%) as opposed to the floristic survey (~4%). This may reflect the increased chance of detecting rare, odd-numbered ploidies with increased sampling. The prevalence of studies that detect odd-numbered ploidies in sexual species suggests that intercytotype hybridization may be more widespread than initially thought. The low frequencies of such hybrids (Husband and Schemske 1998; Baack 2004; Trávníček et al. 2010, 2011a; but see Trávníček et al. 2011b) may reflect the dependence of polyploidy formation on rare unreduced gametes and the low relative fitness of these ploidies.

16.3.2 Cytotype Distribution

The spatial relationships between ploidy states within species can be categorized as sympatric, parapatric or allopatric, depending on whether they are geographically intermixed, adjacent or disjunct, respectively. Examples of all three forms have been described in the literature. One of the most extreme examples of an allopatric distribution involves *Nymphoides indica*. This pantropical species is diploid in the Old World but tetraploid in the New World (Ornduff 1970). In contrast diploid and tetraploid cytotypes of the endemic *Galax urceolata* have a sympatric distribution throughout much of the species' geographic range in the Appalachian Mountains (Nesom 1983; Burton and Husband 1999). An example of parapatry is *Chamerion angustifolium*, which has diploids in northern latitudes and tetraploids in southern latitudes in North America and a narrow zone of sympatry along the southern limit of the boreal forest and in the Rocky Mountains (Husband and Schemske 1998). Similarly, in *Vicia cracca* of Central Europe, diploids in the east give way to tetraploids in the west (Trávníček et al. 2010).

Geographic patterns of polyploidy may arise through two main mechanisms. Sympatric and parapatric distributions can arise when polyploids form *de novo* from local diploids

Table 16.2 Summary of 26 intensive surveys of intraspecific ploidy variation. Ploidy variation was surveyed across 2–336 populations and involved sample sizes between N = 145 and 6,554. Surveys did not always cover the entire geographic range of the species and occasionally mixed-cytotype populations were sampled more extensively. + indicates that presence of the cytotype was reported but the overall frequency was not

			Cytotypes												Populations		
Species	Family	N	2*x*	3*x*	4*x*	5*x*	6*x*	7*x*	8*x*	9*x*	10*x*	12*x*	#	% odd	N	% mixed	Reference
Actinidia chinensis	Actinidiaceae	338	64.2		14.8	0.6	20.4						4	0.6	16 ^{E}	62.5	Li et al. (2010)
Allium oleraceum	Amaryllidaceae	4,347			+	+	+						3		325 ^{P}	23.1	Duchoslav et al. (2010)
Allium przewalskianum	Amaryllidaceae	844	26.3		73.7								2	0.0	62 ^{P}	6.5	Xie-Kui et al. (2008)
Arnica cordifolia	Asteraceae	870		43.4	55.3	1.3							3	44.7	21 ^{E}	57.1	Kao (2008)
Aster amellus	Asteraceae	2,175	34.3	0.1		0.1	65.4			0.1			5	0.3	87 ^{P}	6.9	Mandáková and Münzbergová (2006)
Arrhenatherum elatius	Poaceae	145	72.4	0.7	26.9								3	0.7	2 ^{C}	50.0	Petit et al. (1997)
Centaurea jacea	Asteraceae	541	41.2		58.6		0.2						3	0.0	26 ^{P}	15.4	Hardy et al. (2001)
Chamerion angustifolium	Onagraceae	2,628	53.9	3.3	42.8								3	1.4	51 ^{C}	58.8	Sabara (2008)
Dianthus broteri	Caryophyllaceae	244	25.8	2.0	41.4		11.1					19.7	5	2.0	25 ^{E}	4.0	Balao et al. (2009)
Galax urceolata	Diapensiaceae	1,570	55.0	11.0	34.0								3	34.0	42 ^{P}	59.5	Burton and Husband (1999)
Gymnadenia conopsea s.l.	Orchidaceae	3,581			61.7		1.9		35.6		0.5	0.3	5	0.0	43 ^{P}	58.1	Trávníček et al. (2011a)
Heuchera grossulariifolia	Saxifragaceae	855	87.9	1.4	10.7								3	1.4	14 ^{E}	21.4	Thompson et al. (1997)
Ixeris nakazonei	Asteraceae	592		0.7	17.6	3.0	18.3	0.8	59.6				6	4.5	98 ^{E}	20.4	Denda and Yokota (2004)
Knautia arvensis	Dipsacaceae	4,648	36.95	0.04	64.01	[b]	[b]						3	0.04	279 ^{P}	6.5	Kolář et al. (2009)
Lythrum salicaria	Lythraceae	1,884	4.5	0.9	86.2		8.4						4	0.9	124 ^{P}	2.4	Kubátová et al. (2008)
Melampodium cinereum	Asteraceae	156	51.3	0.6	48.1								3	0.6	75 ^{E}	2.7	Stuessy et al. (2004)
Melampodium leucanthum	Asteraceae	440	77.5	0.2	22.3								3	0.2	188 ^{E}	2.7	Stuessy et al. (2004)
Oxycoccus palustris	Ericaceae	252			17.5	32.1	50.4						3	32.1	11 ^{P}	54.5	Suda (2003)
Parasenecio auriculata	Asteraceae	271	54.2	0.4	45.4								3	0.4	44 ^{P}	4.5	Nakagawa (2006)
Pilosella echioides	Asteraceae	4,410	26.6	42.6	28.7	1.9	0.2						5	44.5	46 ^{E}	23.9	Trávníček et al. (2011b)
Pilosella officinarum	Asteraceae	1,059			40.2	36.7	22.8	0.3					4	37.0	336 ^{E}	10.6	Mráz et al. (2008)
Plantago media	Plantaginaceae	569	58.0	0.2	41.8								3	0.2	27 ^{C}	33.3	van Dijk et al. (1992)
Ranunculus adoneus	Ranunculaceae	1,618	42.1	1.2	56.7								3	1.2	38 ^{E}	5.3	Baack (2004)
Senecio carniolicus	Asteraceae	5,033	33.8	0.1	15.6	0.7	49.6	0.1	0.04	0.1			8	1.0	100 ^{P}	45.0	Sonnleitner et al. (2010)
Solidago altissima	Asteraceae	500	65.8		11.6		22.6						3	0.0	16 ^{C}	50.0	Halverson et al. (2008)
Vicia cracca	Fabaceae	6,554	34.5	0.1	65.4								3	0.1	257 ^{P}	7.0	Trávníček et al. (2010)
Mean		1,744											3.7	8.3	90.5	26.6	

^{E} entire range, ^{C} contact zone only, ^{P} part of range (not specifically the contact zone)

[b]Hexa- and pentaploids were also present but they correspond to *K. dipsacifolia* and *K. arvensis* × *dipsacifolia* hybrids, respectively

(primary contact). Alternatively, polyploids may have diverged elsewhere and, through range expansion, have come into contact with more distantly related diploids (secondary contact). These two scenarios cannot be easily distinguished without additional genetic evidence. In primary contact zones, plants of different ploidy are genetically more similar and more closely related to each other than to geographically distant plants of the same ploidy, whereas cytotypes in regions of secondary contact are genetically divergent. A mosaic of both primary and secondary contact zones has recently been suggested for several species (e.g., Rivero-Guerra 2008; Kolář et al. 2009; Trávníček et al. 2010) but only rarely confirmed by molecular techniques (Kolar et al, 2012). Additional genetic information and

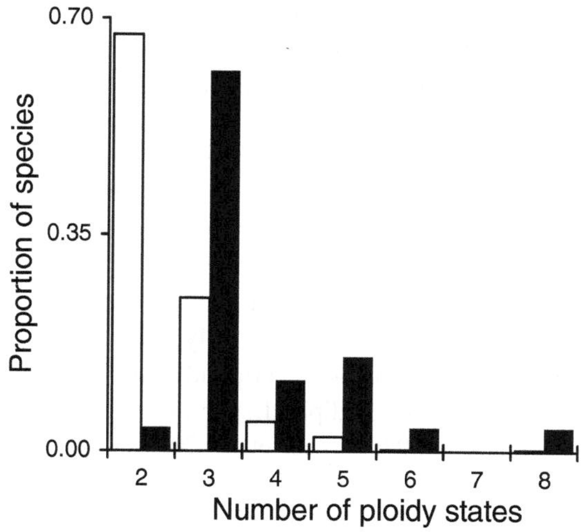

Fig. 16.3 Frequency of species containing two to eight ploidy states in the Californian flora (*open bars*, N = 334; J. Ramsey and B. C. Husband, unpublished) and among 26 arbitrarily selected species with intensive sampling (*solid bars*; studies listed in Table 16.2). Note the large difference in the proportion of species with only two ploidy states observed between the two data sets

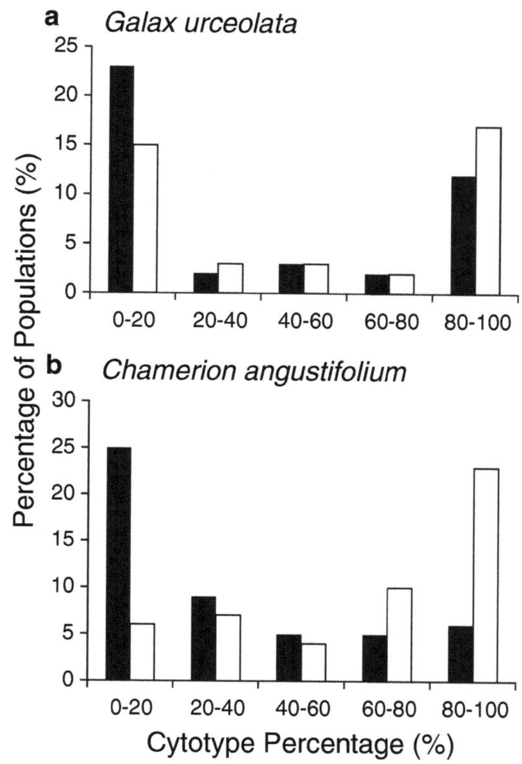

Fig. 16.4 Distribution of diploid (*black bars*) and tetraploid (*white bars*) relative frequencies among (**a**) 40 populations of *Galax urceolata* in southeastern United States (From Burton and Husband 1999) and (**b**) 51 populations of *Chamerion angustifolium* in the Rocky Mountains, Canada (Sabara 2008). Populations were located within the sympatric zones of each species

reciprocal transplants would be required to determine what maintains these distributions and specifically to distinguish the role of ecological differentiation and environmental sorting versus selection against hybrids (tension zone).

16.3.3 Cytotype Frequencies Within Populations

High-throughput ploidy analyses such as flow cytometry are now enabling intensive population-level surveys, which have improved detection of rare ploidies, mixed populations and estimates of cytotype frequencies within populations. The incidence of mixed-ploidy populations is of particular interest because it provides insight into the mechanisms of origin and coexistence of polyploids. Global estimates, however, are still uncertain because of limited numbers of studies, heterogeneity in sampling procedures, and complexities of the geographic structure of species. Based on intensive geographic surveys of 26 mixed-ploidy angiosperm species (from 15 families), the proportion of populations with multiple ploidies per species was variable and ranged from 2.4% to 62.5%, with the mean of 26.6% (Table 16.2). One example is the perennial herbaceous plant, *Chamerion angustifolium*. Within the contact zone in the Rocky Mountains, 59% of populations surveyed contained both ploidies and several of those had triploid hybrids (Husband and Sabara 2004; Sabara 2008). In *Gymnadenia conopsea*, up to five different cytotypes can coexist within natural populations (Trávníček et al. 2011a). We note that variation in mixed-ploidy populations differs among species in part due to variation in the degree of sympatry among cytotypes and the degree to which sampling was restricted to the contact zone. Estimates of ploidy mixture are therefore likely to be overestimates of the species-wide frequency of mixed-ploidy populations.

How does the incidence of mixed-ploidy populations compare to theoretical expectations? Theoretical studies predict that within zones of sympatry, mixed-ploidy populations (and coexistence generally) should be relatively uncommon and evolutionary unstable because of frequency-dependent selection against rare cytotypes (Levin 1975; Husband 2000). This is because, under random mating, most potential mates of the rare cytotype are of the alternate ploidy, yielding unfit hybrids (Felber 1991). This prediction, however, has rarely been tested in natural populations. Within the contact zone of *Chamerion angustifolium*, the ploidy frequency distribution matches the expected U shaped pattern. That is, populations are more likely to be dominated by one cytotype (usually tetraploids, in this case) than have even frequencies (Fig. 16.4; Husband and Sabara 2004; Sabara 2008). Similarly, in *Galax urceolata*, evenly mixed-ploidy populations within the zone of sympatry are relatively uncommon. Most populations are dominated by one cytotype, occasionally diploids but usually tetraploids (Fig. 16.4; Burton and Husband 1999).

The presence of mixed-ploidy populations in many published studies raises the question of what allows cytotypes to coexist within populations. Theory suggests that multiple cytotypes may persist if they are prezygotically isolated by spatial or reproductive segregation (Fowler and Levin 1984; Rodriguez 1996) or if the rare cytotype forms recurrently (Felber 1991; Bever and Felber 1992). Evidence for both processes have been reported. Spatial segregation of cytotypes due to either ecological differentiation or environmentally-independent processes such as founder effects or limited seed dispersal (Kolář et al. 2009) is frequently observed (Bretagnolle and Thompson 1996; Felber-Girard et al. 1996; Duchoslav et al. 2010; Sonnleitner et al. 2010) and widely considered a key factor in the coexistence of different ploidies. Polyploidization is often accompanied by changes in morphology or physiology, resulting in ecological differentiation between diploids and their polyploid derivatives (Levin 2002). Habitat segregation often occurs at different spatial scales. On a coarse grain, spatial arrangement of different cytotypes may range from near-complete niche differentiation (e.g., Felber-Girard et al. 1996) to apparent sympatry, which may disappear as the scale of observation becomes finer (Trávníček et al. 2011a, b). Although spatial segregation is widely observed, more information is needed on the influence of this separation on patterns of prezygotic mating, which is necessary to overcome minority cytotype exclusion (Husband and Schemske 1995; Husband and Sabara 2004).

In the absence of strong spatial isolation, coexistence may be favoured by various reproductive barriers that prevent hybridization between neighbouring cytotypes (Husband and Sabara 2004). Such barriers can exist at any number of steps in the reproductive process. Studies have found differences between ploidy states with respect to insect visitor composition (Segraves and Thompson 1999) or insect foraging patterns (Kennedy et al. 2006), which diminish pollen exchange between cytotypes. This isolation can be reinforced by flowering asynchrony, associated with differences in cell cycles and development rates between ploidies (Bretagnolle and Thompson 1996; Jersáková et al. 2010). Reproductive isolation acting between pollination and fertilization (gametophytic selection) may also be a common mechanism to prevent hybridization (Howard et al. 1998). Reduced germination on stigmata, reduced pollen tube growth or decreased ovule fertilization of heterospecific pollen may result in conspecific gamete precedence (Howard 1999). As demonstrated for *Chamerion angustifolium*, conspecific pollen precedence can be strong and asymmetrical, and may evolve quickly as a direct consequence of polyploidization (Husband et al. 2002; Baldwin and Husband 2011). Higher rates of self-fertilization in polyploids (Barringer 2007; Husband et al. 2008) compared with diploids due to the breakdown of self-incompatibility systems (Stone 2002) and reduced inbreeding depression (Husband and Schemske 1996) can also reduce extra-ploidy mating. More studies that quantify the individual and collective strength of these barriers are needed to understand their role in promoting coexistence of multiple ploidies.

Coexistence will also be promoted by the recurrent formation of polyploids within diploid populations (Felber 1991), although in this case polyploids are usually maintained at low frequencies. Triploids have been reported at low levels in several diploid species (e.g., Bureš et al. 2004; Koutecký 2007; Slovák et al. 2009; Dušková et al. 2010). Although there is evidence that polyploidy has repeatedly evolved in some taxa (Soltis and Soltis 1999), it is unlikely that the frequency is high enough to fully compensate for the mating disadvantage experienced by rare cytotypes (Ramsey and Schemske 1998). Until now, there is only one case (Pannonian populations of *Hieracium echioides*) in which complex ploidy mixtures have clearly been maintained by a sufficient frequency of polyploid origins (Trávníček et al. 2011b). Diploid-polyploid coexistence may be further enhanced in species with dominant mechanisms of clonal reproduction (Eckert et al. 2003; Joly and Bruneau 2004; Kao 2007), allowing them to bypass or further diminish the mating disadvantage experienced by sexual species (Levin 1975).

Coexistence is particularly common in species in which diploid-polyploid hybrids are produced (Table 16.2). In these cases, coexistence may occur because the hybrids are partially fertile and serve as a regular source of higher ploidy states through backcrosses to diploid progenitors (Bever and Felber 1992; Husband 2004). Coexistence is evolutionarily stable particularly when the hybrid polyploid (most often $3x$) produces monoploid and polyploid (unreduced) gametes, which can result in the production of both diploid and polyploid progeny. Few direct tests of this possibility exist for natural populations (although see Husband 2004). Alternatively, mixed populations in these species may simply be transient stages in the evolutionary transition from diploid to polyploid. In the absence of triploids, the evolutionary transition from diploid to tetraploid is very rapid once the conditions for the spread of tetraploids are met, indicating that mixed-ploidy transition states will be rare on the landscape (Fig. 16.5). However, the same transition with fertile triploids present is more complex (Fig. 16.5). Mixed populations will be more common and will take several forms. Based on simulations (Fig. 16.6), populations often enter a brief stage of $2x/3x$ followed by longer periods in $2x/3x/4x$, $3x/2x/4x$, $4x/3x/2x$, and $4x/3x$ stages before becoming fixed for $4x$. Two species in which triploids are present show high frequencies of mixed-ploidy populations and patterns of cytotype structure consistent with this transition. *Galax urceolata* contains a modest number of $2x/3x$ (15%) and $3x/$

Fig. 16.5 Transition states in the evolution of tetraploid populations from diploids in (**a**) absence of triploids, (**b**) presence of triploids. *Bold arrows* indicate where the presence of polyploids can accelerate their own establishment in a positive feedback. *Dotted arrows* indicate that these stages may occur as stable states under some conditions

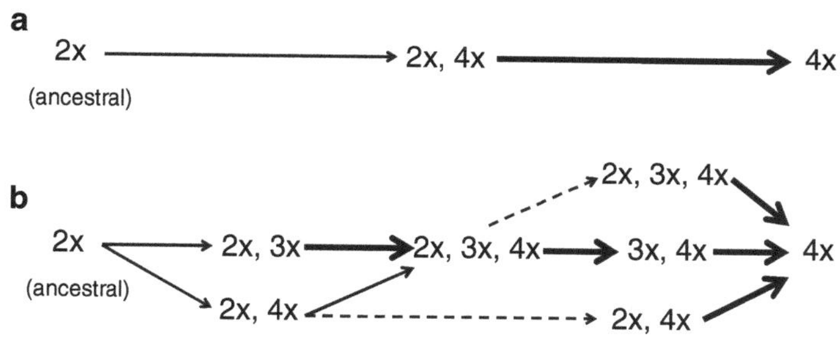

4*x* (5%) populations, but the majority of mixed populations are 2*x*/3*x*/4*x* (27.5%) (Burton and Husband 1999). In *Chamerion angustifolium*, 1.9% of populations were 2*x*/3*x*, no populations contained 3*x*/4*x* and 12.5% were 2*x*/3*x*/4*x* (Sabara 2008). Unfortunately few species have been examined within this evolutionary framework.

16.4 Mode of Origin (Autopolyploid versus Allopolyploid)

Polyploids are generally classified as being one of two types: autopolyploids and allopolyploids (Kihara and Ono 1926). Autopolyploids are formed through the multiplication of genomes within species, either through the union of genomes from different individuals or a single individual. Allopolyploids represent those formed through the union of chromosome sets from different species either by hybridization between autopolyploids or chromosome doubling of diploid hybrids. Distinguishing these forms has never been straightforward since by this definition, allopolyploids can vary widely with respect to the homology of parental chromosome sets and patterns of inheritance (Ramsey and Schemske 1998). With respect to the genetic similarity among genomes, polyploids vary along a continuum from homogeneous (autopolyploidy), to partially divergent (segmental allopolyploidy) and highly divergent genomes (allopolyploidy) (Stebbins 1947). Typically, autopolyploids are associated with multivalent formation and polysomic inheritance, whereas allopolyploids exhibit bivalents and disomic inheritance (Stebbins 1947, 1950; Jackson and Casey 1982; Ramsey and Schemske 1998), although exceptions are not uncommon. The inconsistencies in the genetic attributes of auto- and allopolyploids create difficulties in classifying polyploids unambiguously. This is further exacerbated by the fact that diploid progenitors are often not known or may be extinct. Phylogenetic reconstructions of taxonomic groups are offering a method for delineating them that is in parallel to the definition currently used (Rousseau-Gueutin et al. 2009).

Plant biologists have long disagreed over the relative frequency and evolutionary importance of auto- and allopolyploids. Historically, autopolyploids have been viewed as less common and less important evolutionarily. This perception has prevailed from the early 1900s (Müntzing 1936), as a result of the success in synthesizing new allopolyploids or recreating *de novo* existing polyploid species. Further, researchers such as Wettstein (1927) and others argued that autopolyploids were not viable and would eventually revert to diploids. In contrast, Blakeslee and Avery (1919), and Jorgensen (1928) have argued that autopolyploids can be viable under certain circumstances but are likely to be rare in nature and necessarily associated with apomixis and vegetative reproduction. Stebbins (1950) and Grant (1981) have also argued that autopolyploids should be rare as a result of abnormal chromosome pairing and chromosomal segregation leading to infertility of gametes and inviability of offspring. Ramsey and Schemske (1998) estimated key parameters in the formation of polyploids and found that the rate of formation is likely to be higher in hybrids, thus providing an additional argument for the prevalence of allopolyploids.

The relative frequencies of auto- and allopolyploids remain unresolved, not only because of the difficulty in classifying polyploids but also because of difficulties detecting them. Techniques for characterizing patterns of genetic inheritance were not readily available until the adoption of isozyme analyses in evolutionary biology. Moreover, the close morphological similarities between autopolyploids and their diploid progenitors have made some taxonomists reluctant to recognize them as distinct species (Soltis et al. 2007). Thus, most autopolyploids are hidden as cryptic species within diploid taxa. Nevertheless, over time, as detection improves, the general consensus is that autopolyploids may be found to be more common than once thought (Soltis and Soltis 1993, 1999; Tate et al. 2005).

16.5 Polyploidy and Life History

16.5.1 Growth Form

Current evidence suggests that the frequency of polyploidy differs among different life forms. The proportion of polyploids is generally highest among perennial herbaceous

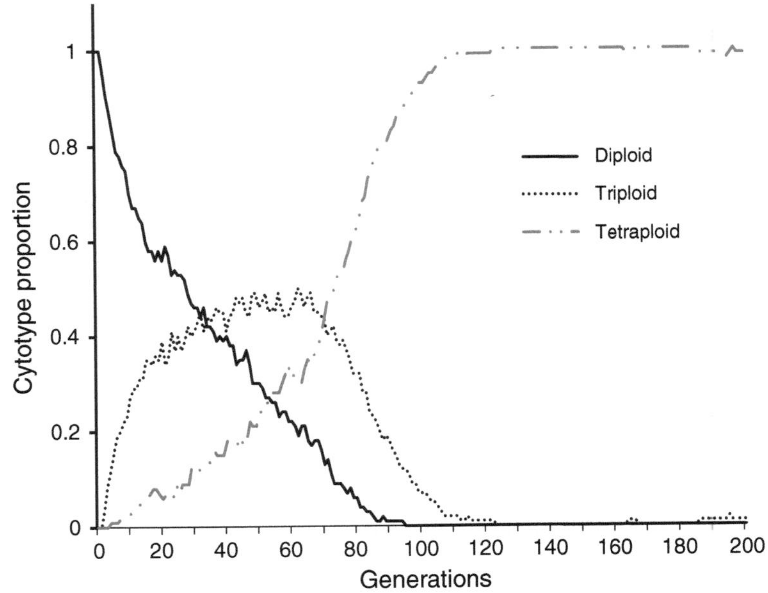

Fig. 16.6 Change in frequency of diploids, triploids and tetraploids during a simulated transition from a diploid population to tetraploid. In this individual-based simulation, diploids and tetraploids had equal fitness but triploid fitness was at 10%. Diploids produced 0.03 unreduced gametes per generation. Mixed populations persist through most of the 130 generations required for tetraploids to become fixed

plants, compared to annuals, bulbs and woody species (Stebbins 1938). This was confirmed in several early vegetation surveys in Canada (Clark et al. 1909), Switzerland (Gustafsson 1948), and Pakistan (Baquar 1976), which revealed that perennials were more likely to be polyploid and annuals more likely to be diploid. In addition, Otto and Whitton (2000) report that the high rate of polyploidy in dicots is driven by polyploidization in herbaceous species and not woody ones. However, little experimental research beyond comparative surveys has explored the basis for these patterns. It has been suggested that reduced frequency of polyploidy in trees may be the result of the impact of cell size on the formation of wood fibres, as with gymnosperms, while polyploidy in bulbs may be constrained by large monoploid genome sizes (Otto and Whitton 2000). The low frequency of polyploidy in annuals may be because of strong selection against polyploid establishment due to the frequency-dependence of homoploid mating and the importance of rapid development associated with colonizing species (Leitch and Bennett 2007). Until analyses control for phylogenetic trends we cannot be sure that this pattern is consistent across groups. In fact, Vamosi and Dickinson (2006) recently found no association between ploidy and herbaceous habit in the Rosaceae family.

16.5.2 Reproductive Systems

Since the early twentieth century, plant biologists have debated the association between polyploidy and reproductive systems (e.g., Ernst 1918; Gustafsson 1947; Stebbins 1950). Since then, particular views have been perpetuated but without a full systematic analysis. Here we examine the relationship between polyploidy and two related aspects of reproduction: asexual reproduction (including clonality and apomixis) and self-fertilization.

16.5.2.1 Asexual Reproduction

Wettstein (1927) hypothesized that polyploidy was only possible in plants with apomixis or vegetative reproduction. Gustafsson (1948) was the first to survey different floras for the incidence of asexual reproduction in polyploid plants and he found very strong trends. Surveys of polyploidy in animals have also provided evidence for the association between polyploidy and asexual reproduction. In fact, a leading hypothesis for the deficiency of polyploidy in animals is the scarcity of asexual reproduction (Schultz 1969) and in support of this is the observation that polyploidy is present among animal groups capable of asexual reproduction such as molluscs (Burch 1964), flatworms (White 1954), and insects (Lokki and Saura 1980). Thus, the association between asexuality and polyploidy is consistent between plant and animal kingdoms.

Apomixis and Polyploidy

Many plants that reproduce by apomixis (embryos that form asexually) are polyploid (Stebbins 1941). In fact there are many examples of mixed-ploidy species, such as *Pilosella officinarum* (Mráz et al. 2008), that are comprised of apomictic polyploids and sexual diploids. Polyploidy and apomixis may co-occur because they share common developmental pathways. Both traits may be induced after hybridization, involve the production of unreduced gametes, and restore fertility after hybridization. Although progress has been made in identifying the genetic determinants of apomixis in many cases (Nogler 1984), understanding when these traits are selected and how they facilitate polyploid formation and establishment has not been fully explored.

A reasonable explanation for the link between apomixis and polyploidy is that the production of unreduced gametes is required for both phenomena. In gametophytic apomixis, the most common form of apomixis in polyploids (Stebbins 1950), unreduced egg cells develop into functional embryos (parthenogenesis) that are genetically identical to the maternal plant. Similarly, the most likely mode of polyploid formation is via fertilization involving at least one unreduced gamete (Ramsey and Schemske 1998). Unreduced gametes may also facilitate the transmission of the genes coding for apomixis. This could occur if the genes are lethal in the haploid form (Nogler 1984) or if the genes are associated with deleterious recessive mutations as in *Hieracium* (Bicknell et al. 2000) and as inferred in *Taraxacum* (van Dijk 2003).

The concurrent evolution of apomixis and polyploidy may also be associated with their connection to hybridization. Both traits can be induced after hybridization and may be selected to restore fertility after hybridization. Ernst (1918) first proposed that hybridization could trigger the production of both polyploidy and apomixis. This could occur since diploid hybrids suffer from numerous meiotic disturbances, and such dysfunction can lead to the production of unreduced gametes. The frequency of unreduced gametes has been shown to be higher in hybrids than non-hybrid diploids (Ramsey and Schemske 2002) and this leads to the testable prediction that apomixis is more common in allopolyploids than autopolyploids (Whitton et al. 2008), or in offspring from inter-ploidy crosses than from single cytotype crosses. Apomixis and polyploidy may also both be favoured to restore fertility after hybridization or in other cases when sexual reproduction is less effective (e.g., mate limitation and/or when there is a high risk of inbreeding depression). Even if apomixis arises equally in diploid and polyploid populations, it may only become fixed when sexual reproduction is maladaptive (Stebbins 1950; Dickinson et al. 2007).

Vegetative Reproduction (Clonality)

As mentioned previously (Sect. 16.5.1), polyploids are more likely to be perennial and diploids more likely to be annual. This difference becomes more pronounced in clonal, or "root-wandering" perennials, which are about twice as likely to be polyploid than annual (reviewed by Gustafsson 1948). There are three main explanations for the correlation between polyploidy and clonality, the relative importance of which has not been fully assessed.

First, genome duplication may directly trigger the development of modes of vegetative reproduction that weren't present in the diploid progenitor (Müntzing 1936) or may intensify the extent of clonality (Stebbins 1950). This effect might be expected if genome duplication results in an increased number of meristems per plant followed by differentiation into meristems with a diverse range of functions.

Second, asexual reproduction may be prevalent in polyploids because diploid species with clonal reproduction are more apt to facilitate the formation and establishment of polyploid populations (Stebbins 1938; Bretagnolle et al. 1998; Otto and Whitton 2000). This is because clonality enables the polyploid to persist in the face of the strong mating disadvantage experienced by newly formed polyploids (i.e., minority cytotype exclusion: Levin 1975; Otto and Whitton 2000) and increases the availability of polyploid mates.

Lastly, vegetative reproduction may be intensified relative to diploids through selection after the genome duplication event (Stebbins 1950). A disproportionate increase in clonality of polyploids might be favoured by selection in mixed-ploidy populations to reduce the effects of hybridization and minority cytotype exclusion. Currently, however, there are few empirical tests of differences in the degree of asexual reproduction between diploid and polyploid relatives. One such example is *Campanula patula* in which the proportion of vegetative reproduction via runners (and plant longevity) increases from diploids to octoploids (F. Rooks, H. Látalová, J. Suda and F. Krahulec, unpublished). There is a need for comparisons of vegetative reproduction among diploids, initial polyploids and established polyploids within and among sister taxa to isolate the effects of genome duplication, preadaptation and selection on the evolution of vegetative reproduction in polyploids.

16.5.2.2 Polyploidy and Mating System

The association between polyploidy and mating system in angiosperms is supported by two apparently conflicting patterns. On one hand, there is a significant tendency for polyploids to be more self-fertilizing than diploids (Barringer 2007; Husband et al. 2008). On the other, polyploidy has been shown to be associated with gender dimorphism and hence outcrossing in several genera (Miller and Venable 2000). The linkage to self-fertilization was originally identified by Grant (1956) and Stebbins (1957) who observed a higher incidence of allopolyploids among lists of selfing species. More recently, Barringer (2007) conducted the first phylogenetically informed analysis of variation in outcrossing rate (t) estimates using genetic markers to determine the proportion of progeny derived from cross-fertilization. Based on 235 taxa, he identified 32 phylogenetically independent contrasts in which increased ploidy was associated with a decrease in outcrossing rate (and hence increase in self-fertilization). Husband et al. (2008) used 10 diploid-polyploid congeneric species pairs to address the same question. Like Barringer, they found that polyploids were significantly more selfing (or less outcrossing; $t = 0.23$) than diploids ($t = 0.51$). They also discovered that increased selfing (lower outcrossing) was primarily associated with allopolyploids ($t = 0.20$), not autopolyploids ($t = 0.64$). This pattern corroborates Stebbins' (1957) original observations, although broader sampling is needed to confirm the difference between types of polyploids across angiosperms.

The correlation between ploidy and selfing has been predicted on multiple grounds. Grant (1956) and Stebbins (1957) suggested, and Ramsey and Schemske (1998) demonstrated, that polyploid formation is more likely via selfing than outcrossing, especially in allopolyploids (34 times more likely). Once formed, the establishment of polyploid populations is also aided by selfing since the strong negative fitness costs of between-ploidy mating are avoided (Levin 1975). In addition, most theory argues that the fitness costs of selfing (i.e., inbreeding depression) should be lower in polyploids (Lande and Schemske 1985; Hedrick 1987; Ronfort 1999; Rausch and Morgan 2005; but see Busbice and Wilsie 1966), making it easier for selfing to evolve. These predictions are supported to different degrees by a limited number of empirical studies (Husband and Schemske 1997; Rosquist 2001; Barringer and Geber 2008).

In contrast to this body of evidence, Miller and Venable (2000) showed that, in multiple plant families, there was a significant association between increased polyploidy and the shift from hermaphroditism to gender dimorphism. This pattern suggests that, in fact, outcrossing can be favoured in some polyploids (or polyploidy is favoured in outcrossed populations). Miller and Venable (2000) have proposed that polyploidy creates the conditions for the spread of male sterile mutants in hermaphrodite populations by (1) causing the loss of self-incompatibility, (2) increasing the rate of self-fertilization, and (3) expressing sufficient inbreeding depression in hermaphrodites to favour the establishment of females. This hypothesis requires inbreeding depression to be strong in new polyploids, a prediction that is not supported by observations of synthesized polyploids (Husband et al. 2008). However, it is possible that inbreeding depression rises in tetraploids with successive generations of selfing, as recently observed by Ozimec and Husband (2011). Clearly, additional work on a variety of species is required to reconcile the strongly opposed patterns of mating in diploids and polyploids and to understand the evolutionary forces driving outcrossing.

16.6 Geographic Distribution, Range Size and Ecological Adaptation

A major observation made by plant biologists over the last century concerns the association between polyploidy and the geographic and ecological range limits of species (Levin 2002). A species' geographic range, the area within which an organism occurs for part of its life, is the result of many complex interacting forces, including the history of colonization and the contemporary interactions between the organisms and its environments. Polyploidy is considered an important genetic determinant of a species' range. Indeed, arguably, this single genetic event may represent the largest influence on the ecological or geographic range of a species.

In general, plant biologists have postulated that: (1) polyploid ranges will differ from those of the diploids, (2) polyploid species will have larger geographic ranges, and (3) the ranges for polyploid species will include more extreme environments such as high latitudes and altitudes, than diploid species (Hagerup 1927; Stebbins 1950, 1985; Otto and Whitton 2000; Levin 2002). The differences in geographic range between polyploids and diploids have been explained through a variety of different mechanisms (Müntzing 1936; Stebbins 1950). For example, polyploid individuals are expected to have greater breadth of ecological tolerance, or plasticity, due to higher heterozygozity and allelic diversity (genetic or biochemical buffering) or greater adaptive potential due to higher population genetic diversity (Otto and Whitton 2000). They may also be less vulnerable, at least initially, to the maladaptive effects of gene flow from other diploid populations (Kirkpatrick and Barton 1997). Finally, polyploids may differ phenotypically from diploids in a systematic way, as a result of the direct effects of genome duplication, particularly cell size. On this basis, it has been argued that polyploids will be more cold tolerant, drought tolerant or tolerant of disturbance (Hagerup 1927). The effects of these different mechanisms are rarely disentangled but they might be expected to influence species ranges and ecological amplitude in different ways.

The strength of the link between distribution and ploidy is as varied as the kinds of evidence used to assess it. In the first half of the twentieth century, the dominant evidence was in the form of floristic and taxonomic case studies (Löve and Löve 1943; Clausen et al. 1945) that examined whether the frequency of polyploids was higher in extreme environments, be it high latitude (Tischler 1935; Löve and Löve 1943; Packer 1969), altitude (Sokolovskaya and Strelkova 1940), or saline habitats (Shimotomai 1933; Tischler 1937). In general, these studies led to the conclusion that polyploidy is more common in extreme environments, which was often attributed to greater cold-hardiness and tolerance to long day photoperiods than diploids. In recent years, high-throughput analyses of polyploidy have been extended into the arctic (Brochmann et al. 2004) and less-studied regions at lower latitudes (Popp et al. 2008; Suda et al. 2009; Dušková et al. 2010); similar latitudinal comparisons have been conducted in mosses (Kuta and Przywara 1997). In general, the high incidence of polyploidy at high latitude holds reasonably well (Suda et al. 2009).

The use of floristic approaches for testing the effect of polyploidy on distribution has been criticized for a number of reasons. First, the comparison of floras is confounded with different biogeographical and phylogenetic histories (Gustafsson 1947, 1948). The prevalence of polyploids in northern floras may be directly related to the species that colonize those areas and their characteristics (e.g., perenniality, evergreenness) rather than selection favouring polyploids *per se*. Secondly, there is much inconsistency in the geographic position of diploids and polyploids within individual taxonomic groups as well as in their cold hardiness, when tested experimentally (Bowden 1940; Gustafsson 1947; Nielsen 1947). Finally, geographic regions have not received uniform effort when screening for polyploids, particularly subtropical and tropical regions. What has been lacking until recently is a broad comparison of range attributes of a large number of species, ideally where the range attributes of polyploid species were compared to sister diploid species or at least those of recent common ancestry.

Three studies have compared the range attributes of diploids and polyploids in broad surveys. Petit and Thompson (1999) compared the ecological range limits of diploids and polyploids in the Pyrenees. They found no evidence that polyploids occurred over a wider range of ecological environments than diploids. Similarly, Stebbins and Dawe (1987) found little effect of polyploidy on the percentage of species that were widespread in European genera. In a recent analysis, Martin and Husband (2009) compared the extent and mean geographic and ecological ranges for diploid and polyploid taxa within 144 North American plant genera. They found no difference in the geographic area of the range between ploidy levels and no evidence that the centroid of the range of polyploids was shifted further north than diploids. Interestingly, they did observe that range overlap between congeneric diploids and polyploids was greater than between two congeneric diploids. This requires further examination but could reflect the ability to coexist because of strong ecological or reproductive isolation. In addition, comparisons of the breadth, maximum or minimum range values for precipitation or temperature showed no consistent differences between ploidy levels (Martin and Husband 2009).

Potentially contrary to this evidence is a parallel literature that has used comparative evidence to show that the incidence of polyploidy increases with invasiveness and decreases with rarity (Pandit 2006; Pandit et al. 2011). The most recent study, based on an analysis of 640 endangered species and 81 invasive species, found that rare species are most likely to be diploid, whereas invasive species are more likely to be polyploid. To the extent that invasive species tend to be widespread, this pattern may conflict with the lack of association between range size and ploidy. However, there may be no conflict as definitions of invasive/rare take many factors into account, only one of which is geographic range size, and may also be influenced by variation in political designations and lack of uniformity of information on which designations are based.

The literature to date does not imply that polyploidization has no effect on ecological tolerances or species ranges. Indeed, there is a rich literature corroborating the effect of genome duplication on habit, phenotype and physiology (Levin 2002). Polyploids frequently differ in geographic range from diploid relatives. However, the data show, in general, that differences among polyploids and their congeners are not systematic in direction and may not be greater than the differences among homoploid congeners. Nevertheless, the question remains as to what drives the shifts in range size between individual diploids and polyploid relatives. Are these shifts a direct result of genome duplication or are they the product of evolutionary divergence operating after duplication? Work on the ecological attributes of newly synthesized polyploids will be particularly insightful in this regard. For example, research on the distribution of diploid and tetraploid *Chamerion angustifolium* (fireweed) supports the idea that range shifts are not achieved as a direct result of the duplication (Sabara 2008). In fireweed, diploids are more often found at higher altitudes and latitudes than tetraploids. Reciprocal transplants show that populations are weakly differentiated and adapted with respect to elevation. However, in terms of traits such as drought physiology and reproductive morphology, neopolyploids resemble diploids more than extant tetraploids (Sabara 2008; Maherali et al. 2009), suggesting that genome duplication is not the sole contributor to the patterns of ecological divergence.

16.7 Conclusions and Future Directions

Our knowledge of the incidence of polyploidy is dominated by a limited number of long-standing patterns first identified by the pioneers in this field. These patterns describe variation in polyploidy as a function of taxonomic group, life history traits and geography. In recent years, additional evidence has accrued using broader taxonomic perspectives and new genetic or phylogenetic analyses, but how have our perceptions of polyploidy and its significance changed?

Although our perception of the dominance of polyploidy in plants has remained relatively constant, some notable changes to the patterns have occurred. With higher throughput analyses, we have seen a steady increase in the frequency of polyploidy, with almost no group untouched, and an increase in the incidence of autopolyploidy and intraspecific ploidy variation. Combined with recent genomic comparisons, which confirm a role of genome duplication in early plant evolution, we can only conclude that polyploidy has influenced almost all extant species at some stage in their evolutionary history. This observation has led to more significant questions such as: does polyploidy contribute to evolution in a way that diploids do not (Otto and Whitton 2000)? The incidence of polyploidy, combined with a phylogenetic perspective, has been equivocal on its role in species diversification, but for the first time we have estimates of the rate of polyploid speciation (Wood et al. 2009). With respect to geography, the dogma that polyploids have larger ranges and are more common at high latitudes and extreme environments has become less clear and possibly even misleading since phylogenetic control and broad comparative analyses have been used. Similarly, patterns of reproductive biology among polyploids are variable and even conflict with common dogma.

The original goal of studying the incidence of polyploidy was to identify key traits or attributes that may offer the universal key to polyploid significance. However, as more evidence and more powerful analyses accumulate, the strength of particular associations is not always clearer. In fact, if there is one lesson emerging, it may be that polyploidy does not have a single cause or unifying advantage. Rather the impact of polyploidy on ecological and evolutionary function is consistent only in that it has an effect, even if the specific direction is unpredictable. This suggests that it may be time to redirect our focus towards the context in which polyploidy operates. This approach shifts the emphasis from asking whether polyploids will differ from diploids to what will best predict the effects of genome duplication in any one circumstance. An understanding of phenotypes at the level of gene interactions and function will be essential for building this predictive framework.

To continue to advance our understanding of the patterns of occurrence of polyploidy in plants, several avenues of research would be profitable. In some less intensively studied taxonomic groups (e.g., Glaucophyta), there is still a need for additional surveys of ploidy. Flow cytometry and whole genome sequencing should help to achieve this. In addition, it is clear that certain traits such as those related to the formation of polyploids (e.g., unreduced gametes) have not been explored in natural populations fully enough to draw any inferences (Bretagnolle and Thompson 1995). In many cases, however, the data are now widely available in the literature and in accessible databases (Bennett and Leitch 2010). The challenge now will be to re-evaluate our long-accepted patterns within a phylogenetic framework. This will provide a stronger measure of the phylogenetic contributions to polyploid formation and establishment but will also allow us to test patterns with more rigour and generality (e.g., Vamosi and Dickinson 2006; Wood et al. 2009). Lastly, experimental research will be critical for corroborating our hypotheses with more direct approaches. For example, the use of synthetic polyploids (neopolyploids), in conjunction with physiological, genetic and genomic tools, will be instrumental for evaluating what of these many associations are the direct effects of genome duplication as opposed to selection operating afterwards (Bretagnolle and Lumaret 1995; Ramsey and Schemske 2002), and more generally what are the evolutionary processes underlying the patterns of polyploid incidence in plants.

References

Adl SM, Simpson AG, Farmer MA, Andersen RA, Anderson OR, Barta JR, Bowser SS, Brugerolle G, Fensome RA, Fredericq S, James TY, Karpov S, Kugrens P, Krug J, Lane CE, Lewis LA, Lodge J, Lynn DH, Mann DG, McCourt RM, Mendoza L, Moestrup O, Mozley-Standridge SE, Nerad TA, Shearer CA, Smirnov AV, Spiegel FW, Taylor MF (2005) The new higher level classification of eukaryotes with emphasis on the taxonomy of protists. J Eukaryot Microbiol 52:399–451

Ahuja MR (2005) Polyploidy in gymnosperms: revisited. Silvae Genet 54:59–69

Anderson JB, Sirjusingh C, Ricker N (2004) Haploidy, diploidy and evolution of antifungal drug resistance in *Saccharomyces cerevisiae*. Genetics 168:1915–1923

Baack EJ (2004) Cytotype segregation on regional and microgeographic scales in snow buttercups (*Ranunculus adoneus*: Ranunculaceae). Am J Bot 91:1782–1788

Balao F, Casimiro-Soriguer R, Talavera M, Herrera J, Talavera S (2009) Distribution and diversity of cytotypes in *Dianthus broteri* as evidenced by genome size variations. Ann Bot 104:965–973

Baldwin SJ, Husband BC (2011) Genome duplication and conspecific pollen precedence. P Roy Soc Lond B Bio 278:2011–2017

Baquar SR (1976) Polyploidy in the flora of Pakistan in relation to latitude, life form, and taxonomic groups. Taxon 25:621–627

Barker MS (2013) Karyotype and genome evolution in pteridophytes. In: Leitch IJ, Greilhuber J, Doležel J, Wendel JF (eds) Plant genome diversity, vol 2, Physical structure, behaviour and evolution of plant genomes. Wien, pp 245–253

Barringer BC (2007) Polyploidy and self-fertilization in flowering plants. Am J Bot 94:1527–1533

Barringer BC, Geber MA (2008) Mating system and ploidy influence levels of inbreeding depression in *Clarkia* (Onagraceae). Evolution 62:1040–1051

Becker B, Marin B (2009) Streptophyte algae and the origin of embryophytes. Ann Bot 103:999–1004

Bennert W, Lubienski M, Körner S, Steinberg M (2005) Triploidy in *Equisetum* subgenus *Hippochaete* (Equisetaceae, Pteridophyta). Ann Bot 95:807–815

Bennett MD, Leitch IJ (2005) Plant genome size research: a field in focus. Ann Bot 95:1–6

Bennett MD, Leitch IJ (2010) Plant DNA C-values database (release 5.0, Dec 2010). http://data.kew.org/cvalues/

Bever JD, Felber F (1992) The theoretical population genetics of autopolyploidy. Oxford Surv Evol Biol 8:185–217

Bicknell RA, Borst NK, Koltunow AM (2000) Monogenic inheritance of apomixis in two *Hieracium* species with distinct developmental mechanisms. Heredity 84:228–237

Blakeslee AF, Avery BT (1919) Mutations in the Jimson weed. J Hered 10:111–120

Bowden WM (1940) Diploidy, polyploidy, and winter hardiness relationships in the flowering plants. Am J Bot 27:357–371

Bretagnolle F, Lumaret R (1995) Bilateral polyploidization in *Dactylis glomerata* L. subsp. *lusitanica*: occurrence, morphological and genetic characteristics of first polyploids. Euphytica 84:197–207

Bretagnolle F, Thompson JD (1995) Gametes with the somatic chromosome number: mechanisms of their formation and role in the evolution of autopolyploid plants. New Phytol 129:1–22

Bretagnolle F, Thompson JD (1996) An experimental study of ecological differences in winter growth between sympatric diploid and autotetraploid *Dactylis glomerata*. J Ecol 84:343–351

Bretagnolle F, Felber F, Calame F (1998) La polyploïdie chez les plantes. Bot Helv 108:5–37

Brochmann C, Brysting AK, Alsos IG, Borgen L, Grundt HH, Scheen AC, Elven R (2004) Polyploidy in arctic plants. Biol J Linn Soc 82:521–536

Brownsey PJ, Lovis JD (1987) Chromosome numbers for the New Zealand species of *Psilotum* and *Tmesipteris*, and the phylogenetic relationships of the Psilotales. New Zealand J Bot 25:439–454

Burch JB (1964) Cytological studies of Planorbidae (Gastropoda: Basommatophora). I. The African subgenus *Bulinus* s. s. Malacologia 1:387–400

Bureš P, Yi-Feng W, Horová L, Suda J (2004) Genome size variation in Central European species of *Cirsium* (Compositae) and their natural hybrids. Ann Bot 94:353–363

Burton TL, Husband BC (1999) Population cytotype structure in the polyploid *Galax urceolata* (Diapensiaceae). Heredity 82:381–390

Busbice TH, Wilsie CP (1966) Inbreeding depression and heterosis in autotetraploids with application to *Medicago sativa* L. Euphytica 15:52–67

Clark GH, Fletcher J, Criddle N (1909) Farm weeds of Canada, 2nd edn. Department of Agriculture Seed Branch, Ottawa

Clausen J, Keck DD, Hiesey WM (1945) Experimental studies on the nature of species. II. Plant evolution through amphiploidy and autoploidy, with examples from the Madiinae, vol 564. Carnegie Inst, Washington, Carnegie Inst Wash

Cole KM (1990) Chromosomes. In: Cole KM, Sheath RG (eds) Biology of the red algae. Cambridge University Press, Cambridge, pp 71–101

Courties C, Perasso R, Chrétiennot-Dinet M-J, Gouy M, Guillou L, Troussellier M (1998) Phylogenetic analysis and genome size of *Ostreococcus tauri* (Chlorophyta, Prasinophyceae). J Phycol 34:844–849

De Wet JMJ (1980) Origins of polyploids. In: Lewis WH (ed) Polyploidy: biological relevance. Plenum Press, New York, pp 3–16

Denda T, Yokota M (2004) Cytogeography of *Ixeris nakazonei* (Asteraceae, Lactuceae) in the Ryukyu Archipelago of Japan and Taiwan. J Plant Res 117:3–11

Dickinson TA, Lo E, Talent N (2007) Polyploidy, reproductive biology, and Rosaceae: understanding evolution and making classifications. Plant Syst Evol 266:59–78

Duchoslav M, Šafářová L, Krahulec F (2010) Complex distribution patterns, ecology and coexistence of ploidy levels of *Allium oleraceum* (Alliaceae) in the Czech Republic. Ann Bot 105: 719–735

Dušková E, Kolář F, Sklenář P, Rauchová J, Kubešová M, Fér T, Suda J, Marhold K (2010) Genome correlates with growth form, habitat and phylogeny in the Andean genus *Lasiocephalus* (Asteraceae). Preslia 82:127–148

Eckert CG, Lui K, Bronson K, Corradini P, Bruneau A (2003) Population genetic consequences of extreme variation in sexual and clonal reproduction in an aquatic plant. Mol Ecol 12:331–334

Ernst A (1918) Bastardierung als Ursache der Apogamie im Pflanzenreich. Gustav Fischer, Jena

Fawcett JA, Van de Peer Y, Maere S (2013) Significance and biological consequences of polyploidization in land plants. In: Leitch IJ, Greilhuber J, Doležel J, Wendel JF (eds) Plant genome diversity, vol 2, Physical structure, behaviour and evolution of plant genomes. Springer-Verlag, Wien, pp 277–293

Felber F (1991) Establishment of a tetraploid cytotype in a diploid population. J Evol Biol 4:195–207

Felber-Girard M, Felber F, Buttler A (1996) Habitat differentiation in a narrow hybrid zone between diploid and tetraploid *Anthoxanthum alpinum*. New Phytol 133:531–540

Fowler NL, Levin DA (1984) Ecological constraints on the establishment of a novel polyploid in competition with its diploid progenitor. Am Nat 124:703–711

Fritsch R (1991) Index to bryophyte chromosome counts. Bryophytorum Bibliotheca 40:1–352

Goldblatt P (1980) Polyploidy in angiosperms: monocotyledons. In: Lewis WH (ed) Polyploidy: biological relevance. Plenum Press, New York, pp 219–240

Grant V (1956) The influence of breeding habit on the outcome of natural hybridization in plants. Am Nat 90:319–322

Grant V (1963) The origin of adaptations. Columbia University Press, New York

Grant V (1981) Plant speciation. Columbia University Press, New York

Gregory TR, Mable BK (2005) Polyploidy in animals. In: Gregory TR (ed) The evolution of the genome. Elsevier, San Diego, pp 585–675

Grif VG (2000) Some aspects of plant karyology and karyosystematics. Int Rev Cytol 196:131–175

Grimsley N, Péquin B, Bachy C, Moreau H, Piganeau G (2010) Cryptic sex in the smallest eukaryotic marine green alga. Mol Biol Evol 27:47–54

Gustafsson A (1947) Apomixis in higher plants. Acta Univ Lund Kungl Fysiogr Sallsk Handl 42–44:1–370

Gustafsson A (1948) Polyploidy, life-form, and vegetative reproduction. Hereditas 34:1–22

Hagerup O (1927) *Empetrum hermaphroditum* (Lange) Hagerup, a new tetraploid bisexual species. Dansk Bot Ark 5:1–17

Halverson K, Heard SB, Nason JD, Stireman JO (2008) Origins, distribution, and local co-occurrence of polyploid cytotypes in *Solidago altissima* (Asteraceae). Am J Bot 95:50–58

Hardy OJ, De Loose M, Vekemans X, Meerts P (2001) Patterns of allozymic variation in diploid and tetraploid *Centaurea jacea* at different spatial scales. Evolution 55:943–954

Haufler CH (1987) Electrophoresis is modifying our concepts of evolution in homosporous pteridophytes. Am J Bot 74:953–966

Hedrick PW (1987) Gametic disequilibrium measures: proceed with caution. Genetics 117:331–341

Hijmans RJ, Gavrilenko T, Stephenson S, Bamberg J, Salas A, Spooner DM (2007) Geographical and environmental range expansion through polyploidy in wild potatoes (*Solanum* section *Petota*). Global Ecol Biogeogr 16:485–495

Hirayoshi I, Nakamura Y (1943) Chromosome number of *Sequoia sempervirens*. Bot Zool 2:73–75

Hoshaw RW, McCourt RM (1988) The Zygnemataceae (Chlorophyta): a twenty-year update of research. Phycologia 27:511–548

Howard DJ (1999) Conspecific sperm and pollen precedence and speciation. Annu Rev Ecol Syst 30:109–132

Howard DJ, Reece M, Gregory PJ, Chu J, Cain ML (1998) The evolution of barriers to fertilization between closely related organisms. In: Howard DJ, Berlocher SH (eds) Endless forms: species and speciation. Oxford University Press, New York, pp 279–288

Husband BC (2000) Constraints on polyploidy evolution: a test of the minority cytotype exclusion principle. Proc Roy Soc Lond B Bio 267:217–233

Husband BC (2004) The role of triploids in the evolutionary dynamics of mixed-ploidy populations. Biol J Linn Soc 82:537–546

Husband BC, Sabara HA (2004) Reproductive isolation between autotetraploids and their diploid progenitors in fireweed, *Chamerion angustifolium* (Onagraceae). Am J Bot 85:1688–1694

Husband BC, Schemske DW (1995) Magnitude and timing of inbreeding depression in a diploid population of *Epilobium angustifolium* (Onagraceae). Heredity 75:206–215

Husband BC, Schemske DW (1996) Evolution of the magnitude and timing of inbreeding depression in plants. Evolution 50:54–70

Husband BC, Schemske DW (1997) The effect of inbreeding in diploid and tetraploid populations of *Epilobium angustifolium* (Onagraceae): implications for the genetic basis of inbreeding depression. Evolution 51:737–746

Husband BC, Schemske DW (1998) Cytotype distribution at a diploid-tetraploid contact zone in *Chamerion* (*Epilobium*) *angustifolium* (Onagraceae). Am J Bot 85:1688–1694

Husband BC, Schemske DW, Burton TL, Goodwillie C (2002) Pollen competition as a unilateral reproductive barrier between sympatric diploid and tetraploid *Chamerion angustifolium*. Proc Roy Soc Lond B Bio 269:2565–2571

Husband BC, Ozimec B, Martin SL, Pollock L (2008) Mating consequences of genome duplication: current patterns and insights from neopolyploids. Int J Plant Sci 169:195–206

Ickert-Bond SM (2003) Systematics of New World *Ephedra* L. (Ephedraceae): integrating morphological and molecular data. Thesis, Arizona State University, pp 363

Ickert-Bond SM, Wojciechowski MF (2004) Phylogenetic relationships in *Ephedra* (Gnetales): evidence from nuclear and chloroplast DNA sequence data. Syst Bot 29:834–849

Jackson RC (1976) Evolution and systematic significance of polyploidy. Annu Rev Ecol Syst 7:209–234

Jackson RC, Casey J (1982) Cytogenetic analyses of autopolyploids: models and methods for triploids to octoploids. Am J Bot 69:487–501

Jersáková J, Castro S, Sonk N, Milchreit K, Schödelbaurová I, Tolasch T, Dötterl S (2010) Absence of pollinator-mediated premating barriers in mixed-ploidy populations of *Gymnadenia conopsea* s.l. (Orchidaceae). Evol Ecol 24:1199–1218

Johnson MAT, Kenton AY, Bennett MD, Brandham PE (1989) *Voanioala gerardii* has the highest known chromosome number in the monocotyledons. Genome 32:328–333

Joly S, Bruneau A (2004) Evolution of triploidy in *Apios americanum* (Leguminosae) revealed by genealogical analysis of the histone H3-D gene. Evolution 58:284–295

Jorgensen CA (1928) The experimental formation of heteroploid plants in the genus *Solanum*. Genetics 19:133–211

Kao RH (2007) Asexuality and the coexistence of cytotypes. New Phytol 175:764–772

Kao RH (2008) Origins and widespread distribution of co-existing polyploids in *Arnica cordifolia* (Asteraceae). Ann Bot 101:145–152

Kapraun DF (2005) Nuclear DNA content estimates in multicellular green, red and brown algae: phylogenetic considerations. Ann Bot 95:7–44

Kapraun DF (2007) Nuclear DNA content estimates in green algal lineages: chlorophyta and streptophyta. Ann Bot 99:677–701

Karlin EF, Gardner GP, Lukshis K, Boles S, Shaw AJ (2010) Allopolyploidy in *Sphagnum mendocinum* and *S. papillosum* (Sphagnaceae). Bryologist 113:114–119

Keeling PJ (2004) Diversity and evolutionary history of plastids and their hosts. Am J Bot 91:1481–1493

Kennedy BF, Sabara HA, Haydon D, Husband BC (2006) Pollinator-mediated assortative mating in mixed ploidy populations of *Chamerion angustifolium* (Onagraceae). Oecologia 150:398–408

Khandelwal S (1990) Chromosome evolution in the genus *Ophioglossum* L. Bot J Linn Soc 102:205–217

Khoshoo TN (1959) Polyploidy in gymnosperms. Evolution 13:24–39

Khoshoo TN, Ahuja MR (1963) The chromosomes and relationships of *Welwitschia mirabilis*. Chromosoma 14:522–533

Kihara H, Ono T (1926) Chromosomenzahlen und systematische Gruppierung der *Rumex*-Arten. Z Zellforsch Mikr Anat 4:475–481

King GC (1960) The cytology of the desmids: the chromosomes. New Phytol 59:65–72

Kirkpatrick M, Barton NH (1997) Evolution of a species' range. Am Nat 150:1–23

Klekowski EJ, Baker HG (1966) Evolutionary significance of polyploidy in the Pteridophyta. Science 153:305–307

Kolář F, Fér T, Štech M, Trávníček P, Dušková E, Schönswetter P, Suda J (2012) Bringing together evolution on serpentine and polyploidy: spatiotemporal history of the diploid-tetraploid complex of *Knautia arvensis* (Dipsacaceae). PLoS One 7:e39988

Kolář F, Štech M, Trávníček P, Rauchová J, Urfus T, Vít P, Kubešová M, Suda J (2009) Towards resolving the *Knautia arvensis* agg. (Dipsacaceae) puzzle: primary and secondary contact zones and ploidy segregation at landscape and microgeographic scales. Ann Bot 103:963–974

Koutecký P (2007) Morphological and ploidy level variation of *Centaurea phrygia* agg. (Asteraceae) in the Czech Republic, Slovakia and Ukraine. Folia Geobot 42:77–102

Kron P, Suda J, Husband BC (2007) Applications of flow cytometry to evolutionary and population biology. Annu Rev Ecol Evol Syst 38:847–876

Kubátová B, Trávníček P, Bastlová D, Čurn V, Jarolímová V, Suda J (2008) DNA ploidy-level variation in native and invasive populations of *Lythrum salicaria* at a large geographical scale. J Biogeogr 35:167–176

Kuta E, Przywara L (1997) Polyploidy in mosses. Acta Biol Cracov Ser Bot 39:17–26

Lande R, Schemske DW (1985) The evolution of self-fertilization and inbreeding depression in plants. I. Genetic models. Evolution 39:24–40

Leitch IJ, Bennett MD (2007) Genome size and its uses: the impact of flow cytometry. In: Doležel J, Greilhuber J, Suda J (eds) Flow cytometry with plant cells: analysis of genes, chromosomes and genomes. Wiley, Weinheim, 153–176

Leitch IJ, Leitch AR (2013) Genome size diversity and evolution in land plants. In: Leitch IJ, Greilhuber J, Doležel J, Wendel JF (eds) Plant genome diversity, vol 2, Physical structure, behaviour and evolution of plant genomes. Springer-Verlag, Wien, pp 307–322

Levin DA (1975) Minority cytotype exclusion in local plant populations. Taxon 24:35–43

Levin DA (2002) The role of chromosomal change in plant evolution. Oxford University Press, New York

Lewis LA, McCourt RM (2004) Green algae and the origin of land plants. Am J Bot 91:1535–1556

Li D, Liu Y, Zhong C, Huang H (2010) Morphological and cytotype variation of wild kiwifruit (*Actinidia chinensis* complex) along an altitudinal and longitudinal gradient in central-west China. Bot J Linn Soc 164:72–83

Lokki J, Saura A (1980) Genetic polymorphism and evolution in parthenogenetic animals XI. Genetic differentiation in parthenogenetic populations. Hereditas 92:275–282

Löve Á, Löve D (1943) The significance of differences in the distribution of diploids and polyploids. Hereditas 29:145–163

Löve Á, Löve D, Pichi-Sermolli REG (1977) Cytotaxonomical atlas of the Pteridophyta, vol 3, Cytotaxonomical atlases. J. Cramer, Vaduz

Lutz AM (1907) A preliminary note on the chromosomes of *Oenothera lamarckiana* and one of its mutants, *O. gigas*. Science 26:151–152

Mahelka V, Suda J, Jarolímová V, Trávníček P, Krahulec F (2005) Genome size discriminates between closely related taxa *Elytrigia repens* and *E. intermedia* (Poaceae: Triticeae) and their hybrid. Folia Geobot 40:367–384

Maherali H, Walden AE, Husband BC (2009) Genome duplication and the evolution of physiological responses to water stress. New Phytol 184:721–731

Mainx F (1927) Untersuchungen über die Ernährung und Zellteilung bei *Eremosphaera viridis* De Bary. Arch Protistenk 57:1–13

Mandáková T, Münzbergová Z (2006) Distribution and ecology of cytotypes of the *Aster amellus* aggregates in the Czech Republic. Ann Bot 98:845–856

Martin SL, Husband BC (2009) Influence of phylogeny and ploidy on species ranges of North American angiosperms. J Ecol 97:913–922

Masterson J (1994) Stomatal size in fossil plants: evidence for polyploidy in majority of angiosperms. Science 264:421–423

Meyers LA, Levin DA (2006) On the abundance of polyploids in flowering plants. Evolution 60:1198–1206

Miller JS, Venable DL (2000) Polyploidy and the evolution of gender dimorphism in plants. Science 289:2335–2338

Moretti A, Sabato S (1984) Karyotype evolution by centromeric fission in *Zamia* (Cycadales). Plant Syst Evol 146:215–223

Mosquin T, Small E (1971) An example of parallel evolution in *Epilobium* (Onagraceae). Evolution 25:678–682

Mráz P, Šingliarová B, Urfus T, Krahulec F (2008) Cytogeography of *Pilosella officinarum* (Compositae): altitudinal and longitudinal differences in ploidy level distribution in the Czech Republic and Slovakia and the general pattern in Europe. Ann Bot 101:59–71

Muller HJ (1925) Why polyploidy is rarer in animals than in plants. Am Nat 59:346–353

Müntzing A (1936) The evolutionary significance of autopolyploidy. Hereditas 21:236–378

Nakagawa M (2006) Ploidy, geographical distribution and morphological differentiation of *Parasenecio auriculata* (Senecioneae; Asteraceae) in Japan. J Plant Res 119:51–61

Nesom GL (1983) *Galax* (Diapensiaceae): geographic variation in chromosome number. Syst Bot 8:1–14

Nichols HW (1980) Polyploidy in algae. In: Lewis WH (ed) Polyploidy: biological relevance. Plenum Press, New York, pp 151–161

Nielsen EL (1947) Polyploidy and winter survival in *Panicum virgatum* L. J Am Soc Agron 29:822–827

Nogler GA (1984) Gametophytic apomixis. In: Johri BM (ed) Embryology of angiosperms. Springer, Berlin/Heidelberg/New York, pp 475–518

Ornduff R (1970) Cytogeography of *Nymphoides* (Menyanthaceae). Taxon 19:715–719

Orr HA (1990) "Why polyploidy is rare in animals than plants" revisited. Am Nat 136:759–770

Orzechowska M, Siwinska D, Maluszynska J (2010) Molecular cytogenetic analyses of haploid and allopolyploid *Pellia* species. J Bryol 32:113–121

Otto SP, Whitton J (2000) Polyploid incidence and evolution. Annu Rev Genet 34:401–437

Ozimec B, Husband BC (2011) Effect of recurrent selfing on inbreeding depression and mating system evolution in an autopolyploid plant. Evolution 65:2038–2049

Packer JG (1969) Polyploidy in the Canadian arctic archipelago. Arctic Alpine Res 1:15–28

Pandit MK (2006) Continuing the search for pattern among rare plants: are diploid species more likely to be rare? Evol Ecol Res 8:543–552

Pandit MK, Pocock MJO, Kunin WE (2011) Ploidy influences rarity and invasiveness in plants. J Ecol 99:1365–2745

Petit C, Thompson JD (1999) Species diversity and ecological range in relation to ploidy level in the flora of the Pyrenees. Evol Ecol 13:45–66

Petit C, Lesbros P, Ge X, Thompson JD (1997) Variation in flowering phenology and selfing rate across a contact zone between a diploid and tetraploid *Arrhenatherum elatius* (Poaceae). Heredity 79:31–40

Popp M, Gizaw A, Nemomissa S, Suda J, Brochmann C (2008) Colonization and diversification in the African 'sky islands' by Eurasian *Lychnis* L. (Caryophyllaceae). J Biogeogr 35:1016–1029

Price DC, Chan CX, Yoon HS et al. (2012) *Cyanophora paradoxa* genome elucidates origin of photosynthesis in algae and plants. Science 335:843–847

Przywara L, Kuta E (1995) Karyology of bryophytes. Polish Bot Stud 9:1–83

Ramsey J, Schemske DW (1998) Pathways, mechanisms, and rates of polyploid formation in the flowering plants. Annu Rev Ecol Syst 29:467–501

Ramsey J, Schemske DW (2002) Neopolyploidy in flowering plants. Annu Rev Ecol Syst 33:589–639

Rausch JH, Morgan MT (2005) The effect of self-fertilization, inbreeding depression and population size on autopolyploid establishment. Evolution 59:1867–1875

Ricca M, Beecher FW, Boles SB, Temsch E, Greilhuber J, Karlin EF, Shaw AJ (2008) Cytotype variation and allopolyploidy in North American species of the *Sphagnum subsecundum* complex (Sphagnaceae). Am J Bot 95:1606–1620

Rivero-Guerra AO (2008) Cytogenetics, geographical distribution, and pollen fertility of diploid and tetraploid cytotypes of *Santolina pectinata* Lag. (Asteraceae: Anthemideae). Bot J Linn Soc 156:657–667

Rodriguez DJ (1996) A model for the establishment of polyploidy in plants. Am Nat 147:33–46

Ronfort J (1999) The mutation load under tetrasomic inheritance and its consequences for the evolution of the selfing rate in autotetraploid species. Genet Res 74:31–42

Rosquist R (2001) Reproductive biology in diploid *Anthericum ramosum* and tetraploid *A. liliago* (Anthericaceae). Oikos 92:143–152

Rousseau-Gueutin M, Gaston A, Ainouche A, Ainouche ML, Olbricht K, Staudt G, Richard L, Denoyes-Rothan B (2009) Tracking the evolutionary history of polyploidy in *Fragaria* L. (strawberry): new insights from phylogenetic analyses of low-copy nuclear genes. Mol Phyl Evol 51:515–530

Sabara HA (2008) The evolution of reproductive isolation between diploid and tetraploid *Chamerion angustifolium* (Onagraceae). Ph.D. thesis, University of Guelph, Canada

Sarma YSRK (1982) Chromosome numbers in algae. Nucleus 25:66–108

Schranz ME, Osborn TC (2004) De novo variation in life-history traits and responses to growth conditions of resynthesized polyploid *Brassica napus* (Brassicaceae). Am J Bot 91:174–183

Schultz RJ (1969) Hybridization, unisexuality, and polyploidy in the teleost *Poeciliopsis* (Poeciliidae) and other vertebrates. Am Nat 103:605–619

Segraves KA, Thompson JN (1999) Plant polyploidy and pollination: floral traits and insect visits to diploid and tetraploid *Heuchera grossulariifolia*. Evolution 53:1114–1127

Shimotomai N (1933) Zur Karyogenetic der Gattung *Chrysanthemum*. J Sci Hiroshima Univ Ser B Div 2:1–100

Slovák M, Vít P, Urfus T, Suda J (2009) Complex pattern of genome size variation in a polymorphic member of the Asteraceae. J Biogeogr 36:372–384

Smith SA, Beaulieu JM, Donoghue MJ (2009) Mega-phylogeny approach for comparative biology: an alternative to supertree and supermatrix approaches. BMC Evol Biol 9:37

Sokolovskaya AP, Strelkova OS (1940) Karyological investigation of the alpine flora on the main Caucasus range and the problem of geographical distribution of polyploids. Dokl Akad Nauk SSSR 29:415–418

Soltis DE, Soltis PS (1987) Polyploidy and breeding systems in homosporous Pteridophyta: a re-evaluation. Am Nat 130:219–232

Soltis DE, Soltis PS (1988) Are lycopods with high chromosome numbers ancient polyploids? Am J Bot 75:238–247

Soltis DE, Soltis PS (1993) Molecular data and the dynamic nature of polyploidy. Crit Rev Plant Sci 12:243–273

Soltis DE, Soltis PS (1999) Polyploidy: origins of species and genome evolution. Trends Ecol Evol 14:349–351

Soltis PS, Soltis DE (2000) The role of genetic and genomic attributes in the success of polyploids. Proc Natl Acad Sci USA 97:70051–77057

Soltis DE, Soltis PS, Schemske DW, Hancock JF, Thompson JN, Husband BC, Judd WS (2007) Autopolyploidy in angiosperms: have we grossly underestimated the number of species? Taxon 56:13–30

Sonnleitner MR, Flatscher P, García E, Rauchová J, Suda J, Schneeweiss JM, Hülber K, Schönswetter P (2010) Distribution and habitat segregation on different spatial scales among diploid, tetraploid and hexaploid cytotypes of *Senecio carniolicus* (Asteraceae) in the eastern Alps. Ann Bot 106:967–977

Stace CA (2000) Cytology and cytogenetics as a fundamental taxonomic resource for the 20th and 21st centuries. Taxon 49:451–477

Stebbins GL (1938) Cytological characteristics associated with the different growth habits in the dicotyledons. Am J Bot 25:189–198

Stebbins GL (1941) Apomixis in the angiosperms. Bot Rev 7:507–542

Stebbins GL (1947) Types of polyploids: their classification and significance. Adv Genet 1:403–429

Stebbins GL (1950) Variation and evolution in plants. Columbia University Press, New York

Stebbins GL (1957) Self fertilization and population variability in the higher plants. Am Nat 91:337–354

Stebbins GL (1971) Chromosomal evolution in higher plants. Addison-Wesley, London

Stebbins GL (1980) Polyploidy in plants: unsolved problems and prospects. In: Lewis WH (ed) Polyploidy: biological relevance. Plenum Press, New York, pp 495–520

Stebbins GL (1985) Polyploidy, hybridization, and the invasion of new habitats. Ann Missouri Bot Gard 72:824–832

Stebbins GL, Dawe JC (1987) Polyploidy and distribution in the European flora: a reappraisal. Bot Jahrb Syst 108:343–354

Stone JL (2002) Molecular mechanisms underlying the breakdown of gametophytic self-incompatibility. Quart Rev Biol 77:17–32

Štorchová Z, Breneman A, Cande J, Dunn J, Burbank K, O'Toole E, Pellman D (2006) Genome-wide genetic analysis of polyploidy in yeast. Nature 443:541–547

Stuessy TF, Weiss-Schneeweiss H, Keil DJ (2004) Diploid and polyploid cytotype distribution in *Melampodium cinereum* and *M. leucanthum* (Asteraceae, Heliantheae). Am J Bot 91:889–898

Suda J (2003) Sympatric occurrences of various cytotypes of *Vaccinium* sect. *Oxycoccus* (Ericaceae). Nord J Bot 22:593–601

Suda J, Kron P, Husband BC, Trávníček P (2007) Flow cytometry and ploidy: applications in plant systematics, ecology and evolution. In: Doležel J, Greilhuber J, Suda J (eds) Flow cytometry with plant cells: analysis of genes, chromosomes and genomes. Wiley–VCH, Weinheim, pp 103–130

Suda J, Loureiro J, Trávníček P, Rauchová J, Vít P, Urfus T, Kubešová M, Dreyer LL, Oberlander KC, Wester P, Roets F (2009) Flow cytometry and its applications in plant population biology, ecology and biosystematics: new prospects for the Cape flora. S Afr J Bot 75:389

Tate JA, Soltis DE, Soltis PS (2005) Polyploidy in plants. In: Gregory TR (ed) The evolution of the genome. Elsevier, San Diego, pp 371–426

Temsch EM, Greilhuber J, Krisai R (2010) Genome size in liverworts. Preslia 82:63–80

Thompson JD, Lumaret R (1992) The evolutionary dynamics of polyploid plants: origins, establishment and persistence. Trends Ecol Evol 7:302–307

Thompson JN, Cunningham BM, Segraves KA, Althoff DM, Wagner D (1997) Plant polyploidy and insect/plant interactions. Am Nat 150:730–743

Thorpe PH, González-Barrera S, Rothstein R (2007) More is not always better: the genetic constraints of polyploidy. Trends Genet 23:263–266

Tindale MD, Roy SK (2002) A cytotaxonomic survey of the Pteridophyta of Australia. Austral Syst Bot 15:839–937

Tischler G (1935) Die Bedeutung der Polyploidie für die Verbreitung der Angiospermen erläutert an den Arten Schleswig-Holsteins, mit Ausblicken auf andere Florengebiete. Bot Jahrb 67:1–36

Tischler G (1937) Die Halligenflora der Nordsee im Lichte cytologischer Forschung. Cytologia 8:162–170

Trávníček P, Eliášová A, Suda J (2010) The distribution of cytotypes of *Vicia cracca* in Central Europe: the changes that have occurred over the last four decades. Preslia 82:149–163

Trávníček P, Kubátová B, Čurn V, Rauchová J, Krajníková E, Jersáková J, Suda J (2011a) Remarkable coexistence of multiple cytotypes of the *Gymnadenia conopsea* aggregate (the fragrant orchid): evidence from flow cytometry. Ann Bot 107:77–87

Trávníček P, Dočkalová Z, Rosenbaumová R, Kubátová B, Szeląg Z, Chrtek J (2011b) Bridging global and microgeographic scales: ploidy distribution in *Pilosella echioides* (Asteraceae) in Central Europe. Ann Bot 107:443–454

Troìa A (2001) The genus *Isoëtes* L. (Lycophyta, Isoëtaceae): synthesis of karyological data. Webbia 56:201–218

Uhl CH (1978) Chromosomes of Mexican *Sedum* II. Section *Pachysedum*. Rhodora 80:491–512

Vamosi JC, Dickinson TA (2006) Polyploidy and diversification: a phylogenetic investigation in Rosaceae. Int J Plant Sci 167:349–358

van Dijk PJ (2003) Ecological and evolutionary opportunities of apomixis: insights from *Taraxacum* and *Chondrilla*. Philos Trans Roy Soc Lond B Bio 358:1113–1120

van Dijk HM, Hartog M, van Delden W (1992) Single cytotype areas in autopolyploid *Plantago media* L. Bot J Linn Soc 46:315–331

Vida G (1976) The role of polyploidy in evolution. In: Novak VJA, Pacltová I (eds) Evolutionary biology. Czechoslovak Academy of Sciences, Prague, pp 267–304

Voglmayer H (2000) Nuclear DNA amounts in mosses (Musci). Ann Bot 85:531–546

Wettstein F (1927) Die Erscheinung der Heteroploidie, besonders im Pflanzenreich. Ergeb Biol 2:311–356

White MJD (1954) Animal cytology and evolution, 2nd edn. Cambridge University Press, Cambridge

Whitton J, Sears CJ, Baack EJ, Otto SP (2008) The dynamic nature of apomixis in the angiosperms. Int J Plant Sci 169:169–182

Winge Ö (1917) The chromosomes: their number and general importance. Compt Rend Trav Lab Carlsberg 13:131–275

Wodniok S, Brinkmann H, Glöckner G, Heidel AJ, Philippe H, Melkonian M, Becker B (2011) Origin of land plants: do conjugating green algae hold the key? BMC Evol Biol 11:104

Wolfe KH (2001) Yesterday's polyploids and the mystery of diploidization. Nat Rev Genet 2:333–341

Wood TE, Takebayashi N, Barker MS, Mayrose I, Greenspoon PB, Rieseberg LH (2009) The frequency of polyploid speciation in vascular plants. Proc Natl Acad Sci USA 106:13875–13879

Wyatt R, Odrzykoski IJ, Stoneburner A, Bass HW, Galau GA (1988) Allopolyploidy in bryophytes: multiple origins of *Plagiomnium medium*. Proc Natl Acad Sci USA 85:5601–5604

Xie-Kui C, Ao CQ, Zhang Q, Chen LT, Liu JQ (2008) Diploid and tetraploid distribution of *Allium przewalskianum* Regel (Liliaceae) in the Qinghai–Tibetan Plateau and adjacent regions. Caryologia 61:190–198

Zeyl C, Vanderford T, Carter M (2003) An evolutionary advantage of haploidy in large yeast populations. Science 299:555–558

Significance and Biological Consequences of Polyploidization in Land Plant Evolution

17

Jeffrey A. Fawcett, Yves Van de Peer, and Steven Maere

Contents

17.1 Introduction

The abundance of polyploids in nature has been long recognized. However, it is only in the last decade with the advance in genome sequencing that we have begun to appreciate their evolutionary potential. In fact, some researchers considered polyploidy as an 'evolutionary dead end' that cannot result in any further diversification of species (Stebbins 1950). It therefore came as a big surprise that the genome of *Arabidopsis thaliana*, the first plant genome to be sequenced, chosen for its small genome size, had experienced multiple rounds of polyploidization in its evolutionary past (Arabidopsis Genome Initiative 2000). This finding brought about some new questions; when did these Whole Genome Duplications (WGDs) occur? Are there traces of ancient WGDs in other plant genomes? And most importantly, what has been the significance of these polyploidy events in plant evolution?

Subsequent analyses of various other plant species over the past decade, based on whole-genome sequences or large Expressed Sequence Tag (EST) collections, have to a certain extent answered these questions. We now know that most, if not all eudicots are derived from a hexaploid ancestor (Fig. 17.1) (Jaillon et al. 2007; Tang et al. 2008a, b). The ancestor of *Arabidopsis* has undergone two WGDs after the hexaploidization (Jaillon et al. 2007; Ming et al. 2008; Tang et al. 2008a). Other lineages such as those of poplar, tomato, and apple have also each undergone additional WGDs (Tuskan et al. 2006; Tang et al. 2008a; Velasco et al. 2010). The lineage of grass species such as rice, maize, sorghum, and wheat appear to have undergone two WGDs after the divergence of monocots and eudicots, although the timing of the earlier WGD is less clear (Tang et al. 2010). Furthermore, it was recently suggested that one WGD has occurred in the common ancestor of all seed plants, and that another WGD has occurred in the common ancestor of all angiosperms (Jiao et al. 2011). Also, the genome of the moss species *Physcomitrella patens* has experienced a WGD (Rensing et al. 2007; 2013 this

S. Maere (✉)
Department of Plant Systems Biology, VIB, Technologiepark 927, Ghent B-9052, Belgium

Department of Plant Biotechnology and Bioinformatics, Ghent University, Technologiepark 927, Ghent B-9052, Belgium
e-mail: stmae@psb.vib-ugent.be

I.J. Leitch et al. (eds.), *Plant Genome Diversity Volume 2*,
DOI 10.1007/978-3-7091-1160-4_17, © Springer-Verlag Wien 2013

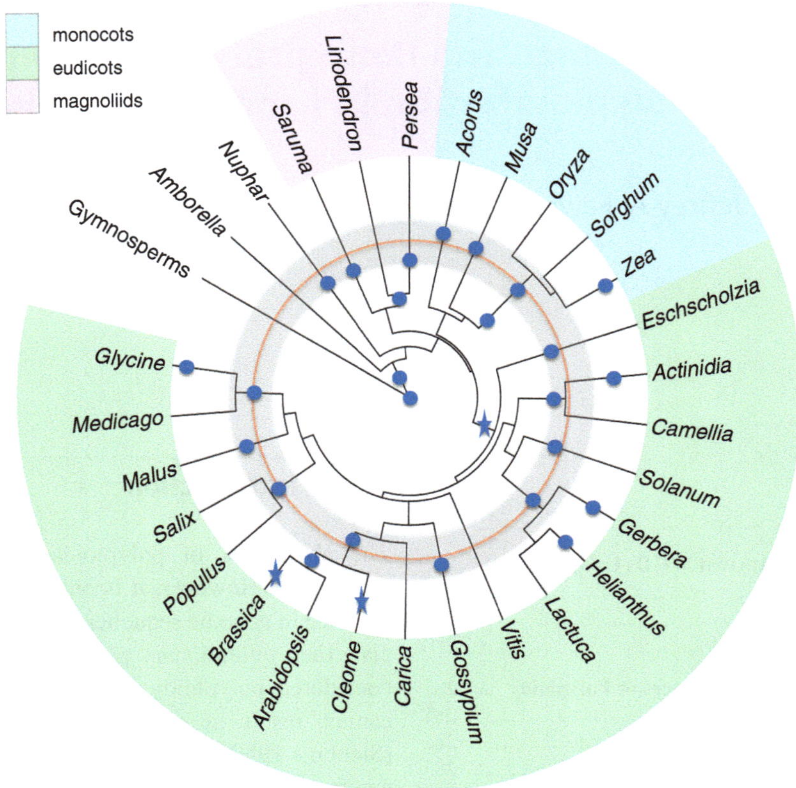

Fig. 17.1 Phylogenetic tree of angiosperm species showing ancient WGDs that have been proposed (Lysák et al. 2005; Cui et al. 2006; Jaillon et al. 2007; Barker et al. 2008, 2009; Lescot et al. 2008; Fawcett et al. 2009; Paterson et al. 2009; Schnable et al. 2009; Schmutz et al. 2010; Shi et al. 2010; Tang et al. 2010; Velasco et al. 2010). Ancient WGDs are observed throughout angiosperms, including in the common ancestor of most (if not all) eudicots. *Blue dots* represent duplication events whereas *blue stars* represent triplication events. Although many genera contain species with different ploidy levels (e.g., *Solanum, Oryza, Gossypium*), these polyploids are not shown. The *red line* represents the Cretaceous - Tertiary (Paleogene) boundary. Although there is some uncertainty associated with the ages of the WGDs, it seems that many of the WGDs occurred close to the time of the K-T extinction event (Fawcett et al. 2009)

volume). Note that the main focus of this chapter will be on angiosperms as little is known about WGDs in plant species outside angiosperms, with the exception of *Physcomitrella*.

'How many percent of plants are (ancient) polyploids?' has been a question that many researchers have been trying to answer (Otto and Whitton 2000). This question is becoming increasingly irrelevant as in fact most angiosperms, or even most seed plants might have a polyploid ancestry (Soltis et al. 2009; Jiao et al. 2011). The significance of polyploidy in the evolution of angiosperms is now undeniable, and a good understanding of polyploidy is vital if we are to understand the evolution of angiosperms. In this chapter, we will first describe how ancient WGDs can be identified in the genomes of extant species, and how the timing of ancient WGDs can be determined. After that, we will describe the molecular biological consequences of polyploidization: what happens to the genes and genomes of polyploid species immediately after polyploidization, and how do the genes and genomes evolve? Finally, we will discuss how the different WGDs might have contributed to the evolutionary success of the descendant plant lineages.

17.2 Detecting and Dating Ancient WGDs Using Genomic Data

17.2.1 Detecting Ancient WGDs

Ancient WGDs can be detected within the genomes of extant species by two different approaches—based on the age distribution of duplicated genes, and based on genomic collinearity. If the species in question had undergone a WGD in its evolutionary past, the genome would contain a large number of duplicated gene pairs of the same age. Thus, if one were to build an age distribution of a large collection of duplicated gene pairs, a large-scale or whole-genome duplication should correspond to a significant peak in the age distribution (Fig. 17.2). The number of synonymous substitutions per synonymous site (K_S) is often used as a proxy for time since duplication, assuming that synonymous substitutions occur at a constant rate across all duplicates. The K_S of duplicate pairs created by the same WGD should be similar. The K_S distribution approach has the advantage of not

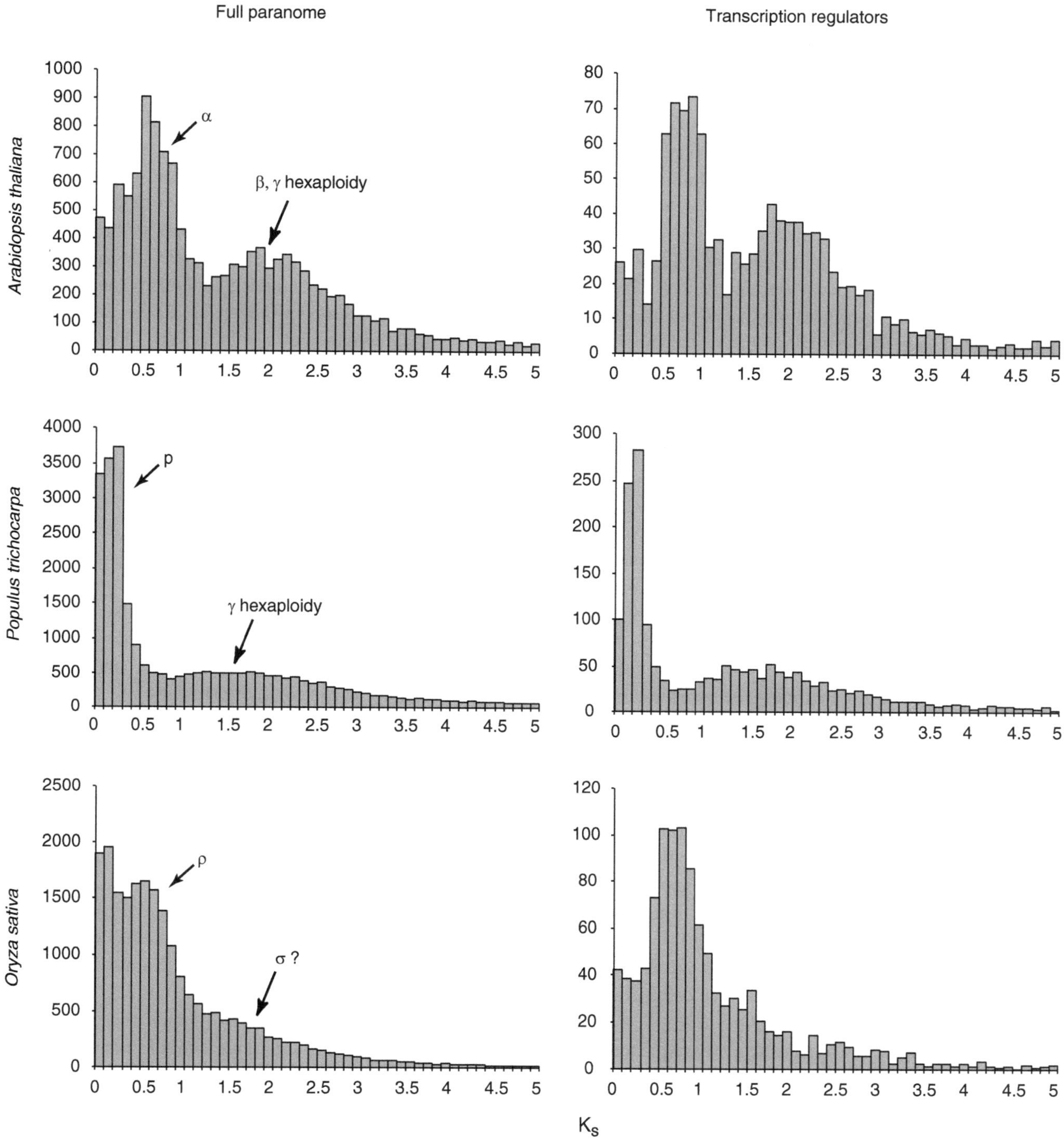

Fig. 17.2 In K_S distributions of duplicated gene pairs, genome duplications (*arrows*) show up as peaks superimposed on a roughly exponentially decaying background of small-scale duplications. Due to increasing variation in K_S with age since duplication, older peaks become progressively broader and may appear merged with other peaks. On the *left* are the K_S distributions of the whole paranome (complete set of duplicated genes) in three plant species. On the *right* are the corresponding distributions for transcription regulatory genes only. Even without rigorous modelling, it is clear from these figures that WGDs contributed comparatively more to the expansion of transcription regulatory gene families than to the expansion of the whole paranome. Distributions were calculated as in Maere et al. (2005). WGD nomenclature follows Tang et al. (2008a, 2010). The older σ WGD proposed to have happened in monocots (Tang et al. 2010) is not clearly visible, and neither are the recently proposed angiosperm and seed plant WGDs (Jiao et al. 2011)

requiring assembled genomic sequences, and many researchers have used it to search for traces of ancient WGDs in EST libraries, leading to the identification of putative WGDs in many lineages such as rice, tomato, soybean, lettuce, and California poppy (Blanc and Wolfe 2004a; Schlueter et al. 2004; Cui et al. 2006; Barker et al. 2008).

One downside of the K_S approach is that if the WGD is too old, the K_S values of the duplicate pairs have a large variance, and the WGD peak may overlap with other peaks or even an artefactual saturation peak in the K_S distribution (Fig. 17.2). In such cases, more sophisticated approaches such as building age distributions based on phylogenetic dating are required (Fawcett et al. 2009; Jiao et al. 2011). WGDs that are too young can also escape detection as the divergence between the duplicates will be low and indistinguishable from the young single gene duplicates or from sequencing errors, which occur frequently in ESTs (Blanc and Wolfe 2004a). It is worth noting that segmental duplication or bursts of single gene duplications can also be responsible for peaks in the age distribution (Blanc and Wolfe 2004a). It is advisable to be cautious when the number of duplicates present in the peak is small and/or if the K_S of the peak is large and perhaps saturated, even if it is statistically significant by some measure.

If the whole genome sequence is available, ancient WGDs can be detected based on genomic collinearity (Van de Peer 2004). Although the similarity between the duplicated genomes decreases as they diverge from each other and many duplicate genes are lost, remnants of genome duplications are often still detectable in the form of duplicated blocks, i.e., sequence stretches containing members of the same paralogous gene pairs in the same (collinear) order, intermingled with single-copy genes that have lost their WGD paralog. Different methods exist to assess the statistical significance of a collinear arrangement of duplicated gene pairs. Including additional genomes in collinearity analyses can increase the resolution of detecting duplicated blocks. Genes that have lost their WGD paralog in one species might still have a paralog in another species, allowing for transitive detection of collinear regions (Vandepoele et al. 2002; Proost et al. 2009). If several paralogous regions appear to be of similar age, i.e., contain paralogous gene pairs with similar K_S, we can assume that they were created by the same WGD event.

Collinearity analyses on various recently sequenced plant genomes have indeed uncovered several instances of ancient WGDs. The genome of rice showed evidence of an ancient WGD that has been confirmed to have occurred in the common ancestor of most grass species such as rice, maize, sorghum, and wheat, ~65–70 Mya, in addition to an older WGD (Paterson et al. 2004; Tang et al. 2010). The genomes of the legumes *Medicago*, *Lotus*, and soybean showed that a WGD occurred before the divergence of these three species (Cannon et al. 2006; Sato et al. 2008; Schmutz et al. 2010). The genomes of poplar and apple both contain evidence of independent WGDs that occurred >60 Mya and >50 Mya, respectively (Tuskan et al. 2006; Velasco et al. 2010).

17.2.2 Dating Ancient WGDs

Once ancient WGDs are identified in a number of species, the next task is to identify the timing of these WGDs. For instance, is an ancient WGD identified in one species shared with another species or specific to that species? Again, genomic collinearity analysis has proven to be very powerful. The detection of collinearity between multiple eudicot genomes, including genomes that appear to have a more ancestral genome structure such as grapevine and papaya, substantially improved our understanding of the timing of ancient WGDs in eudicots (Tang et al. 2008a; Van de Peer et al. 2009b). Grapevine belongs to an early-diverging lineage of rosids and is thus sister to both *Arabidopsis* and poplar (Jansen et al. 2006). It was observed that many regions of the grapevine genome showed homology to two other regions elsewhere in the genome, representing a triplicate structure. This suggests that the grapevine lineage derived from a hexaploid ancestor. In addition, some of these regions showed a one to two relationship to regions in the poplar genome, and a one to four relationship to regions in the *Arabidopsis* genome. This led to the suggestion that the hexaploidization occurred in the common ancestor of grapevine, *Arabidopsis*, and poplar, and that one additional WGD occurred in the lineage leading to poplar, and that two additional WGDs occurred in the lineage of *Arabidopsis* (Jaillon et al. 2007). The genomes of papaya and apple also showed traces of this hexaploidization event, confirming a common hexaploid ancestry of most rosids (Ming et al. 2008; Velasco et al. 2010). The genome of papaya, which is a rosid species more closely related to *Arabidopsis* than to poplar or grapevine, had no traces of additional WGDs after the hexaploidy event, indicating that the two additional WGDs observed in the *Arabidopsis* genome occurred after the divergence of the *Arabidopsis* and papaya lineages (Ming et al. 2008). Comparison of the genomic sequences of grapevine and tomato, an asterid species, suggests that tomato also shares the hexaploidy event (Tang et al. 2008b). This would mean that the hexaploidization pre-dates the divergence of rosids and asterids, and that it would thus be shared by most eudicots.

In the absence of multiple genomic sequences, phylogenetic approaches have been used to identify the timing of WGDs. Suppose that we want to ask whether a WGD that has been detected in the genome of species A is shared with species B or not. If the WGD had occurred in the ancestor of A after the divergence of A and B, the paralogs derived from the WGD (A1 and A2) should be more closely related to each other than to their closest homolog in B (say B1), leading to the topology ((A1,A2),B1). On the other hand, if the WGD had occurred in the common ancestor of A and B, the closest homolog in B should be the ortholog of A1 or A2 and more closely related to A1 or A2 than A1 and A2 are to each other, leading to the topology ((A1,B1),A2) or ((A2,B1),A1). Although this is a useful approach, it can easily yield misleading results because the different substitution rates in different lineages can result in erroneous topologies, and careful choice of species and tree reconstruction is required (Tang et al. 2008b; Van de Peer et al. 2009b; Jiao et al. 2011).

Ideally, one would like to know the timing of any WGD event in terms of 'how many million years ago', rather than 'before the divergence of these species and after the divergence of these species', because this will enable us to place the WGDs in the context of the geological time frame. The simplest approach to dating WGDs is to build an age distribution based on the K_S of all WGD-derived duplicate pairs, and convert the K_S value of the peak of the distribution into time based on estimates of synonymous substitution rates. Although such an approach can provide an adequate approximate in some cases, it is also known to be error-prone for various reasons. For instance, the substitution rate is known to vary across time and species, and estimating the rate for any given species can be very difficult—consider how our understanding of the evolutionary rate of the model species *Arabidopsis thaliana* has totally changed in the past few years (Koch et al. 2000; Jakobsson et al. 2006; Beilstein et al. 2010). In addition, dating older events is especially problematic because the K_S value becomes less reliable and may approach saturation. Recent studies have attempted to date WGD events by building phylogenetic trees with gene families including WGD-derived duplicates and dating the divergence of all duplicated genes using methods that account for rate variation such as Bayesian methods or the penalized likelihood method (Fawcett et al. 2009; Jiao et al. 2011). Such phylogenomic-like approaches will increase in accuracy as genomic data from many different species become available.

17.3 Molecular Biological Consequences of Polyploidy

Although several groups of species contain recently formed polyploids and evidence of ancient WGDs has been found in all angiosperm genomes, the process leading from the formation of polyploids to their establishment is rather complex. First, we will discuss the changes that occur in polyploid genomes immediately or shortly after polyploidization. Recent natural polyploids in genera such as *Brassica*, *Arabidopsis*, cotton, soybean, wheat, tobacco, and *Tragopogon* can provide insights into the early stages of polyploid evolution (Wendel 2000; Kashkush et al. 2003; Comai 2005; Doyle et al. 2008; Jackson and Chen 2010). In addition, synthetic polyploids have been created in many of these species, which allow us to trace the changes that occur immediately after the polyploidization.

17.3.1 Genetic and Epigenetic Changes

Various genetic changes such as indels and rearrangements have been reported to occur shortly after polyploidization in both synthetic and natural polyploids. In synthesized allopolyploids in wheat, elimination of DNA sequences was observed already in the first generation (Ozkan et al. 2001). Although genetic changes were found to be rare in the first generation of ~50 re-synthesized allopolyploid lines of *Brassica napus*, several genetic changes, in particular non-reciprocal transpositions of DNA segments between homoeologous chromosomes were documented during the S2–S5 generations (Gaeta et al. 2007; Lukens et al. 2006). *Tragopogon mirus* is an allopolyploid that has formed repeatedly in nature within the last 80 years (Soltis and Soltis 1999). Loss of DNA has been observed in multiple individuals of this species that are of independent origin (Koh et al. 2010). On the other hand, no large-scale genomic rearrangements were observed in the synthetic *Gossypium* allopolyploids, suggesting that the responses vary among different polyploids (Liu et al. 2001). The genetic changes appear to be a combination of random and non-random changes. In the ~50 re-synthesized allopolyploid lines of *B. napus*, there was a large variation among the different lines of the extent and direction (in which parental genome the changes took place) of the genetic changes. However, the changes were not random in respect to the genomic location, and tended to occur more frequently on chromosomes with large regions of homeology between the two subgenomes (Gaeta et al. 2007).

Changes at the epigenetic level, such as DNA methylation or histone methylation and acetylation, are also likely to play a significant role in newly formed polyploids (Lukens et al. 2006; Wang et al. 2006b). Changes in the pattern of DNA methylation have been especially well-studied. Synthetic polyploids in wheat, *B. napus*, and *Arabidopsis suecica* all showed non-additive methylation patterns with respect to their diploid progenitors, but to varying degrees. No methylation changes were observed in synthetic cotton allopolyploids. Methylation changes have also been documented in natural polyploids in wheat and *Spartina anglica* (Shaked et al. 2001; Salmon et al. 2005). Extensive changes in DNA methylation were also documented in the first generation of the aforementioned synthetic *B. napus* allopolyploids, in contrast to the paucity of genetic changes that took place in the first generation (Lukens et al. 2006). Many of these methylation changes remained fixed in their S5 progeny, although some methylations were reversed and some new changes were observed (Gaeta et al. 2007).

Another documented phenomenon following polyploidy is the activation of transposable elements (TEs). This is most probably related to changes at the epigenetic level that reverse the epigenetic silencing of TEs. Increased transcriptional activity of TEs following polyploidy is thought to be common and has been documented in various polyploids including synthetic allopolyploids in wheat and *Arabidopsis* (Kashkush et al. 2003; Madlung et al. 2005; Parisod et al. 2010). TE amplification after polyploidization appears to be less common, although it has been documented in some cases such as the *Tnt1* retrotransposon in *Nicotiana* synthetic allopolyploids (Petit et al. 2010). This is likely to be because while TE activation is more of an immediate response to the effect of hybridization or polyploidization, the amplification of TEs requires their fixation in new genomic loci, which might be countered by purifying selection due to deleterious effects. Nevertheless, the doubling of the genome might relax the purifying selection against deleterious TE insertions, and reduction in the effective population size following polyploid formation might increase the chance for neutral or slightly deleterious TE insertions to fix through stochastic processes, thereby providing the opportunity for TE amplification (Parisod et al. 2010).

17.3.2 Changes in Expression Pattern

One question of great interest has been how the expression patterns change in homoeologous genes following polyploidization (Hegarty and Hiscock 2008; Jackson and Chen 2010). For instance, are the expression levels of both homoeologous genes maintained at the same levels as in their parental diploid species? How do the expression patterns of homoeologous genes diverge?

It has been shown in various polyploids that the expression levels of homoeologous genes are frequently non-additive, meaning that one or both genes are up- or down-regulated in the polyploid species compared to their parental species. In *Arabidopsis* synthetic polyploids formed by combining *A. arenosa* and *A. thaliana*, 5–38% of the homoeologous gene pairs were non-additively expressed. More than 65% of these genes were down-regulated compared to the parental midpoint, and >94% of the repressed genes were genes that were expressed at higher levels in *A. thaliana* than in *A. arenosa*, indicating extensive suppression of *A. thaliana*-derived homoeologs in the allopolyploid (Wang et al. 2006b). Similar biases where homoeologs from one of the parental species get preferentially silenced have also been observed in cotton and *Brassica* (Chen and Pikaard 1997; Flagel et al. 2008). A recent study quantified the leaf transcriptome size of a *Glycine* tetraploid species and its diploid progenitors. Despite the genome size of the tetraploid being close to the sum (94.3%) of the genome sizes of the two diploid progenitors, the transcriptome size was only 70% of the sum of the transcriptomes of its diploid progenitors, suggesting global downsizing of the transcriptome (Coate and Doyle 2010).

The expression pattern of homoeologous genes can also diverge in a tissue-specific manner. For instance, in some homoeologs in cotton natural or synthetic allopolyploids, the A-genome homoeolog would be silenced in some tissues whereas the D-genome homoeolog would be silenced in other tissues (Adams et al. 2003; Chaudhary et al. 2009). Novel expression patterns have also been observed where a homoeolog is expressed in a tissue in which it was not expressed in the diploid progenitor (Chaudhary et al. 2009). Such expression profile changes are also often observed in F1 diploid hybrids, and it has been suggested that the merging of two different genomes, rather than the doubling, has the largest effect on the expression divergence (Chaudhary et al. 2009). This is not to say that genome doubling does not affect the expression pattern, as shown in *Senecio* (Hegarty et al. 2006). Interestingly, the expression changes of some homoeologs in synthetic allopolyploids mimic the expression changes observed in natural allopolyploids that have in some cases undergone a few million years of evolution (Chaudhary et al. 2009). This suggests that tissue-specific differential silencing that occurs immediately after polyploidization might be maintained for over a million years. A recent study showed that the expression of small RNAs also changes in allopolyploids. It was suggested that the small RNAs might have a role in adjusting the expression levels of mRNAs and maintaining genome and chromatin stability (Ha et al. 2009).

The various genetic and epigenetic changes described above can partly explain the expression divergence that occurs in polyploids. For instance, indels or mutations including TE insertions, or changes in DNA methylation in the upstream regulatory region are likely to alter the expression pattern of the downstream genes. It has indeed been shown that blocking DNA methylation can reactivate silenced genes in polyploids (Lee and Chen 2001). However, expression divergence has been observed when little genetic or epigenetic changes have been detected, and it is thought that the coming together of genes with diverged regulatory sequences plays a large role (Albertin et al. 2006; Chaudhary et al. 2009; Hegarty et al. 2006). For instance, consider two genes A and B in two different species 1 and 2; in the diploid parents, A1 regulates the expression of B1 and A2 regulates the expression of B2, and the expression pattern of B1 and B2 may be different due to divergence between the parent species (at the genetic or epigenetic level). When the two different species form a new polyploid (or a hybrid), the polyploid can potentially gain two new regulatory interactions where A1 regulates B2 and A2 regulates B1, so that B1 and B2 can be regulated by both A1 and A2. Thus, B1 and B2 can have novel expression patterns that were not present in either parental species (Fig. 17.3). Such reuniting of diverged regulatory interactions is likely to be responsible in part for the immediate expression changes and novel

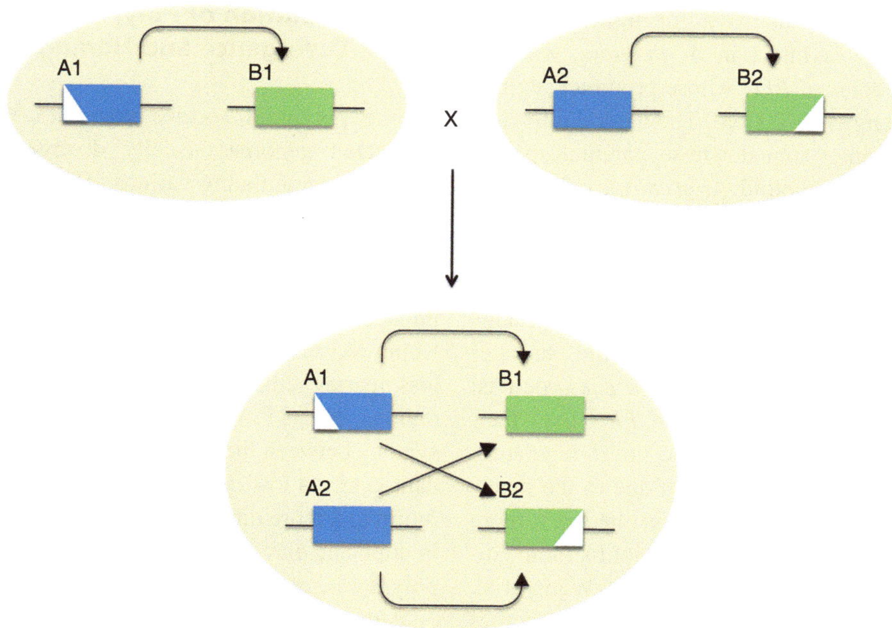

Fig. 17.3 The emergence of novel expression patterns in newly formed polyploids. Gene A regulates the expression of gene B in two different species 1 and 2, but the genes A and B have independently accumulated changes in both species since their divergence. As such, the expression pattern of gene B (B1 and B2) is different in the two species. This can be due to changes at the genetic or epigenetic level in the coding or non-coding regions of A and/or B. As these two species form an allopolyploid, apart from inheriting the regulatory interactions from both parental species (A1–B1 and A2–B2), the polyploid can potentially gain novel regulatory interactions (A1–B2 and A2–B1) immediately. This can result in phenotypes not present in the parental species, which might allow the polyploids to adapt to different environments

expression patterns that are induced by genome merger (Riddle and Birchler 2003).

17.3.3 Adaptive Potential of Novel Polyploids

Polyploidy can be generally considered an abnormal state that has many disruptive effects. Most of the molecular biological consequences described above are usually deleterious (Comai 2005). However, some of these changes might also result in new traits that allow polyploids to adapt to environments that their diploid progenitors could not (Hegarty and Hiscock 2008). In fact, it has been frequently observed that polyploids have colonized novel, often harsh environments (Fawcett and Van de Peer 2010). In *Achillea borealis*, autohexaploids have a fitness advantage over tetraploids in dune habitats. Transplantation experiments showed that neohexaploids already exhibited higher survivorship than tetraploids in dune habitats, indicating that an increase in ploidy level per se can result in a fitness advantage (Ramsey 2011). The adaptive potential of polyploids has been questioned because deleterious mutations will persist longer in a polyploid population and increase the genetic load, whereas the effects of new beneficial mutations might get masked by other alleles at the same locus (Stebbins 1971). However, theory suggests that polyploid populations may adapt faster than diploid populations under certain conditions. For instance, the masking of deleterious mutations can provide polyploids an initial advantage over diploids as long as the deleterious allele is recessive and the masking is stronger in the polyploid than in the diploid (Otto and Whitton 2000). This might contribute to the survival of polyploids shortly after polyploidization by buffering crucial functions (Chapman et al. 2006). Also, new advantageous alleles can spread more rapidly in polyploid populations if the allele is dominant and if the population size is small (Otto and Whitton 2000).

The ability of polyploids to adapt to more extreme environments is probably related more to gene expression changes after polyploidization than merely to copy number changes (Crow and Wagner 2006; Ha et al. 2007; Hegarty and Hiscock 2008; Hegarty et al. 2008; Jackson and Chen 2010). Novel regulatory interactions that emerge due to the hybridization of two genomes can cause immediate changes in gene expression and concomitant changes in phenotype. One example is flowering time variation across *Arabidopsis* diploids and tetraploids. In *Arabidopsis*, the flowering time is largely controlled by the *FRIGIDA* (*FRI*) and *FLOWERING LOCUS C* (*FLC*) genes where *FRI* upregulates *FLC* expression that inhibits early flowering (Johanson et al. 2000). *FRI* is non-functional in the early-flowering ecotypes of *A. thaliana*, whereas the late-flowering autotetraploid *A. arenosa* contains a functional *FRI*. By contrast, the *FLC* gene is intact in *A. thaliana*, whereas *A. arenosa* has two copies of *FLC* whose promoter regions contain deletions (Wang et al. 2006a). The natural allotetraploid *A. suecica*, a winter-annual whose parental species are *A. thaliana* and autotetraploid *A. arenosa*, exhibits an extremely late flowering time, even later than *A. arenosa*. It was found that synthetic F1 allotetraploids containing the genomes of *A. thaliana* and *A. arenosa* already showed a later flowering time than *A. arenosa* autotetraploids. This was attributed to the trans-activation of the stronger *A. thaliana*-derived *FLC* by the functional *A. arenosa*-derived *FRI*, which allowed higher expression of *FLC* than their parental plants. Thus, it appears that the novel regulatory interaction that arose in the allopolyploid due to the divergence between the two parental species resulted in a more extreme phenotype, providing the allopolyploid an immediate adaptive advantage in novel environments (Wang et al. 2006a). More generally, the merger of divergent genomes through polyploidization can lead to transgressive segregation, i.e., the formation of phenotypes that are more extreme than the diploid parent phenotypes, and hybrid vigor (Osborn et al. 2003; Rieseberg et al. 2003, 2007; Crow and Wagner 2006; Hegarty et al. 2008).

Polyploids may also support a wider *range* of transgressive and non-transgressive expression patterns than their diploid counterparts, through an increase in the number of possible allele dosage combinations (Osborn et al. 2003). This may facilitate adaptation to novel environments (Hegarty and Hiscock 2008) if there is sufficient allelic variation in the polyploid population. Allelic variability may be accomplished through gene flow between polyploid populations that originated independently (Soltis and Soltis 1999), through mutations and epigenetic changes that occur immediately after polyploidization, or through polysomic inheritance in autopolyploids (Soltis and Soltis 2000; Osborn et al. 2003)

Epigenetic changes are also likely to be important in the adaptation of polyploids. Epigenetic changes can result in gene expression changes, and are likely to proceed more rapidly than genetic changes, creating phenotypic variation in the early stage of the evolution of polyploids. A recent study on three closely related allopolyploid species of the orchid *Dactylorhiza* suggested that the ecological differentiation and adaptation of the different species were facilitated by methylation changes (Paun et al. 2010). The authors showed that the three species, and also two geographically distinct populations of *D. traunsteineri*, could be clearly differentiated based on their methylation pattern (Paun et al. 2010), even though the genetic divergence between the different allopolyploid species is limited (Pillon et al. 2007). Moreover, the gene expression differences between the allopolyploid species were best explained by differences in their methylation pattern.

17.3.4 Evolution of Polyploid Genomes; Divergence and Homogenization

As the polyploid species continues to evolve, the two duplicated genomes usually diverge further from each other by accumulating various changes at the genetic level. Although epigenetic changes and expression changes might be crucial in determining adaptation and short-term success of polyploids, the genetic changes are likely to have a greater impact on an evolutionary time scale (Edger and Pires 2009; Kejnovsky et al. 2009; Fawcett and Van de Peer 2010). DNA loss, rearrangements, recombination between homoeologous regions, and TE amplification result in the disruption of synteny between the two genomes. A large number of redundant genes is lost or pseudogenized, causing the genome to return to a more diploid state. But the genes that are retained in duplicate play a crucial role in the evolutionary success of polyploids, as we discuss below.

Although it is often assumed that the divergence between paralogous sequences increases over time, this is not always true. Non-reciprocal recombination, or gene conversion can occur between paralogous sequences, which will counteract their divergence. Homogenization by gene conversion has been reported in young polyploids such as cotton and tobacco (Kovarik et al. 2008; Salmon et al. 2010). In addition, extensive gene conversion has been recently reported in duplicates created by the WGD that occurred in the common ancestor of most grass species such as rice and sorghum. In the most extreme case, the short arms' termini of rice chromosomes 11 and 12 show very high similarity throughout the entire 3 Mb, including non-coding regions, and was initially thought to represent a <10 million year old segmental duplication (The Rice Chromosomes 11 and 12 Sequencing Consortia 2005; Wang et al. 2005). However, this duplication was found to be present also in sorghum, which diverged from rice ~50 Mya, and other grass species (Jacquemin et al. 2009; Paterson et al. 2009; Wang et al. 2009). Thus, it appears that this duplicated segment was created by the WGD that occurred in the common ancestor of rice and sorghum, and has been extensively homogenized in rice since then. The corresponding segments in sorghum also appear to have undergone gene conversion, but to a lesser extent than in rice. Gene conversion, like any other mutational process, can have both positive and negative effects (Fawcett and Innan 2011). Although gene conversion reduces the divergence between the paralogous sequences, it simultaneously increases the variation at the haplotype level by creating chimeras of the two paralogous sequences (Fawcett and Innan 2011). It has also been suggested that gene conversion might be important for buffering crucial functions in some WGD duplicates (Chapman et al. 2006) or for maintaining high dosage (Sugino and Innan 2006). In addition, gene conversion is likely to increase the efficiency of selection on duplicated genes (Mano and Innan 2008;

Marais et al. 2010), and it has been reported that the genes undergoing extensive gene conversion in rice are evolving faster (diverging faster from their sorghum orthologs) compared to duplicates that are not undergoing gene conversion (Wang et al. 2009).

17.4 Evolutionary Significance of WGDs in Angiosperms

Most, if not all eudicots derived from a hexaploid ancestor (Tang et al. 2008b; Van de Peer et al. 2009b). All angiosperms might also have a common polyploid ancestry, as well as all seed plants (Jiao et al. 2011). It is also known that most vertebrates derived from a common ancestor that underwent two rounds of WGD (Kasahara 2007). These facts underscore that polyploids inherently have an enormous potential for long-term evolutionary success. Here, we will discuss links between WGDs and two processes that could have contributed to their evolutionary success: species radiation and the generation of evolutionary novelty. Since hard evidence for such links is scarce, we will not limit our discussion to the plant kingdom, but also include evidence from other organisms, mainly vertebrates and teleost fish.

17.4.1 Genome Duplications and Speciation

Genome duplications have repeatedly been linked to species radiations, but the evidence in support of this theory is mostly circumstantial. Examples of putative correlations between genome duplication and increased diversity are found among flowering plants and teleost fish.

Soltis et al. studied the relationship between polyploidy and species richness in angiosperms by comparing the species richness in clades that are ancient polyploids with sister clades that are not (Soltis et al. 2009). In many cases, a strong correlation was found between diversification rates and the occurrence of genome duplications. Whole-genome duplications have been reported in many of the most species-rich plant families, including the Poaceae (>10,000 species), Solanaceae (>3,000 species), Asteraceae (23,000 species), Fabaceae (19,400 species), Lauraceae (>2,000 species) and Brassicaceae (3,700 species) (Cui et al. 2006; Fawcett et al. 2009). However, the phylogenetic position of the WGDs in all of these families is currently insufficiently resolved to causally link genome duplications to radiations (Soltis et al. 2009).

The Actinopterygii or ray-finned fish include more than 25,000 species, the vast majority of which belong to the most derived division, the teleosts. All older, more basal groups of ray-finned fish, namely the Chondrostei (bichirs, sturgeons and paddlefish), and the basal neopterygian orders (gars and bowfin), consist of only a few (~44) extant species. Depending on the source, a fish-specific genome duplication (3R) is estimated to have occurred in the teleost lineage between 226 and 350 Mya (Christoffels et al. 2004; Hoegg et al. 2004; Vandepoele et al. 2004; Hurley et al. 2007). Phylogenetic analyses indicate that 3R separates the species-poor early branching ray-finned fish lineages from the extremely species-rich teleost lineage, suggesting that 3R might be causally related to the teleost radiation. However, fossil evidence suggests that the major teleost radiations did not take place until late in the Cretaceous, more than 150 million years after the youngest 3R age estimate (Hurley et al. 2007). Thus, there seems to be a major time gap between 3R and the radiation of the teleosts, rendering it less likely that genome duplication has been a causative agent in the radiation process. Could the speciation potential of 3R have been kept in store for 150 million years?

17.4.2 Reciprocal Gene Loss and Subfunction Partitioning

Although it has proven difficult to firmly establish causative links between WGD and speciation, there is at least one mechanistic process through which gene duplication may facilitate speciation. WGDs create massive amounts of redundant gene pairs that in the course of evolution return to singleton status through duplicate gene loss or pseudogenization. It can happen that different copies of a duplicated gene pair get lost in different populations, a process referred to as reciprocal gene loss (RGL) or divergent resolution. Mating between individuals from those populations leads to heterozygous F1 progeny with one functional allele at each locus of the duplicated gene. The F2 offspring will inherit a variable number of functional alleles, and 1/16 of the progeny will not inherit any functional alleles. Divergent resolution at 20–30 essential locus pairs could be sufficient to reproductively isolate two populations (Werth and Windham 1991; Lynch and Conery 2000; Semon and Wolfe 2007b). After a genome duplication, divergent resolution could be operating on thousands of duplicate pairs simultaneously, rendering it a very effective speciation mechanism. A similar reproductively isolating process could occur at the level of duplicate subfunctionalization (Postlethwait et al. 2004). If a duplicated gene is multifunctional, different subfunctions could be divergently resolved in different populations and contribute to reproductive incompatibility. The efficiency of this mechanism would of course depend on the prevalence of subfunctionalization after WGD.

Recently, Bikard et al. found evidence that divergent resolution of duplicated genes can indeed be a source of genetic incompatibilities (Bikard et al. 2009). These authors studied crosses between the Col and Cvi accessions of *Arabidopsis thaliana* and came to the conclusion that a homozygous combination of the Col allele of one particular locus with the Cvi allele of another unlinked locus led to arrested embryo development and seed abortion. In Col, these loci contain a paralogous pair of histidinol-phosphate aminotransferases, *HPA1* and *HPA2*. Homozygous *HPA1* deletion is embryo lethal. But in Cvi, the *HPA1* ortholog is missing and the *HPA2* ortholog appears to be the functional allele.

The analysis of complete sequences of pre- and post whole genome duplication species showed that RGL of duplicated genes after WGD is very common. Scannell et al. (2006) showed that after the WGD in ascomycetous yeasts, almost half of the pairs that returned to single-copy status were divergently resolved, suggesting that RGL could have been a major factor in post-WGD speciation. No large-scale studies have been performed so far on plants, but widespread RGL has also been documented in teleost fish. Using the post-WGD genome sequences of *Tetraodon* and zebrafish, and the pre-genome duplication sequences of human and chicken as outgroups, it was estimated that ~1,700 WGD-pairs were divergently resolved since the *Tetraodon*-zebrafish split (Semon and Wolfe 2007b). The studies in yeast and fish also suggest that RGL and/or subfunction partitioning can promote speciation over long periods of time, even tens to hundreds of millions of years after a WGD (Scannell et al. 2006; Semon and Wolfe 2007b), which may help explain how the 3R WGD in teleosts could have facilitated the teleost radiation 150 million years later.

17.4.3 Dosage Balance Effects and the Regulatory Spandrel

WGDs leave a very specific signature in the duplicate retention pattern of certain gene families. In fact, the duplicate retention behaviour of some gene classes after WGD is the exact opposite of their behaviour after small-scale duplications (SSD) (Maere et al. 2005; Freeling 2009). Many transcriptional and developmental regulators, transporters, signal transducers and complex-forming genes appear to be preferentially retained after WGD and do not seem to duplicate easily through SSD. This observation has been made for WGDs in diverse organisms, from yeast (Davis and Petrov 2005) and *Arabidopsis* (Blanc and Wolfe 2004b; Seoighe and Gehring 2004; Maere et al. 2005) to vertebrates (Blomme et al. 2006; Putnam et al. 2008) including teleosts (Blomme et al. 2006; Brunet et al. 2006). The peculiar reciprocal retention patterns of certain gene classes after SSD and WGD can be explained by dosage balance effects (Papp et al. 2003; Birchler et al. 2005). When a dosage-sensitive gene is duplicated on its own, the effects of its increased dosage with respect to its interaction partners or targets would be deleterious, and the duplication would be selected against. But if the gene and its interaction partners/targets are duplicated together, as in WGD, their relative dosage is preserved. Moreover, if after WGD one of the duplicates would be lost, this would create a reverse dosage balance effect, causing dosage sensitive genes to be selectively retained after WGD.

The consequence of these dosage balance effects is that post-WGD organisms are endowed with a regulatory 'spandrel' (Gould and Lewontin 1979), a collection of extra transcription factors, transporters and complexes that may not immediately be useful but that cannot be purged easily from the genome. In the long run, these non-adaptively preserved genes may be co-opted for adaptive innovations. According to Freeling and Thomas (2006) and Freeling (2009) the dosage balance-mediated increase of the regulatory and complex-forming gene repertoire after WGD could even cause a predictable drive toward higher complexity, not unlike the meiotic drive.

17.4.4 WGDs and Evolutionary Novelty

It appears that some genome duplications have indeed been followed by a substantial increase in morphological complexity in the affected organisms. Early WGDs in the angiosperm plant lineage have been invoked to explain the rapid rise and diversification of angiosperms in the Early Cretaceous (145–125 Mya) (Crepet 2000; Otto and Whitton 2000; De Bodt et al. 2005; Otto. 2007; Soltis et al. 2008, 2009; Jiao et al. 2011), linked to the invention of the closed carpel, the emergence of double fertilization and the invention of the flower (Stuessy 2004). Several authors have also suggested a causal link between the 2R duplications and the emergence of the vertebrates, (e.g., Ohno 1970; Aburomia et al. 2003; Holland 2003). The two rounds of genome duplication (2R) in the vertebrates approximately coincide with the development of important morphological innovations, not only in the endoskeleton but also in the nervous system and sensory organs, and endocrine, immune and circulatory systems (Shimeld and Holland 2000; Holland and Chen 2001; Khaner 2007; Holland et al. 2008; Wagner 2008). Post-3R teleosts (Stellwag 2004; Heimberg et al. 2008) also developed some novel morphological features, such as pharyngeal jaws, an oxygen secretion system to inflate the swimbladder (Berenbrink et al. 2005), and complex pigment patterning systems (Mellgren and Johnson 2002; Braasch et al. 2008). Moreover, there are indications that the oxygen secretion system originated independently at least four times in teleosts (Berenbrink et al. 2005), suggesting that these fish shared the potential to develop such a system whereas fish that did not undergo the 3R duplication did not.

It is unclear to what extent genome duplications are involved in the actual *invention* of fundamental innovations.

There are indications that genome duplications may play a more subtle role instead, in refining and elaborating innovations rather than inventing them. In many organisms, fundamental innovations have been followed by the emergence of many variants or elaborations. The invention of the flower for instance was followed by the development of a huge variety of floral forms, specialized pollination syndromes and fruits. Many of the morphological innovations in vertebrates also derive from a single key innovation: the neural crest, a migratory cell population that gives rise to numerous differentiated cell types (Shimeld and Holland 2000; Holland et al. 2008). The evolutionary success of the vertebrates is as much related to the development of various neural crest elaborations as to the invention of the neural crest itself. The complex pigment patterning systems of post-3R teleost fish for instance also find their origin in a further expansion of the neural crest-derived cell type repertoire (Mellgren and Johnson 2002; Braasch et al. 2008).

It is likely that regulatory spandrels formed by WGD may have contributed to the elaboration, if not invention, of the key morphological innovations in angiosperms and vertebrates. In angiosperms, many regulatory gene families that exhibit WGD-mediated expansion, e.g., the Aux/IAA auxin response regulators (Remington et al. 2004) and some MADS-box transcription factor subfamilies (Zahn et al. 2005; Veron et al. 2007), are directly involved in the development of morphological features that are considered angiosperm inventions and derivatives thereof. The same is true for many of the expanded regulatory gene families in vertebrates, such as homeobox genes (Holland and Garcia-Fernandez 1996; Holland 2003; Holland et al. 2008) and neural crest specifier genes (Holland et al. 2008).

There is an interesting parallel to be drawn between the origin of novel gene functions through gene duplication and the origin of innovations through genome duplication. Evolution after single gene duplication has been studied for decades, and many models have been developed to explain how gene duplicates can develop new functions. The classical neofunctionalization model of Ohno (1970) starts from the premise that a duplicate gene is free from selection and that, by genetic drift, it might discover a new beneficial function. It has since become clear that the vast majority of such neutrally evolving duplicates will be lost long before they can ever neofunctionalize. Alternatively, duplicates of multifunctional genes may be preserved through subfunctionalization, i.e., partitioning of the different subfunctions of a gene over the different copies (Force et al. 1999). But subfunctionalization in itself does not lead to novelty. More recently, two models have been proposed that combine aspects of sub- and neofunctionalization, namely the 'Escape from Adaptive Conflict' (EAC) model (Hittinger and Carroll 2007; Des Marais and Rausher 2008) and the 'Innovation, Amplification, Divergence' (IAD) model (Hendrickson et al. 2002; Francino 2005; Bergthorsson et al. 2007). The central idea of these models is that secondary functions of an ancestral gene can get co-opted to a primary role in one of the gene duplicates (Conant and Wolfe 2008). In other words, the 'new' function in the duplicated gene does not arise *de novo* but is already present in a seminal form in the ancestral gene. But the primary and secondary functions in the ancestral gene may be subject to pleiotropic constraints, precluding optimization or elaboration of both functions simultaneously. Gene duplication offers the opportunity to escape from such adaptive conflicts, allowing natural selection to optimize the primary and secondary function independently in different copies.

A similar mechanism might be operating on the level of functional innovation through genome duplication. Analogous to adaptive conflicts in single genes being resolved through gene duplication, adaptive conflicts in gene networks could be resolved through genome duplication. As in the EAC and IAD models of evolution after gene duplication, genome duplications may be instrumental in lifting pleiotropic constraints in molecular networks and perfecting primitive versions of network-level inventions, rather than facilitating innovation *de novo*.

17.4.5 WGDs Increase Evolutionary Potential, and May Do So for a Long Time

What WGDs essentially do is increase the evolutionary potential of an organism. But this potential is not necessarily fulfilled soon (or ever (Semon and Wolfe 2007a)). Circumstances like the availability of niche space are equally important in determining whether or not a post-WGD organism can advantageously use its genomic spandrel. And even under the right conditions, expressing the evolutionary potential of a lineage likely takes time and proceeds through a series of intermediate forms that are later outcompeted by more derived relatives (Knoll and Carroll 1999; Feild and Arens 2007). Indeed, Donoghue and Purnell (2005) have observed that there are no bursts in morphological innovation in post-WGD clades when taking into account extinct lineages. Although a genome duplication can be considered a saltational event in terms of genome evolution, the immediate phenotypical or morphological impact is usually limited. Post-WGD organisms may be considered Goldschmidtian 'hopeful monsters' in the genomic sense, but most often not in the phenotypic sense (Goldschmidt 1940).

Given the importance of opportunity and time in developing evolutionary innovations, a crucial feature of WGDs is that their evolutionary potential can be preserved over long periods of time. Dosage balance effects and functional divergence mechanisms such as subfunctionalization can in principle cause indefinite preservation of duplicate genes, and in particular the regulatory spandrel, in the absence of positive selection. A recent study on the functional divergence of Scn4aa and Scn4ab, a duplicated pair of sodium channels remaining from the teleost-specific genome duplication, confirms that such duplicated genes can neofunctionalize and contribute to morphological innovations more than 100 million years after they were duplicated (Arnegard et al. 2010). The authors investigated whether neofunctionalization of Scn4aa could be directly linked to the origin of adult myogenic electric organs, which happened 100 million years after 3R and on two separate occasions, in African mormyroid and South American gymnotiform fishes. They found that in both lineages, Scn4aa has been co-opted for exclusive expression in myogenic electric organs, and that coinciding with gene co-option, functionally important motifs of Scn4aa evolved under positive selection, suggesting that Scn4aa has been instrumental in the evolution of myogenic electric organs.

17.4.6 Opportunities, Niches and Mass Extinctions

Compared to the high frequency of polyploidization in modern plants (in the order of 1 in 10^5 individuals) (Ramsey and Schemske 1998), the number of preserved ancient genome duplications in plants is low. This pattern is even more conspicuous in vertebrates and yeast (Van de Peer et al. 2009a). Thus, despite the success of some WGD-derived lineages such as the angiosperms, vertebrates, or teleost fish, it appears that genome duplications are seldom successful on long evolutionary timescales. The diversification of a group of species likely depends largely on the ecological opportunities. Even if polyploids possess the potential for the diversification of species and evolutionary novelty, such potential is unlikely to be realized unless an ecological opportunity presents itself. In normal, stable ecosystems, newly formed polyploids are probably not able to compete with the highly adapted occupants of existing niches, including their diploid ancestors (Conant and Wolfe 2007; Feild and Arens 2007; Otto 2007; Hegarty and Hiscock 2008). The number of available niches might be a limiting factor for polyploid lineages to further diversify into species-rich groups. Then, how might the various polyploid ancestors of the extant plant species have managed to survive and diversify? Although it is difficult to know the environmental conditions in which ancient polyploids originated, recent evidence indicates that large-scale ecological catastrophes might have been instrumental in setting the stage for novel or young polyploid lineages to diversify (Fawcett et al. 2009; Van de Peer et al. 2009a). Fawcett et al. (2009) found that many angiosperm plant lineages independently underwent a genome duplication around the Cretaceous-Tertiary (K-T) boundary, ~65 Mya. The K-T event is the most recent large-scale mass extinction of animal and plant species, catalyzed at least in part by one or more catastrophic events such as an asteroid impact on the Yucatan peninsula and increased volcanic activity in India. The plant lineages that underwent WGD around the K-T boundary gave rise to some of the most species-rich angiosperm families such as the Poaceae, Solanaceae, Asteraceae, and Fabaceae. The proposed catalyzing function of mass extinctions in WGD establishment combined with the fact that WGDs increase the evolutionary potential of a lineage suggest somewhat paradoxically that mass extinctions may ultimately lead to more complex plants.

17.5 Conclusions and Perspectives

We are now witnessing a dramatic increase in the availability of whole-genome sequences in plants. As the whole-genome sequence of *A. thaliana* has been instrumental in efforts to understand the biology of *A. thaliana*, a large number of plant genome sequences is likely to help us better understand the molecular basis of the morphological diversity encoded within the different plant genomes. The current morphological diversity is the consequence of a long history of evolutionary and ecological processes, and as we have discussed in this chapter, the evolution of plants, or at least flowering plants, is also the evolution of polyploids. We now know that most eudicots, and possibly all angiosperms are derived from polyploids, and that many of the most species-rich families are also derived from independently formed polyploids (Tang et al. 2008b; Fawcett et al. 2009; Soltis et al. 2009; Van de Peer et al. 2009a). The evolutionary success of various polyploid lineages was probably made possible by their potential for successive rounds of speciation by reciprocal gene loss, and their potential to evolve new functions or morphological features due to the duplication of the entire gene content (in particular the regulatory spandrel) and genetic network (Edger and Pires 2009; Kejnovsky et al. 2009; Van de Peer et al. 2009a). However, concrete examples of such processes are still scarce. We know little about how reciprocal gene loss following polyploidization actually contributed to the species diversification of the various angiosperm lineages. This probably requires a considerable number of genome sequences of species that diverged from each other after polyploidization, which are not available now, but might be in the future. We also do not know much about how the different polyploid-derived lineages achieved specific adaptations and evolved different molecular functions and morphological features.

This requires further understanding of the biology and various molecular aspects of different plant species apart from a few model species such as *Arabidopsis*, with the aid of the genomic information becoming available. It is also important to note that most of our knowledge regarding WGDs in plants comes from angiosperms, with the exception of the moss *Physcomitrella patens* (Rensing et al. 2007; 2013 this volume) and the fern *Ceratopteris richardii* (see Barker 2013, this volume). Examining the impact of WGDs in non-angiosperm plant species will be necessary in order to understand the full extent of the significance of WGDs in the evolution of plants.

Acknowledgments SM is a fellow of the Fund for Scientific Research-Flanders. JAF is a JSPS (Japan Society for the Promotion of Science) postdoctoral fellow.

References

Aburomia R, Khaner O, Sidow A (2003) Functional evolution in the ancestral lineage of vertebrates or when genomic complexity was wagging its morphological tail. J Struct Funct Genomics 3:45–52

Adams KL, Cronn R, Percifield R, Wendel JF (2003) Genes duplicated by polyploidy show unequal contributions to the transcriptome and organ-specific reciprocal silencing. Proc Natl Acad Sci USA 100:4649–4654

Albertin W, Balliau T, Brabant P, ChËvre A-M, Eber F, Malosse C, Thiellement H (2006) Numerous and rapid nonstochastic modifications of gene products in newly synthesized *Brassica napus* allotetraploids. Genetics 173:1101–1113

Arabidopsis Genome Initiative (2000) Analysis of the genome sequence of the flowering plant *Arabidopsis thaliana*. Nature 408:796–815

Arnegard ME, Zwickl DJ, Lu Y, Zakon HH (2010) Old gene duplication facilitates origin and diversification of an innovative communication system—twice. Proc Natl Acad Sci USA 107:22172–22177

Barker MS (2013) Karyotype and genome evolution in pteridophytes. In: Leitch IJ, Greilhuber J, Doležel J, Wendel JF (eds) Plant genome diversity, vol 2, Physical structure, behaviour and evolution of plant genomes. Wien, pp 245–253

Barker MS, Kane NC, Matvienko M, Kozik A, Michelmore RW, Knapp SJ, Rieseberg LH (2008) Multiple paleopolyploidizations during the evolution of the Compositae reveal parallel patterns of duplicate gene retention after millions of years. Mol Biol Evol 25:2445–2455

Barker MS, Vogel H, Schranz ME (2009) Paleopolyploidy in the Brassicales: analyses of the *Cleome* transcriptome elucidate the history of genome duplications in *Arabidopsis* and other Brassicales. Genome Biol Evol 1:391–399

Beilstein MA, Nagalingum NS, Clements MD, Manchester SR, Mathews S (2010) Dating molecular phylogenies indicate a Miocene origin for *Arabidopsis thaliana*. Proc Natl Acad Sci USA 107:18724–18728

Berenbrink M, Koldkjaer P, Kepp O, Cossins AR (2005) Evolution of oxygen secretion in fishes and the emergence of a complex physiological system. Science 307:1752–1757

Bergthorsson U, Andersson DI, Roth JR (2007) Ohno's dilemma: evolution of new genes under continuous selection. Proc Natl Acad Sci USA 104:17004–17009

Bikard D, Patel D, Le Metté C, Giorgi V, Camilleri C, Bennett MJ, Loudet O (2009) Divergent evolution of duplicate genes leads to genetic incompatibilities within *Arabidopsis thaliana*. Science 323:623–626

Birchler JA, Riddle NC, Auger DL, Veitia RA (2005) Dosage balance in gene regulation: biological implications. Trends Genet 21:219–226

Blanc G, Wolfe KH (2004a) Widespread paleopolyploidy in model plant species inferred from age distributions of duplicate genes. Plant Cell 16:1667–1678

Blanc G, Wolfe KH (2004b) Functional divergence of duplicated genes formed by polyploidy during *Arabidopsis* evolution. Plant Cell 16:1679–1691

Blomme T, Vandepoele K, De Bodt S, Simillion C, Maere S, Van de Peer Y (2006) The gain and loss of genes during 600 million years of vertebrate evolution. Genome Biol 7:R43

Braasch I, Volff J-N, Schartl M (2008) The evolution of teleost pigmentation and the fish-specific genome duplication. J Fish Biol 73:1891–1918

Brunet FG, Crollius HR, Paris M, Aury JM, Gibert P, Jaillon O, Laudet V, Robinson-Rechavi M (2006) Gene loss and evolutionary rates following whole-genome duplication in teleost fishes. Mol Biol Evol 23:1808–1816

Cannon SB, Sterck L, Rombauts S, Sato S, Cheung F, Gouzy J, Wang X, Mudge J, Vasdewani J, Schiex T, Spannagl M, Monaghan E, Nicholson C, Humphray SJ, Schoof H, Mayer KF, Rogers J, Quetier F, Oldroyd GE, Debelle F, Cook DR, Retzel EF, Roe BA, Town CD, Tabata S, Van de Peer Y, Young ND (2006) Legume genome evolution viewed through the *Medicago truncatula* and *Lotus japonicus* genomes. Proc Natl Acad Sci USA 103:14959–14964

Chapman BA, Bowers JE, Feltus FA, Paterson AH (2006) Buffering of crucial functions by paleologous duplicated genes may contribute cyclicality to angiosperm genome duplication. Proc Natl Acad Sci USA 103:2730–2735

Chaudhary B, Flagel L, Stupar RM, Udall JA, Verma N, Springer NM, Wendel JF (2009) Reciprocal silencing, transcriptional bias and functional divergence of homeologs in polyploid cotton (*Gossypium*). Genetics 182:503–517

Chen ZJ, Pikaard CS (1997) Transcriptional analysis of nucleolar dominance in polyploid plants: biased expression/silencing of progenitor rRNA genes is developmentally regulated in *Brassica*. Proc Natl Acad Sci USA 94:3442–3447

Christoffels A, Koh EG, Brenner S, Aparicio S, Venkatesh B (2004) Fugu genome analysis provides evidence for a whole-genome duplication early during the evolution of ray-finned fishes. Mol Biol Evol 21:1146–1151

Coate JE, Doyle JJ (2010) Quantifying whole transcriptome size, a prerequisite for understanding transcriptome evolution across species: an example from a plant allopolyploid. Genome Biol Evol 2:534–546

Comai L (2005) The advantages and disadvantages of being polyploid. Nat Rev Genet 6:836–846

Conant GC, Wolfe KH (2007) Increased glycolytic flux as an outcome of whole-genome duplication in yeast. Mol Syst Biol 3:129

Conant GC, Wolfe KH (2008) Turning a hobby into a job: how duplicated genes find new functions. Nat Rev Genet 9:938–950

Crepet WL (2000) Progress in understanding angiosperm history, success, and relationships: Darwin's abominably "perplexing phenomenon". Proc Natl Acad Sci USA 97:12939–12941

Crow KD, Wagner GP (2006) What is the role of genome duplication in the evolution of complexity and diversity? Mol Biol Evol 23:887–892

Cui L, Wall PK, Leebens-Mack JH, Lindsay BG, Soltis DE, Doyle JJ, Soltis PS, Carlson JE, Arumuganathan K, Barakat A, Albert VA, Ma H, dePamphilis CW (2006) Widespread genome duplications throughout the history of flowering plants. Genome Res 16:738–749

Davis JC, Petrov DA (2005) Do disparate mechanisms of duplication add similar genes to the genome? Trends Genet 21:548–551

De Bodt S, Maere S, Van de Peer Y (2005) Genome duplication and the origin of angiosperms. Trends Ecol Evol 20:591–597

Des Marais DL, Rausher MD (2008) Escape from adaptive conflict after duplication in an anthocyanin pathway gene. Nature 454:762–765

Donoghue PCJ, Purnell MA (2005) Genome duplication, extinction and vertebrate evolution. Trends Ecol Evol 20:312–319

Doyle JJ, Flagel LE, Paterson AH, Rapp RA, Soltis DE, Soltis PS, Wendel JF (2008) Evolutionary genetics of genome merger and doubling in plants. Annu Rev Genet 42:443–461

Edger PP, Pires JC (2009) Gene and genome duplications: the impact of dosage-sensitivity on the fate of nuclear genes. Chromosome Res 17:699–717

Fawcett JA, Innan H (2011) Neutral and non-neutral evolution of duplicated genes with gene conversion. Genes 2:191–209

Fawcett JA, Van de Peer Y (2010) Angiosperm polyploids and their road to evolutionary success. Trends Evol Biol 2:e3

Fawcett JA, Maere S, Van de Peer Y (2009) Plants with double genomes might have had a better chance to survive the Cretaceous-Tertiary extinction event. Proc Natl Acad Sci USA 106:5737–5742

Feild TS, Arens NC (2007) The ecophysiology of early angiosperms. Plant Cell Environ 30:291–309

Flagel L, Udall J, Nettleton D, Wendel J (2008) Duplicate gene expression in allopolyploid *Gossypium* reveals two temporally distinct phases of expression evolution. BMC Biol 6:16

Force A, Lynch M, Pickett FB, Amores A, Yan YL, Postlethwait J (1999) Preservation of duplicate genes by complementary, degenerative mutations. Genetics 151:1531–1545

Francino MP (2005) An adaptive radiation model for the origin of new gene functions. Nat Genet 37:573–577

Freeling M (2009) Bias in plant gene content following different sorts of duplication: tandem, whole-genome segmental, or by transposition. Annu Rev Plant Biol 60:433–453

Freeling M, Thomas BC (2006) Gene-balanced duplications, like tetraploidy, provide predictable drive to increase morphological complexity. Genome Res 16:805–814

Gaeta RT, Pires JC, Iniguez-Luy F, Leon E, Osborn TC (2007) Genomic changes in resynthesized *Brassica napus* and their effect on gene expression and phenotype. Plant Cell 19:3403–3417

Goldschmidt R (1940) The material basis of evolution. Yale University Press, New Haven CT

Gould SJ, Lewontin CR (1979) The spandrels of San Marco and the panglossian paradigm: a critique of the adaptationist programme. Proc Roy Soc Lond B Bio 205:581–598

Ha M, Li W-H, Chen ZJ (2007) External factors accelerate expression divergence between duplicate genes. Trends Genet 23:162–166

Ha M, Lu J, Tian L, Ramachandran V, Kasschau KD, Chapman EJ, Carrington JC, Chen X, Wang X-J, Chen ZJ (2009) Small RNAs serve as a genetic buffer against genomic shock in *Arabidopsis* interspecific hybrids and allopolyploids. Proc Natl Acad Sci USA 106:17835–17840

Hegarty MJ, Hiscock SJ (2008) Genomic clues to the evolutionary success of polyploid plants. Curr Biol 18:R435–R444

Hegarty MJ, Barker GL, Wilson ID, Abbott RJ, Edwards KJ, Hiscock SJ (2006) Transcriptome shock after interspecific hybridization in *Senecio* is ameliorated by genome duplication. Curr Biol 16:1652–1659

Hegarty MJ, Barker GL, Brennan AC, Edwards KJ, Abbott RJ, Hiscock SJ (2008) Changes to gene expression associated with hybrid speciation in plants: further insights from transcriptomic studies in *Senecio*. Philos Trans Roy Soc Lond B Bio 363:3055–3069

Heimberg AM, Sempere LF, Moy VN, Donoghue PC, Peterson KJ (2008) MicroRNAs and the advent of vertebrate morphological complexity. Proc Natl Acad Sci USA 105:2946–2950

Hendrickson H, Slechta ES, Bergthorsson U, Andersson DI, Roth JR (2002) Amplification-mutagenesis: evidence that "directed" adaptive mutation and general hypermutability result from growth with a selected gene amplification. Proc Natl Acad Sci USA 99:2164–2169

Hittinger CT, Carroll SB (2007) Gene duplication and the adaptive evolution of a classic genetic switch. Nature 449:677–681

Hoegg S, Brinkmann H, Taylor JS, Meyer A (2004) Phylogenetic timing of the fish-specific genome duplication correlates with the diversification of teleost fish. J Mol Evol 59:190–203

Holland PW (2003) More genes in vertebrates? J Struct Funct Genomics 3:75–84

Holland ND, Chen J (2001) Origin and early evolution of the vertebrates: new insights from advances in molecular biology, anatomy, and palaeontology. Bioessays 23:142–151

Holland PW, Garcia-Fernandez J (1996) Hox genes and chordate evolution. Dev Biol 173:382–395

Holland LZ, Albalat R, Azumi K, Benito-Gutierrez E, Blow MJ, Bronner-Fraser M, Brunet F, Butts T, Candiani S, Dishaw LJ, Ferrier DE, Garcia-Fernandez J, Gibson-Brown JJ, Gissi C, Godzik A, Hallbook F, Hirose D, Hosomichi K, Ikuta T, Inoko H, Kasahara M, Kasamatsu J, Kawashima T, Kimura A, Kobayashi M, Kozmik Z, Kubokawa K, Laudet V, Litman GW, McHardy AC, Meulemans D, Nonaka M, Olinski RP, Pancer Z, Pennacchio LA, Pestarino M, Rast JP, Rigoutsos I, Robinson-Rechavi M, Roch G, Saiga H, Sasakura Y, Satake M, Satou Y, Schubert M, Sherwood N, Shiina T, Takatori N, Tello J, Vopalensky P, Wada S, Xu A, Ye Y, Yoshida K, Yoshizaki F, Yu JK, Zhang Q, Zmasek CM, Putnam NH, Rokhsar DS, Satoh N, Holland PW (2008) The amphioxus genome illuminates vertebrate origins and cephalochordate biology. Genome Res 18:1100–1111

Hurley IA, Mueller RL, Dunn KA, Schmidt EJ, Friedman M, Ho RK, Prince VE, Yang Z, Thomas MG, Coates MI (2007) A new time scale for ray-finned fish evolution. Proc Roy Soc Lond B Bio 274:489–498

Jackson S, Chen ZJ (2010) Genomic and expression plasticity of polyploidy. Curr Opin Plant Biol 13:153–159

Jacquemin J, Laudié M, Cooke R (2009) A recent duplication revisited: phylogenetic analysis reveals an ancestral duplication highly-conserved throughout the *Oryza* genus and beyond. BMC Plant Biol 9:146

Jaillon O, Aury JM, Noel B, Policriti A, Clepet C, Casagrande A, Choisne N, Aubourg S, Vitulo N, Jubin C, Vezzi A, Legeai F, Hugueney P, Dasilva C, Horner D, Mica E, Jublot D, Poulain J, Bruyere C, Billault A, Segurens B, Gouyvenoux M, Ugarte E, Cattonaro F, Anthouard V, Vico V, Del Fabbro C, Alaux M, Di Gaspero G, Dumas V, Felice N, Paillard S, Juman I, Moroldo M, Scalabrin S, Canaguier A, Le Clainche I, Malacrida G, Durand E, Pesole G, Laucou V, Chatelet P, Merdinoglu D, Delledonne M, Pezzotti M, Lecharny A, Scarpelli C, Artiguenave F, Pe ME, Valle G, Morgante M, Caboche M, Adam-Blondon AF, Weissenbach J, Quetier F, Wincker P (2007) The grapevine genome sequence suggests ancestral hexaploidization in major angiosperm phyla. Nature 449:463–467

Jakobsson M, Hagenblad J, Tavaré S, Säll T, Halldén C, Lind-Halldén C, Nordborg M (2006) A unique recent origin of the allotetraploid species *Arabidopsis suecica*: evidence from nuclear DNA markers. Mol Biol Evol 23:1217–1231

Jansen RK, Kaittanis C, Saski C, Lee S-B, Tomkins J, Alverson AJ, Daniell H (2006) Phylogenetic analyses of *Vitis* (Vitaceae) based on complete chloroplast genome sequences: effects of taxon sampling and phylogenetic methods on resolving relationships among rosids. BMC Evol Biol 6:32

Jiao Y, Wickett NJ, Ayyampalayam S, Chanderbali A, Landherr L, Ralph PE, Tomsho LP, Hu Y, Liang H, Soltis PS, Soltis DE, Clifton SW, Schlarbaum SE, Schuster SC, Ma H, Leebens-Mack J,

dePamphilis CW (2011) Ancestral polyploidy in seed plants and angiosperms. Nature 473:97–100
Johanson U, West J, Lister C, Michaels S, Amasino R, Dean C (2000) Molecular analysis of FRIGIDA, a major determinant of natural variation in *Arabidopsis* flowering time. Science 290:344–347
Kasahara M (2007) The 2R hypothesis: an update. Curr Opin Immunol 19:547–552
Kashkush K, Feldman M, Levy AA (2003) Transcriptional activation of retrotransposons alters the expression of adjacent genes in wheat. Nat Genet 33:102–106
Kejnovsky E, Leitch IJ, Leitch AR (2009) Contrasting evolutionary dynamics between angiosperm and mammalian genomes. Trends Ecol Evol 24:572–582
Khaner O (2007) Evolutionary innovations of the vertebrates. Integ Zool 2:60–67
Knoll AH, Carroll SB (1999) Early animal evolution: emerging views from comparative biology and geology. Science 284:2129–2137
Koch MA, Haubold B, Mitchell-Olds T (2000) Comparative evolutionary analysis of chalcone synthase and alcohol dehydrogenase loci in *Arabidopsis*, *Arabis*, and related genera (Brassicaceae). Mol Biol Evol 17:1483–1498
Koh J, Soltis PS, Soltis DE (2010) Homeolog loss and expression changes in natural populations of the recently and repeatedly formed allotetraploid *Tragopogon mirus* (Asteraceae). BMC Genomics 11:97
Kovarik A, Dadejova M, Lim YK, Chase MW, Clarkson JJ, Knapp S, Leitch AR (2008) Evolution of rDNA in *Nicotiana* allopolyploids: a potential link between rDNA homogenization and epigenetics. Ann Bot 101:815–823
Lee HS, Chen ZJ (2001) Protein-coding genes are epigenetically regulated in *Arabidopsis* polyploids. Proc Natl Acad Sci USA 98:6753–6758
Lescot M, Piffanelli P, Ciampi AY, Ruiz M, Blanc G, Leebens-Mack J, da Silva FR, Santos CM, D'Hont A, Garsmeur O, Vilarinhos AD, Kanamori H, Matsumoto T, Ronning CM, Cheung F, Haas BJ, Althoff R, Arbogast T, Hine E, Pappas GJ Jr, Sasaki T, Souza MT Jr, Miller RN, Glaszmann JC, Town CD (2008) Insights into the *Musa* genome: syntenic relationships to rice and between *Musa* species. BMC Genomics 9:58
Liu B, Brubaker CL, Mergeai G, Cronn RC, Wendel JF (2001) Polyploid formation in cotton is not accompanied by rapid genomic changes. Genome 44:321–330
Lukens LN, Pires JC, Leon E, Vogelzang R, Oslach L, Osborn T (2006) Patterns of sequence loss and cytosine methylation within a population of newly resynthesized *Brassica napus* allopolyploids. Plant Physiol 140:336–348
Lynch M, Conery JS (2000) The evolutionary fate and consequences of duplicate genes. Science 290:1151–1155
Lysák MA, Koch MA, Pecinka A, Schubert I (2005) Chromosome triplication found across the tribe Brassiceae. Genome Res 15:516–525
Madlung A, Tyagi AP, Watson B, Jiang H, Kagochi T, Doerge RW, Martienssen R, Comai L (2005) Genomic changes in synthetic *Arabidopsis* polyploids. Plant J 41:221–230
Maere S, De Bodt S, Raes J, Casneuf T, Van Montagu M, Kuiper M, Van de Peer Y (2005) Modeling gene and genome duplications in eukaryotes. Proc Natl Acad Sci USA 102:5454–5459
Mano S, Innan H (2008) The evolutionary rate of duplicated genes under concerted evolution. Genetics 180:493–505
Marais GA, Campos PR, Gordo I (2010) Can intra-Y gene conversion oppose the degeneration of the human Y chromosome? A simulation study. Genome Biol Evol 2:347–357
Mellgren EM, Johnson SL (2002) The evolution of morphological complexity in zebrafish stripes. Trends Genet 18:128–134
Ming R, Hou S, Feng Y, Yu Q, Dionne-Laporte A, Saw JH, Senin P, Wang W, Ly BV, Lewis KL, Salzberg SL, Feng L, Jones MR, Skelton RL, Murray JE, Chen C, Qian W, Shen J, Du P, Eustice M, Tong E, Tang H, Lyons E, Paull RE, Michael TP, Wall K, Rice DW, Albert H, Wang ML, Zhu YJ, Schatz M, Nagarajan N, Acob RA, Guan P, Blas A, Wai CM, Ackerman CM, Ren Y, Liu C, Wang J, Na JK, Shakirov EV, Haas B, Thimmapuram J, Nelson D, Wang X, Bowers JE, Gschwend AR, Delcher AL, Singh R, Suzuki JY, Tripathi S, Neupane K, Wei H, Irikura B, Paidi M, Jiang N, Zhang W, Presting G, Windsor A, Navajas-Perez R, Torres MJ, Feltus FA, Porter B, Li Y, Burroughs AM, Luo MC, Liu L, Christopher DA, Mount SM, Moore PH, Sugimura T, Jiang J, Schuler MA, Friedman V, Mitchell-Olds T, Shippen DE, dePamphilis CW, Palmer JD, Freeling M, Paterson AH, Gonsalves D, Wang L, Alam M (2008) The draft genome of the transgenic tropical fruit tree papaya (*Carica papaya* Linnaeus). Nature 452:991–996
Ohno S (1970) Evolution by gene duplication. Springer, Berlin/Heidelberg/New York
Osborn TC, Pires JC, Birchler JA, Auger DL, Chen ZJ, Lee HS, Comai L, Madlung A, Doerge RW, Colot V, Martienssen RA (2003) Understanding mechanisms of novel gene expression in polyploids. Trends Genet 19:141–147
Otto SP (2007) The evolutionary consequences of polyploidy. Cell 131:452–462
Otto SP, Whitton J (2000) Polyploid incidence and evolution. Annu Rev Genet 34:401–437
Ozkan H, Levy AA, Feldman M (2001) Allopolyploidy-induced rapid genome evolution in the wheat (*Aegilops-Triticum*) group. Plant Cell 13:1735–1747
Papp B, Pal C, Hurst LD (2003) Dosage sensitivity and the evolution of gene families in yeast. Nature 424:194–197
Parisod C, Alix K, Just J, Petit M, Sarilar V, Mhiri C, Ainouche M, Chalhoub B, Grandbastien M-A (2010) Impact of transposable elements on the organization and function of allopolyploid genomes. New Phytol 186:37–45
Paterson AH, Bowers JE, Chapman BA (2004) Ancient polyploidization predating divergence of the cereals, and its consequences for comparative genomics. Proc Natl Acad Sci USA 101:9903–9908
Paterson AH, Bowers JE, Bruggmann R, Dubchak I, Grimwood J, Gundlach H, Haberer G, Hellsten U, Mitros T, Poliakov A, Schmutz J, Spannagl M, Tang H, Wang X, Wicker T, Bharti AK, Chapman J, Feltus FA, Gowik U, Grigoriev IV, Lyons E, Maher CA, Martis M, Narechania A, Otillar RP, Penning BW, Salamov AA, Wang Y, Zhang L, Carpita NC, Freeling M, Gingle AR, Hash CT, Keller B, Klein P, Kresovich S, McCann MC, Ming R, Peterson DG, ur Rahman M, Ware D, Westhoff P, Mayer KFX, Messing J, Rokhsar DS (2009) The *Sorghum bicolor* genome and the diversification of grasses. Nature 457:551–556
Paun O, Bateman RM, Fay MF, Hedrén M, Civeyrel L, Chase MW (2010) Stable epigenetic effects impact adaptation in allopolyploid orchids (*Dactylorhiza*: Orchidaceae). Mol Biol Evol 27:2465–2473
Petit M, Guidat C, Daniel J, Denis E, Montoriol E, Bui QT, Lim KY, Kovarik A, Leitch AR, Grandbastien M-A, Mhiri C (2010) Mobilization of retrotransposons in synthetic allotetraploid tobacco. New Phytol 186:135–147
Pillon Y, Fay MF, Hedrén M, Bateman R, Devey D, Shipunov A, van der Bank M, Chase MW (2007) Evolution and temporal diversification of western European polyploid species complexes in *Dactylorhiza* (Orchidaceae). Taxon 56:1185–1208
Postlethwait J, Amores A, Cresko W, Singer A, Yan Y-L (2004) Subfunction partitioning, the teleost radiation and the annotation of the human genome. Trends Genet 20:481–490
Proost S, Bel MV, Sterck L, Billiau K, Parys TV, Van de Peer Y, Vandepoele K (2009) PLAZA: a comparative genomics resource to study gene and genome evolution in plants. Plant Cell 21:3718–3731
Putnam NH, Butts T, Ferrier DE, Furlong RF, Hellsten U, Kawashima T, Robinson-Rechavi M, Shoguchi E, Terry A, Yu JK, Benito-Gutierrez EL, Dubchak I, Garcia-Fernandez J, Gibson-Brown JJ,

Grigoriev IV, Horton AC, de Jong PJ, Jurka J, Kapitonov VV, Kohara Y, Kuroki Y, Lindquist E, Lucas S, Osoegawa K, Pennacchio LA, Salamov AA, Satou Y, Sauka-Spengler T, Schmutz J, Shin IT, Toyoda A, Bronner-Fraser M, Fujiyama A, Holland LZ, Holland PW, Satoh N, Rokhsar DS (2008) The amphioxus genome and the evolution of the chordate karyotype. Nature 453:1064–1071

Ramsey J (2011) Polyploidy and ecological adaptation in wild yarrow. Proc Natl Acad Sci USA 108:7096–7101

Ramsey J, Schemske DW (1998) Pathways, mechanisms, and rates of polyploid formation in flowering plants. Annu Rev Ecol Syst 29:467–501

Remington DL, Vision TJ, Guilfoyle TJ, Reed JW (2004) Contrasting modes of diversification in the Aux/IAA and ARF gene families. Plant Physiol 135:1738–1752

Rensing SA, Ick J, Fawcett JA, Lang D, Zimmer A, Van de Peer Y, Reski R (2007) An ancient genome duplication contributed to the abundance of metabolic genes in the moss *Physcomitrella patens*. BMC Evol Biol 7:130

Rensing SA, Beike AK, Lang D (2013) Evolutionary importance of generative polyploidy for genome evolution of haploid-dominant land plants. In: Leitch IJ, Greilhuber J, Doležel J, Wendel JF (eds) Plant genome diversity, vol 2, Physical structure, behaviour and evolution of plant genomes. Springer-Verlag, Wien, pp 295–305

Riddle NC, Birchler JA (2003) Effects of reunited diverged regulatory hierarchies in allopolyploids and species hybrids. Trends Genet 19:597–600

Rieseberg LH, Raymond O, Rosenthal DM, Lai Z, Livingstone K, Nakazato T, Durphy JL, Schwarzbach AE, Donovan LA, Lexer C (2003) Major ecological transitions in wild sunflowers facilitated by hybridization. Science 301:1211–1216

Rieseberg LH, Kim SC, Randell RA, Whitney KD, Gross BL, Lexer C, Clay K (2007) Hybridization and the colonization of novel habitats by annual sunflowers. Genetica 129:149–165

Salmon A, Ainouche ML, Wendel JF (2005) Genetic and epigenetic consequences of recent hybridization and polyploidy in *Spartina* (Poaceae). Mol Ecol 14:1163–1175

Salmon A, Flagel L, Ying B, Udall JA, Wendel JF (2010) Homoeologous nonreciprocal recombination in polyploid cotton. New Phytol 186:123–134

Sato S, Nakamura Y, Kaneko T, Asamizu E, Kato T, Nakao M, Sasamoto S, Watanabe A, Ono A, Kawashima K, Fujishiro T, Katoh M, Kohara M, Kishida Y, Minami C, Nakayama S, Nakazaki N, Shimizu Y, Shinpo S, Takahashi C, Wada T, Yamada M, Ohmido N, Hayashi M, Fukui K, Baba T, Nakamichi T, Mori H, Tabata S (2008) Genome structure of the legume, *Lotus japonicus*. DNA Res 15:227–239

Scannell DR, Byrne KP, Gordon JL, Wong S, Wolfe KH (2006) Multiple rounds of speciation associated with reciprocal gene loss in polyploid yeasts. Nature 440:341–345

Schlueter JA, Dixon P, Granger C, Grant D, Clark L, Doyle JJ, Shoemaker RC (2004) Mining EST databases to resolve evolutionary events in major crop species. Genome 47:868–876

Schmutz J, Cannon SB, Schlueter J, Ma J, Mitros T, Nelson W, Hyten DL, Song Q, Thelen JJ, Cheng J, Xu D, Hellsten U, May GD, Yu Y, Sakurai T, Umezawa T, Bhattacharyya MK, Sandhu D, Valliyodan B, Lindquist E, Peto M, Grant D, Shu S, Goodstein D, Barry K, Futrell-Griggs M, Abernathy B, Du J, Tian Z, Zhu L, Gill N, Joshi T, Libault M, Sethuraman A, Zhang X-C, Shinozaki K, Nguyen HT, Wing RA, Cregan P, Specht J, Grimwood J, Rokhsar D, Stacey G, Shoemaker RC, Jackson SA (2010) Genome sequence of the palaeopolyploid soybean. Nature 463:178–183

Schnable PS, Ware D, Fulton RS, Stein JC, Wei F, Pasternak S, Liang C, Zhang J, Fulton L, Graves TA, Minx P, Reily AD, Courtney L, Kruchowski SS, Tomlinson C, Strong C, Delehaunty K, Fronick C, Courtney B, Rock SM, Belter E, Du F, Kim K, Abbott RM, Cotton M, Levy A, Marchetto P, Ochoa K, Jackson SM, Gillam B, Chen W, Yan L, Higginbotham J, Cardenas M, Waligorski J, Applebaum E, Phelps L, Falcone J, Kanchi K, Thane T, Scimone A, Thane N, Henke J, Wang T, Ruppert J, Shah N, Rotter K, Hodges J, Ingenthron E, Cordes M, Kohlberg S, Sgro J, Delgado B, Mead K, Chinwalla A, Leonard S, Crouse K, Collura K, Kudrna D, Currie J, He R, Angelova A, Rajasekar S, Mueller T, Lomeli R, Scara G, Ko A, Delaney K, Wissotski M, Lopez G, Campos D, Braidotti M, Ashley E, Golser W, Kim H, Lee S, Lin J, Dujmic Z, Kim W, Talag J, Zuccolo A, Fan C, Sebastian A, Kramer M, Spiegel L, Nascimento L, Zutavern T, Miller B, Ambroise C, Muller S, Spooner W, Narechania A, Ren L, Wei S, Kumari S, Faga B, Levy MJ, McMahan L, Buren PV, Vaughn MW, Ying K, Yeh C-T, Emrich SJ, Jia Y, Kalyanaraman A, Hsia A-P, Barbazuk WB, Baucom RS, Brutnell TP, Carpita NC, Chaparro C, Chia J-M, Deragon J-M, Estill JC, Fu Y, Jeddeloh JA, Han Y, Lee H, Li P, Lisch DR, Liu S, Liu Z, Nagel DH, McCann MC, SanMiguel P, Myers AM, Nettleton D, Nguyen J, Penning BW, Ponnala L, Schneider KL, Schwartz DC, Sharma A, Soderlund C, Springer NM, Sun Q, Wang H, Waterman M, Westerman R, Wolfgruber TK, Yang L, Yu Y, Zhang L, Zhou S, Zhu Q, Bennetzen JL, Dawe RK, Jiang J, Jiang N, Presting GG, Wessler SR, Aluru S, Martienssen RA, Clifton SW, McCombie WR, Wing RA, Wilson RK (2009) The B73 maize genome: complexity, diversity, and dynamics. Science 326:1112–1115

Semon M, Wolfe KH (2007a) Consequences of genome duplication. Curr Opin Genet Dev 17:505–512

Semon M, Wolfe KH (2007b) Reciprocal gene loss between *Tetraodon* and zebrafish after whole genome duplication in their ancestor. Trends Genet 23:108–112

Seoighe C, Gehring C (2004) Genome duplication led to highly selective expansion of the *Arabidopsis thaliana* proteome. Trends Genet 20:461–464

Shaked H, Kashkush K, Ozkan H, Feldman M, Levy AA (2001) Sequence elimination and cytosine methylation are rapid and reproducible responses of the genome to wide hybridization and allopolyploidy in wheat. Plant Cell 13:1749–1759

Shi T, Huang H, Barker MS (2010) Ancient genome duplications during the evolution of kiwifruit (*Actinidia*) and related Ericales. Ann Bot 106:497–504

Shimeld SM, Holland PW (2000) Vertebrate innovations. Proc Natl Acad Sci USA 97:4449–4452

Soltis DE, Soltis PS (1999) Polyploidy: recurrent formation and genome evolution. Trends Ecol Evol 14:348–352

Soltis PS, Soltis DE (2000) The role of genetic and genomic attributes in the success of polyploids. Proc Natl Acad Sci USA 97:7051–7057

Soltis DE, Bell CD, Kim S, Soltis PS (2008) Origin and early evolution of angiosperms. Ann N Y Acad Sci 1133:3–25

Soltis DE, Albert VA, Leebens-Mack J, Bell CD, Paterson AH, Zheng C, Sankoff D, dePamphilis CW, Wall PK, Soltis PS (2009) Polyploidy and angiosperm diversification. Am J Bot 96:336–348

Stebbins GL (1950) Variation and evolution in plants. Columbia University Press, New York

Stebbins GL (1971) Process of organic evolution. Englewood Cliffs, NJ

Stellwag EJ (2004) Are genome evolution, organism complexity and species diversity linked ? Integr Comp Biol 44:358–365

Stuessy TF (2004) A transitional-combinatorial theory for the origin of angiosperms. Taxon 53:3–16

Sugino RP, Innan H (2006) Selection for more of the same product as a force to enhance concerted evolution of duplicated genes. Trends Genet 22:642–644

Tang H, Bowers JE, Wang X, Ming R, Alam M, Paterson AH (2008a) Synteny and collinearity in plant genomes. Science 320:486–488

Tang H, Wang X, Bowers JE, Ming R, Alam M, Paterson AH (2008b) Unraveling ancient hexaploidy through multiply-aligned angiosperm gene maps. Genome Res 18:1944–1954

Tang H, Bowers JE, Wang X, Paterson AH (2010) Angiosperm genome comparisons reveal early polyploidy in the monocot lineage. Proc Natl Acad Sci USA 107:472–477

The Rice Chromosomes 11 and 12 Sequencing Consortia TRCaS (2005) The sequence of rice chromosomes 11 and 12, rich in disease resistance genes and recent gene duplications. BMC Biol 3:20

Tuskan GA, Difazio S, Jansson S, Bohlmann J, Grigoriev I, Hellsten U, Putnam N, Ralph S, Rombauts S, Salamov A, Schein J, Sterck L, Aerts A, Bhalerao RR, Bhalerao RP, Blaudez D, Boerjan W, Brun A, Brunner A, Busov V, Campbell M, Carlson J, Chalot M, Chapman J, Chen GL, Cooper D, Coutinho PM, Couturier J, Covert S, Cronk Q, Cunningham R, Davis J, Degroeve S, Dejardin A, Depamphilis C, Detter J, Dirks B, Dubchak I, Duplessis S, Ehlting J, Ellis B, Gendler K, Goodstein D, Gribskov M, Grimwood J, Groover A, Gunter L, Hamberger B, Heinze B, Helariutta Y, Henrissat B, Holligan D, Holt R, Huang W, Islam-Faridi N, Jones S, Jones-Rhoades M, Jorgensen R, Joshi C, Kangasjarvi J, Karlsson J, Kelleher C, Kirkpatrick R, Kirst M, Kohler A, Kalluri U, Larimer F, Leebens-Mack J, Leple JC, Locascio P, Lou Y, Lucas S, Martin F, Montanini B, Napoli C, Nelson DR, Nelson C, Nieminen K, Nilsson O, Pereda V, Peter G, Philippe R, Pilate G, Poliakov A, Razumovskaya J, Richardson P, Rinaldi C, Ritland K, Rouze P, Ryaboy D, Schmutz J, Schrader J, Segerman B, Shin H, Siddiqui A, Sterky F, Terry A, Tsai CJ, Uberbacher E, Unneberg P, Vahala J, Wall K, Wessler S, Yang G, Yin T, Douglas C, Marra M, Sandberg G, Van de Peer Y, Rokhsar D (2006) The genome of black cottonwood, *Populus trichocarpa* (Torr. & Gray). Science 313:1596–1604

Van de Peer Y (2004) Computational approaches to unveiling ancient genome duplications. Nat Rev Genet 5:752–763

Van de Peer Y, Fawcett JA, Proost S, Sterck L, Vandepoele K (2009a) The flowering world: a tale of duplications. Trends Plant Sci 14:680–688

Van de Peer Y, Maere S, Meyer A (2009b) The evolutionary significance of ancient genome duplications. Nat Rev Genet 10:725–732

Vandepoele K, Simillion C, Van de Peer Y (2002) Detecting the undetectable: uncovering duplicated segments in *Arabidopsis* by comparison with rice. Trends Genet 18:606–608

Vandepoele K, De Vos W, Taylor JS, Meyer A, Van de Peer Y (2004) Major events in the genome evolution of vertebrates: paranome age and size differ considerably between ray-finned fishes and land vertebrates. Proc Natl Acad Sci USA 101:1638–1643

Velasco R, Zharkikh A, Affourtit J, Dhingra A, Cestaro A, Kalyanaraman A, Fontana P, Bhatnagar SK, Troggio M, Pruss D, Salvi S, Pindo M, Baldi P, Castelletti S, Cavaiuolo M, Coppola G, Costa F, Cova V, Ri AD, Goremykin V, Komjanc M, Longhi S, Magnago P, Malacarne G, Malnoy M, Micheletti D, Moretto M, Perazzolli M, Si-Ammour A, Vezzulli S, Zini E, Eldredge G, Fitzgerald LM, Gutin N, Lanchbury J, Macalma T, Mitchell JT, Reid J, Wardell B, Kodira C, Chen Z, Desany B, Niazi F, Palmer M, Koepke T, Jiwan D, Schaeffer S, Krishnan V, Wu C, Chu VT, King ST, Vick J, Tao Q, Mraz A, Stormo A, Stormo K, Bogden R, Ederle D, Stella A, Vecchietti A, Kater MM, Masiero S, Lasserre P, Lespinasse Y, Allan AC, Bus V, Chagn'e D, Crowhurst RN, Gleave AP, Lavezzo E, Fawcett JA, Proost S, Rouz'e P, Sterck L, Toppo S, Lazzari B, Hellens RP, Durel C-E, Gutin A, Bumgarner RE, Gardiner SE, Skolnick M, Egholm M, Van de Peer Y, Salamini F, Viola R (2010) The genome of the domesticated apple (*Malus domestica* Borkh.). Nat Genet 42:833–839

Veron AS, Kaufmann K, Bornberg-Bauer E (2007) Evidence of interaction network evolution by whole-genome duplications: a case study in MADS-box proteins. Mol Biol Evol 24:670–678

Wagner A (2008) Gene duplications, robustness and evolutionary innovations. Bioessays 30:367–373

Wang X, Shi X, Hao B, Ge S, Luo J (2005) Duplication and DNA segmental loss in the rice genome: implications for diploidization. New Phytol 165:937–946

Wang J, Tian L, Lee H-S, Chen ZJ (2006a) Nonadditive regulation of *FRI* and *FLC* loci mediates flowering-time variation in *Arabidopsis* allopolyploids. Genetics 173:965–974

Wang J, Tian L, Lee H-S, Wei NE, Jiang H, Watson B, Madlung A, Osborn TC, Doerge RW, Comai L, Chen ZJ (2006b) Genomewide nonadditive gene regulation in *Arabidopsis* allotetraploids. Genetics 172:507–517

Wang X, Tang H, Bowers JE, Paterson AH (2009) Comparative inference of illegitimate recombination between rice and sorghum duplicated genes produced by polyploidization. Genome Res 19:1026–1032

Wendel JF (2000) Genome evolution in polyploids. Plant Mol Biol 42:225–249

Werth CR, Windham MD (1991) A model for divergent, allopatric speciation of polyploid pteridophytes resulting from silencing of duplicate-gene expression. Am Nat 137:515–526

Zahn LM, Kong H, Leebens-Mack JH, Kim S, Soltis PS, Landherr LL, Soltis DE, dePamphilis CW, Ma H (2005) The evolution of the *SEPALLATA* subfamily of MADS-box genes: a preangiosperm origin with multiple duplications throughout angiosperm history. Genetics 169:2209–2223

Evolutionary Importance of Generative Polyploidy for Genome Evolution of Haploid-Dominant Land Plants

18

Stefan A. Rensing, Anna K. Beike, and Daniel Lang

Contents

18.1 Introduction

Generative polyploidy is a widespread phenomenon among land plants and has been pivotal to the evolution of land plant genomes (Soltis and Soltis 2009; Van de Peer et al. 2009, Fawcett et al. 2013, this volume). These large-scale or whole genome duplication events (WGD) are widely recognized as driving forces behind diversification and speciation of plant lineages and have often been hypothesized as important factors for the evolution of organismal complexity (Crow and Wagner 2006; Van de Peer et al. 2009). Initial evidence for the latter hypothesis was recently provided by phylogenetic comparative genomics revealing a correlation of WGD with morphological complexity (Lang et al. 2010).

Autopolyploidization, i.e., the doubling of the genome without hybridization, might represent a short-term evolutionary advantage during periods of environmental change that force range shifts (Parisod et al. 2010). Allopolyploidization, i.e., the doubling of the genome in association with hybridization of genetically different chromosome sets, is also regarded as enabling an increased potential for sub- and neofunctionalization of genes and thus speciation (Soltis and Soltis 2009). Allopolyploidization has also been hypothesized to be correlated with periods of environmental upheaval (Fawcett et al. 2009, 2013).

Key to our understanding of the evolutionary importance of WGD events and especially their effects on diversification is the fate of the duplicated chromosomal regions. The diversifying effects of this post-duplication phase encompass large-scale phenomena like subsequent diploidization (or haploidization, in the case of haploid-dominant plants), chromosome loss and rearrangements, as well as small-scale, adaptive processes related to the functionalization and diversification of duplicated gene paralogs; which in the case of WGD are termed paleologs or homoeologs. There are various models describing the fate of paralogs, covered comprehensively by, for example, Innan and Kondrashov (2010).

S.A. Rensing (✉)
Faculty of Biology, University of Freiburg, Schänzle str. 1, Freiburg 79104, Germany
e-mail: stefan.rensing@biologie.uni-freiburg.de

I.J. Leitch et al. (eds.), *Plant Genome Diversity Volume 2*,
DOI 10.1007/978-3-7091-1160-4_18, © Springer-Verlag Wien 2013

It is presumed that most duplicated segments are lost quickly during the early phases of diploidization (Ma and Gustafson 2005). Even if duplicated segments survive the early post-duplication phase and the retained homoeologs are becoming fixed in the population, the most common fate is death by gradual pseudogenization and non-functionalization, except if a selective advantage is introduced. Therefore, central to our understanding of the importance of generative polyploidy is why certain homoeologs are retained and which models can be applied to describe their distinct evolutionary paths which fall somewhere between the extremes of diversification leading to the acquisition of new gene functions and redundancy which is thought to be important for gene dosage (Innan and Kondrashov 2010).

Most of the evolutionary models describing the post-duplication fates of paralogous gene copies have been developed and applied to explain gene duplication in diploid-dominant species, like seed plants that exhibit a heterophasic life cycle with a dominant diploid sporophytic generation. Indeed they are the major focus of the discussion of generative polyploidy in the chapter by Fawcett et al. (2013), this volume. Here we focus on aspects and effects of polyploidization in haploid-dominant plants, like bryophytes, which have a dominant haploid gametophyte and a reduced diploid sporophytic generation. What role does polyploidization play in the evolution of these organisms? Can we find evidence of ancient and recent polyploidization events in the genomes of extant bryophytes? What is the importance of auto- versus allopolyploidy? What is the extent of the paranome, i.e., the genomic fraction of paralogs, within a bryophyte genome? What is the prevalent model for functionalization of paralogs and which kind of functional classes of genes have been retained preferentially?

Bryophytes form a paraphyletic group at the base of the land plant tree of life and comprise liverworts, mosses and hornworts—listed in order of currently assumed divergence from the vascular plant lineage (Qiu et al. 2006; see also Soltis and Soltis 2013, this volume). Although the fossil record of early land plants is scarce, there are fossils indicating that early land plants belonged to the liverwort-lineage (Kenrick and Crane 1997; Wellman et al. 2003). The last common ancestor of vascular (diploid-dominant) and non-vascular (haploid-dominant) land plants was recently dated to have lived about 500 million years ago (Mya) (Lang et al. 2010). As mentioned above, this deep divergence has led to a radically different approach to life on land compared with seed plants, including poikilohydry (i.e., the change of cellular water content in response to water availability in the environment) and the dominance of the gametophytic generation, which results in a haploid state of the genome for the dominant part of the heterophasic life cycle.

It is tempting to speculate that WGD (comprising auto- and allopolyploidizations) represent an even greater potential evolutionary advantage in haploid-dominant organisms (like bryophytes) than in those that spend most of their life cycle in the diploid phase (like seed plants, monilophytes and lycophytes). The presence of an additional copy of each gene and chromosome after a WGD might render a bryophyte more robust against somatic mutations eventually affecting the germ line. In addition, it would have the same potential to evolve new functions as in diploid-dominant organisms. Indeed it has been shown that allopolyploid diploids of the moss *Sphagnum subsecundum* do exhibit greater genetic diversity and linkage disequilibrium than haploids (Shaw et al. 2008).

The estimated frequency of polyploid species differs between the three bryophyte phyla and between authors. For example, Wyatt et al. (1988) estimated that about 79% of mosses, 11% of liverworts and 2% of hornworts are naturally occurring polyploids. Examples include the liverwort *Pellia borealis* (Odrzykoski et al. 1996; Orzechowska et al. 2010), and various moss genera including *Sphagnum* (Shaw et al. 2008; Karlin et al. 2009; Ricca and Shaw 2010), *Plagiomnium* (Wyatt et al. 1988, 1992) and *Physcomitrella*/*Physcomitrium*.

The proposed widespread existence of naturally occurring generative polyploidy within mosses (reviewed by Natcheva and Cronberg 2004), as well as the potential differences compared with diploid-dominant plants reviewed in Husband et al. (2013, this volume), make mosses a particularly interesting subject to study the influence and implications of WGD for the evolution and biology of haploid-dominant organisms.

Heterosis effects of WGD, especially with regard to body size, might explain the sudden increase in size of mosses observed in extant species compared with those embedded in Eocene (45 million year old) amber (Frahm 2010). Indeed, the estimated timing (30–60 Mya) of the WGD event in the model moss *Physcomitrella patens* (Hedw.) Bruch & Schimp. (Rensing et al. 2007) falls in the period of the mass extinction events of the "Grand Coupure" at the end of the Eocene (33 Mya) and of the Cretaceous-Tertiary (K-T) boundary (65 Mya).

Physcomitrella patens has been developed as a model organism over the last two decades, and a well-developed molecular toolbox including efficient gene targeting (Frank et al. 2005; Cove et al. 2009) is now available together with sequenced nuclear and organellar genomes (Sugiura et al. 2003; Terasawa et al. 2007; Rensing et al. 2008). This makes *P. patens*, and the family Funariaceae to which it belongs, ideal candidates to study and demonstrate the effects of polyploidization on haploid genomes.

18.2 History Repeating: Artificial and Natural Polyploidy in the Funariaceae

In the early twentieth century the moss family Funariaceae was extensively studied by von Wettstein in his groundbreaking work on its genetics and the appearance of both

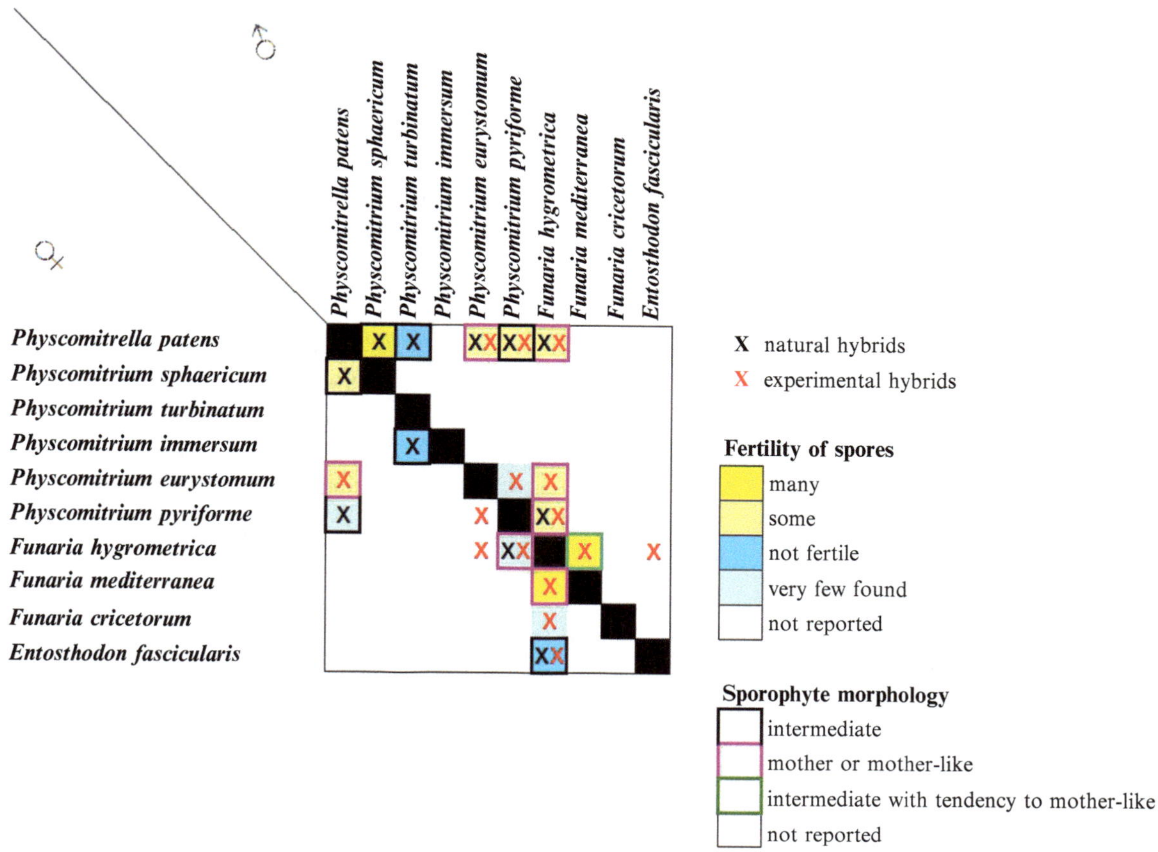

Fig. 18.1 Matrix of natural and experimental hybrid occurrence in the Funariaceae

natural and experimental hybrids (Wettstein 1924, 1932). His results are shown in Fig. 18.1 and demonstrate that hybridization is quite common within this family. If hybrids are fertile and thus able to develop sporophytes, these are usually of intermediate or mother-like morphology and the spores are sometimes few and often not fertile (Fig. 18.1). While *P. patens* seems to represent a good mother line, *Funaria hygrometrica* is a good father line. Interestingly, these two species occupy the opposite ends on the complexity scale of sporophyte morphology; while *F. hygrometrica* features an elaborate sporophyte with a long seta (the stalk on which the spore capsule rests) and operculum (the preformed breaking point for the release of spores), *P. patens* harbours a highly reduced sporophyte without operculum that, due to the extreme reduction of the seta, is embedded into the gametophores that gave rise to it. The reduced sporophyte of *P. patens* was recently proposed to be a secondary reduction of a more complex, potentially operculum-bearing sporophyte in the Funariaceae ancestor (McDaniel et al. 2010). Studies have shown that the chromosomes of *P. patens* are apparently able to pair and recombine with those from several Funariaceae species and usually force the reduced phenotype onto the hybrid offspring (Fig. 18.2). In contrast, while the *F. hygrometrica* gametes are able to fertilize eggs from a range of other Funariaceae, they are not able to pass on the elaborate sporophytes to their hybrid offspring. It might well be that aneuploidization, as has been hypothesized for *P. patens* (Rensing et al. 2007; Beike and Rensing 2010), is the basis of this behaviour.

The hybrid lines mentioned above are usually homoploid hybrids (i.e., lacking a WGD) and as they are haploid no allele can cover for essential functions if the mating partner's genome is altered in that regard. Consequently, a lot of the hybrid offspring are not expected to survive due to severe incompatibility or will be sterile. However, as mentioned earlier, *Physcomitrella* is a paleopolyploid (Rensing et al. 2007) and molecular dating of the WGD event implies ancient polyploidization in the ancestor of the Funariaceae (see below). This may explain, in part, why natural hybrids like *Physcomitrium collenchymatum* or *Physcomitrium eurystomum* are able to survive. Crosses of *P. patens* isolates, that may very well represent cryptic species, usually display a slightly aberrant sporophyte morphology (see Fig. 18.2d and 18.2e, compared with selfing, Fig. 18.2a), but do produce fertile spores, whereas crosses with, for example, *Physcomitrium*, if they occur at all, may result in intermediate sporophyte morphology, as mentioned above (Fig. 18.2b). It should be noted that it is not only

Fig. 18.2 Sporophytes resulting from (**a**) selfing of *Physcomitrella patens* and (**b**) *P. patens* crossed with *Physcomitrium sphaericum* and (**c** and **f**) selfed again. (**d**) and (**e**) show sporophytes from crosses between different isolates of *P. patens*

the hybrid sporophytes arising directly from crossings that display the intermediate morphology, the same is true also for the following generation, i.e., sporophytes eventually formed upon germination of the spores and selfing of the resulting gametangia (Fig. 18.2c, f).

Artificial or experimentally induced autopolyploidization is also frequently observed after transfection of *P. patens* protoplasts (12%; Schween et al. 2005b). Among transformants the polyploid, mostly diploid, plants cannot be distinguished from homoploid, haploid plants by morphological features alone (Schween et al. 2005a).

18.3 Natural Polyploidization and Hybridization Among the Funariaceae

The existence of natural polyploids derived from unreduced spores (diplospory), by reprogramming of vegetative sporophytic cells (e.g., after wounding; apospory) or fusion of sporocytes prior to meiosis (syndiplospory) is known for individual moss plants (Crawford et al. 2009). Also, generative polyploidy at the population or species level has been known for a long time, as outlined above. Nevertheless, the question as to which of the two forms of genome doubling is most prevalent among bryophytes has been an issue of debate. In contrast to allopolyploidy, von Wettstein considered autopolyploidy to be of little or no evolutionary importance for mosses due to the abnormality and cytological instability of the resulting moss strains (reviewed in English by Müntzing 1936). In contrast, in the 1970s it was assumed (Longton 1976; Smith 1979) that most polyploid bryophytes were autopolyploids (with a few exceptions) and thus hybridization followed by genome duplication (i.e., allopolyploidy) was of little significance.

Recent molecular studies on polyploid bryophytes show that allopolyploidy is the rule rather than the exception, and is common in both mosses and liverworts (Natcheva and Cronberg 2004). After hybridization, allopolyploidization may occur in three ways: (1) apospory (i.e., regeneration of diploid gametophytes from the tissues of the hybrid sporophyte), (2) diplospory, or (3) syndiplospory.

Different chromosome counts have been reported for Funariaceae with the most common numbers being 14, 21 and 28. In *P. patens* the chromosome number per haploid cell varies from $n = 14$, $n = 16$ as reported by von Wettstein (1924) and Engel (1968) to $n = 27$ reported by Bryan

(1957) and Reski et al. (1994) with the latter count being recorded for the isolate “Gransden 2004” used for genome sequencing (Rensing et al. 2008). Among true mosses (Bryopsida), the base number of chromosomes is considered to be four, five, six and seven (Frahm 2001). Considering a base number of seven chromosomes among Funariaceae, a model for the genome evolution of *P. patens* can be constructed based on two whole genome duplication events. Starting with the ancient hybridization of two (male and female) parental chromosome sets of $n = 7$, the resulting allopolyploidization and subsequent haploidization (i.e., a loss of complete redundancy by paralog decay and sub- and neofunctionalization) may have led to a haploid, mon- and synoicous species with a chromosome number of $n = 14$. Another (auto- or allo-) polyploidization and haploidization, followed by a putative aneuploidization, would result in the extant haploid *P. patens* strain with 27 chromosomes (Rensing et al. 2007; see also Fig. 3 in Rensing et al. 2009 for illustration). Thus, the last common ancestor of Funariaceae potentially became hermaphroditic through hybridization and polyploidization (Rensing et al. 2007).

Evidence that this might reflect a more general trend among mosses comes from a recent phylogenetic comparative study indicating that polyploid mosses were more likely to be hermaphrodite, thus suggesting that changes in chromosome number through polyploidy could cause changes in the sexual system (Crawford et al. 2009). The selective advantage of hybridization and polyploidization depends on successful reproduction and transmission of genetic information. The evolutionary benefit of a syn- and monoecious moss species over a dioicous species might be the easier way of sexual reproduction through selfing, as male and female gametangia are located on the same plant, and even on the same gametophore. However, inbreeding depression can be a considerable selective disadvantage of monoecy, even if out-crossing among monoecious bryophytes occurs in nature as already described for the synoicous polyploid moss *Plagiomnium medium* (Wyatt et al. 1988, 1992). Interestingly, sporophytic inbreeding depression was noted in the dioicous species *Ceratodon purpureus*, but not in the monoecious, self-fertilizing *Funaria hygrometrica*. Indeed, the low rate of inbreeding depression observed in selfing *F. hygrometrica*, based on capsule and seta length, spore number or capsule mass, may help in maintaining the evolutionary stability of hermaphroditic populations (Taylor et al. 2007). In addition, among bryophytes, the ability and high frequency of vegetative reproduction via fragmentation of gametophytic tissue or gemmae should be considered in this context. A successful vegetative reproduction system can diminish the potentially negative consequences of sexual isolation through polyploidization.

P. patens is a paleopolyploid moss that underwent at least one WGD (Rensing et al. 2007). This ancient polyploidization was identified from an analysis of 2,907 paralogous genes out of 24,845 transcriptomic unigenes. Based on the synonymous substitution rates of homoeologs, the divergence of the paralogous gene copies was dated to 30–60 Mya ago. An average of 45 Mya can therefore be expected for the date of WGD in *P. patens*. According to phylogenetic analyses, the age of the subclass Funariidae has been estimated to be approximately 172 Mya (Newton et al. 2007). Therefore, the detected genome duplication during the Eocene (~45 Mya) most likely occurred after diversification within Funariidae. Although the age of Funariaceae is still unclear, molecular analysis of a conserved domain of MADS-box transcription factors in *F. hygrometrica*, *Funariella curviseta*, *P. eurystomum* and *P. patens* hints at a shared polyploidization history within this family (manuscript in preparation; Zobell et al. (2010)). The analysed species share the same MADS-box transcription factors, which are known to be preferentially retained after duplication events (Veron et al. 2007).

For some time now, bryologists have discussed monoicy as a derived trait over dioicy (Ando 1980). For dioicous mosses reproductive success is strongly dependent on the cytotype frequencies in the population (Crawford et al. 2009), therefore asexual reproduction might be predicted to be more frequent among dioicous than monoecious mosses. In terms of the distribution of sexual reproductive systems among Funariidae, both Encalyptales and Timmiales contain monoecious and dioicous species, and among the Funariales only the family Funariaceae is entirely monoecious. The extension of the phylogenetic analysis of homoeolog-containing gene families, like the MADS-box transcription factors mentioned above, to include additional representatives of the Funariidae will provide further insight concerning the role of WGD events in the establishment of monoicy in mosses.

The genome duplication during the Eocene provides no explanation for the range of chromosome numbers encountered in extant Funariaceae. However, since several Funariaceae show no effective intergeneric breeding barrier, and since some genera, especially *Funaria*, *Physcomitrium* and *Physcomitrella*, grow in comparable habitats in close proximity, a high number of hybrid species can be expected in nature. Thus speciation through hybridization may be caused by reproductive isolation due to allopolyploidy and this may contribute to the range of chromosome numbers observed. It is noted that although some hybrid species from different moss families have been observed in nature, the evolutionary significance and the frequency of hybridization among bryophytes is considered to be widely underestimated at present (Natcheva and Cronberg 2004).

Natural hybrids among Funariaceae have frequently been reported over the last two centuries (Britton 1895; Andrews 1918, 1942; Pettet 1964), and outlined above. Recently, the hybrid origin of *Physcomitrium eurystomum* and *Physcomitrium collenchymatum* was verified based on molecular data and genealogical analyses of six loci, including the ribosomal internal transcribed spacer (ITS), one plastid marker (*atpB-rbcL*) and four protein coding loci. Furthermore, out of a taxon set comprising Funariaceae species from geographically widespread populations in both the Northern and Southern Hemisphere, other *Physcomitrium* species also showed a clear signature of hybridization (i.e., there were discrepancies in the phylogenetic trees of the six chosen loci). Moreover, the amplification of a highly conserved nuclear single copy gene (*BRK1*) from a comparable Funariaceae taxon set, including multiple isolates of four *Physcomitrium* species, has underlined the hybrid origin of *P. eurystomum* and *P. collenchymatum*. *BRK1* is present as multiple paralogs in the respective species, which are recognized through sequence polymorphisms in terms of multiple peaks in the direct sequencing products. Both distinct parental homologs were identified by cloning the PCR products and then sequencing several clones per species isolate. In addition, different isolates of *P. pyriforme* (one of the parental lines giving rise to *P. eurystomum* and *P. collenchymatum*) were shown to contain multiple paralogs of the single copy gene *BRK1*. Flow cytometric measurements revealed all those isolates as recent allopolyploids (neopolyploids), since they have a larger genome size compared to other Funariaceae (manuscript in preparation). Chromosome counts of $n = 9$ to $n = 72$ for *P. pyriforme*, and $n = 9$ to $n = 54$ for *P. eurystomum* (Fritsch 1991) also suggest a high frequency of putative hybrid species generation in the genus *Physcomitrium*. In contrast, there is so far no evidence for a hybrid origin or recent allopolyploidy within *Physcomitrella*. The genomic DNA content of the different isolates is comparable to the well characterized Gransden isolate and the direct sequencing of the single copy gene *BRK1* was unambiguous (manuscript in preparation).

Initially, *Physcomitrella* was subdivided into three species (Tan 1978). *P. readeri* from USA, Australia and Japan (Ochi 1968), *P. magdalenae* from Rwanda and the Democratic Republic of Congo, Africa (Sloover 1975; Mueller 1995) and *P. patens* with a wide distribution in the Northern Hemisphere. Based on variable, but overlapping, phenotypic and key taxonomic characteristics of these species, the taxon *Physcomitrella* was later subdivided into four subspecies: *P. patens* subsp. *patens*, *P. patens* subsp. *magdalenae* de Sloover, *P. patens* subsp. *readeri* C. Muell., and *P. patens* subsp. *californica* Crum & Anderson (Tan 1979). Current molecular data however cannot confirm this latter classification. Instead, phylogenetic analyses of different loci, including plastid, ribosomal and nuclear marker genes, show three distinct lines of *Physcomitrella* which are clearly separated from each other. These data therefore suggest that (1) the genus is polyphyletic and arose at least three times from distinct ancestors within Funariaceae (more or less mirroring the older classification scheme) and (2) the secondary reduction of the sporophyte has occurred independently in each lineage. With regard to these molecular and phylogenetic analyses, a revised classification for the *Physcomitrium*/*Physcomitrella* species complex, following the initial taxonomy (Tan 1978), might be necessary.

18.4 Gene Retention and Gene Family Evolution

Previous studies of WGD events in *P. patens* were based on transcriptome representations (Rensing et al. 2007) and in order to avoid false-positives due to redundant representation of transcripts (by means of several unigenes per locus), nearly identical sequences were removed from such analyses. Repeating the analysis of Rensing et al. (2007) but based on the most recent genomic data gives the results shown in Fig. 18.3 (K_s plot, synonymous substitutions per site for paralog clusters). The first bin (synonymous substitutions per site between 0.0 and 0.1) contains a higher number of paralogs as previously reported (Rensing et al. 2007), i.e., ~40%, exceeding the estimate based on transcriptome data. The WGD (secondary peak) comprises another ~40% of the genes. The latter homoeologs show significant sequence variation and have probably undergone sub- and neofunctionalization (Ohno 1970; Wendel 2000). Mutant analysis of closely related paralogs, usually retained after WGD, has shown in several cases that the proteins are often partly functionally redundant, so that they all need to be down-regulated in order to exhibit a drastic aberrant phenotype, yet they differ, for example, in their expression pattern. Examples of such *P. patens* gene families are *rad51* (Markmann-Mulisch et al. 2002, 2007), *lhcsr* (Alboresi et al. 2010), *rsl* (Menand et al. 2007) and *ftsZ* (Martin et al. 2009a, b) (i.e., knockout mutants of the homoeologs all show partially redundant or additive phenotypes, yet differential expression levels or domains). These observations indicate that both, gene dosage-dependent redundancy and gene subfunctionalization have occurred. More detailed phylogenetic and experimental analysis is necessary to ascertain the contribution of these different types of evolution following polyploidy.

Paralogs from the first bin of K_s plots are usually considered to be the result of local duplication events, such as segmental duplications (Lynch and Conery 2000; Blanc and Wolfe 2004a). Indeed, analysis of the *P. patens* genome has demonstrated that ~1% of the genes are located in tandem arrays, i.e., (often identical) paralogs are in close proximity to each other (Rensing et al. 2008). Currently, we do not know whether such tandemly arrayed genes are the result of recent duplications (which are prone to subsequent paralog loss) or whether they are kept alike by concerted evolution/gene conversion (Wendel 2000; Wang et al. 2007). Even more intriguing is the high proportion of genes that are nearly identical but are not tandemly arranged

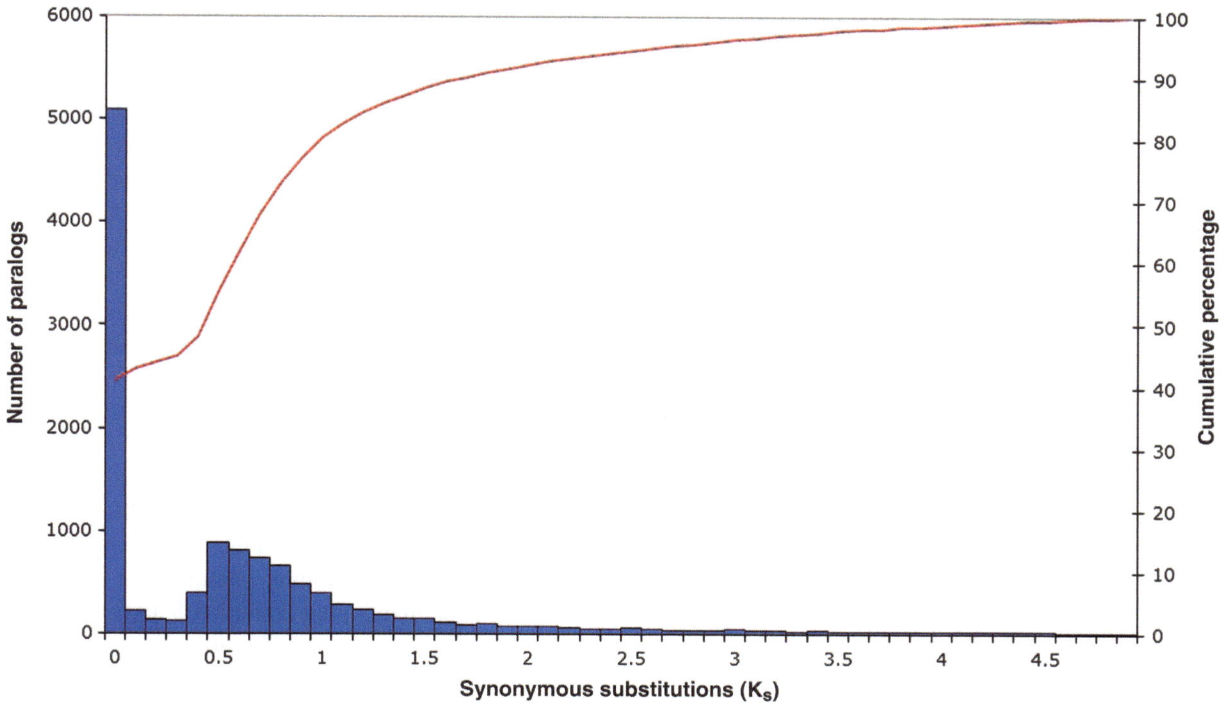

Fig. 18.3 K_s plot of *Physcomitrella patens* paralogs. The plot was generated as previously described (Rensing et al. 2007) but is based on the most recent gene predictions and does not feature a cutoff criterion for nearly identical sequences

(hypothesized from the large first bin, Fig. 18.3). Whether this is due to the still fragmented nature of the genome assembly or reveals a complex pattern of intra- and inter-chromosomal gene conversion has yet to be determined. This pattern might be related to the moss' high rate of DNA repair by homologous recombination and be indicative of concerted evolution of these loci via gene conversion to allow the haploid organism to maintain 'pseudoalleles' of dosage-sensitive or highly expressed genes (Lang et al. 2008). Additional evidence that gene conversion is active in *P. patens* comes from the analysis of the *ftsZ* gene family, where conversion tracts have been shown to maintain functional tubulin/GTPase domains among *ftsZ* homoeologs (Martin et al. 2009b), mirroring the pattern predicted on a larger scale for rice paralogs (Wang et al. 2007).

Enrichment analysis of functional annotations of both paleologs (Rensing et al. 2007) and pseudoalleles (Lang 2008) in *P. patens* revealed an overrepresentation of gene products involved in metabolic processes. This is in clear contrast to what was found previously for seed plant genomes, where the genes involved in signal transduction and transcriptional regulation were the ones that were found to be preferentially retained, e.g., in *Arabidopsis thaliana* (Seoighe and Gehring 2004; Blanc and Wolfe 2004b). This striking difference can at least partly be interpreted in light of Funariaceae ecology. Species belonging to the family are typically short-lived and annual (maximally biennial). This contrasts with the perennial life style that is common in many other moss families. Based on their adopted life strategies, Funariaceae are pioneer plants, classified as fugitives and annual shuttle species, implying they have adapted to open, unshaded and unstable habitats where survival depends on their ability to pass quickly through their life cycle when the conditions are optimal and competition from other species is low. Thus, the observed functional enrichment of metabolic genes might reflect a specific adaptation, ensuring the correct gene dosage important to their particular ecological niche. It might also reflect, for example, the poikilohydric life style of bryophytes in general. Poikilohydry enables bryophytes to resuscitate and restart their entire metabolism almost instantly after suspending it to endure various environmental stresses (e.g., heat, drought, cold, radiation). Indeed, the high abundance of metabolic genes has been found not only in *P. patens* but also in the rehydration transcriptome of *Tortula ruralis* (Oliver et al. 2004). Furthermore, genes in metabolic pathways have recently been found to constitute important fractions of retained paralogs in other species (the protozoon *Paramecium*, and yeast) as well (Gout et al. 2009; van Hoek and Hogeweg 2009).

18.5 Outlook

Physcomitrella patens was initially selected for genome sequencing because it was already well established as a model organism with the full molecular and genetic toolkit available (Frank et al. 2005; Reski and Frank 2005; Quatrano et al. 2007). It is also important from a phylogenetic perspective (Rensing et al. 2002; Cove et al. 2006), bridging the evolutionary gap between sequenced green algae and flowering plants, and thus making it a suitable organism for plant evo-devo studies (e.g., Tanahashi et al. 2005; Menand et al. 2007; Mosquna et al. 2009; Khandelwal et al. 2010).

In the light of what we have outlined above, *P. patens* and its siblings also appears now as a model ideally suited to study genome evolution through polyploidization and hybridization in haploid-dominant organisms. Moreover, due to the extensive range of morphological differences of the sporophyte, ranging from the elaborate *F. hygrometrica* to the vastly reduced *P. patens* morphology, Funariaceae are an excellent model to study gene-phene evolution. The latter is made even more interesting since the reduced sporphyte structure (as observed in *P. patens*) has evolved independently several times and might very well be related to a different mode of propagation (predominantly selfing, spores resting in the ground) in contrast to the cosmopolitan *F. hygrometrica* (which produces numerous small spores suitable for long range dispersal, the species propagating weed-like). The fact that there is a low breeding barrier among the family also enables crossing and associated genetic studies.

For Funariaceae to enter the age of population genomics we will need to re-sequence transcriptomes and genomes of other species and accessions from this family. Indeed, first steps have already been made by genome and transcriptome sequencing of *P. patens* isolates from geologically distinct locations, such as the French Villersexel accession (available through the cosmoss.org genome browser), or are progressing for cryptic species from Africa and Australia. In addition, the transcriptome of *F. hygrometrica* has recently been sequenced, mapped against *P. patens* as reference, and used for comparison of gametophyte and sporophyte biased gene expression in comparison with *Arabidopsis thaliana* (Szovenyi et al. 2010).

P. patens is but the first moss and bryophyte genome, but we certainly need more (Beike and Rensing 2010) in order to detect trends that cannot be hypothesized based on a single species. Fortunately, the genome of *Ceratodon purpureus*, which is already established as a genetic model (Cove and Quatrano 2006), has been granted for sequencing by the US Department of Energy. *C. purpureus* belongs to the Dicranidae whereas *P. patens* is in the Funariidae. The last common ancestor of the two lineages lived approximately 200 Mya (Newton et al. 2007). This offers an evolutionary horizon somewhere between the monocot-eudicot (~90–300 Mya) and angiosperm-gymnosperm (~290–385 Mya) divergence (Zimmer et al. 2007). In the same way that the comparative analysis of the *P. patens* genome has helped to understand the water-to-land-transition of plant life (Rensing et al. 2008), we can expect the *C. purpureus* genome to greatly inform comparative evolutionary studies of later steps in (moss) evolution. The availability of a second moss genome from a different subclass will also aid in discovering genomic traits that are common to mosses or unique to either lineage. Metabolic pathways in general seem to be more versatile and more redundantly encoded in mosses than in seed plants (Oliver et al. 2004; Lang et al. 2005; Rensing et al. 2007, 2008). The genome of a second moss will thus help in determining novel protein functions in this regard, and aid in the detection of metabolic pathways so far unknown among land plants. Also, *C. purpureus* is a dioicous, obligate outbreeder and harbours sex chromosomes (McDaniel 2005), so this will provide a further interesting dimension to compare with the *P. patens* genome that lacks them.

Finally, what about liverworts, the earliest branching division among land plants (Qiu et al. 2006) and sister to mosses: are they different from mosses in terms of genome evolution? The most common chromosome counts among liverworts (Marchantiophyta) are $n = 8$ and $n = 9$ and only a very few species are reported to have higher chromosome counts, like e.g., $n = 16$, 18 or $n = 36$ (Fritsch 1991). The genus *Marchantia* is reported to have a quite uniform number of 8 or 9 chromosomes although there are exceptions i.e., *Marchantia grisea* ($n = 9$ for male, $n = 10$ for female), *M. breviloba* ($n = 18$), *M. globosa* ($n = 18$), and *M. planiloba* ($n = 8, n = 16 + 2$). The genome of *Marchantia polymorpha* is currently being sequenced by the US Department of Energy and many chromosome counts are available for isolates from Japan, Europe, and America. They mainly confirm a chromosome number of eight or nine for this widely distributed species, with the single exceptions of $n = 10$–18 (Fritsch 1991). Thus, genome plasticity as measured by chromosome numbers seems to be significantly lower in liverworts than in mosses. The forthcoming analysis of the *M. polymorpha* genome will reveal whether it is a paleopolyploid. If it is not, as might be expected based on the narrow range of eight or nine chromosomes among liverworts, the genome analyses might reveal some surprises with regard to the evolution of this most ancient group of haploid-dominant plants.

Acknowledgments We are grateful to Stanislav Karnatsevych for conducting literature and experimental research with regard to crossing and hybridization within the Funariaceae.

References

Alboresi A, Gerotto C, Giacometti GM, Bassi R, Morosinotto T (2010) *Physcomitrella patens* mutants affected on heat dissipation clarify the evolution of photoprotection mechanisms upon land colonization. Proc Natl Acad Sci USA 107:11128–11133

Ando H (1980) Evolution of bryophytes in relation to their sexuality. Proc Bryol Soc Jpn 2:129–130

Andrews AL (1918) A new hybrid in *Physcomitrium*. Torreya 18:52–54
Andrews AL (1942) Taxonomic notes II. Another natural hybrid in the Funariaceae. Bryologist 45:176–178
Beike AK, Rensing SA (2010) The *Physcomitrella patens* genome—a first stepping stone towards understanding bryophyte and land plant evolution. Trop Bryol 31:43–49
Blanc G, Wolfe KH (2004a) Widespread paleopolyploidy in model plant species inferred from age distributions of duplicate genes. Plant Cell 16:1667–1678
Blanc G, Wolfe KH (2004b) Functional divergence of duplicated genes formed by polyploidy during *Arabidopsis* evolution. Plant Cell 16:1679–1691
Britton EG (1895) Contributions to American bryology IX. Bull Torrey Bot Club 22:62–68
Bryan VS (1957) Cytotaxonomic studies in the Ephemeraceae and Funariaceae. Bryologist 60:103–126
Cove DJ, Quatrano RS (2006) Agravitropic mutants of the moss *Ceratodon purpureus* do not complement mutants having a reversed gravitropic response. Plant Cell Environ 29:1379–1387
Cove D, Bezanilla M, Harries P, Quatrano R (2006) Mosses as model systems for the study of metabolism and development. Annu Rev Plant Biol 57:497–520
Cove DJ, Perroud P-F, Charron AJ, Mcdaniel SF, Khandelwal A, Quatrano RS (2009) The moss *Physcomitrella patens*: a novel model system for plant development and genomic studies. Cold Spring Harb Protoc 2009: pdb.emo115
Crawford M, Jesson LK, Garnock-Jones PJ (2009) Correlated evolution of sexual system and life-history traits in mosses. Evolution 63:1129–1142
Crow KD, Wagner GP (2006) What is the role of genome duplication in the evolution of complexity and diversity? Mol Biol Evol 23:887–892, In: Proceedings of the SMBE tri-national young investigators' workshop 2005
Engel PP (1968) The induction of biochemical and morphological mutants in the moss *Physcomitrella patens*. Am J Bot 55:438–446
Fawcett JA, Maere S, Van de Peer Y (2009) Plants with double genomes might have had a better chance to survive the Cretaceous-Tertiary extinction event. Proc Natl Acad Sci USA 106:5737–5742
Fawcett JA, Van de Peer Y, Maere S (2013) Significance and biological consequences of polyploidization in land plants. In: Leitch IJ, Greilhuber J, Doležel J, Wendel JF (eds) Plant genome diversity, vol 2, Physical structure, behaviour and evolution of plant genomes. Springer-Verlag, Wien, pp 277–293
Frahm J-P (2001) Biologie der Moose. Spektrum Akademischer Verlag, Heidelberg/Berlin
Frahm J-P (2010) Die Laubmoosflora des Baltischen Bernsteinwaldes. Weißdorn-Verlag, Jena
Frank W, Decker EL, Reski R (2005) Molecular tools to study *Physcomitrella patens*. Plant Biol 7:220–227
Fritsch R (1991) Index to bryophyte chromosome counts. Bryophytorum Bibliotheka vol 40. J. Cramer/Gebrueder Borntraeger, Berlin/Stuttgart
Gout JF, Duret L, Kahn D (2009) Differential retention of metabolic genes following whole-genome duplication. Mol Biol Evol 26:1067–1072
Husband BC, Baldwin SJ, Suda J (2013) The incidence of polyploidy in natural plant populations: major patterns and evolutionary processes. In: Leitch IJ, Greilhuber J, Doležel J, Wendel JF (eds) Plant genome diversity, vol 2, Physical structure, behaviour and evolution of plant genomes. Springer-Verlag, Wien, pp 255–276
Innan H, Kondrashov F (2010) The evolution of gene duplications: classifying and distinguishing between models. Nat Rev Genet 11:97–108
Karlin EF, Boles SB, Ricca M, Temsch EM, Greilhuber J, Shaw AJ (2009) Three-genome mosses: complex double allopolyploid origins for triploid gametophytes in *Sphagnum*. Mol Ecol 18:1439–1454
Kenrick P, Crane PR (1997) The origin and early evolution of plants on land. Nature 389:33–39
Khandelwal A, Cho SH, Marella H, Sakata Y, Perroud PF, Pan A, Quatrano RS (2010) Role of ABA and ABI3 in desiccation tolerance. Science 327:546
Lang D (2008) Comparative genomic and phylogenomic analysis of the moss *Physcomitrella patens* (Hedw.) Bruch & Schimp. Albert-Ludwigs-Universität Freiburg, Freiburg
Lang D, Eisinger J, Reski R, Rensing SA (2005) Representation and high-quality annotation of the *Physcomitrella patens* transcriptome demonstrates a high proportion of proteins involved in metabolism among mosses. Plant Biol 7:228–237
Lang D, Zimmer AD, Rensing SA, Reski R (2008) Exploring plant biodiversity: the *Physcomitrella* genome and beyond. Trends Plant Sci 13:542–549
Lang D, Weiche B, Timmerhaus G, Richardt S, Riano-Pachon DM, Correa LG, Reski R, Mueller-Roeber B, Rensing SA (2010) Genome-wide phylogenetic comparative analysis of plant transcriptional regulation: a timeline of loss, gain, expansion, and correlation with complexity. Genome Biol Evol 2:488–503
Longton RE (1976) Reproductive biology and evolutionary potential in bryophytes. J Hattori Bot Lab 41:205–223
Lynch M, Conery JS (2000) The evolutionary fate and consequences of duplicate genes. Science 290:1151–1155
Ma XF, Gustafson JP (2005) Genome evolution of allopolyploids: a process of cytological and genetic diploidization. Cytogenet Genome Res 109:236–249
Markmann-Mulisch U, Hadi MZ, Koepchen K, Alonso JC, Russo VE, Schell J, Reiss B (2002) The organization of *Physcomitrella patens* RAD51 genes is unique among eukaryotic organisms. Proc Natl Acad Sci USA 99:2959–2964
Markmann-Mulisch U, Wendeler E, Zobell O, Schween G, Steinbiss HH, Reiss B (2007) Differential requirements for RAD51 in *Physcomitrella patens* and *Arabidopsis thaliana* development and DNA damage repair. Plant Cell 19:3080–3089
Martin A, Lang D, Hanke ST, Mueller SJ, Sarnighausen E, Vervliet-Scheebaum M, Reski R (2009a) Targeted gene knockouts reveal overlapping functions of the five *Physcomitrella patens* FtsZ isoforms in chloroplast division, chloroplast shaping, cell patterning, plant development, and gravity sensing. Mol Plant 2:1359–1372
Martin A, Lang D, Heckmann J, Zimmer AD, Vervliet-Scheebaum M, Reski R (2009b) A uniquely high number of ftsZ genes in the moss *Physcomitrella patens*. Plant Biol 11:744–750
McDaniel SF (2005) Genetic correlations do not constrain the evolution of sexual dimorphism in the moss *Ceratodon purpureus*. Evolution 59:2353–2361
McDaniel SF, von Stackelberg M, Richardt S, Quatrano RS, Reski R, Rensing SA (2010) The speciation history of the *Physcomitrium-Physcomitrella* species complex. Evolution 64:217–231
Menand B, Yi K, Jouannic S, Hoffmann L, Ryan E, Linstead P, Schaefer DG, Dolan L (2007) An ancient mechanism controls the development of cells with a rooting function in land plants. Science 316:1477–1480
Mosquna A, Katz A, Decker EL, Rensing SA, Reski R, Ohad N (2009) Regulation of stem cell maintenance by the Polycomb protein FIE has been conserved during land plant evolution. Development 136:2433–2444
Mueller F (1995) Neue und bemerkenswerte Moosfunde aus Zaire. Trop Bryol 10:81–90
Müntzing A (1936) The evolutionary significance of autopolyploidy. Hereditas 21:363–378

Natcheva R, Cronberg N (2004) What do we know about hybridization among bryophytes in nature? Can J Bot 82:1687–1704

Newton AE, Wikström N, Bell N, Forrest LL, Ignatov MS (2007) Dating the diversification of the pleurocarpous mosses. In: Tangney N (ed) Pleurocarpous mosses: systematics and evolution. Systematics Association Special, vol 71. Systematics Association, CRC Press, Boca Raton

Ochi H (1968) A revision of the family Funariaceae (Musci) in Japan and the adjacent regions. Jpn J Bot 20:1–34

Odrzykoski IJ, Chudzinska E, Szweykowski J (1996) The hybrid origin of the polyploid liverwort *Pellia borealis*. Genetica 98:75–86

Ohno S (1970) Evolution by gene duplication. Springer, New York

Oliver MJ, Dowd SE, Zaragoza J, Mauget SA, Payton PR (2004) The rehydration transcriptome of the desiccation-tolerant bryophyte *Tortula ruralis*: transcript classification and analysis. BMC Genomics 5:89

Orzechowska M, Siwinska D, Maluszynska J (2010) Molecular cytogenetic analyses of haploid and allopolyploid *Pellia* species. J Bryol 32:113–121

Parisod C, Holderegger R, Brochmann C (2010) Evolutionary consequences of autopolyploidy. New Phytol 186:5–17

Pettet A (1964) Hybrid sporophytes in Funariaceae. I. Hybrid sporophytes on *Physcomitrella patens* (Hedw.) B. & S., and *Physcomitrium sphaericum* (Schkuhr) Brid. in Britain. Trans Br Bryol Soc 4:642–648

Qiu YL, Li L, Wang B, Chen Z, Knoop V, Groth-Malonek M, Dombrovska O, Lee J, Kent L, Rest J, Estabrook GF, Hendry TA, Taylor DW, Testa CM, Ambros M, Crandall-Stotler B, Duff RJ, Stech M, Frey W, Quandt D, Davis CC (2006) The deepest divergences in land plants inferred from phylogenomic evidence. Proc Natl Acad Sci USA 103:15511–15516

Quatrano RS, McDaniel SF, Khandelwal A, Perroud PF, Cove DJ (2007) *Physcomitrella patens*: mosses enter the genomic age. Curr Opin Plant Biol 10:182–189

Rensing SA, Rombauts S, Van de Peer Y, Reski R (2002) Moss transcriptome and beyond. Trends Plant Sci 7:535–538

Rensing SA, Ick J, Fawcett JA, Lang D, Zimmer A, Van de Peer Y, Reski R (2007) An ancient genome duplication contributed to the abundance of metabolic genes in the moss *Physcomitrella patens*. BMC Evol Biol 7:130

Rensing SA, Lang D, Zimmer AD, Terry A, Salamov A, Shapiro H, Nishiyama T, Perroud PF, Lindquist EA, Kamisugi Y, Tanahashi T, Sakakibara K, Fujita T, Oishi K, Shin IT, Kuroki Y, Toyoda A, Suzuki Y, Hashimoto S, Yamaguchi K, Sugano S, Kohara Y, Fujiyama A, Anterola A, Aoki S, Ashton N, Barbazuk WB, Barker E, Bennetzen JL, Blankenship R, Cho SH, Dutcher SK, Estelle M, Fawcett JA, Gundlach H, Hanada K, Heyl A, Hicks KA, Hughes J, Lohr M, Mayer K, Melkozernov A, Murata T, Nelson DR, Pils B, Prigge M, Reiss B, Renner T, Rombauts S, Rushton PJ, Sanderfoot A, Schween G, Shiu SH, Stueber K, Theodoulou FL, Tu H, Van de Peer Y, Verrier PJ, Waters E, Wood A, Yang L, Cove D, Cuming AC, Hasebe M, Lucas S, Mishler BD, Reski R, Grigoriev IV, Quatrano RS, Boore JL (2008) The *Physcomitrella* genome reveals evolutionary insights into the conquest of land by plants. Science 319:64–69

Rensing SA, Lang D, Zimmer AD (2009) Comparative genomics. In: Knight C, Perroud P-F, Cove D (eds) The moss *Physcomitrella patens*, vol 36. Wiley-Blackwell, Oxford, pp 42–75

Reski R, Frank W (2005) Moss (*Physcomitrella patens*) functional genomics—gene discovery and tool development, with implications for crop plants and human health. Brief Funct Genomic Proteomic 4:48–57

Reski R, Faust M, Wang XH, Wehe M, Abel WO (1994) Genome analysis of the moss *Physcomitrella patens* (Hedw.) B.S.G. Mol Gen Genet 244:352–359

Ricca M, Shaw AJ (2010) Allopolyploidy and homoploid hybridization in the *Sphagnum subsecundum* complex (Sphagnaceae: Bryophyta). Biol J Linn Soc 99:135–151

Schween G, Egener T, Fritzkowsky D, Granado J, Guitton M-C, Hartmann N, Hohe A, Holtorf H, Lang D, Lucht JM, Reinhard C, Rensing SA, Schlink K, Schulte J, Reski R (2005a) Large-scale analysis of 73,329 gene-disrupted *Physcomitrella* mutants: production parameters and mutant phenotypes. Plant Biol 7:238–250

Schween G, Hohe A, Schulte J, Reski R (2005b) Effect of ploidy level on growth, differentiation and phenotype in *Physcomitrella patens*. Bryologist 108:27–35

Seoighe C, Gehring C (2004) Genome duplication led to highly selective expansion of the *Arabidopsis thaliana* genome. Trends Genet 20:461–464

Shaw AJ, Pokorny L, Shaw B, Ricca M, Boles S, Szovenyi P (2008) Genetic structure and genealogy in the *Sphagnum subsecundum* complex (Sphagnaceae: Bryophyta). Mol Phylogenet Evol 49:304–317

Sloover JL (1975) Note de bryologie africaine III. *Physcomitrella magdalenae* sp. nov. Bull Jard Bot Natl Belg/Bull Natl Plant Belg 45:131–135

Smith AJE (1979) Cytogenetics, biosystematics and evolution in the Bryophyta. Adv Bot Res 6:195–276

Soltis PS, Soltis DE (2009) The role of hybridization in plant speciation. Annu Rev Plant Biol 60:561–588

Soltis PS, Soltis DE (2013) Angiosperm phylogeny: A framework for studies of genome evolution. In: Leitch IJ, Greilhuber J, Doležel J, Wendel JF (eds) Plant genome diversity, vol 2, Physical structure, behaviour and evolution of plant genomes. Springer-Verlag, Wien, pp 1–11

Sugiura C, Kobayashi Y, Aoki S, Sugita C, Sugita M (2003) Complete chloroplast DNA sequence of the moss *Physcomitrella patens*: evidence for the loss and relocation of rpoA from the chloroplast to the nucleus. Nucleic Acids Res 31:5324–5331

Szovenyi P, Rensing SA, Lang D, Wray GA, Shaw AJ (2010) Generation-biased gene expression in a bryophyte model system. Mol Biol Evol 2010:20

Tan BC (1978) *Physcomitrella patens* (Musci: Funariaceae) in North America. Bryologist 81:561–567

Tan BC (1979) A new classification for the genus *Physcomitrella* B.S. G. J Hattori Bot Lab 46:327–336

Tanahashi T, Sumikawa N, Kato M, Hasebe M (2005) Diversification of gene function: homologs of the floral regulator FLO/LFY control the first zygotic cell division in the moss *Physcomitrella patens*. Development 132:1727–1736

Taylor PJ, Eppley SM, Jesson LK (2007) Sporophytic inbreeding depression in mosses occurs in a species with separate sexes but not in a species with combined sexes. Am J Bot 94:1853–1859

Terasawa K, Odahara M, Kabeya Y, Kikugawa T, Sekine Y, Fujiwara M, Sato N (2007) The mitochondrial genome of the moss *Physcomitrella patens* sheds new light on mitochondrial evolution in land plants. Mol Biol Evol 24:699–709

Van de Peer Y, Maere S, Meyer A (2009) The evolutionary significance of ancient genome duplications. Nat Rev Genet 10:725–732

van Hoek MJ, Hogeweg P (2009) Metabolic adaptation after whole genome duplication. Mol Biol Evol 26:2441–2453

Veron AS, Kaufmann K, Bornberg-Bauer E (2007) Evidence of interaction network evolution by whole-genome duplications: a case study in MADS-box proteins. Mol Biol Evol 24:670–678

Wang X, Tang H, Bowers JE, Feltus FA, Paterson AH (2007) Extensive concerted evolution of rice paralogs and the road to regaining independence. Genetics 177:1753–1763

Wellman CH, Osterloff PL, Mohiuddin U (2003) Fragments of the earliest land plants. Nature 425:282–285

Wendel JF (2000) Genome evolution in polyploids. Plant Mol Biol 42:225–249

Wettstein F (1924) Gattungskreuzungen bei Moosen. Z Indukt Abstammungs-Vererbungsl 33:253–257

Wettstein F (1932) Genetik. In: Verdoorn F (ed) Manual of bryology. Nijhoff, The Hague, pp 233–272

Wyatt R, Odrzykoski IJ, Stoneburner A, Bass HW, Galau GA (1988) Allopolyploidy in bryophytes: multiple origins of *Plagiomnium medium*. Proc Natl Acad Sci USA 85:5601–5604

Wyatt R, Odrzykoski IJ, Stoneburner A (1992) Isozyme evidence of reticulate evolution in mosses: *Plagiomnium medium* is an allopolyploid of *P. ellipticum* × *P. insigne*. Syst Bot 17:532–550

Zimmer A, Lang D, Richardt S, Frank W, Reski R, Rensing SA (2007) Dating the early evolution of plants: detection and molecular clock analyses of orthologs. Mol Genet Genomics 278:393–402

Zobell O, Faigl W, Saedler H, Munster T (2010) MIKC* MADS-box proteins: conserved regulators of the gametophytic generation of land plants. Mol Biol Evol 27:1201–1211

Genome Size Diversity and Evolution in Land Plants

19

Ilia J. Leitch and Andrew R. Leitch

Contents

19.1 Introduction

The amount of DNA in the nucleus of a cell is commonly referred to as the genome size or C-value and people have been estimating this character in plants and animals for over 60 years. Today, with data available for over 7,000 species (Table 19.1), land plants (embryophytes) are the best studied of the major taxonomic groups of eukaryotes. This chapter provides an overview of what is currently known about the diversity of genome sizes encountered in land plant groups and considers how such diversity might have evolved. The chapter by Greilhuber and Leitch (2013, this volume) will discuss the impact of this diversity on plants in terms of how differences in genome size have an impact at all levels of complexity, from the nucleus to the whole organism. This chapter should also be read with reference to those by Weiss-Schneeweiss and Schneeweiss (2013, this volume) who explore intra- and inter-specific chromosome complexity across angiosperms, including polyploidy and dysploidy, Murray (2013, this volume) who discusses gymnosperm chromosomes, and Barker (2013, this volume) who considers the chromosomes of monilophytes and lycophytes.

19.1.1 Terminology

While the term 'genome size' has been broadly used in the literature, it may refer to different things by different people (Greilhuber et al. 2005). For example, it has been used to refer to the amount of DNA in an unreplicated monoploid chromosome set (*n*) or in a polyploid nucleus where the DNA has been replicated. In an attempt to stabilize the terminology and to bring greater precision when referring to the amount of DNA in different phases of the cell cycle (e.g., G1 and G2), different stages of the life cycle (e.g., haplophase and diplophase), and when comparing the amount of DNA in a basic monoploid chromosome set with the total amount of DNA in the

I.J. Leitch (✉)
Jodrell Laboratory, Royal Botanic Gardens, Kew, Richmond, Surrey TW9 3AB, UK
e-mail: i.leitch@kew.org

I.J. Leitch et al. (eds.), *Plant Genome Diversity Volume 2*,
DOI 10.1007/978-3-7091-1160-4_19, © Springer-Verlag Wien 2013

Table 19.1 Genome size data available in land plants

	No. of species recognised	No. of species with genome size data	Representation (%)	Min. (pg)	Max. (pg)	Mean (pg)	Median (pg)	Mode (pg)	Range (max./min.)
Non-vascular plants									
Liverworts	c. 5,000	43	0.9	0.21	7.97	1.2	0.8	0.76	38-fold
Mosses	c. 12,000	184	1.5	0.17	2.05	0.5	0.4	0.45	12-fold
Hornworts	c. 150	0	–	–	–	–	–	–	–
Vascular plants									
Lycophytes	c. 900	15	1.6	0.086	11.96	1.7	0.1	0.086	139-fold
Monilophytes	c. 11,000	67	0.6	0.77	72.68	14.0	9.5	7.95	94-fold
Seed plants									
Gymnosperms	c. 850	204	24	2.25	36.0	18.6	7.9	10.0	16-fold
Angiosperms	c. 352,000	6,287	1.8	0.065	152.23	5.9	2.5	0.60	2,342-fold

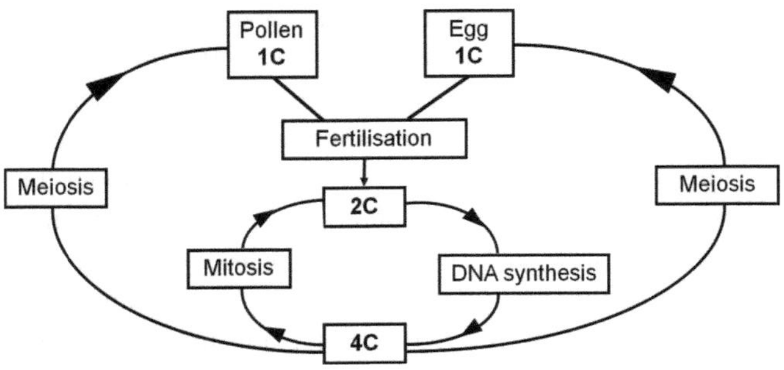

Fig. 19.1 The different levels of DNA amount during the life cycle of an angiosperm or gymnosperm

nucleus of a generative polyploid, Greilhuber et al. (2005) proposed the following terminology which is adopted in the current chapter.

The 'Holoploid genome size' or 'C-value' represents the amount of DNA in the whole chromosome complement of the nucleus with a chromosome number *n*, irrespective of the degree of generative polyploidy.

The 'Monoploid genome size' refers to the amount of DNA in one chromosome set of an organism. It is abbreviated as the Cx-value. In non-polyploid organisms the C-value and Cx-value are the same whereas in polyploids the Cx-value is theoretical, derived by dividing the holoploid C-value by the ploidy level.

In addition, for quantitative comparisons the terms should always be used with a prefix number to indicate the amount of DNA replication (Fig. 19.1). Thus a 1C-value refers to the amount of DNA in an unreplicated, gametic nucleus with a chromosome number *n*.

19.2 Genome Size Data and Diversity Across Land Plants

Land plants first began to diverge around 450–460 million years ago (Mya) during the middle Ordovician (Kenrick and Crane 1997; Rensing et al. 2008) and have since diversified into four major groups: (1) the non-vascular bryophytes which comprise liverworts, mosses and hornworts, (2) the lycophytes, considered to be the first vascular plants to evolve and are sister to the remaining vascular plants (3) the monilophytes comprising marattioid and leptosporangiate ferns, horsetails (*Equisetum*), whisk ferns (*Psilotum, Tmesipteris*) and ophioglossoid ferns (e.g., *Ophioglossum, Botrychium*) and (4) the seed plants comprising angiosperms and gymnosperms. Knowledge of genome sizes in these different groups is uneven and in the past this has hampered studies aiming to compare genome size diversity across land plants. Nevertheless, in recent years there has been a huge increase in the amount of genome size data available (particularly in the angiosperms, Bennett and Leitch 2011) and this provides an opportunity to reassess what it known about genome size diversity across land plants. The following survey is based on genome size data available in the Plant DNA C-values database (release 5.0, December 2010) which contains estimates for 7,058 land plant species together with data not yet compiled into the database.

Fig. 19.2 Chromosomes of (**a**) *Genlisea aurea* (1C = 0.065 pg) and (**b**) *Paris japonica* (1C = 152.23 pg, 2*n* = 40) taken at the same magnification. (Image in (a) from Greilhuber et al. 2006 and (b) from Pellicer et al. 2010)

The new data available have extended the range of genome sizes encountered in land plants at both ends of the scale so C-values now vary nearly 2,400-fold. Both record holders are angiosperms with the smallest reported in the eudicot genus *Genlisea* (*G. aurea* 1C = 0.065 pg, Fig. 19.2a) (Greilhuber et al. 2006; the report of 0.065 pg for *G. margaretae* actually belongs to *G. aurea*; J. Greilhuber, pers. comm.) while the largest to date is in the monocot *Paris japonica* which is an octoploid with 2*n* = 8*x* = 40 and a staggering 1C = 152.23 pg (Pellicer et al. 2010) (Fig. 19.2b).

Each group of land plants is characterized by a distinctive genome size profile (Fig. 19.3) and to understand how these may have evolved requires viewing the data within an evolutionary framework. Fortunately, understanding the relationships between and within the different groups which comprise land plants has increased in recent years and is outlined by Soltis and Soltis (2013, this volume). Indeed, having such a framework is essential for providing insights into the direction of genome size evolution in different groups.

19.2.1 Bryophytes

Traditionally, mosses (Bryophyta s.s.), liverworts (Marchantiophyta) and hornworts (Anthocerotophyta) were grouped together into the bryophytes, comprising the three extant lineages of non-vascular land plants. However, they are no longer considered to form a monophyletic group and instead are recognised as three distinct paraphyletic lineages (Shaw and Renzaglia 2004). Indeed, with increasing amounts of both molecular sequence and structural data, several studies suggest that liverworts are sister to all other land plants followed by mosses, and with hornworts sister to all vascular plants (tracheophytes) (e.g., Qiu et al. 2006; Pena et al. 2008; Qiu 2008). Nevertheless, studies based on other sources of data (e.g., sperm cell morphology) suggest alternative relationships and more data are clearly needed to fully resolve the branching order (Mishler and Kelch 2009).

Sadly there is not a great wealth of genome size data in the bryophytes despite their key evolutionary position at the base of all land plants. In addition, mosses are possibly the second most diverse group of land plants after the angiosperms.

19.2.1.1 Liverworts

Until recently, genome size data were available for only nine liverworts and even some of these values were called into question (Voglmayr 2000; Temsch et al. 2010) due to problems with methodology and material. For example, a recent comparison between genome size estimates obtained using Feulgen densitometry and flow cytometry highlighted how only the later method may be considered reliable due to the rigid cell walls and crystalline deposits in liverworts that interfere with Feulgen staining (Temsch et al. 2010).

Fortunately, the publication by Temsch et al. (2010) of genome size estimates for 43 species in 31 genera (32 foliose, 11 thallose) from 22 families has increased knowledge of genome sizes in liverworts considerably. Although the data are still far from representative given that there are c. 5,000 liverwort species worldwide comprising 83 families and 391 genera (Crandall-Stotler et al. 2009), some insights into genome size diversity can now be gleaned. Overall, 1C-values range 38-fold from 0.211 pg in *Lejeunea cavifolia* (Lejeuneaceae) to 7.966 pg in *Mylia taylorii* (Myliaceae) (Table 19.1) with no significant difference in the ranges encountered in thallose (1C = 0.293–7.966 pg) and foliose (0.211–0.757 pg) liverworts (Temsch et al. 2010).

A histogram of the frequency of species with different genome sizes shows that most liverworts have small genomes (up to around 2 pg; Fig. 19.4a). Only two genera have larger genomes – *Pellia* (Pelliaceae) and *Mylia* (Myliaceae) and their genomes certainly appear as outliers. What is notable about these species is that they belong to different orders – *Pellia* to the Order Pelliales and *Mylia* which is in Jungermanniales (Crandall-Stotler et al. 2009). While there are currently no other genome size estimates for species belonging to the other genus of Order Pelliales (i.e., *Noteroclada*), cytological studies show their chromosomes are considerably smaller, suggesting much smaller genome sizes in this genus (Proskauer 1950). In the Order Jungermanniales several estimates are available and they are all less than 1C = 2.02 pg. Such observations suggest

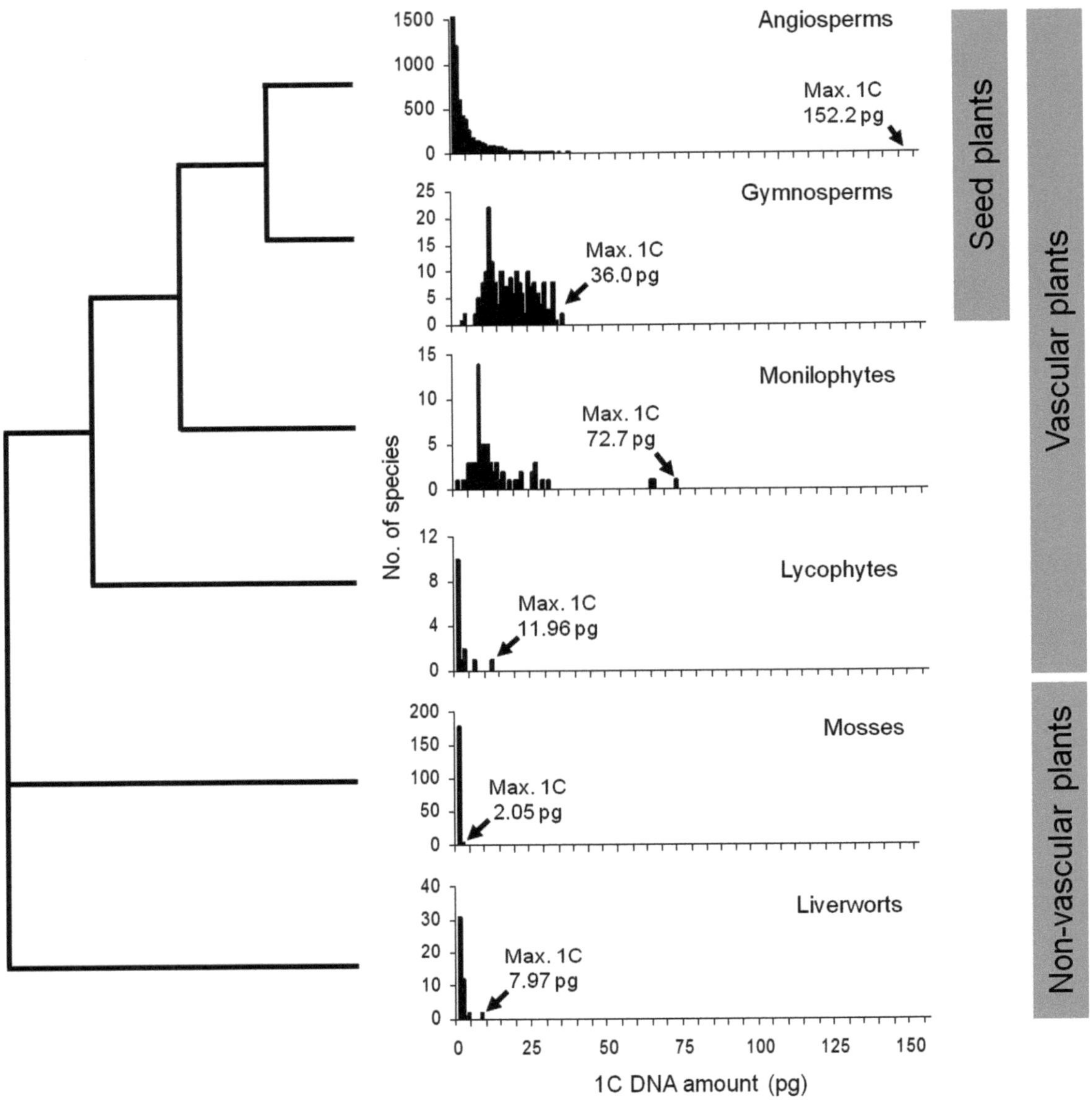

Fig. 19.3 Histograms showing the genome size profiles for each of the major land plant groups with genome size data

that there have been two independent increases in genome size within liverworts. The most extensive increase is in *Mylia taylorii* as this species is reported to be haploid with $n = 9$. In *Pellia*, the largest genome is for a polyploid (*P. borealis* with 1C = 7.4 pg and $n = 18$). Nevertheless, its monoploid genome size (1Cx = 3.7 pg) is similar in size to related haploid species (with 1C = 3.4–3.8 pg) which themselves appear as outliers compared with other liverworts (Fig. 19.4a).

Further large genomes may well be encountered in other species of *Mylia* as *M. anomala* has been reported to have a karyotype comprising eight similarly large chromosomes as observed in *M. taylorii*, and these are said to be amongst the largest known in liverworts (Newton 1987). As for *Pellia*, large genomes may also be found in related species as large chromosomes with lots of heterochromatin (up to 35%) have been observed in *P. neesiana* (Newton 1985, 1987). Large genomes may also be found in *Pallavicinia lyellii* (Pallaviciniaceae) since cytogenetic studies by Zheng and Zhu (2009) showed that this species had the same number ($n = 8$) but larger chromosomes than *Pellia epiphylla*. Sadly, there are currently no genome size data for *Pallavicinia*.

19.2.1.2 Mosses

Mosses are the best represented group of bryophytes in terms of genome size with values for 1.5% of species (Table 19.1). Like liverworts, mosses are characterized by small genomes, but they are considerably less variable, ranging just 12-fold from 0.17 pg in *Holomitrium arboretum* to 2.05 pg in *Mnium marginatum* (Fig. 19.4b). It is also noted that the distribution of DNA amounts is strongly skewed towards the lower end of

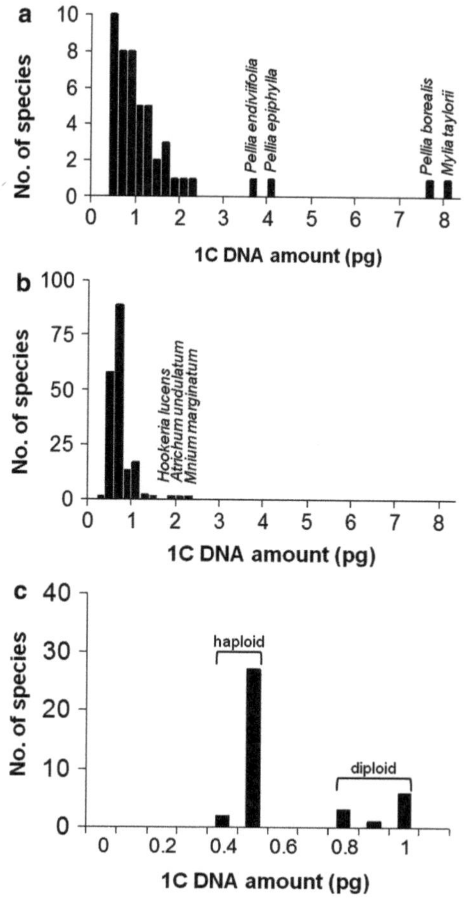

Fig. 19.4 (**a, b**) Histograms showing the distribution of 1C-values in (**a**) 43 liverwort species and (**b**) 184 mosses. (**c**) The distribution of C-values in the gametophytes of 39 species of the moss genus *Sphagnum* showing the two distinct ploidy levels

the genome size range (Fig. 19.4b), so actually the majority of moss genomes show an even narrower range, with 97% of the data varying just five-fold from 1C = 0.2–0.6 pg. It is possible that larger genomes may be uncovered, particularly in species reported to have high chromosome numbers, which may be polyploids (e.g., $n = 72$ in *Physcomitrium pyriforme* and *Leptodictyum riparium* – the highest chromosome number so far reported for a moss; Przywara and Kuta 1995). However, since increasing levels of polyploidy may be accompanied by decreases in chromosome size and genome size, as noted in some liverworts (Bornefeld and Grillenberger 1987; Orzechowska et al. 2010) and some mosses (Abderrahman 2004), whether genomes larger than the largest liverworts will be found in mosses is uncertain.

Among the mosses, data are available for 80 genera and 36 families although 96% of the genera and 75% of the families have five or fewer genome size estimates. Most data are available for *Sphagnum* (the only genus comprising Sphagnaceae) with estimates for 39 species. While C-values here range 0.39–0.95 pg, the data clearly fall into two groups which have been shown to correspond to species whose gametophytes are haploid ($n = 19$) or diploid ($n = 38$) (Temsch et al. 1998; Greilhuber et al. 2003) (Fig. 19.4c). Such studies have highlighted the apparent stability of the monoploid genome size in *Sphagnum* (mean 1Cx = 0.45 pg, S.D. 0.03) although it remains to be determined whether this is typical for other moss genera as insufficient data are currently available.

19.2.1.3 Hornworts

Currently, there are no reliable genome size data for hornworts. Data for two species were reported by Renzaglia et al. (1995) (i.e., *Notothylas orbicularis* 1C = 0.17 pg, *Phaeoceros laevis* 1C = 0.27 pg) and their values are both at the lower end of those found in either mosses or liverworts. However, the Feulgen microdensitometry technique used to measure genome size did not follow best practice recommendations (e.g., unsuitable storage of nuclei prior to analysis, use of animal cells as calibration standards, use of highly condensed sperm nuclei, Temsch et al. 1998; Voglmayr 2000). In addition, the rigid cell walls are likely to prevent access to reagents as noted for liverworts above and hence to unreliable data (Greilhuber pers. comm.).

Nevertheless, very small genomes might be expected given that hornworts are characterized cytologically by possessing low numbers of extremely small chromosomes ($n = 4$–10) and polyploidy is rare or absent (Proskauer 1958, 1967; Kuta and Przywara 2000). Clearly it is imperative that genome size data are obtained for this important group of non-vascular land plants, particularly if they are indeed shown to hold a key phylogenetic position as sister to all vascular plants as suggested by some molecular studies (Qiu et al. 2006).

19.2.2 Lycophytes

Lycophytes which comprise c. 900 species are divided into three lineages (clubmosses and firmosses – Lycopodiaceae, quillworts – Isoetaceae and spikemosses – Selaginellaceae) and have consistently been shown to be monophyletic and sister to all other vascular plants using both molecular and morphological characters (Schneider et al. 2009; Gao et al. 2010). They are therefore considered to hold a key position in plant evolution. Sadly from a genome size perspective they are poorly represented with data for just 15 species although all three families are represented (Fig. 19.5). The smallest genomes (1C = 0.086–0.24 pg; 10 species) are found in *Selaginella* (the only genus comprising Selaginellaceae), which is also noted to be characterized by possessing some of the smallest chromosomes in non-seed vascular plants. In addition, most species are diploid with $2n = 18$ or 20 (Jermy 1967; Takamiya 1993). It therefore seems likely that this family is characterized by a narrow range of small genomes.

In contrast, a larger range of genome sizes is encountered in Isoetaceae (comprising a single genus *Isoetes*) with values

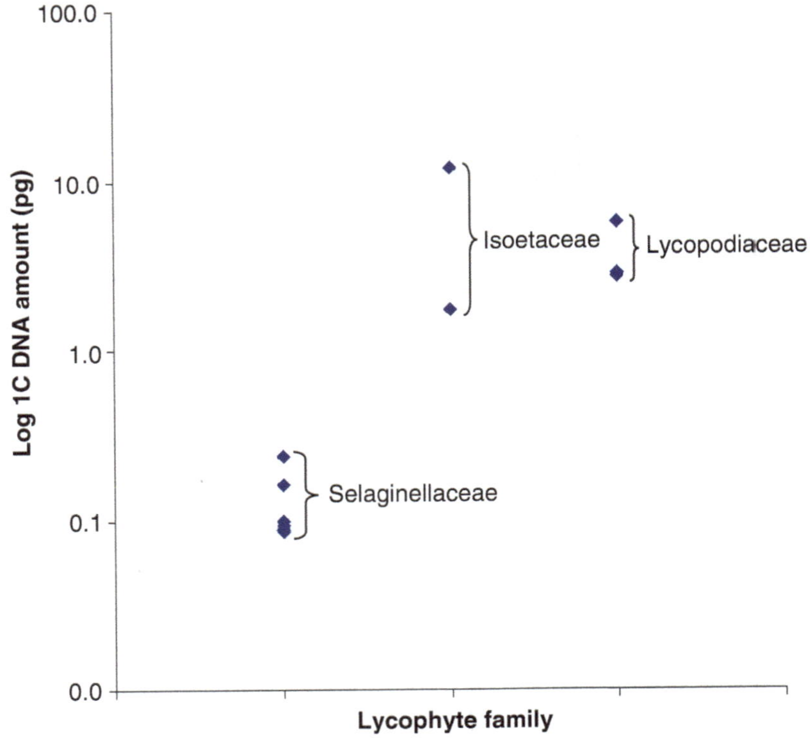

Fig. 19.5 Distribution of genome sizes in the three families which comprise lycophytes. Data available for 10 species in Selaginellaceae, two species in Isoetaceae and three species in Lycopodiaceae

varying seven-fold (i.e., 1C = 1.75 and 11.97 pg), and including the largest genome so far reported for any lycophyte (*Isoetes lacustris*, 1C = 11.97 pg and 2*n* = c. 110, Hanson and Leitch 2002). Even though data are currently available for just two species, this diversity reflects what is known from cytological studies as some of the species have been reported to have the largest chromosomes in lycophytes (Dunlop 1949) while other studies have revealed considerable variation in both chromosome number and size between species (2*n* = 22 – c. 130 and sizes of individual chromosomes from less than 2 to 7–8 μm long, Dunlop 1949; Manton 1950; Troìa 2001; Peruzzi et al. 2003).

Lycopodiaceae have genome sizes ranging from 1C = 2.7–5.7 pg. However these are based on data for just three species and only one chromosome count (i.e., *Lycopodium clavatum, 2n* = 68). Since this group of homosporous lycopods can reach high chromosome numbers (up to 2*n* = c. 510 in *Phylloglossum* (Blackwood 1953) and 2*n* = c. 556 in *Huperzia prolifera* (Tindale and Roy 2002) – the highest for any lycophytes) larger genomes may be encountered as data increase. However, genomes as large as some Isoetaceae seem unlikely in all but *Phylloglossum* since it has been suggested that there are constraints on genome size in groups with biflagellate sperm (i.e., Lycopodiaceae (excluding *Phylloglossum*), Selaginellaceae and all non-vascular plants, Renzaglia et al. 1995). The hypothesis states that for species with only two flagella there is an upper limit to genome size determined by the efficiency with which the flagella can move the sperm from the antheridia to the archegonia to effect fertilization. Due to the positive correlation between genome size and sperm size, an increase in DNA amount would result in larger, less motile sperm, which may be selected against. Multiflagellate sperm, which are more mobile than biflagellate sperm, may not be so impeded and thus genome size is not expected to be under such tight selection pressure. Indeed, as noted above the largest lycophyte genomes so far reported are indeed in Isoetaceae which have c. 20 flagella per sperm (Renzaglia et al. 1995).

19.2.3 Monilophytes

Monilophytes are estimated to comprise c. 11,000 species and data for 67 species show their genomes range 94-fold from 1C = 0.77 pg in the water fern *Azolla microphylla* to 72.68 pg in the whisk fern *Psilotum nudum* (Table 19.1). Yet an analysis of the distribution of genome sizes (Fig. 19.6a) shows that three of the five groups comprising monilophytes (i.e., horsetails, marattioides and leptosporangiate ferns) are characterized by small to medium sized genomes. The very large genomes are restricted to species in the remaining two groups of ferns – whisk ferns and ophioglossoid ferns. Indeed their enormous genomes are clearly outliers, as they are more than twice the size of the next largest monilophyte genome which is in the horsetail *Equistum variegatum* with 1C= 30.3 pg (Fig. 19.6b).

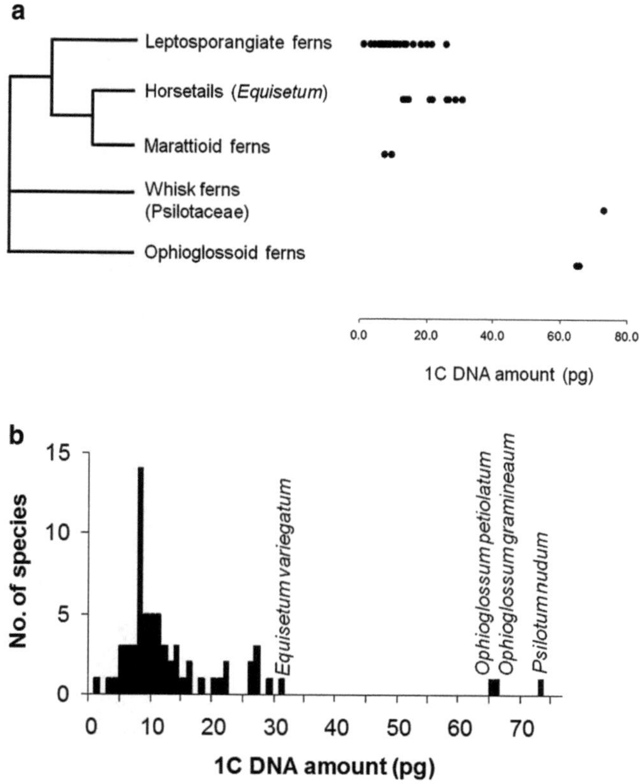

Fig. 19.6 (**a**) Distribution of genome sizes (1C-values) superimposed onto a phylogenetic tree showing relationships between the five major lineages of monilophytes. (**b**) Histogram showing the frequency of genome sizes in 67 monilophyte species

While the monophyly of monilophytes is consistently recovered in phylogenetic studies using molecular data and/or morphological characters from both extant and fossil material (Pryer et al. 2001; Qiu 2008; Schneider et al. 2009), the relationships within the five clades comprising monilophytes are still far from resolved (Schuettpelz and Pryer 2008; Schneider et al. 2009). Yet whisk ferns and ophioglossoid ferns are frequently grouped together and resolved as being distinct from and sister to the remaining monilophytes. Intriguingly, the very large genomes encountered in these two groups appear to have evolved via two different cytological mechanisms. Ophioglossoid ferns such as *Ophioglossum* are characterized by possessing numerous small chromosomes (1.5–4.5 μm in length, Abraham et al. 1962) potentially arising through multiple rounds of polyploidy and/or chromosome fragmentation. Indeed a chromosome count of $2n =$ c. 1,440 which may be 96-ploid has been reported in *O. reticulatum* and is the highest chromosome number ever reported for a plant (Khandelwal 1990). The genome size data available for two species of *Ophioglossum* have counts of $2n =$ c. 720 and c. 960 for *O. gramineum* and *O. petiolatum* respectively. In contrast, the large genomes of the whisk ferns (Psilotaceae – comprising *Psilotum* and *Tmesipteris*) have evolved through increases in chromosome size as cytological data show these genera are characterized by lower numbers (i.e., $2n = 104$, 156, 208, 416) of larger (4.5–18 μm) chromosomes with a maximum ploidy level of $8x$ (Manton 1950; Abraham et al. 1962; Brownsey and Lovis 1987).

At the other end of the scale the smallest monilophyte genome so far reported is in the water fern *Azolla microphylla* which belongs to a derived group of leptosporangiate ferns. Its genome size was estimated to be 1C = 0.77 pg with $2n = 44$ (the lowest number reported for the genus, Stergianou and Fowler 1990; Hanson and Leitch 2002). Currently this is the only genome size estimate for water ferns although *Azolla* is reported to have the smallest chromosomes of any fern (<0.5 μm in *A. pinnata* also with $2n = 44$, Stergianou and Fowler 1990) so very small genomes may well be found in related species. However, bigger genomes are expected in the four other genera comprising water ferns (i.e., *Salvinia, Pilularia, Marsilea* and *Regnellidium*) as cytological studies show they possess larger chromosomes. In *Salvinia* which is considered sister to *Azolla* and represents the other genus of Salvineaceae (Nagalingum et al. 2008) chromosomes were 3–4 μm for species with $2n = 18$ (Tatuno and Takei 1969), while in the three genera comprising Marsileaceae chromosomes varied with 4–8 μm in *Regnellidium* with $2n = 38$ (Loyal 1962), 3–5 μm in *Pilularia* with $2n = 20$ (Large and Braggins 1989) and 1.5–3 μm in *Marsilea* with $2n = 50$ (Lesho 1994).

The best represented group of monilophytes is the horsetails (*Equisitum*) where 10 out of the c. 15 recognized species have genome size estimates. In addition, the phylogenetic relationships and species comprising the two subgenera *Hippochaete* and *Equisetum* are now becoming better resolved (Guillon 2004, 2007) enabling genome size data to be considered within a phylogenetic context. All species (except one reported triploid hybrid) have a chromosome count of $2n = 216$ but genome sizes range from 1C = 12.5–30.3 pg (Fig. 19.7). Nevertheless, when the data are superimposed onto the phylogenetic tree of Guillon (2007) it is clear that they fall into two groups with the species in subgenus *Hippochaete* characterized by larger genomes (1C = 20.7–30.3 pg) than those of subgenus *Equisetum* (1C = 12.5–14.2 pg) (Fig. 19.7). These observations fit with Manton's (1950) cytological observations that species belonging to subgenus *Hippochaete* were characterized by larger chromosomes than those in subgenus *Equisetum*. The recent resolution of the placement of *E. bogotense* within subgenus *Hippochaete* (Guillon 2007) is also consistent with the genome size estimate of this species of 1C = 20.7 pg which is similar to *E. scirpoides*, another early diverging species in the subgenus (Fig. 19.7).

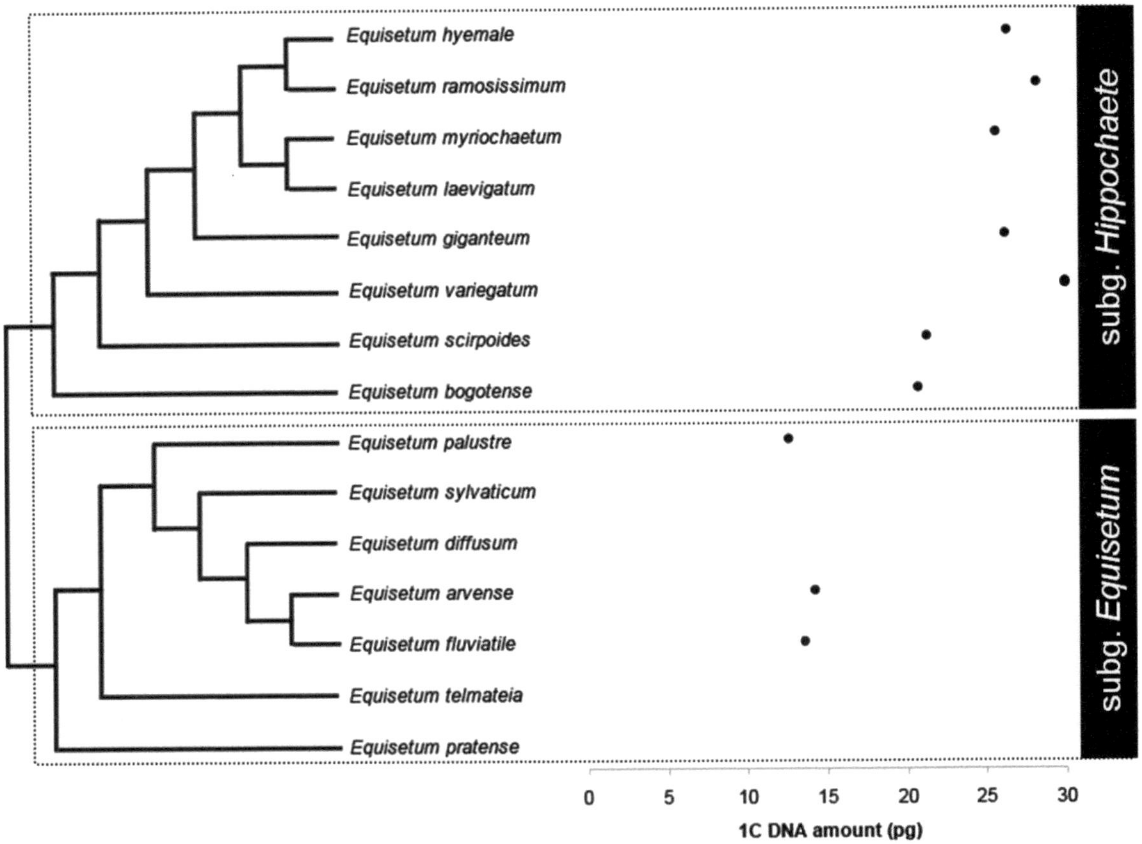

Fig. 19.7 Distribution of genome sizes encountered in *Equisetum* superimposed onto the phylogenetic tree of Guillon (2007)

19.2.4 Seed Plants

Seed plants are estimated to comprise c. 96% of extant vascular plant diversity and are divided into gymnosperms (comprising c. 850 species) and angiosperms with an estimated 352,000 species (Paton et al. 2008).

19.2.4.1 Gymnosperms

Gymnosperms first appeared in the fossil record in the Upper Devonian, c. 350 Mya but while there are arguably 14 gymnosperm groups recognized, only four are represented in extant floras (Bateman et al. 2006). These comprise the cycads (c. 250 species), *Ginkgo* (1 species), Gnetales (c. 50 species) and conifers (c. 550 species) which are typically divided into (1) Pinaceae and (2) the remaining non-Pinaceae conifers. Relationships among the different gymnosperm groups remain enigmatic, especially the position of Gnetales. Molecular data point to affinities with the conifers (e.g., Chaw et al. 2000; Werner et al. 2009), while the analysis of morphological characters provides little support for such a relationship (Palmer et al. 2004) (see also Sect. 14.1 in Murray 2013, this volume and Soltis and Soltis 2013, this volume).

From a genome size perspective, gymnosperms represent the group of land plants with by far the best representation of data. There are C-values for c. 24% of all described species (Table 19.1) including at least one genome size estimate for each of the 13 families recognized. While gymnosperm genomes are, on average, larger than other land plant groups, especially the angiosperms (mean 1C = 18.6 pg compared with angiosperms where mean 1C = 5.9 pg: Table 19.1), gymnosperms remain one of the least variable groups with DNA amounts ranging just 16-fold from 2.25 pg in *Gnetum ula* to 36.0 pg in *Pinus ayacahuite*. In addition, the genome size profile is not strongly skewed towards small DNA amounts, unlike liverworts, mosses, lycophytes and angiosperms (Fig. 19.3).

Cycads, which are often considered to be sister to all other extant gymnosperms, have medium-sized genomes (1C = 12.01–21.1 pg; Fig. 19.8a). While this is based on data for just six species it is consistent with the more extensive chromosomal data showing this group is typically characterized by large chromosomes with a narrow range of numbers ($2n$ = 16–28, e.g., Sax and Beal 1934; Marchant 1968, see also Murray 2013, this volume).

Gnetales comprise three phylogenetically distinct families – Gnetaceae, Ephedraceae and Welwitschiaceae, all of which are monogeneric. They have genomes ranging 8-fold from 1C = 2.25–18.22 pg (Fig. 19.8a). The smallest

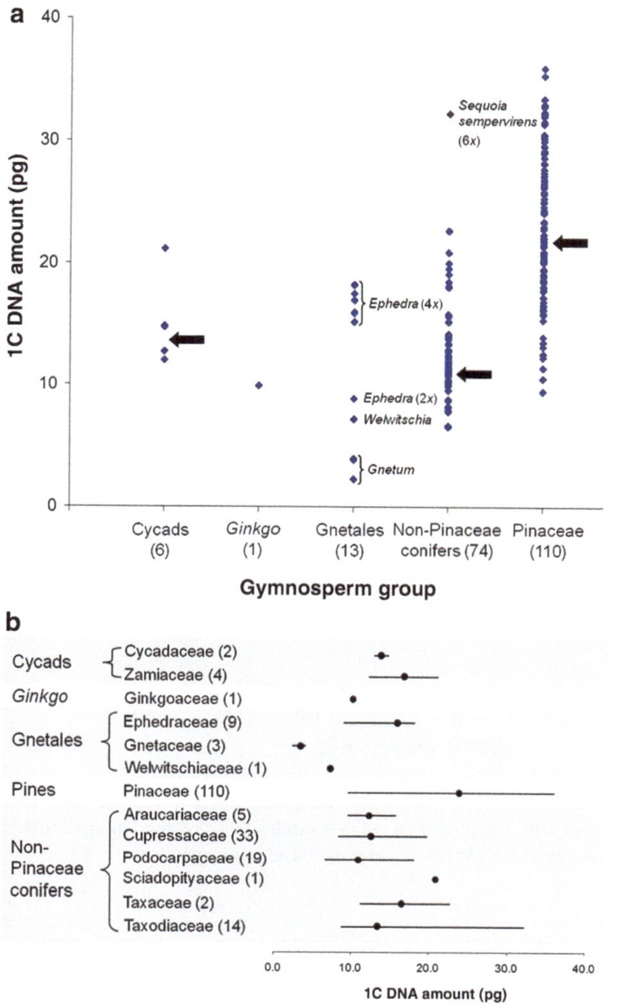

Fig. 19.8 (**a**) Distribution of genome sizes within the four groups of gymnosperms. *Numbers in brackets* correspond to the number of species in each group with genome size data. *Arrows* point to the mean 1C-value for each group (where applicable). (**b**) Comparison of the mean (•) and range of genome sizes in each of the 13 families comprising gymnosperms. Families are arranged into their higher order groups but within each group there is no phylogenetic order. The number in brackets following the family name corresponds to the number of species with genome size data

genomes of any gymnosperm are found in *Gnetum* (Gnetaceae) with 1C = 2.25–3.98 pg and chromosome numbers of $2n = 22$ or 44, representing diploid and tetraploid species. Welwitschiaceae comprises a single species, *Welwitschia mirabilis,* and has an intermediate-sized genome of 1C = 7.2 pg and $2n = 42$. Most variable in Gnetales is *Ephedra* (Ephedraceae) with genome sizes ranging from 1C = 8.8–18.22 pg. This reflects the occurrence of both diploid ($2n = 14$) and tetraploid ($2n = 28$) species with genome size data (Fig. 19.8a). Indeed, although polyploidy is very rare in gymnosperms as a whole, *Ephedra* is distinctive as 44% of species are reported to be polyploid (Delevoryas 1980). (N.B. one hexaploid (*E. distachya* subsp. *distachya*) and two octoploids (*E. gerardiana* and *E. funerea*) have been reported by Ickert-Bond (2003) but no genome size data are currently available for these cytotypes.)

The greatest diversity of genome sizes are found in the conifers with values ranging from 1C = 6.6 to 36.0 pg (Fig. 19.8a and b). However, most of this variation is found in *Pinus* (Pinaceae, 1C = 16.6–36.0 pg) which is the most variable genus of any gymnosperm despite all species having the same chromosome number of $2n = 24$ (see Murray 2013, this volume). The only other notably large genome is that of *Sequoia sempervirens* with 1C = 32.1 pg. Its genome is clearly an outlier compared with the genome sizes encountered in other species of the non-Pinaceae group of conifers (Fig. 19.8a). The large genome of *S. sempervirens* is in part due to polyploidy as cytologically it is a hexaploid with $2n = 6x = 66$, with the highest chromosome number reported to date for any gymnosperm.

19.2.4.2 Angiosperms

Given the astonishing ecological, species and genomic diversity of angiosperms, insights into their genome sizes has long attracted the attention of scientists. Indeed, the first plant to have its genome size measured was an angiosperm *Lilium longiflorum* in 1951 (Ogur et al. 1951). Since then many species have had their genome sizes estimated and there are currently values for 6,287 species in the Plant DNA C-values database (Bennett and Leitch 2010). As already mentioned in Sect. 19.2, the most striking feature of angiosperms is the huge diversity of genome sizes encountered (Figs. 19.2 and 19.3, Table 19.1), considerably greater than any other group of land plants and extending nearly 2,400-fold. In addition, the genome size profile is quite distinct from that of other land plant groups (Fig. 19.3), with most species characterized by having small to very small genomes. Indeed, the modal and median 1C-values for 6,287 species are just 0.6 and 2.5 pg respectively. Species with large to very large genomes are comparatively few and localized in just a few lineages.

Over the years, insights into how this genome size diversity is spread across the phylogenetic tree of angiosperms have been gleaned by superimposing existing data onto phylogenetic trees available at the time of analysis (e.g., Leitch et al. 1998; Soltis et al. 2003; Leitch et al. 2005). The most recent of these was based on data for 4,119 species using the phylogenetic tree published by the Angiosperm Phylogeny Group (APG) in 2003 (APG II 2003). Given a 52% increase in the amount of genome size data available since then, together with an increasingly well resolved phylogenetic tree of the angiosperms (Soltis et al. 2008; APG III 2009), the data have been updated and reanalysed here (Fig. 19.9).

The overall picture, which supports previous studies, shows that all angiosperm groups contain species with small to very small genomes, the extremely large genomes are phylogenetically restricted to some species belonging to

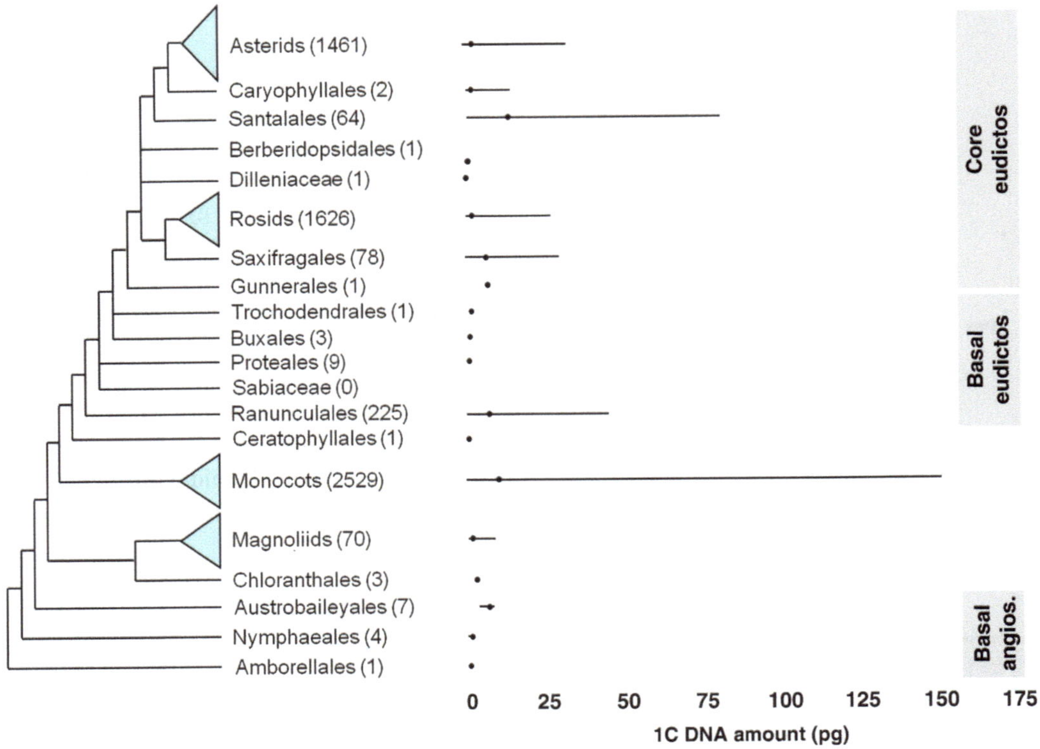

Fig. 19.9 Distribution of genome size across angiosperms superimposed on the phylogenetic tree given in APG III (2009). C-value data for each group are shown as a *line* connecting the minimum and maximum 1C-value with the mean shown as •. The *number in brackets* following the group name gives the number of species with C-value data. Major subdivisions of the angiosperms are shown on the *right* (*Basal angios.* basal angiosperms)

just two groups – the Santalales (within eudicots) and monocots (Fig. 19.9). A closer look shows that even within these groups, species with very large genomes occupy derived positions within the phylogenetic tree (Figs. 19.10 and 19.11).

In Santalales a histogram (Fig. 19.10a) shows that most are characterised by small to medium genomes, the four very large 1C-values (1C = 62.3–82.0 pg) all belong to *Viscum* species and are clearly outliers. Nevertheless, as data representation improves, further similarly large genomes may indeed be found in other genera of Viscaceae as cytological studies have shown that related genera *Notothixos*, *Dendrophthora* and *Phoradendron* contain species with similar numbers of extremely large chromosomes (e.g., Fig. 19.10b, Wiens 1964; Wiens and Barlow 1971). So far the species of *Viscum* which have had their genome sizes estimated have either been diploid with $2n = 20$ (*V. album*), 24 (*V. crassulae*) or 28 (*V. minimum*) or assumed to be based on a similar genome size to the related diploids (i.e., *V. cruciatum*). Although polyploidy is rare in this genus and in the familiy Viscaceae as a whole, a few polyploids of *Viscum* and *Phoradendron* have been reported (Wiens and Barlow 1971) suggesting that considerably larger genomes could be found in this clade of eudicots.

Recent improvements in the understanding of phylogenetic relationships within Santalales by Nickrent et al. (2010) shows that Viscaceae is deeply embedded within the Santalales tree suggesting that the large genomes are derived rather than plesiomorphic for the order. Indeed other genera within Viscaceae are characterised by possessing smaller chromosomes (e.g., *Arceuthobium*), although these are still considerably larger than is typical of many angiosperms (Wiens 1968). An independent increase in genome size may also have occurred in Loranthaceae, another family in Santalales. Although the available genome size for 57 species in this family are all small to medium sized (1C = 1.6–17.5 pg), once again chromosome data suggest the presence of much larger genomes (Wiens 1964). Thus while the early diverging genera of Loranthaceae, *Nuytsia*, *Atkinsonia* and *Gaiadendron*, have relatively small chromosomes and genomes (i.e., *Nuytsia* 1C = 2.9 pg) (Barlow and Wiens 1971; Martin 1983), very large chromosomes have been reported in *Aetanthus*, *Struthanthus* and *Psittacanthus* (Fig. 19.10b, Wiens 1964) all of which belong to tribe Psittacanthinae (*sensu* Nickrent et al. 2010). Indeed, as early as 1946 Covas and Schnack (1946) suggested that *Psittacanthus* possessed the largest chromosomes in the eudicots. As the chromosome numbers of these genera are $2n = 16$ it remains to be seen whether their genomes are larger

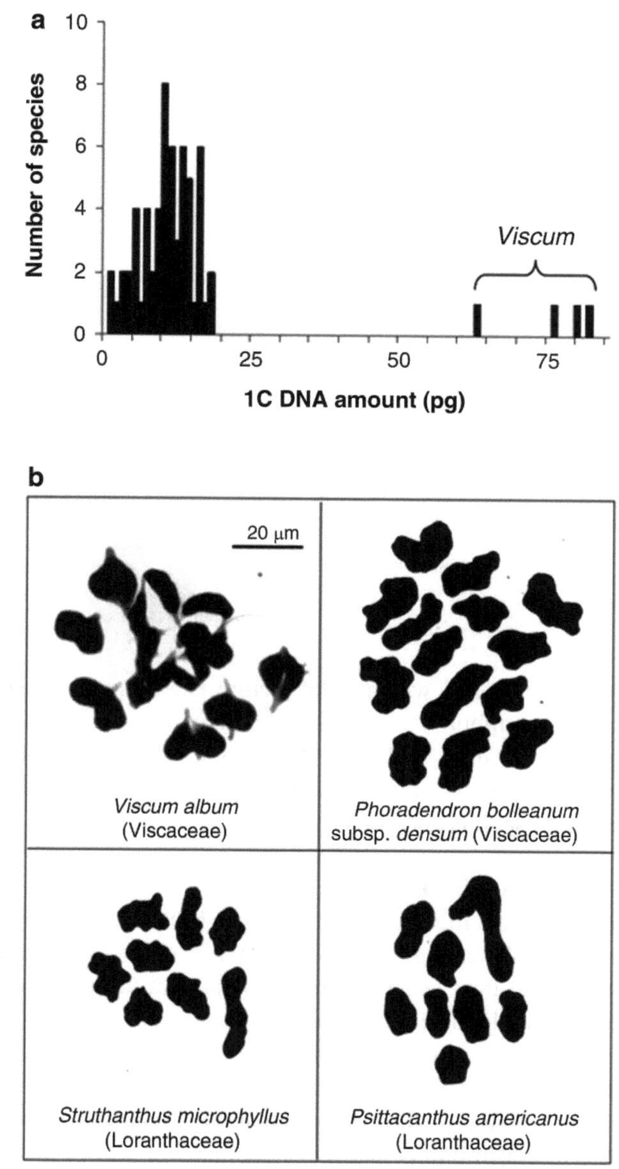

Fig. 19.10 (**a**) Histogram showing the distribution of genome sizes for 64 species of Santalales. (**b**) Meiotic chromosomes of four species in Santalales reported to have large chromosomes, all reproduced at the same magnification (Images for *Phoradendron, Struthanthus* and *Psittacanthus* taken from Wiens 1964 and reproduced with permission)

than those in *Viscum* or indeed *Paris japonica,* the latter with the largest known eukaryote genome (Pellicer et al. 2010).

This pattern is repeated within monocots (Fig. 19.11). Here genome size profiles of all the major groups show most monocots are characterized by small genomes (Leitch et al. 2010). Species with very large genomes are restricted to just a few genera within Asparagales and Liliales, which are well embedded within the monocot tree (Fig. 19.11). The most unusual distribution of genome sizes is found within Liliales where an analysis of 145 species from seven families showed a 100-fold range in genome sizes (1C = 1.53–152.23 pg), and no clear skew in the data towards species with small genomes as observed in most other angiosperm groups (Figs. 19.9 and 19.11). The truly enormous genomes greater than 1C = 100 pg are only found in three genera belonging to Liliales: *Paris* and *Trillium* (Melanthiaceae) and *Fritillaria* (Liliaceae) (Pellicer et al. 2010; Zonneveld 2010).

19.2.5 Mechanisms Responsible for Generating Changes in Genome Size in Different Land Plant Groups

19.2.5.1 Angiosperms

Most of our understanding of the mechanisms responsible for changes in genome size has come from studies of angiosperms (for recent reviews see Kejnovsky et al. 2009; Grover and Wendel 2010). It is recognised that increases in genome size arise predominantly through polyploidy and amplification of non-coding repetitive DNA, especially retrotransposons. These mechanisms are counterbalanced by processes that result in a decrease in genome size. These are less well understood, but have been shown to involve recombination-based processes for example unequal recombination and illegitimate recombination (reviewed by Grover and Wendel 2010).

While polyploidy is widely considered to play a role in generating increases in genome size, its long-term impact on genome size evolution is less clear and may be subject to different selection pressures in the different land plant groups. In angiosperms, while genome size increases immediately following a polyploid event, an extensive survey of genome size data for 3,021 species showed that this is often accompanied over time by 'genome downsizing' (Leitch and Bennett 2004). This is supported by more focused comparative genomic studies which have shown how extensive loss of DNA can take place in specific species. For example, in maize and rice it is estimated that over 50% of the genes have been lost since the last round of polyploidy in their ancestries (e.g., Messing et al. 2004; Wang et al. 2005). These observations suggest that multiple rounds of polyploidy do not inevitably lead to larger genomes (Leitch and Bennett 2004), indeed many angiosperms with very big genomes are not cytological polyploids (e.g., *Fritillaria koidzumiana* $2n = 2x = 24$, 1C = 77.5 pg).

19.2.5.2 Gymnosperms

The mechanisms involved in generating the large genomes typical of many but not all gymnosperms have not been investigated as extensively as angiosperms but preliminary data suggest there are differences. Overall polyploidy, as identified through chromosome numbers is rare in gymnosperms, with only 5% of species estimated to be polyploid, nearly all of which are in *Ephedra* (Delevoryas 1980). Clearly, ongoing polyploidy in extant gymnosperms is not contributing to the large genomes typical of this group of seed plants. Nevertheless, emerging from the increasing

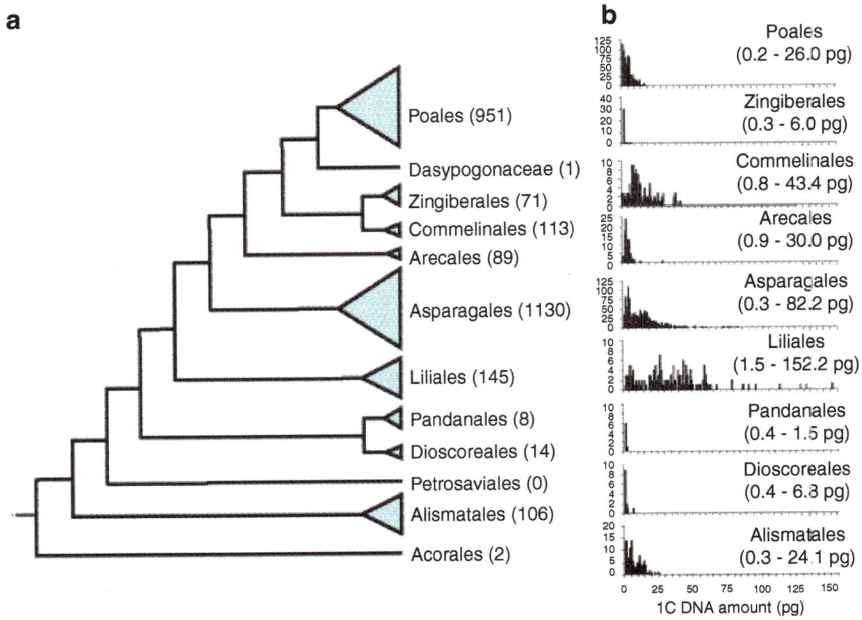

Fig. 19.11 The distribution of genome sizes for 2,530 species of monocots. (**a**) The summary topology of monocots based on Chase et al. (2006) with the *number in brackets* corresponding to the number of species with genome size data. (**b**) Histograms showing the distribution of genome sizes within each order with the range of 1C-values in brackets. (Figure modified from Leitch et al. 2010 and reproduced with permission)

amounts of genomic and transcriptomic sequence data is tentative evidence that at least some gymnosperm groups have undergone whole genome duplications in their evolutionary history (Cui et al. 2006; Barker et al. 2010).

For conifers, retrotransposon amplification is clearly an important contributor to genome expansion. This is based on the increasing amount of data showing that most of their DNA has originated through the amplification of a large diversity of retrotransposons, present in a huge range of copy numbers (Elsik and Williams 2000; Morse et al. 2009). In addition, recent data suggest that many of these retrotransposons are still actively transcribing (Morse et al. 2009; Parchman et al. 2010). Perhaps conifers are less efficient at suppressing retrotransposon amplification than angiosperms (see review by Leitch and Leitch 2012).

Mechanisms of DNA elimination are poorly understood in gymnosperms. A recent study of one retroelement in *Pinus* (Ty-*gypsy* element *Gymny*) suggested there was little recombination between the long terminal repeats (LTRs) of this element (Morse et al. 2009). As this type of recombination has been shown to contribute to DNA loss in angiosperms (Vitte and Panaud 2003; Vitte and Bennetzen 2006), the data may indicate that this mode of genome downsizing is not contributing extensively to eliminating amplified retrotransposons in gymnosperms and may further explain why their genomes are, on average larger than those of angiosperms.

Sadly there are currently little data available for non-conifer gymnosperms especially *Gnetum* which has the smallest genomes (Fig. 19.8) and *Ephedra* where polyploidy is more common. Indeed, investigations into the genomic composition of these non-conifer gymnosperms are essential to provide broader insights into the genomic diversity of gymnosperms and whether the picture emerging from the study of conifers is representative for gymnosperms as a whole. Such data could also shed light on the currently unresolved relationship between Gnetales and the rest of the gymnosperms and angiosperms.

19.2.5.3 Monilophytes

The limited but increasing genomic data becoming available for monilophytes suggest that the genomic organization and evolution of their genomes are different from those of seed plants. Monilophytes, particularly the homosporous ferns, are typically characterized by possessing numerous small chromosomes which are strongly bivalent pairing and undergo relatively little recombination (Nakazato et al., 2008) (see also Barker 2013, this volume)

Given the reported high chromosome numbers in monilophytes, it has been suggested that polyploidy is the predominant mechanism involved in genome size expansion in this group. However, recent studies suggest that polyploidy may not be as prevalent as once thought while in depth genomic studies of the model fern *Ceratopteris richardii* have shown that the evolution of the monilophyte polyploid nucleus may follow a very different trajectory from that in angiosperms.

There is now increasing evidence that the largest clade of homosporous ferns (the polypod ferns) underwent a single whole genome duplication event c. 180 Mya ago (Barker and Wolf 2010). Yet analysis of the genomic structure of *C. richardii* suggests this was not accompanied by extensive chromosome loss or elimination of DNA, as reported in angiosperms. Indeed, linkage map analysis of *C. richardii* shows that its genome has one of the highest proportions of multiple copy genes among plants (Nakazato et al. 2006).

Instead, gene silencing appears to have been the more important diploidization processes operating (Pichersky et al. 1990; Barker and Wolf 2010). These and related studies have led to the suggestion that ferns have a lower rate of whole genome duplication and experience chromosomal and DNA loss at a much slower rate than angiosperms (Barker and Wolf 2010). Together these factors may contribute to explaining the less skewed distribution of genome sizes observed in ferns compared with angiosperms (Fig. 19.3).

Very little is known about the contribution of repetitive DNA especially retrotransposons to genome size diversity in ferns and this seems an area ripe for further study. Limited data have shown that ferns contain the major classes of retrotransposons (Flavell et al. 1992; Voytas et al. 1992) but their contribution to genome size remains unquantified. Brandes et al. (1997) noted that the abundance of Ty-*copia* elements in the leptosporangiate fern *Pteris cretica* was lower than in other land plant groups studied, whereas in the whisk fern *Psilotum triquetrum* not only were high copy numbers estimated to be present but the sequences were also shown to be extremely heterogeneous compared with the 56 other species studies (which included representatives from all the land plant groups, Flavell et al. 1992). Clearly further studies are needed to clarify the nature and contribution of these elements to genome size diversity encountered in monilophytes.

19.2.5.4 Bryophytes

Given the limited genomic data available for hornworts and liverworts the mechanisms responsible for generating the narrow genome sizes observed remain enigmatic. Nevertheless, it is clear that ongoing polyploidy is not a major evolutionary process in these lineages given its rarity in liverworts and particularly hornworts. As to whether paleopolyploidy events played a role in shaping their genomes, this is currently unknown given the lack of genomic sequence data available for these lineages. Nevetheless, insights may soon be gleaned at least for liverworts given that the genome of *Marchantia polymorpha* is currently being sequenced.

In contrast, polyploidy is reported to be common in mosses and the release of the first complete genome sequence of a moss *Physcomitrella patens* (1C = 500 Mb, Rensing et al. 2008) provides at least some molecular insights into its genomic make up and possible factors contributing to the narrow range of small genome sizes encountered in mosses. These studies have shown that the *P. patens* genome is not only a neopolyploid (possibly a triploid with $n = 27$) but also has undergone at least one whole genome duplication c. 30–60 Mya (see also Rensing et al. 2013, this volume). Clearly then polyploidy is an important factor contributing to the genomic make up and genome size of this species. Amplification of retrotransposons has also clearly been occurring given that 48% of its genome is made up of retrotransposons, one of the highest percentages noted for any completely sequenced plant genome to date (Lang et al. 2008). Indeed, evidence of three peaks in retrotransposon activity over the last 3 million years is apparent.

Given the prevalence of polyploidy and retrotransposons, the narrow range of genome sizes in mosses is puzzling and suggests that mechanisms leading to DNA elimination are operating efficiently in moss genomes. In support of this is the nearly complete absence of helitrons in the genome of *P. patens*. These rolling-circle transposons are an ancient class of transposable elements and have been found in all eukaryotic genomes sequenced to date. The presence of just one family of highly conserved helitrons in *P. patens* suggests that this helitron family has been active in the last 3 million years and that the lack of other helitrons is due to their rapid and efficient removal from the genome (Rensing et al. 2008). In addition, the low abundance of complete full length retrotransposons suggests that DNA elimination via recombination is taking place, while the presence of an extensive collection of siRNAs which map to retrotransposons suggests that for most of the time retrotransposons are kept in check by post-transcriptional silencing. Analysis also suggests that local gene duplications do not contribute significantly to the *P. patens* genome as only 1% of protein coding genes were shown to occur in tandem array. This figure is much lower than in angiosperm genomes which have been completely sequenced such as 14% in *Oryza sativa* (International Rice Genome Sequencing Project 2005) and 11% in *Populus trichocarpa* (Tuskan et al. 2006).

Conclusions

Outside of the angiosperms, and to a lesser extent Pinaceae in the gymnosperms, we clearly know very little of the mechanisms shaping genome structure and organization across land plants. Even within these groups we have patchy knowledge. For example, amongst angiosperms we have most in depth understanding of just a few key model or crop species (especially Poaceae and Brassicaceae). However, the limited available data suggest that the mechanisms that shape genomes in land plants (i.e., polyploidy, retroelement mobility and amplification, tandem repeat expansion and contraction, unequal homologous and illegitimate recombination) are likely to occur in all groups, but their relative activities and associated selection pressures may vary significantly.

The most notable feature about genome size diversity in land plants is the frequency of species with small genomes. It is the exception rather than the rule to find truly enormous genomes and these are phylogenetically restricted to just a few angiosperms and ophioglossoid monilophytes.

The small size of most plant genomes suggests an overall strong selection pressure to limit their size. As to the nature of the selection pressures, these may well be different for each land plant group. For example, it has been suggested that there may have been selection for small genomes in the early diverging angiosperms as it enabled them to compete more efficiently with the larger-genomed gymnosperms because of their more rapid development: small genomes correlate with several developmental phenotypic characters such as short minimum generation times and rapid establishment of seedlings (Leitch et al. 2005; see also Greilhuber and Leitch 2013, this volume). In contrast, as noted above (Sect. 19.2.2), selection for a small genome in biflagellate bryophytes and lycophytes may explain the low DNA amounts encountered in these species.

One factor which could be relevant to all land plant groups is the limitation in genome size imposed by the metabolic costs of assimilating and building nucleic acids. Such a constraint on genome size is likely to be particularly acute for primary producers growing in soils limited in the building blocks of DNA (i.e., phosphorous and nitrogen, Hanson et al. 2001; Leitch and Leitch 2008). It is perhaps relevant that gymnosperms which have a larger modal and mean genome size than other land plant groups (Table 19.1) are found in parts of the world that are limited by water rather than nutrients. Typically gymnosperms are specialist of high montane, arctic and sandy soils which are often considered to be relatively nutrient-rich. It is also perhaps notable that the smallest gymnosperm genomes belong to species of *Gnetum,* the only genus of gymnosperm to grow in the relatively nutrient-poor soils of the tropics.

As for the giant genomes of angiosperms and ophioglossoid ferns, these need not necessarily be a consequence of polyploidy, instead a major contributing factor is likely to be the activity of retrotransposons and other non-coding repetitive DNA sequences. It is unknown whether the reason for the massive abundance of repeats is due to a greater activity of the elements or reduced activity of the genome surveillance mechanisms which suppress their activity. Alternatively, it may simply be different selection pressures acting on similar levels of raw variation. With the rapid advances being made in the speed at which sequence data is generated through second and third generation sequencing technologies it is hoped that a phylogenetically diverse range of species will be studied to provide far reaching insights into the genomic basis behind the immense diversity of genome sizes encountered in land plants.

References

Abderrahman SA (2004) Nuclear DNA content of haploid and diploid *Physcomitrium pyriforme* using DAPI staining. Korean J Genet 26:245–250

Abraham A, Ninan CA, Mathew PM (1962) Studies on the cytology and phylogeny of the pteridophytes VII. Observations on one hundred species of South Indian ferns. J Indian Bot Soc 41:339–421

APG II (2003) An update of the Angiosperm Phylogeny Group classification for the orders and families of flowering plants. Bot J Linn Soc 141:399–436

APG III (2009) An update of the Angiosperm Phylogeny Group classification for the orders and families of flowering plants: APG III. Bot J Linn Soc 161:105–121

Barker MS (2013) Karyotype and genome evolution in pteridophytes. In: Leitch IJ, Greilhuber J, Doležel J, Wendel JF (eds) Plant genome diversity, vol 2, Physical structure, behaviour and evolution of plant genomes. Wien, pp 245–253

Barker MS, Wolf PG (2010) Unfurling fern biology in the genomics age. Bioscience 60:177–185

Barker MS, Graham SW, Rieseberg LH (2010) Comparative gymnosperm transcriptomics. Abstract for Botany 2010, http://2010.botany conference.org/engine/search/index.php?func=detail&aid=437. Accessed on 9 July 2012

Barlow BA, Wiens D (1971) The cytogeography of the loranthaceous mistletoes. Taxon 20:291–312

Bateman RM, Hilton J, Rudall PJ (2006) Morphological and molecular phylogenetic context of the angiosperms: contrasting the 'top-down' and 'bottom-up' approaches used to infer the likely characteristics of the first flowers. J Exp Bot 57:3471–3503

Bennett MD, Leitch IJ (2010) Plant DNA C-values database (release 5.0, Dec 2010). http://data.kew.org/cvalues/. Accessed on 9 July 2012

Bennett MD, Leitch IJ (2011) Nuclear DNA amounts in angiosperms – targets, trends and tomorrow. Ann Bot 107:467–590

Blackwood M (1953) Chromosomes of *Phylloglossum drummondii.* Nature 172:591–592

Bornefeld T, Grillenberger C (1987) Some aspects of size in chromosome analysis of liverworts (Marchantiales). Nova Hedwigia 44:91–100

Brandes A, Heslop-Harrison JS, Kamm A, Kubis S, Doudrick RL, Schmidt T (1997) Comparative analysis of the chromosomal and genomic organization of Ty1-copia-like retrotransposons in pteridophytes, gymnosperms and angiosperms. Plant Mol Biol 33:11–21

Brownsey PJ, Lovis JD (1987) Chromosome numbers of the New Zealand species of *Psilotum* and *Tmesipteris*, and the phylogenetic relationships of the Psilotales. New Zealand J Bot 25:439–454

Chase MW, Fay MF, Devey DS, Maurin O, Rønsted N et al (2006) Multigene analyses of monocot relationships: a summary. Aliso 22:63–75

Chaw SM, Parkinson CL, Cheng YC, Vincent TM, Palmer JD (2000) Seed plant phylogeny inferred from all three plant genomes: monophyly of extant gymnosperms and origin of Gnetales from conifers. Proc Natl Acad Sci USA 97:4086–4091

Covas G, Schnack B (1946) Numero de chromosomas en antofitas de la region de Cuya (Republica Argentina). Revista Argent Agron 13:152–166

Crandall-Stotler B, Stotler RE, Long DG (2009) Morphology and classification of the Marchantiophyta. In: Goffinet B, Shaw AJ (eds) Bryophyte biology, 2nd edn. Cambridge University Press, Cambridge, pp 1–54

Cui L, Wall PK, Leebens-Mack JH, Lindsay BG, Soltis DE et al. (2006) Widespread genome duplications throughout the history of flowering plants. Genome Res 16:738–749
Delevoryas T (1980) Polyploidy in gymnosperms. In: Lewis WH (ed) Polyploidy. Plenum Press, New York, pp 215–218
Dunlop DW (1949) Notes on the cytology of some lycopsids. Bull Torrey Bot Club 76:266–277
Elsik CG, Williams CG (2000) Retroelements contribute to the excess low-copy-number DNA in pine. Mol Gen Genet 264:47–55
Flavell AJ, Dunbar E, Anderson R, Pearce SR, Hartley R, Kumar A (1992) Ty1-copia group retrotransposons are ubiquitous and heterogeneous in higher plants. Nucl Acid Res 20:3639–3644
Gao L, Su YJ, Wang T (2010) Plastid genome sequencing, comparative genomics, and phylogenomics: current status and prospects. J Syst Evol 48:77–93
Greilhuber J, Leitch IJ (2013) Genome size and the phenotype. In: Leitch IJ, Greilhuber J, Doležel J, Wendel JF (eds) Plant genome diversity, vol 2, Physical structure, behaviour and evolution of plant genomes. Springer-Verlag, Wien, pp 323–344
Greilhuber J, Sastad SM, Flatberg KI (2003) Ploidy determination in *Sphagnum* samples from Svalbard, Arctic Norway, by DNA image cytometry. J Bryol 25:235–239
Greilhuber J, Doležel J, Lysák MA, Bennett MD (2005) The origin, evolution and proposed stabilization of the terms 'genome size' and 'C-value' to describe nuclear DNA contents. Ann Bot 95:255–260
Greilhuber J, Borsch T, Müller K, Worberg A, Porembski S, Barthlott W (2006) Smallest angiosperm genomes found in Lentibulariaceae with chromosomes of bacterial size. Plant Biol 8:770–777
Grover C, Wendel JF (2010) Recent insights into mechanisms of genome size change in plants. J Bot Article ID 382732, 8 pp. doi:10.1115/2010/382732
Guillon JM (2004) Phylogeny of horsetails (*Equisetum*) based on the chloroplast rps4 gene and adjacent noncoding sequences. Syst Bot 29:251–259
Guillon JM (2007) Molecular phylogeny of horsetails (*Equisetum*) including chloroplast atpB sequences. J Plant Res 120:569–574
Hanson L, Leitch IJ (2002) DNA amounts for five pteridophyte species fill phylogenetic gaps in C-value data. Bot J Linn Soc 140:169–173
Hanson L, McMahon KA, Johnson MAT, Bennett MD (2001) First nuclear DNA C-values for 25 angiosperm families. Ann Bot 87:251–258
Ickert-Bond SM (2003) Systematics of new world Ephedra L. (Ephedraceae): integrating morphological and molecular data. Ph. D. thesis, Arizona State University
International Rice Genome Sequencing Project (2005) The map-based sequence of the rice genome. Nature 436:793–800
Jermy AC (1967) Cytomorphological variation in *Selaginella*. Bot J Linn Soc 60:147–158
Kejnovsky E, Leitch IJ, Leitch AR (2009) Contrasting evolutionary dynamics between angiosperm and mammalian genomes. Trends Ecol Evol 24:572–582
Kenrick P, Crane PR (1997) The origin and early evolution of plants on land. Nature 389:33–39
Khandelwal S (1990) Chromosome evolution in the genus *Ophioglossum* L. Bot J Linn Soc 102:205–217
Kuta E, Przywara L (2000) Cytotaxonomy of bryophytes in the Department of Plant Cytology and Embryology, Institute of Botany, Jagiellonian University, Kracǒw, Poland. IOPB Newsl 32:3–8
Lang D, Zimmer AD, Rensing SA, Reski R (2008) Exploring plant biodiversity: the *Physcomitrella* genome and beyond. Trends Plant Sci 13:542–549
Large MF, Braggins JE (1989) An assessment of characters of taxonomic significance in the genus *Pilularia* (Marsileaceae): with particular reference to *P. americana*, *P. novae-follandiae*, and *P. novae-zelandiae*. New Zealand J Bot 27:481–486
Leitch IJ, Bennett MD (2004) Genome downsizing in polyploid plants. Biol J Linn Soc 82:651–663
Leitch AR, Leitch IJ (2008) Genomic plasticity and the diversity of polyploid plants. Science 320:481–483
Leitch AR, Leitch IJ. (2012) Ecolgoical and genetic factors cause contrasting genome dynamics in seed plants. New Phytol 194:629–646
Leitch IJ, Chase MW, Bennett MD (1998) Phylogenetic analysis of DNA C-values provides evidence for a small ancestral genome size in flowering plants. Ann Bot 82(Suppl A):85–94
Leitch IJ, Soltis DE, Soltis PS, Bennett MD (2005) Evolution of DNA amounts across land plants (Embryophyta). Ann Bot 95:207–217
Leitch IJ, Beaulieu JM, Chase MW, Leitch AR, Fay MF (2010) Genome size dynamics and evolution in monocots. J Bot Article ID 862516, 18 pp. doi:10.1155/2010/862516
Lesho CL (1994) A summary of chromosome numbers in the Marsileaceae, with counts for additional species of Marsilea. Am Fern J 84:121–125
Loyal DS (1962) Abnormal sporelings of *Regnellidium diphyllum* Lindm. and chromosome number of the genus. Res Bull (NS) Panjab Univ 13:25–30
Manton I (1950) Problems of cytology and evolution in the Pteridophyta. Cambridge University Press, Cambridge
Marchant CJ (1968) Chromosome patterns and nuclear phenomena in the cycad families Stangeriaceae and Zamiaceae. Chromosoma 24:100–134
Martin NJ (1983) Nuclear DNA variation in the Australasian Loranthaceae. In: Calder M, Berhnhardt P (eds) Biology of mistletoes. Academic, Sydney, pp 277–293
Messing J, Bharti AK, Karlowski WM, Gundlach H, Kim HR et al (2004) Sequence composition and genome organization of maize. Proc Natl Acad Sci USA 101:14349–14354
Mishler BD, Kelch DG (2009) Phylogenomics and early land plant evolution. In: Goffinet B, Shaw AJ (eds) Bryophyte biology, 2nd edn. Cambridge University Press, Cambridge, pp 175–197
Morse AM, Peterson DG, Islam-Faridi MN, Smith KE, Magbanua Z et al (2009) Evolution of genome size and complexity in *Pinus*. PLoS One 4:e4332
Murray B (2013) Karyotype variation and evolution in gymnosperms. In: Leitch IJ, Greilhuber J, Doležel J, Wendel JF (eds) Plant genome diversity, vol 2, Physical structure, behaviour and evolution of plant genomes. Springer-Verlag, Wien, pp 231–243
Nagalingum NS, Nowak MD, Pryer KM (2008) Assessing phylogenetic relationships in extant heterosporous ferns (Salviniales), with a focus on *Pilularia* and *Salvinia*. Bot J Linn Soc 157:673–685
Nakazato T, Jung MK, Housworth EA, Rieseberg LH, Gastony GJ (2006) Genetic map-based analysis of genome structure in the homosporous fern *Ceratopteris richardii*. Genetics 173:1585–1597
Nakazato T, Barker MS, Rieseberg LH, Gastony GL (2008) Evolution of the nuclear genome of ferns and lycophytes. In: Ranker TA, Haufler CH (eds) Biology and evolution of ferns and lycophytes. Cambridge University Press, Cambridge, UK, pp 175–198
Newton ME (1985) Heterochromatin diversity in two species of *Pellia* (Hepaticae) as revealed by C-, Q-, N- and Hoechst 33258-banding. Chromosoma 92:378–386
Newton ME (1987) Developmental aspects of chromatin condensation in liverworts. Bryologist 90:376–382
Nickrent DL, Malécot V, Vidal-Russell R, Der JP (2010) A revised classification of Santalales. Taxon 59:538–558

Ogur M, Erickson RO, Rosen GU, Sax KB, Holden C (1951) Nucleic acids in relation to cell division in *Lilium longiflorum*. Exp Cell Res 2:73–89

Orzechowska M, Siwinska D, Maluszynska J (2010) Molecular cytogenetic analyses of haploid and allopolyploid *Pellia* species. J Bryol 32:113–121

Palmer JD, Soltis DE, Chase MW (2004) The plant tree of life: an overview and some points of view. Am J Bot 91:1437–1445

Parchman T, Geist K, Grahnen J, Benkman C, Buerkle CA (2010) Transcriptome sequencing in an ecologically important tree species: assembly, annotation, and marker discovery. BMC Genomics 11:180

Paton AJ, Brummitt N, Govaerts R, Harman K, Hinchcliffe S, Allkin B, Lughadha EN (2008) Towards target 1 of the global strategy for plant conservation: a working list of all known plant species – progress and prospects. Taxon 57:602–611

Pellicer J, Fay MF, Leitch IJ (2010) The largest eukaryotic genome of them all? Bot J Linn Soc 164:10–15

Pena MJ, Darvill AG, Eberhard S, York WS, O'Neill MA (2008) Moss and liverwort xyloglucans contain galacturonic acid and are structurally distinct from the xyloglucans synthesized by hornworts and vascular plants. Glycobiology 18:891–904

Peruzzi L, Cesca G, Puntillo D (2003) *Isoetes* (Isoetaceae), *Ophioglossum* and *Botrychium* (Ophioglossaceae) in Calabria (S Italy): more karyological and taxonomical data. Caryologia 56:359–363

Pichersky E, Soltis D, Soltis P (1990) Defective chlorophyll a/b-binding protein genes in the genome of a homosporous fern. Proc Natl Acad Sci USA 87:195–199

Proskauer J (1950) Notes on Hepaticae. I. Bryologist 53:166–172

Proskauer J (1958) Studies on the Anthocerotales. V. Phytomorphology 7:113–135

Proskauer J (1967) The variable ploidy life history of a monoecious self-fertilizing hornwort (Anthercerotales). Am J Bot 54:641–642

Pryer KM, Schneider H, Smith AR, Cranfill R, Wolf PG, Hunt JS, Sipes SD (2001) Horsetails and ferns are a monophyletic group and the closest living relatives to seed plants. Nature 409:618–622

Przywara L, Kuta E (1995) Karyology of bryophytes. Pol Bot Stud 9:1–83

Qiu YL (2008) Phylogeny and evolution of charophytic algae and land plants. J Syst Evol 46:287–306

Qiu YL, Li LB, Wang B, Chen ZD, Knoop V et al (2006) The deepest divergences in land plants inferred from phylogenomic evidence. Proc Natl Acad Sci USA 103:15511–15516

Rensing SA, Lang D, Zimmer AD, Terry A, Salamov A et al (2008) The *Physcomitrella* genome reveals evolutionary insights into the conquest of land by plants. Science 319:64–69

Rensing SA, Beike AK, Lang D (2013) Evolutionary importance of generative polyploidy for genome evolution of haploid-dominant land plants. In: Leitch IJ, Greilhuber J, Doležel J, Wendel JF (eds) Plant genome diversity, vol 2, Physical structure, behaviour and evolution of plant genomes. Springer-Verlag, Wien, pp 295–305

Renzaglia KS, Rasch EM, Pike LM (1995) Estimates of nuclear DNA content in bryophyte sperm cells: phylogenetic considerations. Am J Bot 82:18–25

Sax K, Beal JM (1934) Chromosomes of the Cycadales. J Arnold Arbor 15:255–258

Schneider H, Smith AR, Pryer KM (2009) Is morphology really at odds with molecules in estimating fern phylogeny? Syst Bot 34:455–475

Schuettpelz E, Pryer KM (2008) Fern phylogeny. In: Ranker TA, Haufler CH (eds) Biology and evolution of ferns and lycophytes. Cambridge University Press, Cambridge, pp 395–416

Shaw J, Renzaglia K (2004) Phylogeny and diversification of bryophytes. Am J Bot 91:1557–1581

Soltis PS, Soltis DE (2013) Angiosperm phylogeny: A framework for studies of genome evolution. In: Leitch IJ, Greilhuber J, Doležel J, Wendel JF (eds) Plant genome diversity, vol 2, Physical structure, behaviour and evolution of plant genomes. Springer-Verlag, Wien, pp 1–11

Soltis DE, Soltis PS, Bennett MD, Leitch IJ (2003) Evolution of genome size in the angiosperms. Am J Bot 90:1596–1603

Soltis DE, Bell CD, Kim S, Soltis PS (2008) Origin and early evolution of angiosperms. Ann N Y Acad Sci 1133:3–25

Stergianou KK, Fowler K (1990) Chromosome numbers and taxonomic implications in the fern genus *Azolla* (Azollaceae). Plant Syst Evol 173:223–239

Takamiya M (1993) Comparative karyomorphology and interrelationships of Selaginella in Japan. J Plant Res 106:149–166

Tatuno S, Takei M (1969) Cytological studies of Salviniaceae I. Karyotype of two species in the genus Salvinia. Bot Magazine 82:403–408

Temsch EM, Greilhuber J, Krisai R (1998) Genome size in *Sphagnum* (peat moss). Bot Acta 111:325–330

Temsch EM, Greilhuber J, Krisai R (2010) Genome size in liverworts. Preslia 82:63–80

Tindale MD, Roy SK (2002) A cytotaxonomic survey of the Pteridophyta of Australia. Aust Syst Bot 15:839–937

Troìa A (2001) The genus *Isoëtes* L. (Lycophyta, Isoëtaceae): synthesis of karyological data. Webbia 56:201–218

Tuskan GA, DiFazio S, Jansson S, Bohlmann J, Grigoriev I et al (2006) The genome of black cottonwood, *Populus trichocarpa* (Torr. & Gray). Science 313:1596–1604

Vitte C, Bennetzen JL (2006) Analysis of retrotransposon structural diversity uncovers properties and propensities in angiosperm genome evolution. Proc Natl Acad Sci USA 103:17638–17643

Vitte C, Panaud O (2003) Formation of solo-LTRs through unequal homologous recombination counterbalances amplifications of LTR retrotransposons in rice *Oryza sativa* L. Mol Biol Evol 20:528–540

Voglmayr H (2000) Nuclear DNA amounts in mosses (Musci). Ann Bot 85:531–546

Voytas DF, Cummings MP, Konieczny A, Ausubel FM, Rodermel SR (1992) Copia-like retrotransposons are ubiquitous among plants. Proc Natl Acad Sci USA 89:7124–7128

Wang XY, Shi XL, Hao BL, Ge S, Luo JC (2005) Duplication and DNA segmental loss in the rice genome: implications for diploidization. New Phytol 165:937–946

Weiss-Schneeweiss H, Schneeweiss GM (2013) Karyotype diversity and evolutionary trends in angiosperms. In: Leitch IJ, Greilhuber J, Doležel J, Wendel JF (eds) Plant genome diversity, vol 2, Physical structure, behaviour and evolution of plant genomes. Springer-Verlag, Wien, pp 209–230

Werner T, Braukmann TWA, Kuzmina M, Stefanovic S (2009) Loss of all plastid ndh genes in Gnetales and conifers: extent and evolutionary significance for the seed plant phylogeny. Curr Genet 55:323–337

Wiens D (1964) Chromosome numbers in North American Loranthaceae: (*Arceuthobium*, *Phoradendron*, *Psittacanthus*, *Struthanthus*). Am J Bot 51:1–6

Wiens D (1968) Chromosomal and flowering characteristics in dwarf mistletoes (*Arceuthobium*). Am J Bot 55:325–334

Wiens D, Barlow BA (1971) The cytogeography and relationships of the viscaceous and eremolepidaceous mistletoes. Taxon 20:313–332

Zheng M, Zhu R-L (2009) Karyological studies on some liverworts from China and Singapore. Fieldiana Bot 81:87

Zonneveld BJM (2010) New record holders for maximum genome size in eudicots and monocots. J Bot. doi:10.1155/2010/527357, Article ID 527357

Genome Size and the Phenotype

20

Johann Greilhuber and Ilia J. Leitch

Contents

J. Greilhuber (✉)
Department of Systematic and Evolutionary Botany, Faculty of Life Sciences, University of Vienna, A 1030 Vienna, Austria
e-mail: johann.greilhuber@univie.ac.at

I.J. Leitch et al. (eds.), *Plant Genome Diversity Volume 2*,
DOI 10.1007/978-3-7091-1160-4_20, © Springer-Verlag Wien 2013

20.1 Introduction

Land plant species (Embryophyta) vary more than 2,300-fold in the size of the holoploid genome (C-value) (see Leitch and Leitch 2013, this volume) with the extremes at both ends of the scale contributed by angiosperms. At the lower end we find some species of the carnivorous Lentibulariaceae with ultrasmall genomes, e.g., *Genlisea aurea* with 0.065 pg or 63.5 Mb (1C) (Greilhuber et al. 2006). This is only about 0.4-fold the size of the genome of *Arabidopsis thaliana* with 0.16 pg or 156.5 Mb (1C) (Bennett et al. 2003), a species long considered to be the plant with the smallest reliably determined genome size. At the upper end of the scale stands the octoploid monocot *Paris japonica* (Melanthiaceae) with $2n = 8x = 40$ and 152.23 pg or 148.88 Gb (1C) (Pellicer et al. 2010). Nevertheless, other monocot species such as *Fritillaria davisii* (69.45 pg, 1C; $2n = 2x = 24$, a possible paleotetraploid), *Trillium apetalon* (95.0 pg, 1C; $2n = 4x = 20$), and the dicot tree parasite *Viscum album* (102.9 pg, 1C; $2n = 20$, paleotetraploid?) also rank high on the scale regarding monoploid genome size (i.e., Cx-value = 2C-value divided by ploidy level) (Zonneveld 2010). From these examples it is clear that polyploidy plays only a relatively small role in the origin of the huge genome size differences reported. Instead, it is the accumulation of retrotransposon-like and other repetitive elements in the genomes which are largely responsible for the huge diversity of genome sizes in plants (Bennetzen et al. 2005; Grover and Wendel 2010; see also Kejnovsky et al. 2012 in volume 1). (N.B. Recent studies using flow cytometry to estimate genome size in species with enormous genomes have highlighted how the more traditional approach of estimation using Feulgen densitometry may considerably underestimate genome size at this upper end of the scale (Zonneveld 2010).)

Interest in the diversity of genome sizes has a long history, dating back to the early cytological studies showing enormous variation in the size of cell nuclei and chromosomes between different organisms (Gulliver 1875; Stebbins 1971). However, it was following the discovery of DNA and its presence in the nucleus and chromosomes that studies started to focus on estimating the amount of DNA in cells of different tissues of the same organism and between species (e.g., Boivin et al. 1948; Vendrely and Vendrely 1948; Swift 1950). Such research soon uncovered two notable observations; (1) there was a remarkable constancy in genome size within a species, and (2) there was a surprising variation in genome size between organisms. Indeed, just 15 years after the first genome size for a plant was estimated by Ogur et al. (1951) it soon became clear that there was considerable diversity in genome size even between species and genera. For example, Rothfels et al. (1966) reported a 40-fold range of genome sizes in 22 diploid genera belonging to Ranunculaceae, while Martin and Shanks (1966) noted a five-fold variation within the genus *Vicia*.

The problem during this period (i.e., 1950s – late 1970s) was that it was broadly assumed that the amount of DNA in an organism reflected the number of genes it had, and yet there appeared to be no clear correspondence between genome size and organism complexity (taken as a proxy for gene number). In 1971 Thomas coined the term 'C-value paradox' to reflect this apparent lack of relationship (Thomas 1971).

The solution to the 'C-value paradox' was largely found with the advent of the molecular age, as it soon became clear that most of the DNA in the genome is non-coding. Nevertheless, while the 'paradox' may be considered to have been solved (i.e., genomes vary in size mainly due to differences in the amount of non-coding DNA) there are still many unanswered biological questions remaining (e.g., what types of non-coding DNA sequences predominate in genomes of different sizes? what are the mechanisms involved in generating genome size variation? what are the evolutionary forces driving genome size changes?), leading Gregory (2001) to coin the phrase 'C-value enigma' to reflect this.

Indeed, the biological causes behind genome size differences and the consequences for the organism of having a large or small genome size are topics of a research discipline in its own right (Bennett 1971, 1972; Cavalier-Smith 1985; Bennett 1998; Gregory 2005; Leitch and Bennett 2007). This chapter outlines how genome size is correlated with cell size and cell cycle time (Sect. 20.2), the various hypotheses that have been put forward to explain these correlations (Sect. 20.3) and examples of the impact of genome size variation at the whole plant level (Sect. 20.4).

A necessary premise for the following discussions is to mention the relative constancy of genome size within a species. While many studies have shown that genome size is relatively constant within a species, there have been reports in the literature of extensive intraspecific variation. Many of these studies have subsequently been shown to be artefacts arising from poor technique (discussed by Greilhuber 1998, 2005, 2008), but recent high resolution flow cytometric studies demonstrate that intraspecific and even intrapopulational variation (e.g., Šmarda and Bureš 2006; Šmarda et al. 2007; Slovák et al. 2009; Šmarda et al. 2010) does exist (mostly within 10%, and exceptionally up to 1.37-fold) (reviewed in Šmarda and Bureš 2010). The biological importance of such variation and it potential adaptive significance is discussed in Sect. 20.5.

Another premise is that there will be a certain minimum amount of DNA in a plant genome that is determined by the number of necessary genes and regulatory sequences, together with indispensible sequences which play an architectural role. This amount has been estimated to be about 50 Mb (Bennett and Leitch 2005). Plants such as *Genlisea aurea* are within this order of magnitude. DNA above the required minimum—here we call it *accessory DNA*—may be *inert*, may have a *function*, or may merely have an *effect* (in a similar way as waste may have an effect in human society). Inert DNA is dispensable without disadvantage, DNA with effect may appear dispensable but selectively sensitive, while DNA with function is indispensible or at least selectively strongly sensitive. To separate function from effect is difficult, and unfortunately the possibility of experimenting is very limited. We are therefore currently in the situation where most evidence comes from correlations of genome size with organismal characteristics and environmental parameters, which leaves room for interpretations about the causes involved.

20.2 Basic Facts: The Correlation of Genome Size with Cell Size and Cell Cycle Time

20.2.1 The Nucleo-Cytoplasmic Ratio

Cytologists have been aware of the correlation between nuclear size and cell size in animals and plants for over 130 years, with the first studies by Gulliver in 1875. These observations led to the concept of the 'nucleo-cytoplasmic ratio' (e.g., Strasburger 1893) or the 'karyoplasmic ratio' (e.g., Hertwig 1903) where the ratio of the nuclear volume to that of the cytoplasm is broadly invariant, reflecting the need to balance nuclear and cytoplasmic processes. The correlation was first stated between nuclear volume and cell (cytoplasm) volume, but, as nuclear volume depends on nuclear DNA amount, the correlation also holds between nuclear DNA amount and cell volume, although in a somewhat relaxed manner. It is, of course important to realize that there is a huge range in cell and nuclear size even within the same organism due to cell differentiation (e.g., embryo sac polar nuclei and nucellus nuclei are extremes in close

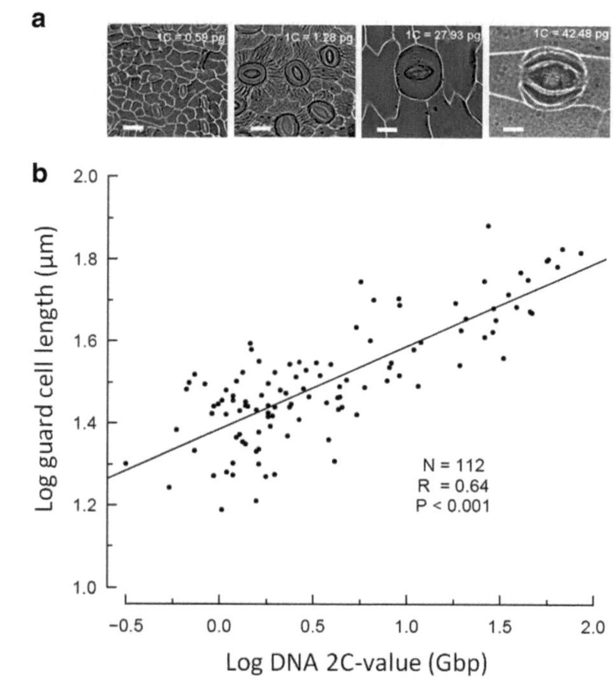

Fig. 20.1 The positive relationship between genome size and guard cell length. (**a**) Images of guard cells from four angiosperm species arranged in increasing C-value (N.B. all images at the same magnification, scale bar = 5 µm). (**b**) Graph plotting the relationship between log 2C DNA amount and log guard cell size for 101 angiosperm species from 34 families. (Images in (**a**) and graph shown in (**b**) are taken from Beaulieu et al. 2008)

proximity, the former decondensed and voluminous, the latter condensed and small) so correlations between cell size and genome size can only be made when comparing the same cell type. So, for example, in animals, the size of nucleate and even enucleate red blood cells show an impressive correlation with genome size (Gregory 2005), while in plants, guard cell size is clearly correlated with C-value across a diverse sample of angiosperms (Beaulieu et al. 2008) (Fig. 20.1). In addition, it is important to note that the best correlations are found among taxonomically close relatives with different genome sizes, since in fairly distantly related species, other developmental factors may come into play and weaken the relationship. For instance, while Bennett (1972) found a significant correlation between pollen size and genome size in 16 species of wind pollinated grass species, Knight et al. (2010) were not able to confirm this in a large scale analysis of 464 angiosperm species belonging to 85 different families.

The clearest correlations between nuclear volume, C-value and cell size are seen in meristematic cells that lack conspicuous vacuoles. In differentiated plant cells additional factors may obscure the correlation. For example, the presence of vacuoles makes quantification of the cytoplamic volume difficult, while cell volume can also be influenced by the particular stage in the cell cycle (replicated or unreplicated, 2C or 4C), and by endopolyploidy (see Maluszynska et al. 2013, this volume). In summary, the size of differentiated organs often does not show a clear relationship with genome size, so that spontaneous polyploids are not always easy to recognize in the field.

One final point regarding the relationship between nuclear volume, C-value and cell size is the observation that, as genome size increases, the amount of (broadly defined) heterochromatin also increases (Flavell 1980). As heterochromatin tends to be more highly methylated and condensed than euchromatin (Houben et al. 2003), the relationship between nuclear volume and genome size may not be linear in species with genomes at the upper end of the range of C-values encountered in plants.

20.2.2 Large Genomes Are Correlated with Long Nuclear Cycles

It has long been appreciated that genome size and cell division rate are linked, with the first studies being conducted in plants by Van't Hof and Sparrow (1963). They studied six diploid angiosperm species growing at 23°C and noted a positive relationship between the duration of mitosis and C-value. Since then there have been numerous other studies although the results have not always been so clear-cut. Thus, while some cell cycle studies suggested that cell cycle length increased by 18–38% every time genome size doubled (Van't Hof 1965; Evans and Rees 1971; Evans et al. 1972), other studies, such as those of Murín (1976), pointed to a much more modest effect of C-value. For example, while *Vicia faba* ($2n = 12$) and *V. sativa* ($2n = 24$), differ seven-fold in C-value (Bennett and Leitch 2010), the cell cycle length for the larger genomed *V. sativa* was only 1.27-fold longer. In part, such contrasting observations may reflect the effect of numerous other factors which can influence cell cycle length independently of genome size (Grif et al. 2002). Indeed, Grif (2000) and Grif et al. (2002) reported that some plants with different genome sizes but grown at their own specific optimum temperature (when the cell cycles in root tips are shortest) had broadly the same cell cycle times. So, for example, the ephemeral *Arabidopsis thaliana* ($2n = 2x = 10$; 0.16 pg, 1C; 22°C) and annual *Avena sativa* ($2n = 6x = 42$; 11.73–13.23 pg, 1C; 25°C) were reported to have identical cell cycle times of 8.5 h despite a 78-fold and 26-fold difference in C-value and Cx-value, respectively. The observations of Grif and colleagues were, however, based on 31 species (22 genera) of different ploidy levels and diploidization status, and some families were better represented than others. In addition, Grif et al. (2002) noted that there could be considerable differences in the cell cycle time for the same species depending on the method used for analysis. Notwithstanding these uncertainties, at 23°C plants with

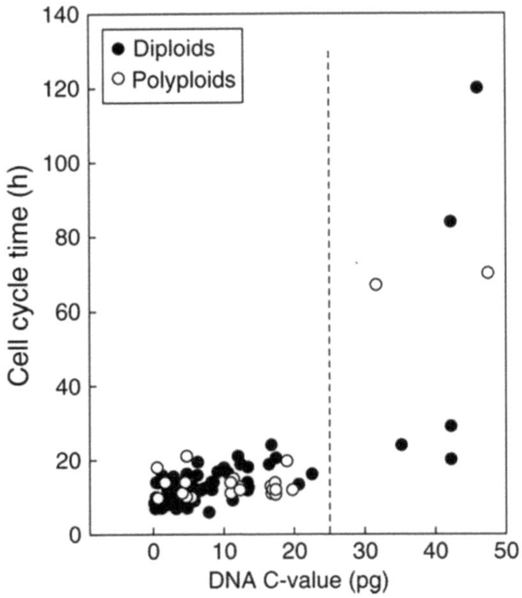

Fig. 20.2 Relationship between DNA C-value (pg) and cell cycle time (h) in root apical meristem cells of 110 diploid and polyploid angiosperms. (Reproduced from Francis et al. 2008)

very large genomes such as *Trillium rhombifolium* ($2n = 6x = 30$; 111.5 pg, 1C) and *Paris quadrifolia* ($2n = 4x = 20$; 60.1 pg, 1C) were shown to have much longer cell cycles of 110 h and 56 h, respectively (Grif 2000; Grif et al. 2002).

In the most comprehensive overview to date, Francis et al. (2008) reported a strong effect of genome size on cell cycle time for 110 species, regardless of ploidy level and life cycle (see Fig. 20.2). The strongest relationship was observed in diploid, perennial monocots with the huge genomes of *Lilium* and *Trillium* having disproportionately long cell cycles. However, it should be noted that the outlying huge genomes of *Trillium* and *Lilium* influenced the line of the regression in favour of diploids, monocots and perennials to the extent that the effect of ploidy may have been obscured. Nevertheless, even if the outlying monocots were excluded from the regression analysis, it was noted that at a given C-value the variation in cell cycle times was considerable, indicating the involvement of other factors affecting cell cycle time.

The influence of polyploidy on cell cycle times is still far from clear, but in part, this is likely to reflect where in the diploidization/polyploidization cycle a particular species falls (see Fawcett et al. 2013, this volume). In neopolyploids cell cycles are not necessarily longer than their corresponding diploids. Indeed, they may be shorter as noted by Bennett and Smith who found that the duration of meiosis in wheat was 42 h for diploids, 30 h for tetraploids and 24 h for hexaploids despite the proportionately higher DNA contents of polyploids (Bennett and Smith 1972). This may arise because the ratio of inert versus genic DNA is balanced. However, in older, more established polyploids that have undergone considerable diploidization, often accompanied by the elimination of non-informative DNA and silencing of some informative DNA, the effect of polyploidy on the relationship between C-values and cell cycle length may become more obscure. Indeed, this may contribute to the lack of a clear effect of polyploidy on cell cycle time in the large analysis by Francis et al. (2008) that included polyploids at all stages of the diploidization/polyploidization cycle.

It has been suggested that phylogeny may play a role in determining the strength of the relationship between cell cycle time and DNA amount. For example, Evans et al. (1972) reported that although the slope of the relationship between C-value and cell cycle time was similar between six monocots and six eudicots (over the range of C-values studied), the cycle time for eudicots was approximately 4 h longer than that of monocots of equivalent C-values, with the increased duration primarily accounted for by a longer G1 phase of the cell cycle. They suggested that this may be due to the denser chromatin packaging observed in the species analysed (Evans et al. 1972). Nevertheless, in the larger study of Francis et al. (2008) a similar, strong effect of phylogeny was not so clear, although a marked increase in cell cycle time in monocots with genomes larger than 25 pg (1C) (i.e., *Lilium* and *Trillium*) was noted. It was suggested that this may reflect the increased packaging of DNA into heterochromatin in species with large genomes, reducing access for the replication machinery and hence leading to a longer DNA synthesis phase and hence longer cell cycle time.

One further factor that has been shown to influence cell cycle times is the presence of B chromosomes, although data are sparse. Ayonoadu and Rees (1968) showed that the presence of B chromosomes increased the duration of the mitotic cycle in *Secale cereale* while Evans et al. (1972) reported that B chromosomes increased both the S-phase and mitotic phase of the cell cycle in three grass species (*Lolium perenne*, *Zea mays* and *Secale cereale*). However, to date the mechanisms responsible are unclear. Evans et al. (1972) proposed that both genotypic effects and/or the late replication of B chromosomes could play a role although they noted that only *Zea mays* and *Secale cereale* were late replicating.

Overall it seems clear that although genome size can considerably influence the length of the cell cycle and indeed will set a minimum time, both genetic (e.g., genotypic differences and variation is gene dosage owing to polyploidy) and abiotic factors (e.g., temperature) can act to increase its duration.

It is perhaps noteworthy that in contrast to eukaryotes, the prokaryotic bacteria show a strong linear correlation between genome size and gene number, but no significant correlation between genome size and doubling time under laboratory conditions, although there is considerable variation in growth rates between taxa (Mira et al. 2001). The reason may be that bacteria do not contain significant amounts of accessory DNA. This may be a consequence of the circular genome they possess, which cannot exceed a certain size to avoid cell division problems. In addition, recombination between direct repeats within the circle may result in the formation of two rings (depending on the resolution of the Holliday junction (s)) and this may be lethal if the two circles fail to segregate properly into daughter cells. In contrast, eukaryotes have linear chromosomes which make the accumulation of accessory DNA possible (Schubert 2011).

20.3 Hypotheses About the Biological Significance of Genome Size Variation

Over the years there have been numerous attempts to try and explain the observed correlation between genome size and cell size and the biological significance of genome size variation. Most of these theories can broadly be divided into three contrasting hypotheses.

20.3.1 Accessory DNA is Mostly of Selfish Origin and Represents Inert Genetic Junk

This hypothesis states that variation in genome size arises purely from mutational events that lead to the differential amplification of non-essential DNA (largely comprising transposable elements (TE) and sometimes referred to as parasitic, selfish or junk DNA). Species with larger genomes thus represent those that are better able to tolerate the 'selfish' accumulation of DNA than those with smaller genomes. Under such a scenario, the relationship between genome size and cell size is considered to be purely coincidental rather than biologically significant (e.g., Doolittle and Sapienza 1980; Orgel et al. 1980; Petrov 2002; Oliver et al. 2007).

The idea of parasitic genetic elements dates back to Östergren (1945), who stated that B chromosomes did not need to have a positive function for the organism in order to exist, they just needed to be able to be self-sufficient, i.e., for their reproduction, like a parasite and able to propagate themselves. Given that mechanisms exist that can accumulate B chromosomes (e.g., preferential distribution into gametes by non-disjunction, see Houben et al. 2013, this volume), one could argue that this is sufficient to explain their existence in a population without the need to invoke an advantage for the organism. Comparable to B chromosomes, TEs are also able to replicate and amplify within the genome (Bennetzen 2005) and so perhaps no other explanation for their existence need be sought. Nevertheless, it is possible to envisage that the neutral accumulation of TEs during the evolution of a species can only continue until developmental constraints (e.g., through excessively long cell cycles or heterochromatization) slow down or stop the reproduction of the organism. In addition, it is now clear that molecular mechanisms exist which are capable of eliminating DNA (see review by Grover and Wendel 2010). These have been shown to operate not only in diploids but also following polyploid formation, leading to genome downsizing (see below, and Fawcett et al. 2013, this volume). Such observations suggest that while selective neutrality could exist within certain limits, large increases in genome size clearly can have certain deleterious effects, which makes a predominantly neutral role of accessory DNA unlikely. Indeed, such theories are also unable to explain why the relationship between genome size and cell size persists following genome size reductions.

20.3.2 The Nucleotype Hypothesis

The correlations of genome size with cell size and cell cycle time form the basis of most hypotheses that ascribe a biological role to genome size, i.e., genome size variation is conceived as being subject to selection rather than representing a neutral character.

Although various researchers in the 1960s suggested that the total mass of DNA played a role in influencing various cell parameters (Commoner 1964; Martin 1966), it was Bennett who coined the term 'nucleotype' to describe "that condition of the nucleus that affects the phenotype independently of the informational content of the DNA" (Bennett 1971, 1972). It should be noted that this definition includes not only the total DNA mass, but also other possibly relevant traits such as GC content (which can effect thermal stability, mutation rate and cost of synthesis; see Šmarda and Bureš 2012, volume 1), and amount and arrangement of satellite DNA (which can influence the extent to which DNA is compacted into heterochromatin). Nevertheless, DNA amount is considered the most important and is so far the only nucleotypic parameter that has been studied in detail.

Bennett (1972) investigated the distribution of C-values in herbaceous angiosperms and observed that the ranges of C-values for each type of life style were different. He noted that perennial herbs varied widely in C-values, whereas annuals exhibited a much narrower range, with C-values restricted to the lower end of the scale. In ephemeral species (germinating and flowering within a few weeks, such as *Arabidopsis thaliana*) only very small genomes were found, with no species having a genome larger than 3.5 pg/1C. Based on the observation that plants with small genomes exhibit faster mitotic and meiotic cell cycles compared with those possessing large genomes (see Sect. 20.2.2 above), Bennett (1972) postulated a causal link between genome size and life style, in which threshold effects were considered to play an important role. He noted that, as genome size increased, this was accompanied by an increase in the minimum generation time (MGT—i.e., the shortest time between germination and the production of the first mature seed) and, at various points this resulted in an inescapable shift in developmental life style. Thus for a species to be an ephemeral it had to possess a genome small enough to enable a MGT of 7 weeks, whereas species with C-values greater than the threshold value for a MGT of 52 weeks (estimated to be about 25 pg /1C) were restricted to being obligate perennials, in part because the time needed for the required number of mitoses and meiosis is too long for an annual lifestyle.

Bennett (1972) recognized that there were many limiting factors which can act to slow down the MGT of a plant (e.g., availability of nutrients, temperature, genes) but they serve only to increase the duration of the life cycle above a certain minimum set by genome size. Thus while species with very small genomes may be ephemeral, annual or perennial, as genome size increases the number of life style options are reduced until, above the c. 25 pg/1C threshold, species are all obligate perennials.

At this point two complications of the nucleotype hypothesis must be mentioned. First, it matters for cell cycle length and thus for life style too, whether the genome is diploid or polyploid. As noted above (Sect. 20.2.2), in *Triticum aestivum* (hexaploid bread wheat), the meiotic cycle was shown to be shorter than for tetraploid and diploid wheats *T. dicoccum* and *T. monococcum* respectively (Bennett and Smith 1972). Moreover, in maize, artificial triploids and tetraploids were shown to have the same cell cycle length as the diploid parent strain (Verma and Lin 1979), and similar results were also reported for diploid and autotetraploid cytotypes of *Avena strigosa* (Yang and Dodson 1970). This absence of an increase in cell cycle length is easily explained in part to be a consequence of everything being multiplied in a polyploid—i.e., not only the DNA quantity but also the machinery to replicate it, transcribe it into mRNA and translate it into proteins. So, it seems to be the *Cx-value*, which influences cell cycle time most strongly. However, in contrast, as far as cell size is concerned, it is noted that the direct and positive correlation between this character and DNA quantity is based on the total DNA amount in the nucleus. So, it is the *C-value* that determines the cell volume at least in meristematic cells (N.B. as noted above, differentiated cells have vacuoles, which introduce a strong element of variation in total cell volume and obscure the relationship between cell size and DNA amount). If developmental speed in terms of body size is adaptively important, the best combination from the nucleotypic viewpoint would seem to be a small Cx-value (fast cell division) plus polyploidy (larger cells).

The second complication is that cell size and cell cycle time can counteract each other in regard to developmental speed, at least under certain circumstances (Price and Bachmann 1976). When fast growth and attained final size are critical (e.g., in understory plants), a larger cell size (through larger genome) can perhaps promote growth better than faster cell divisions (through smaller genome), because the impact of increasing genome size is less on the cell cycle than on the cell volume. This observation is based on early studies which showed that at a constant ploidy level, a doubling of genome size resulted in about a 1.18–1.38-fold increase in cell cycle length (Van't Hof 1965; Evans and Rees 1971; Evans et al. 1972), but a two-fold increase in cell volume (Price et al. 1973; Edwards and Endrizzi 1975). Nevertheless, more data on this relationship are needed.

20.3.3 Nucleoskeletal DNA Hypothesis

In contrast to the nucleotype hypothesis outlined above, Cavalier-Smith proposed that it was cell size which was adaptive and under selection (Cavalier-Smith 1978, 2005). Given that (1) the relationship between cell size and nuclear size is much stronger than that between cell size and genome size, and (2) that cell size is determined primarily via genetic control rather than genome size, Cavalier-Smith (1978) put forward the hypothesis that the role of DNA (including both genic and accessory DNA) in a genome was primarily to act as a nucleoskeleton which determined the nuclear volume (both via DNA amount and the degree of packaging), enabling the ratio between the nuclear and cytoplasmic volume (= nucleo-cytoplasmic ratio) to be optimized. This is considered essential to balance the overall rate of RNA transcription and processing with that of protein synthesis to allow balanced growth of actively dividing and growing cells (as noted in Sect. 20.2.1).

With this hypothesis, the driving force is selection for an optimal cell size; following the maxim "Modify your cell size, the genome will follow!" For example, if a larger cell size became adaptively favorable due to a change in selective forces, Cavalier-Smith envisaged that there would also be positive selection for a corresponding increase in nuclear volume and hence genome size, achieved primarily through increases in the amount of non-coding DNA (although modifications to the folding of DNA can also play a role in determining nuclear volume).

Cavalier-Smith outlined many examples to support his hypothesis (e.g., Cavalier-Smith 2005). One nice illustration is provided by the miniature, enslaved nuclei known as nucleomorphs that are found in two lineages of algae—the cryptomonads and the chlorarachneans (Cavalier-Smith 2002). It is widely accepted that these two algal lineages arose independently via secondary endosymbiosis whereby a eukaryotic alga (a green alga in the case of chlorarachneans and a red alga in cryptomonads) was engulfed by a protozoan (Green 2011). All that remains of the engulfed alga today is the original plasma membrane, the chloroplast (which provides photosynthate to the host cell), and its nucleus (to enable the chloroplast to function). This enslaved nucleus or nucleomorph is surrounded by a typical eukaryotic nuclear membrane, contains chromosomes and multiplies by normal cell division, and yet its genome has undergone a remarkable reduction in size. Indeed, the nucleomorphs have the smallest nuclear genomes so far reported for any eukaryote, comprising just 551 kb for the cryptophyte *Guillardia theta* (Douglas et al. 2001) and 373 kb for the chlorarachnean *Bigelowiella natans* (Gilson et al. 2006).

Despite the independent origins of the nucleomorphs, sequencing of their genomes has revealed similar genomic architecture (e.g., genome divided into three chromosomes c. 100–200 kb long, high gene density, similar proportions of genomic components, and negligible amounts of non-coding DNA) (Gilson et al. 2006). What is so intriguing is that the extreme genome size reduction of the nucleomorph has occurred within the same cell as the main (host) nucleus, which is considerably larger. Indeed, the large range of cell sizes observed in different species of cryptomonads scale with the genome size of the main nucleus, exactly as observed in other eukaryotes (Beaton and Cavalier-Smith 1999). Such an observation is consistent with the nucleoskeletal hypothesis if one envisages that expansion in the main nucleus has arisen through positive selection acting on cell size that has been accompanied by an increase in genome size via amplification of non-coding DNA to provide the nucleoskeleton to balance the nucleocytoplasmic ratio. The fact that selection can reduce genome size in one nucleus (i.e., the nucleomorph) and alter genome size in the main nucleus adaptively within the same cell is considered to provide compelling evidence for a nucleoskeletal role of DNA.

Overall, the distinction between the nucleotype and nucleoskeletal hypotheses is rather subtle and is primarily based on the target on which selection acts. In the nucleotype hypothesis the cell size and/or cell cycle time is optimized via selection for an appropriate genome size (Bennett 1972), whereas in the nucleoskeletal hypothesis the target of selection is cell size, and genome size follows passively, because a concomitantly larger or smaller nucleus allows accumulation or requires deletion of skeletal DNA.

In reality both hypotheses probably apply during adaptive evolutionary processes although the importance of each may vary depending on a multitude of factors. In single celled organisms such as bacteria, unicellular algae, protists, and in organisms with blood cells (animals) it may well be that cell size is most critical. For example, given that larger blood cells have lower surface area to volume ratios and are therefore less efficient for gas exchange than smaller blood cells, it is plausible that selection may act at the level of cell size instead of genome size to optimize gas exchange to meet metabolic demands (Gregory et al. 2009). On the other hand, the cell size argument may be less clearly applicable in land plants, on which Bennett primarily focused with his nucleotype theory. Indeed, it is often difficult to predict where the selective bottleneck is for a plant species. Is it root growth rate, plant biomass, leaf area, reproductive rate, generation time, pollen grain weight, pollen tube growth rate, vessel diameter, stomata size and density, or seed weight and reserve substance content? Certainly, investigations into links between cell size and selection are often complicated by the presence of vacuoles (which make estimations of the cytoplasmic volume difficult) and endopolyploidy (Maluszynska et al. 2013, this volume), which interferes with genome size/cell size correlations as noted above. Even for cell types which show reasonable correlations between cell size and genome size (e.g., guard cell sizes: Beaulieu et al. 2008, Fig. 20.1), taxon-specific factors as well as evolutionary history are likely to play an important role in determining how a particular lineage may respond to prevailing selective forces.

20.3.4 Alternatives to the Nucleotype and Nucleoskeletal Hypotheses

20.3.4.1 Reproduction Mode and Genome Size

Since an annual life history is preferentially associated with inbreeding (Stebbins 1957; Ehrendorfer 1970; Barrett et al. 1997), it is possible that the correlation between low genome size and annual life style noted by Bennett (1972, see Sect. 20.3.2 above) is actually driven by an association between reproductive mode and genome size, with inbreeding species characterized by smaller genomes than outbreeders. Indeed, several authors have addressed this question.

Early studies reporting a correlation between genome size and breeding system include those of Govindaraju and Cullis (1991) who conducted a survey of 176 seed plants, and Labani and Elkington (1987) who investigated *Allium*. Both studies reported smaller genomes in selfing species. However, no correction for phylogenetic relatedness was made in either study. Indeed, the study of Govindaraju and Cullis (1991) included 25 species of distantly related outbreeding gymnosperms with large genomes (e.g., *Pinus*, *Picea*, and *Abies*) and it is possible that they could have influenced the correlations reported. Certainly the results contradict another study by Rees and Jones (1967) who reported the inverse correlation, with inbreeding *Lolium* species characterized by genomes that were 30% larger than their outbreeding relatives.

To test the relationship between genome size and breeding system more fully Albach and Greilhuber recognized the need to use phylogenetically-corrected statistical tests that take evolutionary history into account (Albach and Greilhuber 2004). This is because species do not represent independent samples (as is assumed with most conventional statistics methods). Indeed phylogenetic history has been shown to be an important factor in explaining much of the genome size variation encountered. Using phylogenetic independent contrasts (Felsenstein 1985) and generalized least squares approaches (Pagel 1997, 1999) Albach and Greilhuber (2004) found better statistical support for an association between low genome size with selfing than with annuality in 42 *Veronica* species, with twelve selfing species exhibiting significantly lower Cx-values than outbreeders and three closely related selfing/outcrossing sister species pairs consistently showing the lower Cx-value in the selfer.

Similarly in a phylogenetically-controlled comparison between 14 taxonomically paired outcrossing and highly selfing species comprising eight families including both monocots and eudicots, Wright et al. (2008) found that selfers had significantly smaller monoploid genome sizes (i.e., removing the effect of polyploidy) than outcrossers.

To explain how inbreeding might affect genome size evolution a number of mechanisms have been proposed. For example, given the strong relationship between genome size and amount of transposable elements (TEs) it is possible that changes in the breeding system are accompanied by changes in TE dynamics with inbreeding playing a role by eliminating TEs more efficiently than outbreeders. Indeed computer models have suggested that the smaller genome sizes observed in some inbreeders arise due to a reduced spread of TEs between individuals caused by the lack of outbreeding (Wright and Schoen 1999; Morgan 2001), while the ability of inbreeders to more effectively purge TE insertions with deleterious recessive effects on fitness will also lead to smaller genomes in inbreeders compared with outbreeders (Wright et al. 2008).

Nevertheless, it has also been observed that the evolution of self-fertilization may be associated with a large reduction in the effective rate of recombination and a corresponding decrease in effective population size (Wright et al. 2008). It is therefore possible that the smaller effective population sizes in inbreeders may counteract the effect of TE elimination noted above, due to the lowered efficiency of natural selection to eliminate proliferating TEs (Lynch and Conery 2003). Under such a scenario, one might predict inbreeders would have larger genomes than outbreeders.

In a much larger study analysing over 205 species of angiosperms, Whitney et al. (2010) found only a very weak positive correlation between genome size and breeding system when phylogenetic history was taken into account, suggesting that there is not a strong evolutionary association between these two characters. Nevertheless, they noted the discrepancy between their results and those of Albach and Greilhuber (2004) and suggested it may be a consequence of the different phylogenetic spread of the two datasets and the impact of paleopolyploidy on genome size evolution. Whereas Whitney et al. (2010) analysed 205 species from across the seed plants, which included species at all stages of the diploidization/polyploidization cycle, the study of Albach and Greilhuber was restricted to a single angiosperm genus where the impact of polyploidy was more controlled, perhaps enabling the relationship between the reproductive system and genome size to be more apparent. Clearly more data are needed with analyses conducted at all taxonomic levels to determine to what extent the taxonomic scale is important in revealing the interaction between genome size and reproductive mode.

Certainly these studies highlight the complexity of the relationship between genome size and reproductive mode and suggest that the importance of breeding system on genome size may vary under different conditions. Indeed, with annual inbreeders it is also possible, that in an *a priori* small genome (such as in *Veronica*) the effect on genome size of lethal TE insertions is stronger (as gene density is higher) than in a larger genome (where gene density is lower due to the larger amounts of repetitive DNA). In contrast, in inbreeding taxa with *a priori* larger genomes the effect of TE insertions on genome size and cell cycle length may be more evident and selectively effective. The expectation would thus be (1) in a higher taxon comprising a range of small genomes, the inbreeders might be expected to have the smaller genome sizes (as in *Veronica*), while (2) in a taxon comprising a range of larger genomes, the annuals would have the lower genome sizes. Given such a scenario, it is perhaps plausible to consider that both mechanisms may be operating within angiosperms.

One should also acknowledge that in animals, where selfing is practically unknown, there is clear support for a short life cycle/low genome size/small cell size correlation (Gregory 2005).

20.3.4.2 Genome Size and Phosphate (P) Availability

It has long been recognised that phosphate is an essential macro-nutrient, playing a key role in many cellular components such as nucleic acids (DNA and RNA), membrane lipids and enzymes. Yet despite its abundance in the environment it is not readily accessible to the plant and is often present in such low amounts that it may be considered to be a limiting nutrient for DNA biosynthesis and growth (Raven et al. 2005; Nussaume et al. 2011). Indeed, some plants have evolved specific mechanisms to cope with limiting levels of P. For example, it has been observed that unicellular algae under chronic P-limitation have the capacity to restructure their biochemical make-up by replacing P-rich membrane lipids with sulphur- or nitrogen-containing lipids (Van Mooy et al. 2009). There is also evidence that nutrient limitation may directly affect the DNA composition of plant genomes (Acquisti et al. 2009; Bragg and Wagner 2009).

Given these observations, it has been hypothesized that in P-depleted soils there will be selection for species with reduced genome sizes or metabolic demand for RNA as a way to reduce the biochemical cost of synthesising DNA and/ or RNA, which are both phosphorous–rich molecules. The hypothesis was initially based on the observation that very small genomes were found in some carnivorous plants such as *Drosera* which live in mineral-poor environments (Hanson et al. 2001b) although since then very small genomes have also been noted in other genera growing in nutrient-limited environments such as *Bixa orellana* which occurs in the mineral depleted soils of tropical America (Hanson et al. 2001a). Indeed, the smallest genome so far reported for a plant is found in the carnivorous *Genlisea aurea* (Greilhuber et al. 2006) which grows in tropical nutrient-poor environments on the flat top of granite rocks. Nevertheless, it should be noted that *Genlisea* is a genomically fast-evolving genus (Jobson and Albert 2002; Müller et al. 2006), and so it is possible that genome miniaturization may have arisen as a result of intrinsic genomic factors as a consequence of the carnivorous mode of nutrition rather than due to external selection pressures (Albert et al. 2010).

As an extension of the hypothesis it has also been suggested that polyploidy might be counter-selected under strong P-limitation (Leitch and Bennett 2004; Leitch and Leitch 2008), a hypothesis that finds indirect support from studies indicating that increases in plant ploidy level are often associated with evolutionary reductions in genome size (Leitch and Bennett 2004). Nevertheless, it is clear that further work is necessary to determine the extent to which P-limitations play a role in the establishment and evolution of polyploids in comparison with related diploids. Indeed, cytological studies of plants living in environments with low P-levels such as bogs and heathlands or in some of the highly nutrient-depleted soils of Australia (Beadle 1962) may be particularly informative to determine to what extent P-levels influence the incidence of polyploidy.

20.4 Examples and Explanations

20.4.1 Cell Size, Genome Size and Sperm Mobility in Bryophytes

Genome sizes in bryophytes are much less variable than in tracheophytes (i.e., c. 38-fold in liverworts and 12-fold in mosses, with no data for hornworts—see Leitch and Leitch 2013, this volume). To explain this limited variation in genome size, Renzaglia et al. (1995) proposed that it was because of the nucleotypic effect of genome size on sperm size and mass. This arises because the nucleus contributes most of the cell mass in sperm cells, thus there is a reasonable correlation between sperm cell size and C-value. Since the motility of sperm depends on their size and hence the C-value, it might be envisaged that there would be selection against too much DNA as this would result in larger sperm making them less mobile and hence less efficient at effecting fertilization. Support for the hypothesis comes from the observation that lycophytes with biflagellate sperm are also characterized by very small genomes (e.g., *Selaginella*) whereas lycophytes and gymnosperm groups with multiflagellate sperm (e.g., *Isoetes* and *Phyloglossum* in lycophytes and cycads and *Ginkgo* in the gymnosperms) have larger genomes as might be expected as they will be less constrained in this regard (see Leitch and Leitch 2013; this volume).

20.4.2 The Impact of Genome Size on Phenology

In 1982 Grime and Mowforth put forward an interesting hypothesis concerning the ecological significance of genome size on plant phenology and relating it to the impact of genome size on cell cycle length and cell size (Grime and Mowforth 1982). They observed that in a woodland community of flowering plants in the British Isles, the first species to flower in the year (e.g., snake's head fritillary—*Fritillaria meleagris* and lesser celandine—*Ranunculus ficaria*) were characterized by possessing the largest genomes of the species analysed while as the year progressed, the later flowering species had progressively smaller genomes. The explanation offered by Grime and Mowforth (1982) was based on the relationships between genome size and cell cycle time and between genome size and cell size (see Sect. 20.2). They hypothesized that the plants emerging and flowering very early in the spring had already preformed their organs during the warm period of the preceding season via cell division but without cell expansion. In the following year, with the availability of water after the frost period, a fast expansion of the cells by water uptake could then take place, and given that cell expansion is temperature-independent (unlike cell division), this enabled such plants to grow through cell expansion at a time of year when it was too cold for growth via cell division. Grime (1983) also observed that the rate of leaf expansion was faster for species with larger genomes, thus plants with larger genomes would be able to expand their leaves and hence 'grow' faster than the smaller genomed species, providing an additional advantage of larger genomes for such early growing plants. Those plants flowering later, which did not separate cell division from cell expansion, were limited to growing only once favorable temperatures were reached to enable cell division and were characterized by smaller genomes and hence faster cell cycles. Greilhuber (1995) pointed to the analogous situation in summer-dry habitats, where bulbous and rhizomatous plants with large genomes also thrive.

The same correlation, i.e., increasingly smaller genome sizes in species flowering later in the year, was also noted by Labani and Elkington (1987) in *Allium* and this was confirmed by Baranyi and Greilhuber (1999) for diploids. However, the latter authors failed to confirm the correlation when polyploid *Allium* species were included in the analysis. These results suggest that it is the Cx-value rather than the C-value that matters in this relationship. Although more recently Ohri and Pistrick (2001) failed to find any correlation in yet another extensive analysis of genome size and flowering time in *Allium*, it is currently unclear whether this was due to technical or biological factors. For example, Ohri and Pistrick used the static one-wavelength Feulgen photometry method to estimate genome size in *Allium* and this has been reported to give rise to problematic data (see Greilhuber et al. 2007) and they also did not adjust for ploidy level although their sample included many polyploids. It is clear that further studies using flow cytometry for genome size estimations not only in *Allium* but other genera as well, together with adjustments for ploidy and the inclusion of phylogenetic data will be necessary to determine the full extent to which genome size plays a role in influencing plant phenology.

20.4.3 The Relationship Between Genome Size and Weediness

Weeds are plants invading open arable land and many of them have characteristics of *r*-strategists, i.e., fast reproduction and development, the production of many propagules, and an annual life style being the most common. Given the nucleotypic correlations between genome size (1C and 1Cx-value) and rapid development one might predict that weeds would be characterized by small genomes. To address this question, Bennett et al. (1998) analysed 156 weed species (comprising both British garden weeds and species recognized as important weeds on a global scale) and compared them to 2,685 other non-weed species. They observed that (1) weeds were characterized by a narrower range of genome sizes, limited to the bottom 20% of the range encountered in angiosperms as a whole, (2) the mean C-values and Cx-values were significantly lower in weeds than in the non-weed sample, (3) no weed was observed to have a 1C-value greater than 25 pg and (4) the most aggressive weeds were characterized by the smallest genomes (e.g., water lettuce, *Pistia stratiotes* 1C = 0.3 pg). Further analysis showed that the probability of being a weed decreased with increasing 1C- and 1Cx-values up until the threshold value of 1C = 25 pg was reached, and that the weeds with the largest genomes were predominantly polyploid, in agreement with the observation that cell cycle times in polyploids are often shorter compared to diploids of an equivalent 1C-value (see Sect. 20.2.2 above), although the greater genetic diversity observed in allopolyploids may be an additional factor under selection in weeds. Overall, this study supports the prediction that weeds are characterized by small genomes although it seems likely that selection is acting on correlated factors such as (1) rapid development, (2) fast growth and (3) production of many small and light seeds that are easily dispersible, and it is these evolutionary forces that may be shaping and constraining genome size in weeds.

20.4.4 Genome Size and Invasiveness

Like weediness, to be a successful invasive requires preadaptation or fast adaptation to the new, usually disturbed environments, in which competitive superiority regarding speed of growth is a priority. Once again, cell size and cell cycle time and associated parameters such as seed weight and number,, are considered to be important traits to be a successful invasive, and given the link between genome size and these characters, a small genome size has been suggested to be an important prerequisite for plant invasiveness. Indeed, Rejmánek (1996, 2000) listed a small genome size as one of the eight most important factors contributing to the invasive success of a species, and several studies at the generic level (including the angiosperms *Artemisia* (Garcia et al. 2008), *Briza* (Rejmánek 1996), and gymnosperm *Pinus* (Grotkopp et al. 2004)) and species level (e.g., *Phalaris arundinacea*, Lavergne et al. 2010) have provided support for this.

In a broader analysis, Kubešová et al. (2010) conducted a flow cytometry-based genome size comparison between alien plant species (including both naturalized and invasive species) and their native counterparts in the flora of the Czech Republic. In total, the study analysed 93 species comprising both native and alien species from 70 genera and 32 families and compared their genome sizes with those of congeneric and confamilial non-invading taxa worldwide (using data from the Plant DNA C-values Database). Like the more focused studies mentioned above, this broad scale analysis found that naturalized and invasive plants had significantly smaller genomes (C-values and Cx-values) than their non-invading relatives at both the generic and familial level and indeed, no naturalized species with large genomes were seen. The species with the highest C-value was *Rudbeckia laciniata*, an octoploid with 1C = 15.27 pg and 1Cx = 3.82 pg. A comparative analysis of 401 species of the Californian flora by Knight and Ackerly (2002) also found that the non-native species had significantly smaller genomes than the native species, while an absence of large genomes and overrepresentation of species with small to very small genomes was observed in an analysis of 238 endemics from the oceanic Macaronesian Islands compared with non-Macaronesian relatives (Suda et al. 2005) as well as in a survey of two Pacific archipelagos—Hawaii and Marquesas Islands (Kapralov and Filatov 2011). In addition, a survey of C-values for 3,676 angiosperms by Chen et al. (2010) suggested that both Cx- and C-values had significant effects on plant invasiveness with the exception of trees and species belonging to Fabaceae. Taken together the results do suggest that genome sizes may well be constrained in colonizing/invading plant groups.

Evidence that smaller genomes may indeed be under selection in invasive species comes from a study by Lavergne et al. (2010). They analysed 90 genotypes of the reed canopy grass (*Phalaris arundinaceae*) from N. America and Europe, and observed that the genome size was significantly smaller in the invasive genotypes from N. America compared with the native European genotypes. By modelling the genome size data within a phylogenetic framework they concluded that the smaller genomes observed in the N. American genotypes had been maintained by natural selection during the invasive process. The increased invasive potential was considered, in part, to be due to the more rapid early growth observed in the invasive genotypes arising as a result of the reduction in genome size. Nevertheless, it should be pointed out that a survey of genome sizes in 92 species of Australian *Acacia* comprising both invasive versus non-invasive species failed to find any link between genome size and invasiveness (Gallagher et al. 2011), although such a result was considered to be most likely due to the very small DNA values of *Acacia* species studied. Instead it was suggested that other traits, such as plant height and range size, which were shown to be significantly different between invasive and native species, play a more important role in contributing to the invasive success of some *Acacias* than genome size.

Overall, it seems clear that more comparative data are needed to determine the extent to which a decrease in genome size is a general evolutionary mechanism leading to the evolution of more invasive species. Recent studies have shown that polyploidy (and hence genome size increase, at least initially) also plays a significant role in determining the invasiveness of plants, highlighting the complexity of interactions influencing the ecological traits of a plant (Chen et al. 2010; Pandit et al. 2011; te Beest et al. 2012).

20.4.5 Genome Size and Climate

Studies of genome size have often looked for correlations with various climatic parameters. Of these, many have reported relationships between genome size and elevation above sea level (i.e., altitude) or geographical latitude because they are considered to be associated with parameters such as temperature, precipitation and length of growing season. However, a comparison of results between different studies has shown that the directions of the relationships are not always consistent. For example, Knight et al. (2005) noted that out of 18 studies which looked for a correlation between genome size and latitude, five reported the relationship to be positive, seven found a negative relationship and six failed to find any relationship at all. Similar levels of inconsistencies were also found in studies examining the relationship between genome size and altitude. Given that altitude and latitude are obviously related to climatic factors but not always in the same way, it is perhaps not surprising, that different studies have found different trends. A latitudinal gradient can go from dry to moist or from mesic to (ant) arctic, in the same way as a gradient from low to high elevations. Indeed, it may well be that temperature and precipitation are the more biologically relevant parameters on which selection for genome size acts and these do not vary linearly with altitude or latitude.

Another explanation for the rather confusing picture is that the majority of studies failed to take phylogenetic history into account in their analysis, a point noted by Bancheva and Greilhuber (2006) in their study of *Centaurea*. Here, apparent correlations of genome size with altitude, temperature and precipitation disappeared when phylogenetically closely related species were examined. These observations led to the suggestion that the relationships were pseudo-correlations.

An alternative reason for the inconsistent relationships could arise from technological problems. For example, a previously claimed negative correlation between genome size and latitude for *Glycine max* cultivars reported by Graham et al. (1994) could not be confirmed by Greilhuber and Obermayer (1997). The discrepancy between the two studies was attributed to problems of using external standardization for genome size measurements by Graham et al. (1994). Such an approach has since been shown to lead to serious errors (Greilhuber 1998, 2005) and best practice methods of genome size estimation now emphasize the need to always use internal standardization when estimating genome size (Greilhuber et al. 2007).

On a broader scale, Knight and Ackerly (2002) suggested that the lack of clear relationships between genome size and various climatic variables could have arisen because many of the studies were too narrow in their focus and did not represent the full ecological range of a species (e.g., they did not encompass the full range of latitudes from the tropics to the poles, or altitudes from sea level to mountain tops). In addition, they also recognized that the relationships being studied may not be linear and so the application of linear regression analysis may be inappropriate. For example, they noted the study of Rayburn and Auger (1990) who found the relationship between genome size and altitude in maize (*Zea mays*) was non-linear. From an examination of 23 populations growing at different altitudes, Rayburn and Auger (1990) reported that those growing at sea level and high altitudes had smaller genomes than those growing at intermediate elevations. Similarly, for latitude, Bennett (1987) observed that while genome size of a range of crop species initially increased with increasing latitude, over a certain latitude threshold, species with large genome sizes were progressively excluded, presumably due to selection against species with large genome sizes as the climate became harsher and growing seasons shorter. To overcome these difficulties, Knight and Ackerly (2002) decided to take a novel approach by using quantile regression analysis to analyse genome size diversity in relation to various climatic variables in 401 species across broad environmental gradients of the Californian flora. Their studies showed the relationship between genome size and various environmental parameters was certainly not linear. Thus while species with small genomes were observed across the range of temperatures and precipitation that they encountered (and thus were poorly correlated with genome size), as genome size increased, the correlations became increasingly significant with large genomed species being excluded from the extreme environments with shorter growing seasons (i.e., low or high maximum July temperatures or reduced annual precipitation).

Given these observations one might expect alpine species which experience short growing periods to be characterized by smaller genomes than relatives growing in non-alpine habitats, and yet many have larger genomes. For example, in the study of *Veronica,* Albach and Greilhuber (2004) used phylogenetic independent contrasts (to take phylogenetic history into account), and showed that alpine species had significantly larger 1Cx-values than non-alpine ones. However, a similar analysis based on 1C-values was not significant illustrating the point that for *Veronica* at least it is the cell cycle relevant Cx-values that may be under selection in these habitats. In other species it has been suggested that large genomes may be an advantage in alpine habitats given the observed positive relationship between genome size and frost resistance (MacGillivray and Grime 1995). In addition, it has been suggested that larger genomes may be more readily tolerated in alpine species because the soils are generally more phosphate-rich at higher altitudes (Körner 1989) and thus may not be under such strong selection to limit genome size in order to conserve phosphate which is often in limiting supply in other habitats (see Sect. 20.3.4.2). It should however, be noted that Bennett (1987) argued against large genomes in alpine habitats due to the observed positive correlation between genome size and UV damage (Sparrow and Miksche 1961) and the fact that UV levels in alpine habitats are often higher than at lower elevations.

20.4.5.1 Genome Size and the Impact of Climate Change

Given the observed complexity of interactions between genome size and different climate variables, the question of how plants may respond to climate change is likely to be complex. Yet there are a number of additional studies that are worth mentioning in this respect. For example, Jasienski and Bazzaz (1995) investigated the effect of elevated levels of CO_2 on growth in annual grasses and observed that the extent of the growth response was positively correlated with genome size. In contrast, Grime (1996) reported that under elevated temperatures plants with small genomes showed a greater enhancement of growth compared with species with larger genomes, with the implication that under a scenario of global warming species with small genomes may expand their range more successfully in certain areas (e.g., temperate floras). Nevertheless, this predicted expansion could be held in check in certain areas given the findings of yet another study by MacGillivray and Grime (1995) showing that such potentially responsive species with small genomes are more sensitive to low temperatures and freezing than those with larger genomes (as noted above for alpine species). Thus the potential expansion of species with small genomes under a warming climate would be expected to be reduced in temperate areas experiencing occasional late frosts.

It is clear even from these rather few and limited studies, that predictions of how plants will respond to climate change are fraught with difficulties, yet they do emphasize that genome size does play a role and should be incorporated as a character into models which aim to understand how changes in climate will influence plant distributions in the future.

20.4.6 Genome Size and Pollution

As the world becomes increasingly polluted through human activity, some researchers have questioned whether genome size has any impact on plant survival in polluted environments. Two notable studies are those of Temsch et al. (2010) and Vidic et al. (2009) who used different approaches to investigate a heavy-metal-polluted area in Žerjav, in northern Slovenia known as Dolina Smrty. This "Death Valley" provides an unintended but ideal experimental environment to investigate the question as the soil is heavily polluted with lead from a disused metal smelter and there is a strong gradient of lead pollution with concentrations ranging from 30 g of lead per kg of soil at the site closest to the smelter to 0 g/kg soil at a control site. The studies involved surveying the species diversity and genome sizes of all herbaceous dicots growing in five

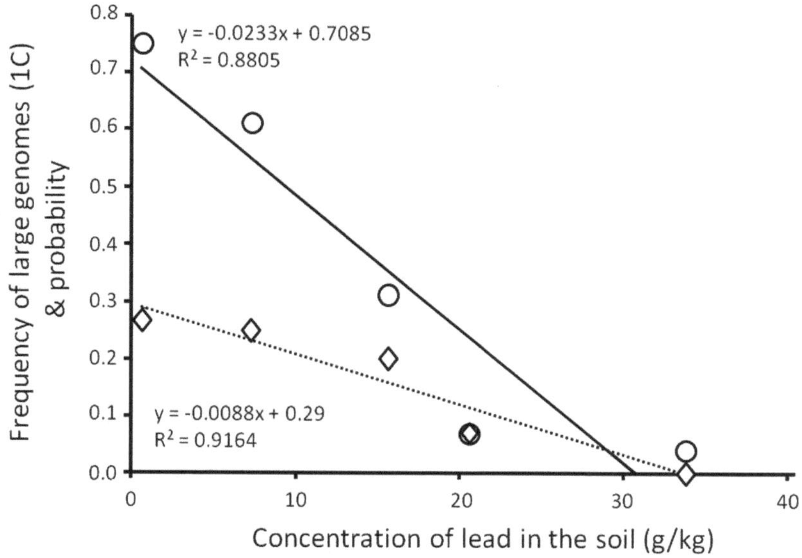

Fig. 20.3 Correlation between concentration of lead in the soil and 1C-values expressed as a percentage of species with large genomes (diamonds) and the probability of finding by chance that number or fewer large genomes (circles) at the five test plots of Dolina Smrty (Žerjav, Slovenia). (Reproduced from Temsch et al. 2010)

study plots and the findings were dramatic. Not only was the number of species progressively reduced with increasing pollution, but the genome sizes of the surviving species were, on average, progressively smaller. The results suggest that as the lead concentration in the soil increases, species with larger genomes are dying out, or are being excluded, compared to those with smaller genomes (Fig. 20.3).

Given that significant mitotic and meiotic disturbances through genotoxic stress have been observed in species with larger genomes from this area (Druškovič 1984), this may explain in part, why species with larger genomes are increasingly excluded as the level of lead pollution rises. Such a correlated sensitivity of genome size with genotoxic stress finds a parallel in the long known higher radiation sensitivity of large genomes observed by Sparrow and Miksche (1961) and may arise because larger genomes have larger and denser nuclei, and thus provide a better target for damage. Similar studies for related pollutants are clearly necessary to determine to what extent such findings may be generalized across the range of pollutants being pumped into the soil through human activity.

20.4.7 The 'Large Genome Constraint Hypothesis'

From the above discussions, it is clear that the impact of genome size on an organism is complex and, in part, is determined by genome size itself, with the consequences often becoming increasingly significant, and options available being more limited as genome size rises. For example, as Bennett (1972) noted, species with small genomes may adopt any life cycle strategy (i.e., they may be ephemeral, annual or perennial) while those with large genomes are restricted to being obligate perennials (see Sect. 20.3.2). This is a clear nucleotypic effect of genome size because even if a species with a large genome contained all the genetic code necessary to be, for example, an ephemeral, the long duration of mitosis and meiosis as a consequence of its genome size would prevent it from completing a sufficient number of cell divisions to grow fast enough to be an ephemeral. Other observations discussed above also suggest that having a large genome restricts the options available to the plant, e.g., restriction to more mesic habitats (Knight and Ackerly 2002), exclusion from being a weed or invasive (Bennett et al. 1998; Suda et al. 2005; Kapralov and Filatov 2011), exclusion from living in a heavily polluted soil (Vidic et al. 2009; Temsch et al. 2010). Studies such as these, together with the observation that most angiosperms have small genomes (see Leitch and Leitch 2013, this volume) despite the prevalence of mechanisms that have the potential to generate large genomes (e.g., polyploidy and retrotransposon amplification), suggest that there may indeed be selection against large genomes. Based on observations such as these, Knight et al. (2005) proposed the 'Large Genome Constraint Hypothesis', suggesting that species with large genomes do pay a cost and may, under certain circumstances, be selected against. In support of their hypothesis they noted how it was often necessary to adopt quantile regression analysis to tease out the nature of the relationships in such non-linear systems where the correlations only become significant at the upper end of the range of genome sizes (as noted in their climatic study—see Sect. 20.4.5).

20.4.7.1 The Large Genome Size Constraint on Speciation Rates and Extinction

Two studies have been conducted to test whether the possession of a large genome constrains the ability of a genus to speciate. Vinogradov (2003) found a negative correlation between the mean genome size of a genus and the number of species in a genus using Spearman's rank correlation analysis, although the correlation was weak ($r = -0.11$, $p < 0.001$). While Knight et al. (2005) also found a weak but significant relationship using slightly different sample data and Spearman's rank correlation analysis, they also tested quantile regression analysis to explore the nature of the relationship. While they found no significant relationship between genome size and diversification rates for species with small genomes, presumably due to the myriad of factors which can influence diversification rates, they did find an increasingly negative trend for larger genomes providing support for the hypothesis that genera with large mean genome sizes are less species rich. Nevertheless, they recognized there were several potential sources of errors in their analysis (e.g., poor representation of species in some genera, the use of mean genome sizes to represent a genus, different evolutionary ages of genera analysed) and concluded that the evidence for a role of genome size in species diversity was, at best, tentative. Indeed, it is also noted that neither analysis was conducted within a phylogenetic framework that could further confound the results.

A recent review by Kraaijeveld (2010) collated studies across eukaryotes (including vertebrates, anamniotes, amniotes, amphibians, reptiles, eutherians and fishes) which had investigated whether there was any relationship between genome size and species diversification in different groups. He concluded that the general pattern seemed to be one in which a higher diversification rate was generally found in taxa with smaller genomes with the only exception being the fishes where a positive relationship was found in all subgroups except pufferfish. Since species diversity represents the product of both speciation and extinction, Kraaijeveld (2010) noted that the absence of species-rich groups in taxa with large genomes could arise either due to constrained speciation rate, or increased extinction rate or both, although distinguishing between these possibilities is currently not possible using available techniques.

In plants there is currently only one comprehensive study that has been conducted looking at the impact of genome size on extinction (Vinogradov 2003). Using data for 3,036 species with both genome size data and information on threat of extinction taken from the United Nations World Conservation Monitoring Centre Database, Vinogradov showed that diploid species with large genomes were significantly more likely to be listed as rare or endangered compared with those with small genomes, and the results were independent of life history type or polyploidy. Although he did not use approaches such as phylogenetic independent contrasts he did also conduct analyses within families that overcame, to some extent, complications arising from phylogenetic issues. However, given the importance of phylogeny in determining the distribution of genome sizes across angiosperms (as previously noted), repeating the analysis using phylogenetically-sensitive approaches is highly recommended, especially since there are now considerably more genome size data available and a greater understanding of phylogenetic relationships across angiosperms.

Given the relationships observed between genome size and many nuclear, cellular and phenotypic characters perhaps one should not be so surprised by the suggestion that species with large genomes may be at greater risk of extinction. As discussed above, the possession of a large genome in many cases does constrain the potential of a species and hence, in a rapidly changing world the more restricted opportunities available may prevent large genomed species from adapting fast enough and hence may be more likely to be threatened with extinction.

20.5 Intrapopulational and Intraspecific Variation of Genome Size: Adaptively Important?

Over the many years of genome size research, scientists have looked for evidence of intraspecific and intrapopulation variation in C-values and there are now over 200 papers examining more than 80 genera which have addressed this question (Šmarda and Bureš 2010). If such variation can be reliably demonstrated, one might hope to find the starting point of the large interspecific variation observed across plants, and eventually to determine if there are ecological correlates to explain why such variation might arise. Indeed, if correlates could be found, then a function (selective role) of genome size at the population level and above (within a species) could be proven.

Despite the numerous investigations that have been made, based on the increased understanding of the strengths and weaknesses of the methods used to estimate genome sizes, it has become clear that many of the Feulgen densitometry-based results should be viewed with caution due to technological shortcomings (unavoidable or otherwise, see Greilhuber 1998, 2005, 2008). Nevertheless, some of the early research looking for the presence of intraspecific variation was conducted with great care, such as the work of Price and co-workers on species belonging to Microseridinae, a N. American tribe within Asteraceae. In one such study, they estimated genome sizes for 222 plants of *Microseris douglasii* from 24 geographically, ecologically and morphologically diverse populations in California and reported up to 1.20-fold variation in DNA amount among individuals within populations and 1.14-fold range in mean DNA values between populations. They also observed differences in DNA amount in the same populations in different years (Price et al. 1981). Overall the results suggested that there was an ecological component to the DNA variation. Plants with larger genomes were limited to clay soils with a higher moisture content than plants with smaller genomes, which typically occupied soils of lower moisture content. Nevertheless, when a further analysis of 210 plants from 10 populations was conducted in subsequent years the expected confirmation of observing higher DNA contents from individuals and populations experiencing increased soil moisture did not hold, highlighting the complexity of the relationship between genome size and environmental parameters (Price et al. 1986). To date, the findings of Price and colleagues have not been confirmed using flow cytometry or investigated with molecular tools. It is therefore unresolved whether there is genuine intraspecific variation in *Microseris* or whether the results are a technological artefact arising from variable amounts of cytosolic inhibitors which can interfere with the quantitative staining of DNA and lead to pseudo-intraspecific variation, as observed for example in sunflowers (Price et al. 2000) and coffee (Noirot et al. 2000; Noirot et al. 2003).

Other studies reporting extensive intraspecific variation from this era include the work of Mowforth and Grime (1989) who noted considerable intrapopulational variation in the grass *Poa annua*, with values between individuals ranging 1.8-fold. Chromosome counts confirmed that all individuals had the same number of $2n = 28$, thereby eliminating the possibility that variation had arisen from the presence of B chromosomes, aneuploidy or polyploidy. In addition, the variation was shown to be positively correlated with cell size and negatively associated with variation in seedling growth rate. The suggestion that there may be selection for different genome sizes attuned to different temporal and climatic niches was proposed to explain the variation (Mowforth and Grime 1989). As for *Microseris douglasii*, there has been no further work on this species and so the results remain unconfirmed. Nevertheless, it seems a perfect candidate to reinvestigate, especially given that many of the reports of intraspecific variation arising during this period have since been reinvestigated and shown to be unreliable such as *Dasypyrum villosum, Glycine max, Arachis hypogaea, Collinsia verna* (reviewed by Greilhuber 1998, 2005).

In recent years it has become clear that even with carefully conducted experiments, Feulgen microdensitometry is no longer considered to be a suitable technique to detect intraspecific variation, given its limitations in the speed and number of samples that can be analysed. Instead, the increasing refinement, accuracy and sensitivity of flow cytometry (assuming best practice methods are followed) means that it has now become the method of choice for detecting intraspecific variation. Flow cytometry is also considerably faster than Feulgen-based approaches enabling the rapid analysis of large numbers of nuclei from many individuals. Best practice techniques for convincingly demonstrating the presence of intraspecific variation have recently been outlined by Šmarda and Bureš (2010) and include: (1) the presence of double peaks in the flow histogram when two individuals differing in genome size are co-chopped and run as a single sample in the flow cytometer, (2) the consistent use of an internal standard for all measurements, (3) demonstrating that the variation is reproducible by repeating the measurements on different days, using different machines and/or different fluorochromes, and (4) testing for the presence of cytosolic inhibitors which have the potential to interfere with the quantitative staining of DNA by the flurochrome. Such developments have now opened up the possibility of investigating the true extent of intraspecific variation for the first time. Indeed, there are now an increasing number of carefully conducted studies which have followed these best practice approaches including the demonstration of double peaks indicative of intraspecific variation (e.g., Pecinka et al. 2006; Šmarda 2006; Šmarda and Bureš 2006; Walker et al. 2006; Leong-Škornickovà et al. 2007; Schönswetter et al. 2007; Suda et al. 2007; Achigan-Dako et al. 2008).

Below several studies are outlined which have sought to document the extent of intraspecific variation in genome size diversity at various spatial and temporal scales and to link this with various environmental factors to provide greater understanding of the forces which might be driving the very first incipient genome size changes which could ultimately lead to the extensive genome size diversity observed across plants.

20.5.1 Evolution Canyon

An area that has been extensively studied in terms of investigating intraspecific variation in genome size is Evolution Canyon I, Lower Nahal Oren, Mount Carmel, in northern Israel (Nevo 2012). The 400 m wide canyon, which dates back to the Plio-Pleistocene era (Nevo et al. 1999), is orientated in an east-west direction and thus presents north and south facing slopes. Both slopes share common geologies and macroclimates but have strongly contrasting microclimates differing in the amount of solar irradiation and aridity. The savannah-like south facing slope (SFS) receives 200–800% more irradiation and is more arid ('Africa-toned') than the forested north facing slope (NFS) which is wetter ('Europe-toned'). Consequently species on the NFS are considered to be under shade stress whereas those on the SFS are under drought stress (Nevo 2012). Given these steep ecological gradients encountered in Evolution Canyon I a number of studies have looked to see whether there is any evidence of intraspecific variation in genome size which can be linked with environmental factors.

20.5.1.1 *Hordeum spontaneum*

Of special relevance are the studies in *Hordeum* species, especially those of *H. spontaneum*, by Schulman and associates, because here genome size, transposable element (TE) variation, and ecology have been considered together. *Hordeum spontaneum* ($2n = 14$) is the wild progenitor of barley, *H. vulgare,* and previous studies have demonstrated the existence of intraspecific variation of up to 1.13-fold in accessions collected from Israel (Kankanpää et al. 1996; Turpeinen et al. 1999). Like other *Hordeum* species, *H. spontaneum* has been shown to contain a transcriptionally-active TE known as *BARE*-1. This is a LTR-retrotransposon belonging to the *copia* family, and is c. 8,932 bp in length (Manninen and Schulman 1993) or 12,088 bp if a 3.14 kb insertion in the 3′ long terminal repeat is present (Suoniemi et al. 1996a).

In a study of *BARE*-1 in different *Hordeum* species, Vicient et al. (1999) analysed nine different accessions of *H. spontaneum* collected from across Israel and showed not only that there was intraspecific variation in genome size between them ($1C = 4.14$–4.68 pg) but that this was correlated with the contribution of *BARE*-1 to the genome, which ranged from 2.8% to 5.8%. Further they noted that the genome size appeared to be higher in accessions collected from desert areas, suggesting a link between genome size, *BARE*-1 copy number and aridity.

To determine whether such a link was also found on the microclimatic scale, Kalendar et al. (2000) conducted a similar analysis to that of Vicient et al. at Evolution Canyon I by sampling individual plants of *H. spontaneum* at six sites across a 300 m transect which traversed the NFS and SFS and differed in height. The study certainly showed that there was considerable variation in the number of full length *BARE*-1 elements, with copy numbers ranging between 8.3 and 22.1×10^3 copies per genome, comprising between 1.8% and 4.7% of the genome. In addition, they found differences in the number of 'solo long terminal repeats' (= solo LTRs) between the sites. Solo LTRs represent 'footprints' of *BARE*-1 elements that have undergone intra-element or possibly intra-chromosomal recombination between the LTRs of the *BARE*-1 element resulting in the elimination of the internal portion of *BARE*-1.

What was most noticeable was the correlation between environmental factors and the copy number of either full length *BARE*-1 elements or solo LTRs. For example, the most arid sites were found to have both the highest copy number of full length elements and the lowest number of solo LTRs. Such results suggest that under stressful conditions imposed by aridity, *BARE*-1 activity may be triggered (possibly via epigenetic modifications) while recombination may be reduced. It may be particularly relevant to note that abscissic acid (ABA) response elements, which are typical of water-stress induced genes, are present within the *BARE*-1 promoter (Suoniemi et al. 1996b) providing further support for the suggestion that water stress may, in part, be responsible for the observed *BARE*-1 proliferation in the most water-stressed sites on the SFS and that *BARE*-1 may have a function for stress resistance and be under selection.

Nevertheless, it is noted that despite differences in copy number of *BARE*-1 between the study sites, there was no evidence of significant intraspecific differences in genome size between the six sites, although pooling the data for NFS and SFS showed that the mean genome sizes between these two slopes were significantly different, with the larger genomes found in the SFS sample. It seems that at this microscale, other repetitive elements within the genome may also be fluctuating in numbers to compensate for the differences in copy number of *BARE*-1 (Kalendar et al. 2000).

20.5.1.2 *Ceratonia siliqua, Lotus peregrinus, Cyclamen persicum*

To investigate to what extent the results of *H. spontaneum* represent a more general response to changes in microclimate at Evolution Canyon I, several additional studies have been conducted on plants with contrasting morphologies and growth characteristics.

The first of these was by Bureš et al. (2004) who investigated the carob tree *Ceratonia siliqua* (Fabaceae). Like *H. spontaneum*, *C. siliqua* is also common in Evolution Canyon I but exhibits a very different life strategy. In contrast to *H. spontaneum,* which is a herbaceous annual grass, *C. siliqua* is a long lived perennial tree. Using flow cytometry Bureš et al. estimated genome sizes in 79 adult trees and 44 seedlings from below the same trees. Overall, genome sizes ranged 1.04-fold, while between the NFS and SFS the difference was considerably less, although still significant, with genome sizes differing just 0.9% (d.f. $=$ 77, $p < 0.001$). As observed for *H. spontaneum*, the south-facing ("African") slope exhibited the larger genomes. Indeed, using multiple regression analysis, the most important factor for genome size variation was shown to be slope orientation (north- or south-facing) ($p = 0.0007$), followed by tree trunk circumference and leaf length ($p = 0.06$). Interestingly, seedlings did not show significant differences in C-values between slopes, suggesting the possibility of microclimatically-mediated selection, with the more stressful arid sites on the SFS favoring, directly or indirectly, larger genomes.

A second study was conducted by Gasmanová et al. (2007), this time on the annual, mostly inbreeding bird's foot trefoil *Lotus peregrinus* (Fabaceae). Genome sizes were estimated in 498 individuals (representing 10 genotypes) collected from six sites along the 300 m transect of Evolution Canyon I. Like previous studies, intraspecific variation was detected with C-values ranging 1.13-fold, and the mean C-value for the arid-stressed SFS (2.549 pg/2C) was shown to be larger than the mean value for the NFS (2.544 pg/2C). Nevertheless, Gasmanová et al. (2007) did not find significant interslope differences in genome size, based on the number of individuals analysed.

Most recently, Pavliček et al. (2008), examined yet another contrasting growth form, a geophyte called the Persian violet (*Cyclamen persicum*, Primulaceae), and once again detected intraspecific variation in genome size, although values were less variable, ranging only 1.06-fold in the 75 samples analysed. They were unable to detect significant differences in genome size between the two slopes. Nevertheless, when the data were combined with previous studies of *C. siliqua*, *L. peregrinus* and a beetle (*Oryzaephilus surinamensis*, Sharaf et al. 2008), in a meta-analysis a positive relationship between drought and genome size was shown to be significant at Evolution Canyon I (Pavliček et al. 2008).

Given the lack of extensive differences in genome size of the plant species studied between the two contrasting microsites of Evolution Canyon I, it remains to be determined to what extent genome size variation may be adaptively significant or whether the subtle, and often non-significant differences between the drier and more stressed SFS and the wetter NFS simply represent correlations rather than causations. To date it is certainly unclear how such small differences in genome size could be visible to selective forces unless selection is operating on one or more aspects of genome size that are not yet recognized as being adaptively relevant (Wendel and Wessler 2000). Indeed, it is likely that different species will be under different selection pressures. In a study of seven leaf characters (e.g., leaf size, leaf thickness) from three woody species in Evolution Canyon I (*Olea europaea*, *Ceratonia siliqua* and *Pistacia lentiscus*), Nevo (2012) reported that all displayed drought resistant adaptations on the SFS compared with the NFS, although the adaptations differed between species. Of most relevance to genome size was the length of the palisade cells (which contributes to leaf thickness, with the thicker leaves characteristic of more drought-adapted plants). The palisade cells were significantly longer in plants of *C. siliqua* from the more arid SFS compared with the NFS. Given the link between genome size and cell size, it is perhaps tempting to speculate that this is one possible place where selection of genome size may be linked to drought adaptation at least in *C. siliqua*. In the other two species, either the positive relationship was only weakly significant (*Pistacia lentiscus*) or the relationship was reversed (i.e., the palisade cells were shorter in the SFS as in *Olea europaea*) or non-significant. Such studies only serve to highlight further the complexity of the interactions and indicate that additional studies are clearly needed.

20.5.2 *Festuca pallens*

So far, intraspecific variation has been clearly demonstrated by double peaks (see above) in four species of the fine-leaved fescues of *Festuca*, i.e., *F. pallens*, *F. polesica*, *F. rupicola* and *F. vaginata* (Šmarda 2006) but the most significant and extensive studies have been made in *F. pallens* which exibits two ploidy levels, diploids with $2n = 14$ and tetraploids with $2n = 28$ (Šmarda and Bureš 2006; Šmarda et al. 2007, 2008, 2010).

The initial investigations of *F. pallens* were conducted at various geographical scales, ranging from the intrapopulation level to surveys across the entire distribution range for the species (Šmarda and Bureš 2006). Such studies uncovered intraspecific differences of nearly 1.2-fold in genome sizes in both diploid and tetraploid cytotypes. At the geographical scale, the larger genomes were observed to predominate in the southeastern part of their range, in contrast to the northwestern populations which were generally characterized by smaller genomes. Such patterns were interpreted as reflecting the glacial history and post-glacial migrations in the region rather than due to selection arising from any environmental factors (Šmarda and Bureš 2006; Šmarda et al. 2007).

Subsequent studies focused on 17 adult plants and 562 progeny and revealed that the range of genome sizes was actually greater in the offspring produced in a single generation than in all the maternal plants studied (Šmarda et al. 2008). Even within a single panicle, the genome sizes of their seedlings were shown to vary up to 1.117-fold. Clearly, in *F. pallens* at least, intraspecific variation is dynamic. Not only can it be generated very rapidly (i.e., in a single generation), but also it is apparent that continuous genome size variation is being added to the population every generation. The question as to whether such variation is under stabilizing selection to minimize variation in established adults was addressed by Šmarda et al. (2010). In a carefully controlled experiment comparing the distribution of genome sizes in a simulated competitive environment, Šmarda et al. demonstrated that the genome size of *F. pallens* is maintained by stabilizing selection for an 'optimal' genome size, although it is weak given the persistent presence of intraspecific variation in the adult plants. Indeed, further studies are needed to determine whether a change in, for example, environmental conditions, would result in a shift of the 'optimal' genome size.

As to how such selection may be operating is currently unclear. Indeed, the lack of observed correlations between genome size and a range of phenotypic characters such as seedling growth, measures of plant fitness, spatial location in the test area, vegetation type or microclimatic conditions (Šmarda et al. 2007, 2008) suggest that, at least in *F. pallens*, selection on genome size is probably indirect and mediated through some trait that interacts directly with the environment (Beaulieu 2010). It would appear therefore that at least during the very initial stages of genome size diversification, genome size may be non-adaptive (Gregory 2001; Bennetzen et al. 2005).

20.6 Conclusions and Future Outlook

From the foregoing discussions, it is clear that there are an impressive array of studies that have reported relationships between genome size and phenotype at the nuclear, cellular and whole plant levels and it is likely that more will be uncovered in the future. Yet it must be borne in mind that in nearly all cases, such relationships remain as correlations with no clear understanding of the molecular and evolutionary forces underpinning the observed nucleotypic effects on the phenotype. Indeed, as is clear from the studies of intraspecific variation (Sect. 20.5), we are still grossly ignorant of how changes in genome size are visible to selection and the mechanisms operating which may lead to shifts in genome size within and between species.

As we enter the next era of genomics, characterized by the overwhelming amount of DNA sequence data being generated by next generation and third generation sequencing technologies, it seems likely that there will be advances in understanding genome size dynamics at the molecular level and the evolutionary forces which may be responsible for generating the diversity of genome sizes observed (e.g., Novák et al. 2010; Kelly and Leitch 2011; Macas et al. 2011; Kelly et al. 2012). Linking this through to the whole plant and phenotypic level will however, require additional understanding and insights from the fields of ecology and physiology as noted by Cavalier-Smith (2005). In addition, the extent to which phylogenetic history plays a role in determining how shifts in genome size at the molecular level are translated into phenotypic effects will be essential. Already there are hints that such phylogenetic effects may be important players. For example, comparisons of genome size diversity across angiosperm families show huge differences between them, with some characterized by extensive variation (e.g., Orchidaceae, Melanthiaceae) while others have much narrower ranges of genome size despite good species representation (e.g., Brassicaceae, Lysák et al. 2009). In addition, the observation that species with truly enormous genomes are phylogenetically clustered to just two orders of monocots (Liliales and Asparagales) and one order of eudicots (Santalales) within the angiosperms (Leitch et al. 2010; Leitch and Leitch 2013) rather than being scattered across angiosperms suggests phylogenetic history may be an important component. The very different genome size profiles observed between angiosperms and gymnosperms (see Leitch and Leitch 2013, this volume) may also, in part, arise from the fact that these two groups of seed plants shared a last common ancestor c. 300 million years ago and have since undergone very different evolutionary trajectories. This is also reflected in their contrasting genome dynamics (Leitch and Leitch 2012).

At the intraspecifc level, we are still grossly ignorant of many aspects of this source of genome size variation such as the factors which effect how often, how quickly and how much intraspecific variation may be generated. In addition, the influence of selection forces, population size and age are also poorly understood and need further investigation (Šmarda and Bureš 2010). Given the improvements in flow cytometry sensitivity, the tools are now at hand to enable at least some of these unknowns to be addressed to give us a more comprehensive view of the role intraspecific variation plays in generating the diversity of genome sizes observed across plants.

Since ultimately the phenotype of a plant will determine where, when and how a plant may grow, it is essential that the role genome size plays in influencing the phenotype is more fully appreciated and understood, especially if we are to predict how plants will respond to the rapidly changing world affected by climate change, increasing CO_2 levels and pollution. As discussed above, the limited studies that have investigated this indicate that genome size may well be important (e.g., see Sects. 20.4.5, 20.4.6). Nevertheless, taken together, they also highlight the complexity of interactions between genome size, phenotype, and a plant's response to environmental change. At the very least, such studies emphasize the need to obtain and incorporate genome size data into computer models and laboratory experiments aiming to understand how plants will respond to the many facets of environmental change induced through human activity.

References

Achigan-Dako E, Fuchs J, Ahanchede A, Blattner F (2008) Flow cytometric analysis in *Lagenaria siceraria* (Cucurbitaceae) indicates correlation of genome size with usage types and growing elevation. Plant Syst Evol 276:9–19

Acquisti C, Kumar S, Elser JJ (2009) Signatures of nitrogen limitation in the elemental composition of the proteins involved in the metabolic apparatus. Proc Roy Soc Lond B Bio 276:2605–2610

Albach DC, Greilhuber J (2004) Genome size variation and evolution in *Veronica*. Ann Bot 94:897–911

Albert VA, Jobson RW, Michael TP, Taylor DJ (2010) The carnivorous bladderwort (*Utricularia*, Lentibulariaceae): a system inflates. J Exp Bot 61:5–9

Ayonoadu U, Rees H (1968) The influence of B-chromosomes on chiasma frequencies in Black Mexican Sweet Corn. Genetica 39:75–81

Bancheva S, Greilhuber J (2006) Genome size in Bulgarian *Centaurea* s.l. (Asteraceae). Plant Syst Evol 257:95–117

Baranyi M, Greilhuber J (1999) Genome size in *Allium*: in quest of reproducible data. Ann Bot 83:687–695

Barrett SCH, Harder LD, Worley AC (1997) The comparative biology of pollination and mating in flowering plants. In: Silvertown J, Franco M, Harper JL (eds) Plant life histories: ecology, phylogeny, and evolution. Cambridge University Press, Cambridge, pp 57–76

Beadle NCW (1962) Soil phophate and the delimitation on plant communitites in eastern Australia II. Ecology 43:281–288

Beaton MJ, Cavalier-Smith T (1999) Eukaryotic non-coding DNA is functional: evidence from the differential scaling of cryptomonad genomes. Proc Roy Soc Lond B Bio 266:2053–2059

Beaulieu JM (2010) Commentary: the right stuff: evidence for an "optimal" genome size in a wild grass population. New Phytol 187:883–885

Beaulieu JM, Leitch IJ, Patel S, Pendharkar A, Knight CA (2008) Genome size is a strong predictor of cell size and stomatal density in angiosperms. New Phytol 179:975–986

Bennett MD (1971) The duration of meiosis. Proc Roy Soc Lond B Bio 178:277–299

Bennett MD (1972) Nuclear DNA content and minimum generation time in herbaceous plants. Proc Roy Soc Lond B Bio 181:109–135

Bennett MD (1987) Variation in genomic form in plants and its ecological implications. New Phytol 106:177–200

Bennett MD (1998) Plant genome values: How much do we know? Proc Natl Acad Sci USA 95:2011–2016

Bennett MD, Leitch IJ (2005) Nuclear DNA amounts in angiosperms—progress, problems and prospects. Ann Bot 95:45–90

Bennett MD, Leitch IJ (2010) Plant DNA C-values database (release 5.0, Dec. 2010). http://data.kew.org/cvalues/

Bennett MD, Smith JB (1972) The effects of polyploidy on meiotic duration and pollen development in cereal anthers. Proc Roy Soc Lond B Bio 181:81–107

Bennett MD, Leitch IJ, Hanson L (1998) DNA amounts in two samples of angiosperm weeds. Ann Bot 82:121–134

Bennett MD, Leitch IJ, Price HJ, Johnston JS (2003) Comparisons with *Caenorhabditis* (~100 Mb) and *Drosophila* (~175 Mb) using flow cytometry show genome size in *Arabidopsis* to be ~157 Mb and thus 25% larger than the *Arabidopsis* Genome Initiative estimate of ~125 Mb. Ann Bot 91:547–557

Bennetzen JL (2005) Transposable elements, gene creation and genome rearrangement in flowering plants. Curr Opin Genet Dev 15:621–627

Bennetzen JL, Ma J, Devos KM (2005) Mechanisms of recent genome size variation in flowering plants. Ann Bot 95:127–132

Boivin A, Vendrely R, Vendrely C (1948) L'acide desoxyribonucleique du noyau cellulaire dépositaire des caracteres hereditaires: arguments d'ordre analytique. C R Acad Sci 226:1061–1063

Bragg JG, Wagner A (2009) Protein material costs: single atoms can make an evolutionary difference. Trends Genet 25:5–8

Bureš P, Pavlíček T, Horová L, Nevo E (2004) Microgeographic genome size differentiation of the carob tree, *Ceratonia siliqua*, at 'Evolution Canyon', Israel. Ann Bot 93:529–535

Cavalier-Smith T (1978) Nuclear volume control by nucleoskeletal DNA, selection for cell volume and cell growth rate, and the solution of the DNA C-value paradox. J Cell Sci 34:247–278

Cavalier-Smith T (1985) Eukaryotic gene numbers, non-coding DNA and genome size. In: Cavalier-Smith T (ed) The evolution of genome size. John Wiley & Sons Ltd, Chichester, pp 69–103

Cavalier-Smith T (2002) Nucleomorphs: enslaved algal nuclei. Curr Opin Microbiol 5:612–619

Cavalier-Smith T (2005) Economy, speed and size matter: evolutionary forces driving nuclear genome miniaturisation and expansion. Ann Bot 95:147–175

Chen GQ, Guo SL, Yin LP (2010) Applying DNA C-values to evaluate invasiveness of angiosperms: validity and limitation. Biol Invasions 12:1335–1348

Commoner B (1964) Roles of deoxyribonucleic acid in inheritance. Nature 202:960–968

Doolittle WF, Sapienza C (1980) Selfish genes, the phenotype paradigm and genome evolution. Nature 284:601–603

Douglas S, Zauner S, Fraunholz M, Beaton M, Penny S et al. (2001) The highly reduced genome of an enslaved algal nucleus. Nature 410:1091–1096

Druškovič B (1984) Uticaj zagadjenja sredine na genetske promene u biljnim populacijama. Thesis, University of Novi Sad, Novi Sad

Edwards GA, Endrizzi JE (1975) Cell size, nuclear size and DNA content relationships in *Gossypium*. Can J Genet Cytol 17:181–186

Ehrendorfer F (1970) Evolutionary patterns and strategies in seed plants. Taxon 19:185–195

Evans GM, Rees H (1971) Mitotic cycles in dicotyledons and monocotyledons. Nature 233:350–351

Evans GM, Rees H, Snell CL, Sun S (1972) The relationship between nuclear DNA amount and the duration of the mitotic cycle. Chromosomes Today 3:24–31

Fawcett JA, Van de Peer Y, Maere S (2013) Significance and biological consequences of polyploidization in land plants. In: Leitch IJ, Greilhuber J, Doležel J, Wendel JF (eds) Plant genome diversity, vol 2, Physical structure, behaviour and evolution of plant genomes. Springer-Verlag, Wien, pp 277–293

Felsenstein J (1985) Phylogenies and the comparative method. Am Nat 125:1–15

Flavell RB (1980) The molecular characterisation and organisation of plant chromosomal DNA sequences. Ann Rev Plant Physiol 31:569–596

Francis D, Davies MS, Barlow PB (2008) A strong nucleotypic effect of DNA C-value on the cell cycle regardless of ploidy level. Ann Bot 101:747–757

Gallagher RV, Leishman MR, Miller JT, Hui C, Richardson DM, Suda J, Trávnícek P (2011) Invasiveness in introduced Australian acacias: the role of species traits and genome size. Diversity Distrib 17:884–897

Garcia S, Canela MA, Garnatje T, McArthur ED, Pellicer J, Sanderson SC, Valles J (2008) Evolutionary and ecological implications of genome size in the North American endemic sagebrushes and allies (*Artemisia*, Asteraceae). Biol J Linn Soc 94:631–649

Gasmanová N, Lebeda A, Dolezalová I, Cohen T, Pavlíček T, Fahima T, Nevo E (2007) Genome size variation of *Lotus peregrinus* at Evolution Canyon I microsite, Lower Nahal Oren, Mt. Carmel, Israel. Acta Biol Cracov Ser Bot 49:39–46

Gilson PR, Su V, Slamovits CH, Reith ME, Keeling PJ, McFadden GI (2006) Complete nucleotide sequence of the chlorarachniophyte nucleomorph: nature's smallest nucleus. Proc Natl Acad Sci USA 103:9566–9571

Govindaraju DR, Cullis CA (1991) Modulation of genome size in plants—the influence of breeding systems and neighborhood size. Evol Trends Plants 5:43–51

Graham MJ, Nickell CD, Rayburn AL (1994) Relationship between genome size and maturity group in soybean. Theor Appl Genet 88:429–432

Green BR (2011) Chloroplast genomes of photosynthetic eukaryotes. Plant J 66:34–44

Gregory TR (2001) Coincidence, coevolution, or causation? DNA content, cell size, and the C-value enigma. Biol Rev 76:65–101

Gregory TR (2005) Genome size evolution in animals. In: Gregory TR (ed) The evolution of the genome. Elsevier, San Diego, pp 3–87

Gregory TR, Andrews CB, McGuire JA, Witt CC (2009) The smallest avian genomes are found in hummingbirds. Proc Roy Soc Lond B Bio 276:3753–3757

Greilhuber J (1995) Chromosomes of the monocotyledons (general aspects). In: Rudall PJ, Cribb PJ, Humphries CJ (eds)

Monocotyledons: systematics and evolution. Whitstable, Royal Botanic Gardens, Kew, pp 379–414

Greilhuber J (1998) Intraspecific variation in genome size: a critical reassessment. Ann Bot 82(suppl A):27–35

Greilhuber J (2005) Intraspecific variation in genome size in angiosperms—identifying its existence. Ann Bot 95:91–98

Greilhuber J (2008) Cytochemistry and C-values—the less well known world of nuclear DNA amounts. Ann Bot 101:791–804

Greilhuber J, Obermayer R (1997) Genome size and maturity group in *Glycine max* (soybean). Heredity 78:547–551

Greilhuber J, Borsch T, Müller K, Worberg A, Porembski S, Barthlott W (2006) Smallest angiosperm genomes found in Lentibulariaceae with chromosomes of bacterial size. Plant Biol 8:770–777

Greilhuber J, Temsch EM, Loureiro J (2007) Nuclear DNA content measurement. In: Doležel J, Greilhuber J, Suda J (eds) Flow cytometry with plant cells. Wiley, Weinheim, pp 67–102

Grif VG (2000) Some aspects of plant karyology and karyosystematics. Int Rev Cytol 196:131–175

Grif VG, Ivanov VB, Machs EM (2002) Cell cycle and its parameters in flowering plants. Tsitologija 44:936–980

Grime JP (1983) Predictions of weed and crop response to climate based upon measurements of nuclear DNA content. Aspects Appl Biol 4:87–98

Grime JP (1996) Testing predictions of the impacts of global change on terrestrial ecosystems. Aspects Appl Biol 45:3–13

Grime JP, Mowforth MA (1982) Variation in genome size—an ecological interpretation. Nature 299:151–153

Grotkopp E, Rejmánek M, Sanderson MJ, Rost TL (2004) Evolution of genome size in Pines (*Pinus*) and its life-history correlates: supertree analyses. Evolution 58:1705–1729

Grover C, Wendel JF (2010) Recent insights into mechanisms of genome size change in plants. J Bot 2010: Article ID 382732 (8 pp)

Gulliver G (1875) Observations on the sizes and shapes of the red corpuscles of the blood of vertebrates, with drawings of them to a uniform scale, and extended and revised tables of measurements. Proc Zool Soc Lond 1875:474–495

Hanson L, McMahon KA, Johnson MAT, Bennett MD (2001a) First nuclear DNA C-values for another 25 angiosperm families. Ann Bot 88:851–858

Hanson L, McMahon KA, Johnson MAT, Bennett MD (2001b) First nuclear DNA C-values for 25 angiosperm families. Ann Bot 87:251–258

Hertwig R (1903) Über Korrelation von Zell- und Kerngröße und ihre Bedeutung für die geschlechtliche Differenzierung und die Teilung der Zelle. Biol Centralbl 23:49–62

Houben A, Demidov D, Gernand D, Meister A, Leach CR, Schubert I (2003) Methylation of histone H3 in euchromatin of plant chromosomes depends on basic nuclear DNA content. Plant J 33:967–973

Houben A, Moghaddam AMB, Klemme S (2013) Biology and evolution of B chromosomes. In: Leitch IJ, Greilhuber J, Doležel J, Wendel JF (eds) Plant genome diversity, vol 2, Physical structure, behaviour and evolution of plant genomes. Wien, pp 149–165

Jasienski M, Bazzaz FA (1995) Genome size and high CO_2. Nature 376:559–560

Jobson RW, Albert VA (2002) Molecular rates parallel diversification contrasts between carnivorous plant sister lineages. Cladistics 18:127–136

Kalendar R, Tanskanen J, Immonen S, Nevo E, Schulman AH (2000) Genome evolution of wild barley (*Hordeum spontaneum*) by *BARE*-1 retrotransposon dynamics in response to sharp microclimatic divergence. Proc Natl Acad Sci USA 97:6603–6607

Kankanpää J, Mannonen L, Schulman AH (1996) The genome sizes of *Hordeum* species show considerable variation. Genome 39:730–735

Kapralov MV, Filatov DA (2011) Does large genome size limit speciation in endemic island floras? J Bot 2011:Article ID 458684 (6 pp)

Kejnovsky E, Hawkins JS, Feschotte C (2012) Plant transposable elements: biology and evolution. In: Wendel JF, Greilhuber J, Doležel J, Leitch IJ (eds) Plant genome diversity, vol 1, Plant genomes, their residents, and their evolutionary dynamics. Springer-Verlag, Wien, pp 17–34

Kelly LJ, Leitch IJ (2011) Exploring giant plant genomes with next-generation sequencing technology. Chromosome Res 19:939–953

Kelly L, Leitch AR, Fay MF, Renny-Byfield S, Pellicer J, Macas J, Leitch IJ (2012) Why size really matters when sequencing plant genomes. Plant Ecol Divers (in press) DOI:10.1080/17550874.2012.716868

Knight CA, Ackerly DD (2002) Variation in nuclear DNA content across environmental gradients: a quantile regression analysis. Ecol Lett 5:66–76

Knight CA, Molinari NA, Petrov DA (2005) The large genome constraint hypothesis: evolution, ecology and phenotype. Ann Bot 95:177–190

Knight CA, Clancy RB, Götzenberger L, Dann L, Beaulieu JM (2010) On the relationship between pollen size and genome size. J Bot 2010: Article ID 612017 (18 pp)

Körner C (1989) The nutritional status of plants from high altitudes. Oecologia 81:379–391

Kraaijeveld K (2010) Genome size and species diversification. Evol Biol 37:227–233

Kubešová M, Moravcová L, Suda J, Jarošik V, Pyšek P (2010) Naturalized plants have smaller genomes than their non-invading relatives: a flow cytometric analysis of the Czech alien flora. Preslia 82:81–96

Labani RM, Elkington TT (1987) Nuclear DNA variation in the genus *Allium* L. (Liliaceae). Heredity 59:119–128

Lavergne S, Muenke NJ, Molofsky J (2010) Genome size reduction can trigger rapid phenotypic evolution in invasive plants. Ann Bot 105:109–116

Leitch IJ, Bennett MD (2004) Genome downsizing in polyploid plants. Biol J Linn Soc 82:651–663

Leitch IJ, Bennett MD (2007) Genome size and its uses: the impact of flow cytometry. In: Doležel J, Greilhuber J, Suda J (eds) Flow cytometry with plant cells. Wiley-VCH, Weinheim, pp 153–176

Leitch AR, Leitch IJ (2008) Perspective—genomic plasticity and the diversity of polyploid plants. Science 320:481–483

Leitch AR, Leitch IJ (2012) Ecological and genetic factors linked to contrasting genome dynamics in seed plants. New Phytol 194:629–646

Leitch IJ, Leitch AR (2013) Genome size diversity and evolution in land plants. In: Leitch IJ, Greilhuber J, Doležel J, Wendel JF (eds) Plant genome diversity, vol 2, Physical structure, behaviour and evolution of plant genomes. Springer-Verlag, Wien, pp 307–322

Leitch IJ, Beaulieu JM, Chase MW, Leitch AR, Fay MF (2010) Genome size dynamics and evolution in monocots. J Bot 2010: Article ID 862516

Leong-Škornickovà J, Šída O, Jarolímová V, Sabu M, Fér T, Trávnícek P, Suda J (2007) Chromosome numbers and genome size variation in Indian species of *Curcuma* (Zingiberaceae). Ann Bot 100:505–526

Lynch M, Conery JS (2003) The origins of genome complexity. Science 302:1401–1404

Lysák MA, Koch MA, Beaulieu JM, Meister A, Leitch IJ (2009) The dynamic ups and downs of genome size evolution in Brassicaceae. Mol Biol Evol 26:85–98

Macas J, Kejnovský E, Neumann P, Novák P, Koblížková A, Vyskot B (2011) Next generation sequencing-based analysis of repetitive

DNA in the model dioceous plant *Silene latifolia*. PLoS One 6:e27335

MacGillivray CW, Grime JP (1995) Genome size predicts frost-resistance in British herbaceous plants—implications for rates of vegetation response to global warming. Funct Ecol 9:320–325

Maluszynska J, Kolano B, Sas-Nowosielska H (2013) Endopolyploidy in plants. In: Leitch IJ, Greilhuber J, Doležel J, Wendel JF (eds) Plant genome diversity, vol 2, Physical structure, behaviour and evolution of plant genomes. Springer-Verlag, Wien, pp 99–119

Manninen I, Schulman AH (1993) *BARE*-1, a copia-like retroelement in barley (*Hordeum vulgare* L.). Plant Mol Biol 22:829–846

Martin PG (1966) Variation in amounts of nucleic acids in cells of different species of higher plants. Exp Cell Res 44:84–94

Martin PG, Shanks R (1966) Does *Vicia faba* have multi-stranded chromosomes. Nature 211:650–651

Mira A, Ochman H, Moran NA (2001) Deletional bias and the evolution of bacterial genomes. Trends Genet 17:589–596

Morgan MT (2001) Transposable element number in mixed mating populations. Genet Res 77:261–275

Mowforth MA, Grime JP (1989) Intra-population variation in nuclear DNA amount, cell size and growth rate in *Poa annua* L. Funct Ecol 3:289–295

Müller KF, Borsch T, Legendre L, Porembski S, Barthlott W (2006) Recent progress in understanding the evolution of carnivorous Lentibulariaceae (Lamiales). Plant Biol 8:748–757

Murín A (1976) Polyploidy and mitotic cycle. Nucleus 19:192–195

Nevo E (2012) "Evolution Canyon", a potential microscale monitor of global warming across life. Proc Natl Acad Sci USA 109:2960–2965

Nevo E, Fragman O, Dafni A, Beiles A (1999) Biodiversity and interslope divergence of vascular plants caused by microclimatic differences at "Evolution Canyon", Lower Nahal Oren, Mount Carmel, Israel. Isr J Plant Sci 47:49–59

Noirot M, Barre P, Louarn J, Duperray C, Hamon S (2000) Nucleus-cytosol interactions—A source of stoichiometric error in flow cytometric estimation of nuclear DNA content in plants. Ann Bot 86:309–316

Noirot M, Barre P, Duperray C, Louarn J, Hamon S (2003) Effects of caffeine and chlorogenic acid on propidium iodide accessibility to DNA: consequences on genome size evaluation in coffee tree. Ann Bot 92:259–264

Novák P, Neumann P, Macas J (2010) Graph-based clustering and characterization of repetitive sequences in next-generation sequencing data. BMC Bioinformatics 11:378

Nussaume L, Kanno S, Javot Hln, Marin E, Nakanishi TM, Thibaud M-C (2011) Phosphate import in plants: focus on the PHT1 transporters. Front Plant Sci 2: Article 83

Ogur M, Erickson RO, Rosen GU, Sax KB, Holden C (1951) Nucleic acids in relation to cell division in *Lilium longiflorum*. Exp Cell Res 2:73–89

Ohri D, Pistrick K (2001) Phenology and genome size variation in *Allium* L.—a tight correlation? Plant Biol 3:654–660

Oliver MJ, Petrov D, Ackerly D, Falkowski P, Schofield OM (2007) The mode and tempo of genome size evolution in eukaryotes. Genome Res 17:594–601

Orgel LE, Crick FH, Sapienza C (1980) Selfish DNA. Nature 288:645–646

Östergren G (1945) Parasitic nature of extra fragment chromosomes. Bot Not 2:157–163

Pagel M (1997) Inferring evolutionary processes from phylogenies. Zool Scr 26:331–348

Pagel M (1999) Inferring the historical patterns of biological evolution. Nature 401:877–884

Pandit MK, Pocock MJO, Kunin WE (2011) Ploidy influences rarity and invasiveness in plants. J Ecol 99:1108–1115

Pavliček T, Bureš P, Horová L, Raskina O, Nevo E (2008) Genome size microscale divergence of *Cyclamen persicum* in Evolution Canyon, Israel. Cent Eur J Biol 3:83–90

Pecinka A, Suchanková P, Lysák MA, Trávnícek B, Doležel J (2006) Nuclear DNA content variation among central European *Koeleria* taxa. Ann Bot 98:117–122

Pellicer J, Fay MF, Leitch IJ (2010) The largest eukaryotic genome of them all? Bot J Linn Soc 164:10–15

Petrov DA (2002) Mutational equilibrium model of genome size evolution. Theor Popul Biol 61:531–544

Price HJ, Bachmann K (1976) Mitotic cycle time and DNA content in annual and perennial Microseridinae (Compositae, Cichoriaceae). Plant Syst Evol 126:323–330

Price HJ, Sparrow AH, Nauman AF (1973) Correlations between nuclear volume, cell volume and DNA content in meristematic cells of herbaceous angiosperms. Experientia 29:1028–1029

Price HJ, Chambers KL, Bachmann K (1981) Geographic and ecological distribution of genomic DNA content variation in *Microseris douglasii* (Asteraceae). Bot Gaz 142:415–426

Price HJ, Chambers KL, Bachmann K, Riggs J (1986) Patterns of mean nuclear DNA content in *Microseris douglasii* (Asteraceae) populations. Bot Gaz 147:496–507

Price HJ, Hodnett G, Johnston JS (2000) Sunflower (*Helianthus annuus*) leaves contain compounds that reduce nuclear propidium iodide fluorescence. Ann Bot 86:929–934

Raven PH, Evert RF, Eichhorn SE (2005) Biology of plants, 7th edn. W.H. Freeman and Company, New York, pp 686

Rayburn AL, Auger JA (1990) Genome size variation in *Zea mays* ssp. *mays* adapted to different altitudes. Theor Appl Genet 79:470–474

Rees H, Jones GH (1967) Chromosome evolution in *Lolium*. Heredity 22:1–18

Rejmánek M (1996) A theory of seed plant invasiveness: the first sketch. Biol Conserv 78:171–181

Rejmánek M (2000) Invasive plants: approaches and predictions. Austral Ecol 25:497–506

Renzaglia KS, Rasch EM, Pike LM (1995) Estimates of nuclear DNA content in bryophyte sperm cells: phylogenetic considerations. Am J Bot 82:18–25

Rothfels K, Sexsmith E, Heimburger M, Krause MO (1966) Chromosome size and DNA content of species of *Anemone* L. and related genera (Ranunculaceae). Chromosoma 20:54–74

Schönswetter P, Suda J, Popp M, Weiss-Schneeweiss H, Brochmann C (2007) Circumpolar phylogeography of *Juncus biglumis* (Juncaceae) inferred from AFLP fingerprints, cpDNA sequences, nuclear DNA content and chromosome numbers. Mol Phylogenet Evol 42:92–103

Schubert I (2011) 'Sex and crime' in evolution—why sexuality was so successful. Genes Genet Syst 86:1–6

Sharaf K, Bureš P, Hovrová L, Pavliček T, Nevo E (2008) Distribution of abundance and genome size variability in the grain beetle *Oryzaephilus surinamensis* (Linnaeus, 1758) (Coleoptera: Silvanidae). Zool Middle East 45:79–89

Slovák M, Vit P, Urfus T, Suda J (2009) Complex pattern of genome size variation in a polymorphic member of the Asteraceae. J Biogeogr 36:372–384

Šmarda P (2006) DNA ploidy levels and intraspecific DNA content variability in Romanian fescues (*Festuca*, Poaceae) measured in fresh and herbarium material. Folia Geobot 41:417–432

Šmarda P, Bureš P (2006) Intraspecific DNA content variability in *Festuca pallens* on different geographical scales and ploidy levels. Ann Bot 98:665–678

Šmarda P, Bureš P (2010) Understanding intraspecific variation in genome size in plants. Preslia 82:41–61

Šmarda P, Bureš P (2012) The variation of base composition in plant genomes. In: Wendel JF, Greilhuber J, Doležel J, Leitch IJ (eds)

Plant genome diversity, vol 1, Plant genomes, their residents, and their evolutionary dynamics. Springer-Verlag, Wien, pp 209–236

Šmarda P, Bureš P, Hovrová L (2007) Random distribution pattern and non-adaptivity of genome size in a highly variable population of *Festuca pallens*. Ann Bot 100:141–150

Šmarda P, Bureš P, Hovrová L, Rotreková O (2008) Intrapopulation genome size dynamics in *Festuca pallens*. Ann Bot 102:599–607

Šmarda P, Horová L, Bureš P, Hralová I, Marková M (2010) Stabilizing selection on genome size in a population of *Festuca pallens* under conditions of intensive intraspecific competition. New Phytol 187:1195–1204

Sparrow AH, Miksche JP (1961) Correlation of nuclear volume and DNA content with higher plant tolerance to chronic radiation. Science 134:282–283

Stebbins GL (1957) Self fertilization and population variability in higher plants. Am Nat 91:337–354

Stebbins GL (1971) Chromosomal evolution in higher plants. Edward Arnold, London

Strasburger E (1893) Über die Wirkungssphäre der Kerne und die Zellgröße. Histol Beitr 5:97–124

Suda J, Kyncl T, Jarolimová V (2005) Genome size variation in Macaronesian angiosperms: forty percent of the Canarian endemic flora completed. Plant Syst Evol 252:215–238

Suda J, Weiss-Schneeweiss H, Tribsch A, Schneeweiss GM, Trávnícek P, Schönswetter P (2007) Complex distribution patterns of di-, tetra-, and hexaploid cytotypes in the European high mountain plant *Senecio carniolicus* (Asteraceae). Am J Bot 94:1391–1401

Suoniemi A, Anamthawat-Jónsson K, Arna T, Schulman AH (1996a) Retrotransposon *BARE*-1 is a major, dispersed component of the barley (*Hordeum vulgare* L.) genome. Plant Mol Biol 30: 1321–1329

Suoniemi A, Narvanto A, Schulman AH (1996b) The *BARE*-1 retrotransposon is transcribed in barley from an LTR promoter active in transient assays. Plant Mol Biol 31:295–306

Swift H (1950) The constancy of desoxyribose nucleic acid in plant nuclei. Proc Natl Acad Sci USA 36:643–654

te Beest M, Le Roux JJ, Richardson DM, Brysting AK, Suda J, Kubesova M, Pysek P (2012) The more the better? The role of polyploidy in facilitating plant invasions. Ann Bot 109:19–45

Temsch EM, Temsch W, Ehrendorfer-Schratt L, Greilhuber J (2010) Heavy metal pollution, selection, and genome size: The species of the Žerjav study revisited with flow cytometry. J Bot 2010: Article ID 596542 (11 pp)

Thomas CA (1971) The genetic organization of chromosomes. Annu Rev Genet 5:237–256

Turpeinen T, Kulmala J, Nevo E (1999) Genome size variation in *Hordeum spontaneum* populations. Genome 42:1094–1099

Van't Hof J (1965) Relationships between mitotic cycle duration, S period duration and average rate of DNA synthesis in root meristem cells of several plants. Exp Cell Res 39:48–58

Van't Hof J, Sparrow AH (1963) A relationship between DNA content, nuclear volume, and minimum mitotic cycle time. Proc Natl Acad Sci USA 49:897–902

Van Mooy BAS, Fredricks HF, Pedler BE, Dyhrman ST, Karl DM et al. (2009) Phytoplankton in the ocean use non-phosphorus lipids in response to phosphorus scarcity. Nature 458:69–72

Vendrely R, Vendrely C (1948) La teneur du noyau cellulaire en acide désoxyribonucléique à travers les organes, les individus et les espèces animales: Techniques et premiers résultats. Experientia 4:434–436

Verma RS, Lin MS (1979) The duration of DNA synthetic (S) period in *Zea mays*: a genetic control. Theor Appl Genet 54:277–282

Vicient CM, Suoniemi A, Anamthawat-Jónsson K, Tanskanen J, Beharav A, Nevo E, Schulman AH (1999) Retrotransposon *BARE*-1 and its role in genome evolution in the genus *Hordeum*. Plant Cell 11:1769–1784

Vidic T, Greilhuber J, Vilhar B, Dermastia M (2009) Selective significance of genome size in a plant community with heavy metal pollution. Ecol Appl 19:1515–1521

Vinogradov AE (2003) Selfish DNA is maladaptive: evidence from the plant Red List. Trends Genet 19:609–614

Walker DJ, Monino I, Correal E (2006) Genome size in *Bituminaria bituminosa* (L.) C.H. Stirton (Fabaceae) populations: separation of "true" differences from environmental effects on DNA determination. Environ Exp Bot 159:104–116

Wendel JF, Wessler SR (2000) Retrotransposon-mediated genome evolution on a local ecological scale. Proc Natl Acad Sci USA 97:6250–6252

Whitney KD, Baack EJ, Hamrick JL, Godt MJW, Barringer BC et al. (2010) A role for nonadaptive processes in plant genome size evolution? Evolution 64:2097–2109

Wright SI, Schoen DJ (1999) Transposon dynamics and the breeding system. Genetica 107:139–148

Wright SI, Ness RW, Foxe JP, Barrett SCH (2008) Genomic consequences of outcrossing and selfing in plants. Int J Plant Sci 169:105–118

Yang DP, Dodson EO (1970) The amounts of nuclear DNA and the duration of DNA synthetic period (S) in related diploid and autotetraploid species of oats. Chromosoma 31:309–320

Zonneveld BJM (2010) New record holders for maximum genome size in eudicots and monocots. J Bot 2010: Article ID 527357

Index

I.J. Leitch et al. (eds.), *Plant Genome Diversity Volume 2*,
DOI 10.1007/978-3-7091-1160-4, © Springer-Verlag Wien 2013

D

E

R

MIX
Papier aus verantwortungsvollen Quellen
Paper from responsible sources
FSC® C105338

If you have any concerns about our products,
you can contact us on
ProductSafety@springernature.com

In case Publisher is established outside the EU,
the EU authorized representative is:
Springer Nature Customer Service Center GmbH
Europaplatz 3, 69115 Heidelberg, Germany

Printed by Libri Plureos GmbH
in Hamburg, Germany